BOILERS, EVAPORATORS, AND CONDENSERS

BOILERS, EVAPORATORS, AND CONDENSERS

Edited by

Sadık Kakaç
Department of Mechanical Engineering
University of Miami
Coral Gables, Florida

A WILEY-INTERSCIENCE PUBLICATION

JOHN WILEY & SONS, INC.

New York • Chichester • Brisbane • Toronto • Singapore

Library of Congress Cataloging in Publication Data:

Boilers, evaporators, and condensers/edited by Sadık Kakaç.
p. cm.
"A Wiley-Interscience publication."
Includes index.
ISBN 0-471-62170-6
1. Steam-boilers. 2. Evaporators. 3. Condensers (Vapors and gases) I. Kakaç, S. (Sadık)

TJ281.C78 1991
621.1'97—dc20 90-22486
CIP

Printed in the United States of America

10 9 8 7 6 5 4 3 2 1

CONTRIBUTORS

A. K. Agrawal, Department of Mechanical Engineering, Clemson University, Clemson, S.C. 29634 U.S.A.

M. J. Albrecht, Fossil Power Division, Babcock & Wilcox Company, Barberton, Ohio 44203

D. Butterworth, Heat Transfer and Fluid Flow Service, Building 392.7 AERE Harwell, Oxfordshire OX11 ORA, United Kingdom

J. G. Collier, Chairman, Nuclear Electric plc, Barnett Way, Barnwood, Gloucester GL4 7RS, United Kingdom

S. Kakaç, Department of Mechanical Engineering, University of Miami, Coral Gables, Florida 33124 U.S.A.

J. B. Kitto, Jr., Research & Development Division, Babcock & Wilcox Co., P.O. Box 385, Alliance, Ohio 44601 U.S.A.

R. Leithner, Institut für Wärme- Und Brennstofftechnik, Franz-Liszt-Str. 35, D-3300 Braunschweig (49), Germany

P. E. Liley, School of Mechanical Engineering, Purdue University, West Lafayette, Indiana 47907 U.S.A.

Z. H. Lin, Energy and Power Engineering Dept., Xi'an Jiaotong University, Xi'an, People's Republic of China

P. J. Marto, Naval Postgraduate School, Monterey, California 93943

R. Oskay, Department of Mechanical Engineering, Middle East Technical University, Ankara, Turkey

M. B. Pate, Department of Mechanical Engineering, Iowa State University, Ames, Iowa 50011 U.S.A.

E. Paykoç, Department of Mechanical Engineering, Middle East Technical University, Ankara, Turkey

J. Taborek, 8002 Ocean Front Drive, Virginia Beach, Virginia 23451

P. B. Whalley, Department of Engineering Science, University of Oxford, Parks Road, Oxford OX1 3PJ, United Kingdom

CONTENTS

Preface xix

1. Introduction 1
S. Kakaç

2. Basic Design Methods of Heat Exchangers 9
S. Kakaç and E. Paykoç

2.1 Introduction 9
2.2 Arrangement of Flow Path in Heat Exchangers 10
2.3 Basic Equations in Design 11
2.4 Overall Heat Transfer Coefficient 14
2.4.1 Order of Magnitude of Thermal Resistances 18
2.5 The LMTD Method for Heat Exchanger Analysis 19
2.5.1 Multipass and Crossflow Heat Exchangers 22
2.6 The ε–NTU Method for Heat Exchanger Analysis 29
2.7 The P–NTU_c Method 42
2.8 The ψ–P Method 44
2.9 Heat Exchanger Design Calculation 48
2.10 Variable Overall Heat Transfer Coefficient 50
2.11 Heat Exchanger Pressure Drop and Pumping Power 52
2.11.1 Tube-Side Pressure Drop 52
2.11.2 Noncircular Cross-Sectional Ducts 55
2.11.3 Shell-Side Pressure Drop 58
2.11.4 Heat Transfer and Pumping-Power Relationship 62
2.12 Summary 65
Nomenclature 65
References 67

3. Forced Convection Correlations for Single-Phase Side of Heat Exchangers 69
S. Kakaç and R. Oskay

3.1 Introduction 69
3.2 Laminar Forced Convection 72
3.2.1 Hydrodynamically Developed and Thermally

Developing Laminar Flow in Smooth Circular Ducts 73
3.2.2 Simultaneously Developing Laminar Flow in Smooth Ducts 74
3.2.3 Laminar Flow Through Concentric Smooth Ducts 75
3.3 The Effect of Variable Physical Properties 77
3.3.1 Laminar Flow of Liquids 78
3.3.2 Laminar Flow of Gases 82
3.4 Turbulent Forced Convection 84
3.4.1 Turbulent Flow in Circular Ducts with Constant Properties 84
3.5 Turbulent Flow in Smooth Straight Noncircular Ducts 88
3.6 The Effect of Variable Physical Properties in Turbulent Forced Convection 88
3.6.1 Turbulent Liquid Flow in Ducts 89
3.6.2 Turbulent Gas Flow in Ducts 89
3.7 Summary 101
Nomenclature 102
References 103

4. Heat Exchanger Fouling **107**
A. K. Agrawal and S. Kakaç

4.1 Introduction 107
4.2 Effects of Fouling 108
4.2.1 Basic Equations 108
4.2.2 Effect of Fouling on Heat Transfer 109
4.2.3 Effect of Fouling on Pressure Drop 113
4.2.4 Cost of Fouling 114
4.3 Aspects of Fouling 115
4.3.1 Categories of Fouling 115
4.3.2 Fundamental Processes of Fouling 117
4.3.3 Prediction of Fouling 119
4.4 Design of Heat Exchangers Subject to Fouling 121
4.4.1 Providing a Fouling Allowance 121
4.4.2 Design Features to Minimize Fouling 132
4.4.3 Design Features to Facilitate Fouling Control 133
4.5 Operation of Heat Exchangers Subject to Fouling 133
4.6 Techniques to Control Fouling 137
4.6.1 Surface Cleaning Techniques 137
4.6.2 Additives 138
Nomenclature 139
References 140

5. Industrial Heat Exchanger Design Practices **143**
J. Taborek

5.1 Introduction 143
5.2 Heat Exchanger Types, Their Characteristics and Selection 143
5.2.1 Shell and Tube 144
5.2.2 Double Pipe or Multitube Hairpin 144
5.2.3 Air-Cooled Heat Exchangers 144
5.2.4 Gasketed Plate Exchangers 144
5.2.5 Matrix and Plate Fin-Tube Exchangers 144
5.2.6 Conclusion 145
5.3 The Strategy of Overall Design Optimization 145
5.3.1 Process Specifications 145
5.3.2 Preliminary Problem Analysis 147
5.3.3 Detailed Thermohydraulic Design 149
5.3.4 Mechanical-Metallurgical Design Aspects 152
5.3.5 Architectural Considerations 152
5.3.6 Maintenance, Operation, and Control Considerations 153
5.4 Shell-and-Tube Heat Exchangers: Characteristics of Constructional Components 153
5.4.1 Shell Types 154
5.4.2 Tube Bundle Types 154
5.4.3 Tube Passes 155
5.4.4 Baffle Types and Geometry 155
5.4.5 Tube Diameter and Tube Length 157
5.4.6 Tube Layout 157
5.5 Comments on Condenser Design 158
5.6 Comments on Reboiler Design 159
5.7 Calculated Example: Butane Cooler 160
5.7.1 Process Specifications 160
5.7.2 Heat Exchanger Type and Fluid Allocation 161
5.7.3 Thermal Profile Analysis: Possible Configurations 161
5.7.4 Selection of Construction Elements 163
5.7.5 Preliminary Estimation of Unit Size 163
5.7.6 Design and Results Evaluation 165
5.8 Design by Computer Programs 167
5.9 Optimization and Expert Systems 168
Nomenclature 169
Appendix 5.1: Step-by-Step Calculations 170
Acknowledgment 176
References 176

6. Fossil-Fuel-Fired Boilers: Fundamentals and Elements 179
J. B. Kitto, Jr. and M. J. Albrecht

6.1 Introduction 179
6.1.1 Background 180
6.1.2 Current Practice 181
6.1.3 Objectives and Overview 181
6.2 Fossil Boiler System 183
6.2.1 Input Requirements and Operating Pressure 183
6.2.2 Power Cycle 184
6.2.3 Types of Boilers 187
6.2.4 System Approach 191
6.3 Major Steam–Water Boiler Components 192
6.3.1 Enclosure Surfaces 192
6.3.2 Superheaters and Reheaters 195
6.3.3 Economizers 199
6.3.4 Steam Temperature Control 200
6.3.5 Steam Drum 200
6.4 Steam–Water System 203
6.4.1 Circulation Methods 204
6.4.2 Boiler Circulation and Flow 206
6.4.3 Furnace Heat Flux Evaluation 208
6.4.4 Circulation Evaluation 216
6.5 Two-Phase Flow Circulation Limiting Criteria 224
6.5.1 Flow Instabilities and General Velocity Limits 224
6.5.2 Heat Transfer and Critical Heat Flux 229
6.5.3 Steam–Water Separation and Drum Capacity 243
6.6 Other Evaluation Factors 252
6.7 Summary 253
Nomenclature 253
References 256
Appendix 6.1: Key Heat Transfer Parameters—Superheater, Reheater, and Economizer 260
Appendix 6.2: Sample Correlations for Two-Phase Multipliers and Void Fraction in Steam–Water Flows 263
Appendix 6.3: Sample Critical Heat Flux (CHF) Correlation 266

7. Once-Through Boilers 277
R. Leithner

7.1 Introduction (Historical Review) 277
7.2 Important Design Criteria in Comparison to Other Systems 283

7.2.1 Main Characteristic Features 284
7.2.2 Pressure Range 286
7.2.3 Operating Modes and Start-Up Period 288
7.2.4 Start-Up Equipment and Problems 296
7.2.5 Evaporator Tube Design 298
7.2.6 Heat Pickup of the Heating Surfaces 304
7.2.7 Differences in Heat Absorption and Flow Resistance in Individual Evaporator Tubes 309
7.2.8 Furnace Wall Design 314
7.2.9 Feed-Water Quality 316
7.2.10 Disturbances 317
7.2.11 Storage Capacity, Load Changes, and Control 317
7.2.12 Unit Capacity, Dimensions, and Design 321

7.3 Special Design Considerations 321
7.3.1 Water Wall Design 321
7.3.2 Steam Preheating Equipment 327
7.3.3 Water Separation 328

7.4 Start-Up Systems and Feed-Water Control 332
7.4.1 Start-Up Systems 332
7.4.2 Feed-Water Control 337

7.5 Examples and Operating Experiences 338
7.5.1 Lignite Fired 600-MW Once-Through Steam Generator 338
7.5.2 Bituminous Coal-Fired 740-MW Once-Through Steam Generator 339
7.5.3 Power Boiler for Supercritical 475-MW Unit 341
7.5.4 Steam Generator Unit for Steam Soak or Steam Drive in Oil Fields 345

7.6 Summary 348
Acknowledgments 349
Nomenclature 349
References 351
Appendix 7.1: Example for Calculating Power Generation Costs 354
Appendix 7.2: Optimal Design of a Recirculation Pump Suction Pipe 357
Appendix 7.3: Steam Generator Energy Balance 359

8. Thermohydraulic Design of Fossil-Fuel-Fired Boiler Components 363
Z. H. Lin

8.1 Introduction 363
8.1.1 Working Principle of a Steam Boiler 363
8.1.2 Main Characteristics of Steam Boilers 366

8.2 Types of Boilers and Construction of Boiler Components 366
8.2.1 Classification of Boilers 366
8.2.2 Construction and Design Problems of Furnaces 366
8.2.3 Construction and Design Problems of Superheaters and Reheaters 382
8.2.4 Construction and Design Problems of Economizers 391
8.2.5 Construction and Design Problems of Air Heaters 394
8.2.6 Construction and Design Problems of Steam Drums 397
8.3 Heat Transfer Calculations of Boiler Components 398
8.3.1 Boiler Efficiency and Weight of Fuel Fired 398
8.3.2 Heat Transfer Calculation of Water-Cooled Furnace 401
8.3.3 Heat Transfer Calculation of Convection Heating Surfaces 406
8.3.4 Procedure for Heat Transfer Calculation of a Boiler 420
8.4 A Numerical Example of the Heat Transfer Calculations of Boiler Components 421
8.5 Steam–Water Systems of Boilers and Circulation Calculations 440
8.5.1 Steam–Water System of Natural-Circulation Boiler and Design Problems 441
8.5.2 Steam–Water System of Controlled-Circulation Boilers and Design Problems 451
8.5.3 Steam–Water System of Once-Through Boilers 456
8.6 A Numerical Example of Boiler Circulation Calculations 457
Nomenclature 466
References 468

9. Nuclear Steam Generators and Waste Heat Boilers **471**
J. G. Collier

9.1 Abstract 471
9.2 Introduction 471
9.3 The Principal Types of Boiler 472
9.3.1 Nuclear Power Plants 472
9.3.2 Waste Heat Boilers 486
9.4 The Thermal and Mechanical Design of Boilers 490
9.4.1 General 490
9.4.2 Primary Side (Unfired Boiler) Design 490
9.4.3 Water-Side (Evaporator) Design 491

9.4.4 An Example: PWR Inverted U-Tube Recirculating Steam Generator 502
9.5 Common Problems in the Operation of Boilers 507
9.5.1 Causes of Steam Generator Problems 507
9.5.2 Worked Solutions 513
9.6 Conclusions 518
Acknowledgment 518
Nomenclature 518
References 519

10. Heat Transfer in Condensation **525**
P. J. Marto

10.1 Introduction 525
10.2 Film Condensation on a Single Horizontal Tube 526
10.2.1 Natural Convection 526
10.2.2 Forced Convection 528
10.3 Film Condensation in Tube Bundles 531
10.3.1 Effect of Condensate Inundation 532
10.3.2 Effect of Vapor Shear 534
10.3.3 Combined Effects of Inundation and Vapor Shear 535
10.3.4 Computer Modeling 539
10.4 Film Condensation Inside Tubes 539
10.4.1 Flow Patterns 539
10.4.2 Condensation in Horizontal Tubes 540
10.4.3 Condensation in Vertical Tubes 544
10.4.4 Condensation in Noncircular Passages 546
10.5 Pressure Drop During Condensation 546
10.5.1 Shell-Side Pressure Drop 546
10.5.2 Pressure Drop Inside Tubes 547
10.6 Condensation Heat Transfer Augmentation 550
10.6.1 Shell-Side Film Condensation Using Integral-Fin Tubes 550
10.6.2 Dropwise Condensation 555
10.7 Condensation of Vapor Mixtures 556
10.7.1 Equilibrium Methods 559
10.7.2 Nonequilibrium Methods 561
Nomenclature 562
References 565

11. Steam Power Plant and Process Condensers **571**
D. Butterworth

11.1 Introduction 571

11.2 Shell-and-Tube Condensers for Process Plant 573
11.2.1 Horizontal Shell-Side Condensers 573
11.2.2 Vertical Shell-Side Condensers 582
11.2.3 Tube-Side Condensers 582
11.2.4 Subcooling in Shell-and-Tube Condensers 583
11.2.5 Choice Between Types 585
11.3 Shell-and-Tube Condensers for Power Plant 585
11.3.1 Steam Turbine Exhaust Condensers 585
11.3.2 Feed-Water Heaters 588
11.4 Plate Exchangers 591
11.5 Spiral Exchangers 591
11.6 Plate-Fin Heat Exchangers 592
11.7 Air-Cooled Heat Exchangers 594
11.8 Direct-Contact Condensers 595
11.9 Thermal Evaluation Methods for Shell-and-Tube Condensers 597
11.9.1 Introduction and Definition of Terms 597
11.9.2 Co-current and Countercurrent Condensers 600
11.9.3 Shell-Side, *E*-Type Condenser with Two Tube-Side Passes 603
11.9.4 Shell-Side, *E*-Type Condenser with Four or More Tube Passes 606
11.9.5 Crossflow Condensers 607
11.9.6 Nonequilibrium Calculation Methods 608
11.9.7 Multidimensional Shell-Side Flows 611
11.10 Thermal Evaluation Method for Direct-Contact Condensers 612
11.10.1 Spray Condensers 612
11.10.2 Tray Condensers 616
11.11 Reasons for Failure of Condenser Operation 616
11.12 Examples 617
11.12.1 Process Condenser 617
11.12.2 Power Condenser 623
Acknowledgment 629
Nomenclature 630
References 631

12. Evaporators and Condensers for Refrigeration and Air-Conditioning Systems **635**
M. B. Pate

12.1 Introduction 635
12.1.1 Background 635
12.1.2 Typical Evaporator Behavior 637

12.1.3 Typical Condenser Behavior 638
12.1.4 Types of Heat Exchangers in Refrigeration and Air-Conditioning Applications 639
12.2 Heat Exchanger Analysis 641
12.2.1 General Equations 641
12.2.2 Lumped Heat Exchanger Analysis Approach 644
12.2.3 Local Heat Transfer Integration Approach 646
12.3 Evaporator Coils 647
12.3.1 Description and Special Considerations 647
12.3.2 In-Tube Refrigerant Evaporation Heat Transfer 650
12.3.3 In-Tube Heat Transfer Augmentation 657
12.3.4 Air-Side Heat Transfer 663
12.3.5 Wet-Coil Heat Transfer 671
12.3.6 Frosted-Coil Heat Transfer 672
12.3.7 Fin Bonding and Thermal Contact Resistance 673
12.4 Condenser Coils 679
12.4.1 Description and Special Considerations 679
12.4.2 Similarities between Condenser and Evaporator Coils 679
12.4.3 In-Tube Refrigerant Condensation Heat Transfer 680
12.4.4 In-Tube Heat Transfer Augmentation 684
12.5 Flooded Evaporators 687
12.5.1 Description and Special Considerations 687
12.5.2 Shell-Side Refrigerant Heat Transfer 688
12.5.3 Shell-Side Heat Transfer Augmentation 694
12.6 Shell-and-Tube Direct Expansion Evaporators 698
12.6.1 Description and Special Considerations 698
12.6.2 In-Tube and Shell-Side Heat Transfer 699
12.7 Shell-and-Tube Condensers 700
12.7.1 Description and Special Considerations 700
12.7.2 Shell-Side Refrigerant Condensation Heat Transfer 701
12.7.3 Shell-Side Heat Transfer Augmentation 704
12.8 Heat Exchanger Design with Alternative Refrigerants 705
Nomenclature 710
References 712

13. Evaporators and Reboilers in the Process and Chemical Industries **717**
P. B. Whalley

13.1 Introduction 717
13.2 Relevance of Upflow and Downflow in Vertical Units 719

13.3 Evaporator Types 721
13.3.1 Horizontal Shell-Side Evaporator 722
13.3.2 Horizontal Falling-Film Evaporator 724
13.3.3 Horizontal Tube-Side Evaporator 725
13.3.4 Short-Tube Vertical Evaporator 726
13.3.5 Long-Tube Vertical Evaporator 729
13.3.6 Climbing-Film Evaporator 730
13.3.7 Vertical Falling-Film Evaporator 730
13.3.8 Agitated Thin Film Evaporator 732
13.3.9 Plate-Type Evaporator 734
13.3.10 Submerged-Combustion Evaporator 736
13.4 Reboiler Types 738
13.4.1 Internal Reboiler 738
13.4.2 Kettle Reboiler 738
13.4.3 Vertical Thermosyphon Reboiler 740
13.4.4 Horizontal Thermosyphon Reboiler 742
13.5 Energy Efficiency in Evaporation 744
13.5.1 Introduction 744
13.5.2 Multiple-Effect Evaporators 744
13.5.3 Vapor Recompression in Evaporation 747
13.5.4 Multistage Flash Evaporator 748
13.6 Heat Transfer and Pressure Drop Problems 751
13.6.1 Initial Sizing of the Unit 751
13.6.2 Two-Phase Vapor–Liquid Pressure Drop 753
13.6.3 Calculation of Natural-Circulation Units 757
13.6.4 Heat Transfer Rates 760
13.6.5 Heat Transfer on the Heating Side 761
13.6.6 Fouling 762
13.6.7 Boiling inside Tubes 762
13.6.8 Boiling outside Tubes 766
13.6.9 Falling-Film Evaporation 767
13.6.10 Agitated-Film Evaporation 768
13.6.11 Mixture Effects 768
13.6.12 Enhanced Surfaces 769
13.7 Possible Problems in the Operation of Evaporators and Reboilers 770
13.7.1 Introduction 770
13.7.2 Corrosion and Erosion 771
13.7.3 Maldistribution 771
13.7.4 Fouling 771
13.7.5 Flow Instability 773
13.7.6 Tube Vibration 774
13.7.7 Flooding 774

13.8 Design Example 774
13.8.1 Further Refinements in the Design 777
Nomenclature 778
References 779

Appendix A. Thermophysical Properties **783**
P. E. Liley
Nomenclature 783
List of Sources of Tables 784
References 785

Table A1. Thermophysical Properties of 113 Fluids at 1 bar, 300 K 786
Table A2. Thermophysical Properties of Liquid and Saturated-Vapor Air 790
Table A3. Thermophysical Properties of Gaseous Air at Atmospheric Pressure 792
Table A4. Thermophysical Properties of Saturated Ammonia (R717) 794
Table A5. Thermophysical Properties of Ammonia (R717) at 1-bar Pressure 795
Table A6. Thermophysical Properties of Saturated Normal Butane (R600) 796
Table A7. Thermophysical Properties of Normal Butane (R600) at Atmospheric Pressure 797
Table A8. Thermophysical Properties of Solid, Saturated-Liquid and Saturated-Vapor Carbon Dioxide 798
Table A9. Thermophysical Properties of Gaseous Carbon Dioxide at 1-bar Pressure 799
Table A10. Thermophysical Properties of Saturated Ethane (R170) 800
Table A11. Thermophysical Properties of Ethane at Atmospheric Pressure 801
Table A12. Thermophysical Properties of Saturated Ethylene (R1150) 802
Table A13. Thermophysical Properties of Ethylene (R1150) at Atmospheric Pressure 803
Table A14. Thermophysical Properties of *n*-Hydrogen (R702) at Atmospheric Pressure 803
Table A15. Thermophysical Properties of Saturated Methane (R50) 804
Table A16. Thermophysical Properties of Methane (R50) at Atmospheric Pressure 804

Table A17. Thermophysical Properties of Nitrogen (R728) at Atmospheric Pressure 805
Table A18. Thermophysical Properties of Oxygen (R732) at Atmospheric Pressure 806
Table A19. Thermophysical Properties of Saturated Normal Propane (R290) 807
Table A20. Thermophysical Properties of Propane (R290) at Atmospheric Pressure 808
Table A21. Thermophysical Properties of Saturated Refrigerant 12 809
Table A22. Thermophysical Properties of Refrigerant 12 at 1-bar Pressure 811
Table A23. Thermophysical Properties of Saturated Refrigerant 22 812
Table A24. Thermophysical Properties of Refrigerant 22 at Atmospheric Pressure 814
Table A25. Thermophysical Properties of Saturated R134a 815
Table A26. Properties of Refrigerant 134a at Atmospheric Pressure 816
Table A27. Thermophysical Properties of Saturated Ice-Water-Steam 817
Table A28. Thermophysical Properties of Steam at 1-bar Pressure 822
Table A29. Thermophysical Properties of Water-Steam at High Pressures 823
Table A30. Thermal Expansion Coefficient $\tilde{\alpha}$ of Water 825
Table A31. Isothermal Compressibility Coefficient β_T of Water 825
Table A32. Thermophysical Properties of Unused Engine Oil 826
Table A33. Conversion Factors 826

Index **829**

PREFACE

Boilers, evaporators, and condensers are vital components of power-producing plants, process chemical industries, and heating, ventilating, air-conditioning, and refrigeration systems. This book focuses on the various types of boilers, condensers, and evaporators, the key elements of these types of heat exchange equipment, their thermohydraulic design, and their design procedures and operational problems. A large number of industries are engaged in designing, constructing, selling, and operating boilers, evaporators, and condensers; in addition, courses are offered at many colleges and universities under various titles concerning these types of heat exchange equipment. There is extensive literature on this subject; however, the information has been widely scattered. Therefore a systematic approach was deemed essential for potential users in both industry and academia, describing the various types of boilers, evaporators, and condensers, detailing their specific fields of application, selection, and thermohydraulic design, and showing the thermal design procedures with numerical examples. It was essential to approach the leading specialists and experts to compile such a book. Engineers in the fields of power, process chemical industries, nuclear engineering, and air conditioning who are dealing with the selection and design of a variety of equipment for vaporization, including steam generators, evaporators, reboilers, condensers, and chillers, will find this book useful. The book has been prepared in a style that will also be suitable for newcomers to the fields of industrial heat transfer, along with senior and/or first-year graduate students in mechanical and chemical engineering, who pursue courses on thermal and fluid design and/or heat exchangers.

The book consists of 13 chapters and an appendix contributed by 15 scientists and experts. It begins with an introduction to two-phase-flow heat exchange equipment in Chapter 1. The basic design methods for rating and sizing problems are given in Chapter 2. In boilers, condensers, and evaporators, in most cases, one side has a single-phase fluid, while the other side has a two-phase fluid. Generally, the single-phase side represents the dominant thermal resistance, particularly with gas and oil flows. Chapter 3 consists of a comprehensive review of the available correlations for laminar and turbulent flow of single-phase Newtonian fluids through circular and noncircular ducts and concludes with recommendations for specific correlations for heat exchanger design. Heat exchanger fouling is discussed in Chapter 4. Guidelines for proper design procedures of heat exchangers are discussed and illustrated with a typical example in Chapter 5. Chapter 6 deals with the various components and the fundamentals of fossil-fuel-fired boilers. Particular emphasis is placed upon the evaluation of large, coal-fired, high-pressure boilers which are prevalent in electric utility power stations. Once-through boilers are introduced in Chapter 7. Chapter 8 reports on the thermohydraulic design and practice of fossil-fuel-fired boilers. Chapter 9 describes the principal types of unfired modern steam riser, concentrating on land-based units in service in the power and process chemical industries and the methods used for the thermal design of the various types along with the common operational problems. Heat transfer in condensation and power condensers and their design are presented in Chapters 10 and 11, respectively. The various types of condensing equipment and the methods available for the design and check-rating of those exchangers are given. Possible explanations regarding the reasons why condensers fail to operate as expected are noted. Chapter 12 introduces design considerations for air-conditioning evaporators and condenser coils. Chapter 13 deals with evaporators and reboilers in the process and chemical industry. This chapter also offers advice on the choice of the best type of evaporators for the many duties encountered in a process plant. More than 30 tables in Appendix A give the thermophysical properties of various fluids, including new refrigerants. A special effort has been made to make each chapter a self-contained unit, so that practicing engineers, professors, and engineering students may be able to select their own chapters of interest. Many references selected from the wealth of sources available are added to each chapter in order to make it a valuable study material for practicing engineers and students of thermal engineering science and applications.

It is hoped that the information given in this book will help students and design engineers to understand fully the basic design principles involved, and that they will be better equipped to develop more sophisticated computer programs to study the effect of various design parameters of interest and to incorporate their results into practice, thus achieving optimization in design.

When I joined the Chair of Heat Technique of the Faculty of Mechanical Engineering at the Technical University of Istanbul as an assistant lecturer in 1955 to replace Dr. Vedat Arpacı (who is now Professor of Mechanical Engineering at the University of Michigan, Ann Arbor), I used to assist Professors F. Narter and N. Aybers in courses on designing boilers and heating, ventilating, refrigeration, and air-conditioning systems. Their ideas, dealing with complex heat transfer engineering problems were so stimulating that I found the subject very fascinating and interesting. Since then, I have been very interested in thermal engineering science and its applications.

I am very pleased to report that when I approached some of the leading scientists and experts on boilers, evaporators, and condensers to prepare a book on this subject, they agreed to contribute to such a vital subject. I am grateful to them for providing excellent material, in such a timely manner within the prescribed length limitations.

Editing of this volume has been finalized during my stay at the Lehrstuhl A für Thermodynamik, Techniche Universität München, FRG, as the receipient of the Senior U.S. Scientists Award from the Alexander von Humboldt Stiftung. I am very grateful to Professor F. Mayinger for providing the facilities of his Chair.

Special thanks are due to the John Wiley editors and staff, in particular, I sincerely appreciate the close cooperation of Mr. Frank Cerra on all matters concerning this book from the start to the final production, and the outstanding editorial work of Ms. Diana Cisek. I also acknowledge the efficient work and excellent cooperation of Mrs. Sarah W. Roesser of Science Typographers and the excellent line drawings prepared by Mr. Ali Akgüneş of the Middle East Technical University, Ankara, Turkey.

Finally, I am very thankful to my wonderful wife Filiz who continuously supported my work.

Every effort has been made by the editor to minimize typographical errors. Each chapter has been independently reviewed by other experts in the field in order to enhance the quality and correctness of the material. If any errors come to the attention of the readers, the editor would greatly appreciate being made aware of them, so that they can be eliminated in subsequent printings. Of course, the editor would also appreciate any additional comments related to any of the chapters.

S. KAKAÇ

January 1991
München, F.R.G.

BOILERS, EVAPORATORS, AND CONDENSERS

CHAPTER 1

INTRODUCTION

S. KAKAÇ
Department of Mechanical Engineering
University of Miami
Coral Gables, Florida 33124

This chapter gives a brief introduction to the following chapters which detail the types, design, and operation of various kinds of boilers, evaporators, and condensers.

Boilers, evaporators, and condensers are two-phase-flow heat exchange equipment; on one side is a boiling or condensing fluid, and on the other side is either a single-phase or two-phase flow. Approximately 60% of all heat exchange equipment used in industrial applications works in a two-phase-flow mode (Fig. 1.1). They are used in the power, process, food industries, and air-conditioning and refrigeration systems. In this book almost every facet of boilers, evaporators, and condensers is considered.

The oldest two-phase-flow heat exchanger used by people was certainly a cooking vessel for preparing meals by boiling water. One of the earliest recorded boilers (130 A.D.), operating on the water-tube principle, supplied steam to Hero's steam engine, a hollow sphere mounted on hollow trunnions that permitted steam to pass into the sphere. The steam was exhausted through two offset nozzles that caused the sphere to revolve, thus providing the world's first steam turbine.

In ancient Egypt, boiling heat transfer is reported to have been used for production through distilling wine. Use in food production was the dominant application of two-phase-flow heat exchangers until the invention of the steam engine by James Watt. The industrial use of boilers (fired-type two-phase-flow heat exchangers) followed later. The production of the first commercial steam engines by Thomas Savery in 1698 and Thomas Newcomer in 1705 provided the impetus for the Industrial Revolution. Since then steam has been extensively used to meet the needs of transportation and industry.

Boilers, Evaporators and Condensers, Edited by Sadik Kakaç
ISBN 0-471-62170-6

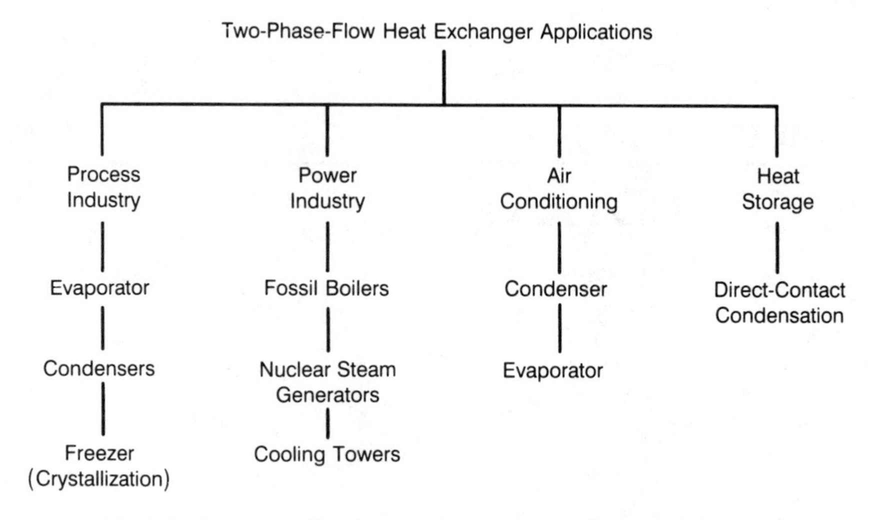

Fig. 1.1. Two-phase-flow heat exchangers according to application.

Early boilers consisted of closed vessels made from sheets of wrought iron and were formed into shapes varying from simple spheres to complex sections; for example, the Waggon boiler of Watt (1788) was shaped as a covered wagon. These vessels were supported by brickwork over a fire which itself was supported on a grade. The working pressure maintained was about 0.7 bar.

In the early boilers, steam was produced at low pressure, typically 1 bar above atmospheric pressure. The trend of maintaining low pressures continued for about a century thereafter. The development of boilers for power plants went through many stages after the first commercially successful steam engine which was patented by Thomas Savery (1698). One of the main steps for the early development of boilers was the introduction of water-tube boilers. One of the early water-tube boilers is shown in Fig. 1.2 [1].

The demand for more powerful engines created a need for boilers that operated at higher pressures, and, as a result, individual boilers were built larger and larger. Eventually, the size of a power plant boiler became so large that existing furnace designs and methods of coal burning, such as stokers, were no longer adequate. This led to the development of pulverized-coal firing and the use of water-cooled furnaces. As a result of these developments, the boiler units used in modern power plants for steam pressure above 1200 psi (80 bar) consist of furnace water-wall tubes, superheaters, and such heat recovery accessories as economizers and air heaters (Fig. 1.3) [1].

In this book emphasis is placed on the problem of modern, high-temperature units of boilers. The development of modern boilers and more efficient condensers for the power industry have represented major milestones in

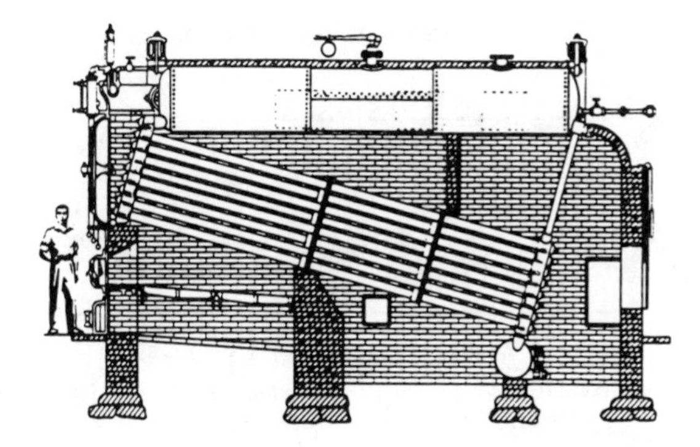

Fig. 1.2. Babcock and Wilcox boiler developed in 1877 [1].

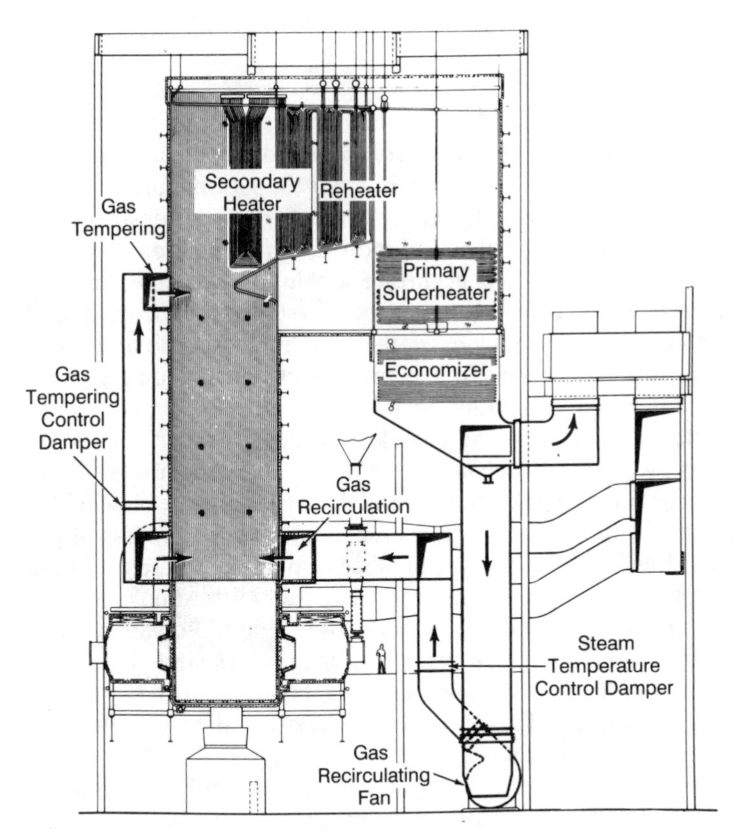

Fig. 1.3. Boiler with water-cooled furnace with heat recovery units [1].

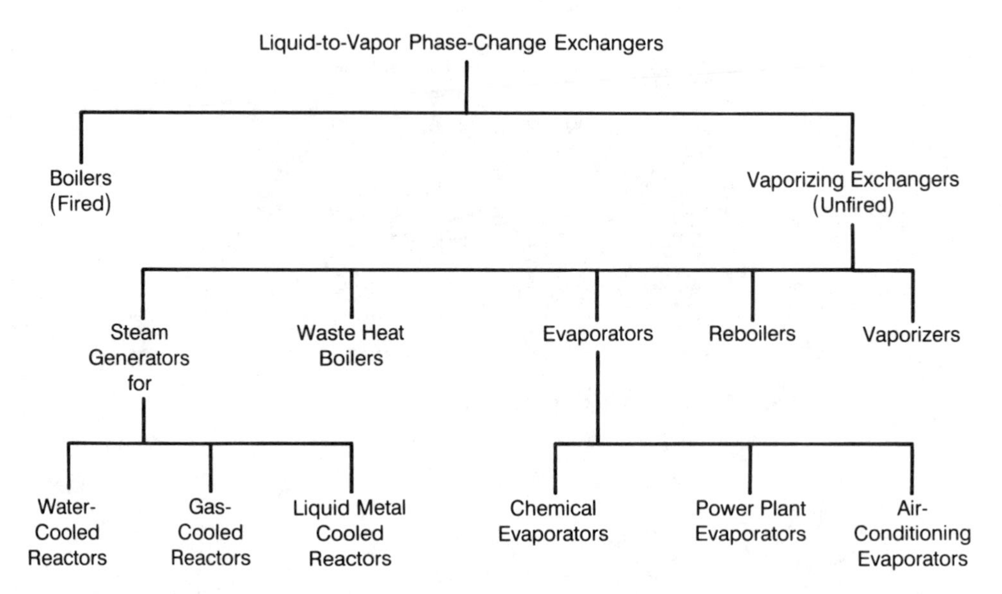

Fig. 1.4. Liquid-to-vapor phase-change exchangers.

engineering. Therefore in the power and process industries, engineers had to design steam generators, evaporators, and condensers to produce steam and to separate a condensable component from the gas mixture.

In the process industry, engineers are concerned with designing equipment to vaporize a liquid. In the chemical industry, the function of an evaporator is to vaporize a liquid or to concentrate a solution by vaporizing part of the solvent. Evaporators may also be used in crystallization processes. Often the solvent is water, but in many cases the solvent is valuable and recovered for reuse. The vaporizers used in the process chemical industry cover a wide range of sizes and applications.

Two-phase-flow equipment includes various types of boilers (fired heat exchangers), vaporizing exchangers (unfired), and condensers (direct-contact and indirect-contact type). These heat exchangers may be classified according to the type of construction, process function, transfer process, flow arrangement, heat transfer mechanisms, number of phases of fluids, and application [2, 3]. The two-phase heat exchangers discussed in this book can be classified according to process function as condensers, liquid-to-vapor phase-change exchangers (boilers and vaporizers), heaters, coolers, and chillers.

Liquid-to-vapor phase-change exchangers are classified as boilers and vaporizing exchangers (Fig. 1.4). Further classification of boilers (fired) and vaporizing exchangers (unfired) is given in Chapters 6, 8, 11, and 13.

Condensers are classified as direct contact and indirect contact (Fig. 1.5), and the detailed classification is given in Chapter 11.

Recently, there has been an impetuous development in the design of evaporators and condensers for power and process engineering, especially in

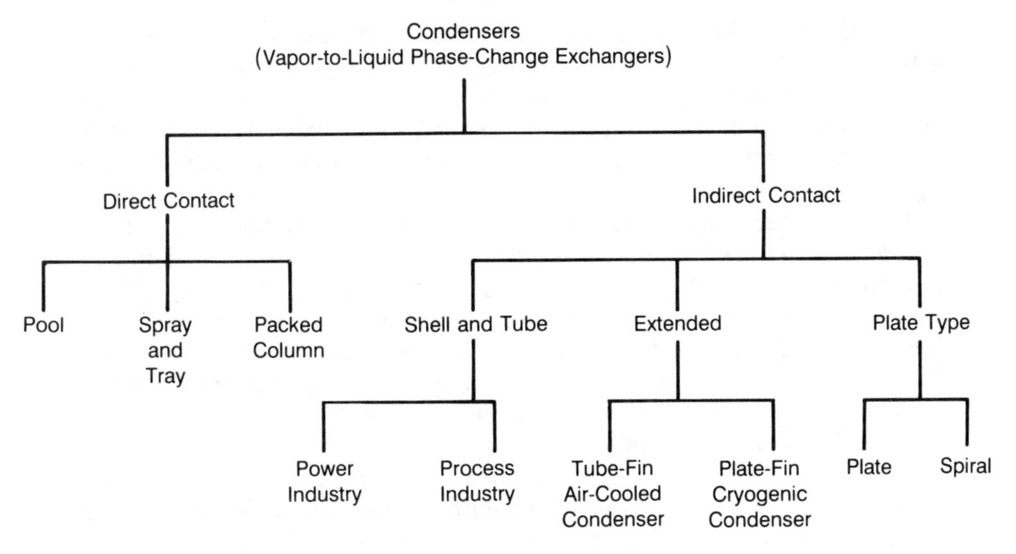

Fig. 1.5. Vapor-to-liquid phase-change exchangers.

power engineering, because of the need for increasing the unit power per station and the demands of nuclear power plants. Until the end of the 1950s, fossil-fired power stations had an electrical output of 50 to 100 MWe per unit; while modern coal-fired power stations are on the order of 700 to 900 MWe [2].

Nuclear power, through the fission of uranium, has provided a further means of producing steam for electricity production. Presently, nuclear reactors provide energy up to 1300 MWe per unit. The increase in electrical output means that the thermal power had to grow from 150 to 4000 MW_{th} per station, which leads to the development of modern boilers, new steam generators, and condensers.

As mentioned before, this book has been written for newcomers to the industrial heat transfer field and mechanical and chemical engineering students, who require a broad introduction to and design information on boilers, condensers, and evaporators and heat exchangers in general.

Basic design methods are reviewed first in Chapter 2. The effects of individual resistances on the overall heat transfer coefficient, temperature distribution in direct-transfer heat exchangers, mean temperature difference LMTD, ε–NTU, P–NTU_c, and ψ–P methods are discussed. Examples of elementary sizing and rating problems are given for a shell-and-tube heat exchanger. Pressure drop relations applicable to the single-phase side of two-phase-flow heat exchangers are also given.

In most applications of boilers, evaporators, and condensers, one side is a single-phase fluid and the other side is a two-phase fluid. One of the important design parameters is calculating the overall heat transfer coefficient. Therefore it is necessary to discover the heat transfer characteristics of the single-phase fluid side in addition to those for the two-phase side. In Chapter 3, a comprehensive review of the available correlations for the laminar and turbulent flow of a single-phase newtonian fluid through circular and noncircular ducts is presented. The effect of property variations is discussed. Important correlations are recommended for heat exchanger design.

Fouling of heat transfer surfaces introduces perhaps the major uncertainty into the design and operation of heat exchange equipment. Chapter 4 introduces the reader to the special problems of design and operation of heat exchange equipment exposed to fouling conditions. The basic concepts of the fouling of heat transfer surfaces are presented, emphasizing important topics that are useful to the designers and users of heat exchange equipment subject to fouling. The impact of fouling on heat transfer and pressure drop is explained by worked examples.

Overall optimization of a particular heat exchanger encompasses a variety of problem areas that involve both quantitative and qualitative information and interpretation of results. Chapter 5 focuses on the logical design modules such as process specifications for design, thermohydraulic design, mechanical–metallurgical design, architectural design, operation, control, and maintenance considerations. The modern strategy of heat exchanger design is a parallel approach to all these pertinent design submodules. Guidelines for a proper design process are discussed and illustrated in a typical worked example. This chapter also briefly introduces the many important problems that may arise in the design and operation of various heat exchange equipment.

Boilers can be broadly classified into two—those heated directly by the combustion of fossil fuels and those heated indirectly by a hot gas or liquid that has received its heat from another energy source. The latter are usually referred to as waste heat boilers. Chapter 6 is concerned with the key elements of steam–water two-phase flow in fossil-fired boilers. Particular emphasis is placed upon the evaluation of large, coal-fired, high-pressure boilers which are prevalent in electric utility power stations. This provides the framework for the common elements with unique characteristics of other steam-generating systems treated as additions or exceptions. The boiler circuitry, furnace absorption, and general circulation calculation are reviewed; the key limiting criteria are addressed including instabilities, heat transfer with particular reference to critical heat flux, steam–water separation, and boiler drum. The focus of the discussion is a modern subcritical pressure, drum-type boiler typically found in the United States, Canada, and the United Kingdom.

Chapter 7 explores selected aspects of once-through boiler design where water is evaporated continuously to dryness in individual boiler tubes. A historical overview is followed by discussion of special design issues, sliding pressure operation, operating and start-up characteristics, and selected two-phase-flow topics. The once-through boiler design typical of units found in

Germany is the basis for this discussion. The classification of boilers is described in Chapters 6 and 8. Once-through boilers involve much more difficult design problems because they must be carefully proportioned to give a good water-flow distribution by avoiding flow instability. Important design criteria in comparison to other systems, special design considerations, start-up systems, feed-water control, and operating experience with various once-through boilers are all described in this chapter.

Chapter 8 considers the construction and design problems of fossil-fuel-fired system boilers and it provides a detailed design process following the boiler design practice of the People's Republic of China (PRC) and, by association, the Soviet Union. Boiler types, the construction of boiler components, and the heat transfer calculations of boiler components are described. Worked examples for heat transfer calculations of boiler components are given. To remove heat from the boiler heating surfaces, it is necessary that proper circulation (the flow of water, steam, or steam–water mixture within the steam boiler) be provided throughout the boiler circuits.

Chapter 9 concerned with the principal types of unfired modern steam generators used in central station nuclear power plants and process chemical industries. This type of heat exchange equipment is called a waste heat boiler. Nuclear steam generators are a special class of waste heat boilers. The methods used for the thermal design of this type of heat exchange equipment are discussed with worked examples, along with the common problems of steam generators encountered during the operation.

Condensation of vapor is an important phenomenon that occurs in numerous heat exchange equipment such as condensers. One of the important steps in the thermal design of condensers is calculating the heat transfer coefficient on the condensing side of the equipment, which is presented in Chapter 10. Calculational methods for the heat transfer coefficient and the pressure drop are presented for a wide variety of surface geometries and flow configurations, including shell-side condensation, in-tube condensation, and direct-contact condensation. Worked examples are provided to illustrate the use of correlations in calculating the condensing-side heat transfer coefficient.

There is a wide variety of equipment available for vapor condensation. Chapter 11 describes the main features of the more important types and discusses their thermal design and check rating; methods available for the design of these exchangers are given, with particular emphasis on the shell and tube which is the most common type. An important calculation step is the determination of the mean temperature difference and mean overall heat transfer coefficient from their local values. This problem is also discussed in some detail with worked examples.

Chapter 12 introduces various types of evaporators and condenser coils in air-conditioning systems. The major factors affecting heat transfer between the refrigerant and air in a plated finned-tube heat exchanger, otherwise known as an evaporator or condenser coil, are discussed. These factors are the inside heat transfer coefficient, the thermal contact resistance between the fin and the tube, and the outside heat transfer coefficient. Correlations are presented to calculate the rate of heat transfer for each of these factors. In-tube augmentation, oil–refrigerant mixture effects, and frosting are also discussed. Tube-circuiting procedures and standards available for rating coils are also presented. Thermal design procedures are illustrated with worked examples.

The main function of an evaporator is to vaporize a liquid or to concentrate a solution by vaporizing part of the solvent. Reboilers are used to vaporize the liquid at the bottom of a distillation column to provide the vapor flow up the column. Chapter 13 describes the main features of different classes of such equipment used in the process and chemical industries. The main applications of various types are briefly discussed. Energy efficiency in evaporation is briefly reviewed, and the possible energy-saving arrangements of multiple-effect evaporation, vapor recompression, and multistage flash evaporation are introduced briefly. The main problems in the heat transfer and pressure drop in evaporators and reboilers are reviewed. Some of the problems encountered during operation are mentioned.

REFERENCES

1. *Steam, Its Generation and Use*. Babcock and Wilcox, New York, 1978.
2. Shah, R. K. (1981) Classification of heat exchangers. In *Heat Exchangers: Thermal-Hydraulic Fundamentals and Design*, S. Kakaç, A. E. Bergles, and F. Mayinger (eds.). Hemisphere, New York.
3. Mayinger, F. (1988) Classification and applications of two-phase flow heat exchangers. In *Two-Phase Flow Heat Exchangers*, S. Kakaç, A. E. Bergles, and E. O. Fernandes (eds.). Kluwer, Dordrecht.

CHAPTER 2

BASIC DESIGN METHODS OF HEAT EXCHANGERS

S. KAKAÇ

Department of Mechanical Engineering
University of Miami
Coral Gables, Florida 33124

E. PAYKOÇ

Department of Mechanical Engineering
Middle East Technical University
Ankara, Turkey

2.1 INTRODUCTION

The most common heat exchanger design problems are rating and sizing problems. In this chapter the basic design methods for two-fluid direct-transfer heat exchangers are reviewed.

A heat exchanger is a device in which heat is transferred from a hot fluid to a cold fluid. In its simplest form, the two fluids mix and leave at an intermediate temperature determined by the conservation of energy. This device is not truly a heat exchanger but rather a mixer. In most applications, the fluids do not mix but transfer heat through a separating wall that takes on a wide variety of geometries. Three categories are normally used to classify heat exchangers: (1) recuperators, (2) regenerators, and (3) direct-contact apparatus.

There are also heat exchangers in which the heat carrier fluid is heated or cooled by means of internal heat sources or sinks.

Recuperators are direct-transfer heat exchangers in which heat transfer occurs between two fluid streams at different temperatures in a space

Boilers, Evaporators and Condensers, Edited by Sadik Kakaç John Wiley & Sons, Inc.
ISBN 0-471-62170-6 ©1991

separated by a thin solid wall (parting sheets or tube walls). Heat transfer occurs by convection from the hotter fluid to the separating wall surface, by conduction through the separating wall, and by convection from the separating wall surface to the cooler fluid. If one of the fluids is a radiating gas, thermal radiation also plays an important role in the heat exchange between the two media. Recuperators include, for example, air heaters, economizers, evaporators, condensers, steam boilers, dry cooling towers, and so on. Regenerators are heat exchangers in which a hot and a cold fluid flow through the same heating surface at certain time intervals. The surface of the regenerator first receives heat energy from the hot fluid; it then releases this thermal energy to the cold fluid. Thus the process of heat transfer is a transient one in a regenerator; that is, the temperatures of the heating wall and of the fluids vary with time during the heat transfer process. Both recuperators and regenerators are surface-type heat exchangers.

In direct-contact heat exchangers, heat is transferred by complete or partial physical mixing of the two streams. Hot and cold fluids entering such an exchanger separately leave together as a single mixed stream. Applications include the jet condenser for water vapor and other vapors using a water spray. Direct-contact heat exchangers also include the cooling towers of thermal and nuclear power stations where heat is transferred through direct contact between the hot and cold immiscible fluids.

Some heat exchangers, such as those of nuclear reactors and electrical heaters, include an internal heat source and a cooling fluid to remove the heat energy liberated in the system. A detailed classification of heat exchangers is given by Shah [1].

In this chapter the basic relationships for sizing and rating two-fluid recuperator-type heat exchangers will be discussed. In most heat exchanger applications, three modes of heat transfer occur in series and result in the continuous temperature change of at least one of the fluid media involved.

2.2 ARRANGEMENT OF FLOW PATH IN HEAT EXCHANGERS

A recuperator-type heat exchanger is classified according to the flow direction of the hot and cold fluid streams and the number of passes made by each fluid as it passes through the heat exchanger. Therefore heat exchangers may have the following patterns of flow: (1) parallel flow with two fluids flowing in the same direction (Fig. 2.1*a*), (2) counterflow with two fluids flowing parallel to one another but in opposite directions (Fig. 2.1*b*), (3) crossflow with two fluids crossing each other (Figs. 2.1*c* and *d*), and (4) mixed flow where both fluids are simultaneously in parallel flow, in counterflow (Figs. 2.2*a* and *b*), and in multipass crossflow (Fig. 2.2*c*). Applications include various shell-and-tube heat exchangers.

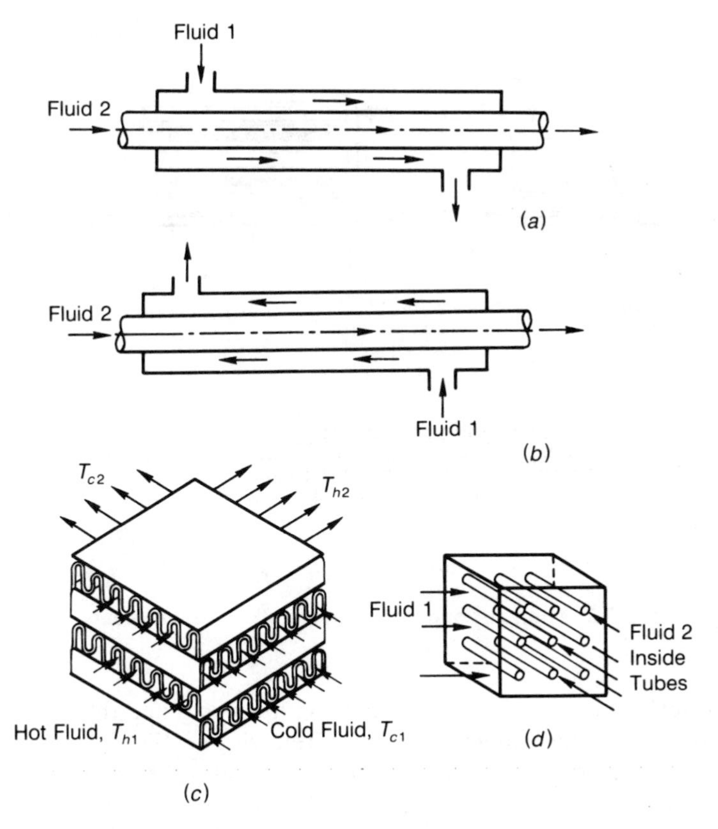

Fig. 2.1. Heat exchanger classification according to flow arrangements.

2.3 BASIC EQUATIONS IN DESIGN

The term *heat exchanger*, although applicable to all four categories listed previously, will be used in this chapter to designate a recuperator in which heat transfer occurs between two fluid streams that do not mix or physically contact each other. Basic heat transfer equations will be outlined for the thermal analysis (sizing and rating calculations) of such heat exchangers. Although complete design of a heat exchanger requires structural and economical considerations in addition to these basic equations, the purpose of the thermal analysis given here will be to determine the heat transfer surface area of the heat exchanger (sizing problem). Performance calculations of a heat exchanger (rating problem) are carried out when the heat exchanger is available, but it is necessary to find the amount of heat transferred, pressure losses, and outlet temperatures of both fluids.

The temperature profiles in usual fluid-to-fluid heat transfer processes, depending on the flow path arrangement, are shown in Fig. 2.3, in which the

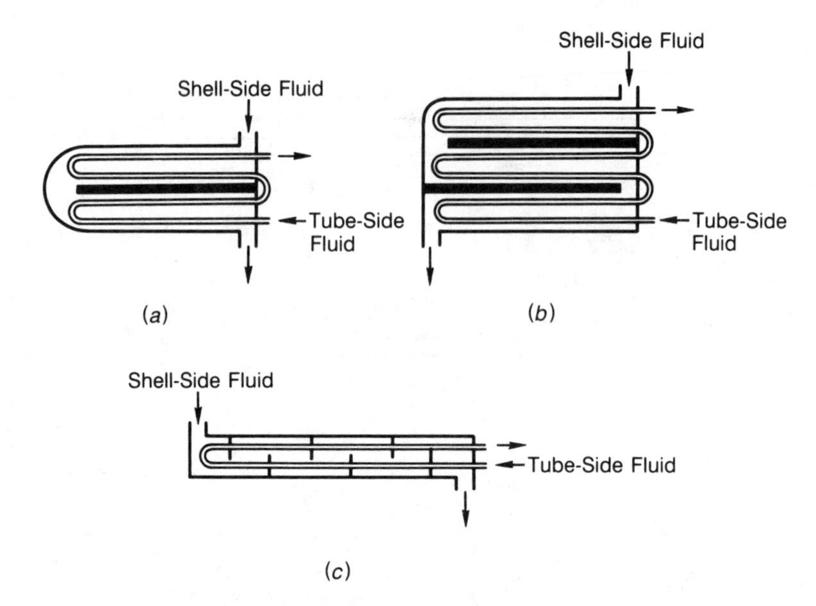

Fig. 2.2. Multipass and multipass crossflow arrangements.

heat transfer surface area A is plotted along the x axis and the temperature of the fluids is plotted along the y axis. Referring to Fig. 2.3, counterflow heat transfer with the two fluids flowing in opposite directions is shown as Fig. 2.3*a*. Parallel-flow heat transfer with the two fluids flowing in the same direction is shown as Fig. 2.3*b*. Heat transfer with the cold fluid at constant temperature (evaporator) is shown as Fig. 2.3*c*. Heat transfer with the hot fluid at constant temperature (condenser) is shown as Fig. 2.3*d*. The nature of the temperature profiles will also depend on the heat capacity ratios ($\dot{m}c_p$) of the fluids and is shown later.

From the first law of thermodynamics for an open system, under steady-state conditions, with negligible potential and kinetic energy changes, the change of enthalpy of one of the fluid streams is (Fig. 2.4)

$$\delta Q = \dot{m}\, di \tag{2.1}$$

where $\dot{m}$ is the rate of mass flow, i is the specific enthalpy, and δQ is the heat transfer rate to the fluid concerned associated with the infinitesimal state change. Integration of Eq. (2.1) gives (Fig. 2.4).

$$Q = \dot{m}(i_2 - i_1) \tag{2.2}$$

where i_1 and i_2 represent the initial and final enthalpies of the fluid stream. Equation (2.2) holds for all processes of Fig. 2.3. Note that δQ is negative for the hot fluid. If there is negligible heat transfer between the exchanger and

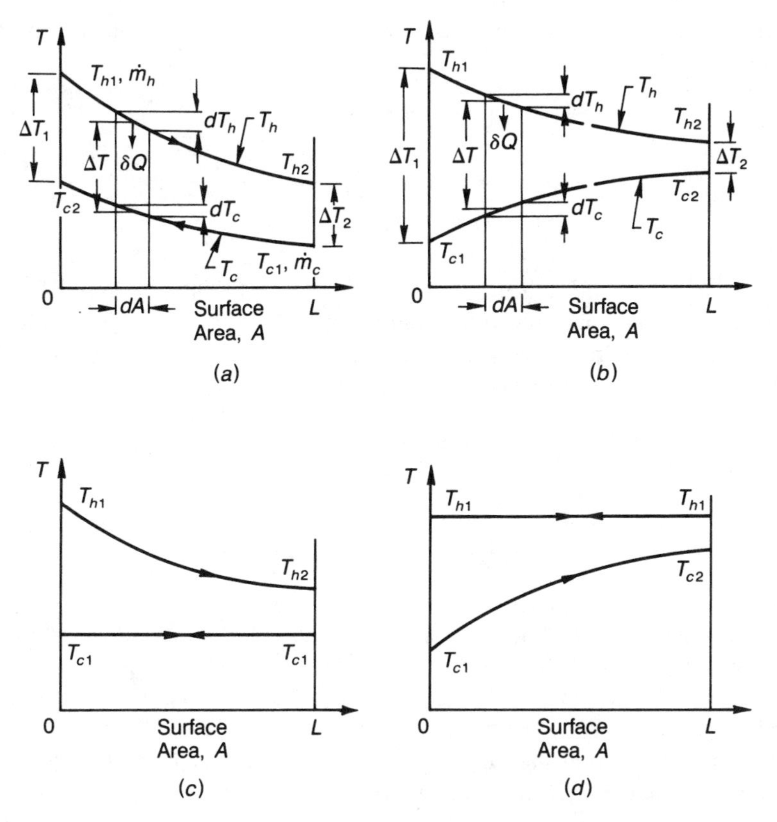

Fig. 2.3. Fluid temperature variation in parallel-flow, counterflow, evaporator, and condenser heat exchangers. (a) Counterflow. (b) Parallel flow. (c) Cold fluid evaporating at constant temperature. (d) Hot fluid condensing at constant temperature.

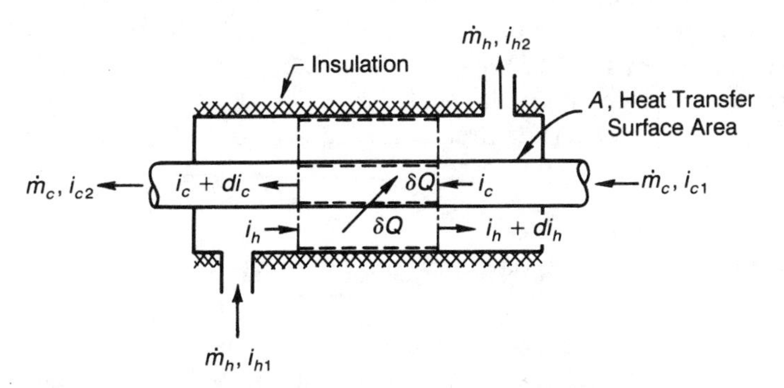

Fig. 2.4. Overall energy balance for the hot and cold fluids of a two-fluid heat exchanger.

its surroundings (adiabatic process), integration of Eq. (2.1) for hot and cold fluids gives

$$Q = \dot{m}_h(i_{h1} - i_{h2}) \tag{2.3}$$

and

$$Q = \dot{m}_c(i_{c2} - i_{c1}) \tag{2.4}$$

The subscripts h and c refer to the hot and cold fluids, where 1 and 2 designate the fluid inlet and outlet conditions. If the fluids do not undergo a phase change and have constant specific heats with $di = c_p\,dT$, then Eqs. (2.3) and (2.4) can be written as

$$Q = (\dot{m}c_p)_h(T_{h1} - T_{h2}) \tag{2.5}$$

and

$$Q = (\dot{m}c_p)_c(T_{c2} - T_{c1}) \tag{2.6}$$

As can be seen from Fig. 2.3, the temperature difference between the hot and cold fluids ($\Delta T = T_h - T_c$) varies with position in the heat exchanger. Therefore, in the heat transfer analysis of heat exchangers, it is convenient to establish an appropriate mean value of the temperature difference between the hot and cold fluids such that the total heat transfer rate Q between the fluids can be determined from the following equation:

$$Q = UA\,\Delta T_m \tag{2.7}$$

where A is the total heat transfer area and U is the average overall heat transfer coefficient based on that area. ΔT_m is a function of T_{h1}, T_{h2}, T_{c1}, and T_{c2}. Therefore a specific form of ΔT_m must be obtained.

Equations (2.5) to (2.7) are the basic equations for the thermal analysis of a heat exchanger under steady-state conditions. If Q, the total heat transfer rate, is known from Eq. (2.5) or (2.6), then Eq. (2.7) is used to calculate the heat transfer surface area A. Therefore it is clear that the problem of calculating the heat transfer area comes down to determining the overall mean temperature difference ΔT_m.

2.4 OVERALL HEAT TRANSFER COEFFICIENT

Heat exchanger walls are usually made of a single material, although the wall may sometimes be bimetallic (steel with aluminum cladding) or coated with a plastic as a protection against corrosion. Most heat exchanger surfaces tend to acquire an additional heat transfer resistance that increases with time.

This may either be a very thin surface oxidation layer, or, at the other extreme, it may be a thick crust deposit, such as that which results from a salt-water coolant in steam condensers. This effect can be taken into consideration by introducing an additional thermal resistance, termed the fouling resistance R_s. Its value depends on the temperature level, fluid velocity, type of surface, and length of service of the heat exchanger [2–4]. Fouling will be discussed in a separate chapter.

In addition, fins are often added to the surfaces exposed to either or both fluids, and, by increasing the surface area, they reduce the resistance to convection heat transfer. The overall heat transfer coefficient for a single smooth and clean plain wall can be written as

$$UA = \frac{1}{R_t} = \frac{1}{\dfrac{1}{h_i A_i} + \dfrac{t}{kA} + \dfrac{1}{h_o A_o}} \tag{2.8}$$

where R_t is the total thermal resistance to heat flow across the surface between the inside and outside flow, t is the thickness of the wall, and h_i and h_o are heat transfer coefficients for inside and outside flow, respectively.

For the unfinned and clean tubular heat exchanger, the overall heat transfer coefficient is given by

$$U_o A_o = U_i A_i = \frac{1}{R_t} = \frac{1}{\left[\dfrac{1}{h_i A_i} + \dfrac{\ln(r_o/r_i)}{2\pi kL} + \dfrac{1}{h_o A_o}\right]} \tag{2.9}$$

If the heat transfer surface is fouled with the accumulation of deposits, this in turn introduces an additional thermal resistance in the path of heat flow. We define a scale coefficient of heat transfer h_s in terms of the thermal resistance R_s of this scale as

$$\frac{\Delta T_s}{Q} = R_s = \frac{1}{Ah_s} \tag{2.10}$$

where the area A is the original heat transfer area of the surface before scaling and ΔT_s is the temperature drop through the scale. $R_f = 1/h_s$ is termed as fouling factor (i.e., unit fouling resistance) which has the unit of $(m^2 \cdot K)/W$. This is discussed in detail in the following chapters and tables are provided for the values of R_f.

We now consider heat transfer across a heat exchanger wall fouled by deposit formation on both the inside and outside surfaces. The total thermal resistance R_t can be expressed as

$$R_t = \frac{1}{UA} = \frac{1}{U_o A_o} = \frac{1}{U_i A_i} = \frac{1}{h_i A_i} + R_w + \frac{R_{fi}}{A_i} + \frac{R_{fo}}{A_o} + \frac{1}{A_o h_o} \tag{2.11}$$

The calculation of an overall heat transfer coefficient depends upon whether it is based on the cold- or hot-side surface area, since $U_o \neq U_i$ if $A_o \neq A_i$. The wall resistance R_w is obtained from the following equations:

$$R_w = \begin{cases} \dfrac{t}{kA} & \text{(for a plane wall)} \quad (2.12a) \\ \dfrac{\ln(r_o/r_i)}{2\pi L k} & \text{(for a tube wall)} \quad (2.12b) \end{cases}$$

A separating wall may be finned differently on each side (Fig. 2.5). On either side, heat transfer takes place from the fins (subscript f in the following equations) as well as from the unfinned portion of the wall (subscript u). Introducing the fin efficiency η_f, the total heat transfer can be expressed as

$$Q = \left(\eta_f A_f h_f + A_u h_u\right) \Delta T \tag{2.13}$$

where ΔT is either $(T_h - T_{w1})$ or $(T_{w2} - T_c)$. The subscripts h and c refer to the hot and cold fluids, respectively.

Taking $h_u = h_f = h$ and rearranging the right-hand side of Eq. (2.13), we get

$$Q = hA\left[1 - \frac{A_f}{A}(1 - \eta_f)\right] \Delta T \tag{2.14}$$

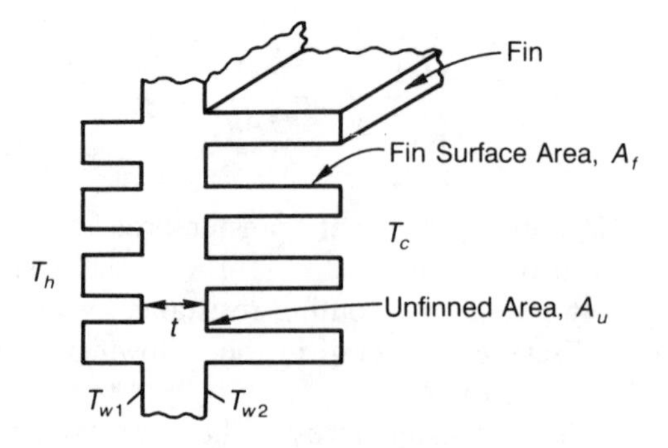

Fig. 2.5. Finned wall.

or

$$Q = \eta_o hA \,\Delta T \tag{2.15}$$

where $\eta_o = [1 - (1 - \eta_f)A_f/A]$ is called the overall surface efficiency and $A = A_u + A_f$.

As can be seen from Eq. (2.15), there will be additional thermal resistances for finned surfaces as $1/\eta_o hA$ on both sides of the finned wall; this is the combined surface resistance of the fin and the unfinned area.

Therefore an overall thermal resistance for the entire wall is then given by

$$R_t = \frac{1}{UA} = \frac{1}{U_o A_o} = \frac{1}{U_i A_i} = \frac{1}{\eta_o h_i A_i} + \frac{R_{fi}}{\eta_o A_i} + R_w + \frac{R_{fo}}{\eta_o A_o} + \frac{1}{\eta_o h_o A_o} \tag{2.16}$$

Contact resistance may be finite between a tube or a plate and the fin. In this case the contact resistance terms on the hot and cold sides are added to Eq. (2.16).

In heat exchanger applications, the overall heat transfer coefficient is usually based on the outer area (cold side or hot side). Then Eq. (2.16) can be represented in terms of the overall heat transfer coefficient based on the outside surface area of the wall as

$$U_o = \frac{1}{\left[\dfrac{A_o}{A_i}\dfrac{1}{\eta_o h_i} + \dfrac{A_o R_{fi}}{\eta_o A_i} + A_o R_w + \dfrac{R_{fo}}{\eta_o} + \dfrac{1}{\eta_o h_o}\right]} \tag{2.17}$$

The expressions or magnitude of η_f for a variety of fin configurations are available in the literature [5]. If a straight or pin fin of length L and uniform thickness is used and an adiabatic tip is assumed, the fin efficiency is given by

$$\eta_f = \frac{\tanh(mL)}{mL} \tag{2.18}$$

where

$$m = \sqrt{\frac{2h}{\delta k_f}} \tag{2.19}$$

where δ is the fin thickness and L is the fin length.

For the unfinned, tubular heat exchangers of Figs. 2.1*a* and *b* and 2.2, Eq. (2.16) reduces to

$$U_o = \frac{1}{\left[\dfrac{r_o}{r_i}\dfrac{1}{h_i} + \dfrac{r_o}{r_i}R_{fi} + \dfrac{r_o \ln(r_o/r_i)}{k} + R_{fo} + \dfrac{1}{h_o}\right]} \tag{2.20}$$

The overall heat transfer coefficient can be determined from knowledge of the inside and outside heat transfer coefficients, fouling factors, and appropriate geometrical parameters.

2.4.1 Order of Magnitude of Thermal Resistances

For a plane wall of thickness t and h_i and h_o on either side with fouling only on one side, Eq. (2.16) becomes

$$\frac{1}{U} = \frac{1}{h_i} + R_{fi} + \frac{t}{k} + \frac{1}{h_o} \tag{2.21}$$

The order of magnitude and range of h for various conditions are given in Table 2.1.

Example 2.1. Determine the overall heat transfer coefficient U for liquid-to-liquid heat transfer through a 0.003-m-thick steel plate [$k = 50$ W/(m · K)] for the following heat transfer coefficients and fouling factor on one side:

$$h_i = 1800 \text{ W/(m}^2\cdot\text{K)} \qquad h_o = 1250 \text{ W/(m}^2\cdot\text{K)}$$

$$R_{fi} = 0.0002 \text{ (m}^2\cdot\text{K)/W}$$

TABLE 2.1 Order of Magnitude of *h*

Fluid	h, W/(m^2 · K)
Gases (natural convection)	5–25
Flowing gases	10–250
Flowing liquids (nonmetal)	100–10,000
Flowing liquid metals	5,000–250,000
Boiling liquids	1,000–250,000
Condensing vapors	1,000–25,000

Substituting into Eq. (2.21), we get

$$\frac{1}{U} = \frac{1}{1800} + 0.0002 + \frac{0.003}{50} + \frac{1}{1250}$$

$$R_t A = 0.00056 + 0.0002 + 0.00006 + 0.0008 = 0.00162 \ (\mathrm{m^2 \cdot K})/\mathrm{W}$$

$$U \cong 617 \ \mathrm{W/(m^2 \cdot K)}$$

In this case none of the resistances is negligible.

Example 2.2. In Example 2.1, replace one of the flowing liquids by a flowing gas [$h_o = 50 \ \mathrm{W/(m^2 \cdot K)}$]:

$$\frac{1}{U} = \frac{1}{1800} + 0.0002 + \frac{0.003}{50} + \frac{1}{50}$$

$$R_t A = 0.00056 + 0.0002 + 0.00006 + 0.02 = 0.02082 \ (\mathrm{m^2 \cdot K})/\mathrm{W}$$

$$U \cong 48 \ \mathrm{W/(m^2 \cdot K)}$$

In this case only the gas side resistance is significant.

Example 2.3. In Example 2.2, replace the flowing liquid by another gas [$h_i = 25 \ \mathrm{W/(m^2 \cdot K)}$]:

$$\frac{1}{U} = \frac{1}{25} + 0.0002 + \frac{0.003}{50} + \frac{1}{50}$$

$$R_t A = 0.04 + 0.0002 + 0.00006 + 0.02 = 0.06026 \ (\mathrm{m^2 \cdot K})/\mathrm{W}$$

$$U \cong 17 \ \mathrm{W/(m^2 \cdot K)}$$

Here the wall and scale resistances are negligible.

2.5 THE LMTD METHOD FOR HEAT EXCHANGER ANALYSIS

In the heat transfer analysis of heat exchangers, the total heat transfer rate Q through the heat exchanger is the quantity of primary interest. Let us consider a simple counterflow or parallel-flow heat exchanger (Figs. 2.3*a* and *b*). The form of ΔT_m in Eq. (2.7) may be determined by applying an energy balance to a differential area element dA in the hot and cold fluids. The temperature of the hot fluid will drop by dT_h. The temperature of the cold fluid will also drop by dT_c over the element dA for counterflow, but it will increase by dT_c for parallel flow if the hot-fluid direction is taken as positive.

Consequently, from the differential forms of Eqs. (2.5) and (2.6) or from Eq. (2.1) for adiabatic, steady-state flow, the energy balance yields

$$\delta Q = -(\dot{m}c_p)_h \, dT_h = \pm(\dot{m}c_p)_c \, dT_c \tag{2.22a}$$

or

$$\delta Q = -C_h \, dT_h = \pm C_c \, dT_c \tag{2.22b}$$

where C_h and C_c are the hot- and cold-fluid heat capacity rates, respectively, and the + refers to parallel flow. The rate of heat transfer δQ from the hot to the cold fluid across the heat transfer area dA may also be expressed as

$$\delta Q = U(T_h - T_c) \, dA \tag{2.23}$$

From Eqs. (2.22) for counterflow, we get

$$d(T_h - T_c) = dT_h - dT_c = \delta Q\left(\frac{1}{C_c} - \frac{1}{C_h}\right) \tag{2.24}$$

Substituting the value of δQ from Eq. (2.23) into Eq. (2.24), we obtain

$$\frac{d(T_h - T_c)}{(T_h - T_c)} = U\left(\frac{1}{C_c} - \frac{1}{C_h}\right) dA \tag{2.25}$$

which, when integrated with constant values of U, C_h, and C_c over the entire length of the heat exchanger, results in

$$\ln \frac{T_{h2} - T_{c1}}{T_{h1} - T_{c2}} = UA\left(\frac{1}{C_c} - \frac{1}{C_h}\right) \tag{2.26a}$$

or

$$T_{h2} - T_{c1} = (T_{h1} - T_{c2})\exp\left[UA\left(\frac{1}{C_c} - \frac{1}{C_h}\right)\right] \tag{2.26b}$$

It is seen that the temperature distribution along the heat exchanger is exponential. Hence in a counterflow heat exchanger the temperature difference $(T_h - T_c)$ decreases in the direction of flow if $C_h < C_c$, but increases if $C_h > C_c$ (Fig. 2.6). The expressions for C_c and C_h can now be obtained from

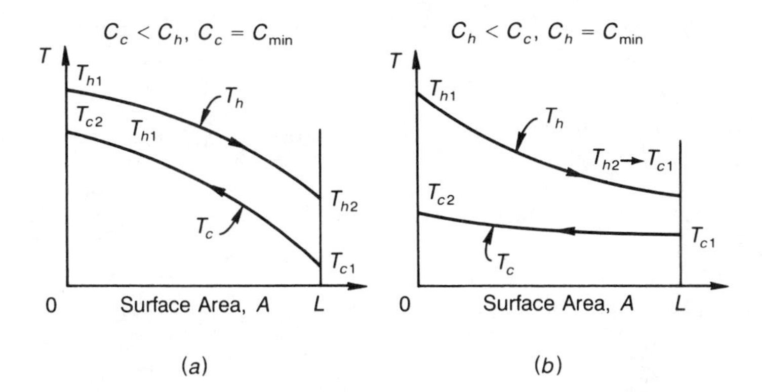

Fig. 2.6. Temperature variation for a counterflow heat exchanger.

Eqs. (2.5) and (2.6) and substituted into Eq. (2.26a). Solving for Q and rearranging, we obtain

$$Q = UA \frac{(T_{h1} - T_{c2}) - (T_{h2} - T_{c1})}{\ln\left(\dfrac{T_{h1} - T_{c2}}{T_{h2} - T_{c1}}\right)} \tag{2.27a}$$

or

$$Q = UA \frac{\Delta T_1 - \Delta T_2}{\ln(\Delta T_1 / \Delta T_2)} \tag{2.27b}$$

where ΔT_1 is the temperature difference between the two fluids at one end of the heat exchanger and ΔT_2 is the temperature difference of the fluids at the other end of the heat exchanger.

Comparison of the preceding expression with Eq. (2.7) reveals that the appropriate average temperature difference between the hot and cold fluids, over the entire length of the heat exchanger, is given by

$$\Delta T_{\mathrm{lm}} = \frac{\Delta T_1 - \Delta T_2}{\ln(\Delta T_1 / \Delta T_2)} \tag{2.28}$$

which is called the log-mean temperature difference (LMTD). Accordingly, the total heat transfer rate between the hot and cold fluids for all single-pass flow arrangements shown in Fig. 2.3 is determined from

$$Q = AU \Delta T_{\mathrm{lm}} \tag{2.29}$$

In the case of counterflow with $(\dot{m}c_p)_h = (\dot{m}c_p)_c$, the quantity ΔT_{lm} is indeterminate since

$$(T_{h1} - T_{h2}) = (T_{c2} - T_{c1}) \qquad \text{and} \qquad \Delta T_1 = \Delta T_2 \tag{2.30}$$

In this case, it can be shown using L'Hospital's rule that $\Delta T_{\text{lm}} = \Delta T_1 = \Delta T_2$ and therefore

$$Q = UA(T_h - T_c) \quad \text{with } (T_h - T_c) = \Delta T_1 = \Delta T_2 \tag{2.31}$$

Starting with Eq. (2.22) for a parallel-flow arrangement, it can be shown that Eq. (2.27b) is also applicable. However, for a parallel-flow heat exchanger, the endpoint temperature differences must now be defined as $\Delta T_1 = (T_{h1} - T_{c1})$ and $\Delta T_2 = (T_{h2} - T_{c2})$.

Note that, for the same inlet and outlet temperatures, the log-mean temperature difference for counterflow exceeds that for parallel flow, $\Delta T_{\text{lm, cf}} > \Delta T_{\text{lm, pf}}$; that is, LMTD represents the maximum temperature potential for heat transfer that can only be obtained in a counterflow exchanger. Hence the surface area required to affect a prescribed heat transfer rate Q is smaller for a counterflow arrangement than that for a parallel-flow arrangement, assuming the same value of U. Also note that T_{c2} can exceed T_{h2} for counterflow but not for parallel flow.

2.5.1 Multipass and Crossflow Heat Exchangers

The LMTD developed previously is not applicable for heat transfer analysis of crossflow and multipass exchangers. The integration of Eq. (2.23) for these flow arrangements results in a form of an integrated mean temperature difference ΔT_m such that

$$Q = UA\,\Delta T_m \tag{2.32}$$

where ΔT_m is the true (or effective) mean temperature difference and it is a complex function of T_{h1}, T_{h2}, T_{c1}, and T_{c2}. Generally this function ΔT_m can be determined analytically in terms of the following quantities [6, 7]:

$$\Delta T_{\text{lm, cf}} = \frac{(T_{h2} - T_{c1}) - (T_{h1} - T_{c2})}{\ln[(T_{h2} - T_{c1})/(T_{h1} - T_{c2})]} \tag{2.33}$$

$$P = \frac{T_{c2} - T_{c1}}{T_{h1} - T_{c1}} = \frac{\Delta T_c}{\Delta T_{\max}} \tag{2.34}$$

and

$$R = \frac{C_c}{C_h} = \frac{T_{h1} - T_{h2}}{T_{c2} - T_{c1}} \tag{2.35}$$

where $\Delta T_{\mathrm{lm,cf}}$ is the log-mean temperature difference for a counterflow arrangement with the same fluid inlet and outlet temperatures. P is a measure of the ratio of the heat actually transferred to the cold fluid to the heat which would be transferred if the same fluid were to be raised to the hot-fluid inlet temperature; therefore P is the temperature effectiveness of the heat exchanger on the cold-fluid side. R is the ratio of the $\dot{m}c_p$ value of the cold fluid to that of the hot fluid and it is called the heat capacity rate ratio (regardless of which fluid is the tube-side or shell-side fluid in the shell-and-tube heat exchanger).

For design purposes, Eq. (2.29) can also be used for multipass and crossflow heat exchangers with a LMTD correction factor F:

$$Q = UAF\,\Delta T_{\mathrm{lm,cf}} \tag{2.36}$$

F is nondimensional; it depends on the temperature effectiveness P, the heat capacity rate ratio R, and the flow arrangement

$$F = \phi(P, R, \text{flow arrangement}) \tag{2.37}$$

The correction factors are available in chart form as prepared by Bowman et al. [6, 7] for practical use for all common multipass shell-and-tube and crossflow heat exchangers and selected results are presented in Figs. 2.7 to 2.11. In calculating P and R to determine F, it is immaterial whether the colder fluid flows through the shell or inside the tubes.

The correction factor F is less than 1 for crossflow and multipass arrangements; it is 1 for a true counterflow heat exchanger. It represents the degree of departure of the true mean temperature difference from the LMTD for a counterflow arrangement.

In a multipass or a crossflow arrangement, the fluid temperature may not be uniform at a particular distance in the exchanger unless the fluid is well mixed along the path length. For example, in crossflow (Fig. 2.12) the hot and cold fluids may enter at uniform temperatures, but if there are channels in the flow path to prevent mixing, the exit temperature distributions will be as shown in Fig. 2.12. If such channels are not present, the fluids may be well mixed along the path length and the exit temperatures are more nearly uniform as in the flow normal to the tube bank in Fig. 1*d*. A similar stratification of temperatures occurs in the shell-and-tube multipass exchanger. A series of baffles may be required if mixing of the shell fluid is to be obtained. Charts are presented for both mixed and unmixed fluids in Figs. 2.13 and 2.14.

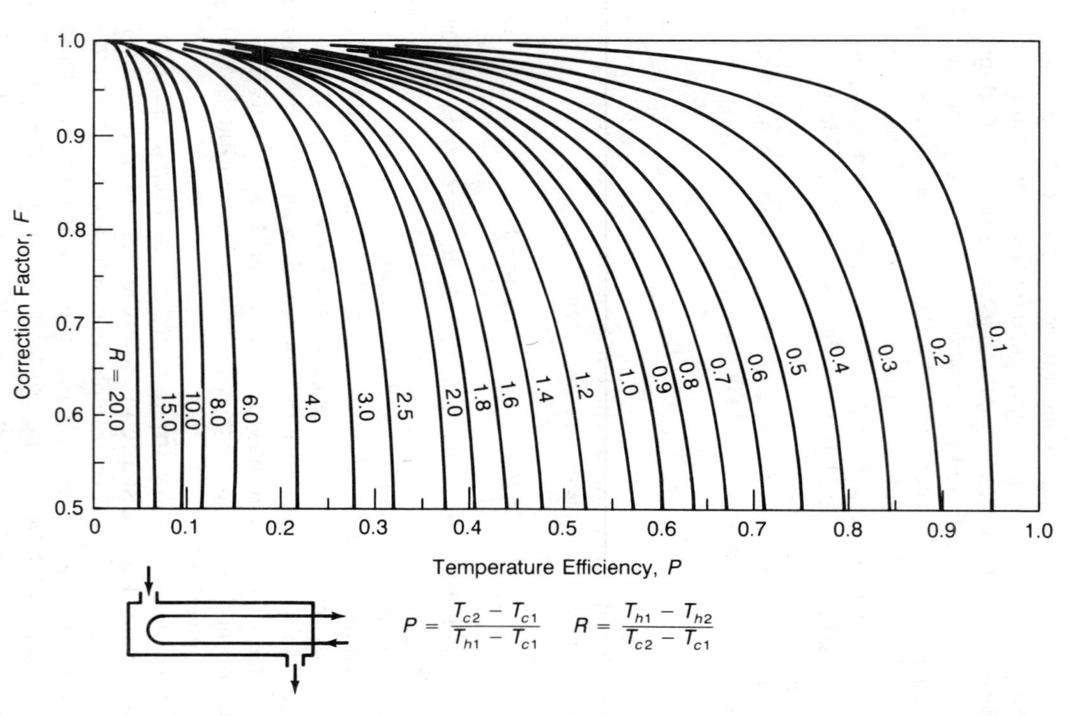

Fig. 2.7. LMTD correction factor F for a shell-and-tube heat exchanger—one shell pass and two or multiple of two tube passes [7].

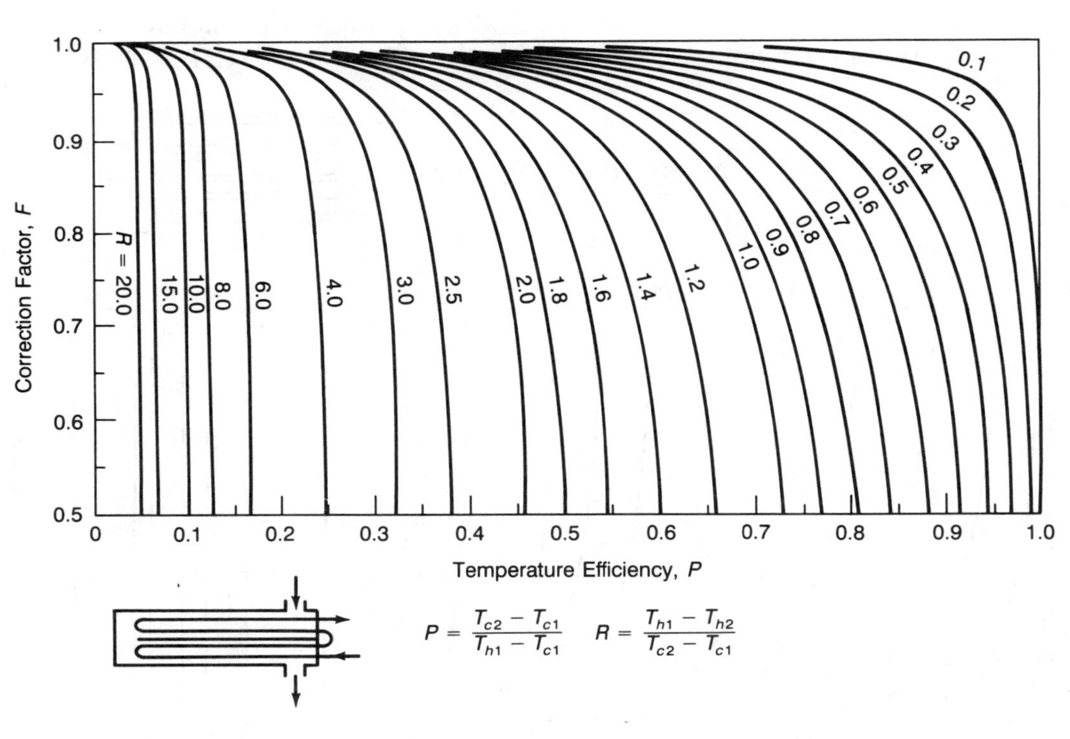

Fig. 2.8. LMTD correction factor F for a shell-and-tube heat exchanger—two shell passes and four or multiple of four tube passes [7].

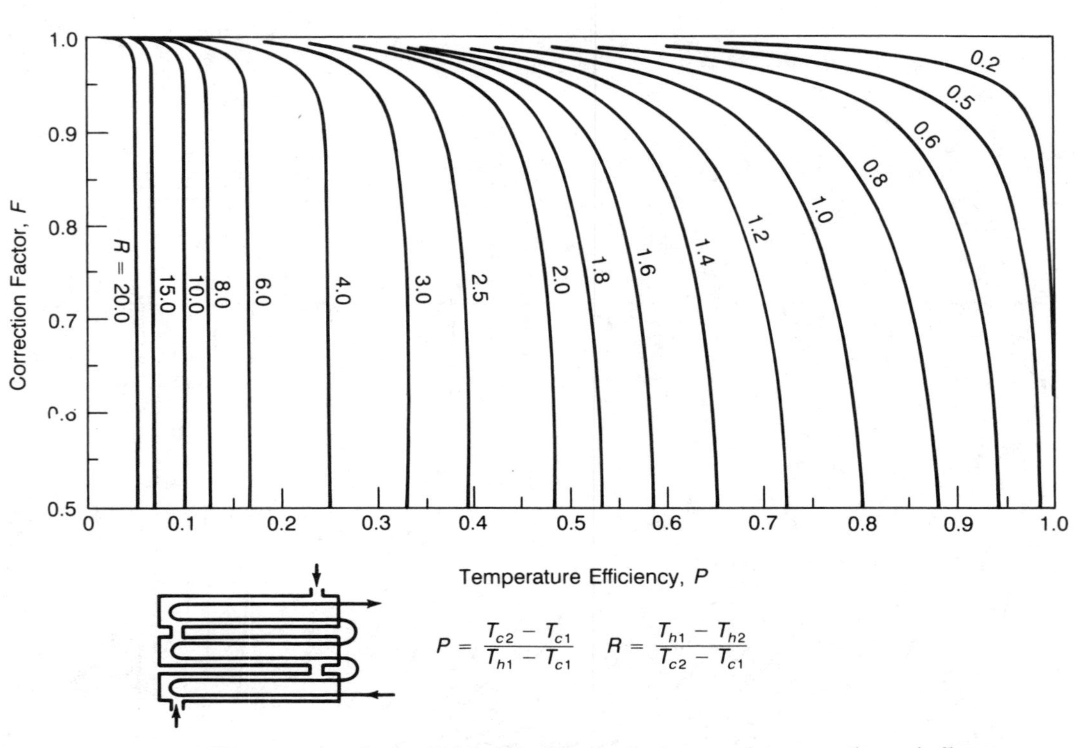

Fig. 2.9. LMTD correction factor F for a shell-and-tube heat exchanger—three shell passes and six or more even number of tube passes [7].

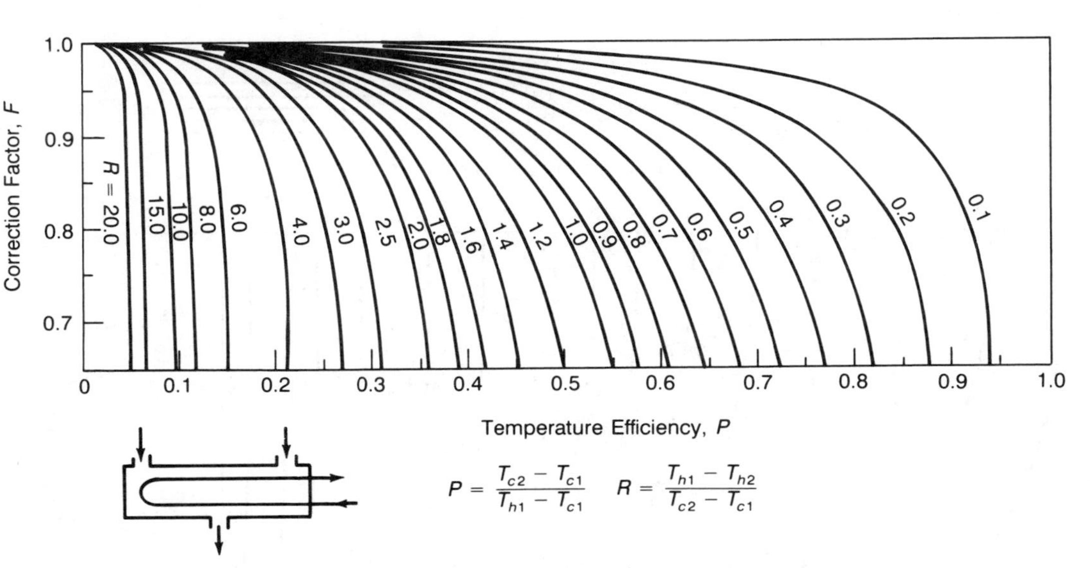

Fig. 2.10. LMTD correction factor F for a divided-flow shell-type heat exchanger—one divided-flow shell pass and even number of tube passes [7].

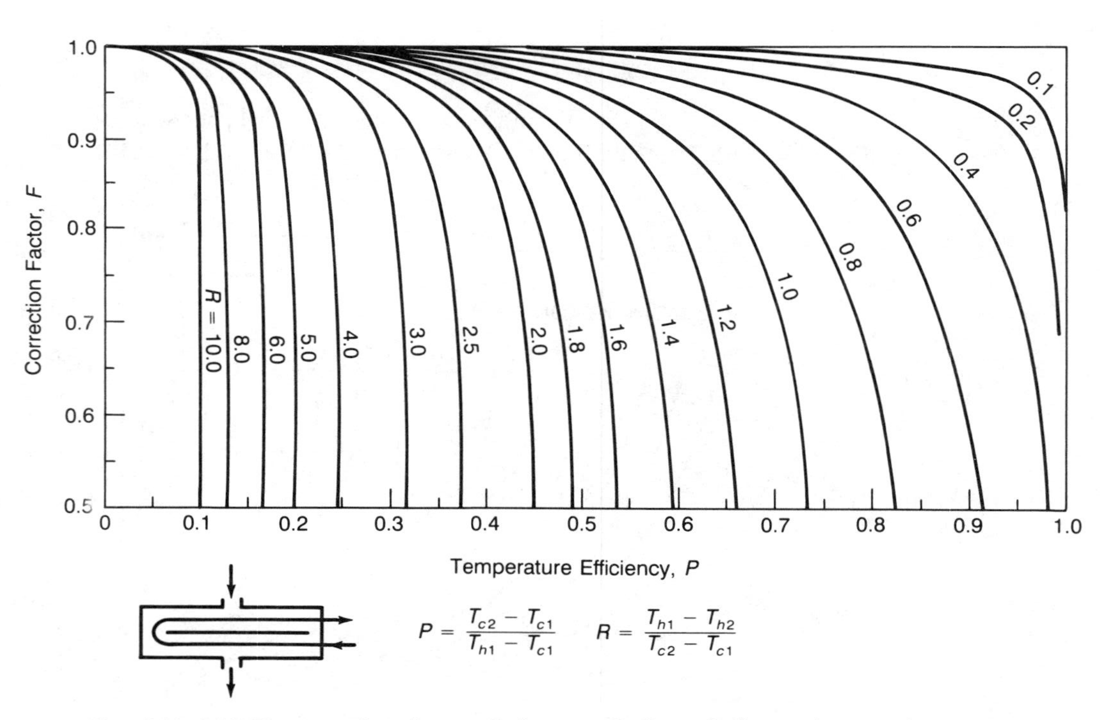

Fig. 2.11. LMTD correction factor F for a split-flow shell-type heat exchanger—one split-flow shell pass and two tube passes [7].

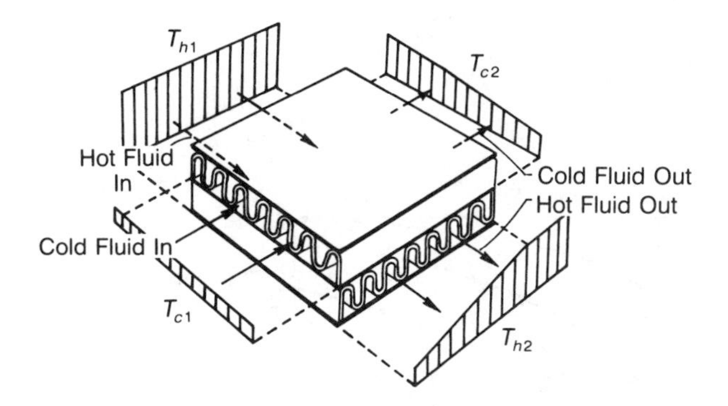

Fig. 2.12. Temperature distribution in a crossflow heat exchanger.

The preceding analysis assumed U to be uniform throughout the heat exchanger. If U is not uniform, the heat exchanger calculations may be made by subdividing the heat exchanger into sections over which U is nearly uniform and by applying the previously developed relations to each subdivision (see Section 2.10).

An important implication of Figs. 2.7 to 2.11, 2.13, and 2.14 is that, if the temperature change of one fluid is negligible, either P or R is 0 and F is 1. Hence heat exchanger behavior is independent of specific configuration. Such would be the case if one of the fluids underwent a phase change. We note from Figs. 2.7 to 2.11, 2.13, and 2.14 that the value of temperature effectiveness P ranges from 0 to 1. The value of R ranges from 0 to ∞, with 0 corresponding to pure vapor condensation, and ∞ to evaporation. It should be noted that a value of F close to 1 does not mean a highly efficient heat exchanger; it means a close approach to the counterflow behavior for comparable operating conditions of flow rates and inlet fluid temperatures.

2.6 THE ε – NTU METHOD FOR HEAT EXCHANGER ANALYSIS

When the inlet or outlet temperatures of the fluid streams are not known, a trial-and-error procedure could be applied for using the LMTD method in the thermal analysis of heat exchangers in order to determine the value of LMTD which will satisfy the requirement that the heat transferred in the heat exchanger [Eq. (2.7)] be equal to the heat convected to the fluid [Eq. (2.5) or (2.6)]. In these cases, to avoid a trial-and-error procedure, the method of the number of transfer units (NTU) based on the concept of a heat exchanger effectiveness may be used. The method is based on the fact that the inlet or exit temperature differences of a heat exchanger are a function of UA/C_c and C_c/C_h [see Eq. (2.25)].

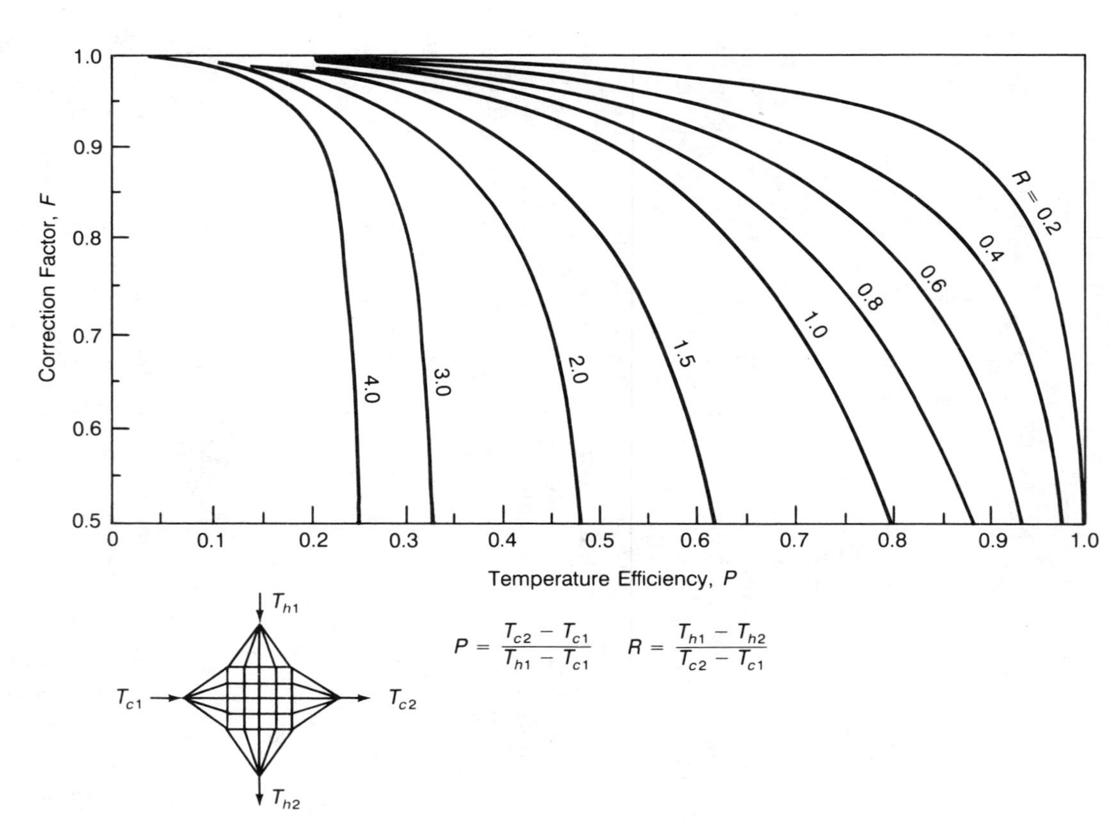

Fig. 2.13. LMTD correction factor F for a crossflow heat exchanger with both fluids unmixed [7].

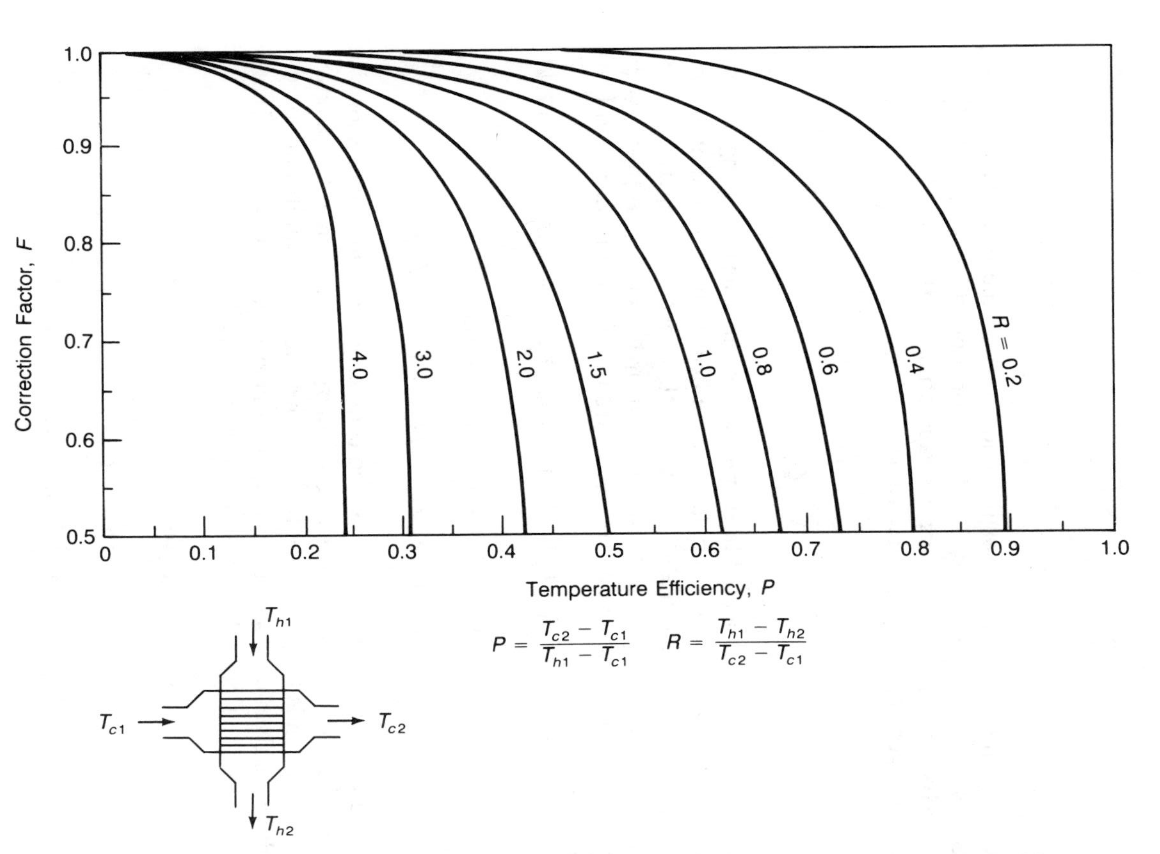

Fig. 2.14. LMTD correction factor F for a single-pass crossflow heat exchanger with one fluid mixed and the other unmixed.

The heat exchanger heat transfer equations such as Eqs. (2.3), (2.4), and (2.26) may be written in dimensionless form resulting in the following dimensionless groups [8].

1. Capacity rate ratio:

$$C^* = \frac{C_{\min}}{C_{\max}} \tag{2.38}$$

where $C_{\min}$ and $C_{\max}$ are the smaller and larger of the two magnitudes C_h and C_c, respectively, and $C^* \leq 1$. $C^* = 0$ corresponds to a finite $C_{\min}$ and $C_{\max}$ approaching ∞ (a condensing or evaporating fluid).

2. Exchanger heat transfer effectiveness:

$$\varepsilon = \frac{Q}{Q_{\max}} \tag{2.39}$$

which is the ratio of the actual heat transfer rate in a heat exchanger to the thermodynamically limited maximum possible heat transfer rate if an infinite heat transfer area were available in a counterflow heat exchanger.

The actual heat transfer is obtained either by the energy given off by the hot fluid or the energy received by the cold fluid, from Eqs. (2.5) and (2.6):

$$Q = (\dot{m}c_p)_h(T_{h1} - T_{h2}) = (\dot{m}c_p)_c(T_{c2} - T_{c1}) \tag{2.40}$$

$$\text{If } C_h > C_c, \text{ then } (T_{h1} - T_{h2}) < (T_{c2} - T_{c1})$$

$$\text{If } C_h < C_c, \text{ then } (T_{h1} - T_{h2}) > (T_{c2} - T_{c1})$$

These are valid for the parallel-flow and counterflow arrangements shown in Fig. 2.3. Therefore the fluid that might undergo the maximum temperature difference, which is the difference between the inlet temperatures of hot and cold fluids, is the fluid having the minimum heat capacity rate $C_{\min}$. Therefore the maximum possible heat transfer is expressed as

$$Q_{\max} = (\dot{m}c_p)_c(T_{h1} - T_{c1}) \quad \text{if } C_c < C_h \tag{2.41a}$$

or

$$Q_{\max} = (\dot{m}c_p)_h(T_{h1} - T_{c1}) \quad \text{if } C_h < C_c \tag{2.41b}$$

which can be obtained with a counterflow heat exchanger (Fig. 2.5). Heat exchanger effectiveness is therefore written as

$$\varepsilon = \frac{C_h(T_{h1} - T_{h2})}{C_{\min}(T_{h1} - T_{c1})} = \frac{C_c(T_{c2} - T_{c1})}{C_{\min}(T_{h1} - T_{c1})} \tag{2.42}$$

The first definition is for $C_h = C_{\min}$ and the second for $C_c = C_{\min}$. Equation (2.42) is valid for all heat exchanger flow arrangements. The value of ε ranges between 0 and 1.

For given ε and $Q_{\max}$, the actual heat transfer rate Q from Eq. (2.39) is

$$Q = \varepsilon(\dot{m}c_p)_{\min}(T_{h1} - T_{c1}) \tag{2.43}$$

If the effectiveness ε of the exchanger is known, Eq. (2.43) provides an explicit expression for the determination of Q.

3. Heat transfer area number:

$$\text{NTU} = \frac{AU}{C_{\min}} = \frac{1}{C_{\min}}\int_A U\,dA \tag{2.44}$$

If U is not constant, the definition of second equality applies. NTU designates the nondimensional heat transfer size of the heat exchanger.

Let us consider a single-pass heat exchanger, assuming $C_c > C_h$, so that $C_h = C_{\min}$ and $C_c = C_{\max}$. With Eq. (2.44), Eq. (2.26b) may be written as

$$T_{h2} - T_{c1} = (T_{h1} - T_{c2})\exp\left[-\text{NTU}\left(\pm 1 - \frac{C_{\min}}{C_{\max}}\right)\right] \tag{2.45}$$

where the + is for counterflow and the − is for parallel flow. With Eqs. (2.5), (2.6), and (2.42), T_{h2} and T_{c2} in Eq. (2.45) can be eliminated and the following expression is obtained for ε for counterflow:

$$\varepsilon = \frac{1 - \exp[-\text{NTU}(1 - C_{\min}/C_{\max})]}{1 - (C_{\min}/C_{\max})\exp[-\text{NTU}(1 - C_{\min}/C_{\max})]} \tag{2.46}$$

If $C_c < C_h$ ($C_c = C_{\min}$, $C_h = C_{\max}$), the result will be the same.

In the case of parallel flow, a similar analysis may be applied to obtain the following expression:

$$\varepsilon = \frac{1 - \exp[-\text{NTU}(1 + C_{\min}/C_{\max})]}{1 + C_{\min}/C_{\max}} \tag{2.47}$$

Two limiting cases are of interest: $C_{\min}/C_{\max}$ equal to 1 or 0. For $C_{\min}/C_{\max} = 1$, Eq. (2.46) is indeterminate, but by applying L'Hospital's rule to Eq. (2.46), the following result is obtained: For $(C_{\min}/C_{\max}) = 1$ for counterflow

$$\varepsilon = \frac{\text{NTU}}{1 + \text{NTU}} \tag{2.48}$$

TABLE 2.2 The ε–NTU Expressions and Limiting Values of ε for $C^* = 1$ and NTU → ∞ for Various Exchanger Flow Arrangements [3, 9]

Flow Arrangement	ε–NTU Formulas
Counterflow	$\varepsilon = \dfrac{1 - \exp[-\text{NTU}(1 - C^*)]}{1 - C^* \exp[-\text{NTU}(1 - C^*)]}$
Parallel Flow	$\varepsilon = \dfrac{1 - \exp[-\text{NTU}(1 + C^*)]}{1 + C^*}$
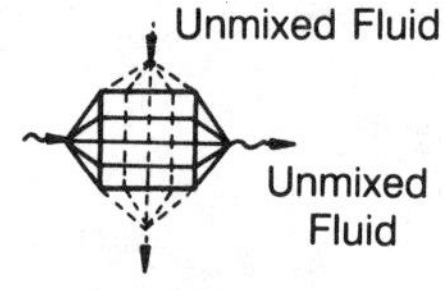 Crossflow, both fluids unmixed	$\varepsilon = 1 - \exp\left[-(1 + C^*)\text{NTU}\right]\left[I_0\left(2\text{NTU}\sqrt{C^*}\right) + \sqrt{C^*}\, I_1\left(2\text{NTU}\sqrt{C^*}\right) - \dfrac{1 - C^*}{C^*} \sum_{n=2}^{\infty} C^{*n/2} I_n\left(2\text{NTU}\sqrt{C^*}\right)\right]$
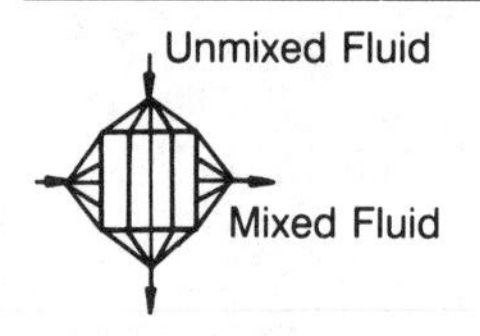 Crossflow, one fluid mixed, other unmixed,	For C_{min} mixed and C_{max} unmixed, $\varepsilon = 1 - \exp[-\{1 - \exp(-\text{NTU} \cdot C^*)\}/C^*]$ For C_{max} mixed and C_{min} unmixed, $\varepsilon = \dfrac{1}{C^*}[1 - \exp\{-C^*[1 - \exp(-\text{NTU})]\}]$
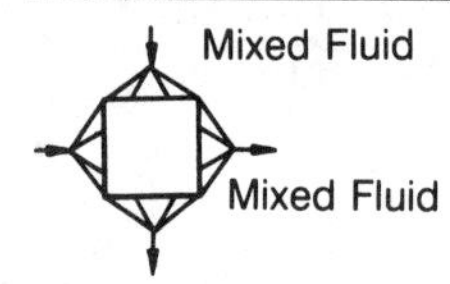 Crossflow, both fluids mixed	$\varepsilon = \dfrac{1}{\dfrac{1}{1 - \exp(-\text{NTU})} + \dfrac{C^*}{1 - \exp(-\text{NTU} \cdot C^*)} - \dfrac{1}{\text{NTU}}}$
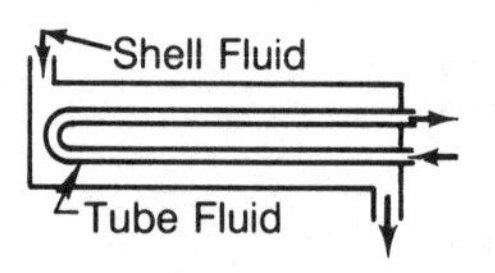 1–2 shell-and-tube exchanger, shell fluid mixed, TEMA *E* shell	$\varepsilon = \dfrac{2}{(1 + C^*) + (1 + C^{*2})^{1/2} \coth(\Gamma/2)}$ where $\Gamma = \text{NTU}[1 + C^{*2}]^{1/2}$ $\coth(\Gamma/2) = (1 + e^{-\Gamma})/(1 - e^{-\Gamma})$
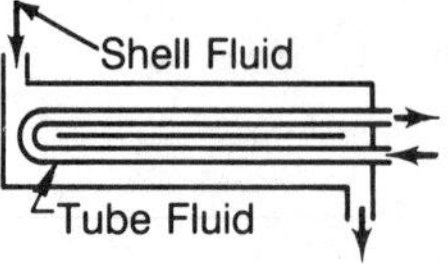 1–2 shell-and-tube exchanger, shell fluid unmixed, TEMA *E* shell	If $C_{min} = C_{tube}$ and $C_{max} = C_{shell}$, $\varepsilon = 1 - \dfrac{2C^* - 1}{2C^* + 1}\left[\dfrac{2C^* + \exp\{-\text{NTU}(C^* + 1/2)\}}{2C^* - \exp\{-\text{NTU}(C^* - 1/2)\}}\right]$ If $C_{min} = C_{shell}$ and $C_{max} = C_{tube}$ $\varepsilon = \dfrac{1}{C^*} - \dfrac{2 - C^*}{C^*(2 + C^*)}\left[\dfrac{2 + C^* \exp\{-\text{NTU}(1 + C^*/2)\}}{2 - C^* \exp\{-\text{NTU}(1 - C^*/2)\}}\right]$

ε–NTU Formulas for $C^* = 1$	Asymptotic Value of ε when NTU $\to \infty$
$\varepsilon = \dfrac{\text{NTU}}{1 + \text{NTU}}$	$\varepsilon = 1$ for all C^*
$\varepsilon = \frac{1}{2}[1 - \exp(-2\text{NTU})]$	$\varepsilon = \dfrac{1}{1 + C^*}$
$\varepsilon = 1 - [I_0(2\text{NTU}) + I_1(2\text{NTU})]e^{-2\text{NTU}}$	$\varepsilon = 1$ for all C^*
$\varepsilon = 1 - \exp[-\{1 - \exp(-\text{NTU})\}]$	For $C_{\min}$ mixed, $\varepsilon = 1 - \exp(-1/C^*)$
$\varepsilon = 1 - \exp[-\{1 - \exp(-\text{NTU})\}]$	For $C_{\max}$ mixed, $\varepsilon = [1 - \exp(-C^*)]/C^*$
$\varepsilon = \dfrac{1}{\dfrac{2}{1 - \exp(-\text{NTU})} - \dfrac{1}{\text{NTU}}}$	$\varepsilon = \dfrac{1}{1 + C^*}$
$\varepsilon = \dfrac{2}{2 + \sqrt{2}\coth(\Gamma/2)}$ where $\Gamma = \sqrt{2}\,\text{NTU}$	$\varepsilon = \dfrac{2}{(1 + C^*) + (1 + C^{*2})^{1/2}}$
If $C_{\min} = C_{\text{tube}}$ and $C^* = 1/2$: $\varepsilon = 1 - \dfrac{1 + e^{-\text{NTU}}}{2 + \text{NTU}}$; If $C^* = 1$: $\varepsilon = 1 - \dfrac{1}{3}\left[\dfrac{2 + \exp(-\frac{3}{2}\text{NTU})}{2 - \exp(-\frac{1}{2}\text{NTU})}\right]$	If $C_{\min} = C_{\text{tube}}$: $\varepsilon = \begin{cases} 2/(1 + 2C^*) & \text{for } C^* \ge 0.5 \\ 1 & \text{for } C^* < 0.5 \end{cases}$; If $C_{\min} = C_{\text{shell}}$: $\varepsilon = \dfrac{2}{2 + C^*}$

TABLE 2.2. (*Continued*)

Flow Arrangement	ε–NTU Formulas
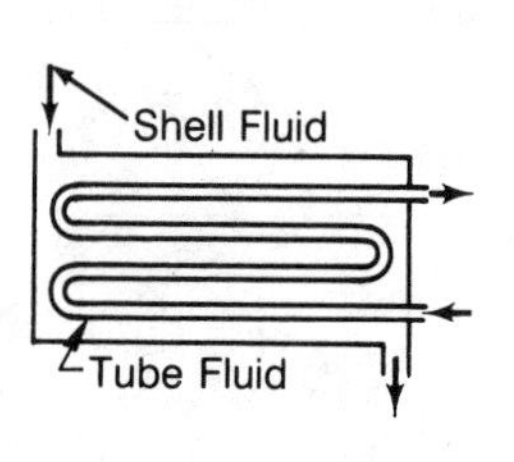 1–4 shell-and-tube exchanger, shell fluid mixed, TEMA E shell	If $C_{\min} = C_{\text{tube}}$ and $C_{\max} = C_{\text{shell}}$ $\varepsilon = 4/\left[2(1 + C^*) + (1 + 4C^{*2})^{1/2}\coth(\Gamma/4) + \tanh(\text{NTU}/4)\right]$ where $\Gamma = \text{NTU}(1 + 4C^{*2})^{1/2}$ If $C_{\min} = C_{\text{shell}}$ and $C_{\max} = C_{\text{tube}}$ $\varepsilon = 4/\left[2(1 + C^*) + (4 + C^{*2})^{1/2}\coth(\Gamma'/4) + C^*\tanh(\text{NTU} \cdot C^*/4)\right]$ where $\Gamma' = \text{NTU}(4 + C^{*2})^{1/2}$
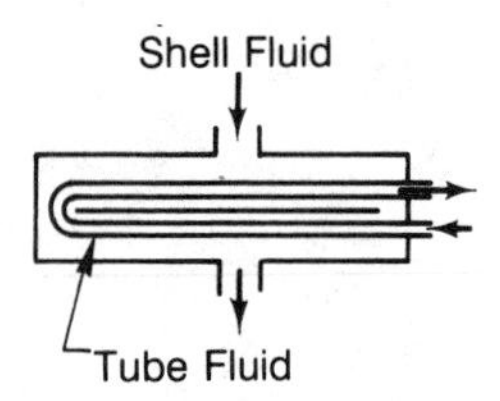 1–2 split-flow exchanger, shell fluid mixed, TEMA G shell	If $C_{\min} = C_{\text{tube}}$ and $C_{\max} = C_{\text{shell}}$ $\varepsilon = \dfrac{(1 + G + 2C^*G) + (2C^* + 1)De^{-\alpha} - e^{-\alpha}}{(1 + G + 2C^*G) + 2C^*(1 - D) + 2C^*De^{-\alpha}}$ where $D = \dfrac{1 - e^{-\alpha}}{2C^* + 1}$, $G = \dfrac{1 - e^{-\beta}}{2C^* - 1} \xrightarrow[C^*=0.5]{} \text{NTU}_t/2$ $\alpha = \frac{1}{4}\text{NTU}(2C^* + 1)$, $\beta = \frac{1}{2}\text{NTU}(2C^* - 1)$ If $C_{\min} = C_{\text{shell}}$ and $C_{\max} = C_{\text{tube}}$, use the preceding formula with C^* replaced by $1/C^*$, NTU replaced by $\text{NTU} \cdot C^*$, and ε replaced by εC^*
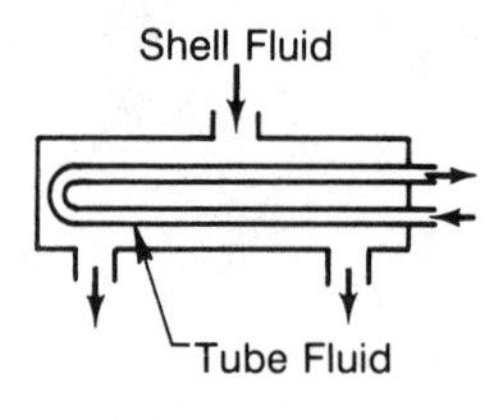 1–2 divided-flow exchanger, shell fluid mixed, TEMA J shell	If $C_{\min} = C_{\text{tube}}$ and $C_{\max} = C_{\text{shell}}$ $\varepsilon = \dfrac{2}{1 + 2C^*\Phi'}$ where $\Phi' = 1 + \gamma\left(\dfrac{1 + \Phi}{1 - \Phi}\right) - 2\gamma\left\{\dfrac{\gamma\Phi + (1 - \Phi)e^{-\text{NTU}C^*(\gamma-1)/2}}{(1 - \Phi)^2 + \gamma(1 - \Phi^2)}\right\}$ $\Phi = \exp(-\text{NTU}C^*\gamma)$, $\gamma = (1 + 4C^{*2})^{1/2}/2C^*$ If $C_{\min} = C_{\text{shell}}$ and $C_{\max} = C_{\text{tube}}$, use the preceding formula with C^* replaced by $1/C^*$, NTU replaced by $\text{NTU}C^*$, and ε replaced by εC^*

ε–NTU Formulas for $C^* = 1$	Asymptotic Value of ε when NTU $\to \infty$
$\varepsilon = \dfrac{4}{4 + \sqrt{5}\coth(\Gamma/4) + \tanh(\text{NTU}/4)}$ where $\Gamma = \sqrt{5}\,\text{NTU}$	If $C_{min} = C_{tube}$ $\varepsilon = \dfrac{4}{2(1 + C^*) + (1 + 4C^{*2})^{1/2} + 1}$
$\varepsilon = \dfrac{4}{4 + \sqrt{5}\coth(\Gamma'/4) + \tanh(\text{NTU}/4)}$ where $\Gamma' = \sqrt{5}\,\text{NTU}$	If $C_{min} = C_{shell}$ $\varepsilon = \dfrac{4}{2(1 + C^*) + (4 + C^{*2})^{1/2} + C^*}$
$\varepsilon = \dfrac{4 - e^{-\text{NTU}/2} - e^{-3\text{NTU}/2}}{4 - 3e^{-\text{NTU}/2} + \frac{2}{3}(2 + 2e^{-3\text{NTU}/4}) - e^{-3\text{NTU}/2}}$	For $C^* > \frac{1}{2}$ $\varepsilon = \dfrac{2C^* + 1}{2C^{*2} + C^* + 1}$ For $C^* \leq \frac{1}{2}$ $\varepsilon = 1$
Same as the preceding formula	For $C^* < 2$ $\varepsilon = \dfrac{C^* + 2}{C^{*2} + C^* + 2}$ For $C^* \geq 2$, $\varepsilon = 1/C^*$
$\varepsilon = \dfrac{2}{1 + 2\Phi'}$ where $\Phi' = 1 + \gamma\left(\dfrac{1 + \Phi}{1 - \Phi}\right) - 2\gamma\left\{\dfrac{\gamma\Phi + (1 - \Phi)e^{-\text{NTU}(\gamma - 1)/2}}{(1 - \Phi)^2 + \gamma(1 - \Phi^2)}\right\}$ $\Phi = \exp(-\text{NTU}\gamma)$, $\gamma = \sqrt{5}/2$	If $C_{min} = C_{tube}$ $\varepsilon = \dfrac{2}{1 + 2C^* + (1 + 4C^{*2})^{1/2}}$
Same as the preceding formula	If $C_{min} = C_{shell}$ $\varepsilon = \dfrac{2}{C^* + 2 + (C^{*2} + 4)^{1/2}}$

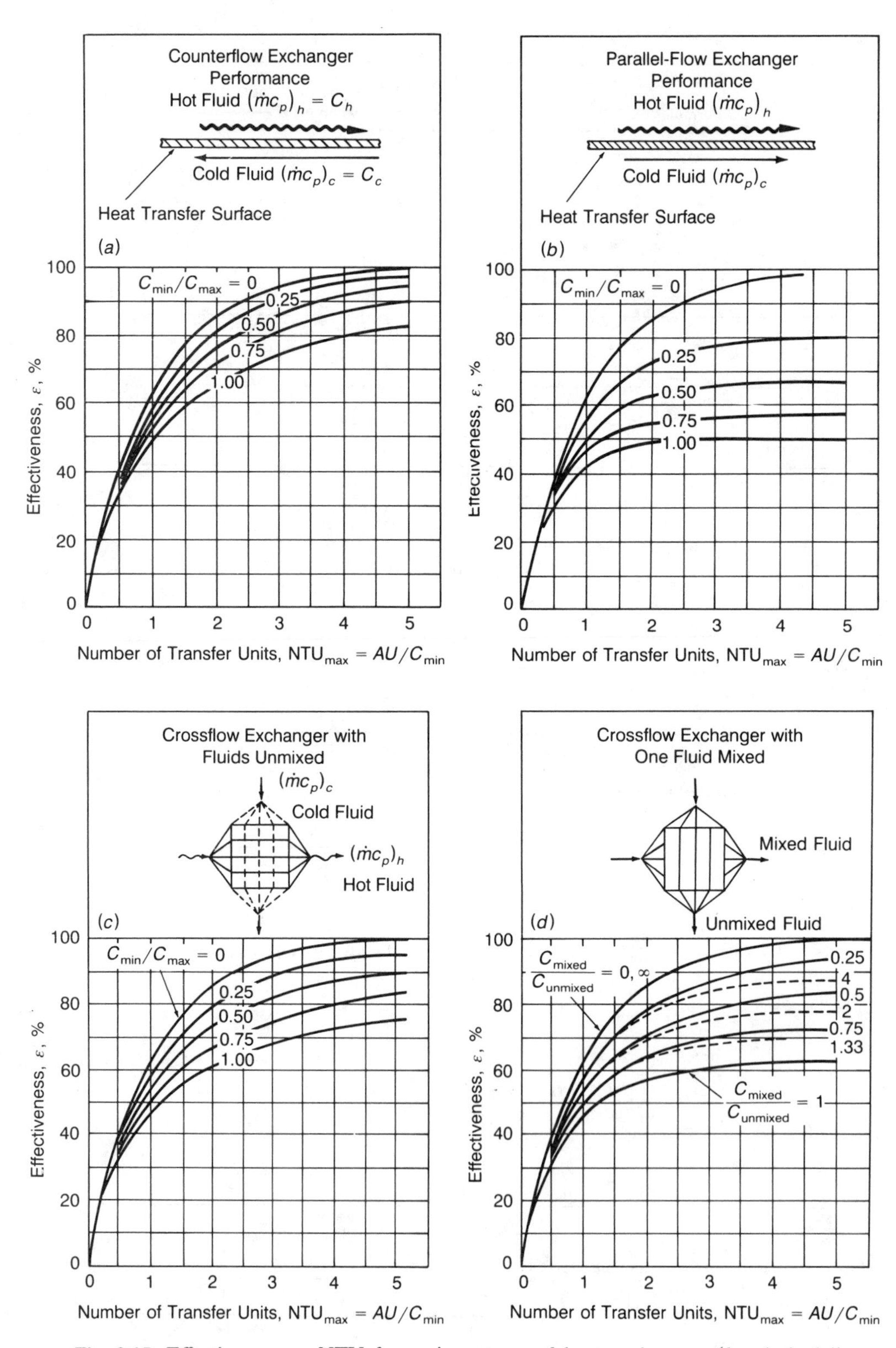

Fig. 2.15. Effectiveness vs. NTU for various types of heat exchangers (for dashed lines $C_{min} = C_{unmixed}$) [8].

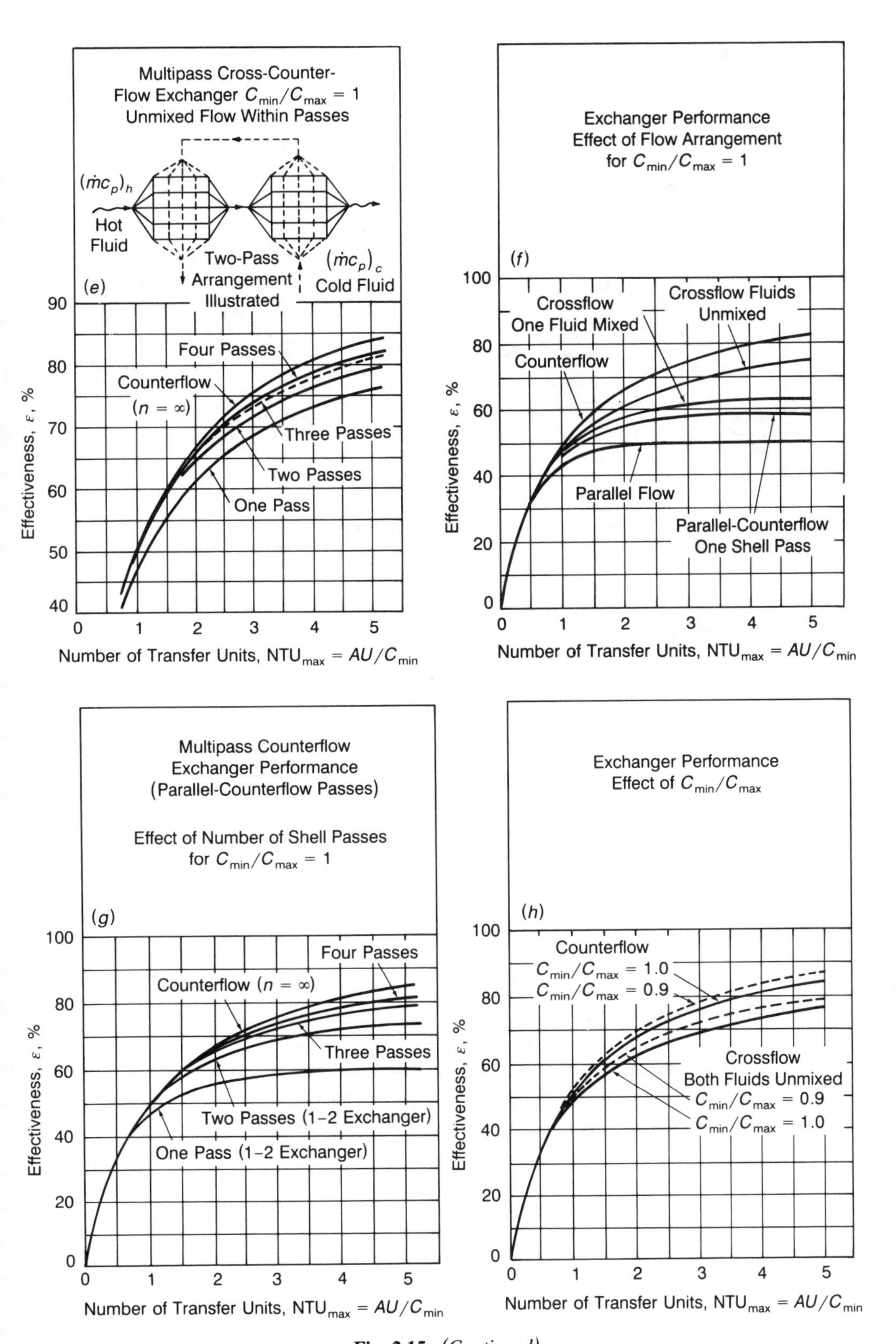

Fig. 2.15. *(Continued)*

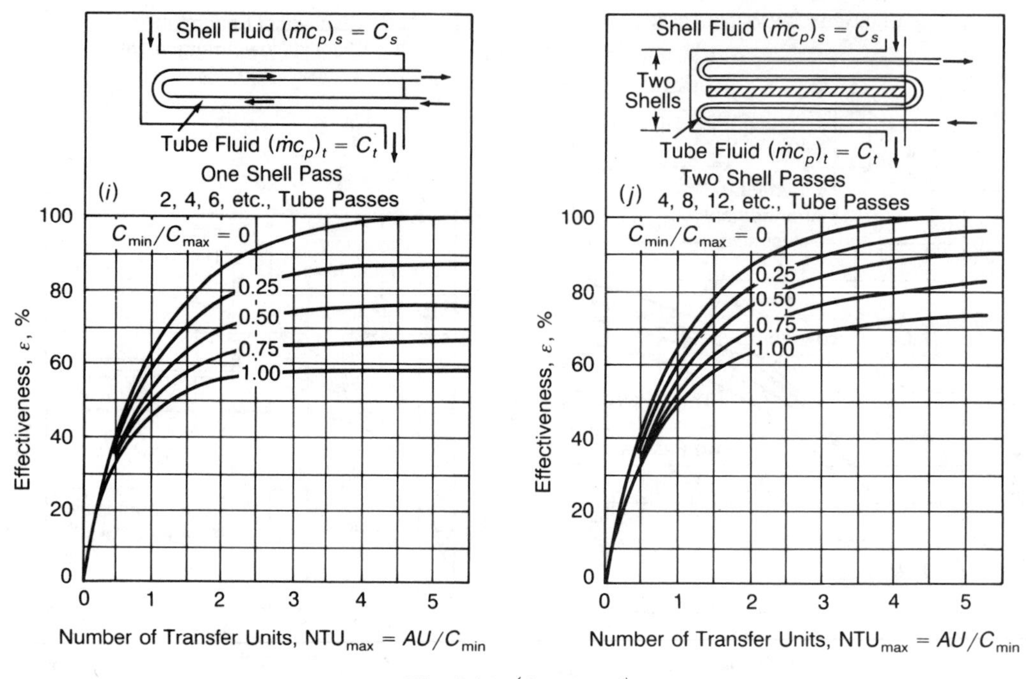

Fig. 2.15. *(Continued)*

and for parallel flow, Eq. (2.47) gives

$$\varepsilon = \tfrac{1}{2}(1 - e^{-2\mathrm{NTU}}) \tag{2.49}$$

For $C_{min}/C_{max} = 0$, as in boilers and condensers (Figs. 2.3*c* and *d*) for parallel flow or counterflow, Eqs. (2.46) and (2.47) become

$$\varepsilon = 1 - e^{-\mathrm{NTU}} \tag{2.50}$$

It is noted from Eqs. (2.46) and (2.47) that

$$\varepsilon = \phi(\mathrm{NTU}, C^*, \text{flow arrangement}) \tag{2.51}$$

Similar expressions have been developed for heat exchangers having other flow arrangements, such as crossflow, multipass, and so forth, and representative results are summarized in Table 2.2 [3, 9, 10].

Some ε–NTU relations are graphically shown in Fig. 2.15 [8]. The following observations may be made by reviewing these figures:

1. The heat exchanger effectiveness ε increases with increasing values of NTU for a specified C^*.
2. The exchanger effectiveness ε increases with decreasing values of C^* for a specified NTU.
3. For $\varepsilon < 40\%$ the capacity rate ratio C^* does not have a significant influence on the exchanger effectiveness.

Because of the asymptotic nature of the ε–NTU curves, a significant increase in NTU, and hence in the heat exchanger size, is required for a small increase in ε at high values of ε.

The counterflow exchanger has the highest exchanger effectiveness ε for specified NTU and C^* values compared to those for all other exchanger flow arrangements. Thus, for a given NTU and C^*, a maximum heat transfer performance is achieved for counterflow: alternately, the heat transfer surface is utilized most efficiently compared to all other flow arrangements.

Example 2.4. A two-pass tube, baffled single-pass shell, shell-and-tube heat exchanger is used as an oil cooler. Cooling water flows through the tubes at 20°C at a rate of 4.082 kg/s. Engine oil enters the shell side at a rate of 10 kg/s. The inlet and outlet temperatures of oil are 65°C and 55°C, respectively. Determine the surface area of the heat exchanger by both the LMTD and ε–NTU methods if the overall heat transfer coefficient based on the outside tube area is 262 W/(m^2 · K). The specific heats of water and oil are 4179 J/(kg · K) and 2047 J/(kg · K), respectively.

Solution: The heat transfer rate Q and LMTD for counterflow will first be calculated. Subsequently, P, R, and the correction factor F will be determined. The heat transfer surface area A will then be determined by the LMTD method using Eq. (2.36). The heat capacity rates for the shell fluid (oil) and the tube fluid are

$$C_h = \left(\dot{m}c_p\right)_h = 10 \times 2047 = 20{,}470 \text{ W/K}$$

$$C_c = \left(\dot{m}c_p\right)_c = 4.082 \times 4179 = 17{,}059 \text{ W/K}$$

From the temperature drop of oil, the heat transfer rate is

$$Q = C_h(T_{h1} - T_{h2}) = 20{,}470 \times (65 - 55) = 204.7 \text{ kW}$$

From the energy balance $Q = C_c(T_{c2} - T_{c1})$, the water outlet temperature is

$$T_{c2} = (204{,}700/17{,}059) + 20 = 32°\text{C}$$

From the definition of ΔT_{lm} in Eq. (2.28) for a counterflow arrangement

$$\Delta T_{\text{lm,cf}} = \frac{33 - 35}{\ln(33/35)} = 34°\text{C}$$

The values of P and R from Eqs. (2.34) and (2.35) are

$$P = \frac{T_{c2} - T_{c1}}{T_{h1} - T_{c1}} = \frac{32 - 20}{65 - 20} = 0.267 \qquad R = \frac{T_{h1} - T_{h2}}{T_{c2} - T_{c1}} = \frac{C_c}{C_h} = 0.833$$

Therefore F from Fig. 2.7 is $F = 0.98$. Thus the heat transfer area from Eq. (2.36) is

$$A = \frac{Q}{UF\,\Delta T_{\mathrm{lm,cf}}} = \frac{204{,}700}{262 \times 0.98 \times 34} = 23.45 \ \mathrm{m}^2$$

In the ε–NTU method, first ε and C^* and subsequently NTU and A will be calculated. In this problem, $C_h > C_c$ and hence $C_c = C_{\min}$:

$$C^* = \frac{C_{\min}}{C_{\max}} = \frac{17{,}059}{20{,}470} = 0.833$$

From the given temperatures, for $C_c = C_{\min}$, Eq. (2.42) gives

$$\varepsilon = \frac{T_{c2} - T_{c1}}{T_{h1} - T_{c1}} = 0.267$$

Now NTU is calculated either from the formula in Table 2.2 or it is found from Fig. 2.15 as NTU = 0.360. Hence

$$A = \frac{C_{\min}}{U} \quad \mathrm{NTU} = \frac{17{,}059 \times 0.360}{262} = 23.44 \ \mathrm{m}^2$$

2.7 THE P–NTU$_c$ METHOD

This method is a variation of the ε–NTU method. As shown in Table 2.2, the ε–NTU relations are different depending upon whether the tube fluid is the $C_{\max}$ or $C_{\min}$ fluid in shell-and-tube heat exchangers. In order to avoid possible confusion about which is the $C_{\min}$ fluid, P is defined as the temperature effectiveness of the heat exchanger on one fluid side, regardless of whether it is the hot side or the cold side. NTU is based on that side's heat capacity rate and R is defined as a ratio of that side's heat capacity rate to that of the other side. Somewhat arbitrarily, that side is chosen as the cold-fluid side. If a distinction is required for the shell side and the tube side instead of hot and cold fluids, the results may also be presented based on the tube side. The P, R, and NTU$_c$ definitions are also valid for the shell side or the tube side as long as they are consistently defined for one side. The definitions of P, R, and NTU are given by Eqs. (2.34), (2.35), and (2.44) where $C_{\min} = C_c$.

Comparing Eqs. (2.34) and (2.42), temperature effectiveness P and exchanger effectiveness ε are related by

$$P = \frac{C_{\min}}{C_c}\varepsilon = \begin{cases} \varepsilon & \text{for } C_c = C_{\min} \\ \varepsilon C^* & \text{for } C_c = C_{\max} \end{cases} \tag{2.52}$$

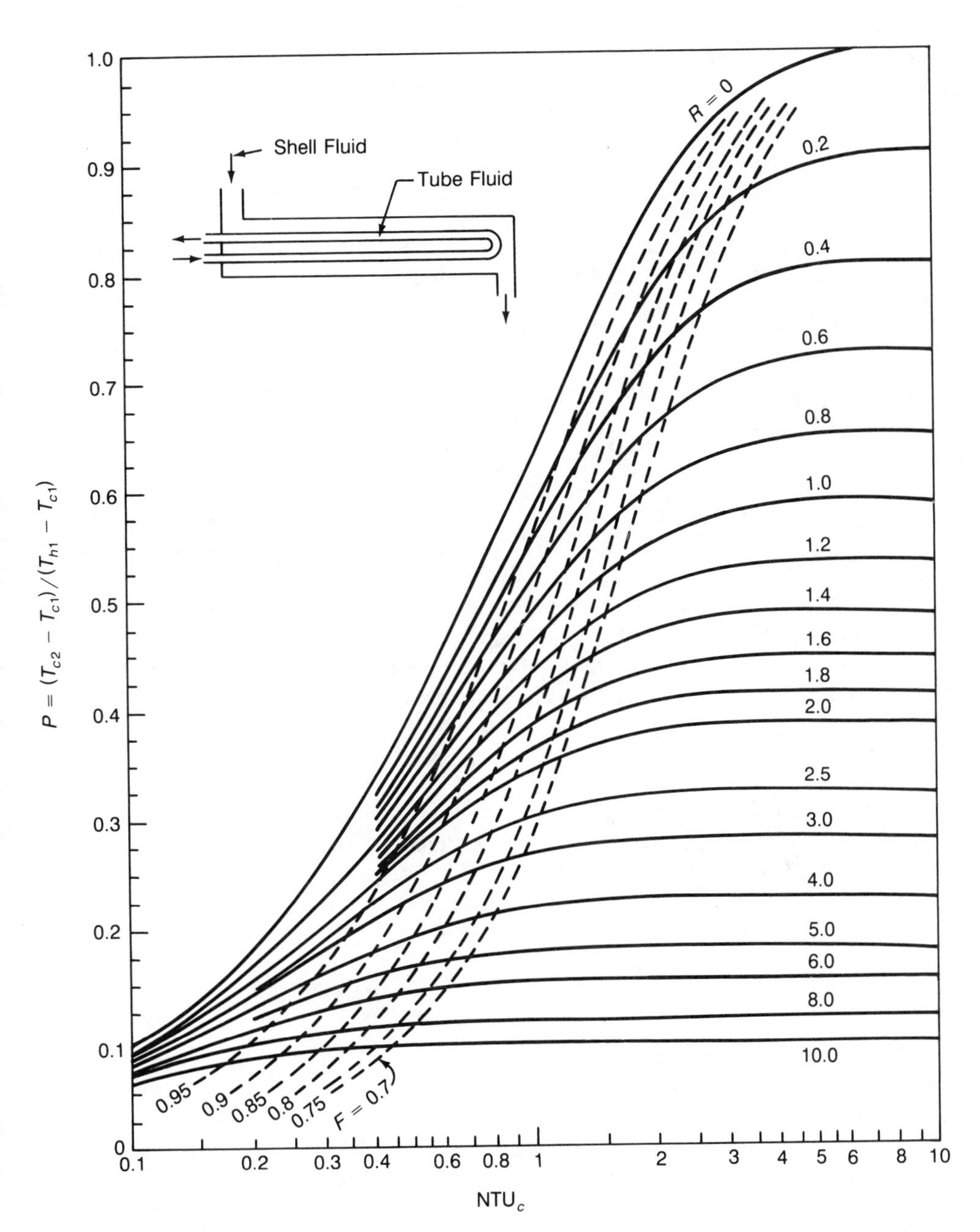

Fig. 2.16. The temperature effectiveness P as a function of NTU$_c$ and R for a 1–2 shell-and-tube heat exchanger with shell fluid mixed [9].

Note that P is always less than or equal to ε. NTU_c is related to NTU based on $C_{\min}$ as

$$\text{NTU}_c = \text{NTU}\frac{C_{\min}}{C_c} = \begin{cases} \text{NTU} & \text{for } C_c = C_{\min} \\ \text{NTU} \cdot C^* & \text{for } C_c = C_{\max} \end{cases} \tag{2.53}$$

Similar to the exchanger effectiveness ε, the temperature effectiveness P is a function of NTU_c, R, and flow arrangement

$$P = \phi(\text{NTU}_c, R, \text{flow arrangement}) \tag{2.54}$$

It should be noted that P, NTU_c, and R are defined consistently based on the cold-side fluid variables; the results are also valid for P, NTU_c, and R all based on the tube-side fluid variables or shell-side fluid variables; they can also be based on the hot-fluid variables.

The origin of the P–NTU_c method is related to shell-and-tube heat exchangers and the most useful NTU_c design range for shell-and-tube exchangers is about 0.2 to 3. The P–NTU_c results are generally presented on a semi-log paper, as shown for example in Fig. 2.16, in order to obtain more accurate graphical values of P or NTU_c.

In the P–NTU_c method, the total heat transfer rate from the hot fluid to the cold fluid in the heat exchanger is expressed as

$$Q = PC_c(T_{h1} - T_{c1}) \tag{2.55}$$

The P–NTU_c relationships can be derived directly for any flow arrangement or can be obtained from the ε–NTU relationship given in Table 2.2, by replacing C^*, ε, and NTU by R, P, and NTU_c, respectively, using Eqs. (2.38), (2.52), and (2.53). For example, for the parallel-flow heat exchanger, the P–NTU_c relationship becomes

$$P = \frac{1 - \exp[-\text{NTU}_c(1 + R)]}{1 + R} \tag{2.56}$$

which is valid for $R = 0$ to $R = \infty$, P is the cold-side (or tube-side) temperature effectiveness and NTU_c is based on C_c (or tube-side C_t).

2.8 THE ψ–P METHOD

As mentioned in Section 2.6, a trial-and-error method is needed for the solution of the rating problem by the LMTD method and for the solution of the sizing problem (for multipass shell-and-tube exchangers) by the ε–NTU method. For the graphical representation of the ε–NTU results, the abscissa ranges from 0 to ∞ (i.e., it is unbounded). A method that combines all the variables of the LMTD and ε–NTU methods and eliminates their limitations for hand calculation has been proposed by Mueller [11]. In this method, a new grouping ψ is introduced. It is a ratio of the true mean temperature difference to the inlet temperature difference of the two fluids

$$\psi = \frac{\Delta T_m}{T_{h1} - T_{c1}} \tag{2.57}$$

It can be shown that ψ is related to ε and NTU as

$$\psi = \frac{\varepsilon}{\text{NTU}} = \frac{P}{\text{NTU}_c} \tag{2.58}$$

The log-mean temperature difference correction factor is defined as [see Eqs. (2.32) and (2.36)]:

$$F = \frac{\Delta T_m}{\Delta T_{\text{lm,cf}}} \tag{2.59}$$

where ΔT_m is the true mean temperature difference. In order to evaluate F, we compare an actual heat exchanger of any flow arrangement of interest with a reference counterflow heat exchanger having the same terminal temperatures. Therefore Eq. (2.59) can also be written as

$$F = \frac{\text{NTU}_{c,\text{cf}}}{\text{NTU}_c} \tag{2.60}$$

where NTU_c represents the actual number of transfer units for a given heat exchanger. The P–NTU_c relationship can be written for counterflow from Table 2.2 as

$$P = \frac{1 - \exp[-\text{NTU}_c(1 - R)]}{1 - R\exp[-\text{NTU}_c(1 - R)]} \tag{2.61}$$

and

$$\text{NTU}_{c,\text{cf}} = \begin{cases} \dfrac{1}{1 - R}\ln\left(\dfrac{1 - RP}{1 - P}\right) & \text{for } R \neq 1 \\[2ex] \dfrac{P}{1 - P} & \text{for } R = 1 \end{cases} \tag{2.62}$$

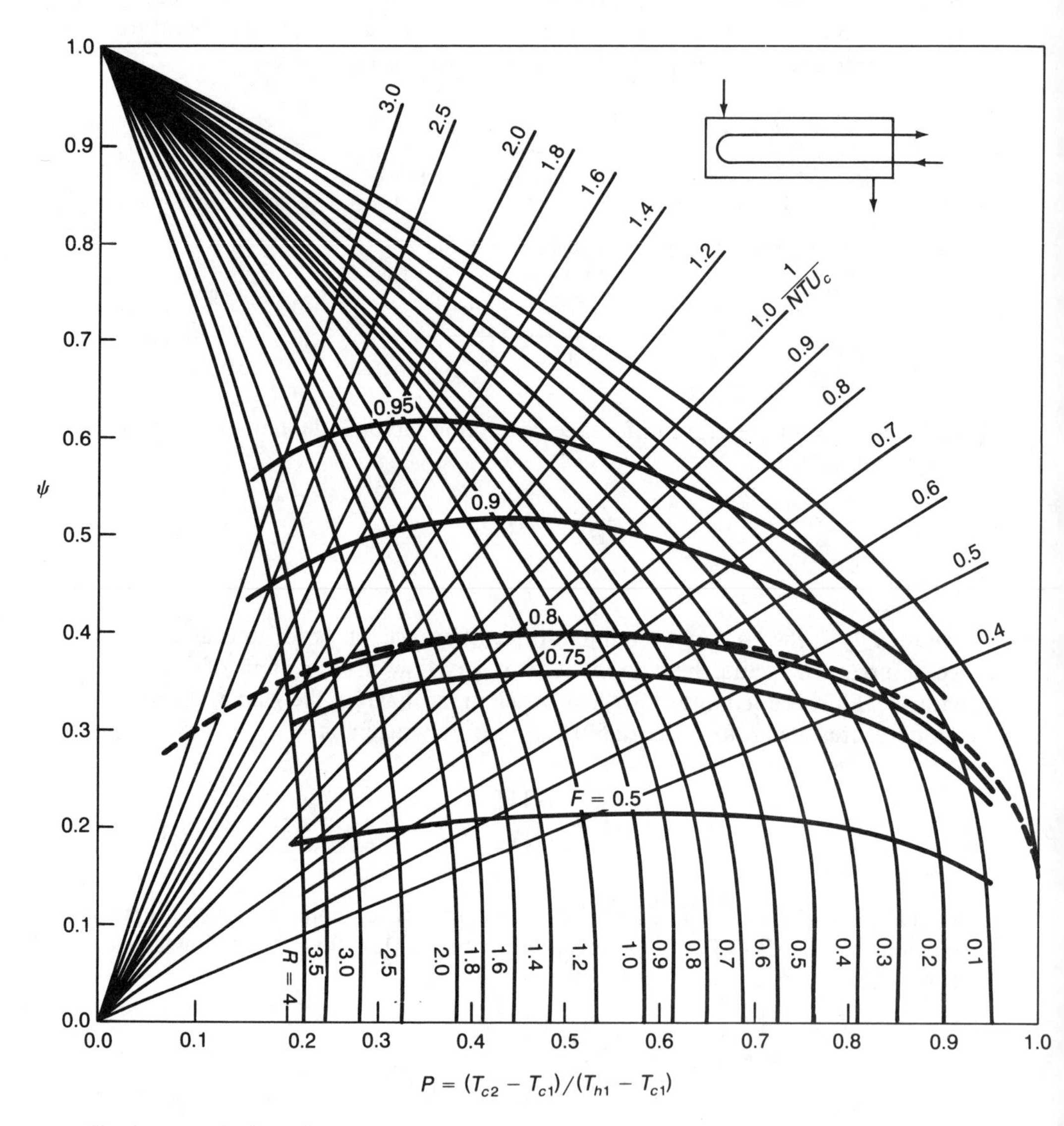

Fig. 2.17. ψ–P chart for a 1–2 shell-and-tube heat exchanger, shell fluid mixed [9].

TABLE 2.3 Working Equations for the LMTD, ε–NTU, P–NTU$_c$, and ψ–P Methods[a]

LMTD	ε–NTU
$Q = UAF\,\Delta T_{\mathrm{lm,cf}}$	$Q = \varepsilon C_{\min}(T_{h1} - T_{c1})$
$\mathrm{LMTD} = \Delta T_{\mathrm{lm,cf}} = \dfrac{\Delta T_1 - \Delta T_2}{\ln(\Delta T_1/\Delta T_2)}$	$\varepsilon = \dfrac{C_h(T_{h1} - T_{h2})}{C_{\min}(T_{h1} - T_{c1})} = \dfrac{C_c(T_{c2} - T_{c1})}{C_{\min}(T_{h1} - T_{c1})}$
$\Delta T_1 = T_{h1} - T_{c2},\ \Delta T_2 = T_{h2} - T_{c1}$	$C^* = \dfrac{C_{\min}}{C_{\max}} = \dfrac{(\dot{m}c_p)_{\min}}{(\dot{m}c_p)_{\max}}$
$P = \dfrac{T_{c2} - T_{c1}}{T_{h1} - T_{c1}},\ R = \dfrac{T_{h1} - T_{h2}}{T_{c2} - T_{c1}}$	$\mathrm{NTU} = \dfrac{UA}{C_{\min}} = \dfrac{1}{C_{\min}}\displaystyle\int_A U\,dA$
$F = \phi(P, R, \text{flow arrangement})$	$\varepsilon = \phi(\mathrm{NTU}, C^*, \text{flow arrangement})$

P–NTU$_c$	ψ–P
$Q = PC_c(T_{h1} - T_{c1})$	$Q = UA\psi(T_{h1} - T_{c1})$
$\mathrm{NTU}_c = \dfrac{UA}{C_c} = \mathrm{NTU}\dfrac{C_{\min}}{C_c}$	$\psi = \dfrac{\Delta T_m}{T_{h1} - T_{c1}}$
$P = \phi(\mathrm{NTU}_c, R, \text{flow arrangement})$	$\psi = \phi(P, R, \text{flow arrangement})$

[a] The definitions of P, R, and NTU$_c$ can also be consistently based on C_t, C_h, or C_s.

Equations (2.58), (2.60), and (2.62) can be combined to relate ψ to F as

$$\psi = \frac{FP(1 - R)}{\ln[(1 - RP)/(1 - P)]} \tag{2.63}$$

In this method the rate of heat transfer can be calculated as

$$Q = UA\psi(T_{h1} - T_{c1}) \tag{2.64}$$

There is no need to calculate ΔT_{lm} since ψ represents the nondimensional ΔT_m. As can be seen from Eq. (2.63), we have

$$\psi = \phi(P, R, \text{flow arrangement}) \tag{2.65}$$

ψ–P charts with R as a parameter have been prepared by Mueller [11]. As an example, the ψ–P chart for a 1–2 shell-and-tube heat exchanger is shown in Fig. 2.17 [9].

Table 2.3 summarizes each of the methods discussed in the preceding sections.

In obtaining expressions for the four basic design methods that we discussed in the preceding sections, Eqs. (2.22) and (2.23) are integrated across the surface area under the following assumptions:

1. The heat exchanger operates under steady-state, steady-flow conditions.
2. Heat transfer to the surroundings is negligible.
3. There is no heat generation in the heat exchanger.
4. In counterflow and parallel-flow heat exchangers, the temperature of each fluid is uniform over every flow cross section; in crossflow heat exchangers each fluid is considered mixed or unmixed at every cross section depending upon the specifications.
5. If there is a phase change in one of the fluid streams flowing through the heat exchanger, phase change occurs at a constant temperature for a single-component fluid at constant pressure.
6. The specific heat at constant pressure is constant for each fluid.
7. Longitudinal heat conduction in the fluid and in the wall are negligible.
8. The overall heat transfer coefficient between the fluids is constant throughout the heat exchanger including the case of phase change.

Assumption 5 is an idealization of a two-phase-flow heat exchanger. Especially, for two-phase flows on both sides, many of the foregoing assumptions are not valid. The design theory of these types of heat exchangers is discussed and practical results are presented in the following chapters.

2.9 HEAT EXCHANGER DESIGN CALCULATION

We have discussed four methods for performing a heat exchanger thermal analysis (Table 2.3). The rating and sizing of heat exchangers are two important problems encountered in the thermal analysis of heat exchangers.

The rating problem is concerned with the determination of the heat transfer rate and the fluid outlet temperatures for prescribed fluid flow rates, inlet temperatures, and the pressure drop for an existing heat exchanger; hence the heat transfer surface area and the flow passage dimensions are available.

The sizing problem, on the other hand, is concerned with the determination of the dimensions of the heat exchanger, that is, selecting an appropriate heat exchanger type and determining the size to meet the specified hot- and cold-fluid inlet and outlet temperatures, flow rates, and pressure drop requirements. For example, if the inlet temperatures and mass flow rates are known and the objective is to design a heat exchanger that will provide a desired value of outlet temperature for one of the fluids, the LMTD method can be used to solve this sizing problem with the following steps:

1. Calculate Q and the unknown outlet temperature from Eqs. (2.5) and (2.6).
2. Calculate ΔT_{lm} from Eq. (2.28) and obtain the correction factor F if necessary.
3. Calculate the overall heat transfer coefficient U.
4. Determine A from Eq. (2.36).

The LMTD method may also be used for rating problems (performance analysis), but computation would be tedious, requiring iteration since the outlet temperatures are not known to calculate the LMTD. In such situations the analysis can be simplified by using the ε–NTU method. The rating analysis with the ε–NTU method will be as follows:

1. Calculate the capacity rate ratio $C^* = C_{\min}/C_{\max}$ and $\mathrm{NTU} = UA/C_{\min}$ from the input data.
2. Determine the effectiveness ε from the appropriate charts or ε–NTU equations for the given heat exchanger and specified flow arrangement.
3. Knowing ε, calculate the total heat transfer rate from Eq. (2.43).
4. Calculate the outlet temperatures from Eqs. (2.5) and (2.6).

The ε–NTU method may also be used for the sizing problem and the procedure will be as follows:

1. Knowing the outlet and inlet temperatures, calculate ε from Eq. (2.42).
2. Calculate the capacity rate ratio $C^* = C_{\min}/C_{\max}$.
3. Calculate the overall heat transfer coefficient U.
4. Knowing ε, C^*, and the flow arrangement, determine NTU from charts or from ε–NTU relations.
5. Knowing NTU, calculate the heat transfer surface area A from Eq. (2.44).

The P–NTU_c method is just a variant of the ε–NTU method. The ψ–P method combines all the variables of the LMTD and ε–NTU methods and eliminates their limitations.

The use of the ε–NTU method is generally preferred in the design of compact heat exchangers for automotive, aircraft, air-conditioning, and other industrial applications where the inlet temperatures of the hot and cold fluids are specified and the heat transfer rates are to be determined. The LMTD and ψ–P methods are traditionally used in the process, power, and petrochemical industries.

2.10 VARIABLE OVERALL HEAT TRANSFER COEFFICIENT

In practical applications, the overall heat transfer coefficient varies along the heat exchanger and it is strongly dependent on the flow Reynolds number, heat transfer surface geometry, and fluid physical properties. Methods to account for specific variations in U are given for counterflow, crossflow, and multipass shell-and-tube heat exchangers.

Figure 2.18 shows typical situations in which the variation of U within a heat exchanger might be very large. The case in which both fluids are changing phase is shown in Fig. 2.18*a*, where there is no sensible heating and cooling; the temperatures simply remain constant throughout. The condenser shown in Fig. 2.18*b* is perhaps more common than the condenser of Fig. 2.3*d*. In the former, the condensing vapor enters at a temperature greater than the saturation temperature and subcooling of the liquid takes place before the hot liquid leaves the exchanger. A corresponding situation, where the cold fluid enters as a liquid and is heated, evaporated, and then superheated, is shown in Fig. 2.18*c*. When the hot fluid consists of both condensable vapor and noncondensable gases, the temperature distribution is more complex as represented in a general way in Fig. 2.18*d*. The difficulty that one faces in designing such a heat exchanger is the continuous variation of U with position within the heat exchanger. If the three parts of the heat exchanger (Figs. 2.18*b* and *c*) had constant values of U, then the heat exchanger could be treated as three different heat exchangers in series. For arbitrary variation of U through the heat exchanger, the exchanger is divided into many segments and a different value of U is then assigned to each

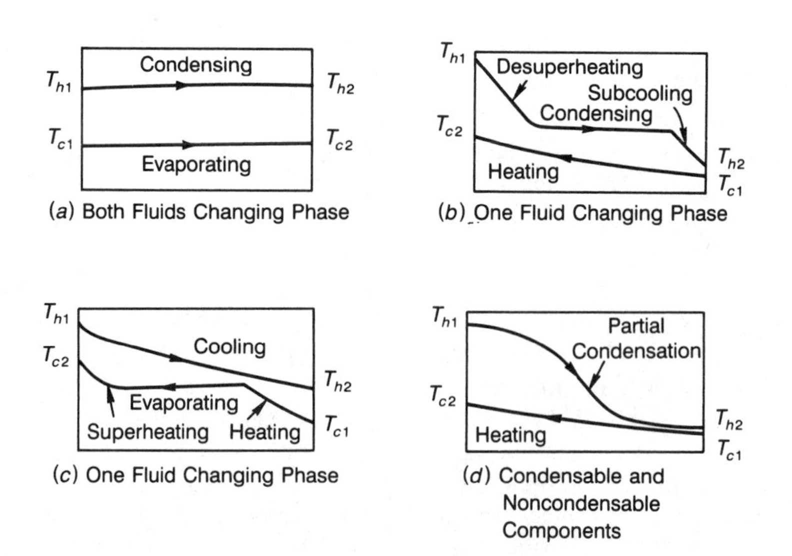

Fig. 2.18. Typical cases of a heat exchanger with variable U.

segment. The analysis is best performed by a numerical or finite-difference method. Let us consider a counterflow double-pipe heat exchanger (Fig. 2.1*b*). The heat exchanger is divided into increments of surface area ΔA_i. For any incremental surface area, the hot- and cold-fluid temperatures are T_{hi} and T_{ci}, respectively, and it will be assumed that the overall heat transfer coefficient can be expressed as a function of these temperatures. Thus

$$U_i = U_i(T_{hi}, T_{ci}) \tag{2.66}$$

The incremental heat transfer in ΔA_i can be calculated from Eq. (2.22):

$$\Delta Q_i = -(\dot{m}c_p)_{hi}(T_{h(i+1)} - T_{hi}) = -(\dot{m}c_p)_{ci}(T_{c(i+1)} - T_{ci}) \tag{2.67}$$

From Eq. (2.23), ΔQ_i is also given by

$$\Delta Q_i = U_i \, \Delta A_i (T_h - T_c)_i \tag{2.68}$$

Equation (2.25) can be written in the finite-difference form as

$$\frac{(T_h - T_c)_{i+1} - (T_h - T_c)_i}{(T_h - T_c)_i} = U_i \left(\frac{1}{C_{ci}} - \frac{1}{C_{hi}} \right) \Delta A_i \tag{2.69}$$

which can be solved for $(T_h - T_c)_{i+1}$:

$$(T_h - T_c)_{i+1} = (T_h - T_c)_i (1 + M_i \, \Delta A_i) \tag{2.70}$$

where

$$M_i = U_i \left(\frac{1}{C_{ci}} - \frac{1}{C_{hi}} \right) \tag{2.71}$$

The numerical analysis can be carried out as follows:

1. Choose a convenient value of ΔA_i for the analysis.
2. Calculate the inner and outer heat transfer coefficients and the value of U for the inlet conditions and through the initial ΔA increment.
3. Calculate the value of ΔQ_i for this increment from Eq. (2.68).
4. Calculate the values of T_h, T_c, and $T_h - T_c$ for the next increment by the use of Eqs. (2.67) and (2.69).

The total heat transfer rate is then calculated from

$$Q = \sum_{i=1}^{n} \Delta Q_i \tag{2.72}$$

For the overall heat transfer coefficient U and ΔT varying linearly with Q, Colburn recommended the following expression to calculate Q [12]:

$$Q = U_m A \, \Delta T_{\mathrm{lm}} = \frac{A(U_2 \, \Delta T_1 - U_1 \, \Delta T_2)}{\ln[(U_2 \, \Delta T_1)/(U_1 \, \Delta T_2)]} \tag{2.73}$$

where U_1 and U_2 are the values of the overall heat transfer coefficients on the ends of the exchanger having temperature differences of ΔT_1 and ΔT_2, respectively.

When both $1/U$ and ΔT vary linearly with Q, Butterworth [13] has shown that

$$Q = U_m A \, \Delta T_{\mathrm{lm}} \tag{2.74}$$

where

$$\frac{1}{U_m} = \frac{1}{U_1}\left[\frac{\Delta T_{\mathrm{lm}} - \Delta T_2}{\Delta T_1 - \Delta T_2}\right] + \frac{1}{U_2}\left[\frac{\Delta T_1 - \Delta T_{\mathrm{lm}}}{\Delta T_1 - \Delta T_2}\right] \tag{2.75}$$

For some condenser applications, Eqs. (2.73) and (2.75) may be applicable (see also chapter 11).

2.11 HEAT EXCHANGER PRESSURE DROP AND PUMPING POWER

The thermal design of heat exchangers is directed to calculate an adequate surface area to handle the thermal duty for the given specifications. Fluid-friction effects in the heat exchanger are equally important. They determine the pressure drop of the fluids flowing in the system, and consequently, the pumping power or fan work input necessary to maintain the flow. Providing for pumps or fans adds to the capital cost and is a major part of the operating cost of the exchanger. Savings in exchanger capital cost achieved by designing a compact unit with high fluid velocities may soon be lost by increased operating costs. The final design and selection of a unit will therefore be influenced just as much by effective use of the permissible pressure drop and the cost of pump or fan power as they are influenced by the temperature distribution and provision of adequate area for heat transfer.

2.11.1 Tube-Side Pressure Drop

In fully developed flow in a tube, the following functional relationship can be written for the frictional pressure drop for either laminar or turbulent flow:

$$\frac{\Delta P}{L} = \phi(u_m, d_i, \rho, \mu, e) \tag{2.76}$$

where the quantity e is a statistical measure of the surface roughness of the tube and has the dimension of length. It is assumed that ΔP is proportional to the length L of the tube. With force F, mass M, length L, and time θ as the fundamental dimensions and u_m, d_i, and ρ as the set of maximum number of quantities, which in themselves cannot form a dimensionless group, the pi theorem leads to

$$\frac{\Delta P}{4(L/d_i)(\rho u_m^2/2)} = \phi\left(\frac{u_m d_i \rho}{\mu}, \frac{e}{d_i}\right) \tag{2.77}$$

where the dimensionless numerical constants 4 and 2 are added for convenience.

The previous dimensionless group involving ΔP has been defined as the Fanning friction factor f:

$$f = \frac{\Delta P}{4(L/d_i)(\rho u_m^2/2)} \tag{2.78}$$

Equation (2.76) becomes

$$f = \phi\left(Re, \frac{e}{d_i}\right) \tag{2.79}$$

Figure 2.19 shows this relationship as deduced by Moody [14] from experimental data for fully developed flow. In the laminar region, existing empirical data on the pressure drop within round pipes can be correlated by a simple relationship between f and Re, independent of the surface roughness

$$f = \frac{16}{Re} \tag{2.80}$$

The transition from laminar to turbulent flow is somewhere in the neighborhood of Re from 2300 to 4000.

The f-versus-Re relation for smooth tubes in turbulent flow has a slight curvature on a log–log plot. A few recommended correlations for turbulent flow in smooth pipes are given in Table 2.4. Two linear approximations shown by the dotted lines in Fig. 2.19 for turbulent flow are

$$f = 0.046 Re^{-0.2} \quad \text{for } 3 \times 10^4 < Re < 10^6 \tag{2.81a}$$

and

$$f = 0.079 Re^{-0.25} \quad \text{for } 4 \times 10^3 < Re < 10^5 \tag{2.81b}$$

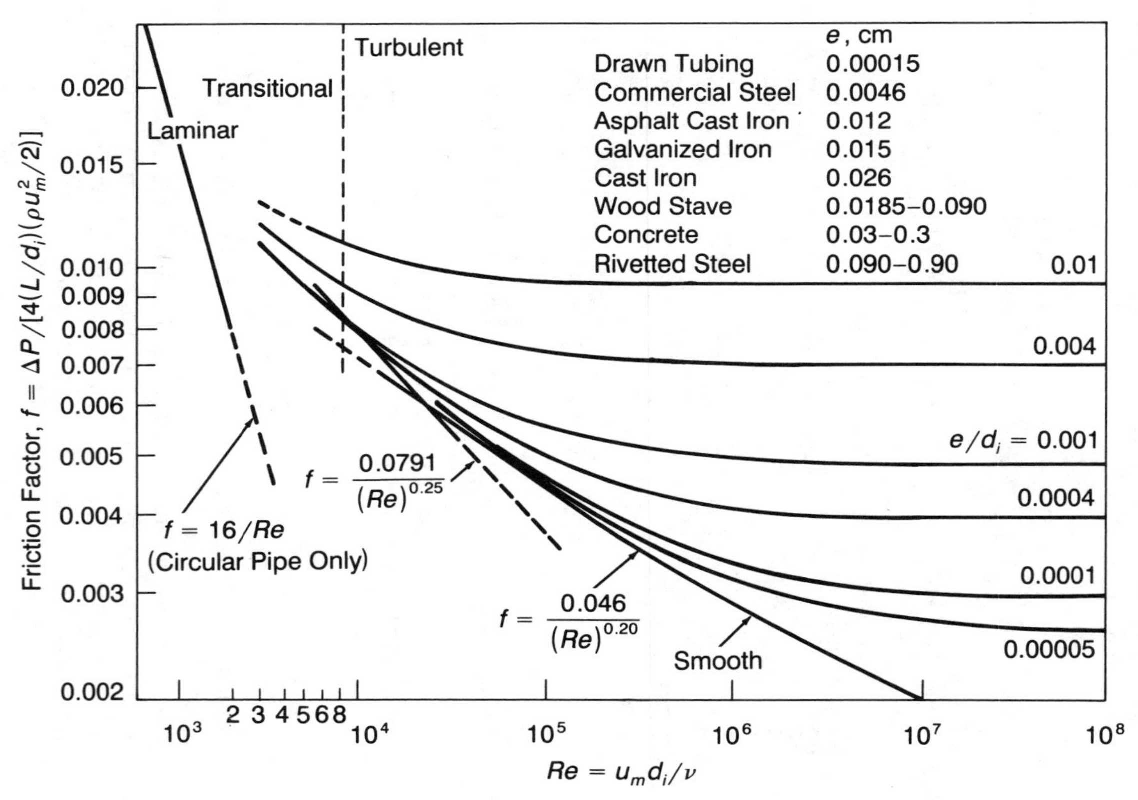

Fig. 2.19. Friction factor for fully developed flow inside a circular duct [14].

TABLE 2.4 Turbulent Flow Isothermal Fanning Friction Factor Correlations for Smooth Circular Ducts [4]

Reference	Correlation[a]	Limitations
1. Blasius	$f = \dfrac{\tau_w}{\rho u_m^2/2} = 0.0791 Re^{-0.25}$	$4 \times 10^3 < Re < 10^5$
2. Drew, Koo, and McAdams	$f = 0.00140 + 0.125 Re^{-0.32}$	$4 \times 10^3 < Re < 5 \times 10^6$
3. Karman and Nikuradse	$\dfrac{1}{\sqrt{f}} = 1.737 \ln(Re\sqrt{f}) - 0.4$ or $\dfrac{1}{\sqrt{f}} = 4 \log_{10}(Re\sqrt{f}) - 0.4$ approximated as $f = (3.64 \log_{10} Re - 3.28)^{-2}$	$4 \times 10^3 < Re < 3 \times 10^6$
4. Approximate Karman–Nikuradse correlation	$f = 0.046 Re^{-0.2}$	$3 \times 10^4 < Re < 10^6$

[a]Properties are evaluated at the bulk temperature.

The friction factor f can be read from the graph, but the correlations for f are useful for computer analysis of heat exchangers. They also show the functional relationship of various quantities.

For fully developed flow in a tube, a simple force balance yields

$$\Delta P \frac{\pi}{4} d_i^2 = \tau_w (\pi d_i L) \tag{2.82}$$

which may be combined with Eq. (2.78) to get an equivalent form for the friction factor, defined as

$$f = \frac{\tau_w}{\rho u_m^2/2} \tag{2.83}$$

2.11.2 Noncircular Cross-Sectional Ducts

A duct of noncircular cross section is not geometrically similar to a circular duct; hence dimensional analysis does not relate the performance of these two geometrical shapes. However, in turbulent flow, f for noncircular cross sections (annular spaces, rectangular, and triangular ducts, etc.) may be evaluated from the data for circular ducts if d_i is replaced by an equivalent diameter (hydraulic diameter) D_e, defined by

$$D_e = \frac{4A}{p_w} = \frac{4(\text{flow area})}{\text{wetted perimeter}} \tag{2.84}$$

Using the equivalent diameter in turbulent flow gives f values within about $\pm 8\%$ of the measured values [15].

The equivalent diameter of an annulus of inner and outer diameters d_i, d_o is

$$D_e = \frac{4(\pi/4)(d_o^2 - d_i^2)}{\pi(d_o + d_i)} = (d_o - d_i) \tag{2.85}$$

For a circular duct, Eq. (2.84) reduces to $D_e = d_i$.

The transition Reynolds number for noncircular ducts $(\rho u_m D_e/\mu)$ is also found to be approximately 2300, as for circular ducts.

For laminar flow, however, the results for noncircular cross sections are not universally correlated. In a thin annulus, the flow has a parabolic distribution perpendicular to the wall and has this same distribution at every circumferential position. If this flow is treated as a flow between two parallel flat plates separated by a distance $2b$, one obtains

$$\frac{\Delta P}{\Delta x} = \frac{12\mu u_m}{b^2} \tag{2.86}$$

Here $D_e = 2b$ and Eq. (2.86) can be written in the form

$$f = \frac{24}{Re} \tag{2.87}$$

with D_e replacing d_i in the definitions of f and Re. This equation is different from Eq. (2.80) which applies to laminar flow in circular ducts.

Flow in a rectangular duct (dimensions $a \times b$) in which $b \ll a$ is similar to this annular flow. For rectangular ducts of other aspect ratios (b/a):

$$f = \frac{16}{\phi Re} \tag{2.88}$$

where

$$D_e = \frac{4ab}{2(a+b)} \tag{2.89}$$

and ϕ is given in Fig. 2.20 [16].

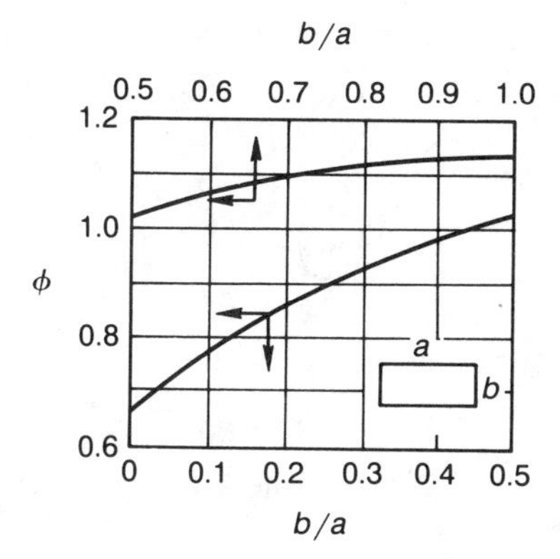

Fig. 2.20. Values of ϕ for rectangular ducts.

For laminar flow in ducts of triangular and trapezoidal cross section, Nikuradse [4] showed that f is approximated by $16/Re$ with D_e given by Eq. (2.84) and transition occurs at approximately $Re = 2300$.

The pressure drop for flow through a duct of length L is generally expressed as

$$\Delta P = 4f \frac{L}{D_e} \frac{\rho u_m^2}{2} \tag{2.90a}$$

or

$$\Delta P = 4f \frac{L}{D_e} \frac{G^2}{2\rho} \tag{2.90b}$$

where $G = u_m \rho$ is referred to as the mass velocity.

The exact form of Eq. (2.90) depends on the flow situation. The pressure drop experienced by a single-phase fluid in the tubes in shell-and-tube heat exchangers must include the effects of all the tubes. The pressure drop for all the tubes can be calculated by

$$\Delta P_t = 4f \frac{LN_p}{D_e} \frac{G^2}{2\rho} \tag{2.91}$$

where N_p is the number of tube passes and $D_e = d_i$. The fluid will experience an additional loss due to sudden expansions and contractions that the tube undergoes during a return. Experiments show that the return pressure loss is given by [17]:

$$\Delta P_r = 4N_p \frac{\rho u_m^2}{2} \tag{2.92}$$

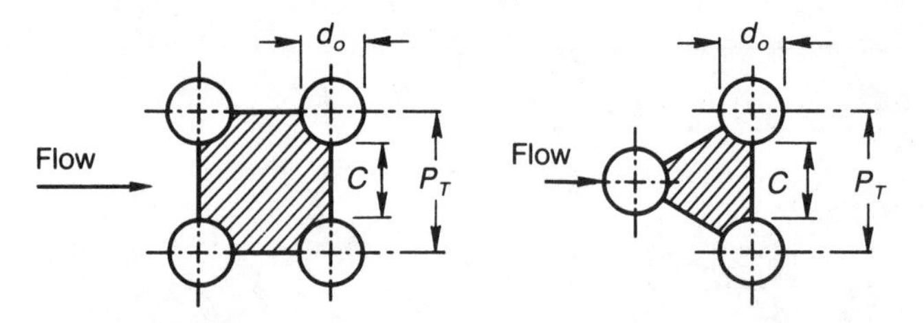

Fig. 2.21. Square and triangular pitch-tube layout.

where N_p is the number of tube passes, which is 2 in a 1–2 heat exchanger and $4N_p$ is the return pressure loss coefficient.

2.11.3 Shell-Side Pressure Drop

In a shell-and-tube heat exchanger with baffles, the fluid velocity fluctuates because of the constricted area between adjacent tubes. The shell width and the number of tubes vary from top to bottom or side to side. There is no true area on the shell side by which the shell-side mass velocity G_s can be computed. For this reason, fictitious values of D_e and G_s can be used. The equivalent diameter (hydraulic diameter) D_e representing shell geometry is calculated along (instead of across) the long axes of the tubes to retain the format of Eq. (2.91). The equivalent diameter is then taken as four times the hydraulic radius obtained by the geometry as layout on the tube sheet. Figure 2.21 shows a square and a triangular pitch layout.

For the square pitch, the area is a square minus the areas of four quarter-circles (the hatched section in Fig. 2.21). The wetted perimeter is the circumference of four quarter-circles. Thus

$$D_e = \frac{4\left(p_T^2 - \pi d_o^2/4\right)}{\pi d_o} \qquad \text{(square pitch)} \tag{2.93}$$

For the triangular pitch

$$D_e = \frac{4\left(p_T^2\sqrt{3}/4 - \pi d_o^2/8\right)}{\pi d_o/2} \qquad \text{(triangular pitch)} \tag{2.94}$$

It will be necessary to calculate a shell-side fluid velocity for a given mass flow rate. This can be done if there is an equation for the shell-side characteristic area. The variables that influence the velocity are the shell inside diameter D_s, the clearance C between adjacent tubes (see Fig. 2.21), the baffle spacing B, and the pitch p_T. The shell characteristic area is given by

$$a_s = \frac{D_s C B}{p_T} \tag{2.95}$$

The shell-side mass velocity is then found with

$$G_s = \frac{\dot{m}}{a_s} \tag{2.96}$$

It should be noted that the shell-side velocity is not constant and the preceding formulation yields an estimate that is useful in calculating the pressure drop and the heat transfer coefficient.

The pressure drop on the shell side depends on the number of crosses of the bundle between baffles. It is proportional to the number of times the fluid crosses the bundle. A correlation has been obtained using the product of distance across the bundle, taken as the inside diameter of the shell D_s, and the number of times the bundle is crossed $(N_B + 1)$, where N_B is the number of baffles. Then the pressure drop is calculated by the following expression [17]:

$$\Delta P_s = f\frac{D_s(N_B + 1)G_s^2}{2\rho D_e \phi_s} \tag{2.97}$$

where D_e is given by Eq. (2.93) or (2.94), and $\phi_s = (\mu_b/\mu_w)^{0.14}$, which will take care of the effect of property variations. The friction factor f for the shell side is given by

$$f = \exp(0.576 - 0.19 \ln Re_s) \tag{2.98}$$

where $400 \leq Re_s\ (= (G_s D_e/\mu) \leq 1 \times 10^6$. The friction factor also accounts for the entrance and exit losses. The additional friction of the shell itself is neglected. If there is no baffle, fluid flow will be along the exchanger and the pressure drop can be calculated as in a tube flow.

The most important parameter affecting pressure drop is the fluid velocity. The pressure drop increases as the square of the velocity. A change in the fluid velocity is therefore more significant than a change in other factors. Doubling the velocity increases the pressure drop by a factor of 4.

The overall pressure drop in the exchanger is the sum of a number of components. The primary loss generally occurs in fluid flow through tubes, in crossflow over the tube bank, or, as in a shell-and-tube heat exchanger, in the combined crossflow and axial flow characterized by repeated passes through the tube bundle in baffled shell-side flow. Secondary losses occur as a result of the sudden contractions and expansions as the fluid enters and leaves the exchanger through the inlet and outlet nozzles or enters and leaves the tube bundles. Reversals in the flow direction in multipass exchangers also increase the pressure drop. The use of augmented surfaces increases the heat transfer coefficient at the expense of increased pressure drop.

In shell-and-tube heat exchangers, the Delaware method is recommended for shell-side thermohydraulic design for single-phase flow. The Delaware method is presented in [18]. The shell-side heat transfer coefficient and pressure drop can be predicted with acceptable accuracy with this method.

The shell-side pressure drop is built up in the Delaware method by summing the pressure drops in crossflow and window sections without leakage or bypass; it is then corrected for a real heat exchanger considering the effect of bypass at the entrance and exit sections, the effects of bypass and leakage in the interior crossflow section, and the effect of leakage in the windows. The total nozzle-to-nozzle shell-side pressure drop is then calculated as the sum of the individual effects. Each of the correction factors can vary over quite wide ranges depending upon the configuration of the heat exchanger.

Equation (2.97) does not take into account the bypass and leakage effects on the total pressure drop calculations. Therefore Eq. (2.97) overestimates the shell-side pressure drop as shown in the following example:

Example 2.5. A single-pass tube, baffled single-pass shell, shell-and-tube heat exchanger is to be used in a hot-water heating system. The water enters the shell side of the heat exchanger at 70°C at the rate of 85,000 kg/hr. The water will be heated by condensing steam inside the tubes at 110°C ($P = 0.143$ MPa). It is proposed to use a heat exchanger that has 17.25 in. ID shell and 0.75 in. OD, 18-BWG tubes. The tubes are laid out on a 0.9375-in. triangular pitch. The length of the heat exchanger is 3 m. The shell contains segmental baffles that are spaced 0.3 m apart. The heat exchanger is designed according to TEMA standards. The number of tubes is 239. Assume fouling factors of 0.000352 $(m^2 \cdot K)/W$ and 0.00009 $(m^2 \cdot K)/W$ for the shell and tube sides, respectively. The shell-side heat transfer coefficient is calculated to be 6974 $W/(m^2 \cdot K)$ and it is assumed that the tube-side heat transfer coefficient is 9000 $W/(m^2 \cdot K)$. The allowable pressure drop on the shell side is 10 psi and the expected outlet temperature of the water is greater than 85°C. The properties of water at 80°C are $c_p = 4197$ J/(kg · K), $\rho = 972$ kg/m^3, $\mu = 0.000352$ Pa · s. Will this heat exchanger be suitable for this application?

Solution: This example may be classified as one requiring a heat exchanger performance calculation. Accordingly, it is expedient to base the calculations on the NTU method.

The overall heat transfer coefficient is obtained as $U_o = 1330$ $W/(m^2 \cdot K)$ from Eq. (2.20):

$$\frac{1}{U_o} = \frac{r_o}{r_i}\frac{1}{h_i} + \frac{r_o}{r_i}R_{fi} + \frac{r_o \ln(r_o/r_i)}{k} + R_{fo} + \frac{1}{h_o}$$

where

$$d_o = 0.01905 \text{ m } (= 0.75 \text{ in.})$$

$$d_i = 0.01656 \text{ m } (= 0.652 \text{ in.})$$

$$k = 53 \text{ W/(m} \cdot \text{K)}$$

The total heat transfer area of the heat exchanger is

$$A_o = n\pi\, d_o L = 239 \times \pi \times 0.01905 \times 3 = 42.91 \text{ m}^2$$

The heat capacity rates are

$$C_c = C_{\min} = \left(\dot{m}c_p\right)_c = \frac{85{,}000 \times 4.197}{3600} = 99.10 \text{ kW/K}$$

$$C^* = \frac{C_{\min}}{C_{\max}} = 0$$

The number of transfer units is

$$\text{NTU} = \frac{U_o A_o}{C_{\min}} = \frac{1330 \times 42.91}{99{,}100} = 0.576$$

From Fig. 2.15 or from Eq. (2.50), the heat exchanger effectiveness is found to be

$$\varepsilon = 0.438$$

It is now a simple matter to determine the outlet temperature of the heating water from Eq. (2.42):

$$\varepsilon = \frac{T_{c2} - T_{c1}}{T_{h1} - T_{c1}} \qquad 0.438 = \frac{T_{c2} - 70}{110 - 70} \qquad T_{c2} = 87.5°\text{C}$$

The pressure drop on the shell side can be calculated from Eq. (2.97). For this equation

$$D_s = 0.4381 \text{ m} \qquad N_B = 9 \qquad p_T = 0.02381 \text{ m} \qquad C = p_T - d_o = 0.004762 \text{ m}$$

The shell characteristic area from Eq. (2.95) is

$$a_s = \frac{D_s C B}{p_T} = \frac{0.4381 \times 0.004762 \times 0.3}{0.02381} = 0.02629 \text{ m}^2$$

The shell-side mass velocity from Eq. (2.96) is

$$G_s = \frac{\dot{m}}{a_s} = \frac{85{,}000}{3600 \times 0.02629} = 898.1 \text{ kg/(m}^2 \cdot \text{s)}$$

The shell equivalent diameter (hydraulic diameter) is calculated from Eq. (2.94):

$$D_e = \frac{4\left(p_T^2\sqrt{3}/4 - \pi d_o^2/8\right)}{\pi d_o/2}$$

$$= \frac{4 \times \left(0.02381^2 \times \sqrt{3}/4 - \pi \times 0.01905^2/8\right)}{\pi \times 0.01905/2}$$

$$= 0.01376 \text{ m}$$

The shell-side Reynolds number is

$$Re_s = \frac{G_s D_e}{\mu} = \frac{898.1 \times 0.01376}{0.000352} = 35{,}107$$

The friction factor for the shell side is given by Eq. (2.98):

$$f = \exp(0.576 - 0.19 \ln Re_s) = 0.2435$$

The shell-side pressure drop is then calculated as

$$\Delta P_s = f\frac{D_s(N_B + 1)G_s^2}{2\rho D_e \phi_s} = 0.2435 \times \frac{0.4381 \times (9 + 1) \times 898.1^2}{2 \times 972 \times 0.01376 \times 1} = 32.16 \text{ kPa}$$

It was assumed that the properties are constant and $\phi_s = 1$.

The Delaware method gives the total nozzle-to-nozzle shell-side pressure drop as 15.67 kPa. Therefore Eq. (2.97) overestimates the shell-side pressure drop by a factor of 2.05 for this configuration and this heat exchanger will serve the purpose.

2.11.4 Heat Transfer and Pumping-Power Relationship

The fluid pumping power $\tilde{P}$ is proportional to the pressure drop in the fluid across a heat exchanger. It is given by

$$\tilde{P} = \frac{\dot{m}\,\Delta P}{\rho \eta_p} \tag{2.99}$$

where η_p is the pump or fan efficiency.

Frequently, the cost in terms of increased fluid friction requires an input of pumping power greater than the realized benefit of increased heat transfer.

In the design of heat exchangers involving high-density fluids, the fluid pumping-power requirement is usually quite small relative to the heat transfer rate, and thus the friction power expenditure (i.e., pressure drop) has hardly any influence on the design. However, for gases and low-density fluids and also for very high viscosity fluids, the pressure drop is always of equal importance to the heat transfer rate and it has a strong influence in the design of such heat exchangers.

Let us consider the single-phase side passage of a heat exchanger where the flow is turbulent and the surface is smooth. The forced convection correlation and the friction coefficient can be expressed as

$$StPr^{2/3} = \phi_h(Re) \tag{2.100}$$

and

$$f = \phi_f(Re) \tag{2.101}$$

Equation (2.100) gives the heat transfer coefficient

$$h = (\mu c_p) Pr^{-2/3} \frac{Re}{D_e} \phi_h \tag{2.102}$$

h can be expressed in $W/(m^2 \cdot K)$ and it can be interpreted as the heat transfer power per unit surface area.

If the pressure drop through the passage is ΔP and the associated heat transfer surface area is A, the pumping power per unit heat transfer area (W/m^2) is given by

$$\frac{\tilde{P}}{A} = \frac{\Delta P \dot{m}}{\rho \eta_p} \frac{1}{A} \tag{2.103}$$

By substituting ΔP from Eq. (2.90) into Eq. (2.103) and noting that $D_e = 4A_c/p_w$,

$$\frac{\tilde{P}}{A} = 8\left(\frac{\mu^3}{\eta_p \rho^2}\right)\left(\frac{1}{D_e}\right)^3 Re^3 \phi_f \tag{2.104}$$

If it is assumed for simplicity that the friction coefficient is given by Eq. (2.81a) and the Colburn analogy is applicable, then

$$\phi_f = 0.046 Re^{-0.2} \tag{2.105}$$

$$\phi_h = 0.023 Re^{-0.2} \tag{2.106}$$

TABLE 2.5 Pumping-Power Expenditure for Various Fluid Conditions (η_p = 80%, D_e = 0.0241 m) [19g]

Fluid Conditions	Power Expenditure, W/m^2
Water at 300 K	
h = 3850 W/(m^2 · K)	3.85
Ammonia at 500 K, atmospheric pressure	
h = 100 W/(m^2 · K)	29.1
h = 248 W/(m^2 · K)	697
Engine oil at 300 K	
h = 250 W/(m^2 · K)	0.270×10^4
h = 500 W/(m^2 · K)	4.3396×10^4
h = 1200 W/(m^2 · K)	92.94×10^4

Equations (2.105) and (2.106) approximate the typical characteristics of fully developed turbulent flow in smooth tubes. Substituting these relations into Eqs. (2.102) and (2.104) and combining them to eliminate the Reynolds number, the pumping power per unit heat transfer area (W/m^2) is obtained as

$$\frac{\tilde{P}}{A} = \frac{C h^{3.5} \mu^{1.83} (D_e)^{0.5}}{k^{2.33} c_p^{1.17} \rho^2 \eta_p} \tag{2.107}$$

where $C = 1.2465 \times 10^4$.

As can be seen from Eq. (2.107), the pumping power depends strongly on fluid properties, as well as on the equivalent diameter of the flow passage. Some important conclusions can be drawn from Eq. (2.107) (see Table 2.5):

1. With a high-density fluid such as a liquid, the heat exchanger surface can be operated at large values of h without excessive pumping-power requirements.
2. A gas with its very low density results in high values of pumping power for even very moderate values of the heat transfer coefficient.
3. A large value of viscosity causes the friction power to be large even though the density may be high. Thus heat exchangers using oils must be designed for relatively low values of h in order to hold the pumping power within acceptable limits.
4. The thermal conductivity k also has a very strong influence and therefore, for liquid metals with very large values of thermal conductivity, the pumping power is seldom of significance.
5. Small values of equivalent diameter D_e tend to minimize the pumping power.

2.12 SUMMARY

In this chapter the basic design methods for two-fluid direct-transfer heat exchangers are reviewed. The LMTD, ε–NTU, P–NTU$_c$, and ψ–P methods are briefly discussed and basic relationships are introduced. Pressure drop relations applicable for certain types of heat exchangers are also reviewed. Some of the assumptions made in this chapter restrict the analysis to single-phase flow on both sides or on one side with dominating thermal resistance. For two-phase flows on both sides, many of the foregoing assumptions are not valid. The design theory of a specific two-phase-flow heat exchanger is presented elsewhere in this book.

NOMENCLATURE

A	total heat transfer area on one side of a recuperator, m^2
A_c	cross-sectional area of a heat exchanger passage, m^2
A_f	fin surface area on one side of a heat exchanger, m^2
A_u	unfinned surface area on one side of a heat exchanger, m^2
a	dimension of rectangular duct, m
a_s	shell characteristic area, m^2
B	baffle spacing, m
b	dimension of rectangular duct, m
b	half-distance between parallel plates, m
C	clearance between adjacent tubes, m
C	constant, in Eq. (2.107)
C	flow stream heat capacity rate, $\dot{m}c_p$, W/K
C_{max}	maximum of C_c and C_h, W/K
C_{min}	minimum of C_c and C_h, W/K
C^*	heat capacity rate ratio, C_{min}/C_{max}
c_p	specific heat at constant pressure, $J/(kg \cdot K)$
D_e	equivalent diameter (hydraulic diameter) of flow passage, $4A/p_w$, m
d_i	tube inside diameter, m
d_o	tube outside diameter, m
e	tube surface roughness, m
F	LMTD correction factor
f	Fanning friction factor, defined by Eq. (2.78)
G	mass velocity, $kg/(m^2 \cdot s)$
h	heat transfer coefficient, $W/(m^2 \cdot K)$
i	specific enthalpy, J/kg
k	thermal conductivity, $W/(m \cdot K)$
L	length of the heat exchanger, m
M_i	parameter, defined by Eq. (2.71), $1/m^2$
$\dot{m}$	fluid mass flow rate, kg/s

N_B	number of baffles
N_p	number of tube passes
NTU	number of heat transfer units based on C_{min}, UA/C_{min}
n	number of incremental areas
P	pressure, Pa
ΔP	pressure drop, Pa
P	temperature effectiveness, defined by Eq. (2.34)
$\tilde{P}$	fluid pumping power, W
Pr	Prandtl number, $\mu c_p/k$
p_w	wetted perimeter, m
p_T	pitch, m
Q	heat transfer rate, W
R	heat capacity rate ratio, defined by Eq. (2.35)
R	thermal resistance, $(m^2 \cdot K)/W$
R_f	fouling factor, $(m^2 \cdot K)/W$
Re	Reynolds number based on the equivalent diameter, $\rho u_m D_e/\mu$
r	tube radius, m
St	Stanton number, $h/\rho c_p u_m$
T	temperature, K
T_c	cold-fluid temperature, K
T_h	hot-fluid temperature, K
ΔT	local temperature difference between two fluids, K
ΔT_{lm}	log-mean temperature difference, defined by Eq. (2.28), K
ΔT_m	true mean temperature difference, defined by Eq. (2.7), K
t	wall thickness, m
U	overall heat transfer coefficient, $W/(m^2 \cdot K)$
u_m	fluid mean velocity, m/s

Greek Symbols

Δ	difference
δ	fin thickness, m
ε	heat exchanger effectiveness, defined by Eq. (2.39)
η_f	fin efficiency
η_o	extended surface efficiency, defined by Eq. (2.15)
μ	dynamic viscosity, Pa · s
ν	kinematic viscosity, m^2/s
ρ	fluid density, kg/m^3
τ_w	wall shear stress, Pa
ϕ	parameter, function of
ψ	ratio of true mean ΔT to inlet ΔT, defined by Eq. (2.57)

Subscripts

c	cold fluid
cf	counterflow

f	fin, finned, friction
h	hot fluid, heat transfer
i	inner, inside
m	mean
max	maximum
min	minimum
o	outer, outside, overall
p	pump
pf	parallel flow
r	return
s	shell, scale
t	tube, thermal, total
u	unfinned
w	wall
x	local
1	inlet
2	outlet

REFERENCES

1. Shah, R. K. (1981) Classification of heat exchangers. In *Heat Exchangers: Thermal-Hydraulic Fundamentals and Design*, S. Kakaç, A. E. Bergles, and F. Mayinger (eds.), pp. 9–46. Hemisphere, New York.
2. Chenoweth, J. M., and Impagliazzo, M. (eds.) (1981) *Fouling in Hot Exchange Equipment*, ASME Symposium Volume HTD-17. ASME, New York.
3. Kakaç, S., Shah, R. K., and Bergles, A. E. (eds.) (1981) *Low Reynolds Number Flow Heat Exchangers*, pp. 21–72. Hemisphere, New York.
4. Kakaç, S., Shah, R. K., and Aung, W. (eds.) (1987) *Handbook of Single Phase Convective Heat Transfer*, Chapters 4 and 18. Wiley, New York.
5. Kern, D. Q., and Kraus, A. D. (1972) *Extended Surface Heat Transfer*. McGraw-Hill, New York.
6. Bowman, R. A., Mueller, A. C., and Nagle, W. M. (1940) Mean temperature difference in design. *Trans. ASME* **62** 283–294.
7. *Standard of the Tubular Exchange Manufacturers Association* (1978) 6th ed. Tubular Exchanger Manufacturers Association (TEMA), New York.
8. Kays, W. M., and London, A. L. (1984) *Compact Heat Exchangers*, 3rd ed. McGraw-Hill, New York.
9. Shah, R. K., and Mueller, A. C. (1985) Heat exchangers. In *Handbook of Heat Transfer Applications*, W. M. Rohsenow, J. P. Hartnett, and E. N. Ganić (eds.), Chapter 4. McGraw-Hill, New York.
10. Kays, W. M., London, A. L., and Johnson, D. W. (1951) *Gas Turbine Plant Heat Exchangers*. ASME, New York.
11. Mueller, A. C. (1967) New charts for true mean temperature difference in heat exchangers. *AIChE* Paper 10, Ninth Nat. Heat Transfer Conference, Seattle.

12. Colburn, A. P. (1933) Mean temperature difference and heat transfer coefficient in liquid heat exchangers. *Ind. Eng. Chem.* **25** 873–877.
13. Butterworth, D. (1981) Condensers: Thermohydraulic design. In *Heat Exchangers: Thermal-Hydraulic Fundamentals and Design*, S. Kakaç, A. E. Bergles, and F. Mayinger (eds.), pp. 647–679. Hemisphere, New York.
14. Moody, L. F. (1944) Friction factor for pipe flow. *Trans. ASME* **66** 671–684.
15. Brundrett, E. (1979) Modified hydraulic diameter for turbulent flow. In *Turbulent Forced Convection in Channels and Bundles*, S. Kakaç and D. B. Spalding (eds.), Vol. 1, pp. 361–367. Hemisphere, New York.
16. McAdams, W. H. (1954) *Heat Transmission*, 3rd ed. McGraw-Hill, New York.
17. Kern, D. Q. (1950) *Process Heat Transfer*. McGraw-Hill, New York.
18. Bell, K. (1981) Delaware method for shell side design. In *Heat Exchangers: Thermal-Hydraulic Fundamentals and Design*, S. Kakaç, A. E. Bergles, and F. Mayinger (eds.), pp. 581–618. Hemisphere, New York.
19. Kakaç, S., Bergles, A. E., and Fernandes, E. O. (eds.) (1990) *Two-Phase Flow Heat Exchangers*, pp. 29–80. Kluwer, Dordrecht.

CHAPTER 3

FORCED CONVECTION CORRELATIONS FOR SINGLE-PHASE SIDE OF HEAT EXCHANGERS

S. KAKAÇ

Department of Mechanical Engineering
University of Miami
Coral Gables, Florida 33124

R. OSKAY

Department of Mechanical Engineering
Middle East Technical University
Ankara, Turkey

3.1 INTRODUCTION

In many two-phase-flow heat exchangers such as boilers, steam generators, power condensers, air conditioning evaporators, and condensers, one side has single-phase fluid while the other side has two-phase flow. Generally, the single-phase side represents higher thermal resistance, particularly with gas or oil flow. In this chapter a comprehensive review is made of the available correlations for laminar and turbulent flow of single-phase newtonian fluid through circular and noncircular ducts with and without the effect of property variations. A large number of experimental and analytical correlations are available for the heat transfer coefficient and the flow friction factor for laminar and turbulent flow through ducts. In this chapter recommended correlations for the single-phase side of heat exchangers are given. Condensation heat transfer is discussed in Chapters 10 and 12. Design information for the boiling side is discussed in Chapters 6, 8, 12, and 13.

Laminar and turbulent forced convection correlations for single-phase fluids represent an important class of heat transfer solutions for heat ex-

Boilers, Evaporators and Condensers, Edited by Sadik Kakaç
ISBN 0-471-62170-6

changer applications. When a viscous fluid flows in a duct, a boundary layer will form along the duct. The boundary layer gradually fills the entire duct and the flow then is said to be fully developed. The distance at which the velocity becomes fully developed is called the hydrodynamic or velocity entrance length (L_{he}). Theoretically, the approach to the fully developed velocity profile is asymptotic and it is therefore impossible to describe a definite location where the boundary layer completely fills the duct.

If the walls of the duct are heated or cooled, then a thermal boundary layer will also develop along the duct. At a certain point downstream, one can talk about the fully developed temperature profile where the thickness of the thermal boundary layer is approximately equal to $d/2$. The distance at which the temperature profile becomes fully developed is called the thermal entrance length (L_{te}).

If heating starts from the inlet of the duct, then both the velocity profile and the temperature profile develop simultaneously. The associated heat transfer problem is referred to as the combined hydrodynamic and thermal entry length problem or simultaneously developing region problem. Therefore there are four types of duct flows with heating, namely, fully developed, hydrodynamically developing, thermally developing, and simultaneously developing, and the design correlations should be selected accordingly.

The rate of development of the velocity and temperature profiles in the combined entrance region depends on the fluid Prandtl number ($Pr = \nu/\alpha$). For high Prandtl number fluids, such as oils, even though the velocity and temperature profiles are uniform at the tube entrance, the velocity profile is established much more rapidly than the temperature profile. In contrast, for very low Prandtl number fluids, such as liquid metals, the temperature profile is established much more rapidly than the velocity profile. However, for Prandtl numbers about 1, as for gases, the temperature and velocity profiles develop at a similar rate simultaneously along the duct, starting from uniform temperature and uniform velocity at the duct entrance.

For the limiting case of $Pr \to \infty$, the velocity profile is developed before the temperature profile starts developing. For the other limiting case of $Pr = 0$, the velocity profile never develops and remains uniform while the temperature profile is developing. The idealized $Pr \to \infty$ and 0 cases are good approximations for highly viscous fluids and liquid metals, respectively.

When fluids flow at very low velocities, the fluid particles move in definite paths called streamlines. This type of flow is called laminar flow. There is no component of fluid velocity normal to the duct axis in the fully developed region. Depending on the roughness of the circular duct inlet and inside surface, fully developed laminar flow will be obtained up to $Re_d \leq 2300$ within the duct length L longer than the hydrodynamic entry length L_{he}; however if $L < L_{he}$, developing laminar flow would exist over the entire duct length. The hydrodynamic and thermal entrance lengths for laminar flow inside conduits have been given in [1, 2]. The hydrodynamic entrance length

TABLE 3.1 Hydrodynamic Entrance Length L_{he} and Thermal Entrance Length L_{te} for Laminar Flow inside Ducts[a]

		$\dfrac{L_{te}/D_h}{Pe}$	
Geometry	$\dfrac{L_{he}/D_h}{Re}$	Constant Wall Temperature	Constant Wall Heat Flux
	0.056	0.033	0.043
	0.011	0.008	0.012
$a/b = 0.25$	0.075	0.054	0.042
0.50	0.085	0.049	0.057
1.0	0.09	0.041	0.066

[a] Based on the results reported in [1, 2]. The thermal entry lengths are for the hydrodynamically developed, thermally developing flow conditions.

L_{he}, for laminar flow inside ducts of various cross sections based on the definition discussed previously, is presented in Table 3.1. Included in this table are the thermal entrance lengths for constant wall temperature and constant wall heat flux boundary conditions for thermally developing, hydrodynamically developed flow. In Table 3.1 the Reynolds number is based on the hydraulic diameter D_h.

If the velocity of the fluid is gradually increased, there will be a point where the laminar flow becomes unstable in the presence of small disturbances and the fluid no longer flows along parallel lines (streamlines), but by a series of eddies that result in a complete mixing of the entire flow field. This type of flow is called turbulent flow. The Reynolds number at which the flow changes from laminar to turbulent is referred to as the critical (value of) Reynolds number. The critical Reynolds number in circular ducts is between 2100 and 2300. Although the value of the critical Reynolds number depends on the duct cross-sectional geometry and surface roughness, for particular applications it can be assumed that the transition from laminar to turbulent flow in noncircular ducts will also take place between $Re_{cr} = 2100\text{–}2300$ when the hydraulic diameter of the duct, which is defined as four times the cross-sectional (flow) area A_c divided by the wetted perimeter P of the duct, is used in calculating the Reynolds number.

At a Reynolds number $Re > 10^4$, the flow is completely turbulent. Between the lower and upper limits lies the transition zone from laminar to turbulent flow. Therefore fully turbulent flow in a duct occurs at a Reynolds number $Re \geq 10^4$.

The heat flux between the duct wall and a fluid flowing inside the duct can be calculated at any position along the duct by

$$\frac{Q}{A} = h_x(T_w - T_b)_x \tag{3.1}$$

where h_x is called the local heat transfer coefficient or film coefficient and is defined on the inner surface of the duct wall by using the convective boundary condition

$$h_x = \frac{-k(\partial T/\partial y)_w}{(T_w - T_b)_x} \tag{3.2}$$

where k is the thermal conductivity of the fluid, T is the temperature distribution in the fluid, and T_w and T_b are the wall and the fluid bulk temperatures, respectively. Then the local Nusselt number is calculated from

$$Nu_x = \frac{h_x d}{k} = \frac{-d(\partial T/\partial y)_w}{(T_w - T_b)_x} \tag{3.3}$$

The fluid bulk temperature T_b, also referred to as the "mixing cup" or flow average temperature, is defined as

$$T_b = \frac{1}{A_c u_m} \int_{A_c} uT\, dA_c \tag{3.4}$$

where u_m is the mean velocity of the fluid and u and T are, respectively, the velocity and temperature profiles of the flow at position x along the duct.

The local heat transfer coefficient is utilized in calculating the axial (in the x direction) variation of the duct wall temperature or the local heat flux. In design problems, it is necessary to calculate the total heat transfer rate over the total (entire) length of a duct using a mean value of the heat transfer coefficient based on the mean value of the Nusselt number defined as

$$Nu = \frac{1}{L} \int_o^L Nu_x\, dx \tag{3.5}$$

3.2 LAMINAR FORCED CONVECTION

Laminar duct flow is generally encountered in compact heat exchangers, cryogenic cooling systems, heating or cooling of heavy (highly viscous) fluids such as oils, and in many other applications. Different investigators performed extensive experimental and theoretical studies with various fluids for numerous duct geometries and under different wall and entrance conditions. As a result, they formulated relations for the Nusselt number versus the Reynolds and Prandtl numbers for a wide range of these dimensionless groups. Shah and London [1] and Shah and Bhatti [2] have compiled the laminar flow solutions.

Laminar flow can be obtained for a specified mass velocity $G = \rho u_m$ for (1) low hydraulic diameter D_h of the flow passage or (2) high fluid viscosity μ. Flow passages with small hydraulic diameter are encountered in compact heat exchangers since they result in large surface area per unit volume of the exchanger. The internal flow of oils and other liquids with high viscosity in noncompact heat exchangers is generally of a laminar nature.

3.2.1 Hydrodynamically Developed and Thermally Developing Laminar Flow in Smooth Circular Ducts

The well-known Nusselt–Graetz problem for heat transfer to an incompressible fluid with constant properties flowing through a circular duct with constant wall temperature boundary conditions and fully developed laminar velocity profile was solved numerically by several investigators [1, 2]. The asymptotes of the mean Nusselt number for a circular duct of the length L are

$$Nu_T = 1.61\left(\frac{Pe_b d}{L}\right)^{1/3} \qquad \text{for } \frac{Pe_b d}{L} > 10^3 \tag{3.6}$$

and

$$Nu_T = 3.66 \qquad \text{for } \frac{Pe_b d}{L} < 10^2 \tag{3.7}$$

The superposition of two asymptotes for the mean Nusselt number derived by Schlünder [3] gives sufficiently good results for most of the practical cases:

$$Nu_T = \left[3.66^3 + 1.61^3\left(\frac{Pe_b d}{L}\right)\right]^{1/3} \tag{3.8}$$

An empirical correlation has also been developed by Hausen [4] for laminar flow in the thermal entrance region of circular duct at constant wall temperature and is given as

$$Nu_T = 3.66 + \frac{0.19(Pe_b d/L)^{0.8}}{1 + 0.117(Pe_b d/L)^{0.467}} \tag{3.9}$$

The results of Eqs. (3.8) and (3.9) are comparable to each other. These equations may be used for the laminar flow of gases and liquids in the range $0.1 < Pe_b d/L < 10^4$. Axial conduction effects must be considered at $Pe_b d/L < 0.1$. All physical properties are evaluated at the fluid bulk mean temperature of T_b, defined as

$$T_b = \frac{T_i + T_o}{2} \tag{3.10}$$

where T_i and T_o are the bulk temperatures of the fluid at the inlet and exit of the duct, respectively.

The asymptotic mean Nusselt numbers in circular tubes with constant wall heat flux boundary conditions are [1]:

$$Nu_H = 1.953\left(\frac{Pe_b d}{L}\right)^{1/3} \qquad \text{for } \frac{Pe_b d}{L} > 10^2 \tag{3.11}$$

and

$$Nu_H = 4.36 \qquad \text{for } \frac{Pe_b d}{L} < 10 \tag{3.12}$$

The fluid properties are evaluated at the mean bulk temperature T_b as defined by Eq. (3.10).

The results given by Eqs. (3.7) and (3.12) represent the dimensionless heat transfer coefficients for laminar forced convection inside a circular duct in the hydrodynamically and thermally developed regions under constant wall temperature and constant wall heat flux boundary conditions, respectively.

3.2.2 Simultaneously Developing Laminar Flow in Smooth Ducts

When heat transfer starts as soon as the fluid enters a duct, the velocity and temperature profiles start developing simultaneously. The analysis of the temperature distribution in the flow, and hence of the heat transfer between

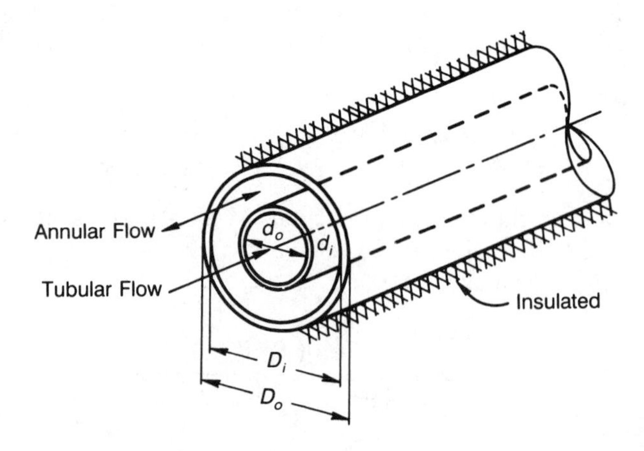

Fig. 3.1. Concentric tube annulus (41).

the fluid and the duct wall, for such situations is more complex because the velocity distribution varies in the axial direction as well as normal to it. Heat transfer problems involving simultaneously developing flow have been mostly solved by numerical methods for various duct cross sections. A comprehensive review of such solutions are given by Shah and Bhatti [2] and Kakaç [5].

Shah and London [1] and Shah and Bhatti [2] presented the numerical values of the mean Nusselt number for this region. In the case of a short duct length, Nu values are represented by the asymptotic equation of Pohlhausen [6] for simultaneously developing flow over a flat plate; for a circular duct, this equation becomes

$$Nu_T = 0.664\left(\frac{Pe_b d}{L}\right)^{1/2} Pr_b^{-1/6} \tag{3.13}$$

The range of validity is $0.5 < Pr_b < 500$ and $Pe_b d/L > 10^3$.

For most engineering applications with short circular ducts ($d/L > 0.1$), it is recommended that, whichever of Eqs. (3.8), (3.9), and (3.13) gives the highest Nusselt number, be used.

3.2.3 Laminar Flow through Concentric Smooth Ducts

Correlations for concentric annular ducts are very important in heat exchanger applications. The simplest form of a two-fluid heat exchanger is a double-pipe heat exchanger made up of two concentric circular tubes (Fig. 3.1). One fluid flows inside the inner tube while the other flows through the annular passage. Heat is usually transferred through the wall of the inner tube while the outer wall of the annular duct is insulated. The heat transfer coefficient in the annular duct depends on the ratio of the diameters (D_i/d_o) because of the shape of the velocity profile.

The hydraulic (equivalent) diameter approach is the simplest method to calculate the heat transfer and the pressure drop in the annulus. In this approach, the equivalent diameter of annulus D_h is substituted instead of the tube diameter in internal flow correlations:

$$D_h = 4\frac{\text{net free-flow area}}{\text{wetted (or heat transfer) perimeter}} \tag{3.14}$$

This approximation is acceptable for heat transfer and pressure drop calculations. The validity of the hydraulic diameter approach has been substantiated by the results of experiments performed with finned annuli [7].

The total wetted perimeter of the annulus for pressure drop calculations is given by

$$P_w = \pi(D_i + d_o) \tag{3.15}$$

and the heat transfer perimeter of the annulus can be calculated by

$$P_h = \pi d_o \tag{3.16}$$

The only difference between P_w and P_h is D_i, which is the inner diameter of the shell (outer tube) of the annulus. This difference is due to the fluid friction on the inner surface of the shell; however, this is not the case for the heat transfer perimeter since the heat transfer takes place only through the walls of the inner tube. The net free-flow area of the annulus is given by

$$A_C = \frac{\pi\left(D_i^2 - d_o^2\right)}{4} \tag{3.17}$$

The hydraulic diameter based on the total wetted perimeter for pressure drop calculation is

$$D_h = \frac{4A_C}{P_w} \tag{3.18}$$

and the hydraulic diameter based on the heat transfer perimeter is given by Eq. (3.19), which is hereafter called the equivalent diameter.

$$D_e = \frac{4A_C}{P_h} \tag{3.19}$$

The Reynolds number, Graetz number, and the ratio d/L are to be calculated with D_h. D_e is used to calculate the heat transfer coefficient from the Nusselt number and to evaluate the Grashof number. Slightly higher heat transfer coefficients arise when D_h is used instead of D_e for heat transfer calculations.

For the constant wall temperature boundary condition, Stephan [8] has developed a heat transfer correlation based on Eq. (3.9). The Nusselt number for hydrodynamically developed laminar flow in the thermal entrance region of an isothermal annulus, the outer wall of which is insulated, may be calculated by the following correlation:

$$Nu_T = Nu_\infty + \left[1 + 0.14\left(\frac{d_o}{D_i}\right)^{-1/2}\right]\frac{0.19(Pe_b D_h/L)^{0.8}}{1 + 0.117(Pe_b D_h/L)^{0.467}} \tag{3.20}$$

where Nu_∞ is the Nusselt number for fully developed flow.

A detailed review of laminar convective heat transfer in ducts for various hydrodynamic and thermal boundary conditions is given in [2].

3.3 THE EFFECT OF VARIABLE PHYSICAL PROPERTIES

When the previously mentioned correlations are applied to practical heat transfer problems with large temperature differences between the surface and the fluid, the constant-property assumption could cause significant errors, since the transport properties of the most fluids vary with temperature, which influence the variation of velocity and temperature through the boundary layer or over the flow cross section of a duct.

For practical applications, a reliable and appropriate correlation based on the constant-property assumption can be modified and/or corrected so that it may be used when the variable-property effect becomes important.

Two methods of correcting constant-property correlations for the variable-property effect have been employed: namely the reference temperature method and the property ratio method. In the former, a characteristic temperature is chosen at which the properties appearing in nondimensional groups are evaluated so that the constant-property results at that temperature may be used to consider the variable-property behavior; in the latter, all properties are taken at the bulk temperature and then all variable-property effects are lumped into a function of the ratio of one property evaluated at the wall (surface) temperature to that property evaluated at the bulk temperature. Some correlations may involve a modification or combination of these two methods.

For liquids, the variation of viscosity is responsible for most of the property effects. Therefore the variable-property Nusselt numbers and friction factors in the property ratio method for liquids are correlated by

$$\frac{Nu}{Nu_{\mathrm{cp}}} = \left(\frac{\mu_b}{\mu_w}\right)^n \tag{3.21a}$$

$$\frac{f}{f_{\mathrm{cp}}} = \left(\frac{\mu_b}{\mu_w}\right)^m \tag{3.21b}$$

where μ_b is the viscosity evaluated at the bulk mean temperature, μ_w is the viscosity evaluated at the wall temperature, and cp refers to the constant-property solution. The friction coefficient usually employed is the so-called Fanning friction factor based on the wall shear rather than the pressure drop.

For gases, the viscosity, thermal conductivity, and density vary with the absolute temperature. Therefore, in the property ratio method, temperature corrections of the following forms are found to be adequate in practical applications for the temperature-dependent property effects in gases:

$$\frac{Nu}{Nu_{cp}} = \left(\frac{T_w}{T_b}\right)^n \tag{3.22a}$$

$$\frac{f}{f_{cp}} = \left(\frac{T_w}{T_b}\right)^m \tag{3.22b}$$

where T_b and T_w are the absolute bulk mean and wall temperatures, respectively.

It must be noted that the constant-property portion of the specific correlation is evaluated in terms of the parameters and conditions defined by its author(s).

Extensive theoretical and experimental investigations on convective heat transfer of fluids with variable properties have been reported in the literature to obtain the values of the exponents n and m which will be cited in the following sections of this chapter.

3.3.1 Laminar Flow of Liquids

Deissler [9] carried out a numerical analysis as described previously for laminar flow through a circular duct at constant heat flux boundary conditions for liquid viscosity variation with temperature given by

$$\frac{\mu}{\mu_w} = \left(\frac{T}{T_w}\right)^{-1.6} \tag{3.23}$$

and obtained $n = 0.14$ to be used with Eq. (3.21a). This has been used widely to correlate experimental data for laminar flow for $Pr > 0.6$.

Deissler [9] also obtained $m = -0.58$ for heating and $m = -0.50$ for cooling of liquids to be used with Eq. (3.21b).

Yang [10] obtained the solution for both constant wall heat flux and constant wall temperature boundary conditions by assuming a viscosity dependence of a liquid on temperature as

$$\frac{\mu}{\mu_w} = \left[1 + A\left(\frac{T_w - T}{T_w - T_i}\right)\right]^{-1} \tag{3.24}$$

where A is a constant. His predictions for both constant wall heat flux and constant wall temperature boundary conditions were correlated with $n = 0.11$ in Eq. (3.21a), and he concluded that the effect of thermal boundary conditions is small and the influence on the friction coefficient is very substantial. He also found that the correction for variable properties is the same for developing and developed regions.

A simple empirical correlation has been proposed by Seider and Tate [11] to predict the mean Nusselt number for laminar flow in a circular duct at constant wall temperature

$$Nu_T = 1.86\left(\frac{Pe_b d}{L}\right)^{1/3}\left(\frac{\mu_b}{\mu_w}\right)^{0.14} \tag{3.25}$$

which is valid for smooth tubes, $0.48 < Pr_b < 16{,}700$, and $0.0044 < (\mu_b/\mu_w) < 9.75$. This correlation has been recommended by Whitaker [12] for values of

$$\left(\frac{Pe_b d}{L}\right)^{1/3}\left(\frac{\mu_b}{\mu_w}\right)^{0.14} \geq 2 \tag{3.26}$$

All physical properties are evaluated at the fluid bulk mean temperature except μ_w, which is evaluated at the wall temperature.

It is not surprising that alternative correlations have been proposed for specific fluids. Oskay and Kakaç [13] performed experimental studies with mineral oil in laminar flow through a circular duct under constant wall heat flux boundary conditions in the range of $0.8 \times 10^3 < Re_b < 1.8 \times 10^3$ and $1 < (T_w/T_b) < 3$ and suggested that the viscosity ratio exponent for Nu should be increased to 0.152 for mineral oil.

Kuznetsova [14] conducted experiments with transformer oil and fuel oil in the range of $400 < Re_b < 1900$ and $170 < Pr_b < 640$ and recommended

$$Nu_b = 1.23\left(\frac{Pe_b d}{L}\right)^{0.4}\left(\frac{\mu_b}{\mu_w}\right)^{1/6} \tag{3.27}$$

Test [15] conducted an analytical and experimental study on the heat transfer and fluid friction of laminar flow in a circular duct for liquids with temperature-dependent viscosity. The analytical approach is a numerical

TABLE 3.2 Laminar Forced Convection Correlations in Smooth Straight Circular Ducts[a]

Number	Reference	Correlation	Limitations and Remarks
1	Nusselt and Graetz [1, 2]	$Nu_T = 1.61(Pe_b d/L)^{1/3}$ $Nu_T = 3.66$	$Pe_b d/L > 10^3$, constant wall temperature $Pe_b d/L < 10^2$, fully developed flow in a circular duct, constant wall temperature
2	Schlünder [3]	$Nu_T = [(3.66)^3 + (1.61)^3 Pe_b d/L]^{1/3}$	Superposition of two asymptotes given in case 1 for the mean Nusselt number. $0.1 < Pe_b d/L < 10^4$
3	Hausen [4]	$Nu_T = 3.66 + \dfrac{0.19(Pe_b d/L)^{0.8}}{1 + 0.117(Pe_b d/L)^{0.467}}$	Thermal entrance region, constant wall temperature. $0.1 < Pe_b d/L < 10^4$
4	Nusselt and Graetz [1, 2]	$Nu_T = 1.953(Pe_b d/L)^{1/3}$ $Nu_H = 4.36$	$Pe_b d/L > 10^2$, constant heat flux $Pe_b d/L < 10$, fully developed flow in a circular duct, constant heat flux
5	Pohlhausen [6]	$Nu_T = 0.664 \dfrac{1}{(Pr)^{1/6}} \left(Pe_b \dfrac{d}{L} \right)^{1/2}$	$Pe_b d/L > 10^3$, $0.5 < Pr < 500$, simultaneously developing flow

6	Stephan [8]	$Nu_T = Nu + \phi\left(\frac{d_o}{D_i}\right)\frac{0.19(PeD_{h/L})^{0.8}}{1 + 0.117(PeD_{h/L})^{0.467}}$	Circular annular duct, constant wall temperature, thermal entrance region
		$\phi(d_o/D_i) = 1 + 0.14(d_o/D_i)^{-1/2}$	Outer wall is insulated, heat transfer through the inner wall
		$\phi(d_o/D_i) = 1 + 0.14(d_o/D_i)^{0.1}$	Heat transfer through outer and inner wall
7	Sieder and Tate [11]	$Nu_T = 1.86(Re_b Pr_b d/L)^{1/3}(\mu_b/\mu_w)^{0.14}$	Thermal entrance region, constant wall temperature, $0.48 < Pr_b < 16{,}700$, $4.4 \times 10^{-3} < (\mu_b/\mu_w) < 9.75$, $(Re_b Pr_b d/L)^{1/3}(\mu_b/\mu_w)^{0.14} > 2$
8	Oskay and Kakaç [13]	$Nu_H = 1.86(Re_b Pr_b d/L)^{1/3}(\mu_b/\mu_w)^{0.152}$	Thermal entrance region, constant wall heat flux, for oils $0.8 \times 10^3 < Re_b < 1.8 \times 10^3$, $1 < (T_w/T_b) < 3$
9	Kuznetsova [14]	$Nu_H = 1.23(Re_b Pr_b d/L)^{0.4}(\mu_b/\mu_w)^{1/6}$	Thermal entrance region, constant heat flux, $400 < Re_b < 1900$, $170 < Pr_b < 640$, for oils
10	Test [15]	$Nu_b = 1.4(Re_b Pr_b d/L)^{1/3}(\mu_b/\mu_w)^n$	Thermal entrance region, $n = 0.05$ for heating liquids, $n = \frac{1}{3}$ for cooling liquids

[a] Unless otherwise stated, fluid properties are evaluated at the bulk mean fluid temperature, $T_b = (T_i + T_o)/2$.

solution of the continuity, momentum, and energy equations. The experimental approach involves the use of a hot-wire technique for determination of the velocity profiles. He obtained the following correlation for the local Nusselt number:

$$Nu_b = 1.4\left(\frac{Pe_b d}{L}\right)^{1/3}\left(\frac{\mu_b}{\mu_w}\right)^n \tag{3.28}$$

where

$$n = \begin{cases} 0.05 & \text{for heating} \\ \frac{1}{3} & \text{for cooling liquids} \end{cases}$$

He also obtained the friction factor as

$$f = \frac{16}{Re}\frac{1}{0.89}\left(\frac{\mu_b}{\mu_w}\right)^{0.2} \tag{3.29}$$

Equations (3.25) and (3.28) should not be applied to extremely long ducts.

3.3.2 Laminar Flow of Gases

The first reasonably complete solution for laminar heat transfer of a gas flowing in a tube with temperature-dependent properties was developed by Worsøe-Schmidt [16]. He solved the governing equations with a finite-difference technique for fully developed gas flow through a circular tube. Heating and cooling with a constant surface temperature and heating with a constant heat flux are considered. In this solution, the radial velocity is included. He concluded that near the entrance, and also well downstream, the results can be satisfactorily correlated for heating $1 < (T_w/T_b) < 3$ by $n = 0$, $m = 1.00$, and for cooling $0.5 < (T_w/T_b) < 1$ by $n = 0$, $m = 0.81$.

Laminar forced convection and fluid flow in ducts have been studied extensively, and numerous results are available for circular and noncircular ducts under various boundary conditions. These results have been compiled by Shah and London [1] and Shah and Bhatti [2]. The laminar forced convection correlations discussed in previous sections are summarized in Table 3.2. The constant-property correlations can be corrected for the variable physical properties by the use of Table 3.3 in which the exponents m and n are summarized. For fully developed laminar flow, $n = 0.14$ is generally recommended for heating liquids.

TABLE 3.3 Exponents n and m Associated with Eqs. (3.21) and (3.22) for Laminar Forced Convection through Circular Ducts, $Pr > 0.5$

Number	Reference	Fluid	Condition	n	m^a	Limitations
1	Deissler [9]	Liquid	Laminar, heating	0.14	−0.58	Fully developed flow,
		Liquid	Laminar, cooling	0.14	−0.50	q''_w = const, $Pr > 0.6$, $\mu/\mu_w = (T/T_w)^{-1.6}$
2	Yang [10]	Liquid	Laminar, heating	0.11	—	Developing and fully developed regions of a circular duct, T_w = const, q''_w = const
3	Worsøe-Schmidt [16]	Gas	Laminar, heating	0	1.00	Developing and fully developed regions, q''_w = const, T_w = const, $1 < (T_w/T_b) < 3$
		Gas	Laminar, cooling	0	0.81	T_w = const, $0.5 < (T_w/T_b) < 1$

[a] Fanning friction factor f is defined as $f = 2\tau_w/(\rho u_m^2)$ and for hydrodynamically developed isothermal laminar flow as $f = 16/Re$.

3.4 TURBULENT FORCED CONVECTION

Extensive experimental and theoretical efforts have been made to obtain the solutions for turbulent forced convection and flow friction problems in ducts because of their frequent occurrence and application in heat transfer engineering. A compilation of such solutions and correlations for circular and noncircular ducts has been summarized by Bhatti and Shah [17]. There are a large number of correlations available in the literature for the fully developed turbulent flow of single-phase newtonian fluids in smooth, straight circular ducts with constant and temperature-dependent physical properties. The objective of this section is to highlight some of the existing correlations to be used in the design of heat exchange equipment and to emphasize the conditions or limitations imposed on the applicability of these correlations.

3.4.1 Turbulent Flow in Circular Ducts with Constant Properties

Extensive efforts have been made to obtain empirical correlations that either represent a best-fit curve to the experimental data or have the constant in the theoretical equations adjusted to best fit the experimental data. An example of the latter is the correlation given by Petukhov and Popov [18]. Their theoretical calculations for the case of fully developed turbulent flow with constant properties in a circular tube with constant heat flux boundary conditions yielded the following correlation, which is based on the three-layer turbulent boundary layer model with constants adjusted to match the experimental data:

$$Nu_b = \frac{(f/2)\,Re_b Pr_b}{(1 + 13.6f) + \left(11.7 + 1.8 Pr_b^{-1/3}\right)(f/2)^{1/2}\left(Pr_b^{2/3} - 1\right)} \tag{3.30}$$

where

$$f = (3.64 \log Re_b - 3.28)^{-2} \tag{3.31}$$

and is defined as $f = \tau_w / \frac{1}{2}\rho u_m^2$.

Equation (3.30) is applicable for fully developed turbulent flow in the range $10^4 < Re_b < 5 \times 10^5$ and $0.5 < Pr_b < 2000$ with 1% error, and in the range $5 \times 10^5 < Re_b < 5 \times 10^6$ and $200 < Pr_b < 2000$ with 1% to 2% error. Equation (3.30) is also applicable to rough tubes. A simpler correlation has also been given by Petukhov and Kirillov as reported in [19] as

$$Nu_b = \frac{(f/2) Re_b Pr_b}{1.07 + 12.7(f/2)^{1/2}(Pr_b^{2/3} - 1)} \tag{3.32}$$

Equation (3.32) predicts the results in the range $10^4 < Re_b < 5 \times 10^6$ and $0.5 < Pr_b < 200$ with 5% to 6% error, and in the range $0.5 < Pr_b < 2000$ with 10% error.

Webb [20] has examined a range of data for fully developed turbulent flow in smooth tubes; he concluded that the relation developed by Petukhov and Popov, given previously, provides the best agreement with the measurements. Sleicher and Rouse [21] correlated analytical and experimental results for the range $0.1 < Pr_b < 10^4$ and $10^4 < Re_b < 10^6$, obtaining

$$Nu_b = 5 + 0.015 Re_b^m Pr_b^n \tag{3.33}$$

with

$$m = 0.88 - \frac{0.24}{4 + Pr_b}$$

$$n = \tfrac{1}{3} + 0.5 \exp(-0.6 Pr_b)$$

Equations (3.30), (3.32), and (3.33) are not applicable in the transition region. Gnielinski [22] further modified the Petukhov–Kirillov correlation by comparing it with the experimental data so that the correlation covers a lower Reynolds number range. Gnielinski recommended the following correlation:

$$Nu_b = \frac{(f/2)(Re_b - 1000) Pr_b}{1 + 12.7(f/2)^{1/2}(Pr_b^{2/3} - 1)} \tag{3.34}$$

where

$$f = (1.58 \ln Re_b - 3.28)^{-2} \tag{3.35}$$

The effect of thermal boundary conditions is almost negligible in turbulent forced convection [24]; therefore the empirical correlations given in Table 3.4 can be used for both constant wall temperature and constant wall heat flux boundary conditions.

TABLE 3.4 Correlations for Fully Developed Turbulent Forced Convection through a Circular Duct with Constant Properties

Number	Reference	Correlation[a]	Remarks and Limitations
1	Prandtl [23, 24]	$Nu_b = \dfrac{(f/2)Re_b Pr_b}{1 + 8.7(f/2)^{1/2}(Pr_b - 1)}$	Based on three-layer turbulent boundary layer model. $Pr > 0.5$
2	McAdams [25]	$Nu_b = 0.021 Re_b^{0.8} Pr_b^{0.4}$	Based on data for common gases; recommended for Prandtl numbers ≈ 0.7
3	Petukhov and Kirillov [19]	$Nu_b = \dfrac{(f/2)Re_b Pr_b}{1.07 + 12.7(f/2)^{1/2}(Pr_b^{2/3} - 1)}$	Based on three-layer model with constants adjusted to match experimental data. $0.5 < Pr_b < 2000$, $10^4 < Re_b < 5 \times 10^6$
4	Webb [20]	$Nu_b = \dfrac{(f/2)Re_b Pr_b}{1.07 + 9(f/2)^{1/2}(Pr_b - 1)Pr_b^{+1/4}}$ $f = (1.58 \ln Re_b - 3.28)^{-2}$	Theoretically based. Webb found case 3 better at high Pr and this one the same at other Pr

5	Sleicher and Rouse [21]	$Nu_b = 5 + 0.015 Re_b^m Pr_b^n$ $m = 0.88 - 0.24/(4 + Pr_b)$ $n = 1/3 + 0.5\exp(-0.6 Pr_b)$	Based on numerical results obtained for $0.1 < Pr_b < 10^4$, $10^4 < Re_b < 10^6$. Within 10% of case 6 for $Re_b > 10^4$.
		$Nu_b = 5 + 0.012 Re_b^{0.83}(Pr_b + 0.29)$	Simplified correlation for gases, $0.6 < Pr_b < 0.9$
6	Gnielinski [22]	$Nu_b = \dfrac{(f/2)(Re_b - 1000) Pr_b}{1 + 12.7(f/2)^{1/2}\left(Pr_b^{2/3} - 1\right)}$ $f = (1.58 \ln Re_b - 3.28)^{-2}$	Modification of case 3 to fit experimental data at low Re ($2300 < Re_b < 10^4$). Valid for $2300 < Re_b < 5 \times 10^6$ and $0.5 < Pr_b < 2000$
		$Nu_b = 0.0214(Re_b^{0.8} - 100) Pr_b^{0.4}$	Simplified correlation for $0.5 < Pr < 1.5$. Agrees with case 4 within -6% and $+4\%$
		$Nu_b = 0.012(Re_b^{0.87} - 280) Pr_b^{0.4}$	Simplified correlation for $1.5 < Pr < 500$. Agrees with case 4 within -10% and $+0\%$ for $3 \times 10^3 < Re_b < 10^6$
7	Kays and Crawford [23]	$Nu_b = 0.022 Re_b^{0.8} Pr_b^{0.5}$	Modified Dittus–Boelter correlation for gases ($Pr \approx 0.5$–1.0). Agrees with case 6 within 0% to 4% for $Re_b \geq 5000$

[a] Properties are evaluated at bulk temperatures.

3.5 TURBULENT FLOW IN SMOOTH STRAIGHT NONCIRCULAR DUCTS

The heat transfer and friction coefficients for turbulent flow in noncircular ducts are compiled in [17]. A common practice is to employ the hydraulic diameter in the circular duct correlations to predict Nu and f for the turbulent flow in noncircular ducts. For most of the noncircular smooth ducts, the accurate constant-property experimental friction factors are within $\pm 10\%$ of those predicted using the smooth circular duct correlation with hydraulic (equivalent) diameter D_h instead of circular duct diameter d. The constant-property experimental Nusselt numbers are also within $\pm 10\%$ to 15% except for some sharp-cornered and narrow channels. This order of accuracy is adequate for the overall heat transfer coefficient and the pressure drop calculations in most of the practical design problems.

Many attempts have been reported in the literature to arrive at a universal characteristic dimension for internal turbulent flows that would correlate the constant-property friction factors and Nusselt numbers for all noncircular ducts [28–30]. It must be emphasized that any improvement made by these attempts is only a few percent, and therefore the circular duct correlations may be adequate for many engineering applications.

The correlations given in Table 3.4 do not account for entrance effects occurring in short ducts. Gnielinski [3] recommends the entrance correction factor derived by Hausen [27] to obtain the Nusselt number for short ducts from the following correlation:

$$Nu_L = Nu_\infty \left[1 + \left(\frac{d}{L} \right)^{2/3} \right] \tag{3.36}$$

where Nu_∞ represents the fully developed Nusselt numbers calculated from the correlations given in Table 3.4. It should be noted that the entrance length depends on the Reynolds and Prandtl numbers and the thermal boundary condition. Thus Eq. (3.36) should be used cautiously.

3.6 THE EFFECT OF VARIABLE PHYSICAL PROPERTIES IN TURBULENT FORCED CONVECTION

When there is a large difference between the duct wall and fluid bulk temperatures, heating and cooling influence the heat transfer and the fluid friction in turbulent duct flow because of the distortion of turbulent transport mechanisms, in addition to the variation of fluid properties with temperature as for laminar flow.

3.6.1 Turbulent Liquid Flow in Ducts

Petukhov [19] reviewed the status of heat transfer and wall friction in fully developed turbulent pipe flow with both constant and variable physical properties.

To choose the correct value of n in Eq. (3.21a), the heat transfer experimental data corresponding to heating and cooling for several liquids over a wide range of values (μ_w/μ_b) were collected by Petukhov [19]. He found that the data are well correlated by

$$\frac{\mu_w}{\mu_b} < 1 \qquad n = 0.11 \qquad \text{for heating liquids} \tag{3.37}$$

$$\frac{\mu_w}{\mu_b} > 1 \qquad n = 0.25 \qquad \text{for cooling liquids} \tag{3.38}$$

which are applicable for fully developed turbulent flow in the range $10^4 < Re_b < 5 \times 10^6$, $2 < Pr_b < 140$, and $0.08 < (\mu_w/\mu_b) < 40$. The value of Nu_{cp} in Eq. (3.21a) is calculated from Eq. (3.30) or (3.32).

The value of Nu_{cp} can also be calculated from the correlations listed in Table 3.4.

Petukhov [19] collected data from various investigators for the variable viscosity influence on friction in water for both heating and cooling and suggested the following correlations for the friction factor:

$$\frac{\mu_w}{\mu_b} < 1 \qquad \frac{f}{f_{cp}} = \frac{1}{6}\left(7 - \frac{\mu_b}{\mu_w}\right) \qquad \text{for heating liquids} \tag{3.39}$$

$$\frac{\mu_w}{\mu_b} > 1 \qquad \frac{f}{f_{cp}} = \left(\frac{\mu_w}{b}\right)^{0.24} \qquad \text{for cooling liquids} \tag{3.40}$$

The friction factor for an isothermal (constant-property) flow f_{cp} can be calculated by the use of Table 3.5 or directly from Eq. (3.31) for the range $0.35 < (\mu_w/\mu_b) < 2$, $10^4 < Re_b < 23 \times 10^4$, and $1.3 < Pr_b < 10$.

3.6.2 Turbulent Gas Flow in Ducts

The heat transfer and friction coefficients for turbulent fully developed gas flow in a circular duct were obtained theoretically by Petukhov and Popov [18] by assuming physical properties ρ, c_p, k, and μ as given functions of temperature. This analysis is valid only for small subsonic velocities, since the variations of density with pressure and heat dissipation in the flow were neglected. The eddy diffusivity of momentum was extended to the case of variable properties. The turbulent Prandtl number was taken to be 1 (i.e.,

TABLE 3.5 Turbulent Flow Isothermal Fanning Friction Factor Correlations for Smooth Circular Ducts [26]

Number	Reference[a]	Correlation[b]	Remarks and Limitations
1	Blasius	$f = \tau_w / 1/2\ \rho u_m^2 = 0.0791 Re^{-1/4}$	This approximate explicit equation agrees with case 3 within $\pm 2.5\%$. $4 \times 10^3 < Re < 10^5$
2	Drew, Koo, and McAdams	$f = 0.00140 + 0.125 Re^{-0.32}$	This correlation agrees with case 3 within -0.5% and $+3\%$. $4 \times 10^3 < Re < 5 \times 10^6$
3	von Kármán and Nikuradse	$1/\sqrt{f} = 1.737 \ln(Re\sqrt{f}) - 0.4$ or $1/\sqrt{f} = 4 \log(Re\sqrt{f}) - 0.4$	von Kármán's theoretical equation with the constants adjusted to best fit Nikuradse's experimental data. Also referred to as the Prandtl correlation. Should be valid for very high values of Re. $4 \times 10^3 < Re < 3 \times 10^6$
		approximated as $f = (3.64 \log Re - 3.28)^{-2}$ $f \quad 0.046 Re^{-1/4}$	This approximate explicit equation agrees with the preceding within -0.4% and $+2.2\%$ for $3 \times 10^4 < Re < 10^6$
4	Flonenko	$f = 1/(1.58 \ln Re - 3.28)^2$	Agrees with case 3 within $\pm 0.5\%$ for $3 \times 10^4 < Re < 10^7$ and within $\pm 1.8\%$ at $Re = 10^4$. $10^4 < Re < 5 \times 10^5$
5	Techo, Tickner, and James	$1/f = \left(1.7372 \ln \dfrac{Re}{1.964 \ln Re - 3.8215}\right)^2$	An explicit form of case 3; agrees with it within $\pm 0.1\%$. $10^4 < Re < 2.5 \times 10^8$

[a] Properties are evaluated at bulk temperatures.
[b] Cited in [17, 23, 24, 26].

TABLE 3.6 Exponents n and m Associated with Eqs. (3.21) and (3.22) for Turbulent Forced Convection through Circular Ducts

Number	Reference	Fluid	Condition	n	m	Limitations
1	Petukhov [19]	Liquid	Turbulent heating	0.11	—	$10^4 < Re_b < 1.25 \times 10^5$, $2 < Pr_b < 140$, $0.08 < \mu_w/\mu_b < 1$
		Liquid	Turbulent cooling	0.25	—	$1 < \mu_w/\mu_b < 40$
		Liquid	Turbulent heating	—	Eq. (3.39)	$10^4 < Re_b < 23 \times 10^4$, $1.3 < Pr_b < 10^4$
					or -0.25	$0.35 < \mu_w/\mu_b < 1$
		Liquid	Turbulent cooling	—	-0.24	$1 < \mu_w/\mu_b < 2$
2	Petukhov and Popov [18]	Gas	Turbulent heating	-0.47	—	$10^4 < Re_b < 4.3 \times 10^6$, $1 < T_w/T_b < 3.1$
		Gas	Turbulent cooling	-0.36	—	$0.37 < T_w/T_b < 1$
		Gas	Turbulent heating	—	-0.52	$14 \times 10^4 < Re_w^* \le 10^6$, $1 < T_w/T_b < 3.7$
		Gas	Turbulent cooling	—	-0.38	$0.37 < T_w/T_b < 1$
3	Perkins and Worsøe-Schmidt [31]	Gas	Turbulent heating	—	-0.264	$1 \le T_w/T_b \le 4$
4	McElligot et al. [32]	Gas	Turbulent heating	—	-0.1	$1 < T_w/T_b < 2.4$

TABLE 3.7 Turbulent Forced Convection Correlations in Circular Ducts for Liquids with Variable Properties

Number	Reference	Correlation	Comments and Limitations
1	Colburn [33]	$St_b Pr_f^{2/3} = 0.023 Re_f^{-0.2}$	$L/d > 60$, $Pr_b > 0.6$, $T_f = (T_b + T_w)/2$; inadequate for large $(T_w - T_b)$
2	Sieder and Tate [11]	$Nu_b = 0.023 Re_b^{0.8} Pr_b^{1/3} \left(\dfrac{\mu_b}{\mu_w}\right)^{0.14}$	$L/d > 60$, $Pr_b > 0.6$, for moderate $(T_w - T_b)$
3	Petukhov and Kirillov [19]	$Nu_b = \dfrac{(f/8) Re_b Pr_b}{1.07 + 12.7\sqrt{f/8}\,(Pr_b^{2/3} - 1)} \left(\dfrac{\mu_b}{\mu_w}\right)^n$	$L/d > 60$, $0.08 < \mu_w/\mu_b < 40$, $10^4 < Re_b < 5 \times 10^6$, $2 < Pr_b < 140$, $f = (1.82 \log Re_b - 1.64)^{-2}$, $n = 0.11$ (heating), $n = 0.25$ (cooling)
4	Hufschmidt et al. [34]	$Nu_b = \dfrac{(f/8) Re_b Pr_b}{1.07 + 12.7\sqrt{f/8}\,(Pr_b^{2/3} - 1)} \left(\dfrac{Pr_b}{Pr_w}\right)^{0.11}$	Water, $2 \times 10^4 < Re_b < 6.4 \times 10^5$, $2 < Pr_b < 5.5$, $f = (1.82 \log Re_b - 1.64)^{-2}$, $0.1 < Pr_b/Pr_w < 10$
5	Yakovlev [35]	$Nu_b = 0.0277 Re_b^{0.8} Pr_b^{0.36} \left(\dfrac{Pr_b}{Pr_w}\right)^{0.11}$	Fully developed conditions. The use of the Prandtl group was first suggested by the author in 1960

6	Oskay and Kakaç [13]	$Nu_b = 0.023 Re_b^{0.8} Pr_b^{0.4} \left(\frac{\mu_b}{\mu_w}\right)^{0.262}$	Water, $L/d > 10$, $1.2 \times 10^4 < Re_b < 4 \times 10^4$
		$Nu_b = 0.023 Re_b^{0.8} Pr_b^{0.4} \left(\frac{\mu_b}{\mu_w}\right)^{0.487}$	30% glycerine–water mixture $L/d > 10$, $0.89 \times 10^4 < Re_b < 2.0 \times 10^4$
7	Hausen [36]	$Nu_b = 0.0235(Re_b^{0.8} - 230)(1.8 Pr_b^{0.3} - 0.8) \times \left[1 + \left(\frac{d}{L}\right)^{2/3}\right]\left(\frac{\mu}{\mu_w}\right)^{0.14}$	Altered form of equation presented in 1959 [4]
8	Sleicher and Rouse [21]	$Nu_b = 5 + 0.015 Re_f^m Pr_w^n$ $m = 0.88 - 0.24/(4 + Pr_w)$ $n = \frac{1}{3} + 0.5e^{-0.6Pr_w}$	$L/d > 60$, $0.1 < Pr_b < 10^5$, $10^4 < Re_b < 10^6$
		$Nu_b = 0.015 Re_f^{0.88} Pr_w^{1/3}$	$Pr_b > 50$
		$Nu_b = 4.8 + 0.015 Re_f^{0.85} Pr_w^{0.93}$	$Pr_b < 0.1$, uniform wall temperature
		$Nu_b = 6.3 + 0.0167 Re_f^{0.85} Pr_w^{0.93}$	$Pr_b < 0.1$, uniform wall heat flux

TABLE 3.8 Turbulent Forced Convection Correlations in Circular Ducts for Gases with Variable Properties

Number	Reference	Correlation	Gas	Comments and Limitations
1	Humble et al. [37]	$Nu_b = 0.023 Re_b^{0.8} Pr_b^{0.4} \left(\frac{T_w}{T_b}\right)^n$	Air	$30 < L/d < 120$, $7 \times 10^3 < Re_b < 3 \times 10^5$, $0.46 < T_w/T_b < 3.5$
		$T_w/T_b < 1$ $n = 0$ (cooling) $T_w/T_b > 1$ $n = -0.55$ (heating)		
2	Bialokoz and Saunders [19]	$Nu_b = 0.022 Re_b^{0.8} Pr_b^{0.4} \left(\frac{T_w}{T_b}\right)^{-0.5}$	Air	$29 < L/d < 72$, $1.24 \times 10^5 < Re_b < 4.35 \times 10^5$, $1.1 < T_w/T_b < 1.73$
3	Barnes and Jackson [38]	$Nu_b = 0.023 Re_b^{0.8} Pr_b^{0.4} \left(\frac{T_w}{T_b}\right)^n$	Air, helium, carbon dioxide	$1.2 < T_w/T_b < 2.2$, $4 \times 10^3 < Re_b < 6 \times 10^4$, $L/d > 60$
		$n = -0.4$ for air, $n = -0.185$ for helium, $n = -0.27$ for carbon dioxide		
4	McElligot et al. [32]	$Nu_b = 0.021 Re_b^{0.8} Pr_b^{0.4} \left(\frac{T_w}{T_b}\right)^{-0.5}$	Air, helium, nitrogen	$L/d > 30$, $1 < T_w/T_b < 2.5$, $1.5 \times 10^4 < Re_{ib} < 2.33 \times 10^5$,
		$Nu_b = 0.021 Re_b^{0.8} Pr_b^{0.4} \left(\frac{T_w}{T_b}\right)^{-0.5} \times \left[1 + \left(\frac{L}{d}\right)^{-0.7}\right]$		$L/d > 5$, local values

5	Perkins and Worsøe-Schmidt [31]	$Nu_b = 0.024 Re_b^{0.8} Pr_b^{0.4} \left(\frac{T_w}{T_b}\right)^{-0.7}$	Nitrogen	$L/d > 40$, $1.24 < T_w/T_b < 7.54$, $18.3 \times 10^3 < Re_{ib} < 2.8 \times 10^5$.
		$Nu_w = 0.023 Re_w^{0.8} Pr_w^{0.4}$		Properties evaluated at wall temperature, $L/d > 24$,
		$Nu_b = 0.024 Re_b^{0.8} Pr_b^{0.4} \left(\frac{T_w}{T_b}\right)^{-0.7} \times \left[1 + \left(\frac{L}{d}\right)^{-0.7} \left(\frac{T_w}{T_b}\right)^{0.7}\right]$		$1.2 \leq L/d \leq 144$
6	Petukov et al. [19]	$Nu_b = 0.021 Re_b^{0.8} Pr_b^{0.4} \left(\frac{T_w}{T_b}\right)^{n}$	Nitrogen	$80 < L/d < 100$, $13 \times 10^3 < Re_b < 3 \times 10^5$,
		$n = -\left(0.9 \log \frac{T_w}{T_b} + 0.205\right)$		$1 < T_w/T_b < 6$
7	Sleicher and Rouse [21]	$Nu_b = 5 + 0.012 Re_f^{0.83} (Pr_w + 0.29)$		For gases, $0.6 < Pr_b < 0.9$
8	Gnielinski [3]	$Nu_b = 0.0214 (Re_b^{0.8} - 100) Pr_b^{0.4} \left(\frac{T_b}{T_w}\right)^{0.45} \times \left[1 + \left(\frac{d}{L}\right)^{2/3}\right]$	Air, helium, carbon dioxide	$0.5 < Pr_b < 1.5$, for heating of gases. The author collected the data from the literature. Second for $1.5 < Pr_b < 500$
		$Nu_b = 0.012 (Re_b^{0.87} - 280) Pr^{0.4} \left(\frac{T_b}{T_w}\right)^{0.4} \times \left[1 + \left(\frac{d}{L}\right)^{2/3}\right]$		
9	Dalle-Donne and Bowditch [39]	$Nu_b = 0.022 Re_b^{0.8} Pr_b^{0.4} \left(\frac{T_w}{T_b}\right)^{-\lvert 0.29 + 0.0019 L/d \rvert}$	Air, helium	$10^4 < Re_b < 10^5$, $18 < L/d < 316$

$\epsilon_H = \epsilon_M$). The analyses were carried out for hydrogen and air for the following range of parameters: $0.37 < (T_w/T_b) < 3.1$ and $10^4 < Re_b < 4.3 \times 10^6$ for air, and $0.37 < (T_w/T_b) < 3.7$ and $10^4 < Re_b < 5.8 \times 10^6$ for hydrogen. The analytical results are correlated by Eq. (3.22a), where Nu_{cp} is given by Eq. (3.30) or (3.31), and the following values for n are obtained:

$$\frac{T_w}{T_b} < 1 \qquad n = -0.36 \qquad \text{for cooling gases} \tag{3.41}$$

$$\frac{T_w}{T_b} > 1 \qquad n = -\left[0.3 \log\left(\frac{T_w}{T_b}\right) + 0.36\right] \qquad \text{for heating gases} \tag{3.42}$$

With these values for n, Eq. (3.22a) describes the solution for air and hydrogen within an accuracy of $\pm 4\%$. For simplicity, one can take n to be constant for heating as $n = -0.47$; then Eq. (3.22a) describes the solution for air and hydrogen within $+6\%$. These results have also been confirmed experimentally and can be used for practical calculations when $1 < (T_w/T_b) < 4$.

A large number of experimental studies are available in the literature for the heat transfer between the tube wall and the gas flow with large temperature differences and temperature-dependent physical properties. The majority of the work deals with gas heating at constant wall temperature in a circular duct; experimental studies on gas cooling are limited.

The results of heat transfer measurements at large temperature differences between the work and the gas flow are usually presented as

$$Nu = CRe_b^{0.8} Pr_b^{0.4} \left(\frac{T_w}{T_b}\right)^n \tag{3.43}$$

For fully developed temperature and velocity profiles (i.e., $L/d < 60$), C becomes constant and n becomes independent of L/d.

A number of heat transfer correlations have been developed for variable-property fully developed turbulent liquid and gas flow in a circular duct, some of which are also summarized in Tables 3.6 to 3.8.

Comprehensive information and correlations for the convective heat transfer and friction factor in noncircular curved ducts and coils, in crossflow arrangements, over rod bundles, in various fittings and liquid metals are given in [40]. The comparison of the important correlations for forced convection in ducts is also given in [41].

Example 3.1. Air at 40°C flows through a heated pipe section with a velocity of 6 m/s. The length and diameter of the pipe are 300 cm and 2.54 cm, respectively. The average pipe wall temperature is 300°C. Determine the average heat transfer coefficient.

Solution: Since the wall temperature is so much greater than the initial air temperature, variable-property flow must be considered. From the Appendix, the properties of air at $T_b = 40°C$ are

$$\rho = 1.128\ \text{kg/m}^3 \qquad c_p = 1005.3\ \text{J/(kg}\cdot\text{K)}$$

$$k = 0.0267\ \text{W/(m}\cdot\text{K)} \qquad \mu = 1.912 \times 10^{-5}\ (\text{N}\cdot\text{s})/\text{m}^2$$

$$Pr = 0.719$$

The inside heat transfer coefficient can be obtained from knowledge of the flow regime, that is, the Reynolds number,

$$Re_b = \frac{\rho U_m d_i}{\mu} = \frac{1.128 \times 6 \times 0.0254}{1.912 \times 10^{-5}} = 8991$$

Hence the flow in the tube is turbulent. On the other hand, $L/d = 3/0.0254 = 118 > 60$, fully developed conditions are assumed. Since $Pr > 0.6$, we can use one of the correlations given in Table 3.4. Hence Gnielinsky's correlation, Eq. (3.34), with constant properties,

$$Nu_b = \frac{(f/2)(Re_b - 1000)Pr_b}{1 + 12.7(f/2)(Pr^{2/3} - 1)}$$

may be used to determine the Nusselt number.

$$f = (1.58 \ln Re - 3.28)^{-2}$$

$$= [1.58 \ln(8991) - 3.28]^{-2} = 0.00811$$

$$Nu_b = \frac{hd}{k} = \frac{(0.00811/2)(8991 - 1000)(0.719)}{1 + 12.7(0.00811/2)^{0.5}(0.719^{2/3} - 1)} = 27.712$$

$$h = Nu_b \frac{k}{d} = \frac{27.72 \times 0.0267}{0.0254} = 29.14\ \text{W/(m}^2\cdot\text{K)}$$

The heat transfer coefficient with variable properties can be calculated from Eq. (3.22a), where n is given in Table 3.6 as $n = -0.47$:

$$Nu_b = Nu_{\text{cp}} \left(\frac{T_w}{T_b} \right)^{-0.47}$$

$$= 27.712 \left(\frac{573}{313} \right)^{-0.47} = 20.856$$

Then

$$h = \frac{Nu_b k}{d} = \frac{(20.856)(0.0267)}{0.0254} = 21.9239 \text{ W/(m}^2 \cdot \text{K)}$$

As can be seen in the case of a gas with temperature-dependent properties, heating a gas decreases the heat transfer coefficient. The heat transfer in the tube is estimated to be $h = 21.9239 \text{ W/(m}^2 \cdot \text{K)}$

Example 3.2. Determine the total heat transfer coefficient at 30 cm from the inlet of a heat exchanger where engine oil flows through the tubes which have a diameter of 0.5 in. Oil flows with a velocity of 0.5 m/s and at a bulk temperature of 30°C while the local tube wall temperature is 60°C.

Solution: From the Appendix, the properties of engine oil at $T_b = 30°\text{C}$ are

$$\rho = 882.3 \text{ kg/m}^3 \qquad c_p = 1922 \text{ J/(kg} \cdot \text{K)}$$

$$\mu = 0.416 \text{ (N} \cdot \text{s)/m}^2 \qquad k = 0.144 \text{ W/(m} \cdot \text{K)}$$

$$Pr = 5550 \qquad \mu_w = 0.074 \text{ (N} \cdot \text{s)/m}^2$$

The heat transfer coefficient may be obtained from knowledge of the Reynolds number

$$Re_b = \frac{\rho U_m d_i}{\mu} = \frac{882.3 \times 0.5 \times 0.0127}{0.416} = 13.47$$

Since $Re < 2300$, the flow inside the tube is laminar. We can calculate the heat transfer coefficient from the Sieler and Tate correlation, Eq. (3.25),

$$Nu_T = 1.86(RePr)^{1/3}\left(\frac{d}{L}\right)^{1/3}\left(\frac{\mu_b}{\mu_w}\right)^{0.14}$$

as long as the following conditions are satisfied:

$$\left(\frac{\mu_b}{\mu_w}\right) = \left(\frac{0.416}{0.074}\right) = 5.62 < 9.75$$

$$\left(RePr\frac{d}{L}\right)^{1/3}\left(\frac{\mu_b}{\mu_w}\right)^{0.14} = \left[\frac{13.47 \times 5550 \times 0.0127}{0.3}\right]^{1/3}\left(\frac{0.416}{0.074}\right)^{0.14} = 18.7 > 2$$

Therefore the preceding correlation is applicable:

$$Nu_T = 1.86 \times 18.7 = 34.8$$

$$h = \frac{Nu_H k}{d_i} = \frac{34.8 \times 0.144}{0.0172} = 394.6 \text{ W/(m}^2 \cdot \text{K)}$$

The Nusselt–Graetz correlation given by Eq. (3.11) which is applicable with constant heat flux boundary conditions can also be used since

$$Re_b Pr_b = 5550 \times 13.47 = 3164 > 100$$

$$Nu_H = 1.953\left(Re_b \frac{d}{L}\right) = 1.953 \times 5550 \times 13.47 = 28.67$$

$$h = \frac{Nu_H k}{d_i} = \frac{28.67 \times 0.144}{0.0127} = 325\ \mathrm{W/(m^2 \cdot K)}$$

The Nusselt–Graetz correlation gives a more conservative answer.

Example 3.3. Water flowing at 5000 kg/hr will be heated from 20 to 35°C by hot water at 140°C. A 15°C hot water temperature drop is allowed. A number of 15-ft (4.5-m) hairpins of 3 in. (ID = 3.068 in., OD = 3.5 in.) by 2 in. (ID = 2.067 in., OD = 2.375 in.) double-pipe heat exchanger with annuli and pipes each connected in series will be used. Hot water flows through the inner tube. Calculate: (a) the heat transfer coefficient in the inner tube and (b) the heat transfer coefficient inside the annulus; the outside of the annulus is insulated against heat loss.

Solution: (a) We first calculate the Reynolds number to determine if the flow is laminar or turbulent, and then select the proper correlation to calculate the heat transfer coefficient. From the Appendix, the properties of hot water at $T_b = 132.5$°C are

$$\rho = 932.4\ \mathrm{kg/m^3} \qquad c_p = 4268.1\ \mathrm{J/kg \cdot K}$$

$$k = 0.688\ \mathrm{W/(m^2 \cdot K)} \qquad \mu = 0.208 \times 10^{-3}\ \mathrm{(N \cdot s)/m^2}$$

$$Pr = 1.29$$

We now make an energy balance to calculate the hot-water mass flow rate:

$$(\dot{m}c_p)_h T_h = (\dot{m}c_p)_c T_c$$

$$\dot{m}_h = \frac{\dot{m}_c c_{pc}}{c_{ph}} = \frac{(5000/3600)(4179)}{4268.1} = 1.360\ \mathrm{kg/s}$$

where $c_{pc} = 4179$ J/(kg · K) is at $T_b = 27.5$°C:

$$Re_b = \frac{\rho U_m d_i}{\mu} = \frac{4\dot{m}}{\pi_\mu d_i} = \frac{4 \times 1.36}{\pi \times 0.0525 \times 0.208 \times 10^{-3}} = 158{,}572$$

Therefore the flow is turbulent and we can select a correlation from Table 3.4. The Petukhov–Kirillov correlation is used here:

$$Nu_b = \frac{(f/2)Re_b Pr_b}{1.07 + 12.7(f/2)^{0.5}(Pr^{2/3} - 1)}$$

where

$$f = (1.58 \ln Re - 3.28)^{-2}$$

$$= [1.58 \ln(158{,}572) - 3.28]^{-2} = 0.00409$$

$$Nu = \frac{(4.09 \times 10^{-3}/2)(158{,}572 \times 1.29)}{1.07 + 12.7(4.09 \times 10^{-3}/2)^{0.5}(1.29^{2/3} - 1)} = 355.6$$

$$h_i = \frac{Nu_b k}{d_i} = \frac{355.6 \times 0.688}{2.067 \times 2.54 \times 10^{-2}} = 4660.36 \text{ W/(m}^2 \cdot \text{K)}$$

The effect of property variations can be found from Eq. (3.21a) with $n = 0.25$ for cooling of a liquid in turbulent flow (Table 3.4).

(b) Calculate the heat transfer coefficient in the annulus. From the Appendix, the properties of cold water at $T_b = 27.5°\text{C}$ are

$$\rho = 996.8 \text{ kg/m}^3 \qquad c_p = 4179 \text{ J/(kg} \cdot \text{K)}$$

$$k = 0.614 \text{ W/(m} \cdot \text{K)} \qquad \mu = 0.846 \times 10^{-3} \text{ (N} \cdot \text{s)/m}^2$$

$$Pr = 5.77$$

The hydraulic diameter of the annulus from Eq. (3.17) is

$$D_i - d_o = (3.068 - 2.375) \times 2.54 \times 10^{-2} = 0.0176 \text{ m}$$

$$Re = \frac{4D_h m_c}{\pi\mu(D_i^2 - d_o^2)} = \frac{4(0.0176)(5000/3600)}{\pi \times (846 \times 10^{-6})(0.002432)} = 15{,}125.13$$

Therefore the flow inside the annulus is turbulent. One of the correlations can be selected from the tables. The Gnielinski correlation is used here. It should be noted that for the annulus, the Nusselt number should be based on the hydraulic diameter (or equivalent diameter) calculated from Eq. (3.19):

$$D_e = \frac{4A_c}{P_h} = \frac{4[\pi/4(D_i^2 - d_o^2)]}{\pi d_o}$$

$$= \frac{[(0.0779^2) - (0.0603^2)]}{0.0603} = 0.0403 \text{ m}$$

$$Nu_b = \frac{(f/2)(Re_b - 1000)Pr_b}{1.07 + 12.7(f/2)^{1/2}(Pr^{2/3} - 1)}$$

$$f = (1.58 \ln Re_b - 3.28)^{-2}$$

$$= [1.58 \ln(15{,}125.13) - 3.28]^{-2}$$

$$= 0.00703$$

$$Nu_b = \frac{(0.00703/2)(15{,}125.13 - 1000)(5.77)}{1.07 + 12.7(0.00703/2)(5.77^{2/3} - 1)} = 104.5$$

$$h_o = \frac{Nu_b k}{D_e} = \frac{2043.3 \times 0.614}{0.0403} = 1593.54 \text{ W}/(\text{m}^2 \cdot \text{K})$$

3.7 SUMMARY

Important and reliable correlations, for newtonian fluids in single-phase laminar and turbulent flow through ducts have been summarized, which can be used in the design of heat transfer equipment.

The tables cover the recommended specific correlations for laminar forced convection through ducts with constant and variable fluid properties. Table 3.3 provides exponents m and n associated with Eqs. (3.21) and (3.22) for laminar forced convection in circular ducts. By the use of this table, the effect of variable properties in laminar flow is incorporated by the property ratio method.

Turbulent forced convection correlations for fully developed flow through a circular duct with constant properties are summarized in Table 3.4. Gnielinski, Petukhov and Kirillov, Webb, Sleicher, and Rouse correlations are recommended for constant-property Nusselt number evaluation for gases and liquids, and the entrance correction factor is given by Eq. (3.36). Recommended turbulent flow isothermal Fanning friction factor correlations for smooth circular ducts are listed in Table 3.5. The correlations given in Tables 3.4, 3.5, 3.7, and 3.8 can also be utilized for turbulent flow in smooth straight noncircular ducts for engineering applications by the use of the hydraulic diameter concept for heat transfer and pressure drop calculations as discussed in Section 3.2.3. Except for sharp-cornered and/or very irregular duct cross sections, the fully developed turbulent Nusselt number and friction factor vary from their actual values within $\pm 15\%$ and $\pm 10\%$, respectively, when the hydraulic diameter is used in circular duct correlations.

When there is a large difference between the wall and fluid bulk temperatures, the influence of variable fluid properties on turbulent forced convection and pressure drop in circular ducts are taken into account by using the exponents m and n given in Table 3.6 with Eqs. (3.21) and (3.22). The correlations for turbulent flow of liquids and gases with variable properties in circular ducts are also summarized in Tables 3.7 and 3.8.

NOMENCLATURE

- A constant
- A_c net free-flow cross-sectional area, m^2
- c_p specific heat at constant pressure, J/(kg · K)
- c_v specific heat at constant volume, J/(kg · K)
- D_e equivalent diameter for heat transfer, $4A_c/P_h$, m
- D_i inner diameter of a circular annulus, m
- D_h hydraulic diameter for pressure drop, $4A_c/P_w$, m
- d circular duct diameter, m
- f Fanning friction factor, $\tau_w / \frac{1}{2}\rho u_m^2$
- G fluid mass velocity, ρu_m, kg/(m^2 · s)
- h average heat transfer coefficient, W/(m^2 · K)
- h_x local heat transfer coefficient, W/(m^2 · K)
- k thermal conductivity of fluid, W/(m · K)
- L distance along the duct, m
- m exponent, Eqs. (3.21b) and (3.22b)
- $\dot{m}$ mass flow rate, kg/s
- Nu average Nusselt number, hd/k
- n exponent, Eqs. (3.21a) and (3.22a)
- L_{he} hydrodynamic entrance length, m
- L_{te} thermal entrance length, m
- Pe Péclet number, $RePr$
- Pr Prandtl number, $c_p\mu/k = \alpha/\nu$
- Re Reynolds number, $\rho u_m d/\mu$, $\rho u_m D_h/\mu$
- T temperature, °C, K
- T_f film temperature, $(T_w + T_b)/2$, °C, K
- u velocity component in axial direction, m/s
- u_m mean axial velocity, m/s
- x cartesian coordinate, axial distance, m
- y cartesian coordinate, distance normal to the surface, m

Greek Symbols

α thermal diffusivity of fluid, m^2/s
μ dynamic viscosity of fluid, $Pa \cdot s$
ν kinematic viscosity of fluid, m^2/s
ρ density of fluid, kg/m^3
τ_w shear stress at the wall, Pa

Subscripts

a arithmetic mean
b bulk fluid condition or properties evaluated at bulk mean temperature
cp constant property
e equivalent
f film fluid condition or properties evaluated at film temperature
H constant heat flux boundary condition
l laminar
i inlet condition
o outlet condition or outer
r reference condition
T constant temperature boundary condition
t turbulent
w wall condition or wetted
x local value at distance x
∞ fully developed condition

REFERENCES

1. Shah, R. K., and London, A. L. (1978) *Laminar Forced Convection in Ducts*. Academic, New York.
2. Shah, R. K., and Bhatti, M. S. (1987) Laminar convective heat transfer in ducts. In *Handbook of Single-Phase Convective Heat Transfer*, S. Kakaç, R. K. Shah, and W. Aung (eds.), pp. 3.1–3.137. Wiley, New York.
3. Schlünder, E. U. (ed.) (1983) *Heat Exchanger Design Handbook*, pp. 2.5.1–2.5.13. Hemisphere, New York.
4. Hausen, H. (1959) Neue Gleichungen für die Warmeübertragung bei freier oder erzwungener Strömung. *Allg. Waermetech.* **9** 75–79.
5. Kakaç, S. (1985) Laminar forced convection in the combined entrance region of ducts. In *Natural Convection: Fundamentals and Applications*, S. Kakaç, W. Aung, and R. Viskanta (eds.), pp. 165–204. Hemisphere, New York.
6. Pohlhausen, E. (1921) Der Warmeaustausch Zwischen festen Körpern und Flüssigkeiten mit Kleiner Reibung und Kleiner Warmeleitung. *Z. Angew. Math. Mech.* **1** 115–121.

7. Delorenzo, B., and Anderson, E. D. (1945) Heat transfer and pressure drop of liquids in double pipe fintube exchangers. *Trans. ASME* **67** 697.
8. Stephan, K. (1959) Warmeübergang und Druckabfall beinichtausgebildeter Laminar Störmung in Rohren und evenen Spalten. *Chem. Ing. Tech.* **31** 773–778.
9. Deissler, R. G. (1951) Analytical investigation of fully developed laminar flow in tubes with heat transfer with fluid properties variable along the radius. NACA TN 2410.
10. Yang, K. T. (1962) Laminar forced convection of liquids in tubes with variable viscosity. *J. Heat Transfer* **84** 353–362.
11. Sieder, E. N., and Tate, G. E. (1936) Heat transfer and pressure drop of liquids in tubes. *Ind. Eng. Chem.* **28** 1429–1453.
12. Whitaker, S. (1972) Forced convection heat-transfer correlations for flow in pipes, past flat plates, single cylinders, single spheres, and flow in packed beds and tube bundles. *AIChE J.* **18** 361–371.
13. Oskay, R., and Kakaç, S. (1973) Effect of viscosity variations on turbulent and laminar forced convection in pipes. *METU J. Pure Appl. Sci.* **6** 211–230.
14. Kuznetsova, V. V. (1972) Convective heat transfer with flow of a viscous liquid in a horizontal tube (in Russian). *Teploenergetika* **19**(5) 84.
15. Test, F. L. (1968) Laminar flow heat transfer and fluid flow for liquids with a temperature dependent viscosity. *J. Heat Transfer* **90** 385–393.
16. Worsøe-Schmidt, P. M. (1966) Heat transfer and friction for laminar flow of helium and carbon dioxide in a circular tube at high heating rate. *Int. J. Heat Mass Transfer* **9** 1291–1295.
17. Bhatti, M. S., and Shah, R. K. (1987) Turbulent forced convection in ducts. In *Handbook of Single-Phase Convective Heat Transfer*, S. Kakaç, R. K. Shah, and W. Aung (eds.), pp. 4.1–4.166. Wiley, New York.
18. Petukhov, B. S., and Popov, V. N. (1963) Theoretical calculation of heat exchange and frictional resistance in turbulent flow in tubes of incompressible fluid with variable physical properties. *High Temperature* **1**(1) 69–83.
19. Petukhov, B. S. (1970) Heat transfer and friction in turbulent pipe flow with variable physical properties. In *Advances in Heat Transfer*, J. P. Hartnett and T. V. Irvine (eds.), Vol. 6, pp. 504–564. Academic, New York.
20. Webb, R. I. (1971) A critical evaluation of analytical solutions and Reynolds analogy equations for heat and mass transfer in smooth tubes. *Warme und Staffübertragung* **4** 197–204.
21. Sleicher, C. A., and Rouse, M. W. (1975) A convenient correlation for heat transfer to constant and variable property fluids in turbulent pipe flow. *Int. J. Heat Mass Transfer* **18** 677–683.
22. Gnielinski, V. (1976) New equations for heat and mass transfer in turbulent pipe and channel flow. *Int. Chem. Eng.* **16** 359–368.
23. Kays, W. M., and Crawford, M. E. (1981) *Convective Heat and Mass Transfer*, 2nd ed. McGraw-Hill, New York.
24. Kakaç, S., and Yener, Y. (1980) *Convective Heat Transfer*. METU Publication 65, Ankara, Turkey; distributed by Hemisphere, New York.
25. McAdams, W. H. (1954) *Heat Transmission*, 3rd ed., McGraw-Hill, New York.

26. Kakaç, S. (1987) The effects of temperature-dependent fluid properties on convective heat transfer. In *Handbook of Single-Phase Convective Heat Transfer*, S. Kakaç, R. K. Shah and W. Aung, (eds.), pp. 18.1–18.92. Wiley, New York.
27. Hausen, H. (1943) Darstellung des Warmeüberganges in Rohren durch verallgeineinerte Potenzbeziebungen. *Z. Ver. Dtsch. Ing. Beiheft Verfahrenstech.* **4** 91–134.
28. Rehme, K. (1973) A simple method of predicting friction factors of turbulent flow in noncircular channels. *Int. J. Heat Mass Transfer* **16** 933–950.
29. Malak, J., Hejna, J., and Schmid, J. (1975) Pressure losses and heat transfer in noncircular channels with hydraulically smooth walls. *Int. J. Heat Mass Transfer* **18** 139–149.
30. Brundrett, E. (1979) Modified hydraulic diameter. In *Turbulent Forced Convection in Channels and Bundles*, S. Kakaç and D. B. Spalding, (eds.), Vol. 1, pp. 361–367. Hemisphere, New York.
31. Perkins, H. C. and Worsøe-Schmidt, P. (1965) Turbulent heat and momentum transfer for gases in a circular tube at wall to bulk temperature ratios to seven. *Int. J. Heat Mass Transfer* **8** 1011–1031.
32. McElligot, D. M., Magee, P. M., and Leppert, G. (1965) Effect of large temperature gradients on convective heat transfer: The downstream region. *J. Heat Transfer* **87** 67–76.
33. Colburn, A. P. (1933) A method of correlating forced convection heat transfer data and comparison with fluid friction. *Trans. AIChE* **29** 174–210.
34. Hufschmidt, W., Burck, E., and Riebold, W. (1966) Die Bestimmung Örlicher und Warmeübergangs-Zahlen in Rohren bei Hohen Warmestromdichten. *Int. J. Heat Mass Transfer* **9** 539–565.
35. Rogers, D. G. (1980) Forced convection heat transfer in single phase flow of a newtonian fluid in a circular pipe. CSIR Report CENG 322, Pretoria, South Africa.
36. Hausen, H. (1974) Extended equation for heat transfer in tubes at turbulent flow. *Warme und Stoffübertragung* **7** 222–225.
37. Humble, L. V., Lowdermilk, W. H., and Desmon, L. G. (1951) Measurement of average heat transfer and friction coefficients for subsonic flow of air in smooth tubes at high surface and fluid temperature. NACA Report 1020.
38. Barnes, J. F., and Jackson, J. D. (1961) Heat transfer to air, carbon dioxide and helium flowing through smooth circular tubes under conditions of large surface/gas temperature ratio. *J. Mech. Eng. Sci.* **3**(4) 303–314.
39. Dalle-Donne, M., and Bowditch, P. W. (1963) Experimental local heat transfer and friction coefficients for subsonic laminar transitional and turbulent flow of air or helium in a tube at high temperatures. Dragon Project Report 184, Winfirth, Dorchester, Dorset, UK.
40. Kakaç, S., Shah, R. K., and Aung, W. (eds.) (1987) *Handbook of Single-Phase Convective Heat Transfer*. Wiley, New York.
41. Kakaç, S., Bergles, A. E., and Fernandes, E. O. (eds.) (1988) *Two-Phase Flow Heat Exchangers*, pp. 123–158. Kluwer, Dordrecht.

CHAPTER 4

HEAT EXCHANGER FOULING

A. K. AGRAWAL
Department of Mechanical Engineering
Clemson University
Clemson, South Carolina 29634-0921

S. KAKAÇ
Department of Mechanical Engineering
University of Miami
Coral Gables, Florida 33124

4.1 INTRODUCTION

Fouling can be defined as the accumulation of undesirable substances on a surface. In general, the collection and growth of unwanted material results in inferior performance of the surface. Fouling occurs in natural as well as synthetic systems. Arteriosclerosis serves as an example of fouling in the human body wherein the deposit of cholesterol and the proliferation of connective tissues in an artery wall form plaque that grows inward. The resulting blockage or narrowing of arteries places increased demand on the heart.

In the present context the term *fouling* is used specifically to refer to undesirable deposits on the heat exchanger surface. A heat exchanger must affect a desired change in the thermal conditions of the process streams within allowable pressure drops and continue to do so for a specified time period. During operation, the heat transfer surface fouls resulting in increased thermal resistance and often an increase in the pressure drop and pumping power as well. Both of these effects compliment each other in degrading the performance of the heat exchanger. The heat exchanger may deteriorate to the extent that it must be withdrawn from service for replacement or cleaning.

Boilers, Evaporators and Condensers, Edited by Sadik Kakaç
ISBN 0-471-62170-6

Fouling may significantly influence the overall design of a heat exchanger and may determine the amount of material employed for construction. Special operational arrangements may be required to facilitate satisfactory performance between cleaning schedules. Consequently, fouling causes an enormous economic loss as it directly impacts the initial cost, operating cost, and heat exchanger performance.

4.2 EFFECTS OF FOULING

Lower heat transfer and increased pressure drop resulting because of fouling decrease the effectiveness of a heat exchanger. These effects and the basic thermohydraulic aspects of heat exchanger design are discussed in this section.

4.2.1 Basic Equations

Thermal analysis of a heat exchanger is governed by the conservation of energy in that the heat released by the hot fluid stream equals the heat gained by the cold fluid stream. The heat transfer rate Q is related to the geometric and flow parameters of the heat exchanger as

$$Q = UA\,\Delta T_m \tag{4.1}$$

where U is the overall heat transfer coefficient based on the heat transfer surface area A. Since the temperature difference along the heat transfer surface is not constant, an effective mean temperature difference ΔT_m is used. The overall heat transfer coefficient depends on the heat transfer mechanisms on both sides of the separating surface and heat conduction through the surface itself. For a clean plain tubular heat exchanger, the overall heat transfer coefficient, based on the outside surface area of the tube, is given by Eq. (2.9) as

$$U_c = \frac{1}{A_o/A_i h_i + A_o \ln(d_o/d_i)/2\pi kL + 1/h_o} \tag{4.2}$$

where h_i and h_o represent the heat transfer coefficients on the inside and outside of the tube, respectively. The order of magnitude and range of heat transfer coefficients for various flow conditions are given in Table 2.1.

The frictional pressure drop for a single-phase flow in the heat exchanger is usually calculated by

$$\Delta P = 4f\left(\frac{L}{d}\right)\left(\frac{\rho u_m^2}{2}\right) \tag{4.3}$$

where f is the Fanning friction factor. Various graphs and correlations to determine the friction factor for single-phase flow are available in the literature [1, 2].

4.2.2 Effect of Fouling on Heat Transfer

A simple visualization of fouling, shown in Fig. 4.1, depicts fouling buildup on the inside and outside of a circular tube. It is evident that fouling adds an insulating layer to the heat transfer surface. For a plain tubular heat exchanger the overall heat transfer coefficient under fouled conditions U_f can be obtained by adding the inside and outside thermal resistances in Eq. (4.2):

$$U_f = \frac{1}{A_o/A_ih_i + A_oR_{fi}/A_i + A_o\ln(d_o/d_i)/2\pi kL + R_{fo} + 1/h_o} \tag{4.4}$$

The overall heat transfer coefficient for a finned tube (based on the outside surface area) is given by Eq. (2.17). Fouling resistances R_{fi} or R_{fo}, also defined in Chapter 2, are sometimes referred to as "fouling factors." The heat transfer in the unwanted material takes place by conduction.

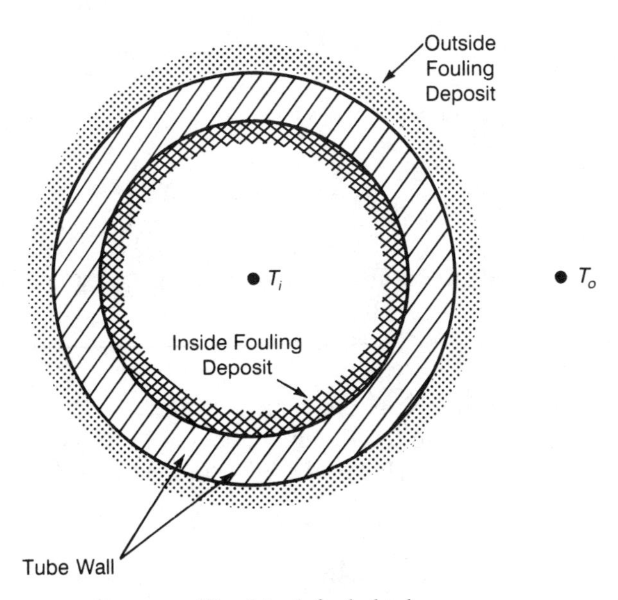

Fig. 4.1. A fouled tube.

Therefore the fouling factor can be related to the fouling thermal conductivity k_f and the fouling thickness t_f as

$$R_f = \frac{t_f}{k_f} \qquad \text{for a plane wall} \tag{4.5a}$$

$$R_f = \frac{d_c \ln(d_c/d_f)}{2k_f} \qquad \text{for a cylindrical wall} \tag{4.5b}$$

The fouling thickness is 0 when the surface is clean. The heat exchanger requires cleaning when the fouling thickness reaches a maximum value often called the design value. In general, the magnitudes of t_f and k_f are unknown since the diversity of applications and operating conditions makes most fouling situations virtually unique. Therefore, in spite of its apparent simplicity, Eq. (4.5) is not very useful in estimating the fouling resistance. It is interesting to note that the tremendous research on single-phase and two-phase heat transfer has markedly reduced the uncertainties in predicting the heat transfer coefficients h_i or h_o. However, the current uncertainty in predictions and/or estimates of the fouling resistances greatly exceeds the uncertainty of the other terms in the overall heat transfer coefficient [Eq. (4.4)].

U_f in Eq. (4.4) can relate to the clean surface overall heat transfer coefficient U_c [given by Eq. (4.2)] as

$$\frac{1}{U_f} = \frac{1}{U_c} + R_{ft} \tag{4.6}$$

where R_{ft} is the total fouling resistance given as

$$R_{ft} = \frac{A_o R_{fi}}{A_i} + R_{fo} \tag{4.7}$$

The heat transfer rate under the fouled conditions Q_f can be expressed as

$$Q_f = U_f A_f \,\Delta T_{mf} \tag{4.8}$$

where the subscript f refers to the fouled conditions. Process conditions usually set the heat duty and fluid temperatures at specified values; that is, $Q_f = Q_c$ and $\Delta T_{mf} = \Delta T_{mc}$. Under these conditions Eqs. (4.1), (4.6), and (4.8) show

$$\frac{A_f}{A_c} = 1 + U_c R_{ft} \tag{4.9}$$

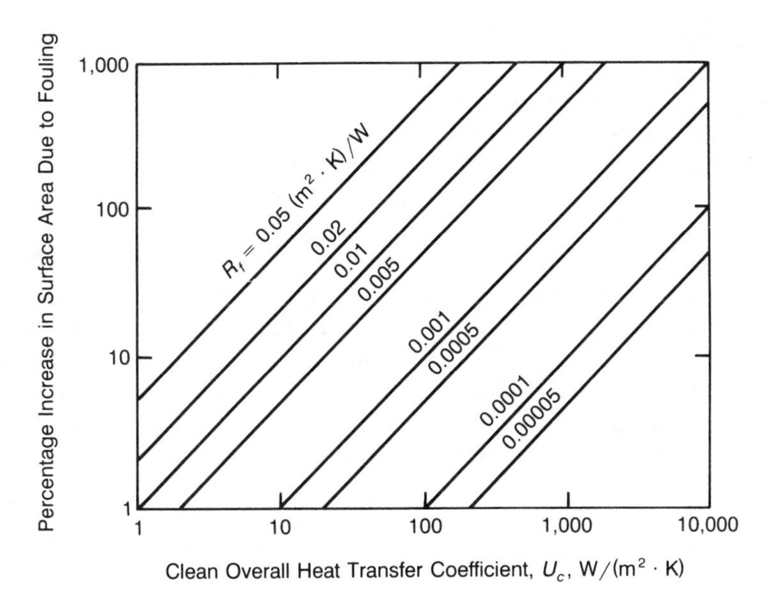

Fig. 4.2. Effect of fouling on surface area.

where A_c is the required surface area if the heat exchanger remains clean. $U_c R_{ft}$ in Eq. (4.9) represents the additional surface area required because of fouling of the heat exchanger. For a range of fouling resistances, Fig. 4.2 shows the percentage of increase in the heat transfer surface area due to fouling. Obviously, the added surface is small if U_c is low [1 to 10 W/(m^2 · K)] even though the total fouling resistance may be high [R_{ft} = 50.0 (m^2 · K/kW]. However, for high U_c [1000 to 10,000 W/(m^2 · K)] even a small fouling resistance [R_{ft} = 0.5(m^2 · K)/kW] results in a substantial increase in the required heat transfer surface area. A 100% increase in the surface area due to fouling alone is not uncommon.

The average total fouling resistances specified in the design of some 750 shell-and-tube heat exchangers, by five different manufacturers, are given in Table 4.1 [3]. Although fouling resistances vary for specific applications, these statistical values reveal present-day industrial trends. It is interesting to note

TABLE 4.1 Specified Average Total Fouling Resistance R_{ft}, (m^2 · K) / kW [3]

Tube Side	Shell Side		
	Vapor	Liquid	Two phase
Vapor	0.37	0.51	0.48
Liquid	0.60	0.79	0.65
Two phase	0.51	0.67	0.51

that the average total fouling resistance with liquid flows, on both the tube and the shell sides of the heat exchanger, is twice the corresponding value with vapor flows. The average total fouling resistances for the other fluid combinations lie in between these two values.

Table 4.2 shows the percentage of increase in the heat transfer surface area for a shell-and-tube heat exchanger with two-phase flow on the shell side. Single-phase or two-phase flow may occur on the tube side. For a given fluid combination the total fouling resistance is taken from Table 4.1. Typical values of the heat transfer coefficients (Table 2.1) are used and the tube wall resistance has been neglected. It is seen that if the heat exchanger involves sensible heating or cooling of gases, fouling does not contribute significantly to increasing the required heat transfer surface area. A high increase of about 25% may take place. However, if a liquid is used for sensible heating or cooling, fouling may substantially increase the required surface area, by a

TABLE 4.2 Added Surface Area for Typical Fluid Combinations

	Tube Side: Gas at High Pressure $h_i = 500$ W/(m$^2 \cdot$ K), $R_{ft} = 0.48$ (m$^2 \cdot$ K)/kW			
	Shell Side (Boiling or Condensation)			
	h_o	U_c	U_f	Percentage Increase in Area
Light organics	1,000	333.3	287.3	16.0
Medium organics	5,000	454.5	373.1	21.8
Steam	10,000	476.2	387.6	22.9
	Tube Side: Liquid Water $h_i = 5{,}000$ W/(m$^2 \cdot$ K), $R_{ft} = 0.65$ (m$^2 \cdot$ K)/kW			
	Shell Side (Boiling or Condensation)			
	h_o	U_c	U_f	Percentage Increase in Area
Light organics	1,000	833.3	540.5	54.2
Medium organics	5,000	2,500.0	952.4	162.5
Water	10,000	3,333.3	1,052.6	216.7
	Tube Side: Vaporizing Water $h_i = 10{,}000$ W/(m$^2 \cdot$ K), $R_{ft} = 0.51$ (m$^2 \cdot$ K)/kW			
	Shell Side (Boiling or Condensation)			
	h_o	U_c	U_f	Percentage Increase in Area
Medium organics	1,000	909.1	621.1	46.4
Water, low pressure	5,000	3,333.3	1,234.6	170.0
Water, high pressure	10,000	5,000.0	1,408.5	255.0

factor of 2 or even 3. Similarly, with two-phase flow on the tube side, fouling may even dictate the overall design of the heat exchanger. Consequently, any attempts to increase the clean surface heat transfer coefficient on either side of the separating wall (including rough or extended surfaces) should be examined carefully by considering the effects of fouling.

4.2.3 Effect of Fouling on Pressure Drop

Interestingly, more heat exchangers are removed from service for cleaning due to excessive pressure drop than for the inability to meet heat transfer requirements. As shown in Fig. 4.1, fouling always results in a finite, although sometimes small, layer. The change resulting in the flow geometry affects the flow field and the pressure drop (hence the pumping power). For example, in a tubular heat exchanger the fouling layer roughens the surface, decreases the inside diameter, and increases the outside diameter of the tubes. The effects of fouling on pressure drop in the shell side are difficult to quantify because of the complex flow passage.

As mentioned previously, the fouling layer decreases the inside diameter and roughens the tube wall resulting in an increase in the pressure drop. Using Eq. (4.3), pressure drops inside a tube under fouled and clean conditions can be related as

$$\frac{\Delta P_f}{\Delta P_c} = \frac{f_f}{f_c}\left(\frac{d_c}{d_f}\right)\left(\frac{u_{mf}}{u_{mc}}\right)^2 \tag{4.10}$$

Assuming that the mass flow rates ($\dot{m} = \rho u_m A_{\text{cr}}$) under clean and fouled conditions are the same, Eq. (4.10) can be written as

$$\frac{\Delta P_f}{\Delta P_c} = \frac{f_f}{f_c}\left(\frac{d_c}{d_f}\right)^5 \tag{4.11}$$

The inside diameter under fouled conditions d_f can be obtained by rearranging Eq. (4.5b):

$$d_f = d_c \exp\left(-\frac{2k_f R_f}{d_c}\right) \tag{4.12a}$$

and the fouling thickness t_f is expressed as

$$t_f = 0.5 d_c \left[1 - \exp\left(-\frac{2k_f R_f}{d_c}\right)\right] \tag{4.12b}$$

TABLE 4.3 Added Pressure Drop (Tube Side) for Typical Fouling Materials

Material	Thermal Conductivity, k W/(m · K) [3]	Fouling[a] Thickness, t mm	Percentage Area Remaining	Percentage Increase in Pressure Drop
Hematite	0.6055	0.24	95.7	11.6
Biofilm	0.7093	0.28	95.0	13.7
Calcite	0.9342	0.37	93.5	18.4
Serpentine	1.0380	0.41	92.8	20.7
Gypsum	1.3148	0.51	90.9	26.9
Magnesium phosphate	2.1625	0.83	85.5	47.9
Calcium sulfate	2.3355	0.90	84.4	52.6
Calcium phosphate	2.5950	0.99	82.9	59.9
Magnetic iron oxide	2.8718	1.09	81.2	68.2
Calcium carbonate	2.9410	1.12	80.8	70.3

[a]Assuming fouling resistance of 0.4 $(m^2 \cdot K)/kW$.

For a specified total fouling resistance, the tube diameter under fouled conditions, can be obtained if the thermal conductivity of the deposits is known. Since the fouling layer consists of several materials, data on thermal conductivity are not readily available. Multiple fouling layers may also lead to nonuniform thermal conductivity. Approximate thermal conductivities of pure materials, constituting fouling deposits, are given in the second column of Table 4.3 [3]. These values have been used to estimate fouling layer thickness in a 25.4 mm OD, 16-*BWG* (22.1 mm ID) tube with a fouling resistance of 0.4 $(m^2 \cdot K)/kW$. Although rarely true, it is assumed that the fouling layer is composed solely of one material. The flow area remaining and the percentage increase in the pressure drop are given in columns 4 and 5, respectively, of Table 4.3. It is seen that for the assumed fouling resistance, the pressure drop increases by up to 70% in some instances. In these calculations it is assumed that fouling does not affect the friction factor (i.e., $f_f = f_c$). Moreover, the increase in the pressure drop because of excess surface area (required to achieve the desired heat transfer under fouled conditions) has not been taken into account.

4.2.4 Cost of Fouling

Fouling of heat transfer equipment introduces an additional cost to the industrial sector. The added cost is in the form of (1) increased capital expenditure, (2) increased maintenance cost, (3) loss of production, and (4) energy losses.

In order to compensate for fouling, the heat transfer area of a heat exchanger is increased. Pumps and fans are oversized to compensate for over-surfacing and the increased pressure drop resulting from reduction in the flow area. Duplicate heat exchangers may have to be installed in order to ensure continuous operation while a fouled heat exchanger is cleaned. High-cost materials such as titanium, stainless steel, or graphite may be required for certain fouling situations. Cleaning equipment may be required for on-line cleaning. All of these items contribute to increasing the capital expenditure.

On-line and off-line cleaning add to the maintenance cost. Fouling increases the normally scheduled time incurred in maintaining and repairing equipment. Loss of production because of operation at reduced capacity or downtime can be costly. Finally, energy losses due to reduction in heat transfer and increase in pumping-power requirements can be a major contributor to the cost of fouling.

The annual costs of fouling and corrosion in U.S. industries, excluding electric utilities, were placed between $3 and $10 billion in 1982 dollars (Garrett-Price et al. [4]). It is clear that the deleterious effects of fouling are extremely costly.

4.3 ASPECTS OF FOULING

A landmark paper by Taborek et al. [5] cited fouling as the major unresolved problem in heat transfer. Since then the great financial burden imposed by fouling on the industrial sector has been recognized. This has resulted in a significant increase in the literature on fouling, and various aspects of the problem have been resolved. The major unresolved problem of 1972 is now the major unsolved problem. This is because the large amount of fouling research has not brought about a significant solution to the prediction and mitigation of fouling.

In the next section some fundamental aspects that help in understanding the types and mechanisms of fouling are discussed. The commonly used methods that aid in developing models to predict fouling are also outlined.

4.3.1 Categories of Fouling

Fouling can be classified a number of different ways. These may include the type of heat transfer service (boiling, condensation), the type of fluid stream (liquid, gas), or the kind of application (refrigeration, power generation). Because of the diversity of process conditions, most fouling situations are virtually unique. However, in order to develop a scientific understanding, it is best to classify fouling according to the principal process that results in it. Such a classification, developed by Epstein [6], has received wide acceptance. Accordingly, fouling is classified into the following categories: particulate, crystallization, corrosion, biofouling, and chemical reaction.

Particulate Fouling The accumulation of solid particles suspended in the process stream onto the heat transfer surface results in particulate fouling. In boilers this may occur when unburnt fuel or ashes are carried over by the combustion gases. Air-cooled condensers are often fouled because of dust deposition. Particles are virtually present in any condenser cooling water. The matter involved may cover a wide range of materials (organic, inorganic) and sizes and shapes (from the submicron to a few millimeters in diameter). Heavy particles settle on a horizontal surface because of gravity. However, other mechanisms may be involved for fine particles to settle onto a heat transfer surface at an inclination.

Crystallization Fouling A common way in which heat exchangers become fouled is through the process of crystallization. Crystallization arises primarily from the presence of dissolved inorganic salts in the process stream which exhibit supersaturation during heating or cooling. Cooling-water systems are often prone to crystal deposition because of the presence of salts such as calcium and magnesium carbonates, silicates, and phosphates. These are inverse solubility salts that precipitate as the cooling water passes through the condenser (i.e., as the water temperature increases). The problem becomes serious if the salt concentration is high. Such a situation may arise, for example, because of accumulation in cooling-water systems with an evaporative cooling tower. The deposits may result in a dense, well-bonded layer referred to as scale or a porous, soft layer described as a soft scale, sludge, or powdery deposit.

Corrosion Fouling A heat transfer surface exposed to a corrosive fluid may react producing corrosion products. These corrosion products can foul the surface provided the pH value of the fluid is not such that it dissolves the corrosion products as they are formed. For example, impurities in fuel like alkali metals, sulfur, and vanadium can cause corrosion in oil-fired boilers. Corrosion is particularly serious on the liquid side. Corrosion products may also be swept away from the surface where they are produced and transported to other parts of the system.

Biofouling Deposition and/or growth of material of a biological origin on a heat transfer surface results in biofouling. Such material may include microorganisms (e.g., bacteria, algae, and molds) and their products result in microbial fouling. In other instances organisms such as seaweed, water weeds, and barnacles form deposits known as macrobial fouling. Both types of biofouling may occur simultaneously. Marine or power plant condensers using seawater are prone to biofouling.

Chemical Reaction Fouling Fouling deposits are formed as a result of chemical reaction(s) within the process stream. Unlike corrosion fouling, the heat transfer surface does not participate in the reaction although it may act as a catalyst. Polymerization, cracking, and coking of hydrocarbons are prime examples.

It must be recognized that most fouling situations involve a number of different types of fouling. Moreover, some of the fouling processes may compliment each other. For example, corrosion of a heat transfer surface promotes particulate fouling. Significant details about each type of fouling are available in the literature. Somerscales and Knudsen [7] and Melo et al. [8] are excellent sources of such information.

4.3.2 Fundamental Processes of Fouling

Even without complications arising from the interaction of two or more categories, fouling is an extremely complex phenomenon. This is primarily due to the large number of variables that affect fouling. To organize thinking about the topic, it will therefore be extremely useful to approach fouling from a fundamental point of view. Accordingly, the fouling mechanisms, referred to as sequential events by Epstein [9], are initiation, transport, attachment, removal, and aging and are described in the following discussion.

Initiation During initiation the surface is conditioned for the fouling that will take place later. Surface temperature, material, finish, roughness, and coatings strongly influence the initial delay induction or incubation period. For example, in crystallization fouling the induction period tends to decrease as the degree of supersaturation increases with respect to the heat transfer surface temperature. For chemical reaction fouling the delay period decreases with increasing temperature because of the acceleration of induction reactions. Surface roughness tends to decrease the delay period [10]. Roughness projections provide additional sites for crystal nucleation thereby promoting crystallization, while grooves provide regions for particulate deposition.

Transport During this phase fouling substances from the bulk fluid are transported to the heat transfer surface. Transport is accomplished by a number of phenomena including diffusion, sedimentation, and thermophoresis. A great deal of information available for each of these phenomena has been applied to study the transport mechanism for various fouling categories [7, 8].

The difference between fouling species, oxygen or reactant concentration in the bulk fluid C_b and that in the fluid adjacent to the heat transfer surface C_s, results in transport by diffusion. The local deposition flux m_d can be written as

$$\dot{m}_d = h_D(C_b - C_s) \tag{4.13}$$

where h_D is the convective mass transfer coefficient. h_D is obtained from the

Sherwood number ($Sh = h_D d/D$) which, in turn, depends on the flow and the geometric parameters.

Because of gravity, particulate matter in a fluid is transported to the inclined or horizontal surface. This phenomenon, known as sedimentation, is important in applications where particles are heavy and fluid velocities are low.

Thermophoresis is the movement of small particles in a fluid stream when a temperature gradient is present. Cold walls attract colloidal particles while hot walls repel these particles. Thermophoresis is important for particles below 5 μm in diameter and becomes dominant at about 0.1 μm.

A number of other processes such as electrophoresis, inertial impaction, and turbulent downsweeps may be present. Theoretical models to study these processes are available in the literature [11–13]. However, application of these models for fouling prediction is often limited by the fact that several of the preceding processes may be involved simultaneously in a particular fouling situation.

Attachment Part of the fouling material transported attaches to the surface. Considerable uncertainty about this process exists. Probabilistic techniques are often used to determine the degree of adherence. Forces acting on the particles as they approach the surface are important in determining attachment. Additionally, properties of the material such as density, size, and surface conditions are important.

Removal Some material is removed from the surface immediately after deposition and some is removed later. In general, shear forces at the interface between the fluid and deposited fouling layer are considered responsible for removal. Shear forces, in turn, depend on the velocity gradients at the surface, the viscosity of the fluid, and surface roughness. Dissolution, erosion, and spalling have been proposed as plausible mechanisms for removal. In dissolution the material exits in ionic form. Erosion, whereby the material exits in particulate form, is affected by fluid velocity, particle size, surface roughness, and bonding of the material. In spalling the material exits as a large mass. Spalling is affected by thermal stress set up in the deposit by the heat transfer process.

Aging Once deposits are laid on the surface, aging begins. The mechanical properties of the deposit can change during this phase because of changes in the crystal or chemical structure for example. Slow poisoning of microorganisms due to corrosion at the surface may weaken the biofouling layer. A chemical reaction taking place at the deposit surface may alter the chemical composition of the deposit and thereby change its mechanical strength.

4.3.3 Prediction of Fouling

The overall result of the processes listed previously is the net deposition of material on the heat transfer surface. Clearly the deposit thickness is time dependent. For heat exchanger design a constant fouling resistance, interpreted as the value reached in a time period after which the heat exchanger will be cleaned, is used. Predicting how fouling progresses over time determines the cleaning cycle. Such information is also required for proper operation of the heat exchanger.

Predictive models are based on the idea that the variation of fouling with time can be expressed as the difference between the deposition rate ϕ_d and removal rate ϕ_r functions [14]:

$$\frac{dR_f}{dt} = \phi_d - \phi_r \tag{4.14}$$

Functions ϕ_d and ϕ_r depend on the events discussed in the previous section. Proper evaluation of these functions will require complete understanding of the phenomena involved during various phases of fouling. Such an approach is not practical since a large number of parameters are involved even in simple fouling situations. Therefore the current predictive models of fouling are semiempirical. From a practical point of view, such an approach is justified since most fouling behavior can be represented by fouling factor–time curves as shown in Fig. 4.3. The shape of these curves relates to the phenomena occurring during the fouling process.

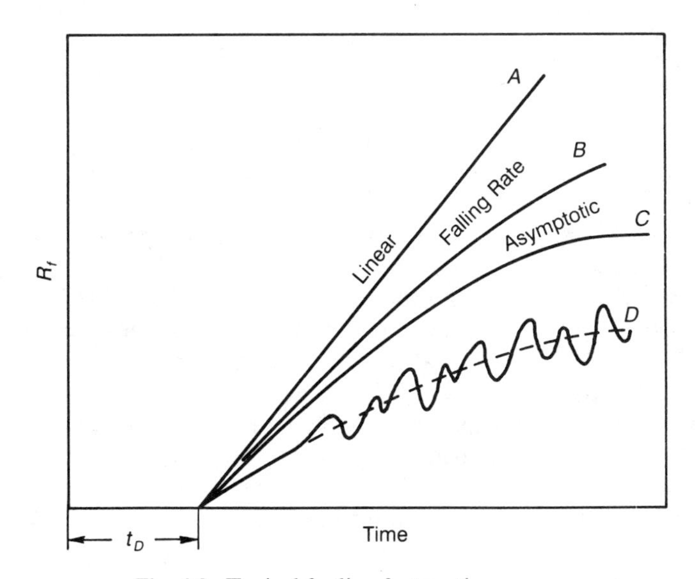

Fig. 4.3. Typical fouling factor–time curve.

If the deposition rate is constant and the removal rate is negligible or if the difference between the deposition and removal rates is constant, the fouling–time curve will be a straight line as shown by curve A in Fig. 4.3. This type of linear fouling is generally represented by tough, hard, adherent deposits. Fouling in such cases will continue to build up unless some type of cleaning is employed. As a special case, if the deposition rate is constant and the removal rate is ignored, Eq. (4.14) can be integrated to yield

$$R_t = \phi_d t \tag{4.15}$$

Equation (4.15), first developed by McCabe and Robinson [15] in 1924, represents the simplest of the fouling models. A widely observed form of fouling is the asymptotic fouling represented by curve C. It results if the deposition rate is constant and the removal rate is proportional to the fouling layer thickness, thus suggesting that the layer's shear strength decreases with time or another mechanism deteriorates the stability of the layer. Such a situation will generally occur if the deposits are soft since they flake easily. The fouling factor in such cases reaches an asymptotic value. Assuming the removal function to be proportional to the fouling resistance ($\phi_r = bR_f$) and the deposition function to be constant ($\phi_d = a$), Eq. (4.14) can be written as

$$\frac{dR_f}{dt} = a - bR_f \tag{4.16}$$

where a and b are constants. The classical Kern–Seaton [16] relation is obtained by integrating Eq. (4.16):

$$R_f = R_f^*(1 - e^{-t/\theta}) \tag{4.17}$$

where R_f^* $(= a/b)$ is the asymptotic fouling factor and θ $(= 1/b)$ is the time constant that indicates how quickly asymptotic fouling conditions are approached.

Falling-rate fouling, shown by curve B, lies between the linear and asymptotic fouling curves. Such behavior may result if the deposition rate is inversely proportional to the fouling thickness. A periodic change in operating conditions results in the sawtooth configuration shown in curve D. This situation is typical of commercial cooling tower water.

Occasionally, particularly for new surfaces, a delay time t_D is observed before deposition occurs. During this time the fouling resistance remains close to 0 while only micronucleation sites are formed. At a certain point in the fouling process, the nucleation sites become so numerous that they combine into an integral blanket resulting in a rapid increase in the fouling rate [5].

A number of semiempirical models have been developed over the years to predict the nature of the fouling curve in a given application. The general

TABLE 4.4 Effect of Parameters on Fouling [17]

Parameter Increased	Deposition Rate	Removal Rate	Asymptotic Fouling
Stickiness	Increases	Decreases	Increases
Surface temperature	Increases	Questionable	Increases
Toughness	Questionable	Decreases	Increases
Roughness	Increases (?)	Increases	Questionable
In-situ corrosion	Increases	Questionable	Increases
Ex-situ corrosion	Increases	Questionable	Increases
Velocity	Decreases	Increases	Decreases

applicability of these models is limited since the various constants or coefficients involved are site dependent and would usually be unknown. Much of the current fouling research is directed toward establishing predictive models. Epstein [6] has tabulated a number of deposition and removal models developed over the past several years. The qualitative effects of increasing certain parameters on the deposition and removal rates and the asymptotic fouling factor are summarized in Table 4.4 [17]. Velocity is the only parameter whose increase causes a reduction in the asymptotic fouling factor even though there may be some exceptions.

4.4 DESIGN OF HEAT EXCHANGERS SUBJECT TO FOULING

Although fouling is time dependent, only a fixed value can be prescribed during the design stage. Therefore the operating characteristics and cleaning schedules of the heat exchanger depend on the design fouling factor. Many heat exchangers operate for long periods without being cleaned while others might require frequent cleaning. Table 4.5, assembled by Garret-Price et al. [18], identifies the type and extent of fouling that may occur in various industry groups. This information is useful in determining the impact of fouling when a particular heat exchanger is designed.

4.4.1 Providing a Fouling Allowance

If fouling is anticipated, provisions should be made during the design stage. A number of different approaches are used to provide an allowance for fouling, all of which result in an excess surface area for heat transfer. Current methods include specifying the fouling resistances, the cleanliness factor, or the percentage over surface.

Fouling Resistance A common practice is to prescribe a fouling resistance or fouling factor [as in Eq. (4.4)] on each side of the surface where fouling is anticipated. The result is a lower overall heat transfer coefficient. Conse-

TABLE 4.5 Fouling of Heat Transfer Surfaces by Industry Groups [18]

Industry Group	Type of Fouling That Occurs in Heat Exchange Equipment	Usual Extent
Food and kindred products	Chemical reaction	Major
	Crystallization (milk processing)	Major
	Biofouling	Medium
	Particulate (gas side) (spray drying)	Minor/major
	Corrosion	Minor
Textile mill products	Particulate (cooling water)	
	Biofouling (cooling water)	
Lumber and wood products including paper and allied products	Crystallization (liquid, cooling water)	Major
	Particulate (process side, cooling water)	Minor
	Biofouling (cooling water)	Minor
	Chemical reaction (process side)	Minor
	Corrosion	Medium
Chemical and allied	Crystallization (process side, cooling)	Medium

quently, excess surface area is provided to achieve the specified heat transfer. The heat exchanger will perform satisfactorily until the specified value of the fouling resistance is reached, after which it must be cleaned. The cleaning interval is expected to coincide with the plant's regular maintenance schedule so that additional shutdowns can be avoided.

It is extremely difficult to predict a specific fouling behavior for most cases since a large number of variables can materially alter the type of fouling and its rate of buildup. Sources of fouling resistances in the literature are rather limited, in part, because of the relatively recent interest in fouling research. Tables found in the standards of the Tubular Exchanger Manufacturers Association (TEMA) [19], reproduced here as Tables 4.6 to 4.10, are probably the most referenced source of fouling factors used in the design of heat exchangers. Unfortunately, the TEMA tables do not cover the large variety of possible process fluids, flow conditions, and heat exchanger configurations. These values allow the exchanger to perform satisfactorily in a designated service for a "reasonable time" between cleaning. The interval between cleaning is not known a priori since it depends on performance of the heat exchanger while it is in service. Quite often sufficient excess area is provided for the exchanger to perform satisfactorily under fouled conditions. Proprietary research data, plant data, and personal or company experience are other sources of fouling resistances.

Cleanliness Factor Another approach that allows for fouling is specifying the cleanliness factor (CF), a term used in the steam power industry. The cleanliness factor relates the overall heat transfer coefficients under fouled

TABLE 4.6 TEMA Design Fouling Resistances for Industrial Fluids, $(m^2 \cdot K)/kW$ [19]

Oils	
Fuel oil #2	0.352
Fuel oil #6	0.881
Transformer oil	0.176
Engine lube oil	0.176
Quench oil	0.705
Gases and vapors	
Manufactured gas	1.761
Engine exhaust gas	1.761
Steam (nonoil bearing)	0.088
Exhaust stream (oil bearing)	0.264–0.352
Refrigerant vapors (oil bearing)	0.352
Compressed air	0.176
Ammonia vapor	0.176
CO_2 vapor	0.176
Chlorine vapor	0.352
Coal flue gas	1.761
Natural gas flue gas	0.881
Liquids	
Molten heat transfer salts	0.088
Refrigerant liquids	0.176
Hydraulic fluid	0.176
Industrial organic heat transfer media	0.352
Ammonia liquid	0.176
Ammonia liquid (oil bearing)	0.528
Calcium chloride solutions	0.528
Sodium chloride solutions	0.528
CO_2 liquid	0.176
Chlorine liquid	0.352
Methanol solutions	0.352
Ethanol solutions	0.352
Ethylene glycol solutions	0.352

and clean conditions as

$$CF = \frac{U_f}{U_c} \tag{4.18}$$

It is apparent from Eq. (4.18) that the cleanliness factor provides a fouling allowance in proportion to the overall heat transfer coefficient under clean conditions. Using Eqs. (4.6) and (4.18), the cleanliness factor and the total

TABLE 4.7 TEMA Design Fouling Resistances for Chemical Processing Streams, $(m^2 \cdot K) / kW$ [19]

Gases and vapors	
Acid gases	0.352–0.528
Solvent vapors	0.176
Stable overhead products	0.176
Liquids	
MEA and DEA solutions	0.352
DEG and TEG solutions	0.352
Stable side draw and bottom product	0.176–0.352
Caustic solutions	0.352
Vegetable oils	0.528

TABLE 4.8 TEMA Design Fouling Resistances for Natural Gas–Gasoline Processing Streams, $(m^2 \cdot K) / kW$ [19]

Gases and vapors	
Natural gas	0.176–0.352
Overhead products	0.176–0.352
Liquids	
Lean oil	0.352
Rich oil	0.176–0.352
Natural gasoline and liquified petroleum gases	0.176–0.352

fouling resistance can be related as

$$R_{ft} = \frac{1 - \text{CF}}{U_c \text{CF}} \tag{4.19a}$$

or

$$\text{CF} = \frac{1}{1 + R_{ft} U_c} \tag{4.19b}$$

Equation (4.19a) has been used to obtain Fig. 4.4, which shows fouling resistance versus clean surface overall heat transfer coefficient curves for various cleanliness factors. Figure 4.4 illustrates that a given CF corresponds to a higher total fouling resistance R_{ft} if the overall heat transfer coefficient U_c is low. Such a trend is desirable for designing steam condensers where U_c is proportional to the velocity. As shown in Table 4.4, lower velocity (hence low U_c) results in increased fouling. Although the cleanliness factor results in favorable trends, the designer is still left with the problem of selecting the appropriate CF for a given application. Typical condenser designs are based

TABLE 4.9 TEMA Design Fouling Resistances for Oil Refinery Streams, $(m^2 \cdot K) / kW$ [19]

Crude and vacuum unit gases and vapors	
Atmospheric tower overhead vapors	0.176
Light naphthas	0.176
Vacuum overhead vapors	0.352
Crude and vacuum liquids	
Crude oil	

	−30 to 120°C Velocity, m/s			120 to 175°C Velocity, m/s		
	< 0.6	0.6–1.2	> 1.2	< 0.6	0.6–1.2	> 1.2
Dry	0.528	0.352	0.352	0.528	0.352	0.352
salt[a]	0.528	0.352	0.352	0.881	0.705	0.705
	175 to 230°C Velocity, m/s			230°C and Over Velocity, m/s		
	< 1.5	0.6–1.2	> 1.2	< 1.5	0.6–1.2	> 1.2
Dry	0.705	0.528	0.528	0.881	0.705	0.705
salt[a]	1.057	0.881	0.881	1.233	1.057	1.057

[a]Assumes desalting at approximately 120°C.

Gasoline	0.352
Naphtha and light distillates	0.352–0.528
Kerosene	0.352–0.528
Light gas oil	0.352–0.528
Heavy gas oil	0.528–0.881
Heavy fuel oils	0.881–1.233
Asphalt and residuum	
Vacuum tower bottoms	1.761
Atmosphere tower bottoms	1.233
Cracking and coking unit streams	
Overhead vapors	0.352
Light cycle oil	0.352–0.528
Heavy cycle oil	0.528–0.705
Light coker gas oil	0.528–0.705
Heavy coker gas oil	0.705–0.881
Bottoms slurry oil (1.4 m/s minimum)	0.528
Light liquid products	0.176
Catalytic Reforming, hydrocracking, and hydrodesulfurization streams	
Reformer charge	0.264
Reformer effluent	0.264
Hydrocracker charge and effluent[b]	0.352
Recycle gas	0.176
Hydrodesulfurization charge and effluent[b]	0.352

TABLE 4.9 *(Continued)*

Overhead vapors	0.176
Liquid product over 50° A.P.I.	0.176
Liquid product 30–50° A.P.I.	0.352

[b]Depending on charge, characteristics, and storage history, charge resistance may be many times this value.

Light ends processing streams	
Overhead vapors and gases	0.176
Liquid products	0.176
Absorption oils	0.352–0.528
Alkylation trace acid streams	0.352
Reboiler streams	0.352–0.528
Lube oil processing streams	
Feed stock	0.352
Solvent feed mix	0.352
Solvent	0.176
Extract[c]	0.528
Raffinate	0.176
Asphalt	0.881
Wax slurries[c]	0.528
Refined lube oil	0.176

[c]Precautions must be taken to prevent wax deposition on cold tube walls.

Visbreaker	
Overhead vapor	0.528
Visbreaker bottoms	1.761
Naphtha hydrotreater	
Feed	0.528
Effluent	0.352
Naphthas	0.352
Overhead vapors	0.264
Catalytic hydro desulfurizer	
Charge	0.705–0.881
Effluent	0.352
Heat transfer separation overhead	0.352
Stripper charge	0.528
Liquid products	0.352
HF alky unit	
Alkylate, deprop. bottoms, main fraction overhead, main fraction feed	0.528
All other process streams	0.352

TABLE 4.10 TEMA Design Fouling Resistances for Water

Temperature of Heating Medium	Up to 115°C		115 to 205°C	
Temperature of Water	50°C		Over 50°C	
	Water Velocity, m/s		Water Velocity, m/s	
	< 0.9	> 0.9	< 0.9	> 0.9
Sea water	0.088	0.088	0.176	0.176
Brackish water	0.352	0.176	0.528	0.352
Cooling tower and artificial spray pond				
Treated makeup	0.176	0.176	0.352	0.352
Untreated	0.528	0.528	0.881	0.705
City or well water	0.176	0.176	0.352	0.352
River water				
Minimum	0.352	0.176	0.528	0.352
Average	0.528	0.352	0.705	0.528
Muddy or silty	0.528	0.352	0.705	0.528
Hard (over 15 grains/gallon)	0.528	0.528	0.881	0.881
Engine jacket	0.176	0.176	0.176	0.176
Distilled or closed cycle				
Condensate	0.088	0.088	0.088	0.088
Treated boiler feed water	0.176	0.088	0.176	0.176
Boiler blowdown	0.352	0.352	0.352	0.352

on a cleanliness factor of 0.80 to 0.85. However, use of the CF for other applications would require careful evaluation.

Percentage over Surface In this approach the designer simply adds a certain percentage of clean surface area to account for fouling. The added surface implicitly fixes the total fouling resistance depending on the clean surface overall heat transfer coefficient. If the heat transfer rates and fluid temperatures under clean and fouled conditions are the same ($Q_f = Q_c$ and $\Delta T_{mf} = \Delta T_{mc}$), the percentage over surface (OS) can be obtained from Eq. (4.9) as

$$\mathrm{OS} = 100\left(\frac{A_f}{A_c} - 1\right) = 100 U_c R_{ft} \tag{4.20}$$

In a shell-and-tube heat exchanger the additional surface can be provided either by increasing the length of tubes or by increasing the number of tubes (hence the shell diameter). The resulting change will also affect the design conditions such as flow velocity, number of cross passes, or baffle spacing.

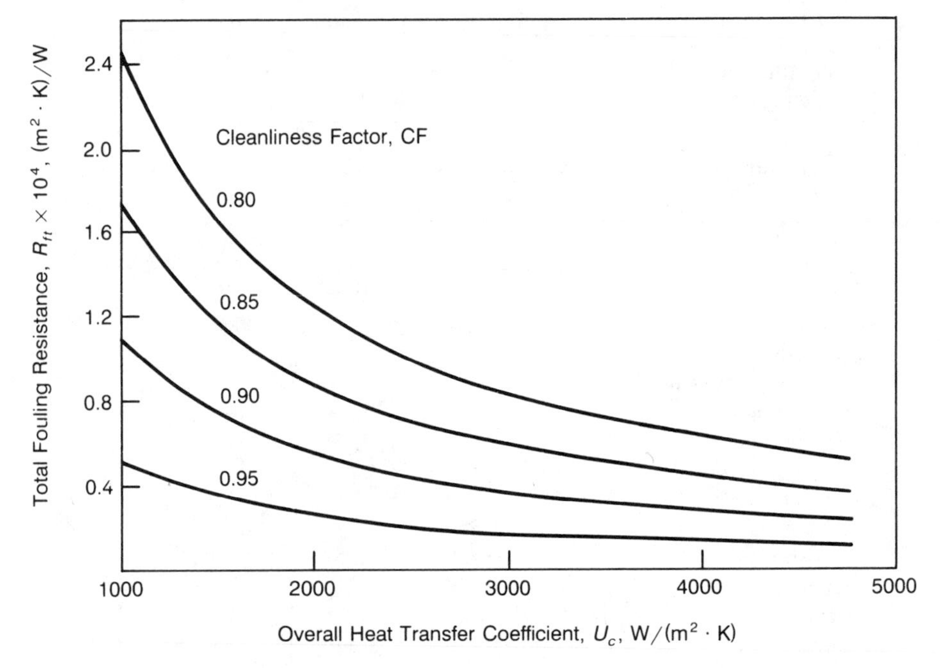

Fig. 4.4. Calculated fouling resistance based on cleanliness factor.

Therefore the new design, with increased surface area, should be re-rated to achieve optimum conditions.

Example 4.1. A double-pipe heat exchanger is used to condense steam at a rate of 113.68 kg/hr at 50°C. Cooling water (brackish water) enters through the tubes at a rate of 0.9 kg/s at 10°C. The tube with 25.4 mm OD, 22.1 mm ID is made of mild steel [$k = 45$ W/(m · K)]. The heat transfer coefficient on the steam side, h_o is 10,000 W/(m² · K).

The exit temperature of the water can be obtained from the heat balance:

$$Q = \dot{m}_o h_{fg} = \dot{m}_i c_p (T_{\text{out}} - T_{\text{in}}) \tag{4.21}$$

with

$$h_{fg} = 2382.7 \text{ kJ/kg} \qquad c_p = 4.18 \text{ kJ/(kg} \cdot \text{K)}$$

$$Q = \frac{113.68}{3600}(2382.7 \times 10^3) = (0.9)(4.18 \times 10^3)(T_{\text{out}} - 10)$$

$$Q = 75{,}240 \text{ W} \qquad T_{\text{out}} = 30°\text{C}$$

Although improved correlations are available [1], for simplicity the heat transfer coefficient on the water side, h_i is obtained from the Dittus–Boelter relation [20]:

$$Nu_i = \frac{h_i d_i}{k} = 0.023 Re^{0.8} Pr^{0.4} \tag{4.22}$$

At the mean temperature of 20°C, the properties of water are

$$\mu = 1007.4 \times 10^{-6}\ (\mathrm{N \cdot s})/\mathrm{m}^2 \qquad \rho = 1000.5\ \mathrm{kg/m}^3$$

$$k = 0.603\ \mathrm{W/(m \cdot K)} \qquad c_p = 4180\ \mathrm{J/(kg \cdot K)}$$

$$Pr = 6.9967$$

$$Re = \frac{\rho u_{mi} d_i}{\mu} \quad \text{or} \quad \frac{4\dot{m}_i}{\pi d_i \mu} = \frac{4 \times 0.9}{\pi \times 0.0221 \times 1007.4 \times 10^{-6}} = 51{,}471$$

Therefore

$$Nu_i = 294.4 \quad \text{or} \quad h_i = 8033\ \mathrm{W/(m^2 \cdot K)} \cong 8000\ \mathrm{W/(m^2 \cdot K)}$$

$$\text{Water-side velocity, } u_{mi} = \frac{\dot{m}_i}{\rho_i A_i} = 2.35\ \mathrm{m/s}$$

Fouling Resistances The inside and outside fouling resistances are obtained from Table 4.10:

$R_{fi} = 0.176\ (\mathrm{m^2 \cdot K})/\mathrm{kW}$ brackish water below 50°C, velocity over 0.9 m/s

$R_{fo} = 0.088\ (\mathrm{m^2 \cdot K})/\mathrm{kW}$ closed cycle condensate

Using Eq. (4.7), the total fouling resistance is obtained as

$$R_{ft} = 0.2903\ (\mathrm{m^2 \cdot K})/\mathrm{kW}$$

The overall heat transfer coefficients and the distribution of thermal resistances under clean and fouled conditions are summarized in Table 4.11. For assumed values of fouling resistances, the heat transfer coefficient, under fouled conditions, is only one-half of that under clean conditions. It is interesting to note that under fouled conditions the heat transfer resistance due to inside tube fouling, dominates all other resistances. In this example, the heat transfer surface area should be increased by 103% to obtain the desired heat transfer rate under fouled conditions. Increasing the surface area by such a magnitude may be expensive. An alternate solution would be to design with lower fouling resistances and arrange for cleaning and/or mitigation techniques to control fouling (see Chapter 5).

TABLE 4.11 Double-Pipe Condenser (Example 4.1)

	Clean	Fouled
Distribution of resistances, %		
Water side		
$[h_i = 8000\ \text{W}/(\text{m}^2 \cdot \text{K})]$	0.8	25.1
Inside fouling	0	35.3 $[R_{fi} = 0.176\ (\text{m}^2 \cdot \text{K})/\text{kW}]$
Tube wall	13.9	6.9
Outside fouling	0	15.3 $[R_{fo} = 0.088\ (\text{m}^2 \cdot \text{K})/\text{kW}]$
Steam side		
$[h_o = 10{,}000\ \text{W}/(\text{m}^2 \cdot \text{K})]$	35.3	17.4
Overall heat transfer coefficient, $\text{W}/(\text{m}^2 \cdot \text{K})$	3534	1744
Mean temperature difference, °C	28.85	28.85
Surface area, m^2	0.7379	1.4952 (103% increase)

Cleanliness Factor If a cleanliness factor of 0.85 (a typical value) is used, U_f can be obtained from Eq. (4.18):

$$U_f = 0.85 \times 3534 = 3004\ \text{W}/(\text{m}^2 \cdot \text{K})$$

Using Eq. (4.19a), the equivalent total fouling resistance is obtained as 0.05 $(\text{m}^2 \cdot \text{K})/\text{kW}$, which is rather small compared to the value obtained from the TEMA standards (Table 4.10). The surface area under fouled conditions is obtained as 0.8682 m^2 corresponding to an increase of only 17.7%. It is clear that the lack of proper fouling data may lead to significantly different heat exchanger designs.

Percentage over Surface If 25% over surface (a typical value) is prescribed, the total fouling resistance is obtained as 0.071 $(\text{m}^2 \cdot \text{K})/\text{kW}$ [Eq. (4.20)], substantially lower than the corresponding value from the TEMA standards (about one-fourth).

TABLE 4.12 Relationship between R_{ft}, CF, and OS [Example 4.1, $U_c = 3534\ \text{W}/(\text{m}^2 \cdot \text{K})$]

R_{ft}, $(\text{m}^2 \cdot \text{K})/\text{kW}$	CF	OS, %
0.05	0.85	17.7
0.10	0.74	35.3
0.15	0.65	53.0
0.20	0.59	70.7
0.25	0.53	88.4
0.30	0.49	106.0
0.35	0.45	123.7
0.40	0.41	141.4

Seemingly different methods that allow for fouling essentially do the same thing; that is, they increase the required heat transfer surface area. As discussed earlier the total fouling resistance, cleanliness factor, and percentage over surface can be related to each other through Eqs. (4.7) and (4.18) to (4.20). The relationship for Example 4.1 is presented in Table 4.12.

Example 4.2. To increase the heat transfer surface area, the plain tube in Example 4.1 is replaced by a low-finned tube (with fins either on the inside or the outside). Fins increase the available surface area by a factor of 2.9.

Assume 100% fin efficiency, identical heat transfer coefficients (h_i and h_o), wall resistances, and inside fouling resistances for plain and finned tubes. Fouling on the shell side can be neglected.

Table 4.13 shows various conditions and results for this example. The heat transfer rate is reflected by the product UA_o/L. A finned tube results in higher heat transfer than the plain tube simply because of increased surface area. Under clean conditions the heat transfer increases by 50% if fins are used on the inside and by 30% if fins are used on the outside. Fins are more effective on the inside because of a relatively lower heat transfer coefficient (hence higher thermal resistance) on the inside. Under fouled conditions, the increase in the heat transfer with fins on the inside and outside of the tube is 88% and 16%, respectively. In this example external fins are not very effective since the heat transfer increases by only 18% even though the outside surface area is increased by 190%. As a general rule, enhancement is effective if done on the side with dominant thermal resistance. It should be recognized that identical fouling resistances have been used for plain and finned tubes. Such an assumption may not be valid if the fins have a significant effect on the fouling characteristics. Moreover, ease of cleaning should also be considered when finned surfaces are used.

TABLE 4.13 Fouling in Plain and Finned Tube (Example 4.2)

Quantity	Plain	Finned Inside	Finned Outside
d_o, mm	25.4	25.4	25.4
d_i, mm	22.1	22.1	22.1
h_i, W/(m$^2 \cdot$ K)	8000	8000	8000
h_o, W/(m$^2 \cdot$ K)	10,000	10,000	10,000
k, W/(m $\cdot$ K)	45	45	45
Area ratio			
A_o/A_{op}	1.0	1.0	2.9
A_i/A_{ip}	1.0	2.9	1.0
A_o/A_i	1.1493	0.3963	3.3333
U_c, W/(m$^2 \cdot$ K)	3534	5296	1586
$U_c A_o/L$, W/(K $\cdot$ m)	282	423	367
Percentage increase	0	50	30
R_{fi}, (m$^2 \cdot$ K)/kW	0.176	0.176	0.176
U_c, W/(m$^2 \cdot$ K)	2061	3867	822
$U_c A_o/L$, W/(K $\cdot$ m)	164	309	190
Percentage increase	0	88	16

4.4.2 Design Features to Minimize Fouling

Fouling may not be avoided in many situations. However, its extent can be minimized by good design practice. If excessive fouling is expected, for example, in geothermal applications, direct-contact heat exchangers should be considered. In shell-and-tube heat exchangers proper allocation of the fluid streams to the tube side and the shell side is very important. In general a fouling-prone fluid stream should be placed on the tube side since cleaning is easier. This is due to ease of cleaning the tubes. Moreover, it is less expensive to provide fouling-resistant material on the tube side.

The designer often has several combinations of operating parameters that will meet the design thermohydraulic requirements. This opportunity should be fully exploited to select parameters that mitigate fouling. In general, higher fluid velocity and lower tube wall temperature impede fouling buildup. A tube-side velocity of 1.8 m/s is a widely accepted figure. Operation above the dew point for acid vapors and above freezing for fluids containing waxes prevents corrosion and freeze fouling from occurring.

In shell-and-tube heat exchangers, the velocity distribution on the shell side is nonuniform because baffles are present. Fouling deposits tend to accumulate in the baffle-to-shell corners where the velocity is low and the flow recirculatory. Figure 4.5 illustrates the observed fouling on the shell side of two small test heat exchangers that differ in their baffle spacing and baffle window size [21]. Both heat exchangers were operated with the same shell-side fluid, crossflow velocity, and surface temperature. Fouling deposits were

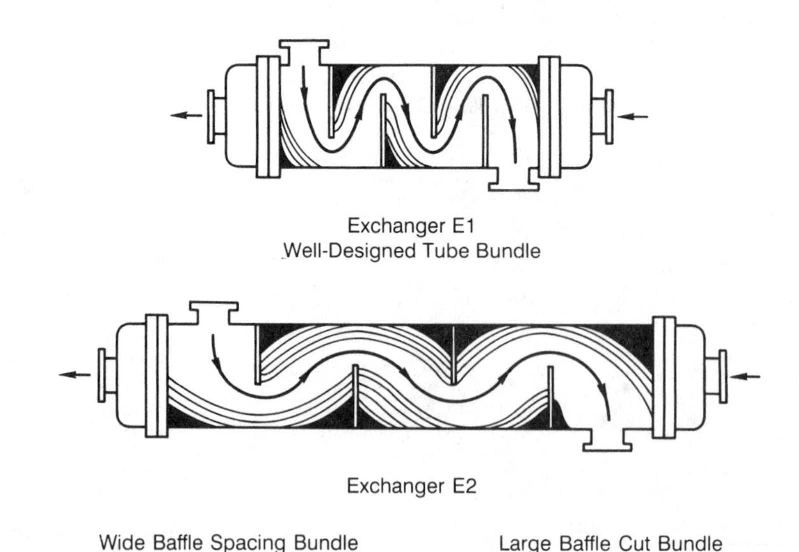

Fig. 4.5. Sketch of observed fouling deposits in two different heat exchanger configurations [17].

heavy in the region of low velocity. Gilmour [22] suggests that baffle cuts of 20% to 25% of the shell diameter are appropriate to minimize fouling.

4.4.3 Design Features to Facilitate Fouling Control

A good design practice alone may not completely eliminate fouling. It should therefore be expected that the heat exchanger will require cleaning at certain time intervals. The cleaning cycle can be extended if on-line cleaning (discussed later) is employed to control fouling. Assurance for continuous cleaning can also result in requiring a smaller fouling allowance. On-line cleaning systems are available and most easily installed when the plant is initially constructed. Unfortunately, selection of the appropriate cleaning procedure is often an "after the fact" decision. For proper operation of the heat exchanger, features that facilitate cleaning, hence better fouling control, should be considered during the design stage.

Horizontal heat exchangers are easier to clean than vertical ones. The G, H, and K types (TEMA standards) are normally oriented horizontally while other types can be oriented horizontally or vertically. A heat exchanger with removable heads and a straight tube would be easy to clean. Space and provision for removing tube bundles need to be available. Valves isolating the exchanger and connecting cleaning service hoses should be incorporated, thus facilitating on-site chemical cleaning.

4.5 OPERATION OF HEAT EXCHANGERS SUBJECT TO FOULING

Specification of excess surface area in the heat exchanger, because of anticipated fouling, leads to operational problems that may accelerate fouling buildup. For example, if the flow rate is maintained at the design value, the heat transfer at the start of the operation will usually be high resulting in undesirable temperature trends. Similarly, to achieve the design value of heat transfer, the flow rate (hence velocity) would be reduced and the temperature increased. As discussed earlier, lower flow velocity and higher surface temperature result in increased fouling. The following example illustrates the effects of fouling on heat exchanger operation under two different conditions.

Example 4.3. A total fouling resistance of 0.176 $(m^2 \cdot K)/kW$ is used to design the condenser in Example 4.1. Consider its operation with (a) the water flow rate maintained at the design value 0.9 kg/s and (b) the heat transfer rate maintained at the design value 75,240 W.

For simplicity, the wall thermal resistance and curvature of the tube are ignored. Therefore the clean surface overall heat transfer coefficient U_c is

$$\frac{1}{U_c} = \frac{1}{8000} + \frac{1}{10{,}000} \quad \text{or} \quad U_c = 4444.4\ \mathrm{W/(m^2 \cdot K)}$$

The fouled surface overall heat transfer coefficient is

$$\frac{1}{U_f} = \frac{1}{4444.4} + 0.000176 \quad \text{or} \quad U_f = 2493.8 \text{ W/(m}^2 \cdot \text{K)}$$

The design surface area is obtained from Eq. (4.8) as

$$Q = 75{,}240 \text{ W}, \qquad U_f = 2493.8 \text{ W/(m}^2 \cdot \text{K)} \qquad \Delta T_{mf} = 28.85°\text{C}$$

Therefore $A_f = 1.0456$ m^2 ($A_c = 0.5868$ m^2, over surface = 78.2%).

The heat exchanger is designed to provide a heat transfer surface area of 1.0456 m^2.

Operation at Design Water Flow Rate If the water flow rate is maintained at the design value, the heat transfer rate will be highest when the surface is clean. A higher heat transfer results in a higher water outlet temperature. The heat transfer rate and the exit temperature can be obtained from the following equations:

$$Q = U_f A_f \, \Delta T_m = \dot{m}_i C_p (T_{\text{out}} - T_{\text{in}}) \tag{4.23}$$

$$\frac{1}{U_f} = \frac{1}{h_i} + \frac{1}{h_o} + R_{ft} \tag{4.24}$$

The known parameters in Eqs. (4.23) and (4.24) are

$$A_f = 1.0456 \text{ m}^2 \qquad \dot{m}_i = 0.9 \text{ kg/s} \qquad c_p = 4180 \text{ J/(kg} \cdot \text{K)}$$

$$T_{\text{in}} = 10°C \qquad h_i = 8000 \text{ W/(m}^2 \cdot \text{K)} \qquad h_o = 10{,}000 \text{ W/(m}^2 \cdot \text{K)}$$

For a specified total fouling resistance, the fouled overall heat transfer coefficient U_f can be obtained from Eq. (4.24). Thereafter, the heat transfer rate Q and the water exit temperature T_{out} can be computed iteratively from Eq. (4.23). Deviations from the design values of the water exit temperature and the heat transfer rate for different fouling resistances are depicted in Fig. 4.6. It has been assumed that the condensing temperature is constant (50°C). Under clean conditions the heat exchanger provides 42% more heat transfer resulting in a water exit temperature 8°C above the design value. As the fouling builds up the heat transfer rate as well as the exit water temperature decrease.

Operation at Constant Heat Transfer Rate Because of anticipated fouling the heat exchanger is designed to provide 78% more surface area at the start of operation. Therefore, to achieve the design heat transfer rate when the heat exchanger is placed in operation under clean conditions, the water

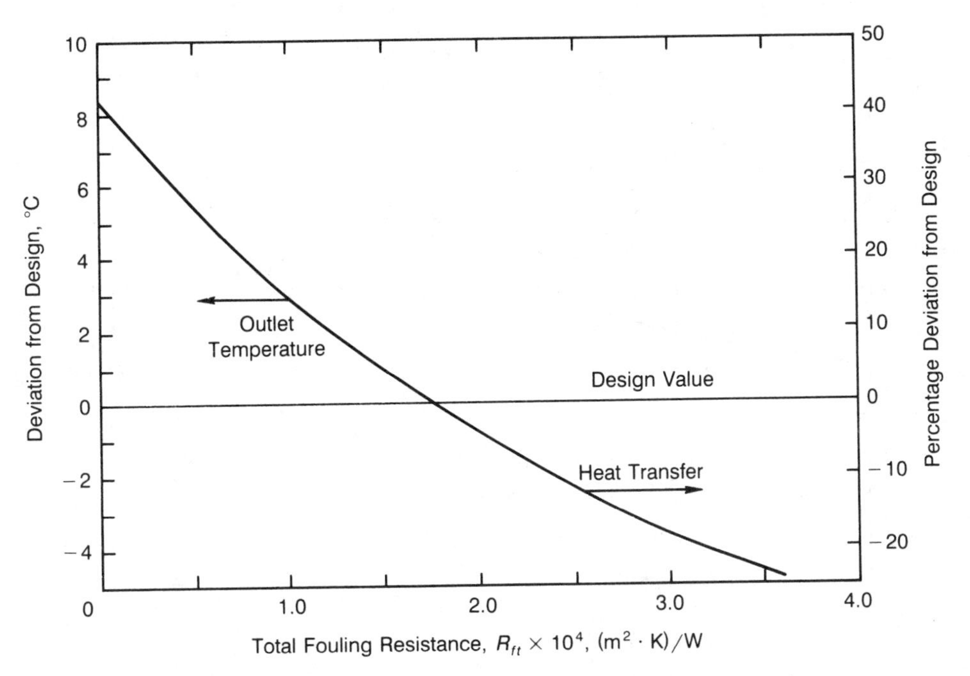

Fig. 4.6. Effect of fouling on outlet temperature and heat transfer.

flow rate (hence the velocity) needs to be adjusted. A common practice (although not recommended) adopted by plant operators is to reduce the mass flow rate. The resulting change in velocity affects the tube-side heat transfer coefficient h_i and the water outlet temperature T_{out}. The water flow rate and the exit temperature to achieve the design heat transfer rate can be computed from Eqs. (4.23) and (4.24). However, the tube-side heat transfer coefficient h_i should be related to the design heat transfer coefficient h_d [= 8000 W/(m² · K)] as

$$\frac{h_i}{h_d} = \left(\frac{\dot{m}_i}{\dot{m}_d}\right)^{0.8} \tag{4.25}$$

The known parameters in Eqs. (4.23) to (4.25) are

$$Q = 75{,}240 \text{ W} \qquad A_f = 1.0456 \text{ m}^2 \qquad c_p = 4180 \text{ J/(kg} \cdot \text{K)}$$

$$T_{in} = 10°\text{C} \qquad h_o = 10{,}000 \text{ W/(m}^2 \cdot \text{K)} \qquad h_d = 8000 \text{ W/(m}^2 \cdot \text{K)}$$

$$\dot{m}_d = 0.9 \text{ kg/s}$$

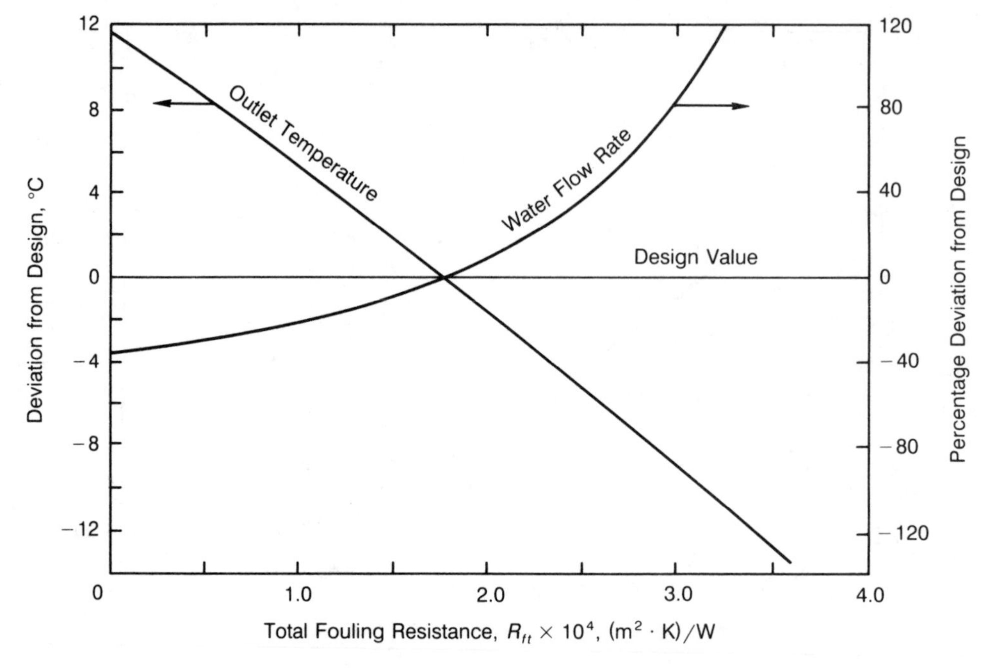

Fig. 4.7. Effect of fouling on outlet temperature and water flow rate.

For a specified R_f, Eqs. (4.23) to (4.25) can be used with a trial-and-error method to obtain $\dot{m}_i$ and T_{out}. The results are shown in Fig. 4.7. When the heat exchanger is placed in operation under clean conditions ($R_f = 0$), 37% less water flow is required and the exit temperature of the water increases by 11.7°C. A higher water temperature as well as a lower water flow rate increase the tendency of the heat transfer surface to foul. This example demonstrates that adding the heat transfer surface area, allowing for anticipated fouling, tends to accelerate the fouling rate.

It would be desirable if, during the initial operating period, the water velocity could be maintained at the design value. This can be accomplished by recirculating some of the water as shown in Fig. 4.8 [17] or by flooding the condenser. The additional capital costs of the pumps and piping required for recirculation should be balanced against the advantage of reducing fouling in the heat exchanger.

Performance of the heat exchanger should be monitored during operation. Based on these observations, a proper cleaning schedule can be established. If proper fouling data are available, the cleaning schedule can be based on a rational economic criterion. Somerscales [23] provides a discussion of some of these techniques.

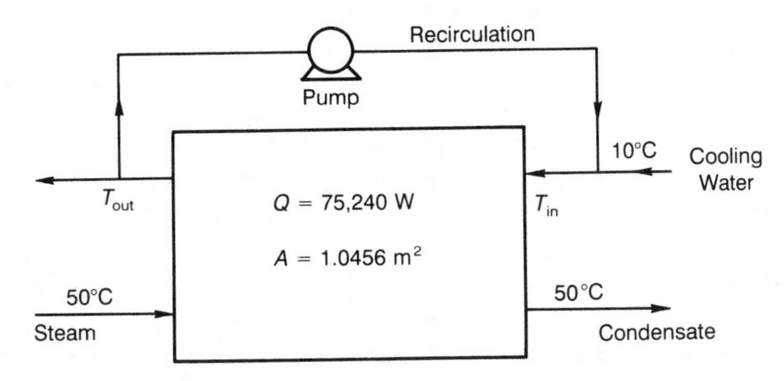

Fig. 4.8. Illustrative example of recirculation.

4.6 TECHNIQUES TO CONTROL FOULING

There are a number of strategies for control of fouling. Additives that act as fouling inhibitors can be used while the heat exchanger is in operation. If it is not possible to stop fouling, it becomes a practical matter to remove it. Surface cleaning can be done either on line or off line. Table 4.14 [21] provides a summary of the various techniques used to control fouling. Following is a discussion of some of these techniques.

4.6.1 Surface Cleaning Techniques

If prior arrangement is made cleaning can be done on line. At other times, off-line cleaning must be used. Cleaning methods can be classified as continuous cleaning or periodic cleaning.

TABLE 4.14 Various Techniques to Control Fouling [21]

On-Line Techniques	Off-Line Techniques
Use and control of appropriate additives	Disassembly and manual cleaning
Inhibitors	
Antiscalants	Lances
Dispursants	Liquid jet
Acids	Steam
	Air jet
On-Line Cleaning	Mechanical cleaning
Sponge balls	Drills
Brushes	Scrapers
Sonic horns	
Soot blowers	
Chains and scrapers	Chemical cleaning
Thermal shock	
Air bumping	

Continuous Cleaning Two of the most common techniques are the sponge-ball and brush systems. The sponge-ball system recirculates rubber balls through a separate loop feeding into the upstream end of the heat exchanger. The system requires extensive installation and therefore is limited to large facilities. The brush system has capture cages at the ends of each tube. It requires a flow-reversal valve which may be expensive.

Periodic Cleaning Fouling deposits may be removed by mechanical or chemical means. The mechanical methods of cleaning include high-pressure water jets, steam, brushes, and water guns. High-pressure water works well for most deposits, but frequently a thin layer of the deposit is not removed resulting in greater affinity for fouling when the bundle is returned to service [5, 24]. High-temperature steam is useful for hydrocarbon deposits. Brushes or lances are scraping devices attached to long rods and sometimes include a water or steam jet for flushing and removing the deposit.

Chemical cleaning is designed to dissolve deposits by a chemical reaction with the cleaning fluid. The advantage of chemical cleaning is that a hard-to-reach area can be cleaned (e.g., finned tubes). However, the solvent selected for chemical cleaning should not corrode the surface.

4.6.2 Additives

Chemical additives are commonly used to minimize fouling. The effect of additives is best understood for water. For various types of fouling Strauss and Puckorious [25] provide the following observations.

Crystallization Fouling Minerals from the water are removed by softening. The solubility of the fouling compounds is increased by using chemicals such as acids and polyphosphates. Crystal modification by chemical additives is used to make deposits easier to remove.

Particulate Fouling Particles are removed mechanically by filtration. Flocculants are used to aid filtration. Dispersants are used to maintain particles in suspension.

Biological Fouling Chemical removal using continuous or periodic injection of chlorine and other biocides is most common.

Corrosion Fouling Additives are used to produce protective films on the metal surface.

A number of additive options may be available depending on the type of fluid and application. Additional information on additives is given by Marner and Suitor [24].

NOMENCLATURE

a, b constants in Eq. (4.16)
A heat transfer surface area, m^2
A_{cr} cross-sectional area, m^2
C_b concentration in the fluid, kg/m^3
c_p specific heat at constant pressure, $J/(kg \cdot K)$
C_s concentration at the heat transfer surface, kg/m^3
CF cleanliness factor defined by Eq. (4.18)
D diffusion coefficient, m^2/s
d tube diameter, m
f Fanning friction factor, dimensionless
h heat transfer coefficient, $W/(m^2 \cdot K)$
h_D mass transfer coefficient, m/s
h_{fg} heat of vaporization, J/kg
k thermal conductivity, $W/(m \cdot K)$
L length of the heat transfer surface, m
$\dot{m}$ fluid mass flow rate, kg/s
$\dot{m}_d$ fouling species deposition flux, $kg/(m^2 \cdot s)$
Nu Nusselt number, hd/k, dimensionless
OS over surface defined in Eq. (4.20), %
ΔP pressure drop, Pa
Pr Prandtl number, $\mu c_p/k$, dimensionless
Q heat transfer rate, W
Re Reynolds number, $\rho u_m d/\mu$, dimensionless
R_f fouling resistance, $(m^2 \cdot K)/kW$
R_{ft} total fouling resistance, $(m^2 \cdot K)/kW$
R_f^* asymptotic fouling resistance, $(m^2 \cdot K)/kW$
Sh Sherwood number, $h_D d/D$, dimensionless
T temperature, °C, K
ΔT_m effective mean temperature difference, °C, K
t time, s
t_f fouling thickness, m
t_D delay time, s
U overall heat transfer coefficient, $W/(m^2 \cdot K)$
u_m average fluid velocity, m/s

Greek Symbols

ρ fluid density, kg/m^3
μ viscosity, $kg/(m \cdot s)$
ϕ_d fouling deposition function, $(m^2 \cdot K)/kJ$
ϕ_r fouling removal function, $(m^2 \cdot K)/kJ$
θ time constant, s

Subscripts

c	clean condition
d	design condition
f	fouled condition
i	inside
in	inlet
o	outside
out	outlet
p	plain tube

REFERENCES

1. Kakaç, S., Shah, R. K., and Aung, W. (eds.) (1987) *Handbook of Single Phase Convective Heat Transfer*, Chapters 3 and 4. Wiley, New York.
2. Kern, D. Q. (1950) *Process Heat Transfer*. McGraw-Hill, New York.
3. Chenoweth, J. M. (1987) Fouling problems in heat exchangers. In *Heat Transfer in High Technology and Power Engineering*, W-J. Yang and Y. Mori (eds.), pp. 406–419. Hemisphere, New York.
4. Garrett-Price, B. A., Smith, S. A., Watts, R. L., Knudsen, J. G., Marner, W. J., and Suitor, J. W. (1985) *Fouling of Heat Exchangers: Characteristics, Costs, Prevention, Control, and Removal*. Noyes, Park Ridge, N.J.
5. Taborek, J., Akoi, T., Ritter, R. B., and Palen, J. W. (1972) Fouling: The major unresolved problem in heat transfer. *Chem. Eng. Prog.* **68** 59–67.
6. Epstein, N. (1978) Fouling in heat exchangers. In *Heat Transfer 1978*, Vol. 6, pp. 235–254. Hemisphere, New York.
7. Somerscales, E. F. C., and Knudsen, J. G. (eds.) (1981) *Fouling of Heat Transfer Equipment*. Hemisphere, New York.
8. Melo, L. F., Bott, T. R., and Bernardo, C. A. (eds.) (1988) *Fouling Science and Technology*. Kluwer, Dordrecht.
9. Epstein, N. (1983) Thinking about heat transfer fouling: A 5 × 5 matrix. *Heat Transfer Eng.* **4** 43–56.
10. Epstein, N. (1981) Fouling in heat exchangers. In *Low Reynolds Number Flow Heat Exchangers*, S. Kakaç, R. K. Shah, and A. E. Bergles (eds.). Hemisphere, New York.
11. Friedlander, S. K. (1977) *Smoke, Dust and Haze*. Wiley, New York.
12. Whitmore, P. J., and Meisen, A. (1977) Estimation of thermo- and diffusiophoretic particle deposition. *Can. J. Chem. Eng.* **55** 279–285.
13. Nishio, G., Kitani, S., and Takahashi, K. (1974) Thermophoretic deposition of aerosol particles in a heat-exchanger pipe. *Ind. Eng. Chem. Proc. Design Dev.* **13** 408–415.
14. Taborek, J., Aoki, T., Ritter, R. B., and Palen, J. W. (1972) Predictive methods for fouling behavior. *Chem. Eng. Progr.* **68** 69–78.

15. McCabe, W. L., and Robinson, C. S. (1924) Evaporator scale formation. *Ind. Eng. Chem.* **16** 478–479.
16. Kern, D. Q., and Seaton, R. E. (1959) A theoretical analysis of thermal surface fouling. *Brit. Chem. Eng.* **4** 258–262.
17. Knudsen, J. G. (1984) Fouling of heat exchangers: Are we solving the problem? *Chem. Eng. Prog.* **80** 63–69.
18. Garret-Price, B. A., Smith, S. A., Watts, R. L., and Knudsen, J. G. (1984) Industrial fouling: Problem characterization, economic assessment, and review of prevention, mitigation, and accommodation techniques. Report PNL-4883, Pacific Northwest Laboratory, Richland, Wash.
19. *Standards of the Tubular Exchanger Manufacturers Association* 7th ed. Tubular Exchanger Manufacturers Association, New York, 1988.
20. Incropera, F. P., and DeWitt, D. P. (1985) *Introduction to Heat Transfer*. Wiley, New York.
21. Chenoweth, J. M. (1988) General design of heat exchangers for fouling conditions. In *Fouling Science and Technology*, L. F. Melo, T. R. Bott, and C. A. Bernardo (eds.), pp. 477–494. Kluwer, Dordrecht.
22. Gilmour, C. H. (1965) No fooling–no fouling. *Chem. Eng. Prog.* **61** 49–54.
23. Somerscales, E. F. C. (1988) Fouling. In *Two-Phase Flow Heat Exchangers: Thermal-Hydraulic Fundamentals and Design*, S. Kakaç, A. E. Bergles, and E. O. Fernandes (eds.), pp. 407–460. Kluwer, Dordrecht.
24. Marner, W. J., and Suitor, J. W. (1987) Fouling with convective heat transfer. In *Handbook of Single-Phase Convective Heat Transfer*, S. Kakaç, R. K. Shah, and W. Aung (eds.), Chapter 21. Wiley, New York.
25. Strauss, S. D., and Puckorious, P. R. (1984) Cooling-water treatment for control of scaling, fouling, and corrosion. *Power* **128** S1–S24.

CHAPTER 5

INDUSTRIAL HEAT EXCHANGER DESIGN PRACTICES

J. TABOREK
Center for Energy Studies
University of Texas
Austin, Texas 78758

5.1 INTRODUCTION

While the thermohydraulic performance of heat exchangers is treated extensively in the literature, the optimum design for industrial applications must reflect several additional criteria, imposed by interrelated design considerations. These include demands, restrictions, and objectives or disciplines such as mechanical; metallurgical; plant-architectural; and operation, fouling, control, and maintenance. This chapter focuses on the interaction of the various, often contradictory aspects of the design process. The principles of such design are illustrated by a detailed calculated example of a process stream cooler, starting from basic specifications and progressing through the suggested steps to the final design and its possible alternatives.

5.2 HEAT EXCHANGER TYPES, THEIR CHARACTERISTICS, AND SELECTION

The selection of heat exchanger type(s) that would be suitable for a given application is the first item in the design process. Often a certain type is virtually predetermined by experience and process conditions, but in a general case the designer would survey the available choices. These are described in most handbooks [2, 4] and are mentioned here only briefly for completeness.

Boilers, Evaporators and Condensers, Edited by Sadik Kakaç
ISBN 0-471-62170-6

5.2.1 Shell and Tube

The universal popularity of shell-and-tube heat exchangers is based on the wide range of applicability, namely,

1. Any temperature and pressure from vacuum to high (material limits)
2. Applications include single phase, condensation, and boiling
3. Size from very small to limits of transportation
4. Pressure drop can be adjusted within an extremely wide range on the shell side through shell type and baffle design
5. Very rugged, but also heavy and bulky (large volume/area)

5.2.2 Double Pipe or Multitube Hairpin

These types were originally used as small-sized classical counterflow heat exchangers; present applications are often of the multitube type with radial or longitudinal finned tubes.

5.2.3 Air-Cooled Heat Exchangers

The scarcity of suitable cooling waters, high cost of their treatments, and environmental considerations increasingly favor the use of air as the final heat sink. The air cooler consists of a finned tube bank (4 to 10 tube rows) and an air blower located before the tube bank (forced draft) or operating in suction (induced draft). Because the heat is discharged to the atmosphere, no limits on the air outlet temperature exist (as is often the case for cooling waters).

5.2.4 Gasketed Plate Exchangers

Originally developed for the food industry because of ease of disassembly and cleaning, modern designs with "herringbone" shaped grooves are applicable in many industrial processes. Some advantages and limitations include the following:

1. Very high coefficients but also high pressure drop; fouling is decreased, but is restricted to small particles
2. Main limitations are pressure (no vacuum, 10 to 15 bar maximum) and temperature (usually 150°C) because of the gasket material
3. Sizes small to very large, with extreme compactness
4. Plate material is usually stainless steel or titanium

5.2.5 Matrix and Plate Fin-Tube Exchangers

A folded sheet metal matrix, usually aluminum, is sandwiched between separating plates to which it is soldered or brazed, forming a system of crossflow passages fastened to headers. While the design is most compact, it does not permit cleaning and is therefore restricted to clean fluids, often gases or cryogenic liquids. The rectangular shape restricts pressures to about 10 bar.

A variation on this design is the so-called plate fin tube, in which one fluid is inside small channels (tubular or rectangular) and gas, often air, is on the plate side. Typical use is for automotive radiators and in refrigeration service with nonfouling fluids.

5.2.6 Conclusion

From this survey, the following criteria of heat exchanger selection stand out as most important:

1. Pressure limitations exclude plate and matrix exchangers above about 10 bar and in vacuum.
2. Temperature above about 150°C excludes plate exchangers because of gasket material limitations.
3. Fouling and cleaning considerations exclude matrix and plate fin-tube exchangers.
4. Low pressure drop limitations will exclude most types except shell-and-tube and air-cooled exchangers (tube side).
5. For applications where the above criteria are of no importance, any design type is applicable and should be considered on merits of cost effectiveness.

However, the shell-and-tube heat exchanger remains as the clearly most versatile type and therefore will be treated here in greater detail.

5.3 THE STRATEGY OF OVERALL DESIGN OPTIMIZATION

The overall optimum design of a particular heat exchanger encompasses a variety of problem areas that involve both quantitative and qualitative information and interpretation of results. We shall focus on the following logical design modules and follow the heat exchanger design process in principle, to demonstrate what is considered the "integral" approach:

1. Process specifications
2. Preliminary problem analysis
3. Detailed thermohydraulic design
4. Mechanical and metallurgical design
5. Architectural design aspects
6. Operational, control, and maintenance considerations

Design methodology requires that the preceding items be solved "in parallel"; that is, the demands of any one discipline must be considered with respect to all others in an interactive process [1]. To illustrate this, let us consider that a particular heat transfer process is suspected of heavy fouling. Fouling results in direct cost through larger surface and possibly more expensive materials, as well as the indirect cost of cleaning with the possible problems of control and production interruption. Therefore all other design aspects will become subservient to the solution or minimization of the fouling problem, and all the following design modules will be involved interactively:

1. Careful selection of the fouling resistance or area over-surface
2. High flow velocity and design geometry considerations to minimize fouling and to facilitate cleaning maintenance
3. Tube material selection to minimize fouling corrosion
4. Architectural placement of the heat exchanger for ease of cleaning "on location," or removal of tube bundle for external cleaning
5. Considerations for proper controls between clean and fouled condition

Other similar examples can be cited, such as design of condensers with noncondensibles, in which the gas removal techniques virtually dictate the selection of unit geometry.

It is not unusual for some of the "preferential" requirements to pose contradictory demands on the overall design. For example, an organic stream would preferably be placed on the shell side, because of a better heat transfer coefficient. However, the stream exhibits fouling, which is easier to accommodate on the tube side. This situation calls for a weighted compromise, frequently requiring the evaluation of alternative designs to optimize the overall effects.

The following sections define in greater detail the problems pertinent to the individual design module areas.

5.3.1 Process Specifications

The major role of the process engineer is to supply all such information to the heat exchanger designer as is needed for proper design, that is:

1. The type of heat exchanger, if such is unconditionally desired (air-cooled heat exchanger, horizontal condenser, etc.)
2. The temperatures and pressures of both streams; acceptable range of seasonal temperature fluctuations (cooling water, air)
3. Flow rates of fluids and composition for mixtures in condensation or boiling
4. The corrosivity of fluids, environmental hazards, and similar

5. The permissible pressure drop (This item is the most important, as it often virtually determines the design.)
6. The fouling resistances of both fluids or area overdesign factor (This specification can have a substantial effect on the design. Careful selection and coordination with the thermohydraulic design, such as the effect of flow velocities and tube wall temperature, is required.)
7. The tube diameter and length (While the process engineer *may* have reasons to specify either or both items, they are often specified with little or no justification and the effects on the heat exchanger design can be detrimental.)

5.3.2 Preliminary Problem Analysis

It is a most useful practice to perform a preliminary analysis of any heat exchanger design problem. This should include:

1. Approximate size. Values of the heat transfer coefficients can be estimated from literature sources such as presented in Table 5.1. From these, together with the mean temperature difference, the exchanger size can be determined in first approximation. This will serve as a guide for subsequent decisions as well as justification for allocation of engineering effort. A small unit with standard operating conditions will require not more than routine attention, while a large unit and/or a complex process, such as condensation with boiling coolant, may need an extensive analysis.
2. The distribution of the thermal resistances under clean and fouled conditions should be determined at this stage from the preceding data. Often one of the resistances will be controlling, requiring special attention.
3. Identification of most restrictive considerations. This will usually determine some design feature and focus the attention toward items identified as needing special attention. Typically, such items will include the following:
   (a) Low pressure drop of gas streams or condensing vapors in vacuum; such streams would preferably be allocated on the shell side, where the designer can control the pressure drop with many options in unit type and baffle selection.
   (b) Thermal profile: for example, is counterflow necessary because of close temperature approach, or can multitube pass design be considered.
   (c) When fouling is a substantial resistance, the design effort must concentrate on defensive measures (flow velocity, material selection, etc.) and ease of cleaning.
   (d) High pressure or corrosivity will place such streams on the tube side because of material cost.

TABLE 5.1 Typical Film Heat Transfer Coefficients and Fouling Resistances for Shell-and-Tube Heat Exchangers

	Fluid Condition	$W/(m^2 \cdot K)$	Fouling Resistance Range, $[kW/(m^2 \cdot K)]^{-1}$
Sensible heat transfer			
Water	Liquid	5,000–7,500	0.1–0.25
Ammonia	Liquid	6,000–8,000	0–0.1
Light organics	Liquid	1,500–2,000	0.1–0.2
Medium organics	Liquid	750–1,500	0.15–0.4
Heavy organics	Liquid		
	Heating	250–750	0.2–1.0
	Cooling	150–400	0.2–1.0
Very heavy organics	Liquid		
	Heating	100–300	0.4–3.0
	Cooling	60–150	0.4–3.0
Gas	1–2 bar abs	80–125	0–0.1
Gas	10 bar abs	250–400	0–0.1
Gas	100 bar abs	500–800	0–0.1
Condensing heat transfer			
Steam, ammonia	No noncondensible	8,000–12,000	0–0.1
Light organics	Pure component, 0.1 bar abs, no noncondensible	2,000–5,000	0–0.1
Light organics	0.1 bar, 4% noncondensible	750–1,000	0–0.1
Medium organics	Pure or narrow condensing range, 1 bar abs	1,500–4,000	0.1–0.3
Heavy organics	Narrow condensing range, 1 bar abs	600–2,000	0.2–0.5
Light multicomponent mixture, all condensible	Medium condensing range, 1 bar abs	1,000–2,500	0–0.2
Medium multicomponent mixture, all condensible	Medium condensing range, 1 bar abs	600–1,500	0.1–0.4
Heavy multicomponent mixture, all condensible	Medium condensing range, 1 bar abs	300–600	0.2–0.8
Vaporizing heat transfer			
Water	Pressure < 5 bar abs, $\Delta T = 25$ K	5,000–10,000	0.1–0.2
Water	Pressure 5–100 bar abs, $\Delta T = 20$ K	4,000–15,000	0.1–0.2
Ammonia	Pressure < 30 bar abs, $\Delta T = 20$ K	3,000–5,000	0.1–0.2
Light organics	Pure component, pressure < 30 bar abs, $\Delta T = 20$ K	2,000–4,000	0.1–0.2
Light organics	Narrow boiling range, pressure 20–150 bar abs, $\Delta T = 15$–20 K	750–3,000	0.1–0.3
Medium organics	Narrow boiling range, pressure < 20 bar abs, $\Delta T_{max} = 15$ K	600–2,500	0.1–0.3
Heavy organics	Narrow boiling range, pressure < 20 bar abs, $\Delta T_{max} = 15$ K	400–1,500	0.2–0.8

4. Identification of probable alternate designs. Some of the previously mentioned items may often pose contradictory demands, requiring compromises, which can usually be identified at this stage. The potential use of enhanced surfaces can also be established.
5. Inadequate or poor specifications. Perhaps the most important item of the preliminary survey is to recognize if the given specifications are suspect of being incorrect or at least requiring additional investigation. Exaggerated fouling, leading to overdesign of more than 50% compared to clean conditions, is a typical example.

5.3.3 Detailed Thermohydraulic Design

The *raison d'être* for the heat exchanger designer is to obtain a global optimum within all the demands and restrictions with respect to initial and operating cost, reliability, and maintenance ease. This will require interactions with the other disciplines of mechanical and metallurgical design and the overall plant architecture, as mentioned earlier. Some representative items are briefly discussed.

1. For shell-and-tube heat exchangers, the selection of the shell construction type and baffle design, together with allocation of the streams to shell and tube side, are the primary decisions. Use is made of the results of the preliminary design estimations; however, these were based on heat transfer performance only, while compliance with the maximum permissible pressure drop as well as full pressure drop utilization must be confirmed through detailed design investigation.

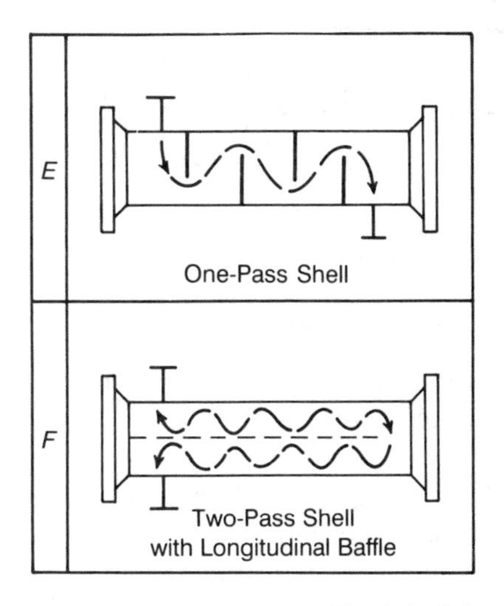

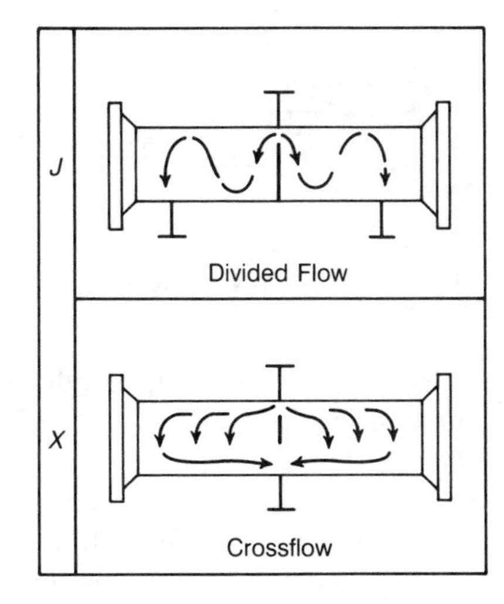

Fig. 5.1. Selected TEMA shell types.

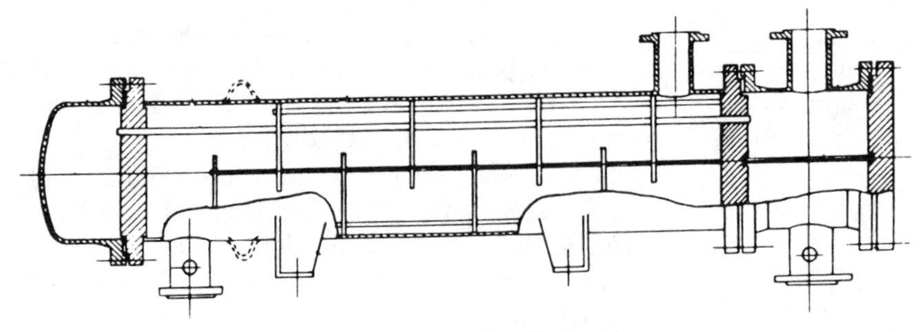

Fixed Tube Sheet Heat Exchanger

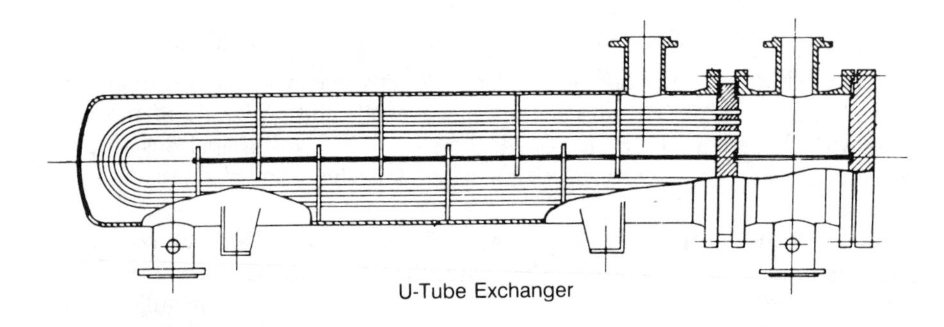

U-Tube Exchanger

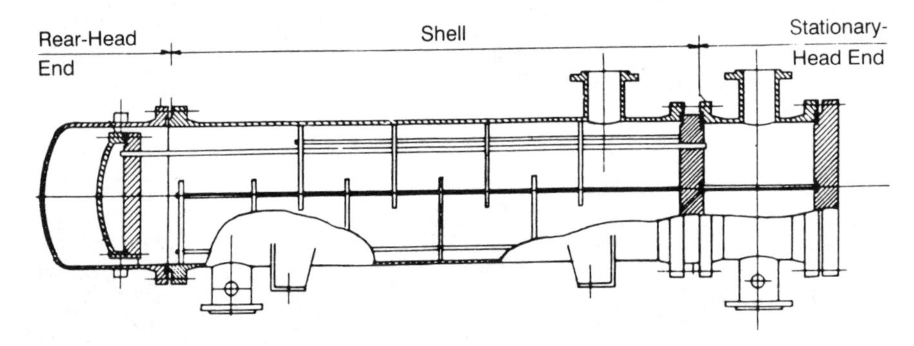

Pull-Through Floating-Head Heat Exchanger

Fig. 5.2. Typical shell-and-tube heat exchanger types.

Shell type and tube bundle construction (Figs. 5.1 and 5.2) will be determined from consideration of three requirements:

(a) The lowest-cost shell is the U tube, followed by fixed tube sheet construction. If fouling is no problem or cleaning can be done by chemical means and flow–temperature conditions permit two or more tube passes, the U tube will be the prime choice. This is typically used for steam-heated designs (steam tube side). If tube-side brush cleaning is required and/or a single tube pass is a consideration because of counterflow, the fixed tube sheet will be preferred, unless excluded by restrictions of thermal expansion.

(b) If the stream temperatures and heat transfer coefficients result in large differences between the temperature of the shell and tubes, the differential thermal expansion must be accommodated as otherwise the tube-to-tube sheet joins would be damaged. Again, the U tube will be the prime choice for cost reasons. Fixed tube sheet design can accommodate only minor expansion stresses. For large differences of wall temperature ($> 50°C$, approximately), "floating head" construction must be used, as described later.

(c) Shell type selection also affects the pressure drop. The TEMA *J* and *X* shells will produce a greatly decreased pressure drop compared to the *E* shell.

2. Selection of possible alternates of constructional elements such as tube diameter and length, tube layout, and pitch, as well as considerations of enhanced surfaces.

3. The specified pressure drop controls to a large extent the design of many heat exchangers, especially in no-phase change. The desired criterion is that the available pressure drop should be utilized as much as possible, as the conversion of pressure drop to heat transfer will result in a smaller unit [2, 5]. Under no circumstances should a design be "pressure drop limited," in which case, the size of a heat exchanger has to be enlarged for the purpose to accommodate pressure drop limits, resulting in heat transfer overdesign. Such design conditions are not readily noted from computer-based designs, which often do not identify intermediate design solutions and the restrictions which force the design into larger sizes. All previously mentioned design elements may enter as optimization parameters, but only few are usually controlled by the computer program logic (e.g., baffle spacing, tube passes), thus requiring human intervention for all others.

4. Pressure drop in peripheral connections, nozzles, and so forth. In some low pressure drop shell designs the pressure drop in the peripherals is a predominant part of the total, requiring careful attention.

5. Provisions for minimization of fouling as well as ease of cleaning. These arrangements can include a variety of items, ranging from the obvious (flow velocity, baffle design, material) to the more complex, such as the quality of water treatment and its fluctuation, fluid storage history, possible contaminants, and so forth. Even more important, each design should be checked for performance under initial clean conditions, in order to highlight the various effects of the specified fouling and to assure that proper controls have been provided.

6. Venting of noncondensible gases can be a dominant design requirement.

7. Tube vibration is a potential problem. However, the use of modern computational methods makes it possible to produce a design virtually free of tube vibration.

These are only examples of the numerous considerations facing the designer of heat exchangers, emphasizing the need for close interaction with the other subset design modules.

5.3.4 Mechanical – Metallurgical Design Aspects

Some of the most costly errors in heat exchanger design are committed in the area of selection of the mechanical and metallurgical design elements, which include [9]:

1. The selection of proper materials for wetted and nonwetted surfaces of the heat exchanger, as well as for the pressure and nonpressure parts. This analysis includes considerations for corrosion, life expectancy, fouling mitigation, and numerous other problems. Not always is the least expensive material the most economical. For example, stainless-steel tubes will increase the initial cost, but may help reduce fouling and eliminate the need for tube replacement.
2. Every heat exchanger design must comply with an often bewildering array of national and international codes, with frequently confusing and contradictory specifications [6].
3. General mechanical design integrity: tube sheet design, tube-to-tube sheet joints, flanges, welds, and the like.
4. Some aspects of tube vibration integrity are intimately connected with thermal design per se (baffle spacing) and may require special construction design provisions (U-tube support plates, flow distributors, etc.).

5.3.5 Architectural Considerations

Architectural considerations include items connected with placing heat exchangers within the overall plant system. Typical problems include the following::

1. Integration of the heat exchanger within the piping system. The cost of piping in a large plant is so considerable that particular design aspects of the heat exchanger may be subordinated. For example, the strategy of piping layout may favor a stream entering and exiting on the same side of a heat exchanger, thus permitting only an even number of tube passes.
2. Vertical versus horizontal orientation. Some heat exchanger processes are orientation dependent, especially boiling and condensation, which are gravity dependent, resulting in preference to vertical orientation.

However, the cost of vertical erection, design for wind and seismic loads, is usually higher than for horizontal. Cleaning on the tube side in place favors horizontal location also, because of ease of access. These considerations may compete with other potentially favorable designs such as condensation in vertical downflow and up-flow boiling.

5.3.6 Maintenance, Operation, and Control Considerations

A large number of items fall into the general category of operation, maintenance, and control. From these we note the following [9]:

1. Heat exchangers designed to fouling conditions will overperform at clean conditions. If streams such as cooling water or air are involved, the heat exchanger is usually designed to 95% of the worst summer conditions, thus absorbing an additional overdesign during the rest of the year. This overperformance is acceptable or desirable only in cases such as power plant condensers, heat recovery units, and similar. In most other cases it has to be either absorbed by downstream units, if acceptable, or eliminated by appropriate controls. This is often difficult and in any case requires careful consideration. Decrease of flow rates of service fluids like cooling water will accomplish the control, but will lead to decreased flow velocity and hence increased fouling. Recirculation of fluids, submergence of part of the heat exchanger surface, and similar measures are resorted to in serious cases.
2. Provisions for thermal expansion between the shell and tubes, which are subjected to different temperatures and are often made out of different materials, is essential to maintain the integrity of tube-to-tube sheet joints and other connecting elements. Leakage and structural damages can develop if this item is not properly attended to by all subdisciplines of design. It must include start-up and shutdown conditions and account for pump failures of individual streams.
3. Provisions for ease of access to the exchanger for cleaning, tube repair, gasket replacement, and general inspection are essential for proper maintenance.

5.4 SHELL-AND-TUBE HEAT EXCHANGERS: CHARACTERISTICS OF CONSTRUCTIONAL COMPONENTS

It is important that the designer of shell-and-tube heat exchangers be familiar with the characteristics of the constructional elements and their effects on the overall performance, as shell-and-tube exchangers are designed on a "custom basis"; that is, every design consists of individually selected components, as described in detail in [4]. This is contrary to many other heat exchanger types, like plate exchangers, which are "modular," meaning that predetermined constructional elements are used.

5.4.1 Shell Types

TEMA standards [3] list a number of shell-and-tube bundle types that are internationally recognized as representing the basic general design types. The most important ones are shown in Fig. 5.1, and their characteristics are briefly described as follows:

E shell: This type is a single-pass shell (inlet and outlet nozzles are on either side) with any number of tube passes, odd or even. With a single-tube pass a nominal counterflow can be obtained. For comparison of pressure drop between the various types, we designate the *E*-shell relative performance by the number 1.

F shell: This two-shell pass unit has a longitudinal dividing partition. It is used when units in series are required, with each shell pass representing one unit. With two tube passes a nominal counterflow is obtained. The pressure drop comparison number is 8, that is, eight times that of the *E* shell for the same shell diameter.

J shell: Fluid entry is centrally located and split into two parts. This shell is used for low pressure drop designs, as the comparison number is 1/8.

X shell: This type has a centrally located fluid entry and outlet, usually with a distributor dome. Crossflow is over the entire length of the tube. Consequently, pressure drop is extremely low. It is used for vacuum condensers and low-pressure gases.

5.4.2 Tube Bundle Types

The most representative tube bundle types are shown in Fig. 5.2. The main design objectives here are to accommodate thermal expansion, to provide for ease of cleaning, or the least expensive construction if other features are of no importance.

U Tube The U tube is the least expensive construction because only one tube sheet is needed. The tube side is difficult to clean by mechanical means because of the sharp U bend. Only an even number of tube passes can be accommodated, but thermal expansion is unlimited.

Fixed Tube Sheet The shell is welded to the tube sheets and there is no access to the outside of the tube bundle for cleaning. This low-cost design option has only limited thermal expansion, which can be somewhat increased by expansion bellows. Cleaning of the tube is easy.

Floating Head Several designs have been developed that permit the tube sheet to "float," that is, to move with thermal expansion. The classic type of pull-through floating head is shown in Fig. 5.2, which permits tube bundle removal with minimum disassembly, as required for heavily fouling units. The cost is high.

5.4.3 Tube Passes

Only the E shell with one tube pass and the F shell with two tube passes result in nominal counterflow and have the highest temperature effectiveness. All multitube passes require a temperature profile correction (factor F), or in some cases, simply cannot deliver the desired temperatures because of temperature cross. In such cases either a single tube pass must be used or, as alternate, multiple units in series (higher cost).

Generally, a larger number of tube passes is used to increase the tube-side flow velocity which results in a higher heat transfer coefficient (within the available pressure drop) and minimizes fouling. However, the pressure drop steps are large, that is, from one to two tube passes the pressure drop rises by a factor of 8. If, for architectural reasons, the tube-side fluid must enter and exit on the same side, an even number of tube passes is mandatory.

5.4.4 Baffle Types and Geometry

The purpose of placing baffles into the tube bundle is (1) to support the tubes for structural rigidity, preventing tube vibration and sagging, and (2) to divert the flow across the bundle, to obtain a higher heat transfer coefficient. The most commonly used baffle types are shown in Fig. 5.3 and described briefly as follows:

Segmental baffles divert the flow most effectively across the tubes. However, the baffle spacing and cut must be chosen very carefully, as otherwise ineffective flow patterns would result, as shown in Fig. 5.4. Optimum baffle spacing is somewhere between 0.4 and 0.6 of the shell diameter, with a baffle cut of 25%.

Double and triple segmental baffles increase the longitudinal flow component, thus lowering the pressure drop, which is approximately 0.5 and 0.3 of the segmental value.

No tubes in the window construction eliminates the tubes that are otherwise supported by only every second baffle, thus minimizing tube vibration. However, fewer tubes can be located in a given shell diameter. Heat transfer effectiveness is high.

Rod or grid baffles are formed by a grid of rod or strip supports. The flow is essentially longitudinal, resulting in very low pressure drops and the highest effectiveness of pressure drop to heat transfer conversion. Because of the usual close baffle spacing, tube vibration danger is virtually eliminated. This construction can also be used effectively for vertical condensers and

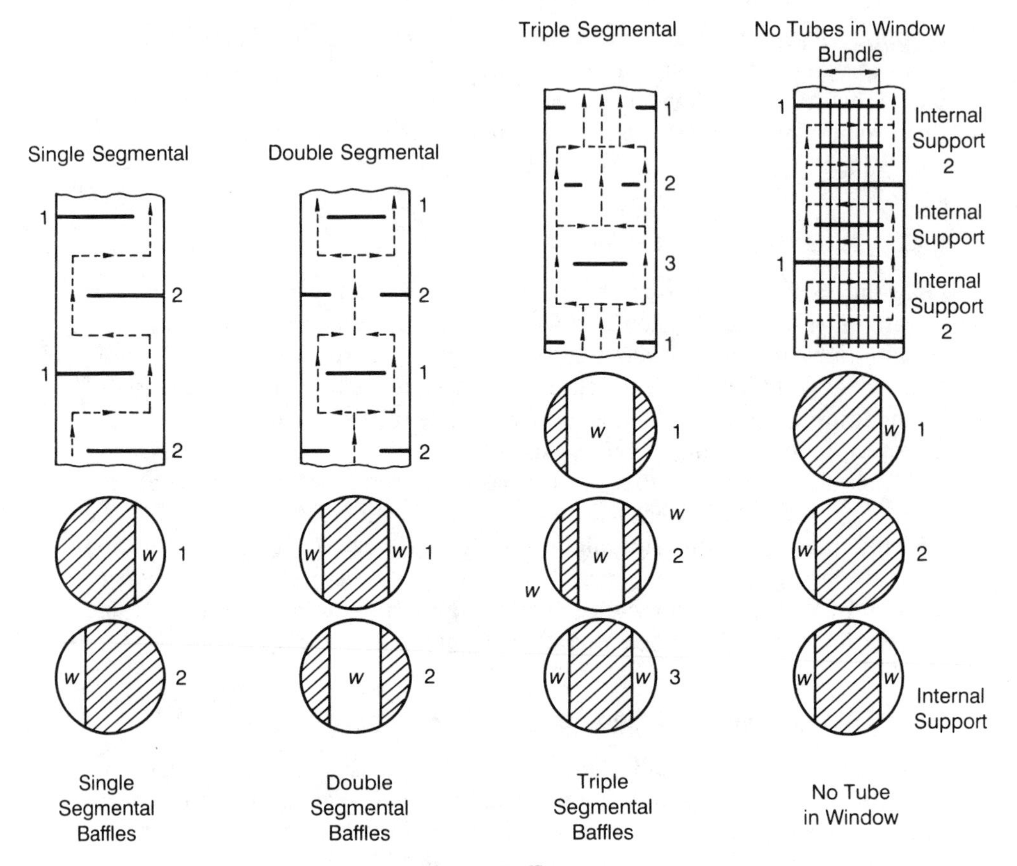

Fig. 5.3. Baffle types.

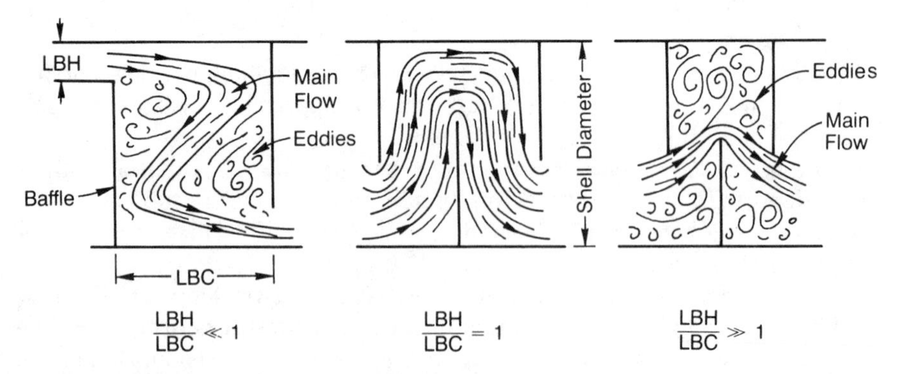

Fig. 5.4. Schematic flow through baffle tube bundles. Effects of baffle cut height (LBH) to baffle spacing (LBC) ratio.

reboilers as there are no "dead" flow areas common to segmental designs (Fig. 5.4).

Disc-and-ring (doughnut) baffles are composed of alternating outer rings and inner discs, which direct the flow radially across the tube field. The potential bundle-to-shell bypass stream is thus eliminated; there are some indications that this baffle type is very effective in pressure drop to heat transfer conversion [5]. At present, these baffles are rarely used in the United States but are very popular in Europe.

5.4.5 Tube Diameter and Tube Length

Small tube diameters (8 to 15 mm OD) are preferred for greater area–volume density, but are limited for cost effectiveness to about 12 mm and for purposes of in-tube cleaning to 20 mm. Larger tube diameters are often required for condensers and boilers for the best performance.

Tube length affects the cost and operation of heat exchangers. Basically, the longer the tube (for any given total surface), the fewer tubes are needed, fewer holes are drilled and the shell diameter decreases, resulting in thinner tube sheets and lower cost. There are, of course, several limits to this general rule, best expressed that the shell-diameter–tube-length ratio should be within limits of about $\frac{1}{5}$ to $\frac{1}{15}$. The maximum tube length is sometimes dictated by architectural layouts and ultimately by transportation to about 20 m.

5.4.6 Tube Layout

Tube layout is characterized by the included angle between tubes, as shown in Fig. 5.5. A layout of 30° results in the greatest tube density and is therefore used, unless other considerations dictate otherwise. For example, clear lanes

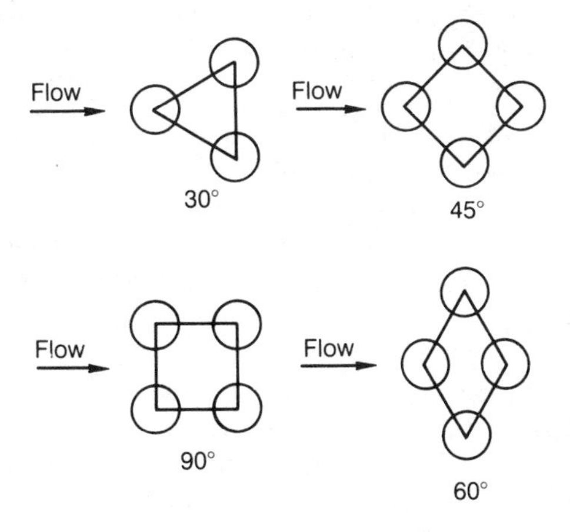

Fig. 5.5. Tube layout angles.

($\frac{1}{4}$ in. or 7 mm) are required for external cleaning, using a square 90° or 45° layout. Tube pitch L_p is usually chosen so that the pitch ratio L_p/D_{to} is between 1.25 and 1.5.

5.5 COMMENTS ON CONDENSER DESIGN

The following general comments should be helpful for design of condensers, selection of construction type, fluid allocation, and other design practices (see also Chapters 10 and 11).

1. *Condensation modes*. Even though dropwise condensation attracted considerable research attention because of the high heat transfer coefficients obtainable, it cannot be relied on to be sustainable for long periods of time on industrial surfaces. Thus all condensers are designed to film condensation principles.
2. *Condensation mechanisms and flow regimes*. Depending on the flow characteristics of the vapor and the condensate, the designer must determine the flow regime applicable along the vapor flow path. At low vapor velocities the so-called "gravity controlled" or Nusselt flow regime exists. At high vapor velocities, the "vapor shear controlled" regime will predominate.
3. *Desuperheating*. Some vapor streams enter the condenser superheated. If the wall temperature is below the dew point, condensation will take place and the desuperheating process will proceed at the rate of the condensation. Otherwise the desuperheating duty must be calculated as a single-phase process.
4. *Subcooling*. It is sometimes desired to subcool the condensate slightly before further processing. This can be accomplished by raising the level of the condensate so that it would be in contact with the cool tubes. For larger subcooling heat applications it is more efficient to allocate a separate unit (see Fig. 11.12).
5. *Constructional considerations*. Some guidelines and practices must be observed for design of well-functioning condensers:
   (a) Vertical in-tube condensation is very effective, but the tube length is limited as it may fill up with condensate. Thus the size of such condensers is restricted, as otherwise large shell diameters would be required.
   (b) Horizontal tube-side condensation is less effective and much more difficult to calculate because of the stratification of the condensate. Positive tube inclination must be used.
   (c) Horizontal shell-side condensation is very popular as it is well predictable, permits use of large surfaces, and the extremely low pressure drop required for vacuum operations can be obtained by

proper unit selection (TEMA *X* shell). Low finned tubes are effective for organic streams and economical, if the condensing side resistance is more than about 60% of the total. Steam cannot be condensed on conventional low finned tubes because the high surface tension of the water retains condensate in the fin valleys.

6. If noncondensible gases are present in vapors, the condensing coefficient will vary an order of magnitude between vapor inlet and outlet. Stepwise calculation is unconditionally required. Furthermore, the unit construction must be suitable for effective removal (venting) of the gases, which would otherwise accumulate and render the condenser inoperative.

5.6 COMMENTS ON REBOILER DESIGN

The following comments may be useful in alerting the designer of shell-and-tube vaporizers to some very crucial problems, especially as applied to the process industry (see also Chapter 13).

1. Nucleate boiling is very sensitive to surface temperature, pressure, and fluid type. Single-component fluids are vaporized only for power, cryogenic, and refrigeration cycles. All other boiling processes involve mixtures of two or many components (e.g., gasoline). The nucleate boiling coefficient decreases drastically if even a small amount (10 mol%) of a second fluid is involved.
2. While nucleate boiling constitutes the basis of all vaporizer calculations, in heat exchangers it is present only in combination with flow (convective) boiling. This makes the industrial boiling process less sensitive to the restrictions of nucleate boiling, especially mixtures.
3. A serious problem in industrial boilers is the need for a steady outflow of vapors. This is easily accomplished in kettle-type and vertical tube-side boilers, where vapors have a natural escape. However, it becomes a problem in shell-side boiling, where vapor pockets can accumulate at the top tube sheet. Boiling in long, horizontal baffled shells exhibits the same tendency toward vapor pocket formation. In either case, the drastic differences in the heat transfer coefficient within the boiling and the vapor regions can generate thermal expansion forces which may cause serious problems. Consequently, it is safer to follow the laws of gravity and design boilers with unrestricted vapor escape upward.
4. The effects of fouling in boiling processes can be absorbed by either hot stream temperature control (steam heating) or by an oversized area. In the latter (usual) case, the process as designed is upset at clean conditions when higher temperature differences will exist. These are generally more difficult to absorb than in no-phase-change operations.

Special controls may be necessary, such as inlet stream control valves used in thermosiphon reboilers, where dryout conditions would render the unit inoperative.

5. Enhanced surfaces are very effective in nucleate boiling, where they promote the onset and enhance the rate of nucleation. Thus low finned tubes would boil at much lower superheat temperatures than plain tubes. Elaborate surfaces composed of sintered materials and simulated cavity surfaces have been developed, which exhibited heat transfer coefficients up to 10 times higher than plain tubes. In cases where nucleate boiling, especially at low temperature differences is predominant (e.g., cryogenics), these surfaces have indeed shown a remarkable performance. When employed under conditions where flow-boiling effects are predominant, the surface enhancement is diminished. However, the surface increase due to finned tubes appears to remain effective even in flow boiling. The designer should be warned that the entire area of enhanced boiling surfaces is extremely competitive and at this time still in the developmental stage; proven experimental data are the only assurance.

5.7 CALCULATED EXAMPLE: BUTANE COOLER

To demonstrate the application of heat exchanger design systematics, we will follow the design of a light organic liquid cooler, from basic specifications through a preliminary estimate to the final product. The serial approach to design, as previously stipulated, will be noted.

5.7.1 Process Specifications

The following data are supplied for thermohydraulic design and reflect the initial input of metallurgical, architectural, and operational requirements.

Hot stream. Liquid *n*-butane at 35 bar pressure and 52.5 kg/s flow rate is to be cooled from an inlet temperature of 113°C to a minimum of 38°C for delivery to a storage tank [C_p = 2960 J/(kg · K)]. From previous plant experiences no fouling of this stream is expected, despite the fact that TEMA fouling tables would suggest 0.00018 (m^2 · K)/W. The pressure drop available is 100 kPa (1 bar).

Coolant. Well-treated cooling tower water [C_p = 4178 J/(kg · K)] is available at 27°C (summer) and 17°C (winter). The outlet temperature should not exceed 45°C because of excessive evaporation in the cooling tower. A fouling resistance of 0.00018 (m^2 · K)/W is suggested together with 25% surface overdesign, whichever is smaller, under the condition that the flow velocity (tube) be kept at approximately 1.5 m/s. A maximum velocity of 3 m/s is also suggested to prevent erosion. A pressure drop of 100 kPa (1 bar) is available.

Construction specifications. A maximum tube length of 10 m is required because of space restrictions. The tube material is 0.5 Cr alloy. Finned tubes are acceptable if so indicated by design. A single-tube pass is acceptable only if substantial advantages can be shown; an even number of tube passes is preferred. A horizontal position is required for ease of cleaning.

5.7.2 Heat Exchanger Type and Fluid Allocation

Because of the high pressure of the butane, shell-and-tube construction is required. Note that otherwise a plate heat exchanger would have been a viable alternative. Water will be placed inside straight $\frac{3}{4}$-in. tubes for ease of cleaning. The allocation of the organic on the shell side is also favorable because of the possible use of finned tubes.

5.7.3 Thermal Profile Analysis: Possible Configurations

The hot outlet temperature of 38°C and the cold outlet temperature of 45°C maximum determine the exchanger configurations to be considered. First we calculate the heat duty from the fully specified hot stream:

$$Q = 52.5 \times 2960 \times (113 - 38) = 11.65 \text{ MW} \tag{5.1}$$

The coolant outlet temperature under fouled conditions is assumed to be 45°C. This assumption requires a water flow rate of

$$\dot{m}_t = \frac{11{,}650{,}000}{4178 \times (45 - 27)} = 155 \text{ kg/s}$$

Three arrangements are theoretically possible (Fig. 5.6):

1. Straight counterflow, 1–1 pass exchanger. This would be the best design if sufficient water velocity can be obtained in a single-tube pass.
2. Two units in series, each with two tube passes. This design is more expensive.
3. Single 1–2 unit, one shell pass, two tube passes (1–2). This design would permit high tube velocity, but a two-tube pass unit, with one pass in parallel with the hot stream, will result in temperature overlap, 45°C versus 38°C for the water outlet. In order to obtain a reasonable LMTD correction factor $F \approx 0.8$, the water outlet temperature should be equal to the hot stream outlet, 38°C. This, in turn, will require a higher water

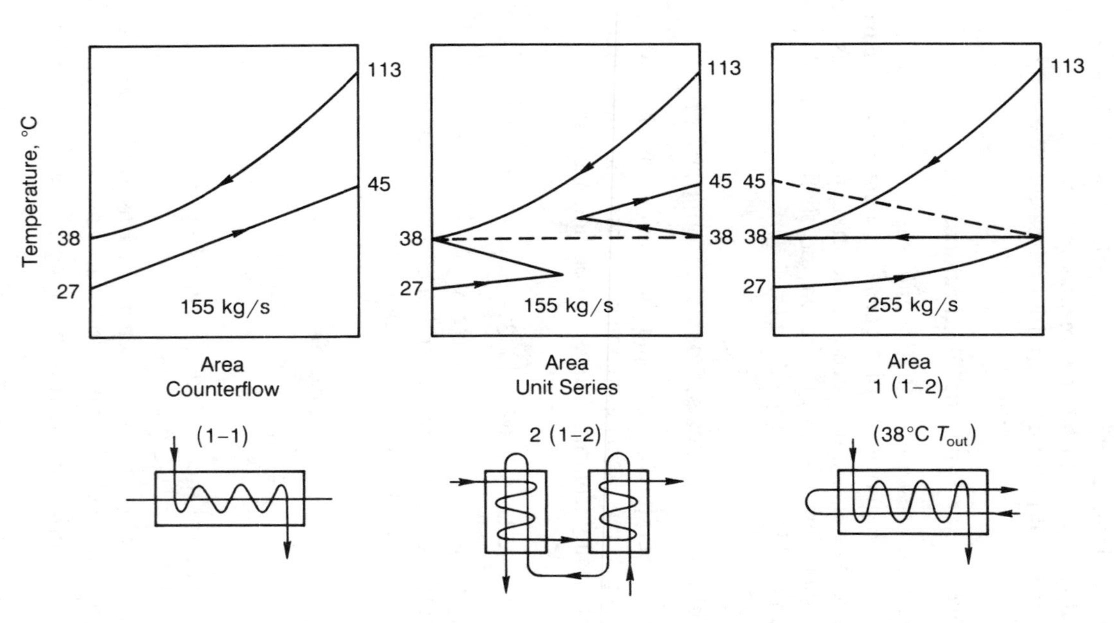

Fig. 5.6. Thermal profile of the three possible solutions.

flow rate

$$\dot{m}_t = \frac{11{,}650{,}000}{4178 \times (38 - 27)} \approx 255 \text{ kg/s}$$

Such a high demand on cooling water would be very expensive (cooling tower and pump capacity) and for this reason this design option may be questionable.

5.7.4 Selection of Construction Elements

We proceed now with the selection of the construction elements as the next step toward design.

1. *Shell type*: A single-pass E shell, fixed tube sheet design can be assumed because the temperature differences are small, but it should be checked later during the mechanical design phase for possible inclusion of expansion bellows.
2. *Tube outside diameter and length*: Assume $\frac{3}{4}$-in. (19-mm) tubes with 16 mm ID for ease of mechanical cleaning. Start with the maximum permissible length of 10 m and adjust if required.
3. *Tube layout*: Start with 30° and a pitch ratio of 1.25; adjust the pitch ratio if required for pressure drop.
4. *Baffles*: Start with segmental, baffle spacing of approximately 0.6 of shell diameter (normally used). However, because of the large fluid flow and rather low permissible pressure drop, we assume 500 mm; baffle cut is set to 25%.
5. *Nozzles*: Assume the nozzle diameter as 25% of shell diameter and check the pressure drop.

5.7.5 Preliminary Estimation of Unit Size

First we estimate the individual heat transfer coefficients. This method is preferable to estimating the overall coefficient as the designer can get a feel for the relative magnitude of the resistances. Tables for the estimation are available in various handbooks; one of the best is shown as Table 5.1, based on [2]. From that table we get for our example:

$$h_{\text{water}} = 7000 \text{ W/(m}^2 \cdot \text{K)} \quad \text{and} \quad h_{\text{light organic}} = 1500 \text{ W/(m}^2 \cdot \text{K)} \tag{5.2}$$

With the specified fouling resistance we can now estimate the overall U:

	$W/(m^2 \cdot K)$	$(m^2 \cdot K)/W$	%
Water	7000	0.00014	15
Water fouling	5556	0.00018	18
n-Butane	1500	0.00067	67

Total resistance 0.00099
Overall coefficient $U = 1000\ W/(m^2 \cdot K)$

Notice the distribution of the resistances, which suggests that the butane is the controlling factor and will dominate future design aspects.

We need to calculate the LMTD from the four given inlet–outlet temperatures.

$$\text{LMTD} = \frac{(T_{h1} - T_{c2}) - (T_{h2} - T_{c1})}{\ln\left(\dfrac{T_{h1} - T_{c2}}{T_{h2} - T_{c1}}\right)} = \frac{(113 - 45) - (38 - 27)}{\ln\left(\dfrac{113 - 45}{38 - 27}\right)} = 31.3°\text{C}$$

Next we calculate the required exchanger area A:

$$A = \frac{11{,}650{,}000}{31.3 \times 1000} = 372\ \text{m}^2 \tag{5.3}$$

The problem now is to convert this area into reasonable dimensions of the first trial unit. The objective is to find a shell diameter D_s which, with a given tube length L, would contain the correct number of tubes N_t of diameter D_{to},

$$A = \pi D_{to} N_t L \tag{5.4}$$

The total number of tubes N_t (i.e., the number of holes in the tube sheet) can be predicted in fair approximation as a function of the shell diameter by taking the shell circle and dividing it by the projected area (in flow direction) of the tube layout pertaining to a single tube A_1 (Fig. 5.5):P

$$N_t = (CTP)\frac{\pi}{4}\frac{(D_s)^2}{A_1} = 0.785\left(\frac{CTP}{CL}\right)\frac{(D_s)^2}{(PR)^2(D_{to})^2} \tag{5.5}$$

where L_p is the tube pitch; PR is the tube pitch ratio $= L_p/D_{to}$; $A_1 = (CL)(L_p)^2$ and $L_p = (PR) \times (D_{to})$, if preferable; CL is the tube layout constant, $= 1.0$ for 90° and 45°; $= 0.87$ for 30° and 60°. The constant CTP accounts for the incomplete coverage of the shell diameter by the tubes, due to necessary clearances between the shell and the outer tube circle and tube omissions due to tube pass lanes for multitube pass designs. Based on the fixed tube sheet (the most common design), the following values are suggested:

One tube pass: $CTP = 0.93$

Two tube passes: $CTP = 0.9$

Four tube passes: $CTP = 0.85$

Equation (5.5) will predict the tube count within 5% for single tube pass, 15-to-25-mm tubes, $PR = 1.25$ to 1.4 and for D_s between 300 and 1000 mm (most common application). The accuracy will decrease somewhat when deviating from this range, but will be fully sufficient for preliminary estimation. More exact estimations are quite complicated and are best documented in [6].

Substituting Eq. (5.4) into Eq. (5.5), we can express the shell diameter D_s as a function of the desired area A, with the tube length L and the tube layout dimensions L_p, PR, and D_{to} as parameters:

$$D_s = 0.637\sqrt{\frac{CL}{CTP}}\left[\frac{(A)(PR)^2(D_{to})}{L}\right]^{1/2} \tag{5.6}$$

Substituting for $A = 372 \text{ m}^2$, $D_{to} = 19$ mm, $L = 10$ m, and $PR = 1.25$, 30° layout, we get $D_s = 647$ mm, rounded off to 650 mm.

5.7.6 Design and Results Evaluation

Summarizing the selected constructional parameters, we get the following initial specification, which can be used as input into a computer rating program or for manual calculations:

Shell diameter = 650 mm

Tube length = 10 m, single tube pass

Tube diameters $D_{to} = 19$ mm, $D_{ti} = 16$ mm

Baffle spacing = 500 mm, baffle cut 25%

Calculations were performed on a simple program [8] using an IBM compatible PC, suitable for educational purposes and based on methods as documented [4]. See also the general comments about computer programs that follow. The results are shown in Table 5.2.

TABLE 5.2 Summary of Calculation Results

Item Case	First Trial (1–1) A1	*n*th Trial (1–1) A2	Two Units (1–2) B
Tube type	Plain	Finned	Finned
Tube diameter, mm	19	19	25
Shell diameter, mm	650	600	675
Tube length, m	10.0	10.0	4.0
Tube pitch ratio	1.25	1.474	1.25
Baffle spacing, mm	500	500	500
Number of tubes	602	366	355
Total area, m^2	355	542[a]	288/unit[a]
Mean temperature difference, °C	31.3	31.3	29.31
Heat transfer coefficient, $W/(m^2 \cdot K)$			
Shell	2,265	1,988[a]	1,950[a]
Tube (inside)	5,998	11,900	11,300
Overall	1,132	737[a]	738[a]
Pressure drop, shell, kPa	118	82	79
Pressure drop, tube, kPa	20	72	104
Flow velocity, tube, m/s	1.26	2.9	2.9
Heat duty required, MW	11.65	11.65	11.65
Heat duty delivered, MW	12.57	12.50	12.47
Safety factor, % (+ or −)			
Fouled–clean	7/30	7/49	6/48
Distribution of resistance, %			
Fouled–clean			
Shell	50–66	37–68[a]	37–68[a]
Tube	22–30	9–16[a]	8–15[a]
Tube fouling	24–0	45–0[a]	45–0[a]
Wall and fin	4–4	9–16[a]	10–17[a]

[a] Designates that value refers to the finned tube area.

However, to demonstrate the details of the calculational process, manual step-by-step calculations by simplified estimation methods were also performed. These are shown in Appendix 5.1, so that the continuity of this analysis would not be interrupted. Note that while the calculational process and sequences are similar to those performed by the computer program, the numerical values from the estimation methods will be somewhat different than those obtained by the more sophisticated procedures in the program. The evaluation of the computer solution cases indicates the following:

Case A1: Single tube pass, plain tubes. The heat duty is achieved, but the shell-side pressure drop is slightly higher than desired. Also, the water velocity is lower (1.26) m/s) than required for fouling prevention (1.5 m/s). The predominant resistance on the organic stream side suggests the use of low finned tubes, used in case A2.

Case A2: Same as case A1 except finned tubes, 19 mm OD, 13.8 mm ID, fin height 1.5 mm, fin thickness 0.5 mm, outside surface 0.15 m^2/m, 90–10 *Cu–Ni*. The tube pitch had to be increased to 28 mm (PR = 1.474) in order to comply with the shell-side pressure drop. The unit is smaller than case A1, 600-mm shell diameter with only 366 tubes versus 602 tubes. All design conditions are satisfied; the water velocity is 2.9 m/s.

Case B: This is the two units in series solution, each with two tube passes. The LMTD correction factor $F = 0.94$ is acceptable, resulting in a mean temperature difference of 29.3°C. The tube diameter had to be changed to 25 mm to satisfy the pressure drop. All other design requirements are also satisfied. However, the cost of the two units will be much higher than for case A2.

5.8 DESIGN BY COMPUTER PROGRAMS

A predominant number of commercial heat exchangers are now designed by computer programs. However, the main advantage of computers is the speed of calculations, while design elements specification, evaluation of results, and eventually alternative design decisions are still made by the engineer. Two types of computer programs are in common use:

1. Performance rating programs, where the geometry of the exchanger is fully specified and the program calculates the thermohydraulic performance. Such results may be the final product if check rating is the objective; alternately, the designer may keep changing selected specifications of geometry or operation, for example, baffle spacing and type, tube length, tube pitch, and so forth, until a satisfactory design is obtained. This method was used in the calculated example presented earlier. It has the advantage that the design engineer is in full control of the design process (interactive design) and is aware of the constraints that need to be modified. A very important application for rating programs is investigation of performance of designs (or existing exchangers) for alternate operating conditions, such as partial loads, seasonal changes in temperature, clean conditions, and so forth.
2. Design programs, requiring that only specification of the heat transfer process and basic elements of the unit geometry be supplied. For shell-and-tube exchangers these are usually tube dimensions, fixed or maximum length, shell and baffle type, and other possible constraints. The program logic then performs calculations with systematic variation of the free parameters (usually shell size, baffle spacing, tube passes) until a configuration is obtained which satisfies the requirements.

However, such a result is rarely the best solution. For example, the designer, after evaluating the result, may suspect that a different shell or baffle type, tube dimension or tube type (finned) would produce better results. The program is rerun until the best solution (or a compromise) is selected. Well-constructed design programs should display intermediate results at each step of the design process, in order to permit a check on the criteria by which the design was generated. For example, a design may be rejected by the computer only because of a minor difference in pressure drop, which a skilled engineer would have accepted, but which may have remained hidden in the program output.

Thus computer programs permit investigation of many alternative designs, a task not practical by hand calculations, but for truly dependable results, careful scrutiny of the results by an experienced designer is still required.

5.9 OPTIMIZATION AND EXPERT SYSTEMS

The early optimization schemes which became popular with the emergence of computers were based on mathematical solutions to an objective function of optimization, usually a complex function of the cost (initial, operating, amortization, etc.), which was hard or impossible to specify. Furthermore, intermediate solutions from the optimization process were usually not available, thus bypassing the crucial engineering judgment. While such optimization systems are not any more used for general design, there will be specific applications with well-defined operational parameters, where optimization logic can be superimposed on the general design logic and produce valid results.

An expert system is a more appropriate variation of the recently introduced concept of artificial intelligence. The following comments are restricted to the application of expert systems in the field of heat exchanger design, and should not be generalized.

Interpreted in this sense, *expert system* means the inclusion of heat exchanger performance rating criteria, which are identified by the predetermined computer program logic (presumably compiled by experts) and acted upon by the program automatically or, preferably, displayed on the screen for actions by the user. This meaning can best be illustrated by an example.

Let us assume that the performance rating of a heat exchanger is available from a competent computer program. The expert system would then analyze the results and identify such problems as may be inconsistent with the prestored rules, codes, or practices in the following categories:

Basic thermohydraulic specifications:

1. Heat duty, pressure drop, flow velocity, and so on
2. Combinations of the preceding
3. Temperature profile effectiveness

Good design practices:

1. Utilization of pressure drop
2. Pressure drop consumed in peripherals (nozzles, etc.)
3. Large parasitic bypass streams, clearances
4. Inefficient baffle design, tube layout, or other inconsistencies

Mechanical design-related subjects:

1. Nozzle velocity and impingement devices
2. Interactions dependent on shell size, tube diameter, and tube length
3. Thermal expansion provisions with respect to specified geometry
4. Analysis of provisions for tube vibration
5. Compliance with codes

Fouling:

1. Flow velocity versus good practice
2. Analysis of performance under clean versus fouled conditions; heat duty, over-surface, outlet temperatures
3. Design geometry elements as such may affect fouling, for example, baffle type, spacing or baffle cut

The ability of the program to identify such items will of course depend on the quality of the program per se and the amount of peripheral information included. Presently available programs are able to supply answers to only a fraction of the previously mentioned problems as "warning messages." It will be up to a new generation of computer programs to provide information of such a nature and quality that expert systems could be used effectively.

NOMENCLATURE

A heat transfer surface area, m^2
A_t tube inside cross-sectional area, m^2
A_1 flow area (projected) of the tube layout, pertaining to one tube, m^2
A_c crossflow area at D_s, m^2
CL tube layout constant
CTP tube count calculation constant
D_s shell inside diameter, mm or m
D_{ti} tube inside diameter, mm or m
D_{to} tube outside diameter, mm or m
k thermal conductivity of the fluid, $W/(m \cdot K)$

L tube length, mm or m
L_p tube pitch, mm or m
$\dot{m}_s$ shell-side flow rate, kg/s
$\dot{m}_t$ tube-side flow rate, kg/s
N_b number of baffles
N_t number of tubes
Nu Nusselt number, hD/k
N_{rc} number of tube rows crossed in exchanger
N_{rc1} number of tube rows crossed in baffle
N_{tc} number of tubes at D_s
PR tube pitch ratio, L_p/D_{to}
Pr Prandtl number, $C_p\mu/k$
Re Reynolds number $= \rho Vd/\mu$
ρ density of fluid, $\mathrm{kg/m^3}$
μ viscosity of fluid, $(\mathrm{N \cdot s})/\mathrm{m^2}$

APPENDIX 5.1: STEP-BY-STEP CALCULATIONS

A sample manual calculation for the first trial of the example, with the following specifications will be demonstrated:

Tube diameter $D_{to} = 19$ mm, $D_{ti} = 16$ mm
Shell-side diameter $D_s = 650$ mm
Tube length $L = 10{,}000$ mm
Tube layout 30°, pitch $L_p = 23.75$ mm
PR (pitch ratio) $= L_p/D_{to} = 1.25$.

Hot fluid: Liquid butane at 35 bar, shell side, flow rate 52.5 kg/s. At 75°C temperature, the properties are as follows: density 504 $\mathrm{kg/m^3}$, $C_p = 2960$ J/(kg · K), viscosity $\mu = 95 \times 10^{-6}$ $(\mathrm{N \cdot s})/\mathrm{m^2}$, Prandtl number $Pr = 3.11$, conductivity $k = 0.09$ W/(m · K).

Cold fluid: Water, tube side, flow rate 155 kg/s; at average temperature 36°C, the properties are as follows: density 1000 $\mathrm{kg/m^3}$, $C_p = 4178$ J/(kg · K), conductivity $k = 0.62$ W/(m · K), viscosity $\mu = 0.00077$ $(\mathrm{N \cdot s})/\mathrm{m^2}$.

The required duty is 11.65 MW and LMTD = 31.3°C.

The estimated performance from Table 5.1 is now to be confirmed and the pressure drop checked that it is within specified limits. Subsequent adjustments must be made if either one is not in the range.

1. Calculate the number of tubes N_t from Eq. (5.5):

$$CTP = 0.93 \qquad CL = 0.87$$

$$N_t = 0.785 \frac{0.93}{0.87} \frac{0.65^2}{1.25^2 \times 0.019^2} = 629$$

The flow area through the tubes is

$$A_t = \frac{\pi}{4}(D_{ti})^2 N_t = \pi \times 0.016^2 \times \frac{629}{4} = 0.1265 \text{ m}^2$$

The flow velocity is

$$V_t = \frac{m_t}{\rho A_t} = \frac{155}{0.1265 \times 1000} = 1.23 \text{ m/s}$$

Note: for the first trial, this is close enough to the desirable 1.5 m/s.

2. Tube-side heat transfer

2.1. Calculate the Reynolds and Prandtl numbers

$$Re_t = \frac{\rho V D_{ti}}{\mu} = \frac{1000 \times 1.23 \times 0.016}{0.00077} = 25{,}558$$

$$Pr_t = \frac{C_p \mu}{k} = \frac{4178 \times 0.00077}{0.62} = 5.19$$

2.2. The heat transfer coefficient by the Dittus–Boelter equation is

$$h = 0.024 \frac{k}{D_{ti}} Re_t^{0.8} Pr_t^{0.4}$$

$$= 0.024 \times \frac{0.62}{0.016} \times 25{,}558^{0.8} \times 5.19^{0.4}$$

$$= 6033 \text{ W/(m}^2 \cdot \text{K)}$$

compared to the estimated value from Table 5.1 of 7000 W/(m^2 · K), which agrees fairly well.

3. Tube-side fluid pressure drop

3.1. The frictional pressure drop Δp_{ft} is

$$\Delta p_{ft} = f \frac{\rho (V_t)^2}{2} \left(\frac{L}{D_{ti}} \right)$$

The friction factor f is calculated from the Filonenko correlation (Darcy definition) as

$$f = 4 \times [1.58 \ln(Re) - 3.28]^{-2} = 0.025$$

$$\Delta p_{ft} = 0.025 \times \frac{1000 \times 1.23^2}{2} \times \frac{10}{0.016} = 11{,}820 \text{ Pa} = 0.12 \text{ bar}$$

3.2. The tube-side pressure drop in a heat exchanger must include the following additional items:

1. Flow expanding from the inlet nozzle into the header and turn (90°) in the header
2. flow contraction into the tubes and expansion from the tubes
3. flow turn (90°) and contraction into the outlet nozzle

The Δp through these components is usually expressed as the momentum change coefficient K in terms of the tube-side velocity

$$K_a = 1.5 \qquad K_b = 1.5 \qquad K_c = 1 \qquad \sum K = 4$$

The total tube-side pressure drop is then

$$\Delta p_t = \Delta p_{ft} + \sum K \frac{\rho (V_t)^2}{2}$$

$$= 11{,}820 + 4 \times \frac{1000 \times 1.23^2}{2} = 14{,}846 \text{ Pa} \approx 15 \text{ kPa}$$

far below the permissible 100 kPa.

4. Shell-side heat transfer

Shell-side calculations are not as straightforward as tube-side calculations, because the shell flow is complex, combining crossflow and baffle window flow, as well as baffle-shell and bundle-shell bypass streams and complex flow patterns, as shown in Fig. 5.4. The calculational method used here is a simplification suitable for manual calculations and educational purposes. A more sophisticated method, based on the Bell–Delaware method, is documented [4] and was also used in the computer program [8] used to generate Table 5.2. This is a nonreiterative method and can be performed by hand-held calculators with moderate difficulties. Higher accuracy is obtained only by the reiterative version of the Tinker stream analysis method, as described in [4], requiring a rather sophisticated computer program.

4.1. Calculate the crossflow area at shell inside diameter $D_s = 650$ mm. The number of tubes at D_s is

$$N_{tc} = \frac{D_s}{L_{tp}} = \frac{650}{23.75} = 27 \text{ (rounded off)}.$$

Next we must assume baffle spacing. Normally 0.4 to 0.6 of D_s would be selected. However, the large flow rate in a long exchanger and tight tube layout ($PR = 1.25$) will probably require large baffle spacing. Assume $L_b =$ 500 mm.

The crossflow area at shell diameter D_s is then

$$A_c = (D_s - N_{tc} D_{to}) L_b$$

$$= (0.65 - 27 \times 0.019) \times 0.5 = 0.0685 \text{ m}^2$$

4.2. Determine the crossflow flow velocity and Reynolds number

$$V_c = \frac{\dot{m}_s}{\rho A_c} = \frac{52.5}{504 \times 0.0685} = 1.52 \text{ m/s}$$

This is a reasonably average value for organic liquids, perhaps somewhat on the high side. This will be tested by the pressure drop calculation.

$$Re_s = \frac{\rho V_s D_{to}}{\mu} = \frac{504 \times 1.52 \times 0.019}{0.000095} = 153{,}000$$

Pr_s is given in input as 3.11.

4.3. The heat transfer coefficient can be calculated from the estimation equation as given by Taborek [4]:

$$Nu = 0.2 Re_s^{0.6} Pr_s^{0.4} = 0.2 \times 153{,}000^{0.6} \times 3.11^{0.4} = 407$$

$$h_s = \frac{Nuk}{D_{to}} = \frac{407 \times 0.09}{0.019} = 1927 \text{ W/(m}^2 \cdot \text{K)}$$

This compares reasonably well with the rough estimate of 1500.

5. Shell-side pressure drop

Estimation of the pressure drop is much more difficult than for heat transfer as baffle turnaround in the window requires rather sophisticated treatment and in general Δp is more sensitive to predict as it is proportional to V^2 compared to heat transfer which is proportional to $V^{0.6}$. There is no quick and simple estimation method for the shell-side pressure drop in the present literature. A new method based on ideal tube bank flow and a simplified estimation of the baffle window pressure drop is presented here.

The general pressure drop formula is based on crossflow velocity at shell diameter D_s, as calculated in section 4.2, $V_c = 1.52$ m/s, and the number of tube rows crossed

$$N_{\text{rc}} = N_{\text{rc1}}(N_b + 1)$$

where N_{rc1} is number of tube rows crossed within one baffle and N_b is number of baffles.

For N_{rc1} we assume that $0.75D_s$ is in effective crossflow, before the window turn. The tube row distance in crossflow direction is L_p for 90° layout and $0.867L_p$ for 30°.

The number of baffles is determined as

$$N_b = \left(\frac{L}{L_b} - 1\right) \quad \text{rounded off to the next lower integer}$$

$$= \frac{10}{0.5} - 1 = 19$$

$$N_{rc1} = \frac{0.75D_s}{0.867L_p}$$

$$= \frac{0.75 \times 0.65}{0.867 \times 0.02375} = 24$$

$$N_{rc} = 24 \times (19 + 1) = 480$$

The shell-side pressure drop is composed of the crossflow part plus the window turnaround:

$$\Delta p_s = f\frac{\rho(V_c)^2}{2}N_{tc} + K_w\frac{\rho(V_c)^2}{2}N_b$$

where K_w is the momentum change coefficient for the baffle window turn, usually assumed as two velocity head, based for simplicity on the crossflow velocity.

The friction factor for crossflow in ideal tube banks can be looked up in many standard references [4] and handbooks. Note that we are using here the Darcy–Blasius definition more common in the literature than the Fanning definition and related as $f_{DB} = 4f_F$. For pitch ratio 1.25 and $Re_s = 153{,}000$, $f_{DB} \approx 0.25$.

The velocity head ($VH = \rho V^2/2$) is determined as

$$VH = \frac{504 \times 1.52^2}{2} = 582 \text{ Pa}$$

Then the pressure drop is

$$\Delta p_s = 0.25 \times 582 \times 480 + 2 \times 582 \times 19 = 91{,}956 \text{ Pa} \approx 92 \text{ kPa}$$

The nozzle pressure drop on the shell side will have much lower relative magnitude than on the tube side, because of the rather high pressure drop in the segmental baffle. We simply add 10% as a safety factor

$$\Delta p_s = 92 \times 1.1 = 101 \text{ kPa}$$

This is close to the desired 100 kPa.

6. The overall clean heat transfer coefficient is then

$$U_c = \left[\frac{1}{h_s} + \frac{1}{h_t} \frac{D_{to}}{D_{ti}} \right]^{-1}$$

$$= \left[\frac{1}{1927} + \frac{1}{6033} \frac{19}{16} \right]^{-1} = 1397 \text{ W}/(\text{m}^2 \cdot \text{K})$$

The tube wall resistance is omitted here for simplification and is approximately 3%.

The overall coefficient under specified fouling is

$$U_f = \left[\frac{1}{U_c} + R_{fi} \left(\frac{D_{to}}{D_{ti}} \right) \right]^{-1} = \left[\frac{1}{1397} + 0.00018 \left(\frac{19}{16} \right) \right]^{-1}$$

$$= 1076 \text{ W}/(\text{m}^2 \cdot \text{K})$$

The difference between U_c and U_f should be no more than about 35%, higher values suggesting fouling overspecification.

In our case the clean versus fouling safety factor is

$$F_f = \frac{U_c}{U_f} = \frac{1397}{1076} = 1.29 \text{ (or 29\%)}$$

The heat duty under the fouled condition is

$$Q_f = U_f A \text{ LMTD} = 1076 \times 372 \times 31.3 = 12.52 \text{ MW}$$

This represents a safety factor between the required and actual duty of

$$F_s = \frac{12.52}{11.65} = 1.07 \text{ (or 7\%)}$$

Closer adjustment by using shorter tubes or smaller shell diameter is in this case hardly possible because of standard increments in tube length and shell diameters.

Of substantial interest to the designer is the distribution of the resistances. This analysis will indicate where the major resistance is—shell, tube, fouling—and gives a clue to the designer where improvements are justified. The calculation is performed as follows (adjusted to outside tube):

$$\text{Total resistance fouled } \frac{1}{U_f} = \frac{1}{1076} = 0.000930$$

$$\text{Shell-side resistance } \frac{1}{h_s} = \frac{1}{1927} = 0.000519$$

$$\text{Tube-side resistance } \frac{1}{h_t}\left(\frac{D_{to}}{D_{ti}}\right) = \left(\frac{1}{6033}\right) \times \left(\frac{19}{16}\right) = 0.000197$$

$$\text{Fouling tube-side resistance } = 0.00018 \times \left(\frac{19}{16}\right) = 0.000214$$

The ratio of individual resistances to the total is calculated as R_i/R_{fo}:

$$\text{Shell side} = \frac{0.000519}{0.000930} = 55.8\%$$

$$\text{Tube side} = \frac{0.000197}{0.000930} = 21.2\%$$

$$\text{Fouling tube side} = \frac{0.000214}{0.000930} = 23.0\%$$

The shell-side resistance is clearly dominating. Use of finned tubes should be investigated.

ACKNOWLEDGMENT

The major part of the work for this chapter was performed during the author's stay at the University of Karlsruhe, Federal Republic of Germany, as recipient of the Senior Scientist Award from the Alexander V. Humboldt Stiftung.

REFERENCES

1. Soler, A. I. (1986) Expert system for design integration: Application to the total design of shell and tube heat exchangers. In *Thermal/Mechanical Heat Exchanger Design—Karl Gardner Memorial Session*, K. P. Singh and S. M. Shankman (eds.), HTD Vol. 64. ASME, New York.
2. Bell, K. J. (1983) In *Heat Exchanger Design Handbook*, E. U. Schlünder (ed.), Vol. 3.1. Hemisphere, New York.

3. *TEMA Standards* (1988), 7th ed. TEMA, Tarrytown, N.Y.
4. Taborek, J. (1983) In *Heat Exchanger Design Handbook* (1983), E. U. Schlünder (ed.), Vol. 3.3. Hemisphere, New York.
5. Taborek, J., and Sharif, A. (1987) Effectiveness of pressure drop to heat transfer conversion for various types of shell side flow. Presented at the ASME/AICHe Nat. Heat Transfer Conf., Pittsburgh (to be published in *Heat Transfer Eng.*).
6. *Heat Exchanger Design Handbook* (1983), E. U. Schlünder (ed.), Vol. 4.2. Hemisphere, New York.
7. Taborek, J. (1988) Strategy of heat exchanger design. In *Two-Phase Flow Heat Exchangers*, S. Kakaç, A. E. Bergles, and E. O. Fernandes (eds.). Kluwer, Dordrecht.
8. Taborek, J., STEX Shell and Tube Program, IBM PC compatible.
9. Yokell, S., *A Working guide to Shell-and-Tube Heat Exchangers*, McGraw-Hill, New York, 1990.

CHAPTER 6

FOSSIL-FUEL-FIRED BOILERS: FUNDAMENTALS AND ELEMENTS

J. B. KITTO, JR.
Research and Development Division
Babcock & Wilcox Company
Alliance, Ohio 44601-2196

M. J. ALBRECHT
Fossil Power Division
Babcock & Wilcox Company
Barberton, Ohio 44203-0351

6.1 INTRODUCTION

Fossil-fuel-fired boilers are perhaps some of the most complex pieces of heat exchange equipment currently supplied—stretching materials and design technologies to their limits. Their basic function is to convert water into steam for electricity generation and process applications. However, they are also being called upon to burn an ever wider variety of fuels, dispose of refuse, enhance oil recovery, recover waste heat, and reduce pollution. Many possible trade-offs can be made in the design of boilers to accommodate local and worldwide variations in application: fuel, reliability, efficiency, environmental protection, customer preferences, and a variety of economic and political factors. As a result, many different approaches to water-tube boiler design have evolved over the past 150 years to meet these diverse needs. Operating pressures, cycling requirements, unit sizes, steam–water circulation options, fuel firing methods, and heat transfer surface arrangements vary widely, even while many of the fundamental technologies remain common to all designs.

Boilers, Evaporators and Condensers, Edited by Sadik Kakaç
ISBN 0-471-62170-6 ©1991 John Wiley & Sons, Inc.

Chapters 6 to 8 present an overview of international and technical boiler design practices, especially as they apply to large utility boilers. This chapter provides a summary of the design process and technology fundamentals, with particular emphasis on steam–water two-phase flow and recirculating boiler design (where water is only partially evaporated each time it passes through the unit). The focus of the discussion in this chapter is a modern subcritical pressure, drum-type boiler typically found in the United States, Canada, and the United Kingdom.

Chapter 7 explores selected aspects of once-through boiler design where water is evaporated continuously to dryness in individual boiler tubes. A historical overview is followed by discussions of special design issues, sliding pressure operation, operating and start-up characteristics, and selected two-phase-flow topics. The once-through boiler design typical of units found in Germany is the basis for this discussion. Both Chapters 6 and 7 provide extensive lists of references from the United States, the United Kingdom, Germany, and Japan.

Finally, Chapter 8 provides a detailed example of the design process following the boiler design practice of the People's Republic of China (PRC) and, by association, the Soviet Union. The relationships between the various design issues are presented in this English summary of the detailed design procedures from the PRC and USSR, which are identified in the references. The example provided combines analytical methods, empirical correlations, and experience-based "rules."

In order to permit each chapter to be relatively complete, there is some limited overlap in the introductory material to each.

6.1.1 Background

Since at least the time of the ancient Greeks and Romans, steam generated from boiling water has been used for a variety of applications to provide heat and power. Initially, boiling water was used for heating applications with an occasionally innovative but mostly ornamental mechanical power use. It was not until the Industrial Revolution with the development of practical steam engines such as those of Savery and Newcomen (circa 1700) [1, 2] that steam and boilers became widely used to generate power for transportation and industry. Today, boilers and the steam they produce generate electricity, heat and cool structures, provide energy to chemical processes, enhance oil recovery, process food, among others.

Boilers are basically enclosed spaces where water can be heated and continuously evaporated to steam. Early designs were little more than empty vessels ("shell" or kettle boilers) to which water could be added, heat externally applied, and steam removed at a pressure slightly above atmospheric (e.g., the Haycock boiler circa 1720). Soon designers learned that large gas-to-water contact areas were needed to generate increasing quantities of steam at higher efficiencies. This led to boiler designs with the combustion products passing through tubes which were surrounded by water:

"fire-tube" type boilers. Eventually, the need for higher pressures and larger capacities led to the introduction of "water-tube" type boilers where water and steam passed through the tubes, which could more easily withstand the higher pressures.

Since that time, boilers have evolved into very large, complex machines which use the most advanced theoretical analyses and advanced materials. New designs are constantly striving for higher efficiency and lower costs. The largest fossil-fired boilers built today operate at supercritical steam pressures [> 22.1 MPa (> 3208 psia)] providing 1260 kg/s (or 10 million lbm/hr) of steam flow at 566°C (1050°F). The steam can be reheated once or twice to 566°C before ultimately being passed to the condenser. These large units produce 1300 MW_e. Fuels have expanded greatly from gas, oil, coal, and wood to include nuclear fission, municipal refuse, oil shale, and biomass, among others. The evolution in the design of fossil-fuel-fired boilers has been led by extensive innovation in theory, design, and materials over the past 150 years. These innovations are far too numerous to cover here in depth. To more fully understand the evolution of boilers and their design, see [3, 4].

6.1.2 Current Practice

Today's fossil-fired boilers are very diverse in design depending on the steam use requirements, fuel, and process needs. Sizes range from 0.1 to over 1260 kg/s (1000 to about 10 million lbm/hr) steam flow. Pressures range from a little over 1 atm to over the critical pressure. However, regardless of the size or application, today's boiler design remains driven by four key factors:

1. Efficiency (boiler and cycle)
2. Reliability
3. Cost
4. Environmental protection

It is these factors that have been combined with specific applications to produce the diversity of designs in service today—from small package boilers used to supply steam in hospitals to the largest electric utility boiler, from an oil-refinery heater to a marine power boiler. However, regardless of the unit size, application, fuel, or design, all of these units share a number of fundamental or key elements upon which the site- and application-specific design is based. This is especially true for the steam–water side of the system which is the focus here.

6.1.3 Objectives and Overview

This chapter provides a general framework for the evaluation of fossil-fuel-fired boilers with particular emphasis on steam–water thermo-hydraulics. In many areas of the complex evaluation process, fundamental relationships have not yet been fully developed or tested. However, a general overall approach to evaluation is provided. Full evaluation will usually rely on field data and empirical models—much of which are proprietary to individual companies—to permit design of reliable, efficient, cost-effective boilers. The detailed evaluation of a boiler presents some of the simplest geometries for two-phase flow evaluation—vertical upward flow of steam and water under constant heat flux conditions beginning with subcooled water. However, full evaluation is greatly complicated by multiple circuits, hundreds of tubes, bends, nonuniform axial and circumferential heating of tubes, and unheated areas.

It would be impractical to provide a complete description of the evaluation of boilers in a short chapter, even when discussing just steam–water flow. The approach taken here is to focus on the key or fundamental elements common to all boilers and to illustrate their application to a 820-MW_e coal-fired utility boiler used in a large electric power station. This typical

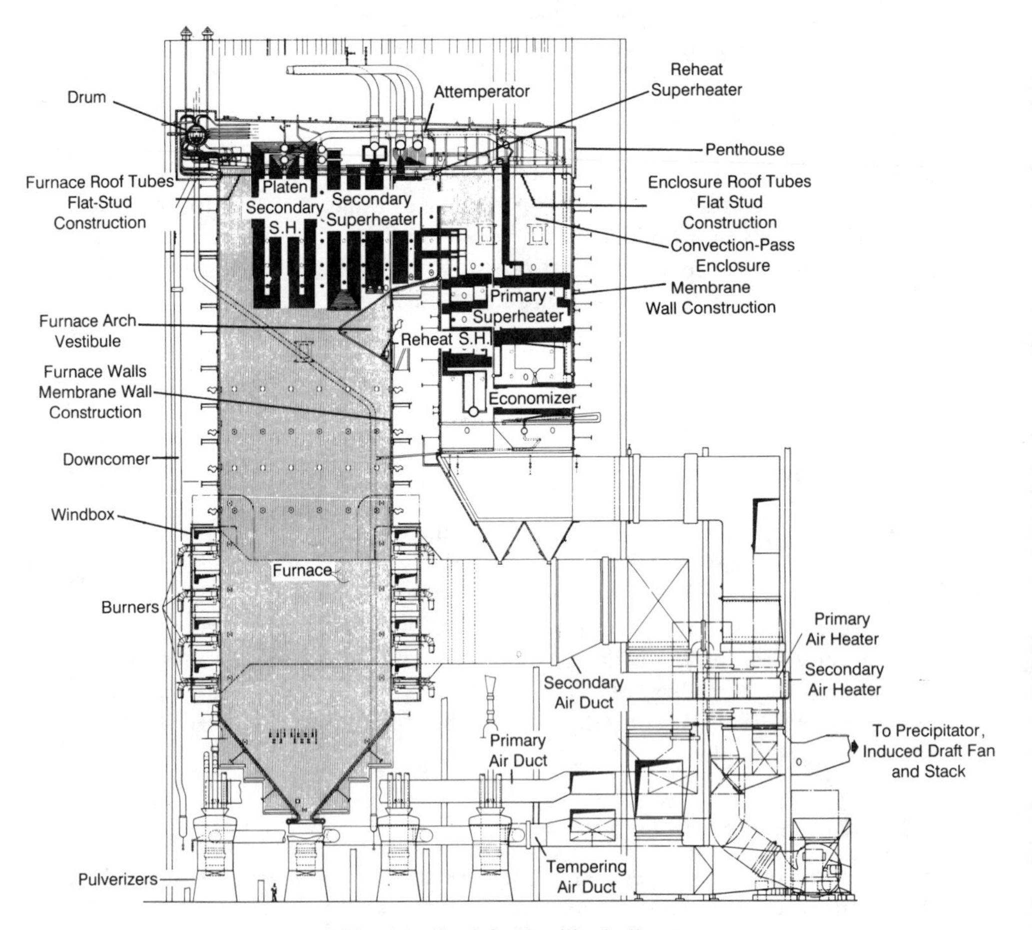

Fig. 6.1. Coal-fired utility boiler.

boiler is shown in Fig. 6.1. The overall evaluation process is presented in more detail in [3–11].

The following sections make up the balance of this chapter. Section 6.2 provides a general discussion of the different types of boilers and how they fit into the overall plant design process. Section 6.3 provides a description and evaluation of each of the major boiler components. Section 6.4 specifically addresses evaluation of the components that are cooled by two-phase flow, while Section 6.5 identifies the limiting criteria. The chapter concludes with Sections 6.6 and 6.7 which identify other selected factors for consideration.

6.2 FOSSIL BOILER SYSTEM

6.2.1 Input Requirements and Operating Pressure

The boiler evaluation begins with the identification of the overall application requirements specified in Table 6.1. These are generally selected in an iterative process balancing initial capital cost, operating costs (especially delivered fuel), steam process needs, and operating experience. Operating steam pressure is a key parameter. For industrial boilers, determination of the outlet steam pressure can be controlled by: (1) process temperature or pressure needs, (2) the need to avoid process fluid leakage into the steam–water system or vice versa, or (3) the maximum allowable metal temperatures to avoid gas-side corrosion or erosion.

For industrial or utility power boilers, outlet steam pressure and temperature are set by the desired power cycle and operating efficiency. Higher temperatures, higher pressures, and the addition of steam reheat sections tend to improve the overall thermal efficiency. However, these must be balanced with the initial capital costs, long-term fuel costs, availability, maintenance costs, unit operating mode (i.e., base versus cycling load), and

TABLE 6.1 Use-Derived Specification Requirements for Boiler Design

Specified Parameter	Comments
Steam use requirements	Flow rates, pressures, temperatures—for utility boilers, the particular power cycle and turbine heat balance
Fuel type and analysis	Combustion characteristics, fouling and slagging characteristics, ash analysis, etc.
Feed-water supply	Source, analysis, and economizer inlet temperature
Pressure drop limits	Gas side
Government regulations	Including emission control requirements
Site-specific factors	Geography, seasonal characteristics
Steam generator use	Cycling, base load, etc.
Customer preferences	Specific design guidelines such as flow conditions, equipment preferences, and steam generator efficiency

other operating costs. Higher pressure and temperature operation tends to increase the initial capital costs. The interaction of all of these factors with the overall power plant design process are addressed in greater depth in [7].

6.2.2 Power Cycle

Even after the last 150 years of power system development, the basic power cycle used in steam–water systems remains the Clausius–Rankine cycle [12, 13] shown in Fig. 6.2 (solid line). Obviously, higher steam outlet temperatures and pressures provide higher thermodynamic efficiencies. This cycle has been modified (at least for utility boilers) to further increase efficiency by adding a section to reheat the steam between turbine stages (dashed line). This version of the cycle is commonly used in utility boilers. Additional cycle efficiency is sometimes obtained through the use of two reheat stages in the turbine–boiler system. "Feed-water heating" (or regenerative heating) has also been added to the cycle to preheat the water before it enters the economizer section of the boiler system. This is used primarily in high-pressure utility boiler designs in order to increase the overall plant thermal efficiency. With reference to boiler evaluation, this significantly increases the temperature of water entering the economizer section of the boiler and limits the potential heat absorption in this section.

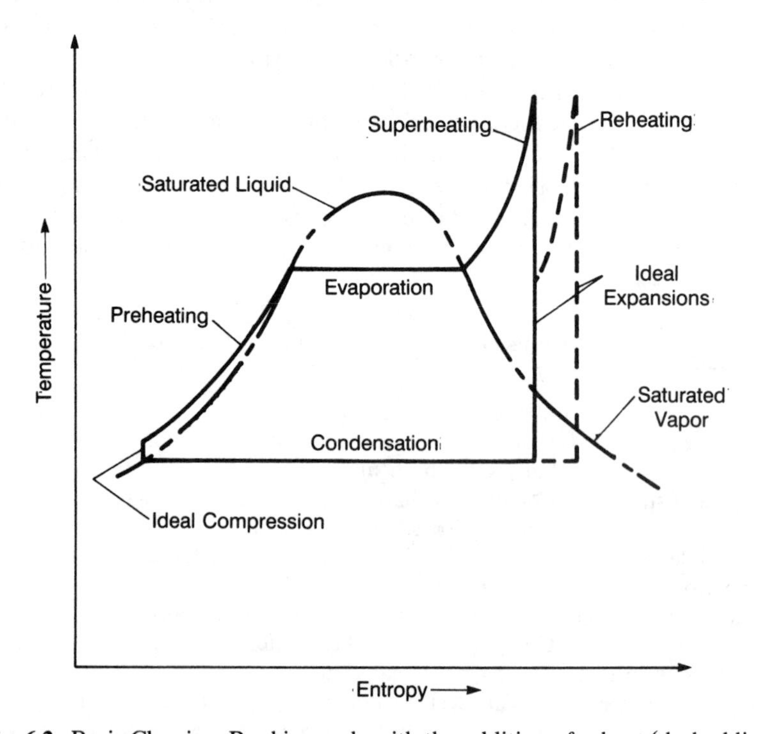

Fig. 6.2. Basic Clausius–Rankine cycle with the addition of reheat (dashed line).

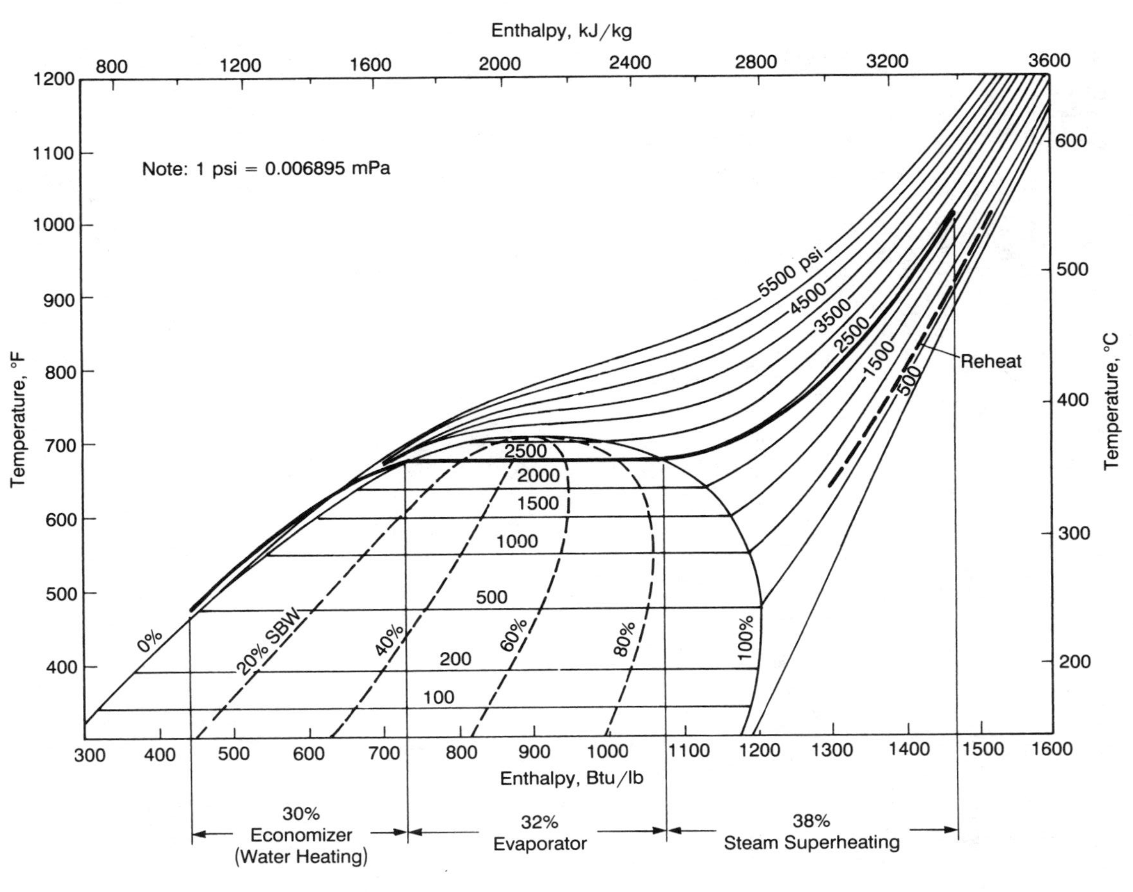

Fig. 6.3. Temperature–enthalpy diagram for boiler heat addition.

From a boiler evaluation perspective, the temperature–enthalpy diagram shown in Fig. 6.3 (for a high-pressure, single reheat unit) provides a useful summary of design information about the unit configuration. As water is converted from subcooled liquid to superheated steam in a typical unit, the relative heat pickup of the economizer (water heating to just below saturation), evaporator, and superheater are 30%, 32%, and 38%, respectively. Reheating the steam (dashed line) increases the total heat absorption by approximately 20% more. For cycles with higher initial operating pressures (usually supercritical), a second steam reheat operation may be added.

Boilers can be designed for subcritical or supercritical pressures. At subcritical pressures, a majority of the furnace enclosure is cooled by two-phase boiling heat transfer phenomena. A small portion of the enclosure will operate at subcooled liquid conditions with the associated liquid convection heat transfer phenomena. Steam–water separation equipment is typically employed to provide saturated (or dry) steam from the evaporator surface to the separate superheater surfaces. Subcritical boilers must be configured to satisfy two-phase-flow and heat transfer limits which assure safe and reliable behavior. Operation is relatively straightforward, compared to supercritical boilers. Industry-accepted water chemistry limits are less stringent, and the

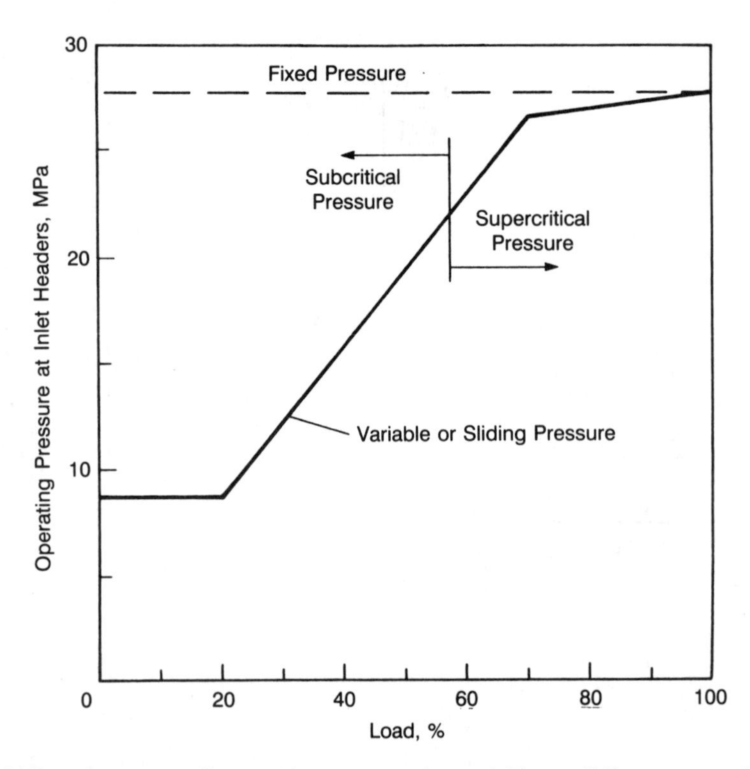

Fig. 6.4. Sample water-wall operating pressure for variable or sliding pressure boiler operation.

steam–water pressure drop inside the tubes is less of an issue on these units. In supercritical constant pressure units, once-through single-phase flow is used to cool the furnace enclosure and must be addressed in the detailed evaluation. In this application, both the steam drum and the internal steam separation equipment can be eliminated. The overall power cycle efficiency is increased but at a cost of higher initial capital, more precise operating requirements, and more stringent water treatment requirements. Finally, variable-pressure boilers where the operating pressure varies with load (see Fig. 6.4) are being installed in Europe and to a lesser degree in the United States [14, 15]. In the latter case, the overall steam–water circuits must be designed for the issues of both subcritical and supercritical pressure operation. As will be discussed later, the operating pressure also sets the types and sizes of heat transfer surfaces to be included in the boiler configuration.

6.2.3 Types of Boilers

Modern boiler equipment consists of a complex configuration of thermohydraulic circuits as shown in Fig. 6.1. Namely:

1. Economizer (feed-water preheating)
2. Evaporator
3. Steam drum (steam–water separation, where needed)
4. Steam superheater
5. Steam reheater (between turbine stages, where used)
6. Steam attemperators (steam temperature control)

They also incorporate firing equipment and are interconnected with fans, controls, pollution control equipment, fuel preparation equipment, and ductwork among other auxiliary equipment in order to provide a complete steam supply system. While the term *boiler* originally referred to the section where evaporation from saturated liquid took place, the terms *boiler* and *steam generator* have come to refer to all of the steam–water components. These components are then optimized for a specific fuel, desired flow rate, and desired steam conditions.

Fuels play an especially significant role in the overall boiler configuration. The variety of fuels used requires a number of different combustion technologies and different convective heat transfer surface configurations to address corrosion, erosion, slagging, fouling and/or cleaning. Even for a single fuel such as coal, a number of different combustion methods may be used: pulverized coal, cyclone, stoker and/or grate, fluidized bed, and so on. Each firing method requires a different water-cooled enclosure shape and size. Figure 6.1 shows a pulverized-fuel-fired utility boiler; Fig. 6.5 shows a typical bark-fired industrial boiler with traveling grate stoker combined with oil- and gas-firing capability. A typical oil- or gas-fired package boiler (preassembled) for industrial and commercial applications is shown in Fig. 6.6.

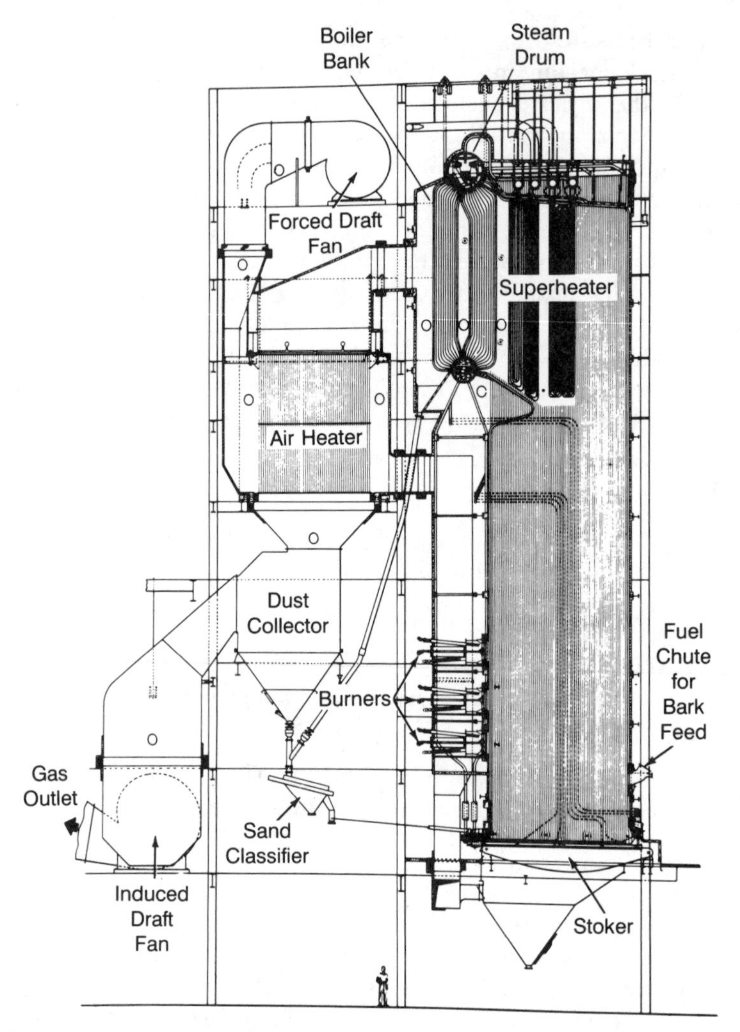

Fig. 6.5. Multifuel industrial boiler with stoker firing (bark, oil, or gas) [3].

Beyond the open combustion furnace, convective tube banks remove additional energy. Where the combustion products (or flue gas) are relatively free of any ash, tighter tube spacings and higher gas velocities are used to minimize the cost and size. Where high ash levels are present in the flue gas, open spacing is used to avoid plugging of the tube bank with ash and to permit cleaning equipment access (these combustion systems and differences are covered in more detail in [3, 5, 10, 11]).

The variety of boiler systems has been classified in a number of ways: end use, type of fuel, firing method, operating pressure, and circulation method, among others (see also Table 8.1 for a more comprehensive list of classifications). From a thermohydraulic perspective, the general breakdown shown in Table 6.2 [16] provides a reasonable starting point.

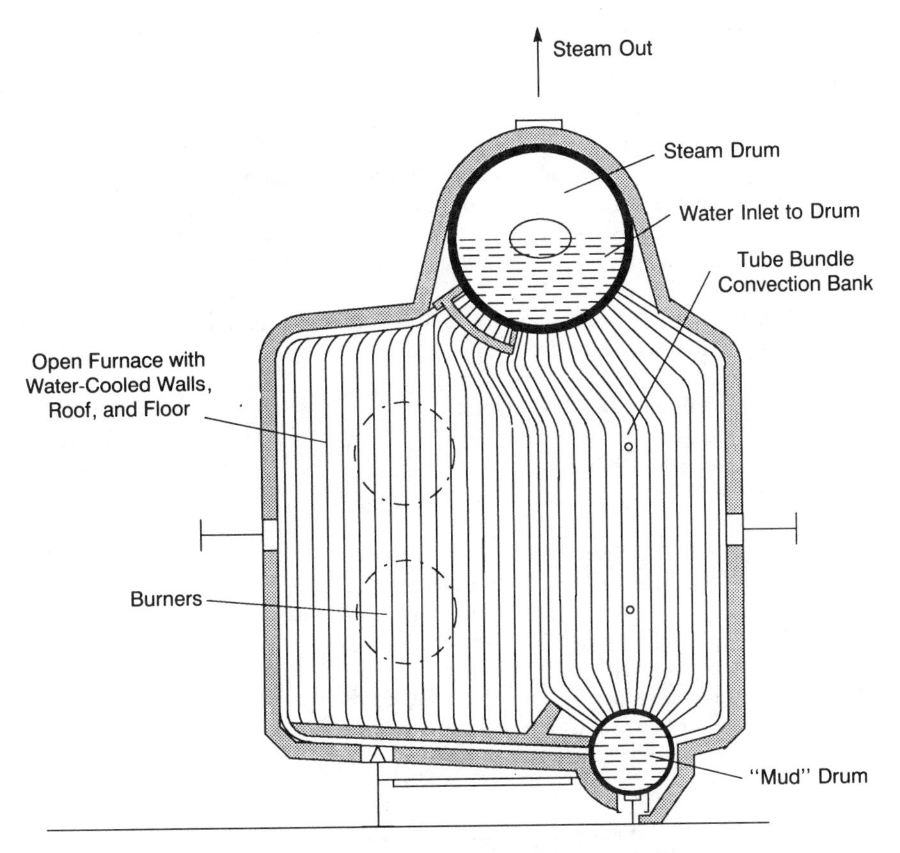

Fig. 6.6. Small oil- and gas-fired package boiler.

Utility boilers (Fig. 6.1) are used for generating electricity in large central power stations and are designed to optimize the overall thermodynamic efficiency at the highest possible availability. Industrial boilers (e.g., Figs. 6.5 and 6.6) are generally used to supply steam to processes or manufacturing activities and thus are designed for: (1) process-controlled (usually low) pressures, (2) high reliability with minimum maintenance, (3) use of available fuels (process waste, if possible), (4) low capital cost, and (5) minimum overall operating cost. A key factor in boiler configurations is the relative amount of energy needed to first evaporate the liquid and then, if necessary, superheat the steam. This controls the quantity of heat transfer surface dedicated to each function and the overall arrangement. This relationship is dependent on pressure and is illustrated in Fig. 6.7. From a thermohydraulic perspective, industrial boilers tend to differ from modern utility boilers in the following general ways:

1. Newer large utility boilers typically operate at high pressures [> 12.4 MPa (> 1800 psia)] for thermodynamic cycle efficiency while industrial boilers typically range from 1.7 to 12.4 MPa (250 to 1800 psia).

TABLE 6.2 Boiler Classifications

- Utility
  - Subcritical pressure
    - Recirculating
      - Nature or thermal or gravity-induced circulation
      - Pumped or forced or controlled circulation
      - Pump-assisted circulation
    - Once through
  - Supercritical pressure
- Industrial (steam supply and/or fuel-use dominant: no reheat)
  - Thermohydraulic basis
    - Recirculating
      - Multiple drum boiler bank
      - Single drum
    - Once through (selected cases)
  - Fuel

Oil	Refuse
Wood–biomass	Byproduct gas
Waste heat	Petroleum coke
Coal	Liquor (chemical recovery)
Gas	Other

  - Firing method

Oil–gas burner	Pulverized coal
Stoker (several types)	Fluidized bed (several types)

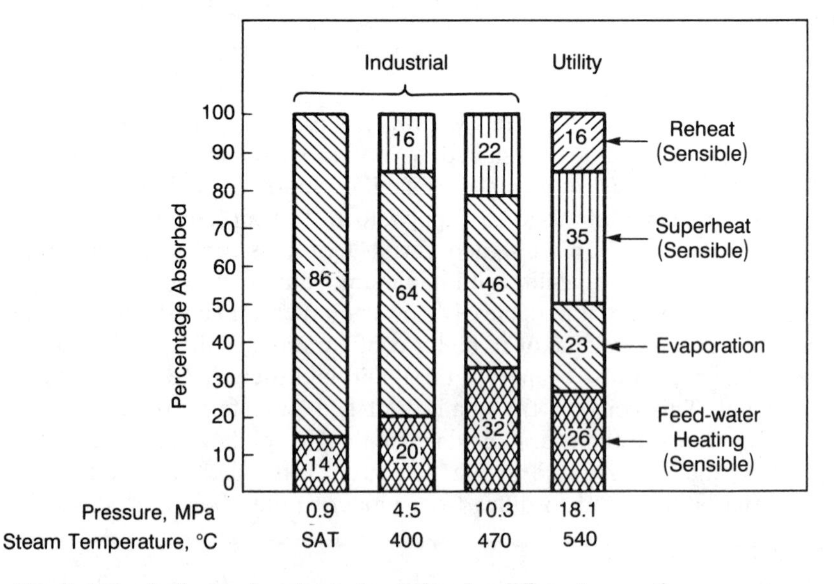

Fig. 6.7. Relative boiler surface heat absorption for different operating pressures.

2. Industrial boilers frequently have separate boiler banks (see Fig. 6.5) to generate steam because the furnace enclosures alone do not provide enough surface area.
3. Reheat of steam is usually not used in industrial units because of higher capital cost and lower economic justification.
4. Pump-assisted circulation is less frequently found in industrial units.

Boiler manufacturers have a relatively limited number of standard designs in order to minimize engineering and fabrication costs. These boilers are usually designed around a number of pre-engineered components to permit maximum flexibility. Most field-erected boilers are custom designed to meet specific fuel and application requirements, although standardized modules are being used more frequently to minimize erection costs.

6.2.4 System Approach

As with most complex engineering problems, there are a variety of boiler evaluation approaches that can be used to meet performance requirements. These include the multiple iterations commonly found in thermal design situations where "real world" complexities and nonlinear interactions prevent a straightforward solution. Boiler evaluation can be approached from two directions: (1) as a steam–water heater or (2) as a flue gas cooler. For the former perspective in a typical boiler, water at about 246°C is converted into 540°C superheated steam. In the latter, the combustion products are cooled from over 1650°C to about 330°C just upstream of the air heater. In most instances, the fuel combustion requirements and gas-side parameters take a leading role in many of the configuration decisions. Thus the overall process focuses on the gas-cooling approach. From the steam flow requirements and thermal power cycle heat balance, the quantity of fuel needed and the enthalpy gain in all boiler sections are established. From here, the process becomes one of adjusting the steam–water circuitry to accommodate the cooling of the gas while checking the thermohydraulic parameters to ensure that key criteria are met. The overall process includes the following general steps:

1. Specifying the steam supply requirements and other parameters (see Table 6.1)
2. Evaluating heat balances and heat absorption by type of surface
3. Performing combustion calculations
4. Configuring the combustion system
5. Configuring the furnace (combustion zone) and convection pass gas-side (temperature limits and material property trade-offs)
6. Specifying the air heater
7. Evaluating the circulation system
8. Verifying overall unit performance

Steps 4, 5, and 7 are of central interest here. For further discussion of the overall process, see Chapter 8.

6.3 MAJOR STEAM–WATER BOILER COMPONENTS

As shown in Fig. 6.1, the steam–water circuitry of a modern high-pressure recirculating (or drum) utility boiler consists of an integrated system of [3, 5, 6, 10]:

1. Enclosure surfaces: (a) furnace: boiling; (b) convection pass: steam cooled or boiling
2. Superheater: primary and secondary
3. Reheater
4. Economizer
5. Attemperator–steam temperature control
6. Drum

Each of these components will now be discussed, including its function, arrangement, size, spacing, thermohydraulic considerations, and materials.

6.3.1 Enclosure Surfaces

The furnace in a large pulverized-coal boiler is a large enclosed space for the combustion of the fuel and for the cooling of the products of combustion prior to their entry into the tube bundles found in the convection pass. Excessive gas temperatures entering these tube banks could lead to unacceptable fouling, slagging, or elevated metal temperatures. Heat transfer to the enclosure walls is basically controlled by radiation. The enclosure walls are cooled by boiling water (subcritical) or high-velocity supercritical pressure water. The convection pass enclosure is composed of the horizontal and vertical-down gas flow passages shown in Fig. 6.1 where most of the superheater, reheater, and economizer surfaces are located. These enclosure surfaces can be water or steam cooled; the heat transfer to the enclosure walls is predominantly controlled by convection. The objective of the water- or steam-cooled wall is to maintain wall metal temperatures within the allowable limits.

The furnace enclosure is usually made of water-cooled tubes in a membrane construction (membrane walls or panels): closely spaced tubes with centerlines slightly larger than the tube outside diameter connected by bars continuously welded to each tube (see Fig. 6.8). Furnace enclosures may also be made from tangent tube construction or closely spaced tubes with a gas-tight seal usually composed of insulation or refractory and lagging. The convection pass enclosure also uses either a membrane or a tangent tube

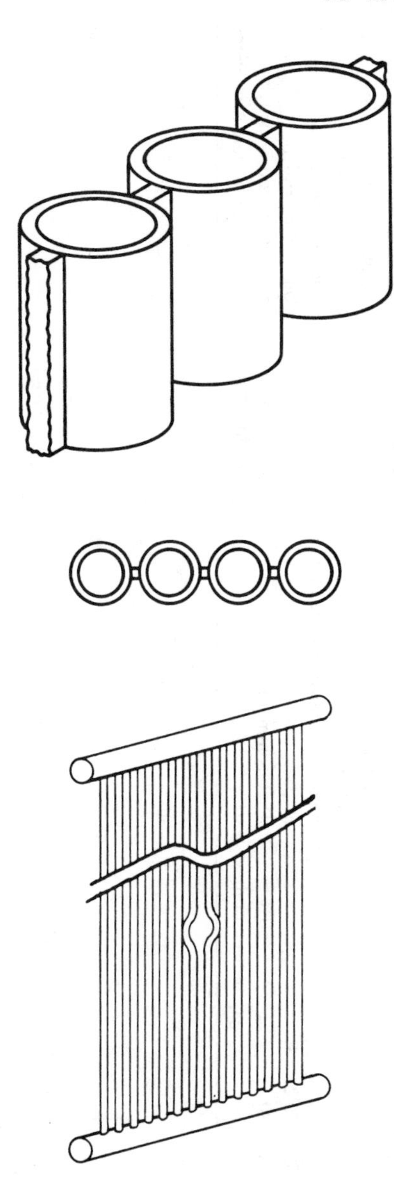

Fig. 6.8. Membrane panel construction.

construction. These tube panels are connected together in parallel flow by inlet and outlet headers. The membrane walls and casing provide a gas (pressure) tight, continuous, rigid construction for the furnace. For a membrane construction, the tube wall and membrane surface are exposed on the inside to the combustion process while insulation and lagging (sheet metal) are provided on the outside to protect the boiler, minimize heat loss, and protect operating personnel. The tubes are usually prefabricated into shippable membrane panel modules for delivery and erection in the field (see

TABLE 6.3 Typical Component Dimensions

Component	Tube Outside Diameter, mm	Tube Centerline, mm	Panel-to-Panel Centerline Spacing, mm	Typical Flue Gas Inlet Temperature, °C	Comments
Furnace water wall					
Thermal circulation	51–76.2	63.5–95.3	—	1650–1900	Membrane wall construction
Pumped circulation	31.75–38.1	44.5–50.8	—	1650–1900	Membrane wall construction
Once through	22.2–34.9	38.1–47.6	—	1650–1900	Membrane wall construction
Furnace division wall	50.8–76.2	60.3–76.2		1650–1900	Membrane wall construction
Superheater	50.8–76.2	—	≥ 1200	—	Radiant platen
			305–610	1005–1215	Convective pendant
			(See economizer comments)	—	Convective horizontal
Reheater	50.8–76.2	—	240	940	Convective pendant
Economizer	44.5–70	44.5–50.8 (gap between tubes)	(See comments)	450–540 (exit 330–370)	Spacing: erosion velocity dependent: in-line arrangement

Fig. 6.8). Openings in these panels are provided for burners, observation doors, slag or ash removal equipment, and gas injection ports. These openings, plus the general boiler arrangement, result in a number of tube shapes and heated–unheated sections that are geometry specific and need to be accounted for in the design. Typical dimensions for enclosure wall tubes are provided in Table 6.3. To minimize costs, manufacturers normally standardize on certain tube sizes, centerline distances, and tube shapes. The thermohydraulics of this portion of the boiler are discussed in detail in Sections 6.4 and 6.5.

6.3.2 Superheaters and Reheaters

Superheaters and reheaters in utility boilers increase the temperature of saturated or near-saturated steam in order to increase the thermodynamic efficiency of the power cycle or to provide the desired process conditions. In general terms, they are simple single-phase heat exchangers with steam flowing inside the tubes and flue gas passing outside the tubes, generally in crossflow. The key criteria in the design of these heat exchangers are:

1. Limiting tube metal temperatures to below acceptable values to meet allowable stress and corrosion–erosion limits
2. Controlling steam outlet temperatures within the specified level over the range of boiler operating conditions
3. Maintaining pressure drop on the steam side within allowable limits (especially for high-pressure subcritical boilers)

The main difference between superheaters and reheaters is operating pressure. In a typical recirculating drum boiler, the outlet pressure of the superheater is 18 MPa while the reheater inlet pressure is only 4 MPa. The volumetric flow rates for the reheater will thus be substantially higher than those of the superheater though the mass flow rate through the reheater is 10% to 15% less than the superheater because of steam extracted from the high-pressure turbine to preheat the feed water.

The mechanical design and location of superheaters and/or reheaters is set by the control range of operation, outlet temperature requirements, overall cycle thermal characteristics, fouling and slagging characteristics of the fuel and cleaning equipment. If high outlet steam temperatures or high absorption (%) are required, some of the heat transfer surface may have to be exposed to radiation from the furnace. There are four general arrangements for superheaters: pendant-platen, pendant, inverted, and horizontal. The locations of these (except inverted) are shown in Fig. 6.9; typical tube sizes and tube spacings are provided in Table 6.3.

The thermohydraulic design of superheaters is a complex trade-off between competing variables including structural and material requirements in

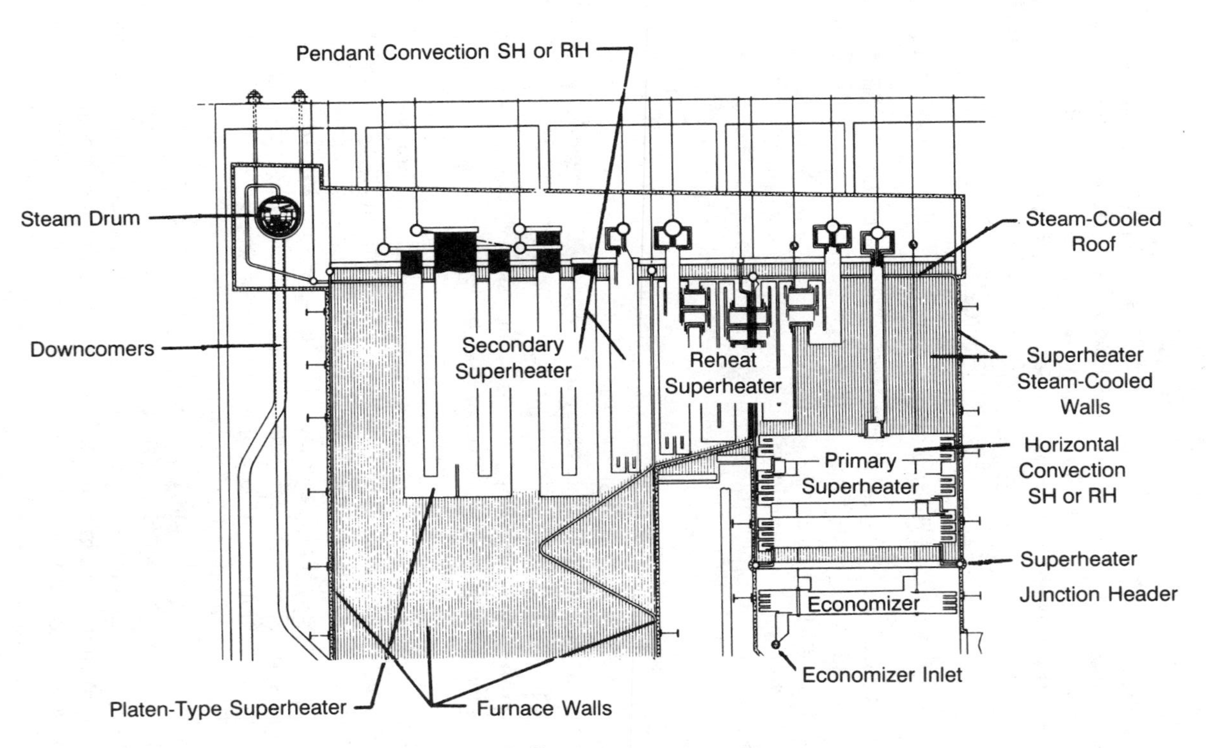

Fig. 6.9. Boiler heat transfer surfaces.

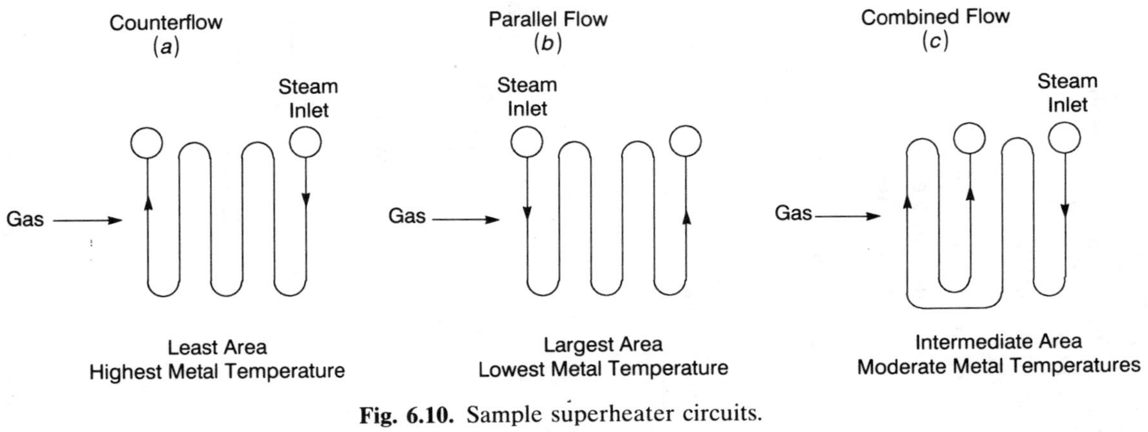

Fig. 6.10. Sample superheater circuits.

order to provide a minimum cost design which meets steam temperature control needs. The key variables that are optimized include:

1. Material cost: surface area, tube thickness, and tubing cost
2. Steam-side pressure drop: limited by possible boiler design conditions, operating cost, and cycle efficiency
3. Gas-side pressure drop: operating cost
4. Tube spacing to handle the expected and worst-case fuel ash deposits (cleaning, fouling factor, erosion limits, etc.)
5. Steam velocities to minimize tube metal temperatures
6. Control of outlet steam temperature

The heat transfer evaluation of superheaters and reheaters is relatively straightforward: single-phase gas flow over a tube bank heating saturated or superheated steam. General heat transfer equations, factors, and assumptions for superheaters and reheaters are provided in Appendix 6.1. Parallel-flow, counterflow, and combination-flow conditions are encountered as the trade-off is made between material cost and heat exchanger efficiency (see

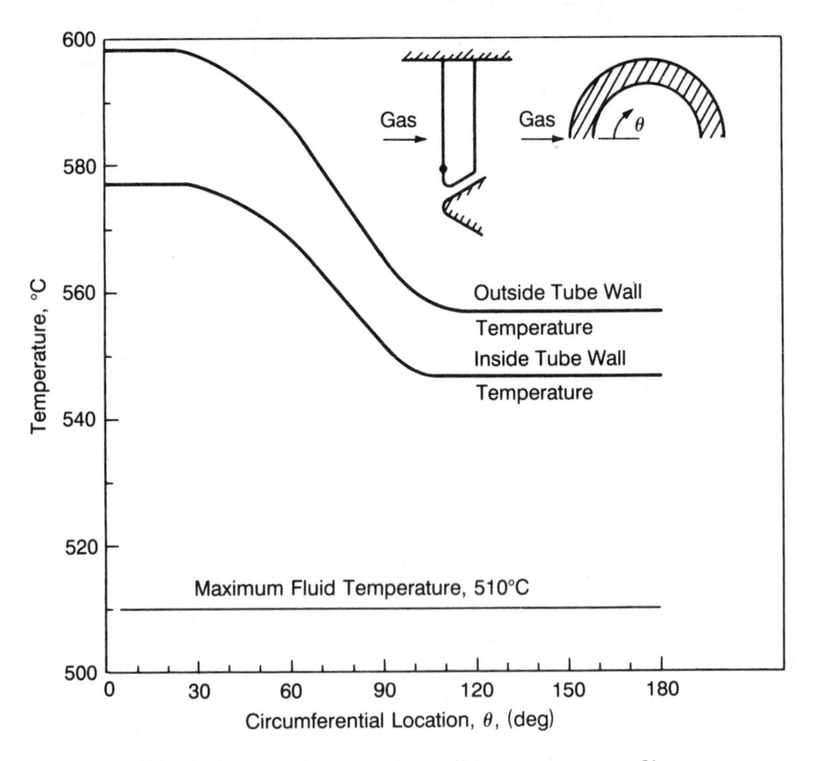

Fig. 6.11. Superheater tube wall temperature profile.

Fig. 6.10). A complicating factor in the evaluation of superheater metal temperatures is radiation from the boiler furnace to the first few tube rows at the furnace exit and the cavities between the tube banks. Nonuniform heating can result in a nonuniform tube wall temperature distribution such as is shown in Fig. 6.11. The additional 40°C could limit the design or require more expensive materials.

The thermohydraulic design of steam reheaters basically follows that of the superheater except, of course, for the lower steam pressure. The flow velocities are typically kept high enough so that the temperature drop across the steam film is 80°C or less. An additional design limitation is the allowable pressure drop. Normally, the total pressure loss through the reheater and the reheat steam piping and fittings is limited to about 10% of the inlet pressure. About half of this is reserved for the connecting piping and fittings with the remainder available for the reheater tubing and headers.

6.3.3 Economizers

Economizers are simple counterflow heat exchangers for recovering additional energy from the combustion products after the superheaters and reheaters but before the air heater, increasing the water temperature after the final regenerative feed-water heater, and minimizing temperature differences between the saturation temperature and the feed-water temperature. The tube bundle is typically an arrangement of parallel horizontal serpentine tubes with both inlet and outlet headers as well as the 180° bends exposed to the flue gas stream. The water flow is usually counter to the flue gas flow. The typical range of tube diameters is provided in Table 6.3. The tube spacing is set to ensure the highest gas velocities which do not exceed the allowable erosion velocities. The bundles have historically been bare tubes configured in an in-line arrangement with appropriate cavities for soot-blower equipment placement. Recently, some extended surface economizers have been used. The economizer location is illustrated in Fig. 6.1. Carbon steel is typically used for this piece of equipment.

From a thermohydraulic standpoint while economizers typically experience crossflow conditions, economizers are evaluated as simple counterflow heat exchangers with the combustion products flowing over the outside of the tubes and subcooled water flowing inside the tubes. The methods for evaluating economizer performance are similar to those for superheaters and reheaters. The following guidelines may also apply:

1. In high-pressure boilers, nonsteaming economizers are frequently used and the maximum water outlet temperature is limited to a temperature that is below the steam saturation temperature.
2. The tube outside metal temperatures are normally fixed in relation to the acid dew-point temperature for the particular products of combus-

tion (temperatures at which acidic combustion product constituents condense).

6.3.4 Steam Temperature Control

The objective of the steam temperature control system is to maintain the superheater and reheater outlet temperatures within a narrow range regardless of changes in boiler load or normal fluctuations in the wide variety of operating variables. Steam temperature reductions of 20°C can reduce the plant heat rate (cycle efficiency) by 1% in high-pressure power boilers (above 12 MPa). Temperature excursions above the nominal design levels may damage superheater, reheater, and turbine components which are temperature limited. Typical operating practice allows fluctuations of ±5 or 6°C in the superheated steam outlet temperature. Factors affecting the superheat control include load, fouling and/or slagging, feed-water change, and burners out of service, among others.

TABLE 6.4 Steam Temperature Control Methods

General Method/Alternate Options
Attemperation (reduce energy per unit mass of steam flow)
Direct-contact spray attemperator (possible locations)
Intermediate between primary and secondary superheater or reheater
Upstream of superheater and reheater
Downstream of superheater and reheater
Indirect attemperator (via heat exchanger)
Steam drum heat exchanger
Separate shell-and-tube heat exchanger
Superheater surface design (see Fig. 6.12)
Gas recirculation (adjust heat absorption in furnace and/or convection pass)
Tempering: gas injected at top of furnace
Recirculation: gas injected at bottom of furnace
Secondary air stream gas injection
Flue gas proportioning (parallel convection passes with controlled flow split) (see Fig. 6.1)
Combustion control (adjust furnace heat absorption)
Tilting burners
Burner rows out of service (BOOS)
Operating procedures
Excess air
Soot-blower control
Feed-water temperature (once-through units)
Separately fired superheater

On-line steam outlet temperature control is achieved by either reducing the energy per unit mass of steam directly by cooling or dilution with water or saturated steam (referred to as attemperation), or changing the relative absorption of heat between reheater, superheater, and furnace. A number of these methods are listed in Table 6.4 and are discussed in more detail in [3, 5, 10]. In large utility boilers, attemperators with direct water or steam injection are typically used for dynamic control because of their rapid response. In most units, they are combined with one or more of the other methods to optimize the temperature control for the overall power cycle. Attemperators are usually installed at the inlet of superheater sections or between superheater sections to control the final superheater outlet metal temperatures.

To a limited degree, superheater sections can be designed and positioned to provide some natural compensation in final superheater outlet tempera-

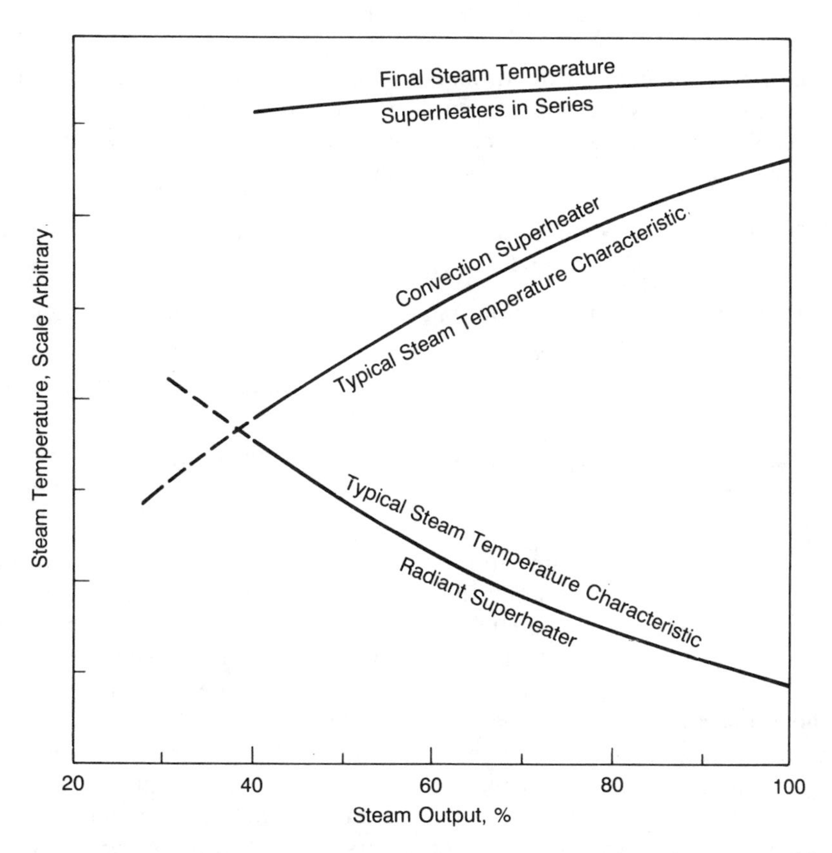

Fig. 6.12. Radiant and convective superheater temperature characteristics [3].

ture which would otherwise vary with load. In a superheater section exposed to the furnace (radiant superheater), heat input is relatively constant with the load. Thus, as steam flow (load) increases, the heat input per pound of steam declines. In a superheater section away from the furnace (i.e., gas convection dominates: convection superheater), the heat input per unit area increases with increasing gas velocity (and load) and steam temperatures climb with increasing load. By matching convection and radiant superheaters, a relatively constant outlet steam temperature can potentially be maintained over a given load range with minimal attemperation (see Fig. 6.12).

6.3.5 Steam Drum

Subcritical recirculating boilers are provided with a large cylindrical pressure vessel or steam drum in which the saturated steam is separated from the two-phase mixture leaving the boiler tubes. These drums can be quite large with diameters ranging from 1 m to several meters and with lengths approaching 30 m. They are fabricated from thick plates that have been rolled into cylinders with hemispherical heads.

The primary function of the drum is to house the equipment necessary to separate the steam–water mixture into saturated steam, which is sent to the steam-cooled surfaces in the boiler, and saturated liquid for recirculation to the furnace circuits. Additional functions include:

1. Mixing the feed water with the saturated liquid after steam separation
2. Mixing the chemicals added to control corrosion
3. Purifying the steam to remove impurities and residual moisture prior to transfer to the superheater
4. Removing a portion of the boiler water to control boiler water chemistry (i.e., blowdown)
5. Containing limited water storage to accommodate some changes in the boiler load level

Typical boiler steam drum internals and their arrangement are shown in Fig. 6.13. In high-pressure utility boilers, the cyclone or centrifugal action caused by the steam–water separators is used universally to separate most of the water from the steam. The reduced density difference between the steam and water at pressures above 12 MPa and the very high flow rates make other methods uneconomical. Scrubber or dryer elements (closely spaced corrugated plates or screens) are then used to remove the remaining moisture and residual impurities prior to steam exit from the drum. Feed-water pipes are installed along the drum length to provide uniform mixing of the feed water with the water discharge from the separators, prior to sending the water to the downcomer outlet tubes or pipes. Manufacturers usually standardize on certain drum diameters and internal arrangements and then vary the drum length to accommodate different size boilers.

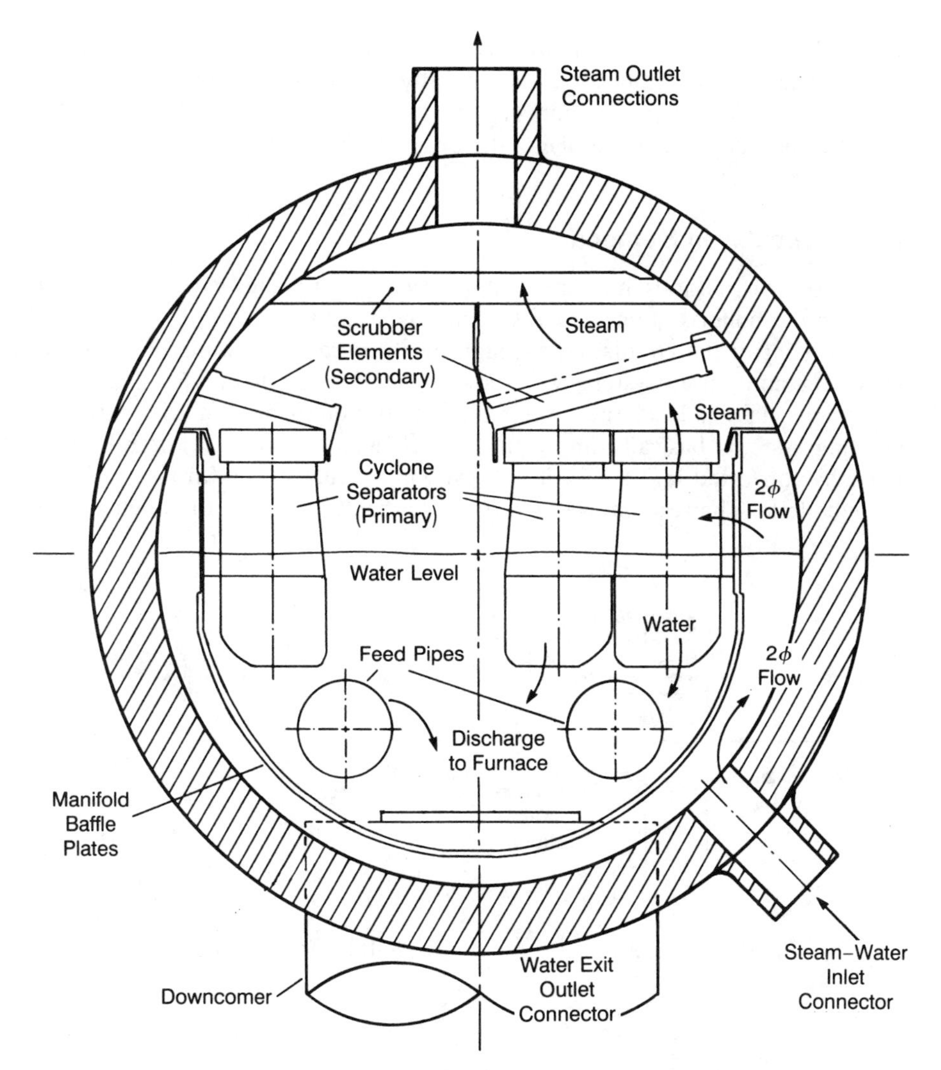

Fig. 6.13. Typical steam drum internals.

6.4 STEAM – WATER SYSTEM

The purpose of the steam–water flow circuitry in fossil-fired boilers is twofold:

1. To generate high-purity superheated steam from subcooled inlet feed water at the specified flow rate, temperature, and pressure
2. To protect metal components from temperature-related failure

The objective of the boiler design is to meet both of these requirements at a minimum cost. The second purpose is particularly important because it establishes many of the evaluation criteria of the major steam–water components. The evaluation becomes an iterative process of component specification, material selection, and criteria verification to set the final boiler configuration capable of supplying the desired steam flows.

6.4.1 Circulation Methods

A number of methods have been developed to circulate water and steam through the boiler system. These systems are usually classified by the means used to control the circulation through the furnace water walls. Five of the most commonly used systems are shown schematically in Fig. 6.14 and may be broadly classified as either "recirculating" or "once-through" types. Recirculating systems basically imply that, at all load levels, the water is only partially converted into steam in the evaporator tubes (i.e., furnace walls in large boilers). The residual water is then recirculated back to the inlets of the

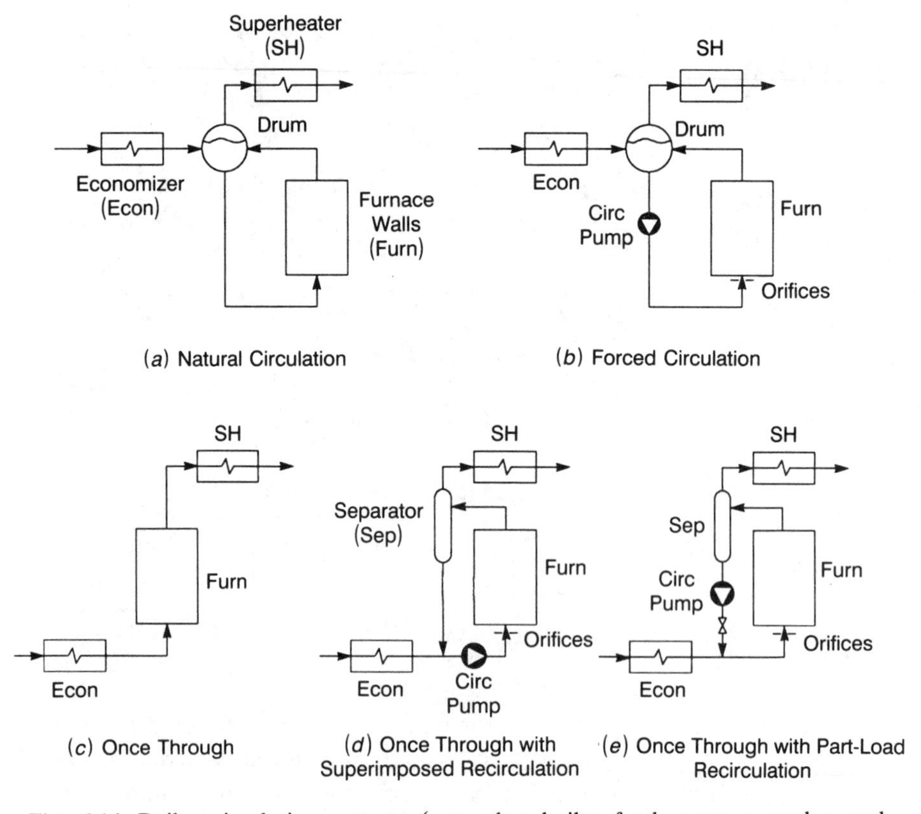

Fig. 6.14. Boiler circulation systems (note that boiler feed pump precedes each system).

evaporator tubes for further heating. In once-through systems, water is boiled completely to dryness (100% steam by weight) in the evaporator circuits during at least full-load operation. The entire evaporator flow is then sent to the superheater. The specific circulating systems include:

Natural (or Thermal) Recirculation (Fig. 6.14*a*) The difference in mean densities between the subcooled water in the downcomer–supply tubes and the heated furnace evaporator tubes (thermal driving head) produces sufficient driving force to overcome the flow resistances in the steam–water circuit. A drum is used to separate the steam–water mixture before the steam is sent to the superheater and the water is returned to the evaporator tubes. Flow is adjusted during the design stage by appropriately sizing the number and diameter of all tubes to provide adequate flow rates for all load conditions. Relatively large diameter tubes are used to minimize friction losses. Flow rates vary directly with heat input (fuel firing rates) and tend to compensate automatically for local heat input upsets. Drums are required for steam–water separation. This system has been used for subcritical pressure boilers.

Pump-Assisted (or Controlled or Forced) Recirculation (Fig. 6.14*b*) To supplement the thermal driving head, one or more pumps are added to the circulation loop to increase the available driving head. Orifices are sized and installed at the entrances to the furnace wall tubes in order to distribute the flow uniformly. The mass flow rate in the furnace tubes is relatively independent of load and local heat flux variations. Drums are required for steam–water separation. This system is used in high-pressure subcritical boilers.

Pure Once-Through Circulation (Fig. 6.14*c*) The boiler feed pump provides the entire driving head to force the water through the economizer, evaporator, and superheater. Water is continuously evaporated to dryness and then superheated without any steam–water separation. This circulation method is applicable to all operating pressures (subcritical and supercritical) although it is not usually used below 8 MPa because the large change in specific volume upon evaporation at low pressures results in excessive pumping power. High-purity water is required since any residual solids will deposit in the boiler when 100% by weight steam conditions are reached. The flow rate is proportional to firing and the furnace wall design is more sensitive to upsets and nonuniform tube-to-tube heating. Special bypass systems are needed to start the units although steam drums are not required. Once-through circulation is sometimes referred to as Bensen or Sulzer monotube systems.

Once-Through with Superimposed Recirculation (Figs. 6.14*d* and *e*) Developed to overcome low-load and start-up limitations of pure once-through designs, these systems permit partial recirculation of fluid to the furnace walls to increase the fluid velocity in these tubes. In constant-pressure supercritical units, part of the furnace flow is continuously mixed with the feed water to increase furnace panel flow except at the highest loads. For subcritical or variable-pressure units at low loads ($<$ 60% to 70% of maximum continuous rating), a limited number of large external steam–water separators (refer to the later discussion about Fig. 6.49) are supplied to recirculate water to the furnace tubes to maintain the tube mass fluxes. Orifices can be used to distribute flow and traditional drums are not required. These systems are applicable to any pressure range.

The selection of the appropriate circulation system for a specific application involves balancing the competing effects of unit pressure, size, planned operating mode, required maneuverability, specific application requirements, and economics, as well as manufacturer and owner philosophies. A more detailed comparison of the different circulation systems is provided in [17].

6.4.2 Boiler Circulation and Flow

Sample System Flow Circuitry A typical steam–water circuit for a large subcritical natural circulation drum boiler (excluding reheater) is shown in Fig. 6.15. Feed water enters the bottom header (A) of the economizer and passes upward in the opposite direction to the flue gas and is collected in an outlet header (B) which can also be located in the flue gas steam. The water then flows through a number of pipes that connect the economizer outlet header to the steam drum. It is sometimes appropriate to run these pipes vertically upward (B–C) through the convection pass to the economizer outlet headers located at the top of the boiler. These tubes are then available to serve as water-cooled supports for the horizontal superheater and reheater when these banks span too great a distance for end support. The feed water is injected into the steam drum (D) where it mixes with the water discharged from the steam–water separators before entering connections to the "downcomer" pipes (D–E) which exit the steam drum.

The water travels around the furnace water-wall circuit to generate steam. The water flows through the downcomer pipes (D–E) to the bottom of the furnace where "supply" tubes (E–F) route the circulating water to the individual lower furnace panel wall headers (F). The water rises through the furnace to an outlet header (G), absorbing energy to become a two-phase mixture. The two-phase mixture passes through the furnace wall outlet headers by means of a number of "riser" tubes (G–D) to be discharged into the drum where the mixture enters the steam–water separators. The steam–water separation equipment returns an essentially steam-free liquid water to the downcomer inlet connections. The residual moisture in the steam that leaves the primary steam separation devices is removed in secondary steam separators and "dry" steam is discharged from the drum to the superheater through a number of drum outlet connections (H–I and H–J).

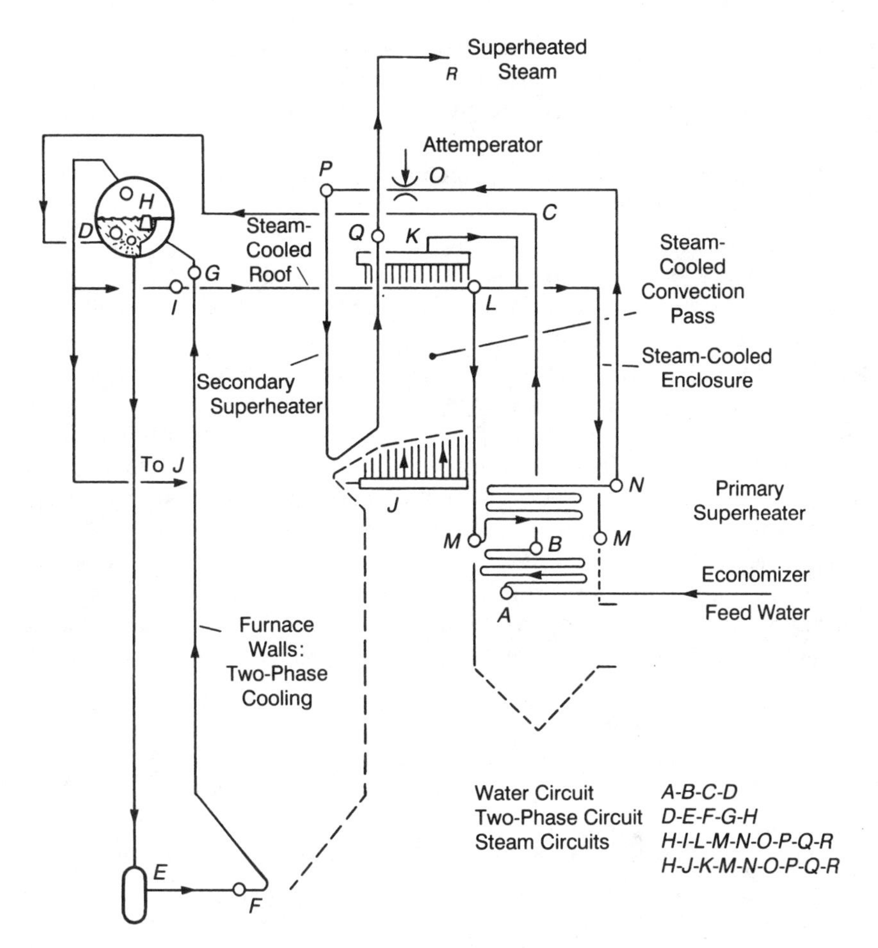

Fig. 6.15. Subcritical natural-circulation boiler steam–water circuitry (reheater excluded here for clarity).

The steam circuitry serves the dual function of cooling the convection pass enclosure and generating the required superheater steam conditions. Steam from the drum passes through multiple connections to a header (I) supplying the roof tubes and, separately, to headers (J) supplying the membrane panels in the horizontal convection pass. The steam flows through these membrane panels to outlet headers (K). Steam from these outlet headers and the roof tube outlet headers then supply the cooling for the vertical convection pass enclosure (L–M). Steam flows downward through these panels and is collected in an outlet header (M) just upstream of the economizer bank.

Steam flow now rises through the primary superheater and discharges through the outlet header (N) and connecting piping equipped with a spray attemperator (O). It then enters the secondary superheater inlet header (P), flowing through the superheater sections to an outlet header (*Q*). A discharge pipe then terminates at a point outside of the boiler enclosure (*R*) where the main steam lines route the steam flow to the control valves and turbine.

Alternate flow circuitry arrangements are possible. Some of these are discussed in Sections 7.2 and 8.2.

Furnace Wall Circuit Evaluation The furnace wall enclosure circuits are perhaps the most critical areas in a boiler. High constant heat flux conditions make uninterrupted cooling of furnace tubes essential. Inadequate cooling can result in rapid overheating, cycling thermal stress failure, or material failures from differential tube expansion. Sufficient conservatism must be engineered into the system to provide adequate cooling even during transient upset conditions. Specified operating parameters must be maintained. Simultaneously, the rated steam-flow conditions must be maintained at the drum outlet. Any of the circulation methods discussed in Section 6.4.1 may be used to cool the furnace water-wall tubes. In evaluating the circulation method selected for a particular situation, a common set of thermohydraulic elements can be used with only minor variations to account for the different systems. The fundamental elements evaluated to ensure proper cooling include:

1. Evaluation parameters:
   applied heat flux distribution
   circulation calculation and balancing of pressure drops
2. Selected limiting criteria:
   instabilities in two-phase flows in parallel circuits
   general velocity limits
   heat transfer rates and critical heat flux
   steam–water separation and drum capacity

The evaluation procedure then becomes an iterative process of selecting standardized components (furnaced membrane panels, drums, headers, etc.) to meet the desired performance limits at the lowest possible cost.

The balance of this section focuses on the evaluation parameters of heat flux distribution and circulation mass flux distribution. Section 6.5 focuses on the limiting criteria.

6.4.3 Furnace Heat Flux Evaluation

Once the system steam parameters, fuel type, and fuel flow rate are set, the evaluation process begins by setting the furnace geometry and assessing the resulting heat flux distribution. The design and thermal evaluation of boiler furnaces is an extremely complex process that does not currently lend itself to exact analytical solutions by theoretical methods alone. Nevertheless, the

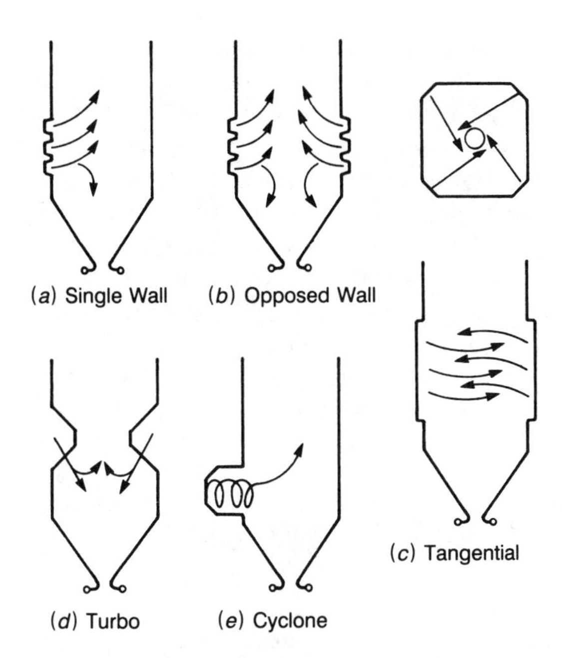

Fig. 6.16. Sample utility boiler furnace configurations.

furnace must be configured to achieve the desired furnace exit gas temperature (FEGT) before the flue gases pass into the closely spaced convection surfaces and the heat flux distribution must be established in order to permit the full evaluation of the furnace wall tube circuits. Several different furnace configurations have been developed to burn the variety of fuels used in large utility boilers (see Fig. 6.16). Manufacturers use a combination of field data, operating experience, plus interpolation and extrapolation based on fundamental relationships to establish proprietary design methods for the application of these furnace configurations. These basic design techniques are now being augmented by advances in computer numerical modeling techniques to permit extrapolation to new, untested geometries and fuels, and to permit parametric optimization of existing designs.

Basic Furnace Configuration Approach The furnace geometry and volume are set by the characteristics of each manufacturer's combustion equipment, fuel input, fouling–slagging characteristics, and the manufacturer's standardized components [3, 5, 18, 19]. The basic approach includes:

1. The furnace depth (burner wall to burner wall in Fig. 6.16*b*) is standardized by manufacturers and is generally dependent on the burner combustion characteristics.

2. Depending on the fuel slagging characteristics (low, medium, high, or severe), the specified value of heat input per unit furnace plan area is set and used to define the furnace width in order to avoid excessive furnace slag accumulation and to minimize pollutant formation.
3. The desired FEGT is set by the ash characteristics to minimize slagging of the convection pass surfaces (< 380 mm centerline spacing). A typical temperature range is 1005 to 1160°C (the more severe slagging characteristics, the lower the exit temperature).
4. Using proprietary procedures, the height of the furnace is defined to provide the desired exit temperature. Depending on the need for superheater or reheater absorption, platen surfaces, furnace division walls, and wing walls may be added to absorb part of the energy to reduce the FEGT to the desired level.

The dramatic impact that fuel ash characteristics and current design practice have on furnace size is shown in Fig. 6.17.

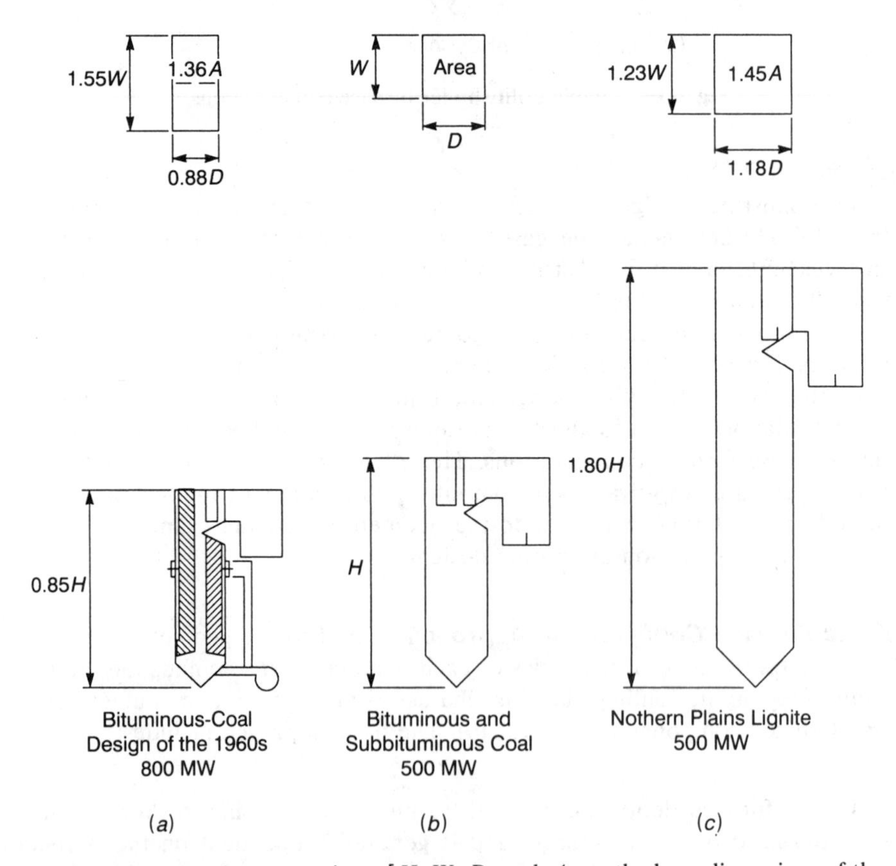

Fig. 6.17. Furnace size comparisons [H, W, D, and A are the base dimensions of the current 500-MW bituminous-coal boiler—diagram (b)].

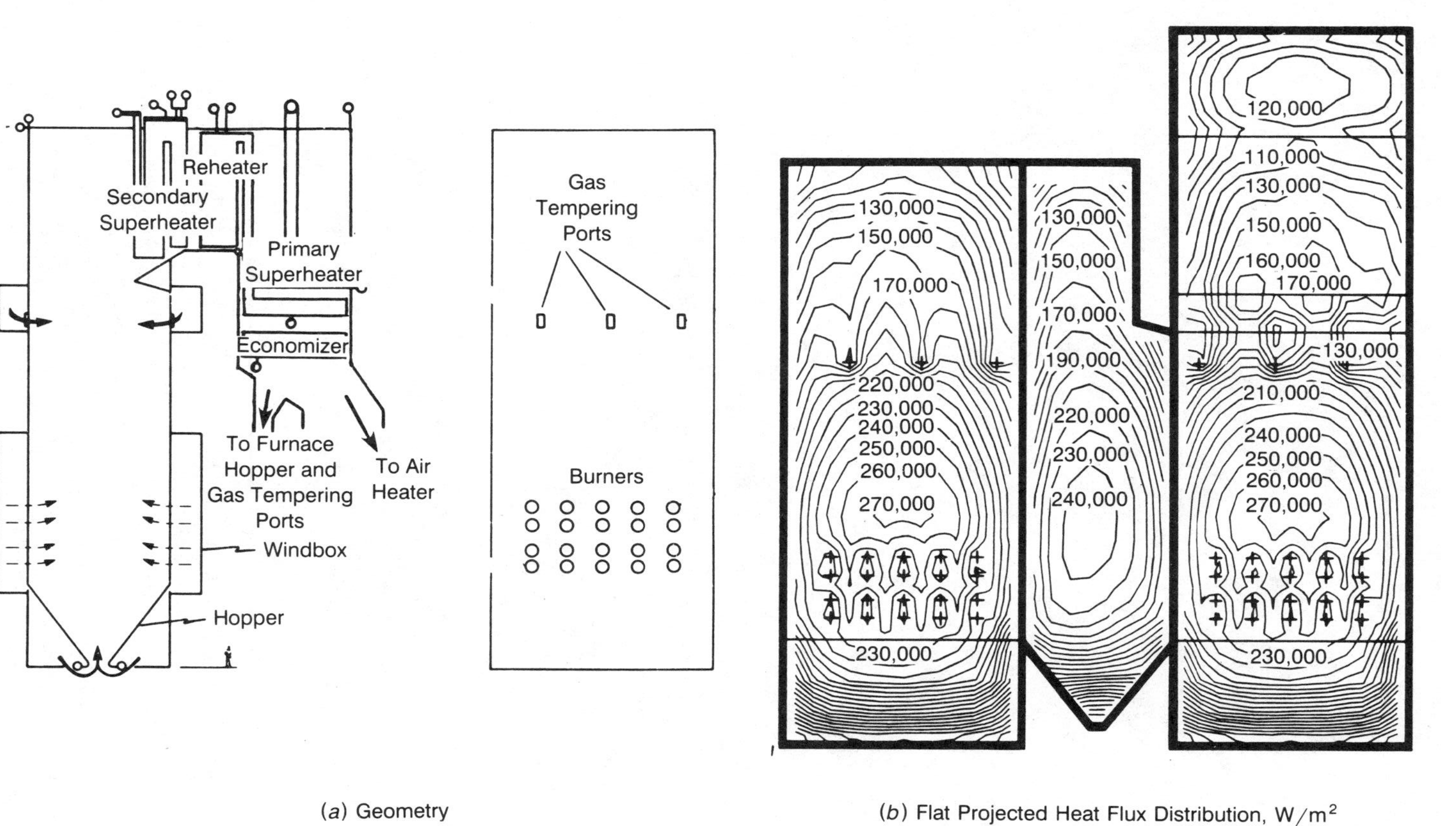

(*a*) Geometry

(*b*) Flat Projected Heat Flux Distribution, W/m^2

Fig. 6.19. Furnace heat flux analysis [20].

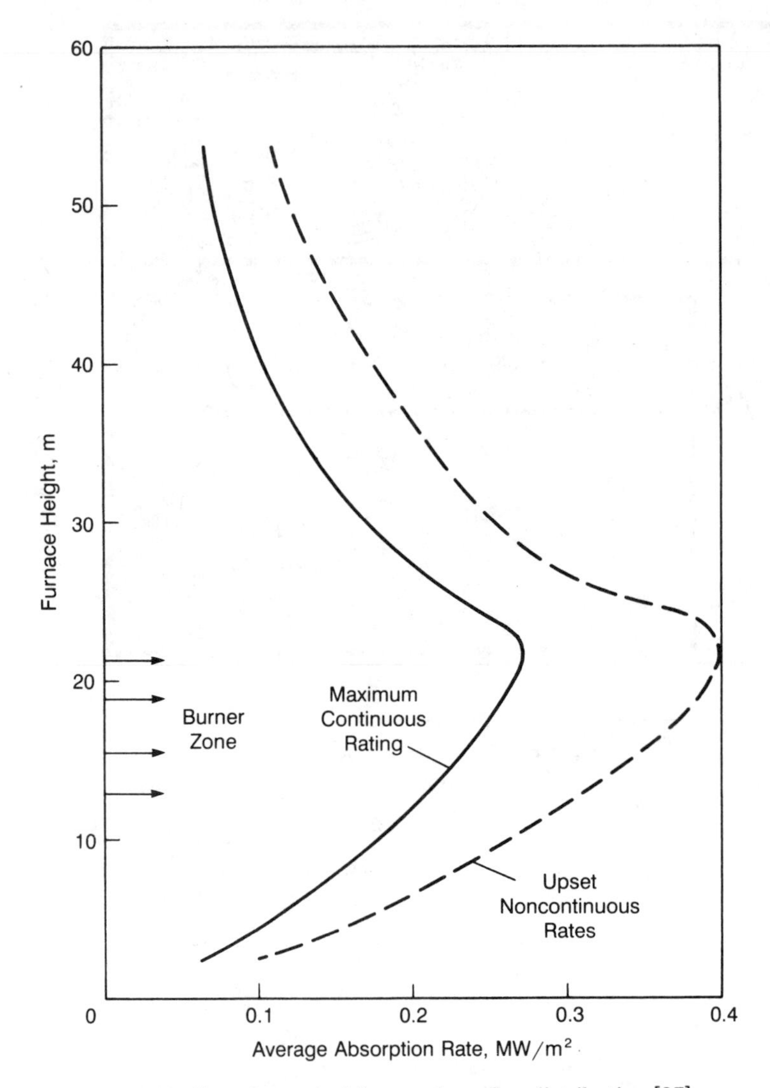

Fig. 6.20. Sample vertical furnace heat flux distribution [27].

insulated to minimize heat loss. The resulting heat flux distribution depends on the tube outside diameter, wall thickness, and centerline spacing, as well as the thickness and materials of the metal bars (web) connecting the tubes. The fluid temperature and inside heat transfer coefficient have secondary effects. This distribution can be evaluated using a variety of finite-difference or finite-element programs plus a uniform flat projected heat flux. A typical distribution is provided in Fig. 6.22.

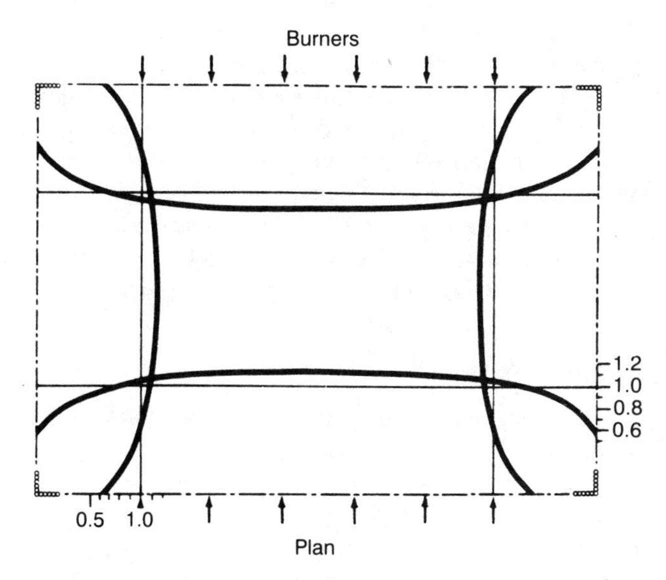

Fig. 6.21. Sample horizontal furnace wall absorption distribution (local to average heat flux ratio R_F) [27].

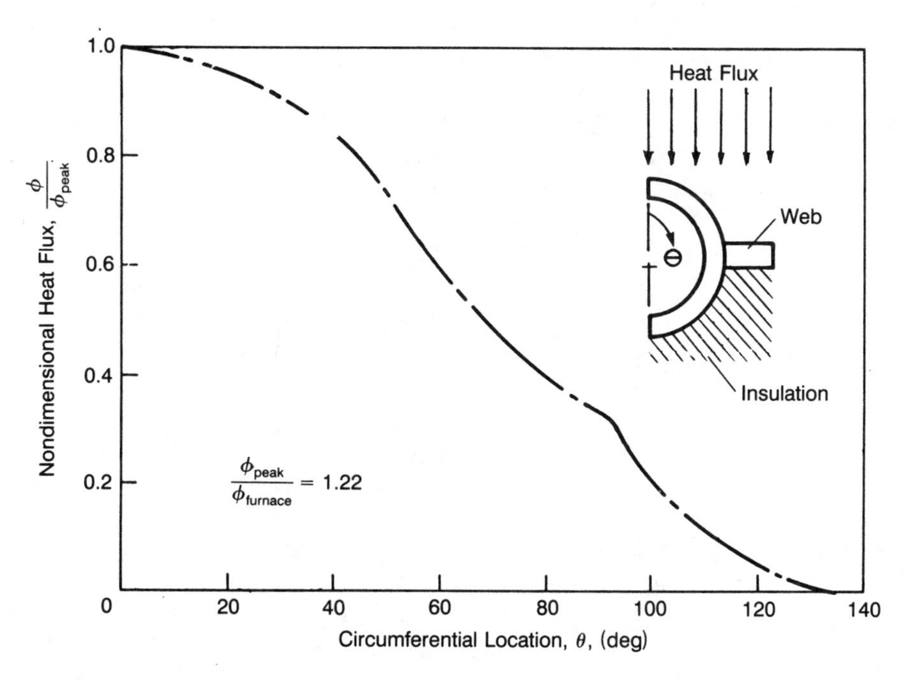

Fig. 6.22. Typical membrane tube circumferential heat flux distribution.

Upset Factors In addition to the reproducible heat flux distribution evaluated in the preceding paragraphs, tests have shown that a noncontinuous variation of absorption rates can be superimposed on the steady patterns. These deviations are caused by unbalanced firing, variations in tube surface condition, differences in slagging, load changes, soot-blower operation, and other variations in unit operation. A typical upset heat flux distribution is shown in Fig. 6.20. These proprietary upset factors are typically a function of vertical–horizontal location, firing method, and fuel and furnace configuration. They are typically set from operating experience.

6.4.4 Circulation Evaluation

General Principles of Natural Circulation The circulation evaluation in a natural circulation boiler provides the flow rates in the furnace wall panels, sizes the various connecting components (supplies, risers, and downcomers), and verifies that the limiting design parameters are met [3, 5, 27]. A simplified circulation loop for natural circulation boiler is shown in Fig. 6.23.

The motion of fluid in this or any circuit can be described by the Navier–Stokes equations. Such a direct solution, however, is too complex to

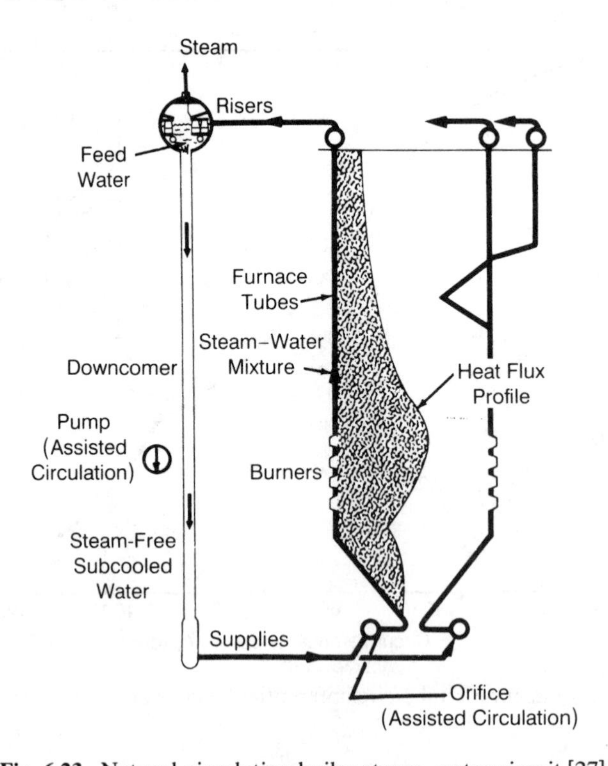

Fig. 6.23. Natural-circulation boiler steam–water circuit [27].

provide a workable approach. In their place, consideration of the hydraulic, thermodynamic, and expansion theories results in the general equation for circulation evaluation which assumes steady, incompressible flow:

$$\sum \text{pressure changes around circulation loop} = 0 \tag{6.1}$$

Alternately, this takes the form:

$$\left(Z\bar{\rho}_d - \int_0^Z \rho(z)\,dz\right) + \Delta P_{\text{pump}} = (\Delta P_{\text{friction}} + \Delta P_{\text{local}} + \Delta P_{\text{acceleration}} + \Delta P_{\text{orifice}}) \tag{6.2}$$

where Z and z are the total and incremental vertical elevation, respectively.

Note that for natural circulation systems, ΔP_{pump} and $\Delta P_{\text{orifice}}$ are set equal to 0 while in pump-assisted systems, these are added to provide additional flow distribution. ΔP_{local} includes the nonrecoverable pressure loss from expansions, contractions, fittings, bends, steam–water separators, and other drum internals.

Explicit solution of Eq. (6.2) for the local flow rates is not currently practical because of the complexity of the flow circuits, the nonuniform heat input, and the nonhomogeneous nature of two-phase flows. As a result, local values for the heat absorption, average density, friction losses, acceleration losses, and local nonrecoverable losses are numerically integrated for the overall flow circuit at a number of flow rates until a flow is identified that satisfies Eq. (6.2).

Pressure Loss: Two-Phase System The local pressure gradient at any location in the system can be represented by

$$-\left(\frac{\delta P}{\delta l}\right) = -\underbrace{\left(\frac{\delta P}{\delta l}F\right)}_{\text{friction}} - \underbrace{\left(\frac{\delta P}{\delta l}A\right)}_{\text{acceleration}} - \underbrace{\left(\frac{\delta P}{\delta l}Z\right)}_{\text{hydrostatic}} + \underbrace{\Delta P_l}_{\text{local losses}} \tag{6.3}$$

Numerical or stepwise integration of this equation around a flow circuit and setting the overall pressure drop equal to 0 (i.e., steady state) provides the system flow rate. Equations for each component are needed for single-phase and two-phase flow.

Using one of several separated flow models for two-phase flow, these components become

$$-\left(\frac{\delta P}{\delta l}F\right) = -\left(\frac{\delta P}{\delta l}F\right)_{LO}\phi_{LO}^2 \tag{6.4}$$

$$-\left(\frac{\delta P}{\delta l}F\right)_{LO} = \frac{f}{D_i}\frac{G^2 v_f}{2g_c} \tag{6.5}$$

$$-\left(\frac{\delta P}{\delta l}A\right) = G^2\frac{\delta}{\delta l}\left(\frac{x^2 v_g}{\alpha} + \frac{(1.0 - x)^2 v_f}{(1.0 - \alpha)}\right) \tag{6.6}$$

$$-\left(\frac{\delta P}{\delta l}Z\right) = g \sin\theta\left(\frac{\alpha}{v_g} + \frac{1.0 - \alpha}{v_f}\right) \quad (\theta = \text{angle from horizontal}) \tag{6.7}$$

$$\Delta P_l = \Phi K\frac{G^2 v_f}{2g_c} \tag{6.8}$$

where Φ is a two-phase multiplier. While ΔP_l usually represents just the irreversible pressure loss in single-phase flows, the complexity of two-phase flows results in the loss ΔP_l typically representing the reversible and irreversible losses for the fitting in the following discussions and in Appendix 6.2 [28–34].

To define each of these local losses, it is necessary to define ϕ_{LO}^2, α, and Φ. Unfortunately, these factors are not well defined because of the complex nature of two-phase flows. Even for the simplest case of two-phase upflow in a vertical tube, a consensus does not currently exist for the two-phase multiplier and void fraction that should be used in different situations:

1. Pre-CHF versus post-CHF flow regimes
2. High versus low velocity
3. High versus low pressure
4. High versus low steam fraction (by weight)
5. Vertical versus horizontal or inclined orientation
6. Small versus large tube diameters
7. Friction versus acceleration versus hydrostatic components

Specific correlations and evaluation approaches can only be used where experimental data under similar conditions provide confidence in the prediction. Collier [28] and more recently Koehler and Kastner [34] provide

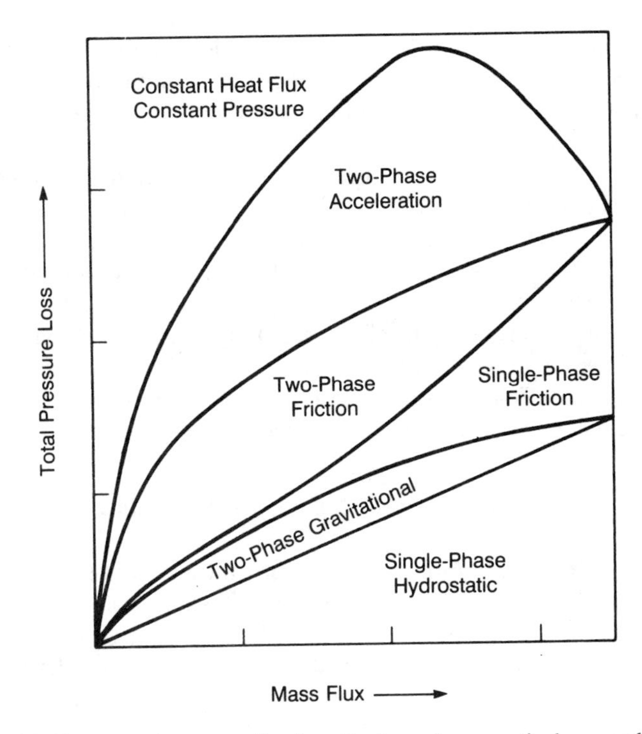

Fig. 6.24. Pressure drop contributions for two-phase vertical upward flow.

overviews of the available correlations for void fraction α and two-phase multiplier ϕ^2_{LO}. The multiplier for fittings and local resistances Φ provides additional uncertainty because the full physical understanding of the behavior of two-phase flows in this area is not yet complete. The work of Chisholm [29–31], Geiger [32], and Idelchik [33] does provide some guidance. Appendix 6.2 provides several sample (but not recommended) correlations covering ϕ^2_{LO}, α, and Φ for consideration.

The relative magnitudes of the hydrostatic, acceleration, and friction pressure losses in a vertical tube with upflow at constant heat flux and pressure are shown in Fig. 6.24.

Circuit Evaluation The furnace volume and configuration are established by the fuel and combustion systems; standardized components are then configured around the required volume. Based on the total steam flow rate and furnace design, the water-wall enclosure tubes are connected into headers of convenient size for assembly and shipment. Engineering standards based on prior experience and engineering analysis set the preliminary size of the drum, drum internals, downcomers, supplies, and risers. The balanced flow for each circuit of the flow system is evaluated for the available pumping head and applied heat flux distribution. For natural circulation systems, the sizes and numbers of the components are adjusted to meet the required criteria. The various load levels and operating conditions at which these analyses are conducted are dependent on the boiler type and projected operating requirements.

Two basic approaches can be used in the circulation evaluation. In the first, the downcomer flow can be used as the driving head to overcome the flow resistance (friction, acceleration, and hydrostatic and local pressure differentials) in the evaporator circuits. Alternately, the difference in density between the downcomer pipe and the evaporator sections can be used as the driving force to overcome the flow resistances in the downcomer (less hydrostatic pressure), supplies, boiler tubes (less hydrostatic pressure), risers, and drum internals. The latter concept can be referred to as the "available ΔP" method where the available ΔP in any section of the boiler circuitry can be defined as

Available ΔP = net driving head − hydraulic resistances

Net driving head = equivalent hydrostatic pressure in an equal vertical length of downcomer

Hydraulic resistance = sum of hydrostatic, acceleration, friction, and local losses in circuit being evaluated

Where natural circulation is used, the "available ΔP" method is frequently useful because it permits sizing of the areas (and thus the flow resistances) of the supplies, risers, and downcomers to balance the flows in all evaporator circuits. The pressure drop versus flow characteristic that is used to balance the evaporator driving head is then the friction and local resistances in the downcomer after appropriate corrections. However, both the "available ΔP" method and downcomer driving head methods are used and will provide equivalent results.

The evaluation process begins by assembling of all the physical geometry information for the preliminary boiler circuits, the applied heat fluxes (refer to an earlier section), and the required steam flow conditions at maximum continuous load. Assumptions are made for a number of flow conditions such as average maximum desired outlet steam quality (percentage steam by weight) based on operating experience. The pressure drop evaluation and circuit balancing processes then proceed. Because of the complex geometry and nonuniform heat flux distribution, the circulation calculation becomes a multistep process of combining elements with common endpoints, geometries, and heat input into subcircuits and establishing the ΔP or "available ΔP" versus flow characteristic for each subcircuit. Figure 6.25 illustrates how tubes can be combined to generate a total flow characteristic. The diagram shows a simple furnace wall panel with a burner opening and assumes that there is no significant flat-projected heat flux variation with width. All of the

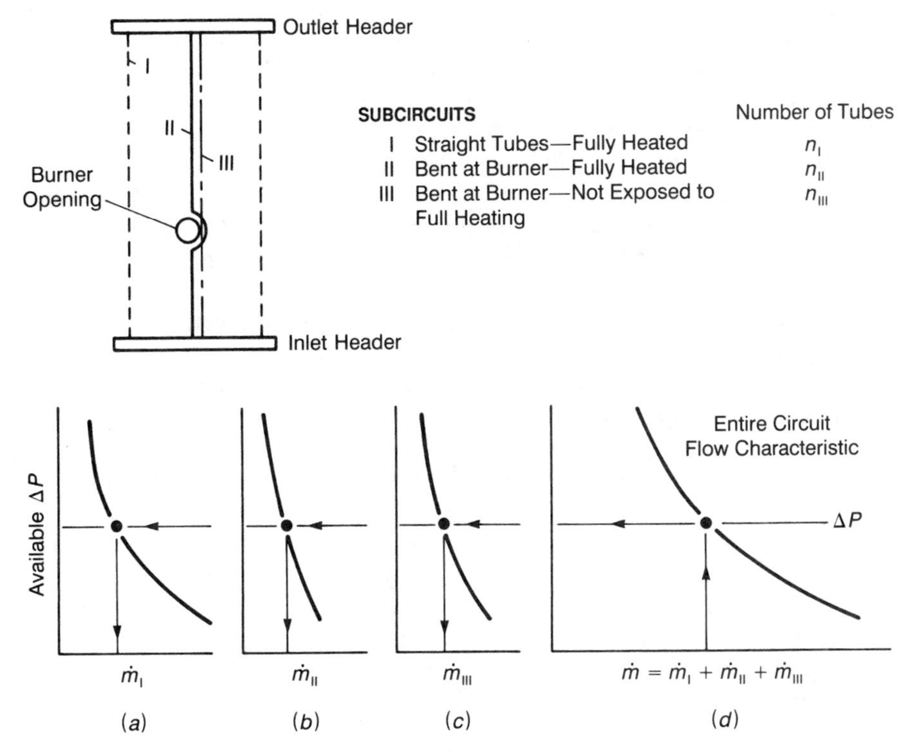

Fig. 6.25. Typical flow balancing using available ΔP approach.

tubes are connected to the same inlet and outlet headers, and thus have similar inlet and outlet pressures (i.e., same ΔP). Because of the physical geometry, there are three different types of tubes of subcircuits as shown: (1) straight fully heated with n_I tubes; (2) fully heated tubes, bent at the burner with n_{II} tubes; and (3) straight with partially heated length with n_{III} tubes. Further subcircuit groups could be defined if there are other significant physical differences or if there is a significant horizontal change in the heat flux distribution. The "available ΔP" (net difference between the hydrostatic head of an equivalent length of downcomer and the tube element flow resistance) is then evaluated as a function of the total flow rate for each group of n similar tubes. This is performed for the applied nonuniform axial heat flux distribution by numerically integrating the pressure losses in each subcircuit. The resulting "available ΔP" versus mass flow rate ($\dot{m}$, kg/hr) curve is shown for each subcircuit in Fig. 6.25—curves (a), (b), and (c). For each value of "available ΔP," the mass flow from all three subcircuits are summed to produce an overall or entire circuit flow characteristic as shown in curve (d) of Fig. 6.25.

These circuits are then combined in a similar fashion to provide flow characteristics for progressively larger units until the whole boiler flow characteristic is evaluated. The final balance is achieved by graphically comparing the pressure drop characteristic of the downcomer or other components with the remaining available pressure drop as shown in Fig. 6.26. The "final available pressure drop" curve for the entire boiler circuit less the downcomer system is plotted. The flow resistance curve (pressure drop less the hydrostatic pressure changes) for the downcomer system (i.e., "remaining flow resistance" curve) is also plotted. The boiler flow rate is then established to be the intersection of the two curves (point A) where the driving force equals the last resistance in the circuit. The flow rate for each individual tube can then be back-calculated. This provides all of the circuit and subcircuit flow rates for comparison to the various limiting criteria.

Given the complexity of this process, computer codes are used to calculate pressure drops, to establish the subcircuit flow characteristics, and to sum the results to provide an overall flow characteristic. The computer models can be designed to solve the flow distribution using the preceding approach or they can iterate until the sum of the pressure drops in the system is equal to 0. Once the system flow distribution is established, the results are compared to the original assumptions with additional iteration steps required if there are any significant deviations. The resulting pressure drop distribution for a 19.7-MPa natural circulation boiler is shown in Fig. 6.27. In this figure the

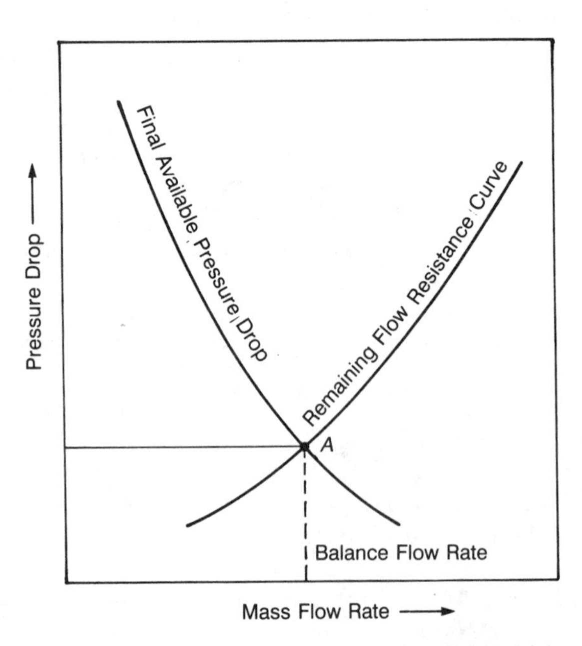

Fig. 6.26. Final natural-circulation system balancing of available driving pressure drop and remaining flow resistance.

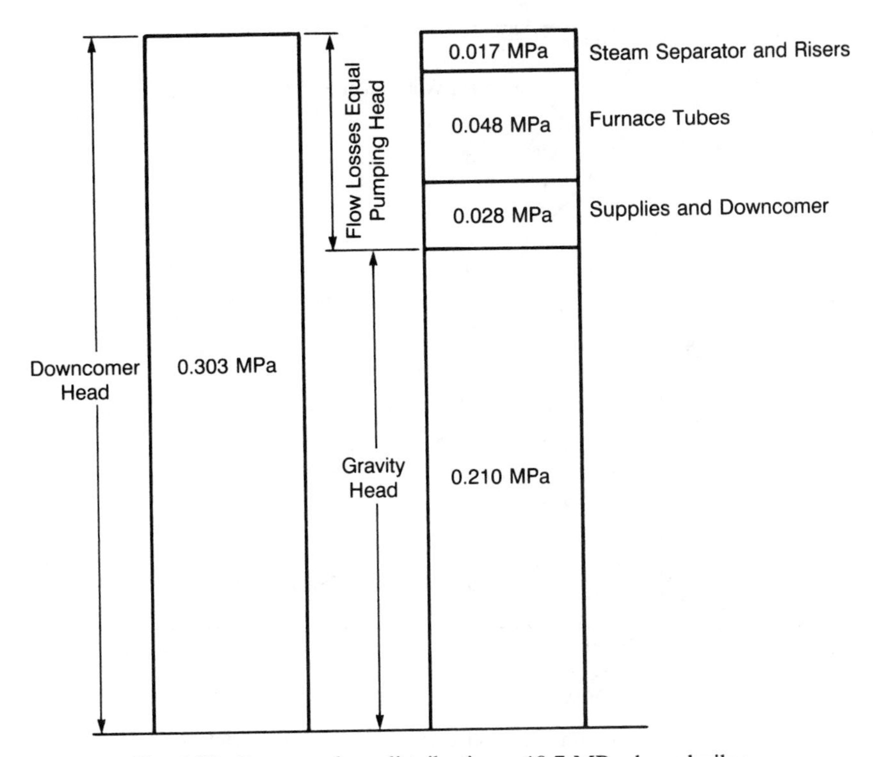

Fig. 6.27. Pressure drop distribution—19.7-MPa drum boiler.

downcomer hydrostatic head (0.303 MPa) is compared to the hydrostatic head in the evaporator tubes and risers to identify the available pumping head. This pumping head is to used to overcome the friction, acceleration, and local pressure losses in the supplies, risers, furnace tubes, steam–water separation equipment, and downcomer. Figure 6.28 shows the same information in the form of the local static pressure in the circulation loop as a function of location (see Fig. 6.23). Beginning in the drum at the drum operating pressure, the local static pressure increases as the subcooled liquid flows toward the bottom of the boiler through the downcomer, reaching a maximum at the downcomer outlet. The local static pressure then falls as the water and then steam–water mixture flows through the supplies, furnace wall inlet headers, evaporator tubes (wall tubes), a furnace wall outlet headers, risers, and the steam–water separators in the steam drum.

Limits Once the circulation rate has been established for a boiler, the operating conditions are checked against several limiting criteria. Among these are: (1) instability and general velocity limits, (2) critical heat flux (CHF), and (3) steam–water separation and drum capacity.

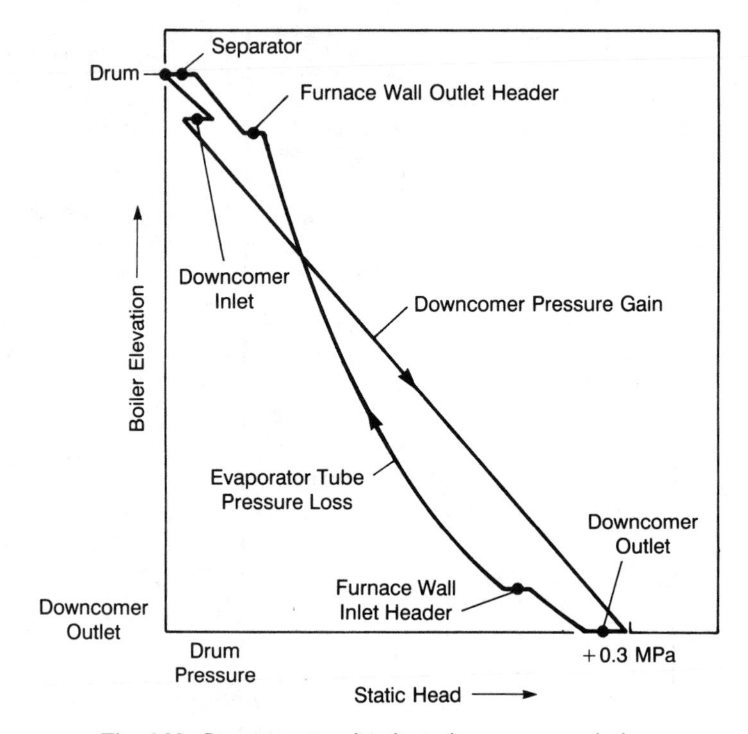

Fig. 6.28. Steam–water circuit static pressure variation.

6.5 TWO-PHASE FLOW CIRCULATION LIMITING CRITERIA

6.5.1 Flow Instabilities and General Velocity Limits

Furnace enclosure steam–water circuit evaluation is partially driven by the desire to generally avoid the possibility of two-phase instabilities [35–42]. Under some operating conditions, the sudden changes in flow direction, reductions in flow rate, and oscillating flows can cause:

1. Unit control problems
2. Complete dryout of the boiling film
3. Departure from nucleate boiling (DNB)
4. Tube metal temperature oscillation and tube wall thermal fatigue
5. Mechanical vibrations with mechanical fatigue
6. Accelerated corrosion attack

Several authors [35–37, 39] have provided detailed overviews of this subject;

TABLE 6.5 Classification of Flow Instability

Class	Type
Static instabilities	
Fundamental (or pure) static instabilities	Flow excursion or Ledinegg instability Boiling crisis
Fundamental relaxation instability	Flow pattern transition instability
Compound relaxation instability	Bumping, geysering, or chugging
Dynamic instabilities	
Fundamental (or pure) dynamic instabilities	Acoustic oscillations Density wave oscillations
Compound dynamic instabilities	Thermal oscillations BWR instability Parallel channel instability
Compound dynamic instability as secondary phenomena	Pressure drop oscillations

thus only a focused review will be provided here. Of the 10 common types of instabilities listed in Table 6.5, three have been found to be of most interest in boiler evaluation: excursive (including Ledinegg) instability, boiling crisis, and density wave oscillations. The first two are static instabilities evaluated using steady-state equations while the last is dynamic in nature requiring the inclusion of time-dependent factors. The boiling crisis or critical heat flux phenomenon is treated separately in the paragraphs dealing with heat transfer and critical heat flux.

Excursive Instability Evaluation The excursive instability is characterized by conditions wherein small perturbations in operating parameters result in a large change in flow rate to a separate steady-state level in multitube panels. Excursive instabilities can be predicted by evaluating the Ledinegg criteria [38]. Instability may occur if the slope of the pressure drop–flow characteristic curve (internal) for the tube becomes less than the slope of the supply (or applied) pressure drop–flow characteristic curve:

$$\left(\frac{\delta \Delta P}{\delta G}\right)_{\text{internal}} \leq \left(\frac{\delta \Delta P}{\delta G}\right)_{\text{applied}} \tag{6.9}$$

The stable and unstable situations are illustrated in Fig. 6.29. As shown in the figure for unstable conditions, if the mass flow drops below point B then the flow rate will continue to fall dramatically because the applied pumping

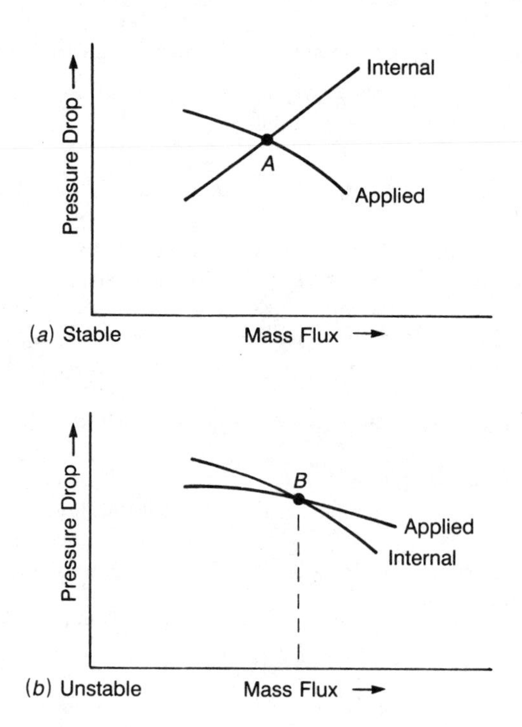

Fig. 6.29. Pressure drop characteristics illustrating Ledinegg instability.

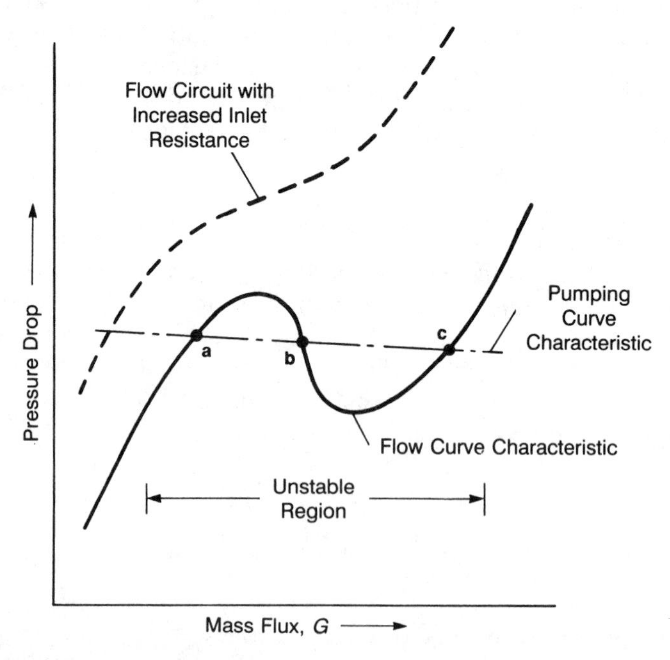

Fig. 6.30. Two-phase pressure drop versus flow curve showing unstable region—constant pressure and heat input.

head is less than the value needed to move the fluid. For slightly higher mass flow rates (higher than point B), a dramatic positive flow excursion will occur because the pumping head exceeds the flow system requirement.

In most systems, the first term is generally positive and the second is generally negative; thus Eq. (6.9) predicts stability. However, in two-phase systems, thermohydraulic conditions may combine to produce a local area where $[(\delta\,\Delta P/\delta G)_{\text{internal}}]$ is negative and the potential for satisfying Eq. (6.9) and observing an instability exists. A heated tube flow characteristic showing a potential region of instability is illustrated in Fig. 6.30 where multiple flow rates can occur for a single applied pressure difference. Operating at point b is unstable with small disturbances resulting in a shift to point a or point c. More intense disturbances could result in flow shifts between points a and c.

For the relatively small subcooling found at the entrance to tube panels in drum boilers and relatively low exit steam qualities, negative slope regions in the ΔP versus G curves are typically not observed for positive flow cases. However, for once-through boilers with high subcooling at the panel inlet and evaporation to dryness, negative slope regions in the upflow portion of the pressure drop characteristic may occur (see also Fig. 6.24). Steps can be taken to avoid operation in any region where the circuit internal $\delta\,\Delta P/\delta G \leq 0$. General effects of boundary condition parameters on the ΔP versus mass flow curves include:

Parameter Increased	Effect on ΔP	Comment
Heat input	Decrease	More stable
Inlet ΔP	Increase	More stable
Pressure	Increase	More stable

Density Wave Instability Density wave instabilities involve kinematic wave propagation phenomena. Regenerative feedback between the flow rate, vapor generation rate, and pressure drop produce self-sustaining alternating waves of higher- and lower-density mixtures that travel through the tube. This instability can occur in single tubes that contain two-phase flows and is of dynamic (time-dependent) behavior.

Density wave oscillations in heated tubes are more frequently encountered at lower pressures. At higher pressures, critical heat flux or burnout is frequently observed before such instabilities have become a significant issue.

Density wave oscillations can be predicted by application of feedback control theory. A number of computer codes have been developed to provide predictions of density wave oscillations. In addition, instability criteria have been developed which used a series of dimensionless parameters to reduce the complexity of the evaluation.

A potentially useful set of dimensionless parameters are those proposed by Friedly and his co-workers [42]. These parameters include

$$Ja = \text{inlet subcooling Jakob number} = \frac{\Delta i_{\text{sub}} v_{fg}}{i_{fg} v_f} \tag{6.10}$$

$$X_F = \text{quality parameter} = \frac{x v_{fg}}{v_f} \tag{6.11}$$

$$R_p = \text{pressure drop ratio} = \frac{\Delta P \text{ exit}}{\Delta P \text{ inlet}} \tag{6.12}$$

Figure 6.31 is a sample stability map using these criteria. The map is the result of the numerical stability evaluation and shows promise for future adaption in stability assessment. Note that the "exit" pressure drop can include the boiler tube, riser, drum, and cyclone pressure losses while the "inlet" pressure drop may include the downcomer and supply pressure differentials.

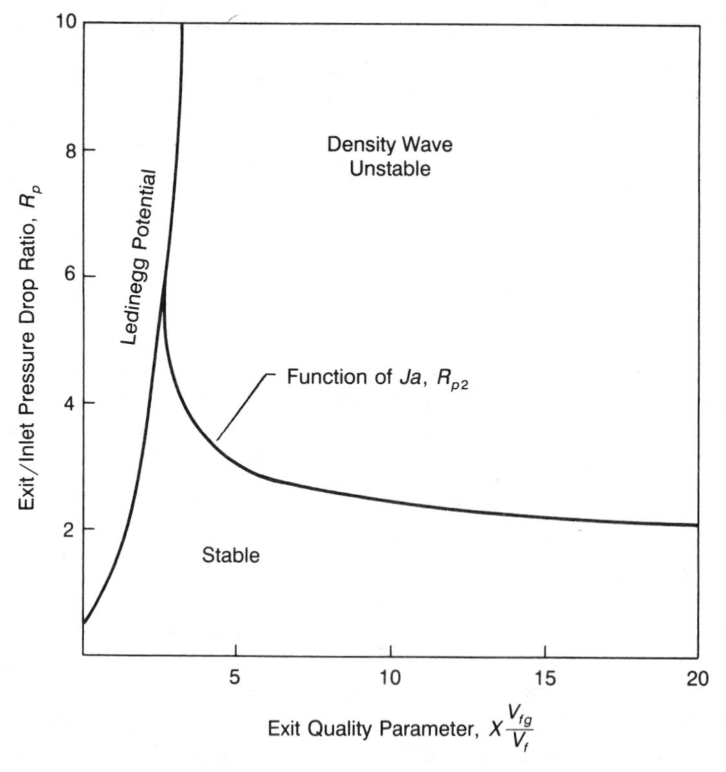

Fig. 6.31. Friedly stability map [42].

Given the complex geometry and heat input in the furnace wall, computer evaluation will be required to define the appropriate operating boundaries. However, with the generally reproducible axial heat flux distribution, the computer-model-generated instability criteria for design may then potentially be recharacterized by a pressure drop ratio (e.g., single-phase to two-phase) and operating pressure for the general design use. The allowable ratio would include a safety margin and could also be adjusted for field experience.

Effects of the operating parameters on the density wave instability include:

Parameter Increased	Change in Stability
Mass flux	Improved
Heat flux	Reduced
Pressure	Improved
Inlet ΔP	Improved
Outlet ΔP	Reduced
Inlet subcooling	Improved

General Velocity Limits In addition to the minimum required mass flux to avoid instabilities, additional minimum mass fluxes (or inlet velocity limits with subcooled liquid) may also be set to address a number of issues [3].

Field and laboratory data are required to set appropriate limits that can be influenced by operating pressure, tube diameter, and heat input. Typical velocity ranges are on the order of: vertical tubes—0.3 to 1.5 m/s; horizontal or inclined tubes—1.5 to 3.0 m/s. Based on experience, higher and lower values may also be used.

6.5.2 Heat Transfer and Critical Heat Flux

Since the overall performance of the furnace flow circuits is controlled to a large extent by the furnace radiation, the objective of the two-phase-flow heat transfer evaluation is to prevent the tube metal from overheating and failing. Boiling heat transfer coefficients are usually high enough so that tube metal temperatures are kept at acceptable levels as long as boiling conditions are maintained in the coal-fired furnace tubes. Thus limits are set by operating conditions that prevent a breakdown of the boiling process: critical heat flux (CHF) or departure from nucleate boiling (DNB).

Convective and Boiling Heat Transfer The heat transfer mechanisms that can occur in fossil boiler smooth bore tubes are defined in Fig. 6.32 and are discussed in [28, 37, 43–47]. Figure 6.32 illustrates a vertical straight

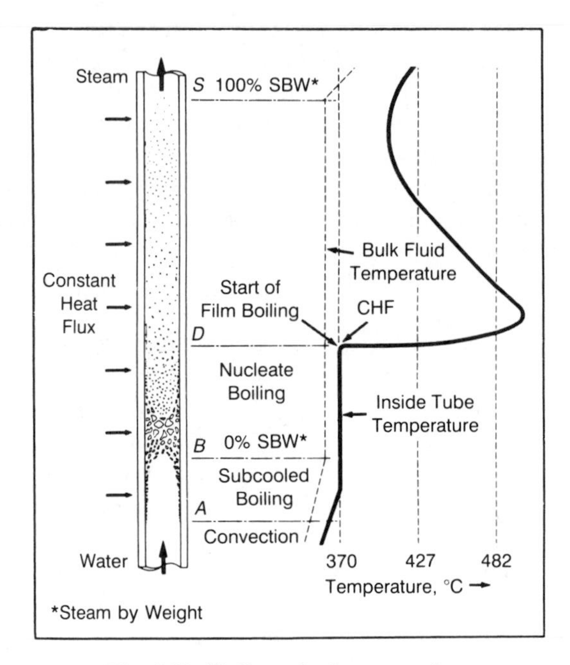

Fig. 6.32. Boiler tube heat transfer.

boiler tube with subcooled inlet temperature conditions and with an applied heat flux. Water is shown evaporating from subcooled conditions to superheated vapor. The inside tube wall temperature and bulk fluid temperature are also shown to the right. The sudden increase in the tube wall temperature shown is a result of the deterioration in boiling heat transfer associated with the CHF phenomenon to be discussed later.

The water entering the furnace tubes is subcooled, and thus a limited amount of forced convection may take place near the tube inlet. Heat transfer coefficients may be evaluated using a variety of techniques—one of which is provided in Appendix 6.1. The convection zone is followed by subcooled, nucleate, and convective boiling regions. For subcooled boiling (up to $x = 0$), a variety of correlations are available to characterize the heat transfer process. Typical are those of Jens and Lottes:

$$(T_w - T_{sat}) = 25\phi^{1/4}e^{-P/62} \tag{6.13}$$

and Thom et al.:

$$(T_w - T_{sat}) = 22.65\phi^{1/2}e^{-P/87} \tag{6.14}$$

where P is the absolute pressure in bar, $\Delta T_{sat} = T_w - T_{sat}$ in °C, and ϕ is the heat flux in MW/m^2. The onset of subcooled boiling and nucleate boiling are discussed at length in [28] along with the evaluation of partial boiling conditions.

Heat transfer in the saturated boiling region occurs by a complex combination of bubble generation at the tube surface (nucleate boiling) and direct evaporation at the steam–water interface (convective boiling). At lower qualities, nucleate boiling dominates while at higher qualities, convective boiling dominates. The most widely used nucleate boiling correlation in this region for any boiling application is that proposed by Chen [46, 47]. While the Chen equation is frequently recommended for use in saturated boiling systems, the additional precision provided is not necessarily required in this application, especially in light of other nonquantified variables such as the effect of tube-side deposits. For general evaluation purposes, most flow boiling correlations provide reasonable approximations for use in the saturated boiling region to characterize the heat transfer process.

Post-CHF Heat Transfer As shown in Fig. 6.33, substantial increases in the tube wall metal temperature are possible if boiling is interrupted by the CHF phenomenon. Since boiler systems are effectively constant heat flux machines, the magnitude of the temperature rise is dependent on the difference between the boiling and post-CHF heat transfer coefficients. The magnitude of the increase will be controlled by the heat flux, mass flux, pressure, and quality at which dryout or CHF occurred. In recirculating boilers, post-CHF conditions are avoided by design. In once-through designs, however, CHF must occur at some location and the associated temperature increase must be accounted for in the design. Since the peak temperature increase occurs just downstream of the dryout location, it is this peak temperature and its location that must be conservatively estimated. Down-

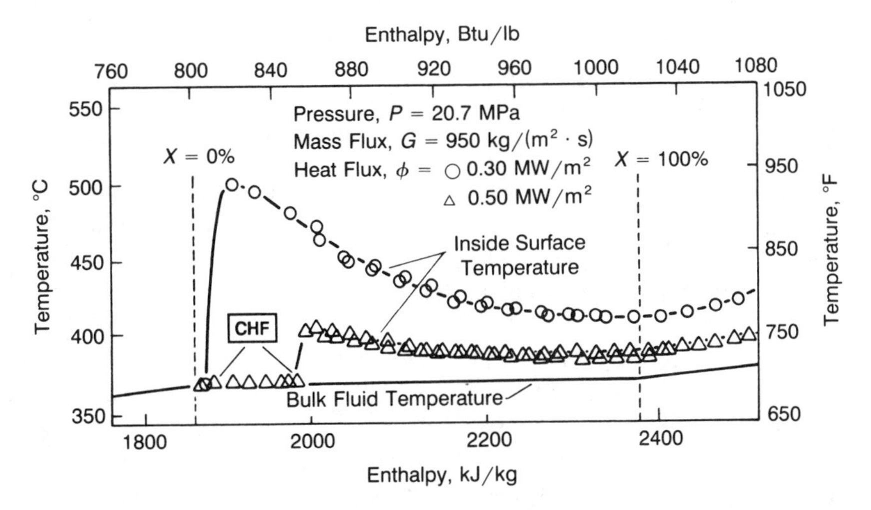

Fig. 6.33. Tube inside surface temperatures for a boiler tube which experiences CHF conditions—two separate heat flux conditions shown.

stream temperatures are of secondary concern. This is discussed more fully in [48–51].

The assessment of post-CHF heat transfer is beyond the scope of this discussion. However, Scruton and Chojnowski [48] provide a comparison of many of the available empirical correlations with high-pressure water data. The key results included:

1. None of the current empirical models provide a good estimate of heat transfer in this region.
2. Groeneveld and Delorme [49] provide the model that best fits the data, but it is very complex.
3. Kimber and Sutton [50] provide a modified equation that is simpler, but less accurate.

For additional discussion of this issue, a recent assessment is provided in [51].

Critical Heat Flux *Objective*. Critical heat flux (CHF) is perhaps the most important design parameter in the furnace wall circuit evaluation. CHF is the term used to denote the set of operating conditions (G, P, ϕ, and x), where the transition from the relatively high heat transfer rates associated with nucleate or forced convective boiling to the lower rates resulting from transition or film boiling occurs (see Fig. 6.32). It is used here to encompass the phenomena frequently referred to as departure from nucleate boiling (DNB), burnout, dryout, boiling crisis, and so forth. The usual objective in the recirculating boiler evaluation is to *avoid* CHF conditions while in once-through boilers the objective is to *predict* their location. In this process, the heat flux profile or level, tube orientation, operating pressure, and inlet enthalpy are usually fixed, leaving mass flux/local quality/inside diameter surface as the more easily adjusted variables.

Overview. The hundreds of experimental and theoretical studies that have been conducted on the CHF phenomenon since the 1950s attest to the complexity of this subject: for selected articles see [37, 52–64]. A number of survey articles have been prepared that provide an overview of this wealth of material with selected papers listed here in the references. A review of these papers leads to the following conclusions:

1. CHF is a very complex combination of thermohydraulic phenomena that are not yet sufficiently well understood to provide a comprehensive theoretical basis. A piecewise approach will probably be necessary to address the range of phenomena observed.
2. Thousands of CHF data and hundreds of correlations define the CHF condition well over *limited* ranges of conditions and geometries. However, some progress is being made to develop more general evaluation

procedures for at least the most studied case—a uniformly heated smooth bore tube in upflow.

Because of this complexity, equipment manufacturers have developed proprietary databanks and associated correlations to accurately define the range of conditions for specific equipment design.

General Model A number of mechanisms have been proposed for the onset of CHF conditions in vertical upflows [52, 57–62]. From a general perspective, these fall into the two categories or regions shown in Fig. 6.34. In this figure the local heat flux (W/m^2) at which CHF conditions [either departure from nucleate boiling (DNB) or film dryout] occur is plotted versus the corresponding average local steam quality (percentage steam by weight). At lower heat fluxes and/or steam qualities, a fully wetted tube wall condition and boiling are maintained. Under slightly subcooled and low steam quality (x) conditions, DNB occurs with the creation of an insulating steam layer adjacent to the tube wall (region I). This may be caused by

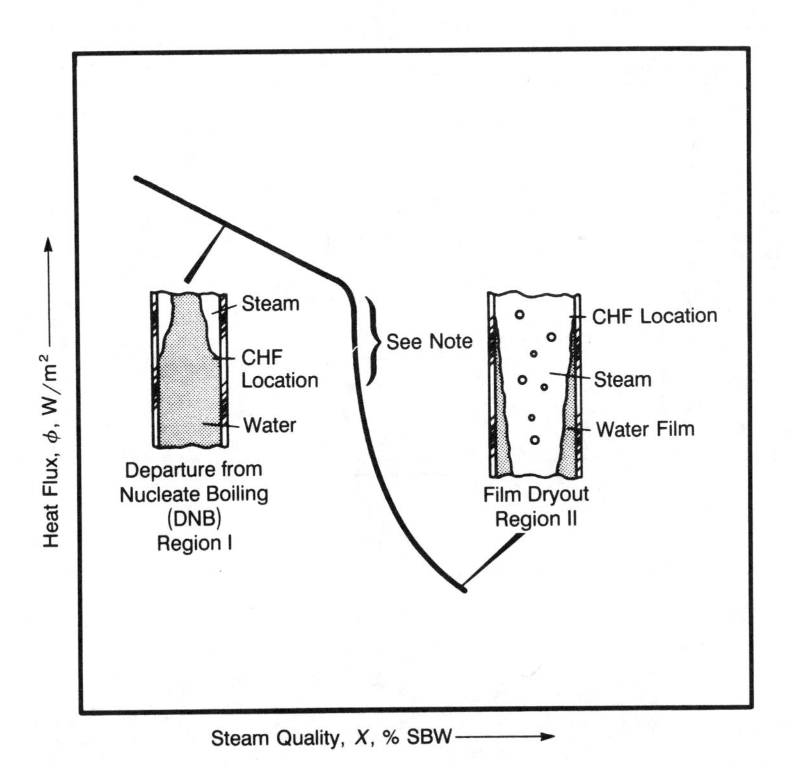

Fig. 6.34. Generalized CHF function for a vertical tube with constant upward flow rate at constant pressure—two separate CHF phenomena shown (note: limiting quality region).

high-vapor generation rates where bubble crowding, bubble shielding, or steam blanketing prevents liquid replenishment at the tube wall. The microlayer under bubbles evaporates resulting in temperature excursions. At high qualities (region II) where annular flow prevails, it is generally accepted that CHF is the result of dryout of the liquid film on the tube wall. The liquid film flow rate is a balance between droplet deposition from the steam core, re-entrainment of liquid from the film, film flow rate, and film evaporation rate. When the film flow rate drops to 0, a temperature excursion occurs. Significantly more complex behavior is observed in horizontal and inclined tubes where a variety of flow patterns play a more important role. An example is shown in Fig. 6.35. Figure 6.35*a* illustrates an idealized CHF

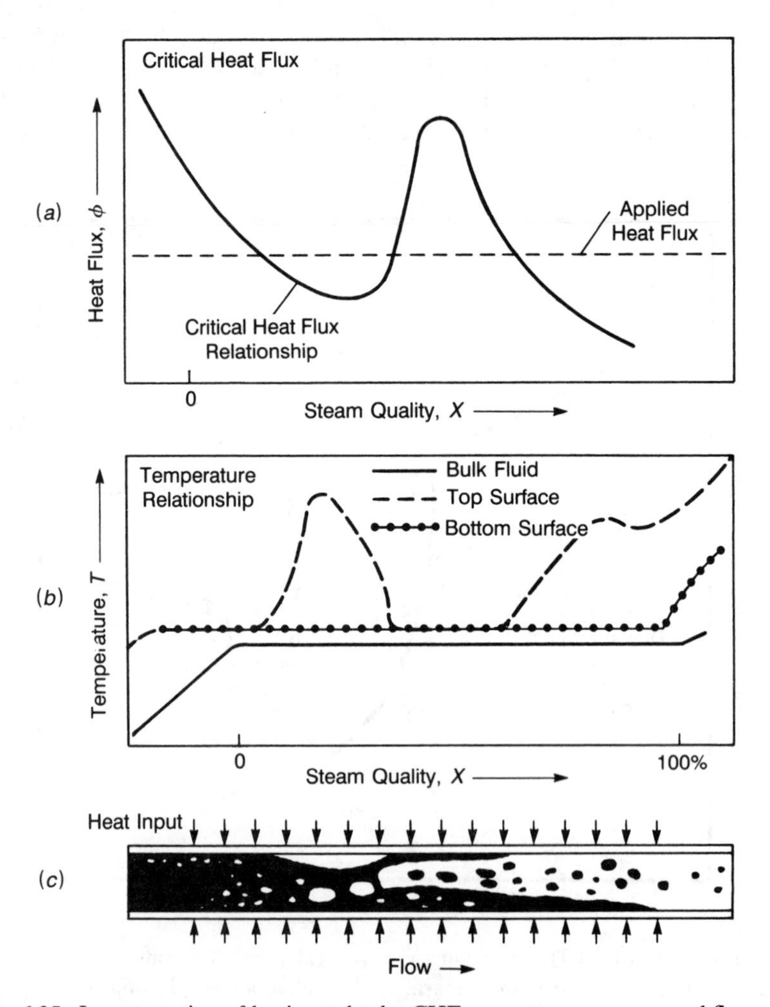

Fig. 6.35. Interpretation of horizontal tube CHF—constant pressure and flow rate.

TABLE 6.6 CHF Parametric Effects

When using the exit quality approach to evaluate CHF conditions, the following effects on the heat flux and quality have been observed. Only the indicated variable changes in each case.

Parameter Increased	Effect on Heat Flux	Figure
Pressure (above 10 MPa)	Reduced	6.36
Mass Flow (Low)	Decrease	6.37
(High)	Increase	6.37
Quality	Decrease	6.37
Angle from vertical	Decrease (G and x dependent)	6.38
Inside diameter	Decrease	—
Length	No significant effect on long tubes	—
Circumferential heat flux ($\phi_{peak}/\phi_{average}$)	Increase (local, x dependent)	6.39
Axial heat flux ($\phi_{peak}/\phi_{average}$)	Decrease (local, heat flux gradient dependent)	—
Internal rifling or multilead ribbed bore tubes	Increase	6.40

versus steam quality curve and the applied axial heat flux of the test section. Figure 6.35*b* illustrates the associated selected temperatures (bulk fluid, top inside tube wall, and bottom inside tube wall) as a function of location. Finally, Fig. 6.35c illustrates a possible flow pattern in the heated section. It should be noted that the sudden increase in CHF at intermediate steam qualities is associated with a transition in flow pattern to annular flow.

Parametric Effects For the upward-flowing water inside tubes, the key variables that affect the CHF phenomena include:

Operating conditions: pressure, mass flux, and steam quality

Physical parameters: inside diameter, inclination, heat flux profile (circumferential/axial), inlet geometry, and tube bore surface (smooth bore or rifled tubes)

The effect of increasing each of the parameters while keeping the other parameters constant [including local quality (x) at CHF] are listed in Table 6.6 with selected effects illustrated in Figs. 6.36 to 6.40. In general for high-pressure utility boilers, increasing pressure (Fig. 6.36), increasing local steam quality (Fig. 6.37), or inclining a tube (Fig. 6.38) tends to decrease the

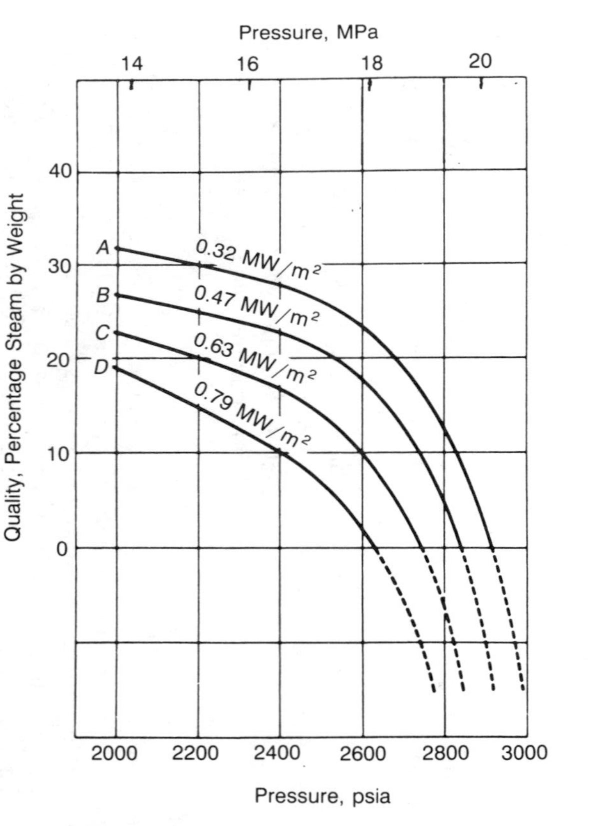

Fig. 6.36. Effect of pressure on CHF [3].

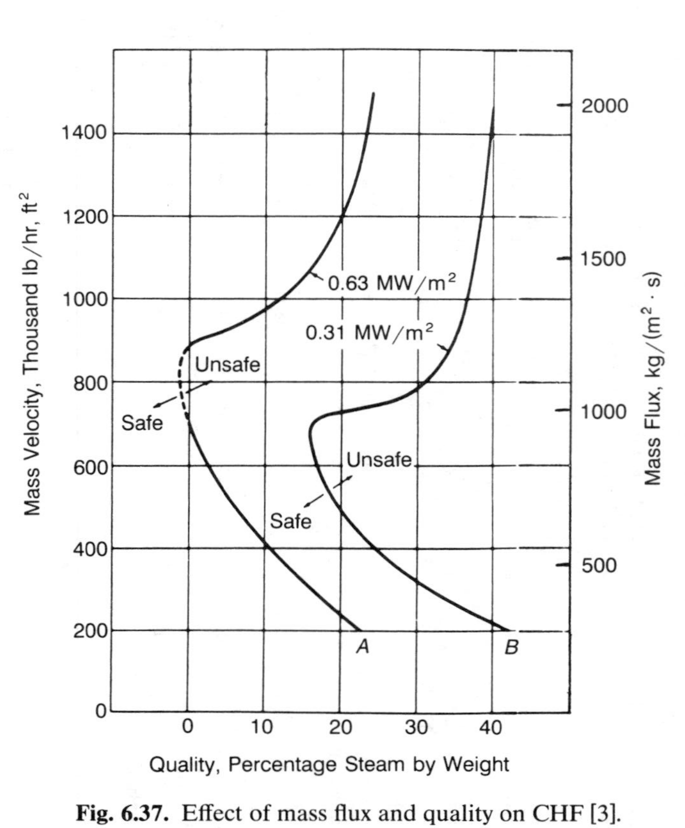

Fig. 6.37. Effect of mass flux and quality on CHF [3].

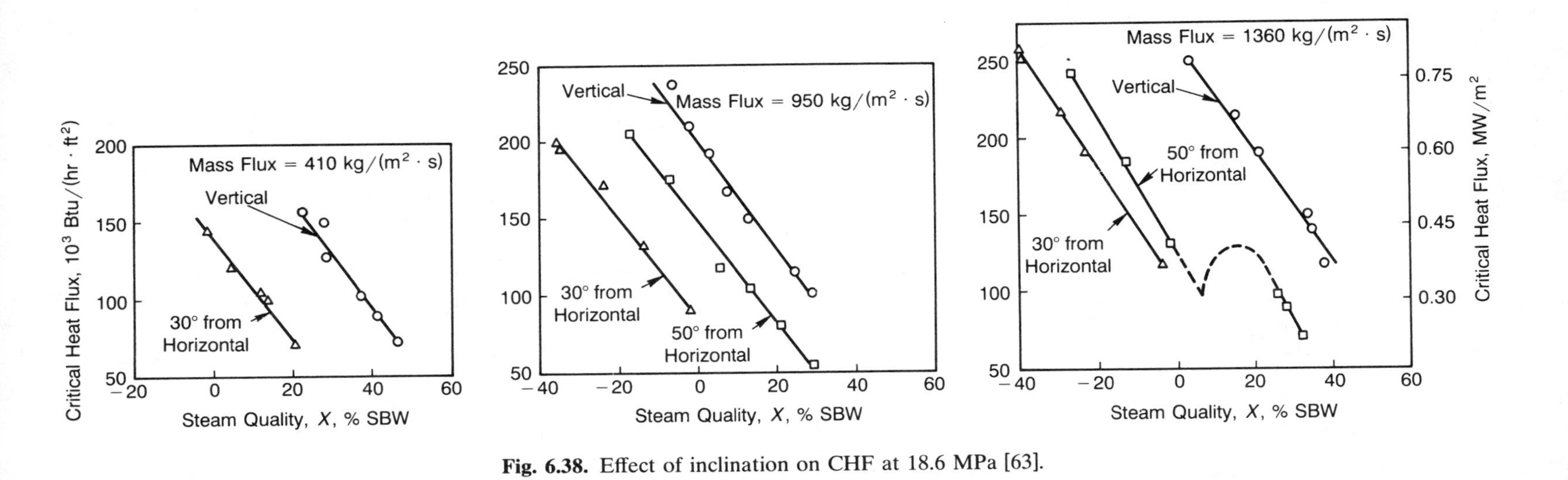

Fig. 6.38. Effect of inclination on CHF at 18.6 MPa [63].

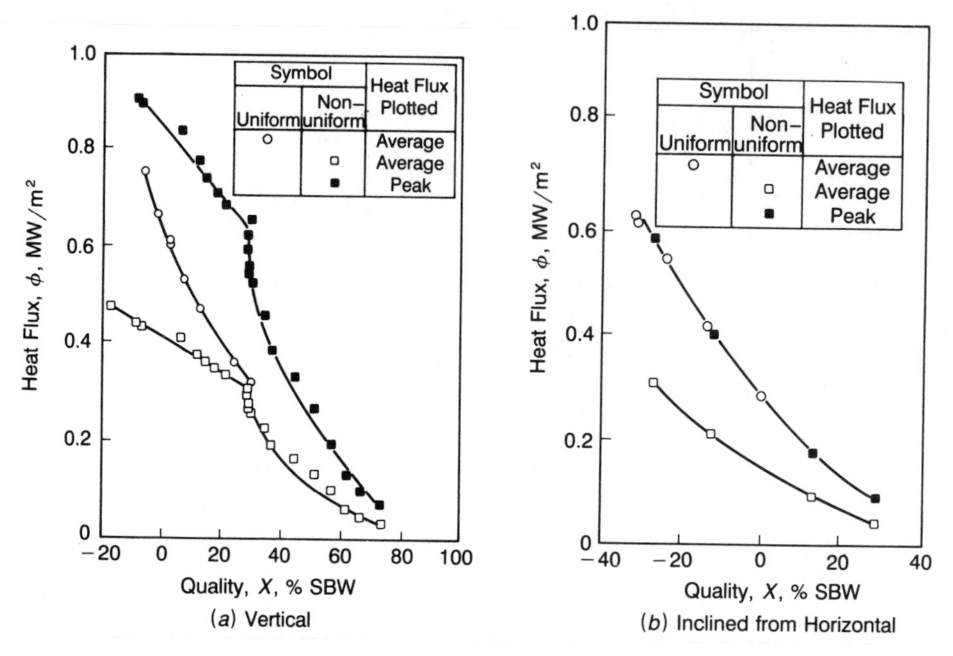

Fig. 6.39. Effect of nonuniform circumferential heating on CHF ($\phi_{peak}/\phi_{ave} \sim 2$) [64].

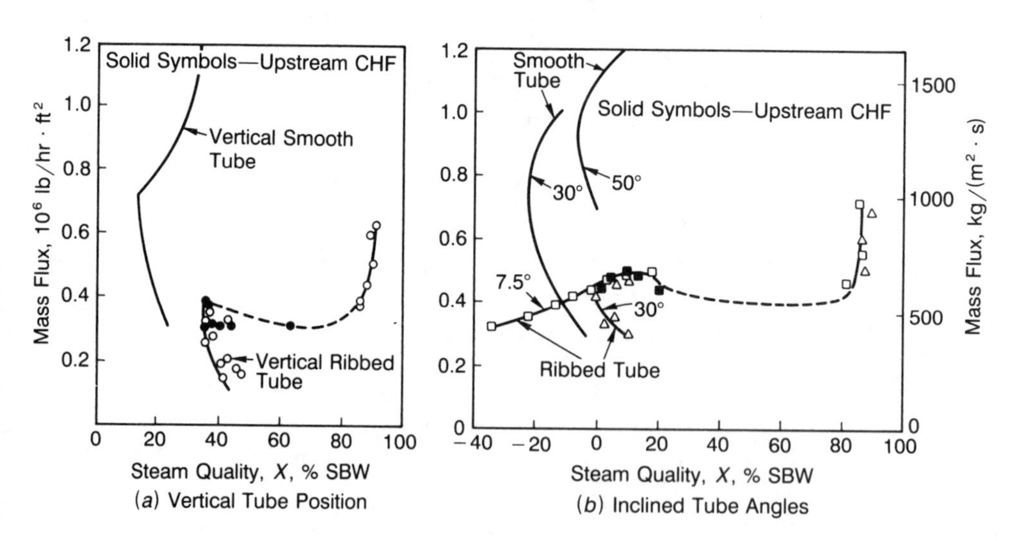

Fig. 6.40. Effect of ribbed bore geometry—18.6 MPa, 0.47 MW/m² [63].

allowable heat flux, while the effect of changing mass flux depends on the initial mass flux level.

Ribbed Tubing The effect of rifling or multilead ribs as shown in Fig. 6.40 is very important in boiler design. The multilead ribbed tube shown schematically in Fig. 6.41 permits much higher steam qualities to occur in the tube circuits for the same boiler load. This permits a reduction in allowable in-tube velocities with the associated reduction in capital and operating costs. This ribbed tube configuration was the result of tests on a large number of devices including twisters, springs, and various grooved, ribbed, and corrugated tubes to improve CHF performance. The multilead spiral rib was found to have the most satisfactory overall performance, balancing CHF improvement, increased pressure drop, and other effects. The ribs generate a swirl flow resulting in a centrifugal action which forces the water to the tube wall and retards reentrainment of the liquid. The steam blanketing and film dryout at CHF conditions are thus prevented until substantially higher steam qualities are reached, as shown in Fig. 6.40. From these figures, an additional

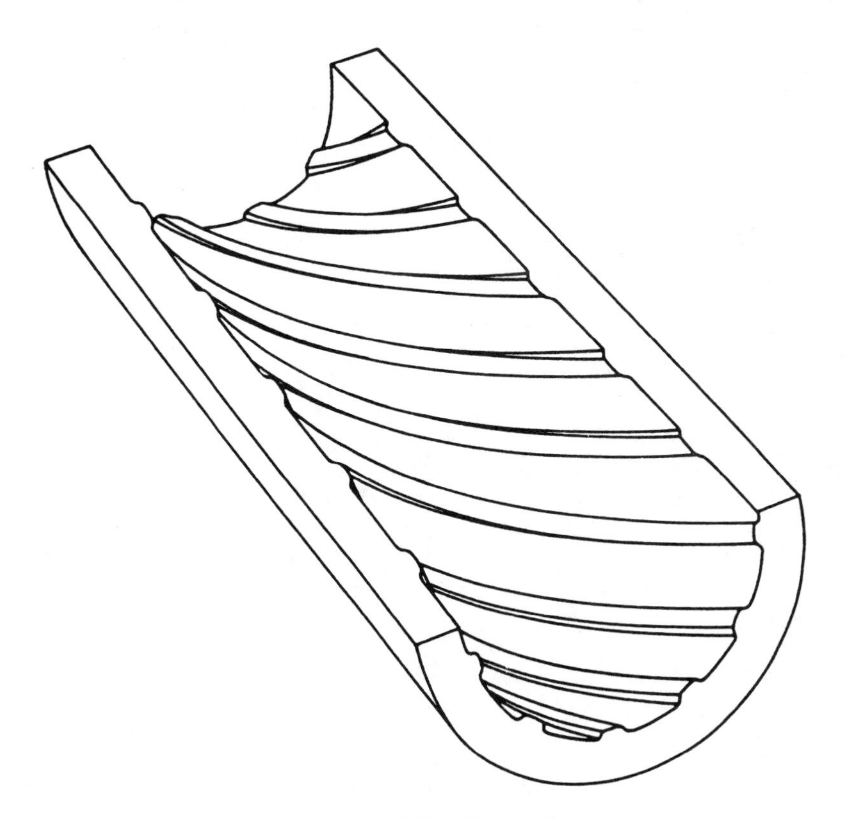

Fig. 6.41. Ribbed bore tube.

TABLE 6.7 Typical CHF Correlation Methods (Vertical Uniformly Heated Tubes)

Method	Author	Reference	Parameter Range	Accuracy
Equations—flow and heat transfer regime implicit	Bowring	71	$0.2 \leq P \leq 19.0$ MPa $2 \leq D_i \leq 45$ mm $0.15 \leq L_T \leq 3.7$ m $136 \leq G \leq 18{,}600$ kg/(m$^2 \cdot$ s) (water)	RMS 7% Base data
Equations—flow and heat transfer regime explicit	Katto	67	$0.05 \leq P \leq 20.3$ MPa $1 \leq D_i \leq 38$ mm $0.01 \leq L \leq 8.8$ m $5 \leq L/D \leq 880$ $0.0003 \leq \rho_g/\rho_f \leq 0.41$ (water included)	—
Graphical—flow and heat transfer regime explicit	Shah	68	$0.03 \leq P \leq 21$ MPa $0.0012 \leq P/P_c \leq 0.94$ $6 \leq G \leq 24{,}300$ kg/(m$^2 \cdot$ s) $-260 \leq x_{CHF} \leq 96\%$ (water included)	—
Tabular methods	Groeneveld	66	$0.1 \leq P \leq 20.0$ MPa $1 \leq D_i \leq 92$ mm $0.08 \leq L \leq 6$ m $-15 \leq x \leq 100\%$ $30 \leq G \leq 7500$ kg/(m$^2 \cdot$ s) (water)	14,401 data points 40.6% ±10% 66.5% ±20% 92.35% ± 50%
	USSR standard tables	72	$3 \leq P \leq 19.6$ MPa $4 \leq D \leq 16$ mm $750 \leq G \leq 5000$ kg/(m$^2 \cdot$ s) (water)	—

evaluation parameter is needed if the advantages of ribbed tubes are to be used—minimum mass flux to generate swirling flow.

Critical Heat Flux Evaluations A wide variety of correlations exist for the estimation of the CHF condition in smooth bore tubes with uniform heating [65–72]. These include mathematical correlations that do not explicitly differentiate between flow–heat transfer regimes, mathematical correlations based on flow–heat transfer regimes, graphical methods, and tabular methods. Five of the candidate correlations are identified in Table 6.7.

These correlations tend to fall into one of three approaches:

1. Local conditions: CHF $= f$ (local values including x); these need to be corrected for upstream conditions and nonuniform axial heating.
2. Quality boiling length: unique combination of boiling length and quality at CHF define dryout condition; averages heat input over boiling length.
3. Power: CHF $= f$ (inlet conditions and geometry).

From an evaluation standpoint, correlations must ultimately provide local parameters that identify CHF conditions, including specific location. Thus correlations such as those of Groeneveld may ultimately provide more useful information than other forms.

Groeneveld's correlation for vertical tubes takes the form:

$$\mathrm{CHF}_{\mathrm{LOCAL}} = \mathrm{CHF}(P, G, x) \cdot K_1 \cdot K_2 \cdot K_3 \cdot K_4 \cdot K_5 \cdot K_6 \qquad (6.15)$$

where CHF(P, G, x) is interpolated from the tables for $D_i = 0.008$ m provided in Appendix 6.3

$$K_1 = \text{diameter correction} = \left(\frac{0.008}{D_i}\right)^{1/3} \qquad \text{for } 0.002 \le D_i \le 0.016 \text{ m}$$

$$= 0.79 \qquad \text{for } D_i > 0.016 \text{ m}$$

$$K_2 = \text{bundle correction factor} = 1.0$$

$$K_3 = \text{grid spacer factor} = 1.0$$

K_4 = heated length factor = 1.0 for long boiler tube conditions or

$$\alpha = \frac{x}{x + (\rho_g/\rho_f)(1.0 - x)}$$

$$K_4 = \exp\left(\frac{De^{2\alpha}}{L}\right)$$

K_5 = axial flux distribution factor

$$= \frac{\phi_{\text{BLA}}}{\phi_{\text{LOCAL}}} \quad \text{(BLA indicates boiling length average) for } x > 0$$

$$= 1.0 \qquad \text{for } x < 0$$

K_6 = flow factor (upflow or downflow: 1.0 for vertical upflow)

This correlation by Groeneveld, as well as the one by Katto, cannot yet be recommended for use in evaluation. When these correlations are compared to large-diameter tubular data between 14.5 and 20.7 MPa, they tend to overpredict the experimental data by 10% to 100% in some cases. These differences probably represent the low availability of published data with diameters greater than or equal to 37 mm and an oversimplified diameter correction factor. Groeneveld also provides for linear interpolation between the wide range of 15.0 and 20.0 MPa which may introduce errors. However, it does form the basis for future refinement. In practice, boiler manufacturers depend on extensive, proprietary CHF databases and correlations covering their specific operating variable ranges.

A number of criteria are used to assess the available critical heat flux safety margins in a particular tube at different elevations. These include:

$$\text{CHF ratio} = \frac{\text{minimum heat flux at CHF conditions}}{\text{upset heat flux}}$$

$$\text{Flow ratio} = \frac{\text{minimum design mass flux}}{\text{mass flux at CHF conditions}}$$

$$\text{Quality margin} = \text{CHF quality (\% SBW)} - \text{maximum design quality (\%SBW)}$$

The CHF ratio is illustrated in Fig. 6.42 for a smooth tube (ϕ_B/ϕ_A) and a ribbed bore tube (ϕ_C/ϕ_A), and is used to assess the margin of safety in furnace designs. It indicates the increase in local heat input which can be tolerated before the onset of CHF conditions. The "upset heat flux" is the

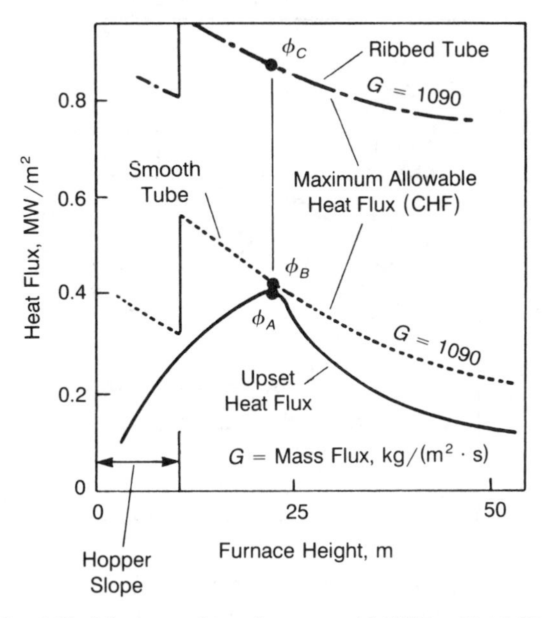

Fig. 6.42. Maximum heat flux to avoid CHF—19.3 MPa.

highest local heat flux that can be experienced at a given boiler panel wall location. The flow ratio provides a measure of the tolerable flow reduction before the onset of CHF (see Fig. 6.43). The smooth tube flow ratio is G_A/G_B; the ribbed bore tube flow ratio is G_A/G_C.

6.5.3 Steam–Water Separation and Drum Capacity

The objectives of the steam–water separation equipment and drum in recirculating boilers include:

1. Removal of moisture from the steam to prevent thermal damage to the superheater
2. Removal of steam from the water entering the downcomer to minimize any decrease in the effective thermal driving force
3. Prevention of the carry-over of solids and contaminants into the superheater and turbine where they can form damaging deposits

General Steam–Water Separation Fundamentals In low steam flow situations, natural steam separation by gravity at the steam–water interface in a drum without the use of other internals is potentially sufficient depending on the drum size, steam flow rate, and the application or end use of the steam. Possible limit curves for this arrangement based on data and analysis are provided in Fig. 6.44. These curves assume a relatively uniform distribution of inlet flow over the drum length.

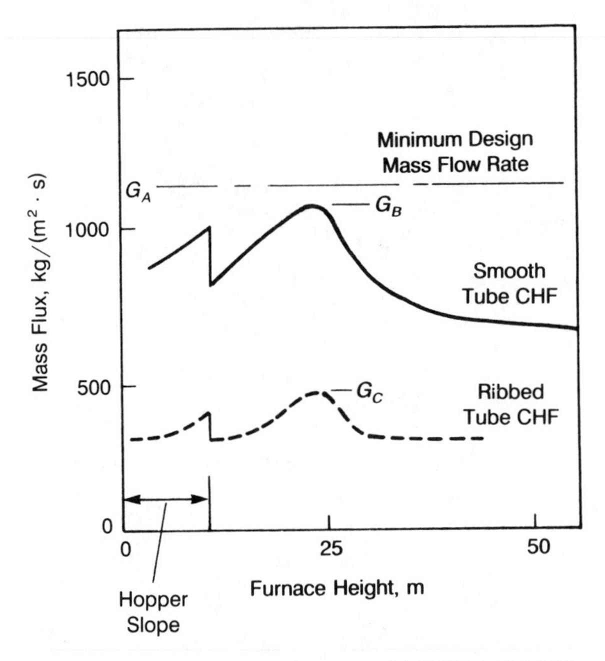

Fig. 6.43. Minimum mass flux to avoid CHF—19.3 MPa.

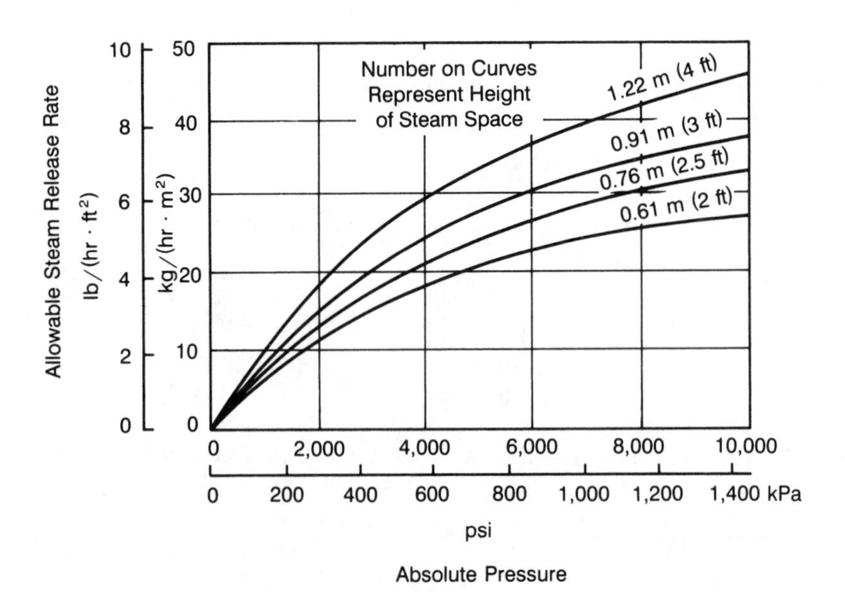

Fig. 6.44. Natural steam–water separation curves for 0.25% moisture [77].

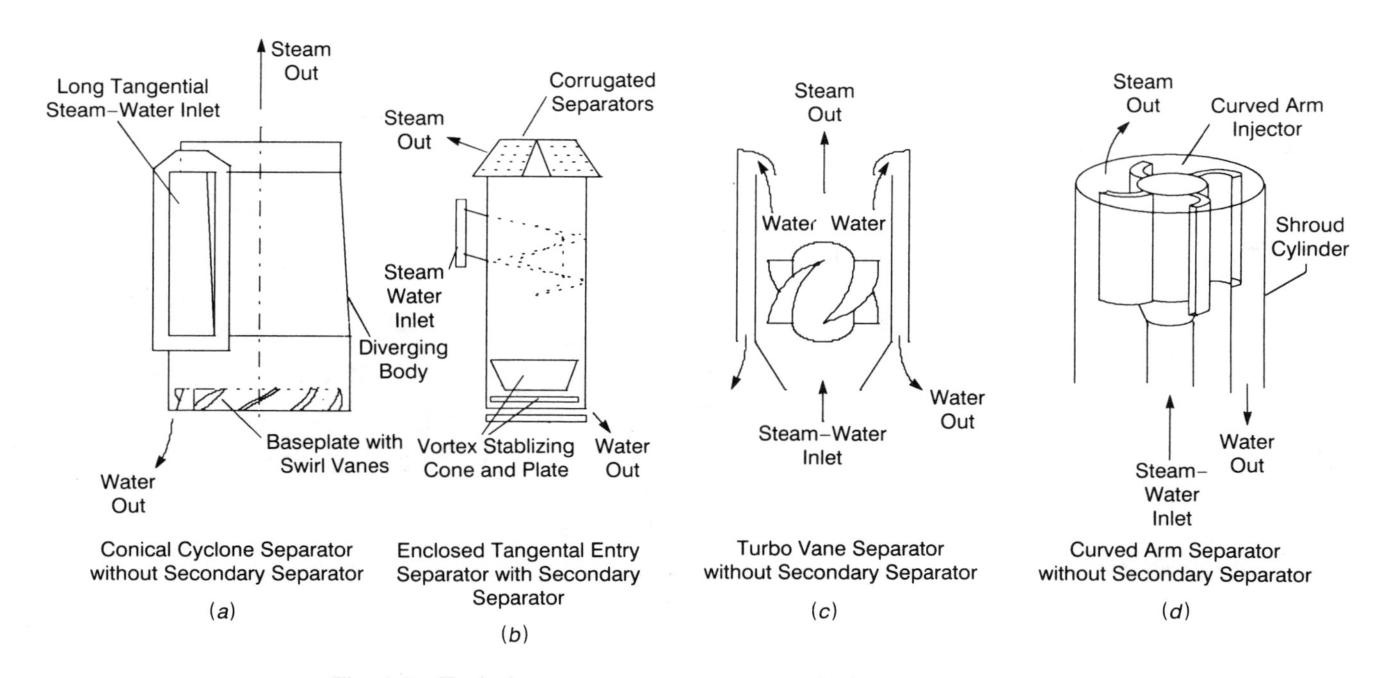

Fig. 6.45. Typical steam–water separators for high-pressure boilers.

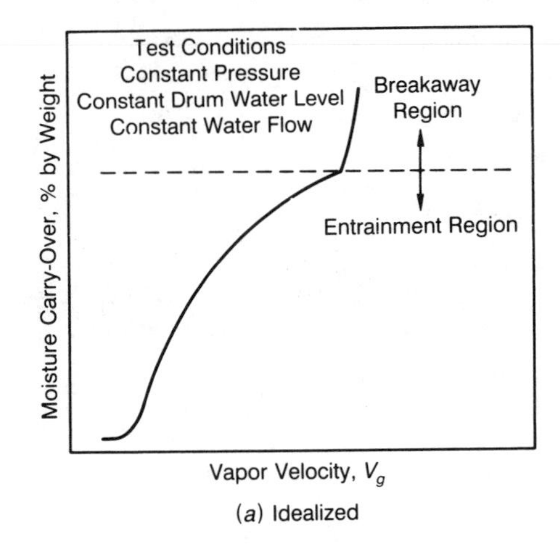

(*a*) Idealized

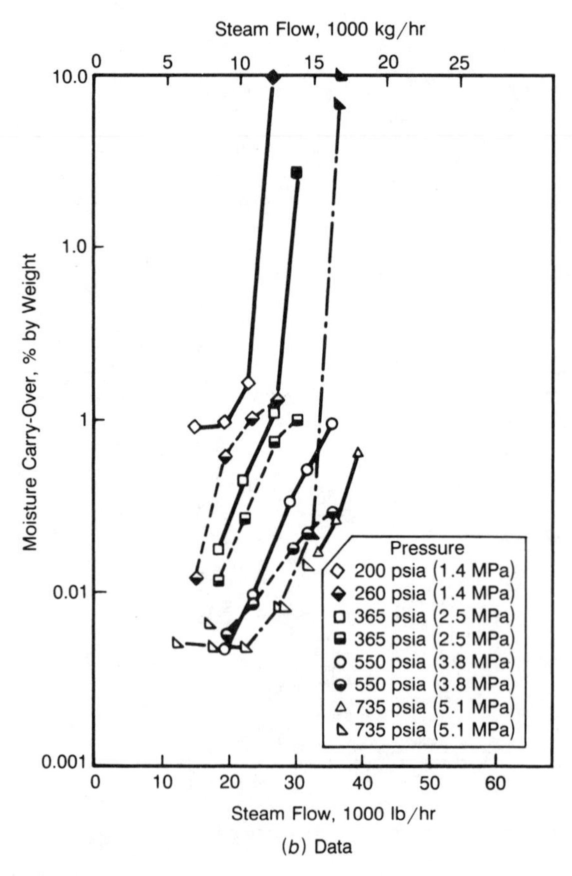

(*b*) Data

Fig. 6.46. Moisture carry-over curves for a steam–water separator [74].

In most large power boiler applications, such natural separation techniques are far too inefficient on a volumetric basis for economical use. For these applications, proprietary design steam separation equipment has been developed with steam purity limits of less than 0.25% carry-over [3, 5, 73–81]. A typical arrangement of these drum internals for a larger boiler is illustrated in Fig. 6.13. As shown, the steam–water separation equipment consists of basically two elements: (1) primary separators where the bulk of the steam and water are separated and (2) secondary separators or scrubbers, which remove the residual moisture from the steam. The primary separators usually use centrifugal force to separate the water. These will be discussed at length in a following paragraph. The secondary scrubbers use large surface-area-to-volume geometries such as closely spaced corrugated plates or wire meshes to provide a tortuous path for droplet impaction to be used to remove the last significant levels of moisture present. These scrubbers are usually limited by critical mass flow velocity, which sets an upper limit on the flow per unit cross-section area. These limits plus any pressure drop are determined experimentally for specific geometries.

Centrifugal force or radial acceleration is used almost universally for steam–water separation in large high-pressure boilers. Four different separator designs are shown in Fig. 6.45. The overall performance of this type of equipment is influenced by flow rate, pressure, length, aperture sizes, water level, inlet steam quality (mass fraction), interior finish, and drum arrangement. While performance data for these specific designs are proprietary, the general trends are shown in Figs. 6.46 to 6.48.

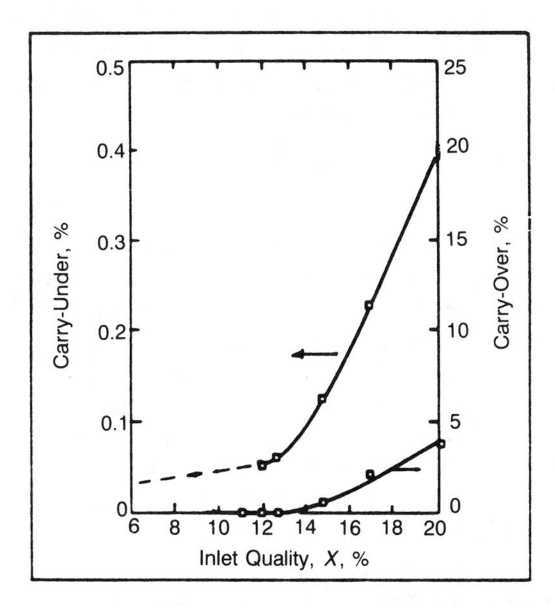

Fig. 6.47. Effect of inlet steam quality on steam–water separation in a steam separator [77].

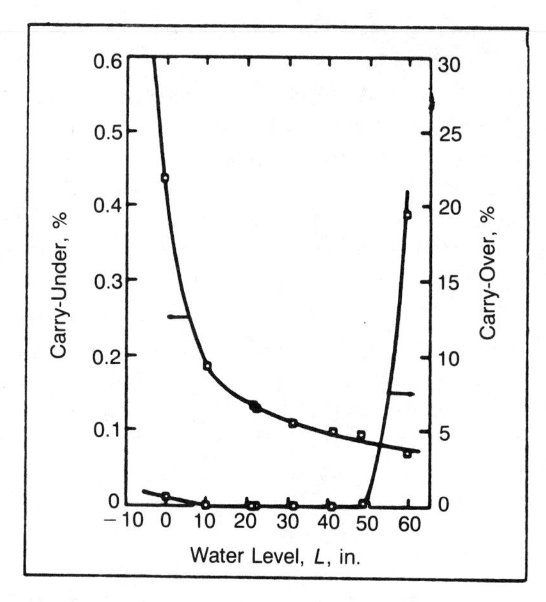

Fig. 6.48. Effect of water level on steam–water separation [77].

1. Moisture carry-over with steam: increases gradually with steam flow until a breakaway point is reached where a sudden rise in carry-over occurs; increases with water level rise until flooding occurs (a typical operating range is ± 0.15 to 0.20 m but this is dependent on the type of primary separation equipment); increases with steam quality.
2. Carry-under of steam with water: declines with increasing water level; declines with decreasing inlet steam quality.
3. Pressure drop $(P_{\text{inlet}} - P_{\text{drum}})$: increases with mass flow and steam quality.

Recently, the use of steam–water separation equipment on variable operating pressure boilers and once-through boilers has become attractive to facilitate low-load operation and start-up and/or shutdown. In place of a large horizontal drum found on recirculating boilers, one or more small-diameter vertical (or integral) steam separators are used. The centrifugal steam–water separation is designed directly into the vertical cylinder that also serves the same purposes as the drum. An example of such a vertical separator is shown in Fig. 6.49. These are typically connected directly to the top of the downcomer pipes. When not needed for steady-state or near-full-load operation, these units can be bypassed.

Steam Separator Evaluation The theoretical analyses available so far have not yet provided a completely satisfactory framework for evaluating the separator equipment performance, as would be expected for this complex

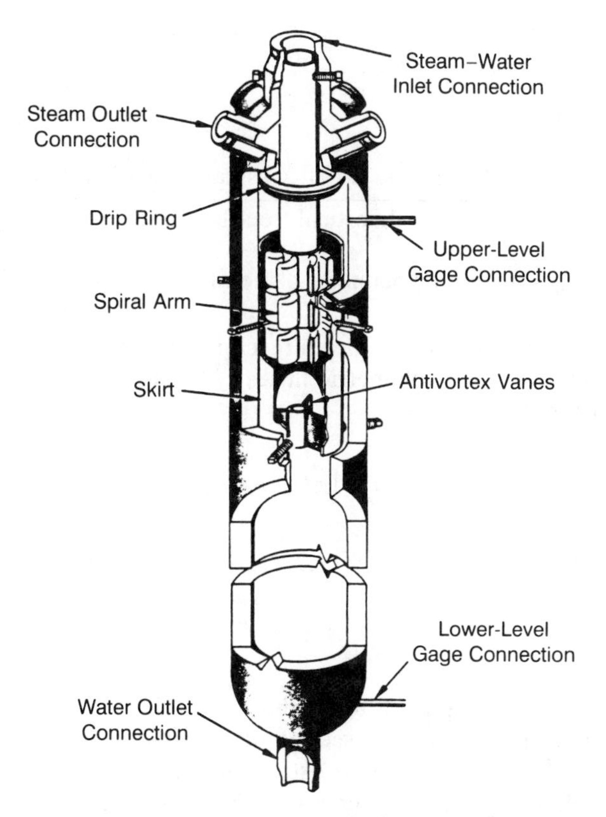

Fig. 6.49. Integral steam–water separator [81] (*Courtesy Foster Wheeler*).

situation. However, the following have been identified as the key nondimensional parameters in evaluating performance:

$$\text{Modified Froude number } (Fr_m) = \frac{\rho_m}{(\rho_f - \rho_g)} \frac{V_s^2}{gD_h} \tag{6.16}$$

$$\text{Weber number } (We) = \frac{\rho_m V_s^2 D_h}{\sigma} \tag{6.17}$$

$$\text{Density ratio } (DR) = \frac{\rho_f - \rho_g}{\rho_g} \tag{6.18}$$

Reynolds number (Re) (bubble and total flow) (6.19)

$$Re = \frac{\rho_m V_s D_h}{\mu}$$

where

$$\rho_m = (xv_{fg} + v_f)^{-1.0} \tag{6.20}$$

$$V_s = \frac{(\text{steam mass flow}) \times (\text{steam specific volume})}{\text{vertical free-flow area}} \tag{6.21}$$

$$D_h = \text{hydraulic diameter of free-flow area} \tag{6.22}$$

In addition, the breakaway moisture carry-over has been found to be a function of the modified volumetric fluxes (j_g^* and j_f^*) for steam and liquid as shown in Fig. 6.50:

$$j_g^* = j_g \rho_g^{1/2}[gD_h(\rho_f - \rho_g)]^{-1/2} \tag{6.23}$$

$$j_f^* = j_f \rho_f^{1/2}[gD_h(\rho_f - \rho_g)]^{-1/2} \tag{6.24}$$

$$C = j_g^{*1/2} + Mj_f^{*1/2} \tag{6.25}$$

where C and M are evaluated for each physical design.

As indicated in the previous section on circulation, the pressure drop of two-phase flow through a fitting is extremely complex. A first approximation to help correlate the data for separators is to use the homogeneous model

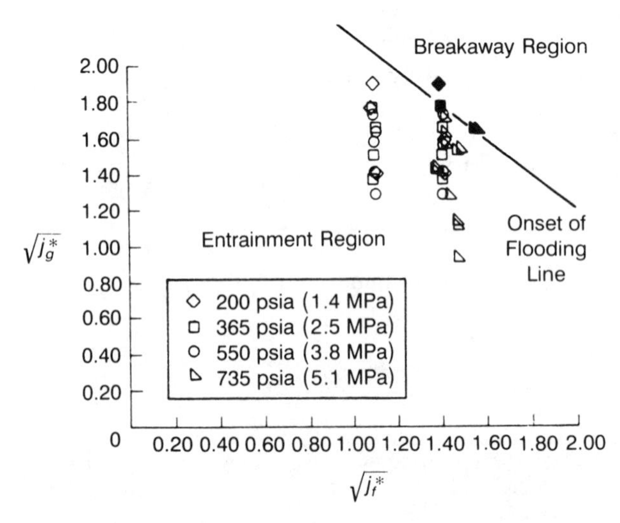

Fig. 6.50. Steam–water separator flooding curve [74].

two-phase multiplier times a loss coefficient [see Eq. (6.8)]:

$$\Delta P_{\text{separator}} = K_{ss}\Phi\frac{G^2 v_f}{2g_c} \tag{6.26}$$

where

$$\Phi = \left(1.0 + \left(\frac{v_g - v_f}{v_f}\right)x\right) \tag{6.27}$$

$$K_{ss} = f(P) \qquad \text{(unique for each physical steam separator design)} \tag{6.28}$$

In applying steam–water separators to a particular design, the objectives are: (1) to select the appropriate number of steam–water separators that will provide less than the specified steam carry-over and still obtain minimum carry-under at the total steam mass flow rate and (2) to define the pressure drop in the separator. Two functional relationships are required—one for pressure drop and one for maximum steam flow capability. Based on the foregoing discussion, these correlations might take the form:

$$\Delta P = f(K_{ss}, x, P, G) \tag{6.29}$$

$$m_s(\max) = f(Fr_m, We, DR, j_g^*, j_f^*) \tag{6.30}$$

where Fr_m is a modified Froude number, We is the Weber number, DR is the liquid-to-vapor density ratio, and j_g^* and j_f^* are the volumetric fluxes for vapor and liquid defined in Eqs. (6.23) and (6.24), respectively.

The maximum steam flow per separator [m_s(max)] defines the minimum number of standard units required and the ΔP is used in the circulation calculations. Given the highly unique nature of each separator design and the current theoretical state of the art, the performance levels should be experimentally determined preferably under full-scale, full-flow, and full-pressure conditions. The functional relationship for m_s(max) changes with different primary steam separator geometries for each manufacturer.

Drum Capability Given the flow limitations of the standardized steam–water separation equipment, the boiler drum is sized to accommodate the number of separators necessary for the largest expected boiler load (maximum steam flow rate). The drum diameter, in incremental steps, and length are adjusted to meet the space requirements at a minimum cost. A typical cross section is shown in Fig. 6.13.

An evaluation limit is the maximum steam "carry-under" into the downcomer. Carry-under or transport of steam into the downcomers is not desirable since it reduces the available thermal pumping force by reducing the density at the top of the downcomer. Carry-under performance is a function of physical arrangement (both furnace connections and steam drum internal arrangements), operating (drum) pressure, feed-water enthalpy, free-water surface area, drum water level, and separator efficiency. Empirical correction factors for specific designs are developed and used in the circulation calculations to account for the steam voidage entering the downcomers. The steam is eventually completely condensed after it travels some distance into the downcomer. However, the average density in the top portion of the downcomer is still lower than thermal equilibrium would indicate.

6.6 OTHER EVALUATION FACTORS

The range of additional evaluation factors that influence the boiler design and more specifically the steam–water system is quite extensive and a comprehensive summary goes beyond the scope of the current work. Some of these factors with particular reference to thermohydraulics include:

1. *Steam–water mixing*: Steam preheating of water and water attemperation of steam are used extensively; appropriate components need to be carefully designed.
2. *Steam–water mixture distribution*: In some instances, it is necessary to collect, mix, and then redistribute two-phase mixtures evenly to avoid problems in downstream components.
3. *Effects of deposits*: Fouling deposits occur inside all boiler tubes. They influence heat transfer to the water, provide sites for chemical concentration, and can negatively impact pressure drop and CHF conditions. Ripple deposits are especially important in supercritical pressure boilers.
4. *Water chemistry*: Virtually all boilers employ some form of internal water treatment. The chemicals added and those that arrive with the feed water can influence steam–water separation, heat transfer rate, CHF phenomena, and local corrosion.
5. *Start-up and operation*: Boilers only operate at full load part of the time. More commonly, load changes occur frequently including: hot overnight shutdowns, weekend shutdowns, periodic downtime for maintenance, emergency shutdowns because of mechanical problems, as well as regular cycling–peaking power duty. Special start-up procedures and bypass systems are needed for these kinds of services. The start-up rate of the boiler and load change rates are closely tied to material stress limits in selected boiler components such as the drum and outlet superheated steam headers. The control systems and strategies can be quite complex.
6. *Supercritical fluid heat transfer*: Heat transfer to supercritical pressure water near the critical point poses additional challenges. Rapid fluid property changes make special heat transfer rate evaluations necessary.

Additionally, although true two-phase flow does not exist, a pseudocritical film boiling phenomenon similar to the degradation of heat transfer at CHF can occur.

7. *Inclined tubes*: Steam and/or water in horizontal or inclined tubes pose a particular problem in avoiding steam–water separation and stratified flows. Such a situation could lead to tube over-heating, chemical deposition, and corrosion.

6.7 SUMMARY

The foregoing discussion has provided only the briefest framework for the initial understanding of fossil boiler thermohydraulic evaluation. Modern utility boilers and the power plants to which they supply steam represent some of the most complex machines that have been developed. Evaluation techniques and material properties are stretched to their economic limits and further advances will continue to be made to meet the challenges of increased efficiency and reliability at a lower cost. However, these utility boilers do share with the smallest utility boiler a core group of two-phase-flow fundamentals covering heat transfer and fluid flow that can be used to understand the overall evaluation process. These fundamentals have been presented here in at least their broadest terms with references provided for further study. However, full evaluation of even the two-phase thermohydraulics still requires extensive field experience and empirical data to permit safe, cost-effective boiler designs.

NOMENCLATURE

a	S_1/D_o, dimensionless
A	Surface area, m^2
b	S_2/D_o, dimensionless
B	Constant or coefficient, as noted
C	Constant or coefficient, as noted
C_o	Drift flux parameter
c_p	Specific heat at constant pressure, J/(kg · K)
D	Tube diameter, m
D_h	Free-flow area hydraulic diameter, 4(minimum free-flow area)/wetted perimeter, m
DP	Circuit composite pressure loss, Pa
DR	Density ratio, dimensionless
f	Friction factor, dimensionless

E	Parameter, as noted
F	Correction factors, as noted
Fr_m	Modified Froude number, dimensionless
g	Gravitational acceleration, m/s^2
g_c	Proportionality constant in Newton's second law of motion = 1
G	Mass flux, $kg/(m^2 \cdot s)$
h	Local heat transfer coefficient, $W/(m^2 \cdot K)$
H	Pumping head, consistent units, Pa, or parameter, as noted
i	Enthalpy per unit mass, J/kg
j	Superficial volumetric flux, Gx/ρ_f or $G(1-x)/\rho_g$, $m^3/s/m^2$
Ja	Subcooling Jakob number, $\Delta i_{sub} v_{fg}/i_{fg}/v_f$, dimensionless
k	Thermal conductivity, $W/(m \cdot K)$
K	Loss coefficient, dimensionless
K_{ss}	Separator loss coefficient, dimensionless
K_1–K_6	Groeneveld correction coefficients, dimensionless
l, L	Length, m
LMTD	Log mean temperature difference, K
M	Steam–water separator breakaway constant
m	Mass flow rate, kg/s
n	Number of tubes
Nu	Nusselt number, hD/k, dimensionless
P	Pressure, Pa
Pr	Prandtl number, $c_p\mu/k$, dimensionless
Q	Heat transfer rate, W
R	Mean bend radius, m
Re	Reynolds number, $\rho DV/\mu$, dimensionless
Re^*	Modified Reynolds number, see Eq. (6.1-11) in Appendix 6.1, dimensionless
R_f	Heat transfer fouling factor, $(m^2 \cdot K)/W$
R_p	Friedly pressure drop ratio, dimensionless
S^*	Area ratio, dimensionless
S_1	Horizontal pitch between tubes perpendicular to flow, m
S_2	Vertical pitch between tubes parallel to flow, m
T	Temperature, °C
U	Overall heat transfer coefficient, $W/(m^2 \cdot K)$
v	Specific volume, m^3/kg
V	Velocity, m/s
V_s	Separator superficial steam velocity, m/s
W_{rel}	Relative velocity, m/s
We	Weber number, $\rho_m V_s^2 D_h/\sigma$
x	Steam quality by weight (mass vapor quality)
X_F	Friedly quality parameter, dimensionless
X	Martinelli parameter (tt), $((1.0-x)/x)^{0.9}(v_f/v_g)^{0.5}(\mu_f/\mu_g)^{0.1}$
Z	Vertical elevation or height, m

Greek Symbols

α	Local cross-section void fraction
Δi_{sub}	Subcooling enthalpy, J/kg
ΔP	General pressure drop, Pa
ΔP_a	Pressure drop from fluid acceleration, Pa
ΔP_f	Pressure drop from friction, Pa
ΔP_{local} (ΔP_l)	Pressure drop from local discontinuities, Pa
ΔP_z	Pressure drop from change in elevation (hydrostatic), Pa
θ	Angle (inclination or around tube circumference)
μ	Dynamic viscosity, kg/(m · s)
ρ ($\bar{\rho}$)	Density, kg/m^3 ($\bar{\rho}$ mean value)
σ	Surface tension, N/m
ϕ	Heat flux, MW/m^2
ϕ_{LO}^2	Two-phase multiplier fluid all as a liquid, dimensionless
Φ	Two-phase multiplier for discontinuities, dimensionless
ψ	$1.0 - (\pi/4a)$, dimensionless

Subscripts

b	Bulk fluid condition
c	Convection term
d	Downcomer property
f	Fluid
furnace	Furnace value
fg	Difference between saturated liquid and saturated vapor
g	Gas
GO	All fluid as a gas or vapor
i	Inside tube
in	Inlet condition
l	Local loss component
L	Laminar or liquid
LO	All fluid as a liquid
m	Metal or homogeneous mean
o	Outside tube
out	Outlet condition
peak	Local maximum value
r	Radiation term
s	Steam separator
sat	Saturated condition
T	Turbulent
w	Wall condition

REFERENCES

1. Dickenson, H. W. (1963) *Short History of the Steam Engine*, 2nd ed. F. Cass, London.
2. Rolt, L. T. C. (1977) *The Steam Engine of Thomas Newcomen*, Science History Publications, New York.
3. *Steam: Its Generation and Use*, 39th ed. (1978) The Babcock & Wilcox Company, New York.
4. Axtman, W. H., Mosher, R. N., and Bahn, C. R. (eds.) (1988) *The American Boiler Industry: A Century of Innovation*, American Boiler Manufacturers Association, Arlington, Va.
5. Singer, J. G. (ed.) (1981) *Combustion: Fossil Power Systems*, 3rd ed. Combustion Engineering Inc., Windsor, Conn.
6. El-Wakil, H. M. (1984) *Power Plant Technology*, pp. 72–172. McGraw-Hill, New York.
7. Aschner, F. S. (1977) *Planning Fundamentals of Thermal Power Plants*. Wiley, New York.
8. Baumeister, T., Avallone, E. A., and Baumeister, T., III (eds.) (1981) *Mark's Standard Handbook for Mechanical Engineers*, 8th ed. McGraw-Hill, New York.
9. Goodall, P. M. (ed.) (1980) *The Efficient Use of Steam*. IPC Science & Technology Press, Surry, UK.
10. Elliot, T. C. (ed.) (1989) *Standard Handbook of Power Plant Engineering*. McGraw-Hill, New York.
11. Boilers and auxiliary equipment. *Power* **132**(6) B-1–B-138.
12. Rankine, W. J. M. (1908) *A Manual of the Steam Engine and Other Prime Movers*, revised by W. J. Miller. Griffon, London.
13. Schmidt, E. (1936) *Thermodynamics: Principles and Applications to Engineering*. Translation by J. Kestin, Oxford University Press, London, pp. 228–231, 1949.
14. Armor, A., et al. (eds.) (1988) *Improved Coal-Fired Power Plants*. Electric Power Research Institute (Report CS-5581-SR), Palo Alto, Calif.
15. Armor, A., et al. (eds.) (1988) *Proc. Second Int. Conf. on Improved Coal Fired Power Plants (1988)*. Electric Power Research Institute, Palo Alto, Calif.
16. Kitto, J. B., and Albrecht, M. J. (1988) Elements of two-phase flow in fossil boilers. In *Two-Phase Flow Heat Exchangers*, S. Kakaç, A. E. Bergles, and E. O. Fernandes, (eds.), pp. 495–552. Kluwer, Dordrecht.
17. Merz, J. (1988) Design considerations for fossil-fuel fired steam generators in favor of the once-through system. In *Two-Phase Flow Heat Exchangers*, S. Kakaç, A. E. Bergles, and E. O. Fernandes (eds.), pp. 595–617. Kluwer, Dordrecht.
18. Smith, V. L. (1975) Coal firing and industrial boiler design–the modern approach. ASME Paper 75-IPWR-14.
19. Barsin, J. A. (1979) Boiler design considerations. *Proc. Coal Combustion Technology Conf*. Pasadena, Calif.
20. Fiveland, W. A., and Wessel, R. A. (1988) Numerical model for predicting performance of three-dimensional pulverized-fuel-fired furnaces. *ASME J. Eng. Gas Turbines and Power* **110** 117–126.

21. Scruton, B., Gibb, J., and Chojnowski, B. (1985) Conventional power station boilers: Assessment of limiting thermal conditions for furnace-wall tubes. *CEGB Res.* pp. 3–11, April.
22. Blokh, A. G. (1988) *Heat Transfer in Steam Boiler Furnaces.* Hemisphere, Washington, D.C.
23. Hottel, H. C., and Sarofim, A. F. (1967) *Radiation Heat Transfer.* McGraw-Hill, New York.
24. So, R. M. C., Whitelaw, J. H., and Mongia, H. C. (eds.) (1986) *Calculation of Turbulent Reactive Flows.* ASME, New York.
25. Presser, A., and Lilley, D. G. (1987) *Heat Transfer in Furnaces.* ASME, New York.
26. Pai, B. R., Michelfelder, S., and Spalding, D. B. (1978) Prediction of furnace heat transfer with a three-dimensional mathematical model. *Int. J. Heat and Mass Transfer* **21** 571–580.
27. Wiener, M. (1977) The latest developments in natural circulation boiler design. *Proc. American Power Conf.* **39** 336–348.
28. Collier, J. G. (1981) Introduction to two-phase flow problems in the power industry. In *Two-Phase Flow and Heat Transfer in the Power and Process Industries*, A. E. Bergles, et al. (eds.), pp. 226–255, 573–579. Washington, D.C.
29. Chisholm, D. (1973) Research note: Void fraction during two-phase flow. *J. Mech. Eng. Sci.* **15**(3) 225–236.
30. Chisholm, D. (1980) Two-phase flow in bends. *Int. J. Multiphase Flow* **6** 363–367.
31. Chisholm, D. (1983) Gas–liquid flow in pipeline systems. In *Handbook of Fluids in Motion*, N. P. Ceremisinoff and R. Gupta (eds.), pp. 483–513. Butterworth, Boston.
32. Geiger, G. E. (1964) Sudden contraction losses in single- and two-phase flow. Ph.D. Thesis, University of Pittsburgh.
33. Idelchik, I. E. (1986) *Handbook of Hydraulic Resistance*, 2nd ed. Hemisphere, Washington, D.C.
34. Koehler, W., and Kastner, W. (1988) Two-phase pressure drop in boiler tubes. In *Two-Phase Flow Heat Exchangers*, S. Kakaç, A. E. Bergles, and E. O. Fernandes (eds.), pp. 575–593. Kluwer, Dordrecht.
35. Bouré, J. A., Bergles, A. E., and Tong, L. S. (1973) Review of two-phase instability. *Nuc. Eng. Des.* **25** 165–192.
36. Bergles, A. E. (1981) Instabilities in two-phase systems. In *Two-Phase Flow and Heat Transfer in the Power and Process Industries*, A. Bergles, et al. (eds.), pp. 383–422. Hemisphere, Washington, D.C.
37. Butterworth, D., and Hewitt, G. F. (eds.) (1978) *Two-Phase Flow and Heat Transfer*, pp. 343–393. Oxford University Press, Oxford.
38. Ledinegg, M. (1938) Instability of flow during natural and forced circulation. *Die Warme* **61**(8) 1938 (AEC-tr-1861, 1954).
39. Kakaç, S., and Veziroglu, T. N. (1985) Two-phase flow instabilities in boiling systems: Summary and review. In *Advances in Two-Phase Flow and Heat Transfer*, S. Kakaç and M. Ishii (eds.), Vol. 2, pp. 577–667. Kluwer, Dordrecht.

40. Gurgenci, H. et al. (1986) Pressure drop and density wave thresholds in boiling channels. ASME Paper 86-WA/HT-73.
41. Nakanishi, S., and Kaji, M. (1986) An approximation method for construction of a stability map of density wave oscillations. *Nuc. Eng. Des.* **95** 55–64.
42. Friedly, J. C., Akinjiola, P. O., and Robertson, J. M. (1979) Flow oscillation in boiling channels. *AIChE Symp. Ser. 189* **75** 204–217.
43. Jens, W. H., and Lottes, P. A. (1951) Analysis of heat transfer burnout, pressure drop and density data for high-pressure water. Report ANL-4627, U.S. Government.
44. Thom, J. R. S., Walker, W. M., and Fallon, T. A., and Reising, G. F. S. (1965) Boiling in subcooled water during flow up heated tubes of annuli. Paper presented at the Symposium on Boiling Heat Transfer in Steam Generating Units and Heat Exchangers, Manchester, UK, IMechE Paper 6, September.
45. Davis, E. J., and Anderson, G. H. (1966) The incipience of nucleate boiling in forced convection flow. *AIChE J.* **12**(4) 774–786.
46. Chen, J. C. (1966) Correlation for boiling heat transfer to saturated fluids in convective flow. *Ind. Eng. Chem. Proc. Des. Dev.* **5** 322–329.
47. Elelstein, S., Perex, A. J., and Chen, J. C. (1984) Analytical representation of convective boiling functions. *AIChE J.* **30** 840–841.
48. Scruton, B., and Chojnowski, B. (1982) Post dryout heat transfer for steam water flowing in vertical tubes at high pressure. *Heat Transfer 1982.* Hemisphere, Washington, D.C.
49. Groeneveld, D. C., and Delorme, G. G. J. (1976) Prediction of thermal nonequilibrium in post dryout regime. *Nucl. Eng. Des.* **36** 17–26.
50. Kimber, G. R., and Sutton, C. (1979) Comparison of post dryout heat transfer correlations with experimental data, UKAEA Report AEEW-R1266.
51. Koehler, W., and Kastner, W. (1988) Post CHF heat transfer in boiler tubes. In *Two-Phase Flow Heat Exchangers*, S. Kakaç, A. E. Bergles, and E. O. Fernandes (eds.), pp. 553–573. Kluwer, Dordrecht.
52. Hewitt, G. F. (1978) Critical heat flux in flow boiling. *Heat Transfer 1978*, Vol. 6, pp. 143–171. Hemisphere, Washington, D.C.
53. Butterworth, D., and Schock, R. A. W. (1982) Flow boiling, *Heat Transfer 1982*, Vol. 1, pp. 11–30. Hemisphere, Washington, D.C.
54. Katto, Y. (1986) Critical heat flux in boiling. *Heat Transfer 1986*, Vol. 1, pp. 171–180. Hemisphere, Washington, D.C.
55. Katto, Y. (1983) Critical heat flux in forced convective flow. *Proc. ASME–JSME Thermal Engineering Conf.*, Vol. 3, pp. 1–10. ASME, New York.
56. Tong, L. S. (1972) *Boiling Crisis and Critical Heat Flux.* U.S. Atomic Energy Commission (TID-25887), Washington, D.C.
57. Collier, J. G. (1980) *Convective Boiling and Condensation*, 2nd ed., pp. 236–300. McGraw-Hill, London.
58. Tong, L. S., and Hewitt, G. F. (1972) Overall viewpoint of flow boiling CHF mechanism. ASME Paper 72-HT-54.
59. Bergles, A. E. (1979) Burnout in boiling heat transfer. II: subcooled and low quality forced convection system. *Nuclear Safety* **20** 671–689.

60. Bergles, A. E. (1979) Burnout in boiling heat transfer. III: High quality systems. *Nuclear Safety* **20** 671–689.

61. Govan, A. H. (1987) Modeling of dryout in vertical upflow. UKAEA Report AERE-R12590, April.

62. Kitto, J. B. (1980) Critical heat flux and the limiting quality phenomenon. *Heat Transfer–Orlando 1980*, AIChE Symp. Ser. 199, pp. 57–78.

63. Watson, G. B., Lee, R. A., and Wiener, M. (1974) Critical heat flux in inclined and vertical smooth and ribbed tubes. *Proc. Fifth Int. Heat Transfer Conf.*, Vol. 4, pp. 275–279.

64. Kitto, J. B., and Wiener, M. (1982) Effects of nonuniform circumferential heating and inclination on critical heat flux in smooth and ribbed bore tubes. *Heat Transfer 1982*, Vol. 4, pp. 297–302. Hemisphere, Washington, D.C.

65. Groeneveld, D. C., and Rousseau, J. C. (1983) CHF and post-CHF heat transfer: An assessment of prediction methods and recommendations for reactor safety codes. In *The Advances in Two-Phase Flow and Heat Transfer*, S. Kakaç and M. Ishii (eds.). Martinus Nijhoff, The Hague.

66. Groeneveld, D. C., Cheng, S. C., and Doan, T. (1986) 1986 AECL-UO critical heat flux lookup table. *Heat Transfer Engineering* **7**(1 and 2) 46–62.

67. Katto, Y., and Ohno, H. (1984) An improved version of the generalized correlation of critical heat flux for the forced convective boiling in uniformly heated vertical tubes. *Int. J. Heat Mass Transfer* **27**(9) 1641–1648.

68. Shah, M. M. (1979) A generalized graphical method for predicting CHF in uniformly heated vertical tubes. *Int. J. Heat Mass Transfer* **22** 557–568.

69. Belyakov, I. I., Smirnov, S. I., and Romanov, D. F. (1983) Investigation of deterioration in the heat transfer in uniformly heated large diameter tubes during vertical motion of the heat transfer medium. *Energomashinostroenie* **3** 10–13.

70. Chojnowski, B., and Wilson, P. M. (1974) Critical heat flux for large diameter steam generator tubes with circumferential variable and uniform heating. *Proc. Fifth Int. Heat Transfer Conf.*, Vol. 4, pp. 260–262.

71. Bowring, R. W. (1972) A simple but accurate round tube, uniform heat flux dryout correlation over pressure range 0.7–17.0 MN/m^2 (100–2500 psia). UKAEA Report AEEW-R-789.

72. Tabular data for calculating burnout when boiling water in uniformly heated round tubes (1976). *Teploenergetica*, **23**(9) 90–92. Translation in *Thermal Engineering*, pp. 77–79, September 1977.

73. Carsen, W. R., and Williams, H. K. (1980) Method of reducing carryover and reducing pressure drop through steam generators. Electric Power Research Institute (Report NP-1607), Palo Alto, Calif.

74. Carter, H. R., and Prueter, W. P. (1980) Evaluation and correlation of the effects of operating conditions on the moisture carryover performance of centrifugal steam water separators. *Proc. Symp. Polyphase Flow and Transport Technology*. ASME, New York.

75. Eaton, A. M., Prueter, W. P., and Wall, J. R. (1985) A study of geometric scaling of curved-arm primary steam–water separators. *Heat Transfer—Denver 1985*. AIChE Symp. Ser. 245.

76. Millington, B. C. (1981) A background to cyclonic separation of steam from water in the power generation industry. Report No. ME/81/18, Dept. of Mech. Eng., Univ. of Southampton, Southampton (UK).
77. Coulter, E. (1989) Moisture separation and steam washing. In *Water Technology for Thermal Power Systems*, P. Cohen (ed.), pp. 10-1–10-43. ASME, New York.
78. Gardner, G. C., Crow, I. G., and Neller, P. H. (1973) Carry-under performance of drums in high-pressure circulation boilers. *Proc. Inst. Mechanical Engineers* **187**(14) 207–214.
79. Thomas, R. M. (1980) Rules for modelling the steady-state carry-under performance of boiler drums using Freon-12. *ASME Special Publication* HTD Vol. 14, pp. 27–36. ASME, New York.
80. Chen, X., and He, X. (1986) An experimental study on the separator on subcritical low circulation rate boiler. In *Two-Phase Flow and Heat Transfer: China–U.S. Progress*, X. Chen and T. N. Veziroglu (eds.), pp. 453–467. Hemisphere, Washington, D.C.
81. Gorzegno, W. P. et al. (1988) Supercritical once-through unit design trends. In *Improved Coal Fired Power Plants*, A. Armor, et al. (eds.), pp. 2-95–2.128. Electric Power Research Institute (Report CS-5581-SR), Palo Alto, Calif.

APPENDIX 6.1: KEY HEAT TRANSFER PARAMETERS—SUPERHEATER, REHEATER, AND ECONOMIZER

Overall Heat Transfer Coefficient (see also Chapters 2 and 3)

$$\frac{1.0}{U_o} = R_{fo} + \frac{1.0}{h_o} + \frac{D_o \ln(D_o/D_i)}{2k_m} + \frac{D_o}{D_i}\frac{1.0}{h_i} + R_{fi}\frac{D_o}{D_i} \quad (6.1\text{-}1)$$

Outside Heat Transfer Coefficient (In-Line—Smooth Tube: Gnielinski Formulation) [1]

$$h_o = h_c + h_r \quad (6.1\text{-}2)$$

$$h_c = \left(\frac{2}{\pi}\frac{k_f}{D_o}\right) F_A F_d \left(0.3 + \sqrt{Nu_L^2 + Nu_T^2}\right) \quad (6.1\text{-}3)$$

$$h_r = \text{intertube radiation conductance [2]} \quad (6.1\text{-}4)$$

$$F_d = (\text{see Fig. 6.1-1}) \quad (6.1\text{-}5)$$

$$F_A = 1.0 + \left[\left(\frac{0.7}{\psi^{1.5}}\right)\left(\frac{b/a - 0.3}{(b/a + 0.7)^2}\right)\right] \quad (6.1\text{-}6)$$

$$Nu_L = 0.664\sqrt{Re^*}\, Pr^{1/3} \quad (6.1\text{-}7)$$

$$Nu_T = \frac{0.037Re^{*0.8}Pr}{1.0 + 2.443Re^{*-0.1}(Pr^{2/3} - 1.0)} \tag{6.1-8}$$

$$a = \frac{S_1}{D_o} \quad \text{(where } S_1 = \text{horizontal pitch perpendicular to flow)} \tag{6.1-9}$$

$$b = \frac{S_2}{D_o} \quad \text{(where } S_2 = \text{pitch in direction of flow)} \tag{6.1-10}$$

$$Re^* = \left(\frac{G}{\psi\mu}\right)\left(\frac{\pi D_o}{2}\right)$$

$$\text{(where } G = \text{mass flux in empty convection pass cross section)} \tag{6.1-11}$$

$$\psi = 1.0 - \frac{\pi}{4a} \tag{6.1-12}$$

$$Pr = \left(\frac{c_p\mu}{k}\right) \tag{6.1-13}$$

All properties are evaluated at the average bulk fluid temperature:

$$T_b = \frac{T_{\text{in}} + T_{\text{out}}}{2}$$

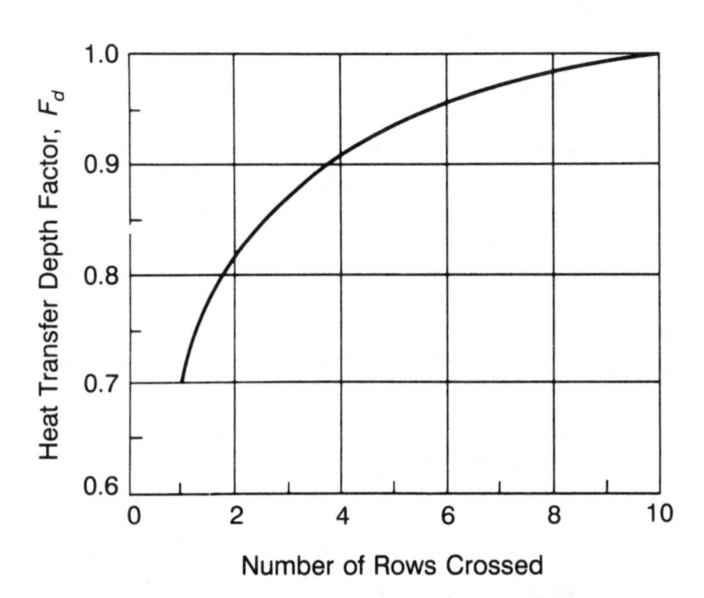

Fig. 6.1-1. Tube bundle convection heat transfer correction factor [2].

Inside Heat Transfer Coefficient for Turbulent Flow (Heating only—Smooth Straight Tube: (Petukhov–Kirillov Correlation [3]):

$$h_i = \left(\frac{k_f}{D_i}\right) \frac{(f/8)RePr}{1.07 + 12.7(f/8)^{1/2}(Pr^{2/3} - 1.0)} \left(\frac{\mu_b}{\mu_w}\right)^{0.11} \tag{6.1-14}$$

$$f = (1.82 \log_{10} Re - 1.64)^{-2} \tag{6.1-15}$$

$$Re = \frac{GD_i}{\mu_f} \tag{6.1-16}$$

$$Pr = \left(\frac{c_p \mu}{k}\right)_f \tag{6.1-17}$$

All properties are evaluated at T_b except for μ_b at the bulk fluid temperature and μ_w at the wall temperature. See also Chapter 3 for additional correlations for forced convection through ducts, and the conditions under which they are applicable.

Fouling Factors—R_{fo} and R_{fi}: Experience Dependent (see Chapter 4)

Total Heat Transfer (see Chapter 2)

$$Q = U_o \cdot A_o \cdot \text{LMTD} \cdot F_{CF} \tag{6.1-18}$$

$$\text{LMTD} = \text{log mean temperature difference} \tag{6.1-19}$$

$$F_{CF} = \text{the LMTD crossflow correction factor for most serpentine bundle arrangements is approximately equal to 1.0} \tag{6.1-20}$$

NOTE: additional nomenclature is defined in the main text.

REFERENCES

1. Schlünder, E. U. (ed.) (1985) *Heat Exchanger Design Handbook*, Vols. 2 and 3 (2.5.3). Hemisphere, Washington, D.C.
2. *Steam*: *Its Generation and Use*, 39th ed. The Babcock & Wilcox Company, New York.
3. Petukhov, B. S. (1970) Heat Transfer and friction in turbulent pipe flow with variable physical properties. In *Advances in Heat Transfer*, J. P. Hartnett and T. F. Irvine (eds.), pp. 504–564. Academic, New York.

APPENDIX 6.2: SAMPLE CORRELATIONS FOR TWO-PHASE MULTIPLIERS AND VOID FRACTION IN STEAM–WATER FLOWS

Pressure Loss Coefficients — Two-Phase — Fittings (Φ) (includes reversible and irreversible losses)

The loss coefficient for two-phase flow in fittings has not been well established [1–5]. Three cases are of interest: loss in a bend, loss in a sudden expansion (into an outlet header), and loss in a sudden contraction (out of a header). For the case of bends, the following correlation has been formulated for 90° bends:

$$\Phi_B = 1.0 + \left(\frac{\rho_f}{\rho_g} - 1.0\right)\left(Bx(1.0 - x) + x^2\right) \tag{6.2-1}$$

$$B = 1.0 + \frac{2.2}{K_{\text{LO}}\left(2 + \dfrac{R}{D_i}\right)} \qquad \text{(Chisholm parameter [1])} \tag{6.2-2}$$

$$K = K_{\text{LO}} \tag{6.2-3}$$

where R equals the radius of the bend, D_i equals the inside tube diameter, and K_{LO} is the loss coefficient for the entire flow as a liquid (see [5]).

For sudden expansions, an approach that has been suggested for the two-phase multiplier and loss coefficient includes:

$$\Phi = \left(1.0 + \frac{C}{X} + \frac{1.0}{X^2}\right) \tag{6.2-4}$$

$$K = 2S^*(1.0 - S^*)(1.0 - x)^2 \tag{6.2-5}$$

where

$$S^* = \frac{\text{small area}}{\text{large area}} \tag{6.2-6}$$

$$C = \left[1.0 - 0.5\left(\frac{v_g - v_f}{v_g}\right)^{1/2}\right]\left[\left(\frac{v_g}{v_f}\right)^{1/2} + \left(\frac{v_f}{v_g}\right)^{1/2}\right] \tag{6.2-7}$$

$$X = \left(\frac{1.0 - x}{x}\right)^{0.9}\left(\frac{v_f}{v_g}\right)^{0.5}\left(\frac{\mu_f}{\mu_g}\right)^{0.1} \tag{6.2-8}$$

For sudden contractions, a possible approach is the application of the homogeneous multiplier:

$$\Phi = \left(1.0 + \frac{(v_g - v_f)}{v_f} x\right) \tag{6.2-9}$$

$$K = (1.0 + \text{single-phase loss coefficient}) \tag{6.2-10}$$

Actual values of the entrance and exit loss coefficients depend on specific tube-to-header geometries. Accurate coefficient values have to be obtained from experiments.

Pressure Loss Coefficients and Void Fraction for Smooth Tubes (pre-CHF)

Given the uncertainties and caution stated in the main text, the following correlations [6–8] for void fraction and two-phase multipliers are presented for preliminary consideration. For void fraction, either a homogeneous void fraction or a drift flux model is typically used. The homogeneous void fraction (used in the acceleration pressure drop term [7], for example) is:

$$\alpha = \frac{v_g x}{v_f(1.0 - x) + v_g x} \tag{6.2-11}$$

For cases where phase slip needs to be accounted for (hydrostatic head, for example), the equation of Rouhani (taken from [7]: version of Zuber–Findlay drift flux model [9]) is sometimes used for void fraction, although it does not adequately address some mass flow effects:

$$\alpha = \frac{x/\rho_g}{C_o\left(x/\rho_g + (1.0 - x)/\rho_f\right) + W_{\text{rel}}/G} \tag{6.2-12}$$

$$C_o = 1.0 + 0.12(1.0 - x) \tag{6.2-13}$$

$$W_{\text{rel}} = 1.18\left(g\sigma(\rho_f - \rho_g)\right)^{1/4}(1.0 - x)/\rho_f^{1/2}$$

$$= \text{relative velocity} \tag{6.2-14}$$

Other correlations for void fraction are listed in [6, 8].

As with void fraction correlations, no two-phase multiplier correlation provides consistent and accurate predictions over a wide range of parameters. One empirical correlation frequently cited for vertical and inclined boiler tubes is that of Friedel [7, 8]:

$$\phi_{LO}^2 = E + \frac{3.23FH}{Fr^{*0.045}We^{*0.035}} \tag{6.2-15}$$

$$E = (1.0 - x)^2 + x^2\left(\frac{\rho_f f_{GO}}{\rho_g f_{LO}}\right) \tag{6.2-16}$$

$$F = x^{0.78}(1.0 - x)^{0.24} \tag{6.2-17}$$

$$H = \left(\frac{\rho_f}{\rho_g}\right)^{0.91}\left(\frac{\mu_g}{\mu_f}\right)^{0.19}\left(1.0 - \frac{\mu_g}{\mu_f}\right)^{0.7} \quad \text{(property parameter)} \tag{6.2-18}$$

$$Fr^* = \frac{G^2}{gD\rho_H^2} \tag{6.2-19}$$

$$We^* = \frac{G^2 D}{\rho_H \sigma} \tag{6.2-20}$$

$$\rho_H = \left(xv_g + (1.0 - x)v_f\right)^{-1.0} \tag{6.2-21}$$

where f_{GO} and f_{LO} are the friction factors for all of the fluid as a gas and as a liquid, respectively. As with the void fraction, *no* two-phase multiplier correlation can be selected for all applications. Significant deviations can occur without a database covering the general operating conditions of interest.

REFERENCES

1. Chisholm, D. (1973) Research note: Void fraction during two-phase flow. *J. Mech. Eng. Sci.* **15**(3) 225–236.
2. Chisholm, D. (1980) Two-phase flow in bends. *Int. J. Multiphase Flow* **6** 363–367.
3. Chisholm, D. (1983) Gas–Liquid flow in pipeline systems. In *Handbook of Fluids in Motion*, N. P. Ceremisinoff and R. Gupta (eds.), pp. 483–513. Butterworth, Boston.
4. Geiger, G. E. (1964) Sudden contraction losses in single- and two-phase flow. Ph.D. Thesis, University of Pittsburgh.

5. Idelchik, I. E. (1986) *Handbook of Hydraulic Resistance*, 2nd ed. Hemisphere, Washington, D.C.
6. Collier, J. G. (1981) Introduction to two-phase flow problems in the power industry. In *Two-Phase Flow and Heat Transfer in the Power and Process Industries*, A. E. Bergles, et al. (eds.), pp. 226–255, 573–579. Hemisphere, Washington, D.C.
7. Koehler, W., and Kastner, W. (1988) Two-phase pressure drop in boiler tubes. In *Two-Phase Flow Heat Exchangers*, S. Kakaç, A. E. Bergles, and E. O. Fernandes, (eds.), pp. 575–593. Kluwer, Dordrecht.
8. Schlűnder, E. U. (ed.) (1985) *Heat Exchanger Design Handbook*, Vols. 2 and 3. Hemisphere, Washington, D.C.
9. Zuber, N., and Findlay, J. A. (1965) Average volumetric concentration in two-phase flows. *J. Heat Transfer* **87** 453–468.

APPENDIX 6.3: SAMPLE CRITICAL HEAT FLUX (CHF) CORRELATION

As indicted earlier in the main text of this chapter, *none* of the existing publicly available critical heat flux (CHF) correlations can be recommended for general use at this time. For large-diameter tubes and pressures over 14.5 MPa, even the best correlations can *overpredict* the actual CHF data by 10% to 100%. To address this complex situation, equipment manufacturers have developed proprietary databases and correlations for specific conditions.

A variety of correlations have been proposed for CHF in smooth tubes. Some of the better known correlations are listed in Table 6.3-1. For illustrative purposes only, the method proposed by Groeneveld [1–4] is supplied here. As was indicated in Section 6.5.2, Groeneveld's correlation for CHF takes the form of a lookup table (included here as Table 6.3-2). This table identifies CHF values as a function of pressure, mass flux, and local steam

TABLE 6.3-1 Selected CHF Correlations

Number	Reference	Author	Year	Type
1	1–4	Groeneveld	1988	Tabular methods with corrections
2	5	Katto	1984	Multiple explicit equations—heat transfer and flow regime dependent
3	6	Shah	1979	Graphical—heat transfer and flow regime dependent
4	7	Bowring	1972	Explicit equations
5	8	USSR standard method	1976	Tabular methods with corrections
6	9	Doroshchuk	1975	Single equation—less reliable at high qualities and large tube diameters

quality. These values apply to a single specific geometry:

- Long vertical smooth bore tube
- 0.008 m inside diameter
- Uniform circumferential and axial heating
- Upward co-current flow

For conditions not explicitly listed in the look-up table, interpolation is recommended by Groeneveld to obtain the heat flux for CHF conditions [CHF(P, G, x)]. This base value can be adjusted for other geometry and heating situations by multiplying it by a series of correlation factors K_1 through K_6:

$$\mathrm{CHF_{LOCAL}} = \mathrm{CHF}(P, G, x) \cdot K_1 \cdot K_2 \cdot K_3 \cdot K_4 \cdot K_5 \cdot K_6 \quad (6.3\text{-}1)$$

For tubular sections with steam–water flows, Eq. (6.15) in Section 6.5.2 provides the correlation factor equations.

REFERENCES

1. Groeneveld, D. C., and Rousseau, J. C. (1983) CHF and post-CHF heat transfer: An assessment of prediction methods and recommendations for reactor safety codes. In *The Advances in Two-Phase Flow and Heat Transfer*, S. Kakaç and M. Ishii (eds.). Martinus Nijhoff, The Hague.
2. Groeneveld, D. C., Cheng, S. C., and Doan, T. (1986) 1986 AECL-UO critical heat flux lookup table. *Heat Transfer Eng.* **7**(1 and 2) 46–62.
3. Groeneveld, D. C. (1982) A general CHF prediction method for water suitable for reactor accident analysis. Centre d'Etudes Nucléaires de Grenoble, Report DRE/STT/SETRE/82-2-E/DGR, Grenoble.
4. Groeneveld, D. C., and Smoek, C. W. (1986) A comprehensive examination of heat transfer correlations suitable for reactor safety analysis. In *Multiphase Science and Technology*, pp. 181–274. Hemisphere, Washington, D.C.
5. Katto, Y., and Ohno, H. (1984) An improved version of the generalized correlation of critical heat flux for the forced convective boiling in uniformly heated vertical tubes. *Int. J. Heat Mass Transfer* **27**(9) 1641–1648.
6. Shah, M. M. (1979) A generalized graphical method for predicting CHF in uniformly heated vertical tubes. *Int. J. Heat Mass Transfer* **22** 557–568.
7. Bowring, R. W. (1972) A simple but accurate round tube, uniform heat flux dryout correlation over pressure range 0.7–17.0 MN/m^2 (100–2500 psia). UKAEA Report AEEW-R-789.
8. Tabular data for calculating burnout when boiling water in uniformly heated round tubes. *Teploenergetica* **23**(9) 90–92. Translation in *Thermal Eng.* 77–79, September 1977.
9. Doroshchuk, V. E., Levitan, L. L., and Lantsman, F. P. (1975) Recommendations for calculating burnout in a round tube with uniform heat release. *Thermal Eng.* **22**(12) 77–80.

TABLE 6.3-2 Groeneveld CHF Look-Up Table (kW/m²) (Appendix 6.3 References 1 to 4)

Pressure kPa	Mass Flux G kg/m² · s	−0.50	−0.40	−0.30	−0.20	−0.15	−0.10	−0.05	0.00
100	0					3,930	3,430	2,045	1,200
100	50					4,008	3,500	2,400	1,700
100	100					4,800	4,200	3,100	2,200
100	200					5,600	4,900	3,800	2,600
100	300					6,400	5,600	4,500	3,000
100	500					7,200	6,300	5,200	3,600
100	750					8,000	7,000	5,900	5,500
100	1,000					8,800	7,600	6,600	4,800
100	1,500					9,600	8,200	7,000	5,500
100	2,000					10,000	8,800	7,700	6,000
100	3,000					10,500	9,400	8,400	6,500
100	4,000					11,000	10,000	9,000	6,900
100	5,000					11,500	10,600	9,500	8,000
100	7,500					12,000	11,500	11,000	10,500
150	0				6,273	4,000	3,500	2,400	1,350
150	50				6,344	4,100	3,600	2,700	1,900
150	100				6,397	4,800	4,200	3,200	2,400
150	200				6,488	5,400	4,800	3,500	2,700
150	300				6,571	6,000	5,400	4,000	3,000
150	500				6,721	6,600	6,000	4,500	3,700
150	750				7,400	7,200	6,600	5,000	4,300
150	1,000				8,000	7,800	7,200	5,500	5,200
150	1,500				8,600	8,400	7,600	6,000	5,500
150	2,000				9,200	9,000	8,200	6,200	5,700
150	3,000				9,800	9,600	6,240	5,800	4,800
150	4,000				10,200	10,000	6,636	6,000	5,000
150	5,000				10,700	10,500	9,900	9,200	9,145
150	7,500				11,200	11,000	10,400	9,700	9,500
200	0				5,971	4,820	3,700	2,774	1,450
200	50				6,044	4,955	4,000	3,094	2,313
200	100				6,099	5,000	4,100	3,200	2,600
200	200				6,193	5,200	4,300	3,400	2,800
200	300				6,278	5,400	4,450	3,500	3,000
200	500				6,433	5,600	4,650	3,700	3,150
200	750				6,610	5,800	4,850	3,900	3,300
200	1,000				6,775	6,000	5,000	4,000	3,400
200	1,500				7,083	6,200	4,600	3,800	3,300
200	2,000				7,370	6,400	4,500	4,000	3,500
200	3,000				7,905	7,000	5,742	4,287	3,900
200	4,000				8,600	8,500	6,411	4,800	4,500
200	5,000				8,880	6,607	6,120	5,267	4,845
200	7,500				9,990	6,990	6,465	5,581	5,077
300	0				5,631	4,650	3,667	2,683	1,700
300	50				5,712	5,063	4,292	3,296	2,975
300	100				5,772	5,200	4,500	3,500	3,400
300	200				5,877	5,600	4,700	3,600	3,500
300	300				5,971	4,592	3,324	3,300	3,250
300	500				6,142	4,634	3,367	3,300	3,250
300	750				6,338	6,000	5,000	3,900	3,800
300	1,000				6,521	6,200	5,200	4,000	3,900
300	1,500				6,862	6,400	4,154	4,100	4,000
300	2,000				7,180	6,600	4,300	4,200	4,100

Quality 0.05	0.10	0.15	0.20	0.25	0.30	0.40	0.50	0.60	0.70	0.80	0.90
1,000	980	800	790	780	710	486	470	454	339	225	89
1,600	1,500	1,300	1,250	1,204	1,126	899	850	800	501	355	350
2,100	2,000	1,450	1,000	1,000	955	950	950	900	660	430	335
2,550	2,500	1,500	1,400	1,300	1,246	1,200	1,124	1,100	854	474	460
3,000	3,000	2,500	2,400	2,300	2,300	2,168	1,721	1,448	818	481	450
3,550	3,500	3,200	3,000	2,500	2,358	2,350	2,301	1,220	700	651	440
5,000	4,700	4,249	4,006	3,104	3,081	2,933	2,407	1,317	911	700	400
4,600	4,406	3,800	3,600	3,300	3,300	3,250	2,503	1,552	1,100	600	300
4,800	4,600	4,500	4,451	4,200	4,000	3,911	3,105	2,601	1,100	600	300
5,400	5,300	5,200	5,100	5,050	5,000	4,900	4,549	2,900	1,100	600	300
6,300	6,000	5,800	5,700	5,400	5,300	5,200	3,900	2,700	1,100	600	300
6,785	6,650	6,100	5,300	5,000	4,600	4,000	3,000	2,000	1,000	500	300
7,875	7,600	7,400	5,200	4,400	3,800	3,000	2,500	1,500	900	500	300
9,233	9,000	7,500	5,000	4,000	3,000	2,200	2,000	1,200	800	400	300
1,200	1,100	1,000	950	930	913	588	550	544	375	250	116
1,800	1,700	1,500	1,450	1,416	1,329	1,002	900	873	539	419	400
2,200	2,000	1,900	1,800	1,400	1,093	1,000	950	950	941	701	450
2,600	2,500	2,100	1,600	1,400	1,200	1,200	1,166	1,150	1,100	1,000	500
2,900	2,800	2,500	2,400	1,500	1,400	1,340	1,300	1,200	1,200	800	500
3,600	3,500	3,100	2,700	1,600	1,500	1,500	1,400	1,200	1,100	670	450
4,150	4,000	3,500	2,800	1,400	1,399	1,300	1,200	1,100	1,000	650	400
4,500	3,985	3,400	3,200	1,200	1,105	1,100	1,100	1,000	1,000	600	350
4,000	3,800	3,300	3,100	2,400	2,000	1,500	1,300	1,200	1,100	550	300
4,700	4,300	4,100	4,000	3,500	3,000	2,500	2,000	1,700	1,500	550	300
4,400	4,300	4,200	4,100	3,800	3,300	3,250	3,200	2,500	1,000	500	300
4,600	4,500	4,300	4,200	3,600	2,800	2,800	2,600	2,000	1,000	450	300
7,405	6,500	5,700	4,000	3,100	2,600	2,300	2,000	1,500	900	450	300
8,881	8,000	6,500	5,000	3,000	2,200	1,800	1,600	1,000	600	400	300
277	262	248	233	219	204	175	146	117	87	58	29
1,794	1,641	1,562	1,483	1,405	1,326	1,244	1,044	650	600	574	571
2,300	2,100	2,000	1,900	1,300	1,200	1,109	800	767	750	750	516
2,500	2,300	2,150	1,400	1,297	1,200	1,180	1,100	1,000	1,000	1,000	514
2,700	2,400	2,200	2,150	2,100	2,000	1,829	1,497	904	417	400	350
2,900	2,500	2,400	2,300	2,250	2,226	1,759	1,388	659	250	200	200
3,000	2,650	2,500	2,314	2,159	1,818	1,590	1,463	1,000	600	450	300
3,000	2,600	2,500	2,134	1,939	1,800	1,700	1,600	1,000	600	450	300
3,000	2,600	2,500	2,350	2,169	1,800	1,700	1,500	1,000	600	450	300
3,000	2,600	2,500	2,150	2,050	1,800	1,700	1,400	1,000	600	450	300
3,000	2,600	2,500	2,150	2,050	1,800	1,700	1,400	1,000	600	450	300
3,000	2,600	2,500	2,150	2,050	1,800	1,700	1,400	1,000	600	450	300
3,000	2,600	2,500	2,150	2,050	1,800	1,700	1,400	1,000	600	450	300
3,000	2,600	2,500	2,150	2,050	1,800	1,700	1,400	1,000	600	450	300
325	308	291	274	257	240	205	171	137	103	68	34
2,481	2,327	2,248	2,168	2,014	1,860	1,551	1,067	535	500	500	450
3,200	3,000	2,900	2,800	2,600	2,400	1,160	900	870	850	800	504
3,200	2,900	2,800	1,600	1,463	1,400	1,400	1,376	1,300	1,200	1,181	569
3,200	2,800	2,700	2,600	2,500	2,400	2,176	1,785	1,043	438	400	362
3,200	2,700	2,600	2,600	2,600	2,500	1,946	1,341	811	200	200	200
3,200	2,800	2,700	2,586	2,475	2,402	1,526	868	490	418	250	200
3,200	2,600	2,500	2,395	2,208	2,100	1,800	1,500	1,200	700	400	200
3,200	2,600	2,600	2,538	2,360	2,100	1,800	1,500	1,200	700	400	200
3,200	2,600	2,500	2,294	2,200	2,100	1,800	1,500	1,200	700	400	200

TABLE 6.3-2 *(Continued)*

Pressure kPa	Mass Flux G kg/m² · s	−0.50	−0.40	−0.30	−0.20	−0.15	−0.10	−0.05	0.00
300	3,000				8,000	7,778	5,690	4,800	4,706
300	4,000				8,600	8,443	6,302	4,518	4,400
300	5,000				8,852	5,419	4,851	3,967	3,525
300	7,500				10,082	5,816	5,198	4,236	3,630
450	0				5,363	4,540	4,222	2,000	1,968
450	50				5,454	5,035	4,627	2,330	2,250
450	100				5,522	5,200	4,100	3,400	2,900
450	200				5,639	5,600	4,150	3,400	3,000
450	300				5,745	4,145	2,896	2,850	2,825
450	500				5,939	4,187	2,939	2,900	2,850
450	750				6,159	6,000	4,300	3,400	3,100
450	1,000				6,365	6,200	4,400	3,400	3,200
450	1,500				6,749	6,400	4,145	3,060	3,000
450	2,000				7,108	6,600	3,521	3,400	3,300
450	3,000				7,775	7,500	5,752	4,146	3,716
450	4,000				8,399	8,000	6,210	4,292	4,076
450	5,000				8,991	6,619	6,002	5,440	3,500
450	7,500				10,376	6,878	6,271	5,707	3,600
700	0			6,548	5,153	4,482	4,264	2,059	1,488
700	50			6,652	5,260	5,021	4,669	2,488	2,300
700	100			6,730	5,339	5,200	4,200	3,500	3,000
700	200			6,865	5,477	5,300	4,250	3,600	3,200
700	300			6,986	5,601	5,400	4,300	3,700	3,400
700	500			7,208	5,827	5,700	4,400	3,800	3,500
700	750			7,460	6,085	6,000	4,600	3,900	3,600
700	1,000			7,696	6,326	6,200	5,060	4,094	3,600
700	1,500			8,136	6,775	6,400	4,991	4,025	3,600
700	2,000			8,547	7,195	6,600	5,200	4,400	3,600
700	3,000			9,312	7,976	6,800	5,400	4,600	3,600
700	4,000			10,026	8,705	7,000	5,600	4,800	3,600
700	5,000			10,705	9,397	7,200	5,800	4,900	3,600
700	7,500			12,294	11,018	7,400	6,000	5,000	3,600
1,000	0			8,092	6,551	4,480	3,866	3,240	2,620
1,000	50			8,195	6,657	5,620	5,092	4,560	4,105
1,000	100			8,271	6,736	6,000	5,500	5,000	4,600
1,000	200			8,403	8,100	8,000	7,500	7,000	6,000
1,000	300			8,522	8,200	8,000	7,500	7,000	6,500
1,000	500			8,738	8,300	8,000	7,500	7,000	6,500
1,000	750			8,985	8,400	8,000	7,500	7,000	6,500
1,000	1,000			9,217	8,500	8,000	7,500	7,000	6,500
1,000	1,500			9,647	8,500	8,000	7,500	7,000	6,500
1,000	2,000			10,049	8,700	8,000	7,500	7,000	6,500
1,000	3,000			10,798	9,362	8,000	7,500	7,000	6,500
1,000	4,000			11,497	10,088	8,000	7,500	7,000	6,500
1,000	5,000			12,161	10,778	8,000	7,500	7,000	6,500
1,000	7,500			13,716	12,391	8,000	7,500	7,000	6,500
1,500	0		8,975	7,719	6,456	4,525	4,007	3,488	2,970
1,500	50		9,088	7,839	6,576	5,781	5,100	4,922	4,200
1,500	100		9,171	7,928	6,666	6,200	5,500	5,400	5,000
1,500	200		9,315	8,083	7,200	7,100	6,900	6,600	6,400
1,500	300		9,445	8,222	7,500	7,400	7,100	6,700	6,500

Quality											
0.05	0.10	0.15	0.20	0.25	0.30	0.40	0.50	0.60	0.70	0.80	0.90
3,200	2,600	2,500	2,294	2,200	2,100	1,800	1,500	1,200	700	400	200
3,200	2,600	2,500	2,294	2,200	2,100	1,800	1,500	1,200	700	400	200
3,200	2,600	2,500	2,294	2,200	2,100	1,800	1,500	1,200	700	400	200
3,200	2,600	2,500	2,300	2,200	2,100	1,800	1,500	1,200	700	400	200
380	360	340	320	300	280	240	200	160	150	120	110
2,195	2,115	2,035	1,993	1,950	1,870	1,860	1,625	600	550	543	500
2,800	2,700	2,600	2,550	2,200	2,000	1,500	1,123	1,112	984	908	595
2,800	2,600	2,550	2,500	2,200	1,700	1,600	1,500	1,400	1,306	890	595
2,800	2,600	2,500	2,450	2,400	2,400	2,300	2,161	1,583	755	447	408
2,800	2,600	2,400	2,400	2,400	2,300	2,245	1,784	1,123	451	250	236
2,800	2,600	2,300	2,150	2,000	1,900	1,355	879	707	467	350	316
2,800	2,600	2,100	1,966	1,786	1,200	700	606	593	500	300	200
2,800	2,600	2,100	2,086	1,500	1,000	700	665	600	500	300	200
2,800	2,600	2,100	1,800	1,500	1,200	1,100	1,000	700	500	300	200
2,800	2,600	2,200	1,800	1,500	1,200	1,100	1,000	700	500	300	200
2,800	2,600	2,200	1,800	1,500	1,200	1,100	1,000	700	500	300	200
2,800	2,600	2,200	1,900	1,500	1,200	1,100	1,000	700	500	300	200
2,800	2,600	2,200	2,000	1,500	1,200	1,100	1,000	700	500	300	200
448	424	401	377	354	330	283	236	189	97	72	70
2,212	2,056	1,825	1,594	1,363	1,133	971	809	500	450	404	400
2,800	2,600	2,300	2,000	1,700	1,400	1,200	1,000	830	820	816	578
3,000	2,900	2,700	2,600	2,500	2,200	2,000	1,800	1,700	1,403	913	649
3,000	2,900	2,700	2,600	2,550	2,500	2,500	2,500	1,892	1,093	628	533
3,000	2,900	2,800	2,700	2,650	2,600	2,600	2,568	1,360	644	385	353
3,000	2,700	2,600	2,500	2,400	2,300	2,284	1,398	680	450	426	420
3,000	2,800	2,700	2,700	2,695	1,600	1,595	914	467	260	250	200
3,000	2,500	2,400	2,300	2,284	1,200	1,166	745	259	199	175	150
3,000	2,500	2,200	1,800	1,500	1,300	1,300	900	250	200	200	150
3,000	2,500	2,200	1,800	1,500	1,300	1,200	900	250	200	200	150
3,000	2,500	2,200	1,800	1,500	1,300	1,200	900	250	200	200	150
3,000	2,500	2,200	1,900	1,500	1,300	1,200	900	250	200	200	150
3,000	2,500	2,200	2,000	1,600	1,300	1,200	900	250	200	200	150
509	483	456	429	402	375	322	268	214	161	107	54
3,427	3,271	3,114	3,032	2,801	2,569	2,330	1,942	1,104	875	814	426
4,400	4,200	4,000	3,900	3,600	3,300	3,000	2,500	1,600	1,547	1,070	659
5,800	5,600	5,400	5,100	4,700	4,400	4,300	3,352	2,455	1,704	1,117	778
6,300	5,600	5,400	5,100	4,700	4,700	4,671	3,958	2,571	1,569	918	711
6,200	5,600	5,400	5,100	4,700	4,652	4,600	3,803	2,087	1,123	603	437
6,100	5,600	5,400	5,000	4,900	4,800	3,940	2,272	1,106	696	527	466
6,000	5,600	5,400	5,369	4,923	4,123	2,905	1,646	888	606	298	218
5,900	5,300	5,000	4,925	4,425	3,325	2,393	1,328	613	256	188	100
5,700	5,200	4,800	4,400	3,600	3,500	3,474	2,807	800	250	150	100
5,500	5,100	4,700	4,200	4,000	3,962	3,858	1,300	750	250	150	100
5,300	5,000	4,500	4,000	3,400	2,800	1,850	1,150	700	250	150	100
5,400	4,900	4,400	3,800	3,300	2,600	1,650	1,000	650	250	150	100
5,500	4,800	4,300	3,600	3,200	2,300	1,500	900	600	250	150	100
584	553	523	492	461	430	369	307	246	184	123	61
4,000	3,513	3,356	3,273	3,115	2,883	2,567	2,327	1,149	1,021	879	426
4,300	4,500	4,300	4,200	4,000	3,700	3,300	3,000	1,600	1,563	1,081	637
6,200	6,000	5,700	5,400	5,000	4,100	3,600	3,411	2,788	2,040	1,265	816
6,300	6,000	5,700	5,300	5,200	5,200	5,110	4,551	3,539	2,240	1,271	855

TABLE 6.3-2 *(Continued)*

Pressure kPa	Mass Flux G $kg/m^2 \cdot s$	−0.50	−0.40	−0.30	−0.20	−0.15	−0.10	−0.05	0.00
1,500	500		9,683	8,477	7,800	7,700	7,200	6,700	6,400
1,500	750		9,954	8,767	8,100	8,000	7,200	6,700	6,300
1,500	1,000		10,207	9,038	8,200	8,000	7,200	6,700	6,250
1,500	1,500		10,678	9,543	8,284	8,000	7,200	6,700	6,250
1,500	2,000		11,119	10,015	8,757	8,000	7,200	6,700	6,200
1,500	3,000		11,939	10,893	9,637	7,529	6,720	6,700	6,100
1,500	4,000		12,705	11,714	10,459	7,645	6,836	6,700	6,100
1,500	5,000		13,433	12,493	11,239	8,500	7,500	6,900	6,300
1,500	7,500		15,137	14,317	13,065	9,000	8,000	7,100	6,300
2,000	0	9,632	8,569	7,495	6,421	4,581	4,130	3,680	3,230
2,000	50	9,749	8,692	7,629	6,550	5,870	5,420	4,970	4,700
2,000	100	9,836	8,784	7,729	6,646	6,300	5,850	5,400	5,200
2,000	200	9,988	8,944	7,902	7,100	7,000	6,900	6,800	6,700
2,000	300	10,124	9,088	8,058	7,450	7,400	7,350	7,200	6,900
2,000	500	10,372	9,349	8,343	7,800	7,700	7,600	7,500	7,200
2,000	750	10,656	9,648	8,668	8,100	8,100	7,800	7,600	7,200
2,000	1,000	10,921	9,928	8,971	8,300	8,200	8,100	7,800	7,714
2,000	1,500	11,414	10,448	9,536	8,700	8,500	8,400	7,800	7,387
2,000	2,000	11,876	10,934	10,064	9,000	8,700	8,500	7,800	7,000
2,000	3,000	12,735	11,840	11,047	9,832	8,944	8,135	7,800	7,000
2,000	4,000	13,537	12,685	11,964	10,713	9,060	8,251	7,000	6,500
2,000	5,000	14,299	13,489	12,836	11,550	10,000	8,500	6,500	5,000
2,000	7,500	16,083	15,369	14,876	13,507	10,300	8,500	5,500	4,800
3,000	0	8,858	8,031	7,207	6,380	4,650	4,290	3,930	3,570
3,000	50	8,993	8,181	7,358	6,518	6,038	5,573	5,108	4,868
3,000	100	9,093	8,291	7,470	6,620	6,500	6,000	5,500	5,300
3,000	200	9,267	8,484	7,665	7,300	7,200	7,100	6,900	6,700
3,000	300	9,424	8,657	7,841	7,600	7,500	7,400	7,352	7,029
3,000	500	9,710	8,973	8,161	8,000	8,000	8,000	7,968	7,645
3,000	750	10,036	9,334	8,527	8,400	8,300	8,100	8,100	7,800
3,000	1,000	10,341	9,671	8,869	8,500	8,300	8,200	8,000	7,711
3,000	1,500	10,909	10,299	9,505	8,800	8,700	8,248	7,907	7,265
3,000	2,000	11,440	10,885	10,099	9,200	9,100	9,000	9,000	8,900
3,000	3,000	12,429	11,977	11,206	10,010	9,500	9,000	8,500	8,000
3,000	4,000	13,351	12,997	12,239	10,947	10,500	9,000	7,027	6,592
3,000	5,000	14,228	13,966	13,221	11,837	10,800	8,505	7,094	5,870
3,000	7,500	16,281	16,234	15,518	13,919	11,200	8,623	7,035	5,246
4,500	0	8,083	7,488	6,895	6,294	4,676	4,400	4,125	3,850
4,500	50	8,250	7,658	7,055	6,432	5,669	5,300	5,006	4,713
4,500	100	8,374	7,784	7,174	6,535	6,000	5,600	5,300	5,000
4,500	200	8,589	8,004	7,381	6,900	6,800	6,700	6,600	6,000
4,500	300	8,783	8,202	7,567	7,100	7,000	6,900	6,490	6,367
4,500	500	9,136	8,562	7,906	7,400	7,300	7,200	7,106	6,783
4,500	750	9,539	8,973	8,294	7,800	7,700	7,100	6,777	6,089
4,500	1,000	9,916	9,357	8,656	7,816	7,700	7,009	6,586	6,110
4,500	1,500	10,617	10,073	9,330	8,500	7,768	5,965	5,534	5,503
4,500	2,000	11,273	10,741	9,959	8,600	6,400	6,382	5,500	5,405
4,500	3,000	12,494	11,986	11,132	8,700	6,299	5,955	5,327	5,000
4,500	4,000	13,633	13,148	12,226	9,215	8,948	6,831	6,216	5,446
4,500	5,000	14,716	14,252	13,266	9,890	8,937	8,000	6,624	5,241
4,500	7,500	17,251	16,835	15,698	10,510	9,913	8,833	7,133	5,109
7,000	0	7,118	6,752	6,381	5,997	4,538	4,348	4,160	3,970
7,000	50	7,306	6,927	6,536	6,127	5,485	5,137	4,790	4,443

Quality											
0.05	0.10	0.15	0.20	0.25	0.30	0.40	0.50	0.60	0.70	0.80	0.90
6,100	5,800	5,500	5,300	4,800	4,600	4,500	4,307	3,103	1,837	956	608
5,900	5,600	5,400	5,200	5,000	4,800	4,531	2,944	1,717	1,082	707	451
5,700	5,600	5,550	5,500	5,400	5,220	3,693	2,081	1,118	818	466	216
5,500	5,300	5,275	5,150	4,849	4,538	2,893	1,440	816	389	233	200
5,500	5,200	4,900	4,043	4,000	3,920	2,578	1,470	1,000	800	400	200
5,600	5,100	4,800	4,700	4,456	4,307	4,009	2,100	1,000	800	400	200
5,600	5,000	4,700	3,800	3,800	3,100	1,900	1,800	1,000	800	400	200
5,600	4,800	4,400	3,600	3,500	2,600	1,700	1,500	1,000	800	400	200
5,500	4,700	4,100	3,600	3,300	2,350	1,600	1,200	1,000	800	400	200
643	609	575	541	507	474	406	338	271	203	135	68
3,800	3,752	3,594	3,510	3,352	3,118	2,801	2,710	1,193	941	927	434
4,900	4,800	4,600	4,500	4,300	4,000	3,600	3,500	1,700	1,601	1,206	762
6,600	6,300	6,000	5,600	5,300	4,800	4,400	3,402	2,741	2,169	1,350	829
6,600	6,600	5,900	5,700	5,500	5,200	4,500	4,400	4,329	2,804	1,516	879
6,600	6,600	5,730	5,348	5,225	5,200	5,191	4,480	3,701	2,446	1,327	772
7,000	6,843	6,405	5,849	5,666	5,300	5,222	3,601	2,349	1,245	789	450
7,073	6,745	6,482	6,202	5,800	5,792	4,283	2,752	1,725	1,012	584	382
6,771	6,388	5,892	4,766	4,687	4,600	3,197	1,729	957	558	316	300
6,500	5,800	5,236	4,471	4,305	4,133	2,221	1,019	481	415	400	300
6,200	5,600	5,240	4,755	4,662	4,459	3,393	1,600	1,200	1,000	400	300
6,000	5,500	5,274	4,779	4,700	4,415	3,215	1,400	1,200	900	400	300
4,800	4,600	4,400	4,267	4,200	4,081	2,400	1,400	1,200	900	600	500
4,600	4,400	4,000	3,483	3,200	2,800	1,700	1,500	1,300	1,100	800	600
919	690	651	613	575	536	460	383	306	240	235	231
4,130	3,997	3,913	3,753	3,594	3,359	3,040	2,721	2,327	340	308	300
5,200	5,100	5,000	4,800	4,600	4,300	3,900	3,500	3,000	1,353	1,058	758
6,600	6,500	6,300	6,000	5,700	5,400	4,800	4,000	3,273	2,203	1,345	819
7,000	7,000	7,000	6,970	6,600	6,260	5,463	4,600	4,538	2,902	1,673	978
7,200	6,500	6,292	5,886	5,724	5,699	5,610	4,283	3,790	2,723	1,694	1,039
7,100	6,695	6,586	6,197	6,104	6,100	5,038	3,594	2,750	1,851	1,228	684
6,945	6,700	6,594	6,432	5,929	5,299	4,140	3,312	2,605	2,201	1,698	809
6,600	6,587	5,751	5,105	4,635	4,505	3,389	2,523	2,100	2,000	1,973	1,208
8,818	7,347	6,162	5,216	4,592	4,164	2,817	2,122	1,209	1,000	1,000	600
7,858	6,203	5,450	4,997	4,486	3,018	2,100	2,015	1,100	1,050	1,000	600
5,973	5,556	5,047	4,701	3,959	2,032	2,000	1,900	1,750	1,350	1,000	600
5,383	5,124	4,591	4,157	3,611	1,742	1,500	1,400	1,350	1,350	1,000	600
4,389	4,067	3,900	3,860	3,188	1,201	1,200	1,200	1,100	1,100	1,000	700
1,362	793	730	687	644	601	515	429	344	304	300	250
3,941	3,573	3,482	3,247	2,711	2,175	1,629	1,232	836	338	300	300
4,800	4,500	4,400	4,100	3,400	2,700	2,000	1,500	1,000	980	866	744
5,800	4,900	4,700	4,600	4,200	3,700	3,200	2,900	2,880	1,645	1,286	916
6,300	5,400	5,300	5,000	4,700	4,600	4,374	3,595	3,365	2,275	1,633	1,100
5,900	5,594	5,229	4,700	4,652	4,093	3,509	3,000	2,920	2,392	1,801	1,192
5,816	5,292	4,841	4,400	4,307	4,100	3,835	2,884	2,611	2,263	1,705	1,075
5,838	5,173	4,473	4,295	4,189	4,100	3,886	3,370	2,679	2,150	2,124	1,531
5,462	5,265	4,728	4,340	4,147	4,047	3,333	2,641	1,744	1,700	1,668	1,608
5,319	5,132	4,722	4,409	4,070	3,564	2,658	2,050	1,446	802	687	200
4,996	4,798	4,500	4,216	3,757	2,817	1,889	1,382	1,018	391	300	224
4,944	4,525	4,180	3,779	3,200	2,481	1,527	939	900	393	309	200
4,586	4,101	3,825	3,541	2,893	2,122	1,366	1,200	1,200	1,100	1,100	900
4,400	4,000	3,900	3,653	2,969	2,355	2,200	2,200	1,800	1,500	1,300	1,000
1,921	1,221	868	761	713	666	570	475	380	350	300	250
3,780	3,380	3,142	2,965	2,578	1,966	1,493	1,169	920	535	500	500

TABLE 6.3-2 *(Continued)*

Pressure kPa	Mass Flux G kg/m$^2\cdot$s	−0.50	−0.40	−0.30	−0.20	−0.15	−0.10	−0.05	0.00
7,000	100	7,446	7,057	6,651	6,223	5,800	5,400	5,000	4,600
7,000	200	7,688	7,284	6,852	6,391	6,200	6,000	5,500	5,093
7,000	300	7,907	7,487	7,032	6,542	6,400	6,300	5,600	5,122
7,000	500	8,305	7,858	7,361	6,817	6,500	6,500	6,000	5,700
7,000	750	8,759	8,282	7,736	7,131	7,100	7,000	6,200	5,500
7,000	1,000	9,184	8,678	8,087	7,425	6,800	6,281	5,758	5,296
7,000	1,500	9,975	9,416	8,740	8,000	7,606	5,366	4,790	4,766
7,000	2,000	10,713	10,104	9,349	6,980	6,100	6,044	4,935	4,728
7,000	3,000	12,089	11,387	10,485	7,859	6,192	5,840	4,995	4,295
7,000	4,000	13,373	12,584	11,544	8,637	8,317	6,334	5,694	4,638
7,000	5,000	14,593	13,721	12,550	9,487	8,696	7,889	6,408	4,682
7,000	7,500	17,447	16,381	14,903	10,262	9,817	8,841	7,042	4,783
10,000	0	4,731	4,566	4,396	4,218	4,144	4,019	3,894	3,770
10,000	50	4,930	4,745	4,551	4,347	4,300	4,230	4,050	4,018
10,000	100	5,078	4,877	4,666	4,500	4,400	4,300	4,200	4,100
10,000	200	5,336	5,108	4,866	4,610	4,600	4,500	4,450	4,400
10,000	300	5,568	5,315	5,046	4,760	4,700	4,262	4,075	4,007
10,000	500	5,600	5,500	5,400	5,344	5,200	5,184	4,454	3,600
10,000	750	5,800	5,600	5,500	4,776	4,650	4,600	4,463	3,768
10,000	1,000	6,924	6,200	5,908	4,873	4,400	4,300	4,254	4,052
10,000	1,500	7,763	6,900	6,476	5,772	4,747	4,678	4,413	3,903
10,000	2,000	8,547	7,982	6,686	6,400	5,776	5,066	4,385	3,955
10,000	3,000	11,650	9,289	8,298	6,700	6,700	5,473	4,476	3,952
10,000	4,000	12,289	10,508	9,489	7,170	7,134	5,995	4,637	3,874
10,000	5,000	13,179	11,414	10,337	8,300	8,223	7,226	5,831	4,235
10,000	7,500	14,766	13,009	11,908	8,703	8,615	7,704	6,241	4,778
15,000	0	3,308	3,249	3,187	3,117	3,103	3,046	2,987	2,930
15,000	50	3,434	3,363	3,287	3,204	3,176	3,087	2,997	2,908
15,000	100	3,528	3,446	3,361	3,269	3,200	3,100	3,000	2,900
15,000	200	3,690	3,592	3,490	3,381	3,200	3,200	3,000	2,500
15,000	300	3,837	3,724	3,606	3,482	3,300	2,900	2,800	2,772
15,000	500	3,800	3,750	3,700	3,569	3,526	3,313	3,007	2,200
15,000	750	3,587	3,400	3,350	2,910	2,700	2,600	2,600	2,510
15,000	1,000	3,200	3,150	3,100	2,980	2,600	2,599	2,300	2,300
15,000	1,500	4,300	4,100	3,947	3,770	3,331	3,246	2,838	2,270
15,000	2,000	5,530	5,091	4,433	4,180	3,670	3,244	2,715	2,391
15,000	3,000	6,904	5,218	5,163	4,629	4,236	3,578	2,967	2,726
15,000	4,000	7,402	5,300	5,100	4,734	4,581	3,906	3,200	2,953
15,000	5,000	8,573	7,329	6,303	5,703	5,700	5,158	4,356	3,567
15,000	7,500	10,140	9,040	7,881	6,401	6,089	5,567	4,749	4,406
20,000	0	1,211	1,204	1,202	1,198	1,107	1,087	1,066	1,045
20,000	50	1,300	1,300	1,294	1,286	1,133	1,101	1,069	1,037
20,000	100	1,400	1,400	1,362	1,351	1,142	1,106	1,070	1,034
20,000	200	1,550	1,500	1,500	1,465	1,142	1,142	1,070	1,034
20,000	300	1,600	1,600	1,587	1,500	1,400	1,300	1,231	1,200
20,000	500	1,665	1,649	1,450	1,400	1,300	1,235	1,100	1,063
20,000	750	1,838	1,819	1,500	1,400	1,300	1,300	1,250	1,250
20,000	1,000	1,580	1,414	1,232	1,200	1,100	1,100	1,000	977
20,000	1,500	2,500	2,249	1,997	1,800	1,750	1,712	1,399	1,174
20,000	2,000	3,261	2,878	2,458	2,098	1,648	1,565	1,500	1,498
20,000	3,000	3,939	3,176	3,017	2,671	2,233	2,000	1,958	1,900
20,000	4,000	4,365	4,000	3,891	2,939	2,576	2,300	2,250	2,235
20,000	5,000	4,041	4,000	3,881	3,732	3,215	2,875	2,580	2,357
20,000	7,500	5,101	5,017	4,900	3,849	3,300	3,228	3,200	3,200

Quality											
0.05	0.10	0.15	0.20	0.25	0.30	0.40	0.50	0.60	0.70	0.80	0.90
4,400	4,100	3,900	3,700	3,200	2,400	1,800	1,400	1,100	1,100	1,000	909
4,800	4,600	4,500	4,400	4,000	3,500	3,000	2,300	1,600	1,300	1,289	1,257
5,000	4,800	4,700	4,400	4,100	4,000	3,500	3,172	2,201	1,937	1,531	1,216
5,600	5,500	5,300	4,482	4,456	3,171	2,539	2,290	2,222	2,066	1,532	952
5,400	5,300	4,928	3,727	3,597	3,296	2,866	2,524	2,342	1,992	1,478	1,167
5,100	4,646	4,057	3,551	3,356	3,340	2,998	2,384	1,771	1,400	1,300	1,100
4,740	4,378	3,896	3,545	3,333	2,956	2,166	1,525	992	900	714	700
4,323	3,685	3,265	3,179	2,917	2,423	1,658	1,023	776	327	189	115
3,832	3,199	2,932	2,905	2,521	1,904	1,049	521	349	213	171	124
3,742	2,929	2,614	2,435	2,085	1,670	851	341	255	234	175	110
3,470	2,600	2,400	2,325	2,000	1,466	846	600	400	300	250	150
3,636	2,941	2,570	2,512	2,200	2,143	1,853	700	500	400	300	200
2,280	1,584	1,181	918	748	698	598	498	399	250	250	200
3,495	2,946	2,545	2,180	1,837	1,524	1,200	950	700	365	300	300
3,500	2,887	2,850	2,835	2,800	2,167	2,081	1,700	1,662	1,123	710	700
2,721	2,500	2,500	2,462	2,422	2,032	1,672	1,600	1,543	1,450	1,432	1,275
3,725	3,400	3,353	3,200	3,164	2,741	2,398	1,826	1,617	1,377	1,125	855
3,500	3,485	3,300	2,859	2,744	2,390	2,161	1,443	1,082	925	723	528
3,500	3,428	3,177	2,870	2,640	2,135	1,683	1,363	1,107	776	491	284
3,713	3,376	3,121	2,827	2,393	1,844	1,061	865	595	395	350	310
3,581	3,166	2,699	2,204	1,873	1,273	871	572	379	350	300	283
3,458	2,768	2,178	1,772	1,387	1,027	666	386	300	250	230	166
3,232	2,374	1,793	1,494	1,259	944	542	267	223	220	209	168
3,110	2,213	1,624	1,378	1,190	846	454	227	200	200	177	160
2,929	2,047	1,579	1,461	1,245	1,128	703	600	500	400	177	160
3,561	2,773	2,277	2,187	2,100	2,000	1,872	1,000	500	400	170	160
2,206	1,731	1,395	1,145	952	798	577	481	385	121	113	110
2,501	2,158	1,999	1,786	1,663	1,549	1,419	795	696	286	262	250
2,400	2,097	2,054	1,905	1,881	1,740	1,664	1,468	1,249	816	544	500
1,636	1,400	1,400	1,389	1,376	1,300	1,275	1,150	1,132	870	643	600
2,770	2,560	2,478	2,201	2,095	1,692	1,488	1,112	816	576	336	300
2,100	2,100	2,050	1,640	1,584	1,300	1,097	634	335	284	236	178
2,196	2,088	1,969	1,541	1,401	1,168	769	517	400	400	400	173
2,183	1,806	1,466	1,229	1,108	930	425	263	242	225	200	173
1,928	1,549	1,264	1,032	850	644	335	220	196	190	180	173
2,104	1,673	1,272	913	718	529	299	222	192	190	180	173
2,288	1,722	1,299	1,017	869	634	431	286	250	225	200	173
2,525	1,906	1,476	1,234	1,111	732	514	324	300	225	200	173
2,841	2,244	1,802	1,577	1,340	1,103	790	760	750	710	400	200
3,800	2,983	2,441	2,309	2,113	1,971	1,562	1,000	950	910	400	200
907	791	692	607	533	467	357	269	200	150	100	50
922	813	762	687	642	598	544	308	264	199	170	146
927	820	785	713	678	642	606	321	285	250	185	150
927	820	800	800	800	750	679	676	675	574	323	300
1,100	1,050	1,000	1,000	900	850	789	657	456	282	183	150
950	915	900	674	657	508	393	325	179	172	150	150
1,219	1,164	1,085	656	541	516	431	328	218	200	200	150
961	801	714	629	620	611	387	254	178	110	100	98
1,054	943	864	730	670	621	375	250	131	110	100	98
1,420	1,354	1,274	1,087	960	793	568	388	195	110	100	98
1,716	1,598	1,461	1,308	1,197	1,001	801	550	189	110	100	98
1,960	1,854	1,679	1,525	1,430	1,159	948	654	189	182	178	143
2,150	2,015	1,745	1,717	1,607	1,301	965	568	268	253	250	214
3,185	2,594	2,500	2,469	2,398	1,902	1,486	1,034	339	325	321	250

CHAPTER 7

ONCE-THROUGH BOILERS

R. LEITHNER

Technical University Braunschweig
Braunschweig, Federal Republic of Germany

7.1 INTRODUCTION (HISTORICAL REVIEW)

Jacob Perkins is reported [1] to be the inventor of the once-through boiler and the supercritical cycle. He was born in Newburyport, Massachusetts, on July 9, 1766, and became an incredibly versatile engineer and scientist, who dealt with central heating plants, machinery for engraving bank notes, steam rockets, and refrigerating machines used to preserve food and make ice. He also proposed propelling a cannonball weighing a ton across the English channel using steam at 50,000 psig (3447 bar) and 1200°F (649°C). In 1822 he built an injection-type boiler, in which *the amount of feed water entering precisely equaled the amount of steam discharged*, composed of a single copper chamber with a 3-in. wall thickness, which was reported to operate at 7500 psig (517 bar). In 1827 he designed a high-pressure boiler consisting of *a continuous tube from the inlet to the outlet* steam chamber discharging to a steam engine cylinder. Jacob Perkins died in London on July 30, 1849. Although his son Angier March Perkins continued his work in high-pressure steam, the engineering concepts of Jacob Perkins were too far ahead of his time and his invention of the once-through boiler at supercritical pressure did not gain any pratical importance. In 1889 Leon Serpollet [1] devised a successful once-through boiler of the flash type for steam automobiles.

As late as 1923 [1] Mark Benson, a Czechoslovakian chemical engineer, who had emigrated to the United States, proposed a turbine to operate at a steam pressure of about 1500 psig (103 bar) though the throttle pressure used at that time in U.S. central stations was 250 psig (17 bar). To avoid problems of fabricating thick-walled drums, Benson designed a once-through boiler

Boilers, Evaporators and Condensers, Edited by Sadik Kakaç
ISBN 0-471-62170-6

operating at critical pressure with a reducing valve ahead of the final superheater. Such power plants, built in England, Belgium, and Germany, had two problems: boiler control and especially the need for demineralized water. The water treatment at that time was insufficient and the remaining impurities in the feed water deposited in the boiler tubes so rapidly that only limited periods of continuous operation could be achieved. During the late 1920s Sulzer Brothers, Ltd. in Winterthur, Switzerland, developed the monotube boiler for operation at subcritical pressure. This concept eliminated the need for the reducing valve, reduced the wall thickness of the evaporator, and lowered the feed pump power. By using a water separator, this concept offered the possibility of flushing the evaporator during operation (chemical blowdown), tolerating a certain degree of impurities in the feed water. Sulzer Brothers also improved the feed water control.

During the 1930s in Europe, the once-through boiler was commercially accepted, whereas in the United States the somewhat lower relative cost of steel and a more rapid development in welding technology allowed large thick-walled (high pressure) drums and lessened the economic advantage of once-through boilers. Thus the broad use of once-through boilers was delayed for another 20 years.

After the development of Philo 6 of the AEP Ohio Power Co. in 1953 as the first commercial supercritical pressure once-through utility boiler in the United States, in 1954 a power generation milestone was announced with the plans to build Eddystone I of Philadelphia Electric Company setting four "firsts" in the power station industry namely: highest capacity (325 MW, 910 t/hr), highest steam pressure (345 bar), highest temperature (650°C, 565°C, 565°C), and highest efficiency. Since that time only the capacity increased up to 4500 t/hr in the United States, up to 2500 t/hr in West Germany, and up to between 1500 and 2500 t/hr as an international standard as shown in Fig. 7.1. In [2], a large number of units with supercritical pressure is listed. Because of the high material costs for austenitic steel and some availability problems in the past, today's standard steam pressure is about 190 bar and the steam temperature about 535°C. One reheater stage and feed-water bled-steam heaters (steam is extracted from the turbine to preheat the feed water thus increasing the cycle efficiency) are provided.

Of course, there are also some special requirements, which make it uneconomical or impossible to use high pressure and high temperatures [4]. Small industrial boilers or boilers for cogeneration of power and heat for district heating and peak plants are examples of such a case. For these boilers it is not economical to select sophisticated parameters with extensive bled-steam feed-water heaters and a reheat system. Together with a back-pressure turbine, the design pressure is normally between 60 and 120 bar, the feed-water temperature is between 120 and 150°C, and the temperature at the superheater outlet is between 400 and 520°C, natural circulation is applied. This leads to a simple mechanical design with a generous furnace as evaporator and an economic arrangement of superheater and economizer.

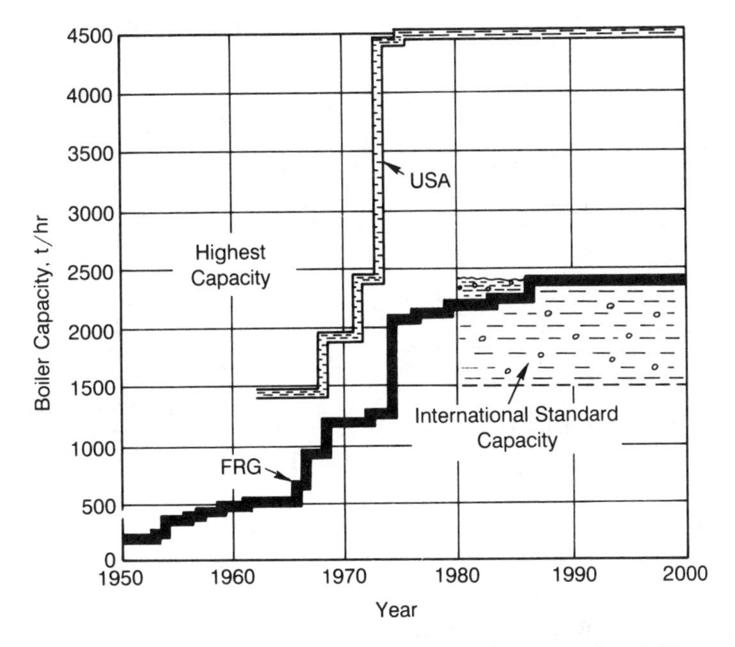

Fig. 7.1. Development of boiler capacity. (Adapted from [2].)

Carbon material is used for all tubes, except the final superheater where a low alloy is necessary to achieve acceptable lifetime (200,000 operating hours).

Other examples are boilers in the chemical industry, where, for example, safety reasons may demand a system pressure below the process pressure to avoid leakage of water and/or steam into the process or the available fuel and its combustion products contain components which allow only a certain (metal) surface temperature to avoid or minimize corrosion. For example, black-liquor-fired boilers in a paper mill allow only a surface metal temperature of about 300°C for the furnace walls leading to a corresponding maximum saturation pressure of 80 bar.

Therefore the selection of the system requires careful consideration of the specific needs, boundary conditions, and customer specifications.

Once-through boilers are preferred for two main reasons:

1. The need for high steam pressure to gain higher cycle efficiency (η_{cyc}) in a water–steam cycle. Figure 7.2 shows the Clausius–Rankine cycle at low and high pressure levels and Fig. 7.3 the further improvement by reheating the steam partly expanded in the high-pressure part of the turbine and preheating the feed water by bled-steam feed-water heating. The cycle becomes more and more similar to the Carnot cycle; the mean upper cycle temperature increases and the mean lower cycle

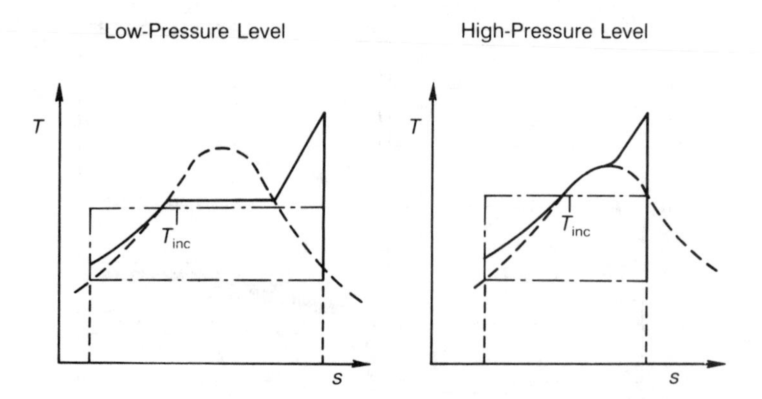

Fig. 7.2. Clausius–Rankine cycle. (With permission of [4].)

temperature decreases resulting in a higher efficiency. Figure 7.4 shows the historical development of the steam pressure, the steam temperatures, heat consumption, and efficiency.

2. The tendency to higher steam capacities because of decreasing specific capital and personnel costs.

Both reasons are connected with the need for drums and greater evaporator tube diameters of circulation boilers, that is, with the higher material costs of these boilers and both reasons influence strongly the cost of power

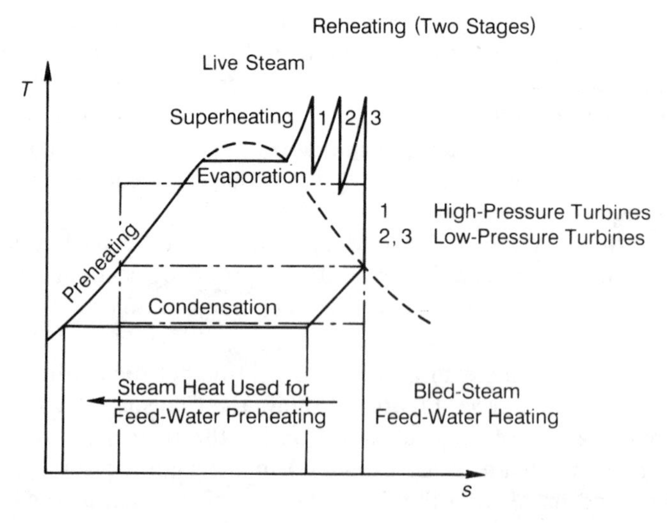

Fig. 7.3. Water–steam cycle with reheaters and feed-water preheaters. (With permission of [4].)

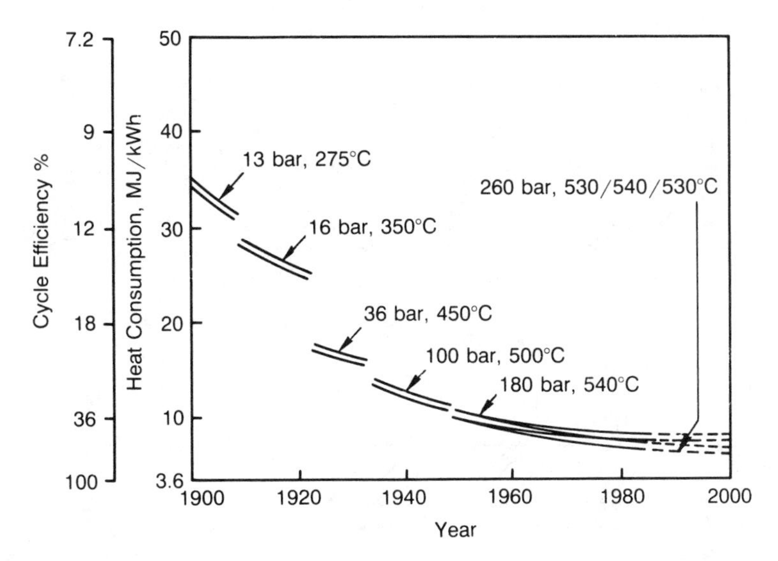

Fig. 7.4. Historical development of steam pressure, temperature, heat consumption, and cycle efficiency. (Adapted from [2].)

generation C_{el}, which can be calculated according to the following equation [5]:

$$C_{el} = \underbrace{\frac{C_{cap} am f_{fc}}{8760 f_{av} f_{us}}}_{\text{investment}} + \underbrace{\frac{C_{fuel} f_{var}}{\eta_{cyc}}}_{\text{operating (fuel) costs}} \tag{7.1}$$

where C_{cap}am is the specific capital cost per year, and $8760 f_{av} f_{us}$ is the equivalent full-load operating hours per year (same total energy output).

One can see from this equation that decreasing specific capital costs (C_{cap}) and increasing cycle efficiency (η_{cyc}) reduce electrical power costs (C_{el}). But, if the increasing efficiency (η_{cyc}) is combined with decreasing availability (f_{av}) and/or increasing specific capital costs (C_{cap}), the electrical power costs (C_{el}) may rise. See also appendix 7.1.

The design, construction, and operation of once-through boilers became easier and easier, because of the progress in water treatment plants, control systems, materials, and the change from brick walls to skin casing and water walls [53] (Fig. 7.5) (which on the other hand was a necessity for higher capacities).

Every time, when the fuel costs rise rapidly, the use of higher steam pressure is discussed and the once-through boiler becomes more economical. Today, an increase in the efficiency would also be desirable because of the reduction of the CO_2 emission and the influence on the global climate.

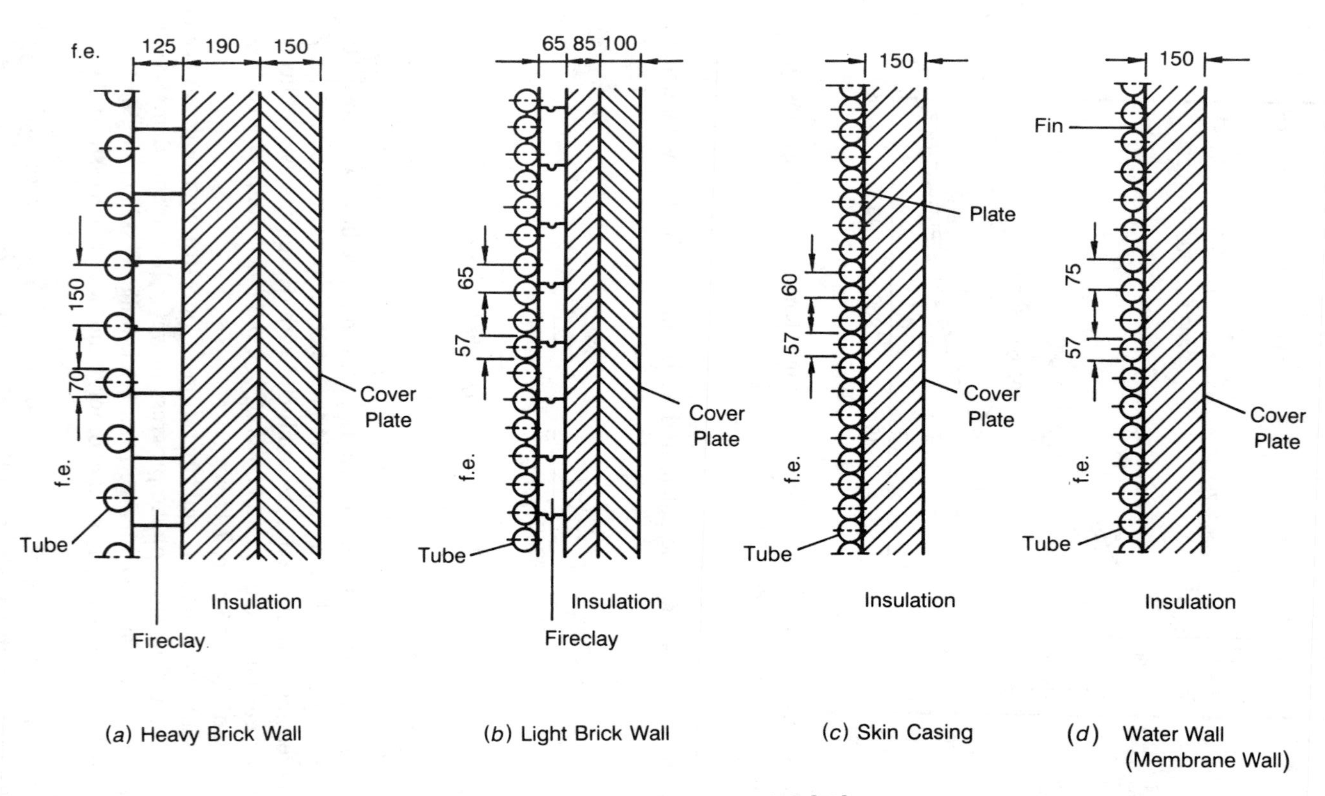

Fig. 7.5. Furnace wall construction [53].

7.2 IMPORTANT DESIGN CRITERIA IN COMPARISON TO OTHER SYSTEMS

The term *once-through* or natural-circulation boiler only refers to the flow in the evaporator. The economizers, superheaters, and reheaters are all operated in the once-through mode even in a natural-circulation boiler. Usually, the evaporator tubes form the furnace walls, mainly for two reasons:

1. The nearly constant temperature (also for supercritical pressure there is a region where the temperature changes very little at nearly constant pressure, when the enthalpy is increased) allows welded water walls with minimum stress.
2. The good heat transfer of nucleate boiling provides low material temperatures in spite of the high radiation heat flux.

The need for high transfer rates for safe operation dictates one of two operating modes:

1. Increase of heat transfer coefficients by increasing mass flux and designing the tubes for the appropriate temperature. For a given steam generation rate (feed-water flow), this solution leads to a smaller evaporator tube diameter, a smaller number of evaporator tubes in

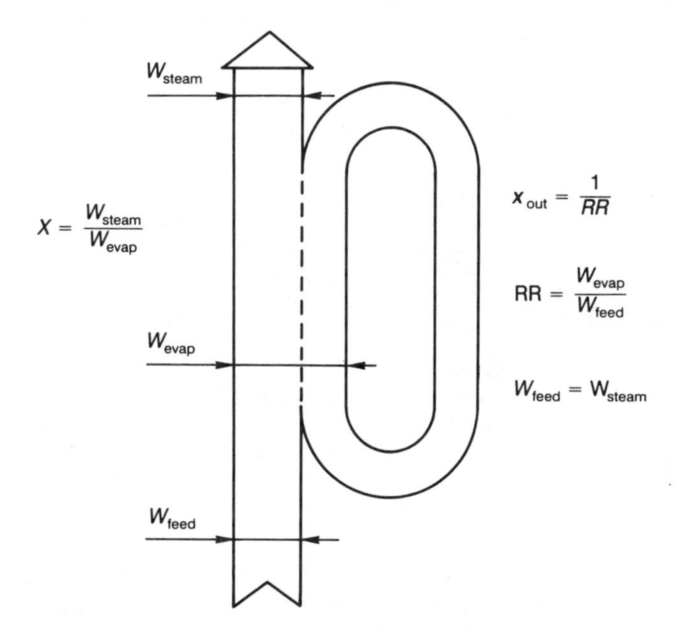

Fig. 7.6. Evaporator recirculation ratio and steam quality at the evaporator outlet.

parallel, and higher pressure drops in the evaporator. This is the once-through boiler solution.

2. Avoidance of departure of nucleate boiling DNB or dryout DO (see Chapter 6 and Section 7.2.5) by flow recirculation and in this way keeping the steam quality low (see Fig. 7.6). This is the evaporator operating mode of all circulation boilers.

7.2.1 Main Characteristic Features

Figure 7.7 shows the flow schemes of five evaporator systems. Basically, there are the two systems according to the two previously mentioned evaporator

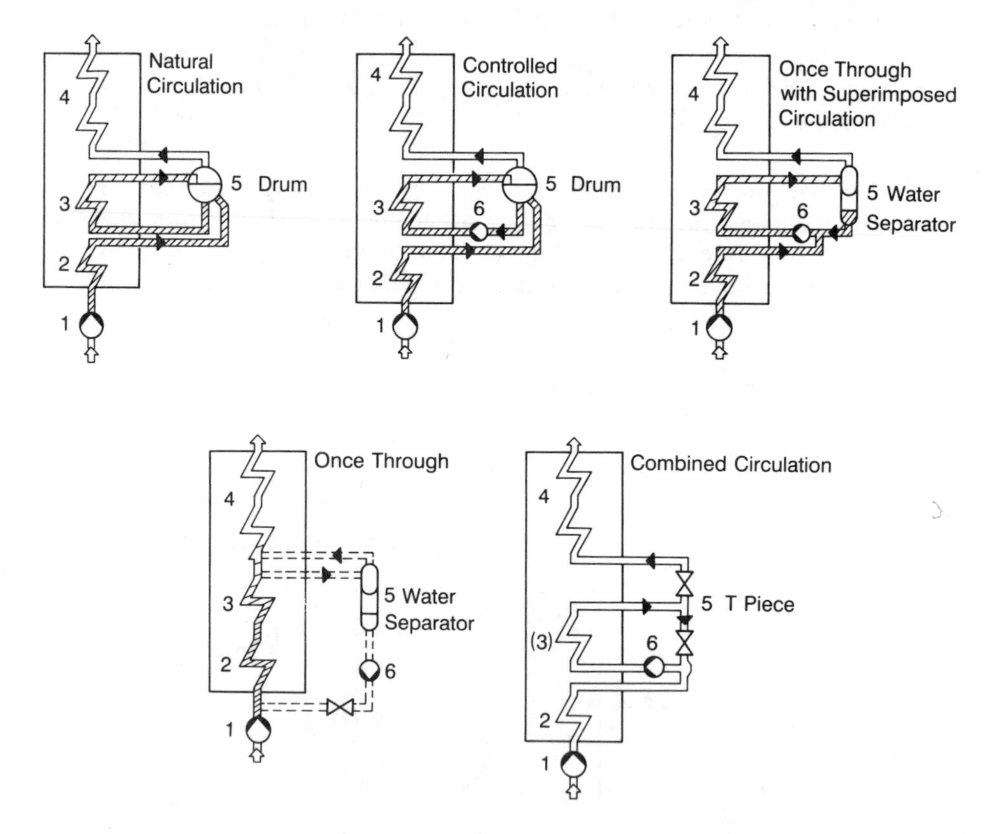

Fig. 7.7. Steam-generating systems. (Adapted from [4].)

operating modes:

1. Once-through system consisting theoretically simply of one tube, in which the water is preheated, evaporated, and superheated, while it flows through. In a real steam generator, this tube is divided into several heating surfaces with many tubes in parallel, headers (collectors) at the inlet and outlet, and pipes connecting the headers of the heating surfaces. These heating surfaces are: economizer (for preheating), evaporator (usually for building the furnace and boiler walls), and superheaters. (Of course, reheaters are also used, but are the same in any system.) The arrangement can be seen in Figs. 7.7, 7.48, 7.51 to 7.56, 7.60, and 7.61.
2. Recirculation systems (see also Chapter 6, especially Section 6.4).

The latter has four subsystems:

1. Natural circulation [the driving force is provided by gravity because of the different weights of the water in the downcomers from the drum and the water–steam mixture in the heated riser tubes forming the walls of the furnace (water walls, membrane walls)] and three types of pump-assisted circulation systems, namely
2. Controlled circulation with a pump assisting natural circulation (allowing higher pressures with less difference in specific weight between water and the steam–water mixture of a certain steam quality), a drum, and a recirculation ratio (see Fig. 7.6) above 2.
3. Once through with superimposed circulation with a pump as in controlled circulation, but with a (vertical) water separator vessel (smaller diameter) instead of a (horizontal) drum and recirculation ratios below 2.
4. Combined circulation with a pump as in controlled circulation and once through with superimposed circulation, but with a reducing valve ahead of the superheater so that the "evaporator" can be operated at all loads at supercritical pressure and the drum or water separator are replaced by a simple T piece. DNB and DO is avoided by supercritical pressure and a sufficient heat transfer coefficient is provided by flow recirculation usually at loads below 70%. Once-through operation is usually applied at full load.

The two systems have two main characteristic features:

1. In a once-through system, the liquid–vapor phase transition point is variable, whereas for circulation systems, the transition point is fixed in the water separator or drum, respectively.
2. In the once-through system, the evaporator pressure drop is supplemented by the feed-water pump. In the natural-circulation system, the

evaporator pressure drop is overcome by the density difference between the water and the steam–water mixture of the downcomers and risers, respectively, and in the other recirculation systems the natural circulation is assisted by a pump.

7.2.2 Pressure Range

Figure 7.8 shows the pressure ranges of the different steam generator systems.

Once-Through Steam Generators In a once-through boiler, the pressure range is not limited by the evaporative system but by the stresses to which furnace walls and main steam headers are subjected. There is also a usual economical lower limit for once-through boilers at about 80 bar because of the high pressure drop in the evaporator.

Steam Generators with Recirculation in the Evaporator

Once-Through Steam Generators with Superimposed Circulation This system is capable of subcritical operation (up to a maximum pressure of

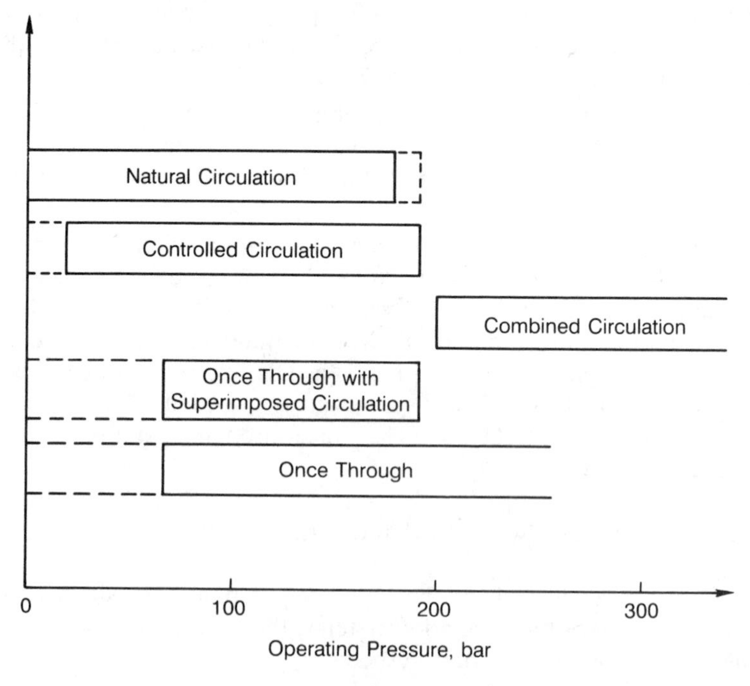

Fig. 7.8. Applicable pressure range of steam generators. (Adapted from [4].)

approximately 200 bar in the water separator, evaporator tubes are designed for DNB or DO) and in principle also of supercritical operation like a CE–combined-circulation steam generator without throttle valves before the superheaters.

In the past it was commonly accepted that the critical pressure range should be avoided because of the deficient accuracy of the water level measurement despite the high enthalpy difference between saturated water and saturated steam. Therefore it was assumed that the feed-water level controller and the feed-water temperature controller would not render a satisfactory control.

At the Mannheim Power Station, Unit 18, a special feed-water control was tested that allows the steam generator to be operated in the critical pressure range [3].

The lower-pressure limit usually at about 80 bar has economic reasons caused by the relative high-pressure drop in the evaporator.

Combined-Circulation Steam Generator The combined-circulation steam generator is operated at higher loads at supercritical pressure. The supercritical pressure can be provided in the evaporator also during part load with sliding pressure operation (see Section 7.2.3) and start-up by throttle valves before the superheaters.

Controlled-Circulation Steam Generator The controlled-circulation steam generators are usually operated up to a maximum pressure of approximately 200 bar in the drum. At high pressures the risk of departure from nucleate boiling (DNB) or dryout (DO) increases. To avoid this, higher recirculation ratios (RR > 2 at full load) compared with the once-through boiler with superimposed recirculation (RR < 2 at full load) are used.

Natural-Circulation Steam Generator Natural circulation in the evaporator system (see Chapter 6, especially Sections 6.3.5 and 6.4) already starts when the water in the furnace wall tubes (evaporator tubes, risers) is hotter than in the downcomers, if the furnace tubes and their connecting tubes to the drum are arranged below the water level in the drum. The water in the downcomers has a lower temperature because of the heat losses of the downcomers, the time lag between risers and downcomers, and (after evaporation started) the mixing of the recirculated water with the feed water. Also at supercritical pressure a natural circulation is possible. Nevertheless, the application of the natural-circulation system is usually limited to a maximum pressure of approximately 180 bar in the drum [6, 7]. This pressure limit is based on the fact that the recirculation ratio RR for a typically designed circulation system (of course, the tube diameters, height, heat input, etc., also influence this RR) decreases as the pressure increases, for example, usually at 100 bar RR $= 10$ at 180 bar RR $= 6.5$. This phenomenon entails the risk of DNB or

DO (see Section 7.2.5) because a lower RR means a lower mass flux and a higher steam quality.

In recent years natural-circulation boilers operating with drum pressures around 20 MPa are in use. The design of such boilers has to concentrate particularly on the correct calculation of the heat transfer and the design temperature of the evaporator tubes (see Section 7.2.5).

Because of the pressure drop of the water-separating internals in the drum and the high pressure drop in the superheater, which is necessary to keep differences in steam temperature between individual superheater tubes low when large flue gas temperature maldistribution occurs, a relatively low superheater outlet pressure is available. In large furnaces fired with coal of low specific heating value and with asymmetrical furnace heat input, it is necessary to split the superheater–reheater steam paths into two or four parallel paths. Crossovers are required at each desuperheating stage. The pressure drop over integral piping (long run, bends) significantly contributes to the already high pressure drop over the superheater stages which is required for good flow distribution [4].

7.2.3 Operating Modes and Start-Up Period

From the turbine power equation

$$P_T = W_{\text{steam}}(i_i - i_o) = W_{\text{steam}}(i_i - i_{o,\text{ad}})\eta_i \tag{7.2}$$

one can see that the turbine power output is mainly controlled by the live steam flow as the turbine inlet and outlet temperatures (and thus approximately the enthalpies) are usually kept constant to avoid reductions in efficiency. The mass flow through a turbine admission valve or through a turbine (as it represents for this purpose only an opening with supercritical pressure ratio) is simply proportional to the product of the flow area A_v and the pressure p before the valve or turbine, respectively,

$$\frac{W_{\text{steam}}}{W_{\text{steam}_o}} = \frac{A_v}{A_{v_o}} \frac{p}{p_o} \tag{7.3}$$

Therefore the turbine power can be controlled by controlling the valve opening and/or by controlling the pressure before the valve or turbine.

There are four operating modes used today as shown in Fig. 7.9.

Natural sliding pressure operation does not need turbine control valves. The turbine power is controlled by the firing controller and has the longest delay (Fig. 7.10). At fixed pressure operation mode, the turbine power is controlled by the turbine valve and the resulting pressure deviations are controlled by the firing rate. The only difference between the fixed pressure and controlled sliding pressure operation modes is that in controlled sliding

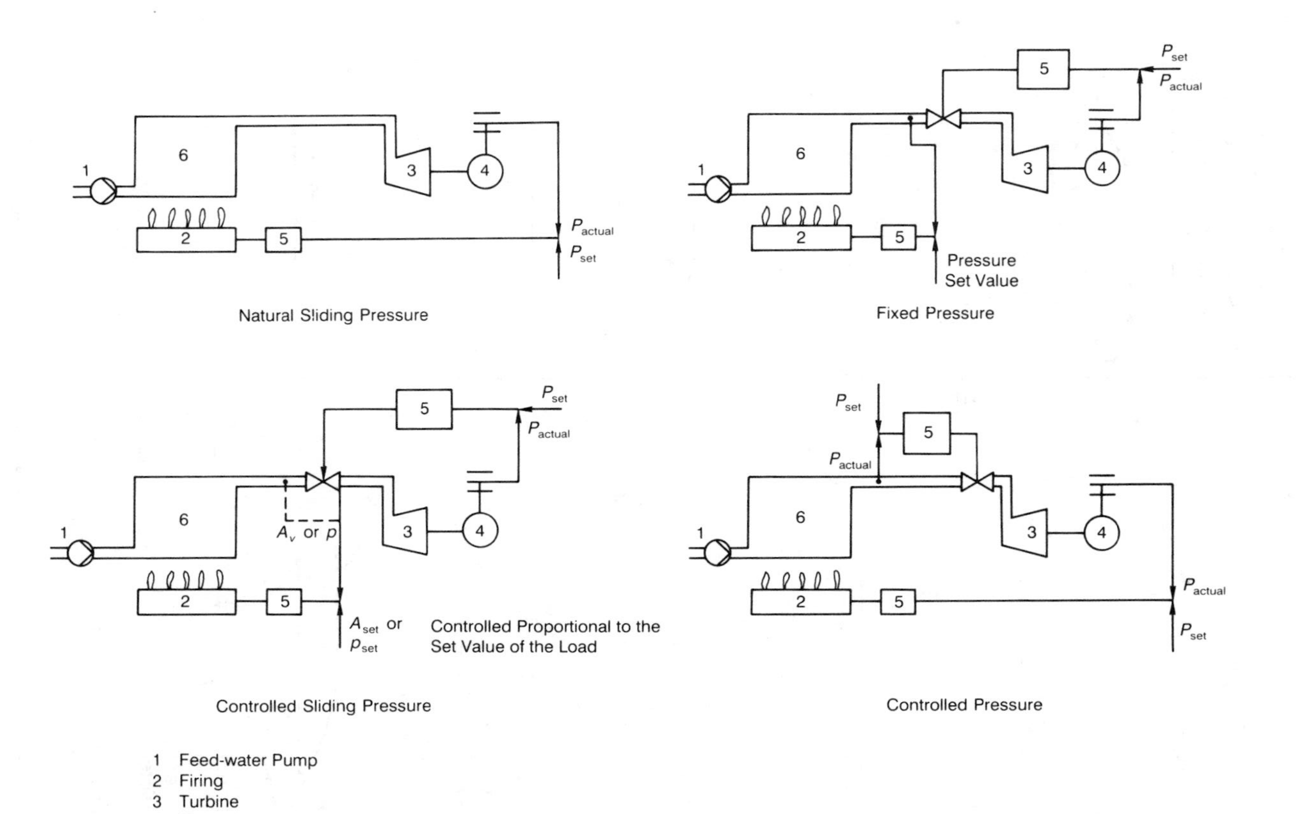

Fig. 7.9. Operating modes.

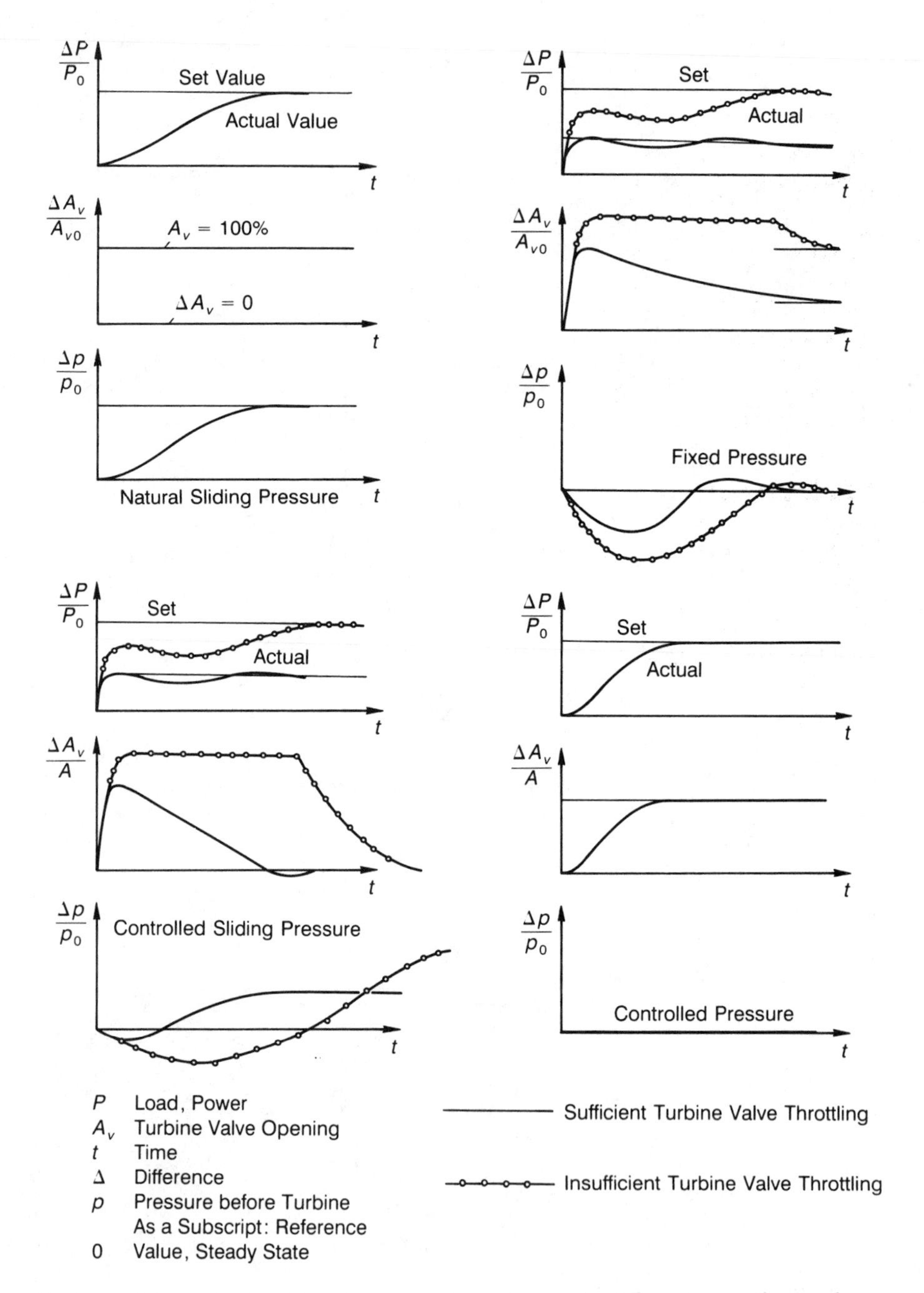

Fig. 7.10. Schematic time histories of step load changes in different operating modes.

pressure operation the set value for the pressure (or the turbine valve throttling) is not fixed but proportional to the set value for the boiler load. As long as the turbine throttling is sufficient, both are equally fast. If the throttling is insufficient, that is, the turbine valve opens fully during the load change, there is a changeover from fast fixed pressure to slow natural sliding operation (Fig. 7.10). At controlled pressure operation mode, the pressure in the boiler is kept constant by the turbine valves, load changes are controlled by the firing rate, and the delay is between sliding and fixed pressure operation mode (Fig. 7.10).

In principle, all steam generator systems are capable of operating at controlled, fixed, natural sliding, or controlled sliding pressure. At least in modern practice steam generators are started up in sliding pressure operation mode.

The boiler pressure drop at fixed pressure operation is proportional to the square of the load, whereas the boiler pressure drop at natural sliding pressure operation is directly proportional to the load due to the fact that the mean steam density is approximately proportional to the pressure; that is, in natural sliding pressure operation it is proportional to the load.

In once-through steam generators with superimposed-circulation and natural-circulation steam generators, however, the admissible rate of pressure change (i.e., rate of load changes at sliding pressure operation) is mostly conditioned by the thermal stresses in the water separator (cyclone) and drum which—as compared to the cyclones in once-through units of the same rating—are larger in diameter, resulting in thicker walls and lower admissible temperature transients under identical pressure, temperature, and material conditions. The other bottlenecks are the headers, which will be discussed later in this chapter.

This limitation applies both to load reduction and load increase [49]. The preceding components are located in the saturated steam range. By approximation, the saturation temperature T_s, in °C, at a pressure p, in bar, is

$$T_s \approx 100\sqrt[4]{p} \tag{7.4}$$

and the saturation temperature variation as a function of pressure variation is

$$\frac{dT_s}{dp} \approx 25p^{-3/4} \tag{7.5}$$

According to the German boiler code TRD 301 [10], Annex 1, the admissible rate of temperature change, in K/s, at a pressure of 1 bar and at operating pressure shall be v_{T1} and v_{T2}, respectively (within certain limits v_{T1} or v_{T2} can be selected).

Neglecting the non-quasistationary conditions that exist in the initial and final phase of a temperature and pressure increase, the required time t, in s, can be calculated provided that the temperature increase occurs at the admissible temperature transient v_T, as applicable. Hence

$$t = \int_{p_1}^{p_2} \frac{dT_s}{dp} \frac{dt}{dT} \, dp \tag{7.6}$$

where dT_s/dp is substituted by Equation (7.5). According to TRD 301, Annex 1, the following shall be applicable for dt/dT:

$$\frac{dT}{dt} = v_T = v_{T1} + \frac{p - p_1}{p_2 - p_1}(v_{T2} - v_{T1}) \tag{7.7}$$

This can be solved as

$$t = \frac{25}{\sqrt{2}} \frac{1}{c_1 c_4 c_3^3} \left\{ \ln \frac{c_4^2 + c_3\sqrt{2c_4} + c_3^2}{c_4^2 - c_3\sqrt{2c_4} + c_3^2} + 2 \arctan \frac{c_3\sqrt{2c_4}}{c_3^2 - c_4^2} \right\}_{c_{41}}^{c_{42}} \tag{7.8}$$

where

$$c_1 = \frac{v_{T2} - v_{T1}}{p_2 - p_1} \tag{7.9}$$

$$c_2 = v_{T2} - c_1 p_1 \tag{7.10}$$

$$c_3^4 = \frac{c_2}{c_1} \tag{7.11}$$

$$c_4 = \sqrt[4]{p} \qquad c_{41} = \sqrt[4]{p_1} \qquad c_{42} = \sqrt[4]{p_2} \tag{7.12}$$

Table 7.1 gives the operating temperature, inside diameter, wall thickness, materials, admissible rate of temperature change, admissible temperature differences, and the time t for different steam generating systems as required for a pressure increase from 1 bar, 2 bar, 5 bar, 10 bar and from 30% to 100% operating pressure.

It becomes evident that the start-up period of drums at operating pressures below 140 bar is almost identical with that of water separators (cyclones) at 190-bar operating pressure. At operating pressures of 190 bar, however, drums show a slow start-up characteristic (see Table 7.1).

Table 7.2 gives the start-up periods for the main steam headers calculated on the assumption that the main steam temperature varies proportionally to the saturation temperature, the maximum main steam temperature being

TABLE 7.1 Start-Up Time for Drums and Cyclones for Different Design Pressures (with Permission of [49])

	Start-Up Time, min				
	1	2	3	4	5
Pressure part	Drum	Drum	Drum	Cyclone	Cyclone
Material	WB 36	WB 36	WB 36	WB 36	13 CrMo 44
100% pressure, bar	72.8	141	185	192	286
Inner diameter, mm	1620	1532	1680	585	420
Wall thickness, mm	41	84	116	46	74.3
a.t.d.[a]					
At 0 bar ΔT_I K	−8.1	−9.5	−8.3	−9.8	−15.4
At 100% bar, ΔT_E K	−71.1	−64.9	−67.4	−63.5	−50.0
a.t.t.[b]					
At 0 bar, v_{TI} K/min	8.7	2.3	1.0	7.0	4.0
At 100% bar, v_{TE}, K/min	75.8	15.4	7.7	45.2	12.8
t, 1 bar, 100%p, min	9.74	56.5	137.7	21.42	55.6
t, 2 bar, 100%p, min	7.85	48.6	120.2	18.82	50.9
t, 5 bar, 100%p, min	5.24	36.9	92.8	14.83	43.5
t, 10 bar, 100%p, min	3.39	27.4	70.4	11.47	36.8
t, 30%, 100%p, min	1.68	9.5	20.6	3.46	12.0

[a]a.t.d.: allowable wall temperature difference, TRD 301.
[b]a.t.t.: allowable temperature transient of fluid, TRD 301.

TABLE 7.2 Start-Up Time for Live Steam Headers for Different Design Pressures (with Permission of [49])

	Start-Up Time, min			
	6	7	8	9
Material	10 CrMo 910	X20 CrMoV 121	X20 CrMoV 121	X20 CrMoV 121
100% pressure, bar	65	171	172	255
100% temperature, °C	515	540	539	530
Inner diameter, mm	250	280	250	200
Wall thickness, mm	29.9	62.1	52.9	49.4
a.t.d.[a]				
At 0 bar ΔT_I K	−25.4	−23.3	−22.7	−20.2
At 100% bar, ΔT_E, K	−37.9	−43.0	−43.2	−46.9
a.t.t.[b]				
At 0 bar, v_{TI} K/min	35.8	5.1	6.9	7.1
At 100% bar, v_{TE}, K/min	53.3	9.5	13.2	16.5
t, 1 bar, 100%ρ, min	8.21	63.9	46.8	44.0
t, 2 bar, 100%ρ, min	7.26	58.5	42.8	40.5
t, 5 bar, 100%ρ, min	5.75	49.6	36.3	34.9
t, 10 bar, 100%ρ, min	4.39	41.6	30.4	29.8
t, 30%, 100%ρ, min	2.91	18.4	13.3	11.1

[a]a.t.d.: allowable wall temperature difference, TRD 301.
[b]a.t.t.: allowable temperature transient of fluid, TRD 301.

reached at full load. A comparison of Tables 7.1 and 7.2 suggests that the main steam headers may not be capable of following the fast temperature transients of the cyclones which would be a deficit in the fast starting characteristic of the once-through system.

For the main steam headers, an initial temperature of 50°C was used resulting in theoretical pressures of less than 1 bar absolute. This corresponds to the conditions at a vacuum start-up and is helpful in avoiding the problems arising from the temperature step changes occurring at 1 bar due to condensation.

The variation of the saturation temperature is particularly dominating at low pressures (low loads in sliding pressure operation). Thus, in this load (and pressure) range, boilers also operated in controlled sliding pressure mode at higher loads are mostly run in the fixed pressure mode at a lower pressure. In addition, this helps to avoid steaming in the economizer. This is also the reason why auxiliary steam preheating is recommended for all steam generator systems (besides the favorable effect on the firing system, avoiding temperatures in the air preheater falling below the dew point, and saving oil).

A detailed revision of the transient conditions only adds to the advantages of the drum taking into consideration that the TRD 301, Annex 1, overestimates the alternating stress range [11]. The main steamline and the tie bars of an inclined tubed water wall have to be taken into analogous considerations, to avoid a bottleneck for start-up [11] (see Sections 7.2.8 and 7.3.1).

Of course, all circulation boilers except the supercritical combined-circulation boiler can have problems when the pressure is decreasing during load reductions in the sliding pressure operation mode or during step-load increases in the fixed pressure operation mode. If the pressure causes evaporization in the downcomers, the (natural-circulation) flow may also nearly stop in the evaporator tubes (furnace wall) resulting in tube damage by overheating. In once-through boilers with superimposed recirculation or in controlled-circulation boilers, the net positive suction head (NPSH) of the recirculation pump may drop below the minimum value causing the recirculation pump to trip in order to avoid cavitation and/or imbalance and the steam generator to be shut down in order to avoid damage of the water wall tubes.

In once-through steam generators with superimposed circulation, sufficient subcooling is basically given at higher loads because of the admission of water from the economizer so that a trip of the recirculation pump following a pressure decrease would only occur during the first stage of start-up.

The admissible rates of pressure reduction and optimum design of downcomers and/or suction pipes to pumps which allow the fastest pressure reduction without evaporization in the downcomers or suction pipes to pumps, causing evaporator tube damage or pump trips, respectively, are described in the following paragraph [8, 9].

For an equally inclined (angle α to the vertical downward direction) pipe in which water (saturated at the inlet = top) flows downward, one can easily find the differential pressure decrease dp, which totally consumes the difference between the pressure gain due to the hydrostatic head and the friction pressure drop, is

$$dp = \rho g \, dh - \lambda_f \frac{dl_t}{d_i} \frac{\rho v^2}{2} \tag{7.13}$$

The division by the differential time dt, the water needs to flow from one point to the other (differential distance

$$dl_t = dh / \cos \alpha \tag{7.14}$$

$$v = \frac{dl_t}{dt} \tag{7.15}$$

differential pressure dp between these two points) gives the pressure transient, which can be applied without evaporization

$$\frac{dp}{dt} = v \left(\rho g \cos \alpha - \frac{\lambda_f}{d_i} \frac{\rho v^2}{2} \right) \tag{7.16}$$

Of course, the pipe should be vertical to allow the fastest pressure decreases and necessary horizontal or inclined parts should be arranged at the bottom, so that there is already a big hydrostatic head gained.

By setting the differentiation of dp/dt by dv to 0, one can obtain the optimal water velocity v_{opt}, which allows the fastest pressure decreases without evaporization

$$v_{\text{opt}} = \sqrt{\frac{2 d_i \rho \cos \alpha}{3 \lambda_f}} \tag{7.17}$$

Eliminating v in Eq. (7.16) by given mass flow W,

$$W = \rho v \frac{d_i^2 \pi}{4} \tag{7.18}$$

and setting the differentiation of dp/dt by dd_i to 0, one obtains the optimal diameter, which allows the fastest pressure decrease without evaporization

$$d_i = \sqrt[5]{\frac{28 \lambda_f W^2}{\rho^2 g \pi^2 \cos \alpha}} \tag{7.19}$$

Of course, small inner diameters of the downcomers reduce the recirculation mass flow increasing the risk of DNB or DO also during steady state. Smaller diameters of suction pipes to pumps increase the pressure drop and thus the risk of cavitation; therefore a bit larger diameters are preferable. See also Appendix 7.2.

7.2.4 Start-Up Equipment and Problems

Start-Up Equipment Independent of the evaporator systems, steam generators should have high-pressure and low-pressure turbine bypasses to allow boiler operation without the turbine and a steam preheating facility (see also Section 7.4).

Concerning the evaporator section, natural-circulation, controlled-circulation, combined-circulation, and once-through boilers with superimposed circulation do not need any additional equipment besides, as all others, a blowdown (drain) line with discharge valves and flash tank (separator) for discharging the start-up water swell (water release due to starting evaporation) from the water separator or drum into the condenser and/or tank and atmosphere (see Section 7.4.1). The size is according to the evaporator volume, start-up time, heat pickup, and possible overfeeding of the evaporator. In the case of once-through steam generators, a choice is to be made between

1. Low load circulation pump
2. Circulation by feed-water pump through heat exchanger and feed-water tank
3. Whether the minimum evaporator flow is maintained by the feed water and the discharge valves are sized for overfeeding of the evaporator. The latter entails the highest losses in water and heat.

Evaporator Recirculation Pump Arrangement Natural-circulation boilers do not need any evaporator recirculation pump. In order to allow circulation in the natural-circulation system at low heat pickup rates and before steaming (to avoid major local temperature differences between tubes, through which water is flowing, and tubes, in which the water stagnates, or temperature transients, when stagnant water starts circulating), particular care should be taken that from all recirculation systems (furnace and boiler parts) at least part of the overflow tubes enter the drum below the water level and not above the water level.

In controlled-circulation boilers the recirculation pump is arranged in the evaporator loop, the economizer discharges into the drum as in natural-circulation boilers or into a vessel in the downcomers.

In once-through boilers with superimposed circulation, the recirculation pump is also arranged in the evaporator loop, but the water flow from the economizer outlet is mixed with the recirculated saturated water from the water separator in a mixing vessel, thus subcooling the water flowing to the recirculation pump and allowing faster pressure decreases, if the water at the economizer outlet is subcooled.

In combined-circulation boilers, the recirculation pump has the same position as in once-through boilers with superimposed circulation (only the water separator is replaced by a simple T-piece).

In once-through boilers, the recirculation pump is usually arranged after the water separator before the mixing with the feed water, thus, pumping saturated water subcooled only by the hydrostatic head, which allows only slow pressure decreases. But there is also a possibility of arranging the recirculation pump in series with the feed water pump after the mixing vessel, thus allowing faster pressure decreases. The advantages and disadvantages of these recirculation pump arrangements are compared in detail in [3].

Only for once-through boilers is it possible to include the economizer in the recirculation loop (without special economizer recirculation equipment) to avoid flow instability and dew point problems (in the economizer, air heater, and ducts by heating the flue gas in the economizer) during start-up, because at higher loads there is no recirculation and the economizer can pick up heat as usual.

Water–Steam Separation Water–steam separation is necessary for once-through boilers only at part load, for all recirculation boilers besides the combined-circulation boiler over the whole-load range. In once-through steam generators with low-load circulation or superimposed circulation, the separation of water and steam is made by means of cyclones (see Section 7.3.3). It is of no essential importance whether the separation of water and steam and the storing of the water takes place in the same or in separate vessels. The latter pattern offers the possibility of using several vessels in parallel arrangements with reduced diameter and wall thickness, allowing higher admissible temperature transients.

The simplest way for water–steam separation in the drum (natural- and controlled-circulation boilers) is by gravitational separation; for higher pressures, other equipment, especially a greater number of small cyclones, may be installed (see Section 6.3.5).

It is important to reach a low residual moisture, though the main aspects nowadays in this requirement are not the minerals (which may deposit in superheaters and turbines) contained in the water (which is usually demineralized), but rather the fact that any water entrained causes a reduction of the spray water flow.

Economizer Steaming In once-through boilers, the two-phase mixture produced by economizer steaming results in a nonuniform flow distribution through and damage to the evaporator tubes. This can be avoided if one does not use intermediate headers between the economizer and evaporator. This solution, however, apart from the problems arising under design aspects and with regard to drainage, implies the risk of flow instabilities [26]. See also Section 6.5.

In once-through steam generators with superimposed circulation and in controlled-circulation steam generators when the economizer discharges into a vessel in the downcomers, the steam leaving the economizer arrives at the recirculation pump causing it to trip.

Only in natural-circulation and controlled-circulation steam generators, when the economizers discharges into the drum, economizer steaming is not prejudicial since the steam can be separated in the drum. But the economizer has to be designed to allow steaming; steady upward flow is useful and the discharge into the drum has to be designed to separate the water and steam.

In supercritical combined-circulation boilers, steaming in the economizer cannot occur.

7.2.5 Evaporator Tube Design

Evaporator Tube Heat Transfer and Wall Temperatures Recirculation of water through the evaporator decreases the steam quality, and departure of nucleate boiling DNB or dryout DO are thus avoided in natural- and controlled-circulation boilers and wall temperatures are only slightly above the water–steam temperature because of the high heat transfer coefficients of nucleate boiling [7, 12]. Once-through evaporation cannot avoid at least DO, and in once-through boilers with superimposed circulation, DNB or DO is also accepted. In combined-circulation boilers the supercritical pressure makes DNB and DO impossible.

Drescher and Köhler [13] verified by their measurements and by about 3000 measurements from other sources that for DNB the values and equations of Doroshchuk [14] and for DO those of Konkov [15] fit best [19].

DNB occurs for heat fluxes higher than the critical heat flux q''_{cr} [19]:

$$q''_{cr} = 10.3 \times 10^3 - 17.5\left(\frac{p}{p_{cr}}\right) + 8\left(\frac{p}{p_{cr}}\right)^2 \left(\frac{8 \times 10^{-3}}{d_i}\right)^{0.5} \times \left(\frac{W''}{1000}\right)^{0.68} \left(\frac{p}{p_{cr}}\right)^{-1.2x-0.3} e^{-1.5x} \tag{7.20}$$

validity range: 29 bar $\leq p \leq$ 196 bar, 500 kg/(m$^2 \cdot$ s) $\leq W'' \leq$ 5000 kg/(m$^2 \cdot$ s), 0 K $\leq \Delta T_u \leq$ 75 K, 4×10^{-3} m $\leq d_i \leq 25 \times 10^{-3}$ m.

DO occurs for heat fluxes higher than the following critical heat fluxes q''_{cr} [19]:

Pressure range: 4.9–29.4 bar:

$$q''_{cr} = 1.8447 \times 10^8 \dot{x}^{-8} W''^{-2.664} (1000 d_i)^{-0.56} e^{0.1372 p} \tag{7.21}$$

Pressure range: 29.4–98 bar:

$$q''_{cr} = 2.0048 \times 10^{10} \dot{x}^{-8} W''^{-2.664} (1000 d_i)^{-0.56} e^{-0.0204p} \tag{7.22}$$

Pressure range: 98–196 bar:

$$q''_{cr} = 1.1853 \times 10^{12} \dot{x}^{-8} W''^{-2.664} (1000 d_i)^{-0.56} e^{-0.0636p} \tag{7.23}$$

Validity range:

$$200 \text{ kg}/(\text{m}^2 \cdot \text{s}) \le W'' \le 5000 \text{ kg}/(\text{m}^2 \cdot \text{s})$$

If one needs the steam quality for DO, one can use the following equations [19]:

Pressure range: 4.9–29.4 bar:

$$\dot{x}_{cr} = 10.795''^{-0.125} W''^{-0.333} (1000 d_i)^{-0.07} e^{0.01715p} \tag{7.24}$$

Pressure range: 29.4–98 bar:

$$\dot{x}_{cr} = 19.398 q''^{-0.125} W''^{-0.333} (1000 d_i)^{-0.07} e^{-0.00255p} \tag{7.25}$$

Pressure range: 98–196 bar:

$$\dot{x}_{cr} = 32.302 q''^{-0.125} W''^{-0.333} (1000 d_i)^{-0.07} e^{-0.00795p} \tag{7.26}$$

In addition, equations for steam superheating due to thermodynamic unbalance and for heat transfer are given [16, 19]. Furthermore, the difference between the wall temperature of a horizontal and vertical tube at DO are described in [17, 19].

With internally rifled tubes, the steam quality at which film boiling (DNB or DO) occurs is considerably higher than with tubes of plain inside surface. Thus the natural-circulation, controlled-circulation, and once-through with superimposed circulation steam generators can (could) be operated at a lower circulation ratio (higher steam quality) and somewhat higher pressure [20, 21].

Internally rifled tubes could also be adopted for once-through systems to allow for changeover to the recirculation mode under very low mass flux and very low load conditions or to use vertical tubing (see Section 7.3.1).

Figure 7.11 shows the fluid temperature, steam quality, heat transfer coefficient and tube wall temperature versus furnace height of a 650-MW bituminous coal-fired once-through steam generator at 100% load. Figure 7.12 shows that at 45% load DO occurs at 63% steam quality. In recirculation boilers, of course, the curve of the actual local heat flux as a function of the local steam quality would not run into DNB or DO areas.

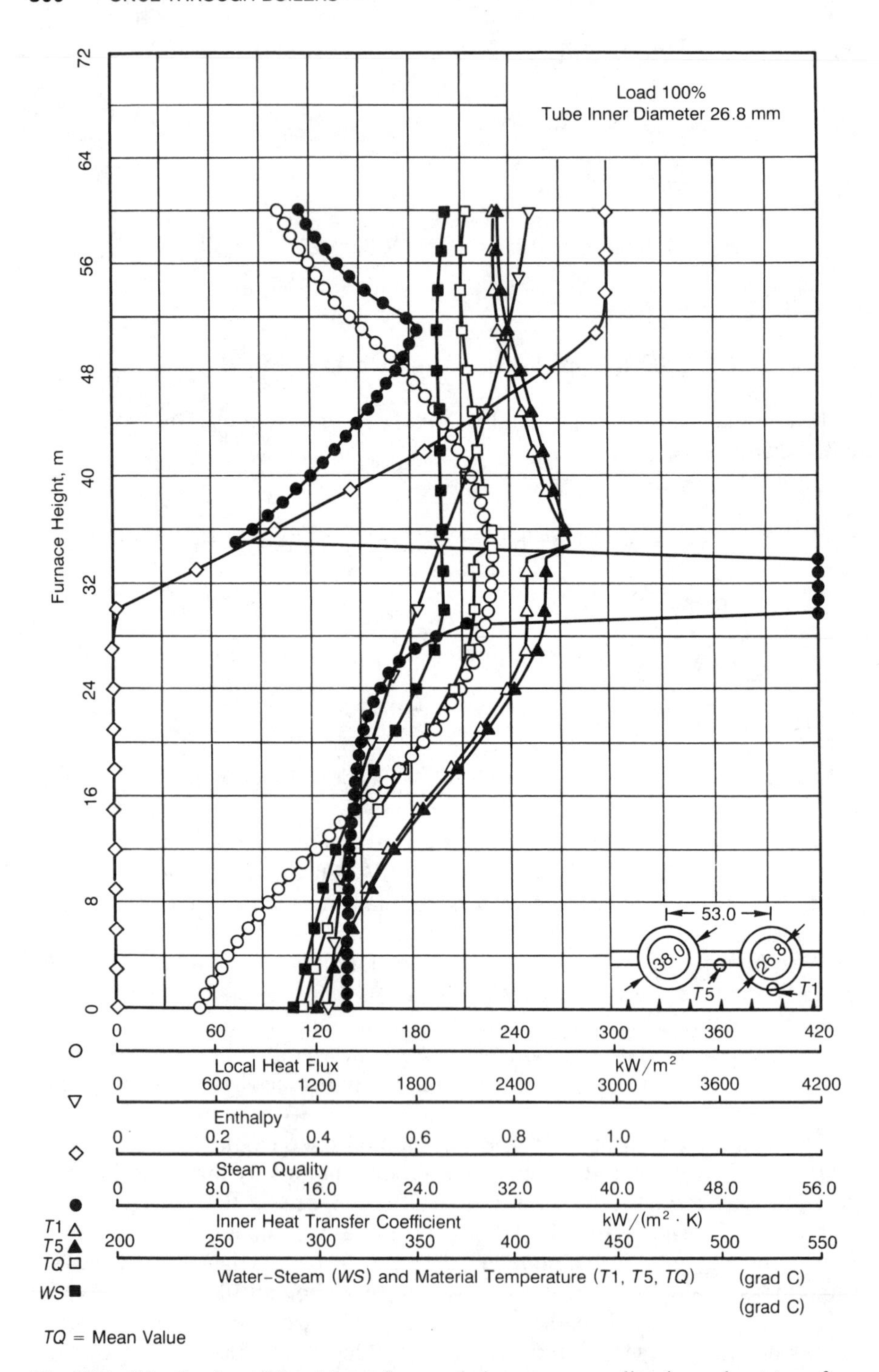

Fig. 7.11. Distribution of local heat flux, enthalpy, steam quality, inner heat transfer coefficient, water–steam, and material temperatures over furnace height of a 680-MW bituminous-coal-fired once-through boiler. (With permission of [49].)

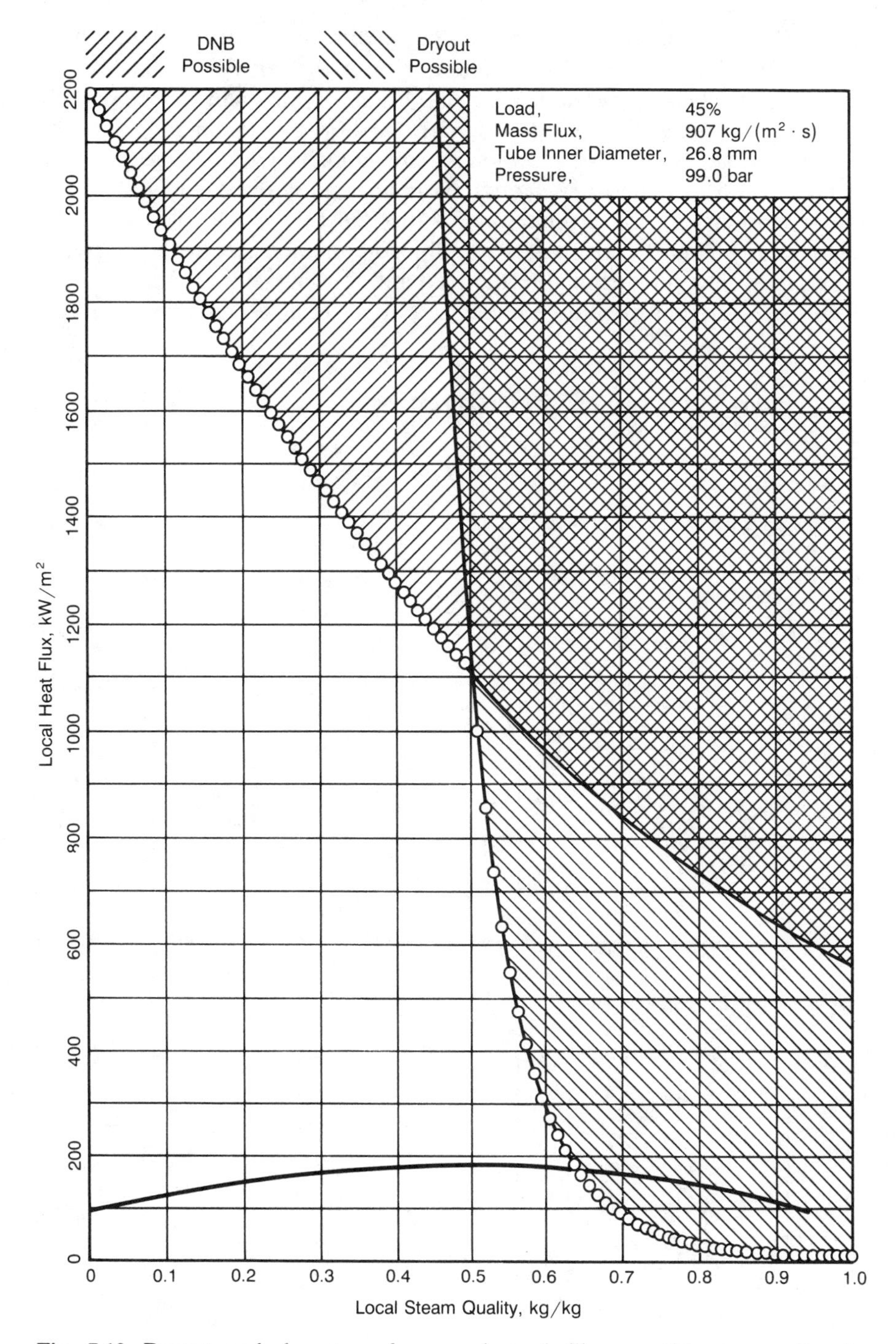

Fig. 7.12. Dryout and departure from nucleate boiling conditions according to Doroshuk and Kon'kov. Actual values of a 680-MW bituminous-coal-fired once-through boiler.

Mass Flux and Pressure Drop in the Evaporator, Ripple Formation, Feed-Water-Pump, and Recirculation Pump Power Consumption In once-through steam generators, the mass flux above the minimum evaporator flow is proportional to the load (Fig. 7.13). Below this minimum value, the mass flow is kept constant by low-load recirculation or overfeeding. Since the local heat flux can almost rise to its maximum value even at low firing rates (oil support firing), the evaporator flow needed to avoid unallowable tube temperatures almost corresponds to full-load conditions. A further increase of the mass flow proportional to the load and thus also of the pressure drop is not necessary and should be minimized by selecting a high minimum load

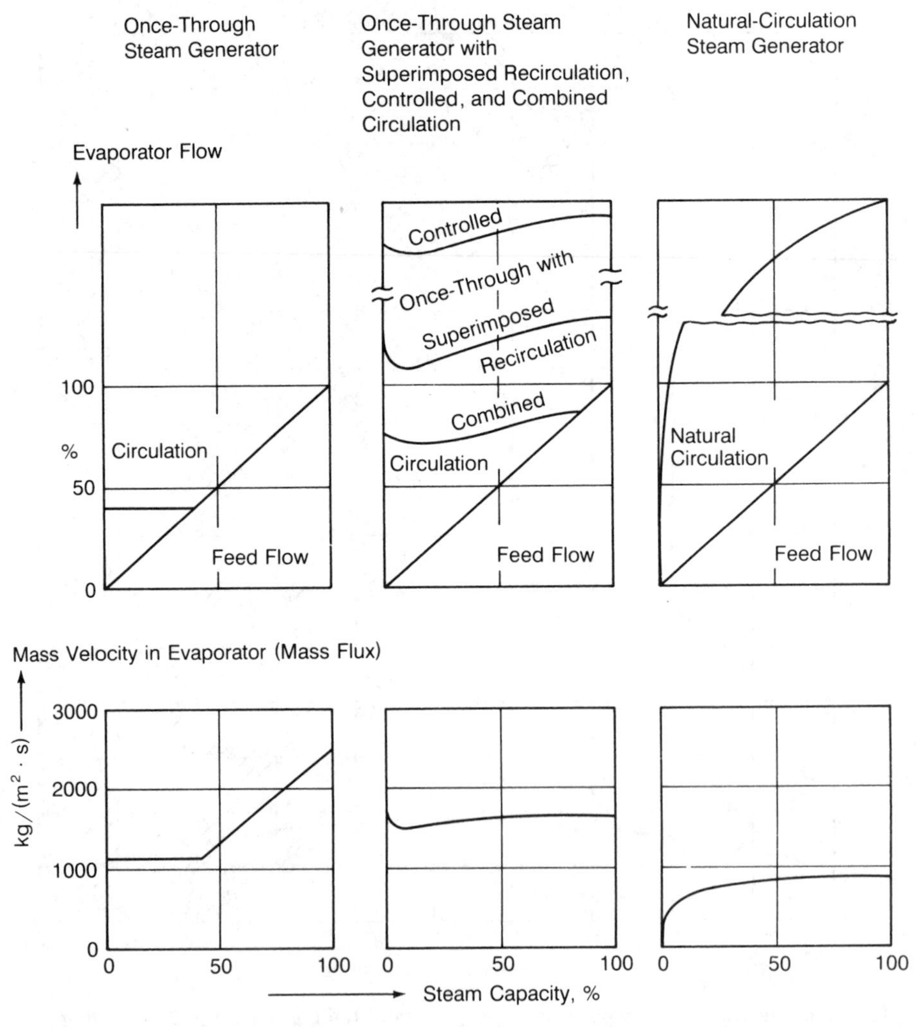

Fig. 7.13. Flow and mass velocity in the evaporator. (With permission of [49].)

for the once-through operation mode. This would economize the feed pump power and allow for design of the parts before the evaporator for lower pressures. In my opinion, the practice to select low minimum loads for once-through operation is based on the following:

1. Operation at low-load recirculation was not commonly accepted (i.e., start-up with overfeeding) in earlier days. Today, planning includes frequent start-ups with the requirement that a low-load circulation system (which can also be used for part-load operation) be available to minimize start-up losses and corrosion and fouling in the economizer and air heater.
2. Problems were encountered when changing over from circulating to once-through operation due to an inadequate control concept as described in Section 7.4.2 and [22].

In the once-through systems with superimposed circulation, the mass flux of the evaporator is almost constant for all loads resulting in ideal conditions as far as pressure loss and cooling of the evaporator tubes are concerned. Yet the required recirculation and feed pump power (only the comparison of the sum of recirculation and feed pump power makes sense) is to be taken into consideration which—as compared to the recirculation and feed-water pump power in once-through steam generators—is higher at part-load and lower at full-load conditions (Fig 7.14). For controlled- and combined-circulation systems the situation should be similar.

In natural-circulation systems, the mass flux increases proportionally to the heat absorption and is with an adequate design basically sufficient. In this

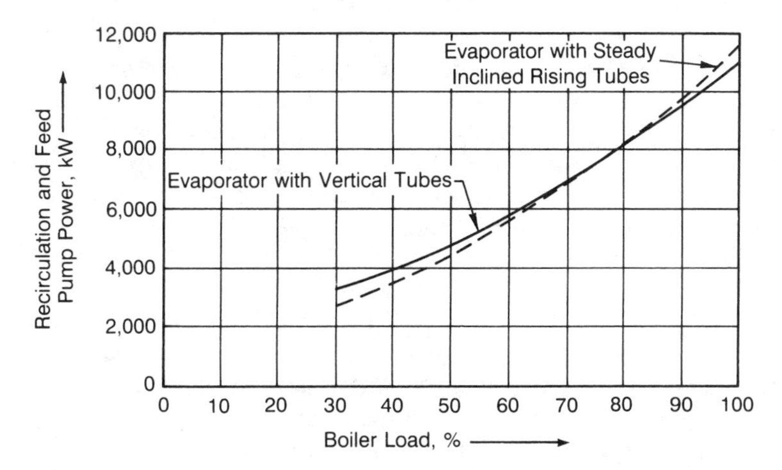

Fig. 7.14. Power consumption of recirculation and feed pump according to evaporator tubing. (With permission of [49].)

case, the evaporator pressure drop can be compensated for without additional pump power. Problems can arise during start-up with the upper burners, that is, nearly no heat input in the lower furnace section and high heat input in the upper furnace section.

In once-through steam generators, ripple roughness has been observed in evaporator tubes causing a marked increase in the pressure drop [23, 24].

In once-through steam generators with superimposed circulation and possibly in natural-circulation boilers, ripple roughness occurs to a small extent only, which, due to the smaller pressure drop of the evaporator, causes a negligible reduction of the circulation ratio. Ripple formation depends on heat flux, mass flux, and the pH value of the feed water. Elevated pH values (greater than 9.3) are counteractive to the formation of ripples but cannot be used without limitation due to their corroding effect on copper tubes which are often used for condensers.

The pressure drop of the two-phase flow in the evaporator is usually calculated [25] as the product of the pressure drop of the same water mass flow and a factor depending on the pressure and steam quality. The results of more recent research concerning the two-phase pressure drop are given in [16, 19]. See also Sections 6.4 and 6.5.

7.2.6 Heat Pickup of the Heating Surfaces

Once-Through Steam Generator Only in the once-through steam generator is the liquid–vapor phase transition point variable. Thus this system is capable of compensating for varying heat absorption between furnace and convective heating surfaces within the range of the admissible material temperatures (Fig. 7.15). At constant spray water flow and other conditions (e.g., constant total heat absorption), superheater steam and material temperatures (except live steam temperature) increase as the heat absorption rate in the (cleaner) furnace increases and decreases as the heat absorption rate in the (slagged) furnace decreases. A further advantage is that the maximum main steam temperature is likely to be reached even at part-load operation. Shifting of the liquid–vapor phase transition point and keeping the live steam temperature constant during part-load operation is subject to limitations if a smooth changeover from once-through to recirculation mode is aimed at (see Section 7.4). Further limitations are the admissible material temperatures and temperature differences of the water walls (see Section 7.3.1).

Once-Through Steam Generator with Superimposed-Circulation, Controlled-Circulation, and Natural-Circulation Steam Generators In once-through steam generators with superimposed-circulation, controlled-circulation, and natural-circulation steam generators, the liquid–vapor phase transition point is located in the water separator and in the drum, respectively. Compensation for varying heat absorption of the furnace and mainte-

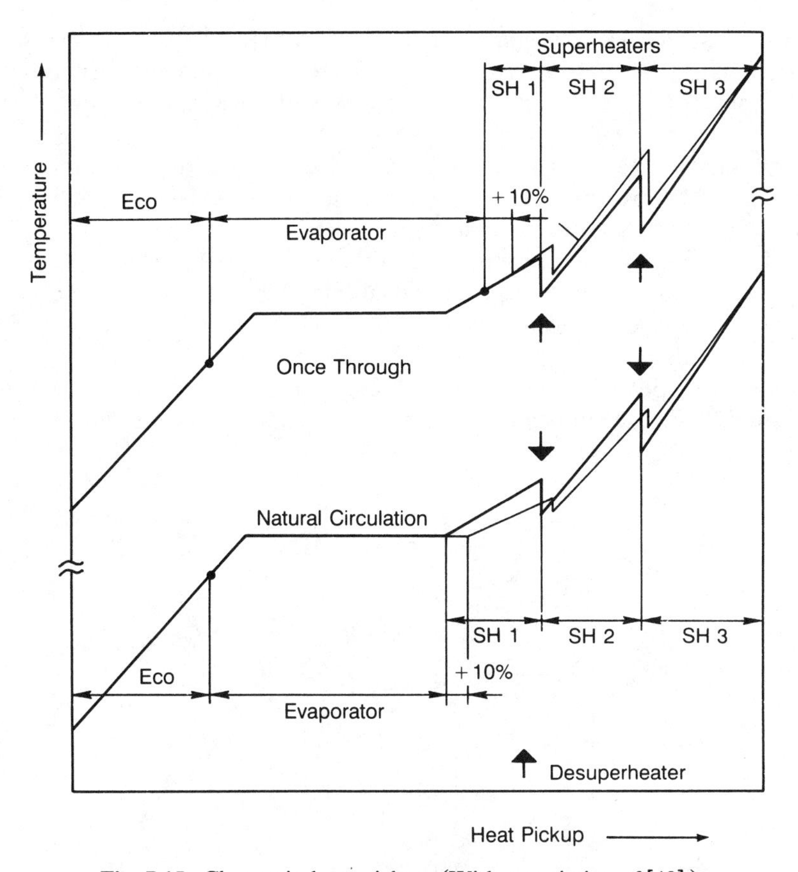

Fig. 7.15. Change in heat pickup. (With permission of [49].)

nance of constant main steam temperatures is only possible by means of spray attemperation unless such compensation is achieved by changing the fuel–air ratio, flue gas recirculation, heat exchangers in the drum, heat absorption of the economizer, or by means of a final evaporative bank or controlled flue gas paths.

At constant other conditions an increase in the furnace heat absorption rate increases the feed-water flow and reduces the spray water flow and superheater material temperatures, whereas a decrease in the furnace heat absorption rate causes a decrease in the feed-water flow and an increase of the spray water flow and of material temperatures before spray attempertors (see Fig. 7.15). At part load the relative furnace heat absorption increases, increasing relatively the feed-water flow and decreasing the spray water flow and/or main steam temperature. This implies that the superheaters be generously designed often in addition to one or more of the already mentioned other possibilities or equipment to change heat absorption rates.

Furnace Outlet Temperature In the case where no radiant wall superheater and/or flue gas recirculation are desired, the minimum furnace outlet temperature is fixed in steam generators with fixed liquid–vapor phase transition point.

The relative heat input to preheat and evaporate the feed water is a function of the pressure, of superheating and reheating (Figs. 7.16 and 7.17). Higher operating pressures, temperatures, and more reheating and preheating (i.e., higher cycle efficiency) require lower relative evaporation heat absorption and evaporative surface, resulting in higher furnace outlet temperatures; however, when another fuel is used, the air ratio is changed or flue gas is recirculated.

In an attempt to determine those furnace outlet temperatures at which no radiant wall superheaters and flue gas recirculation are required, a study was

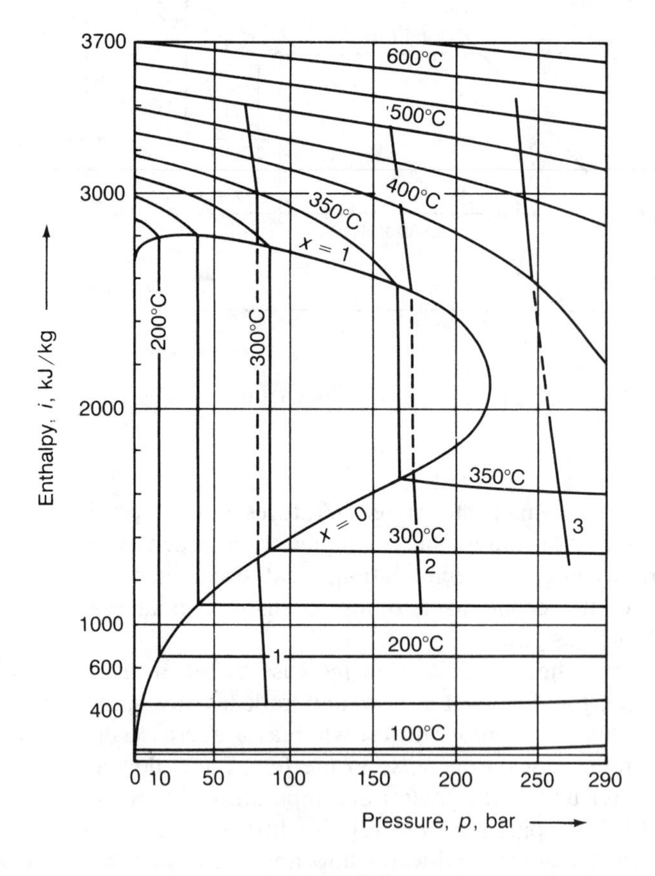

Fig. 7.16. Change of conditions at different pressure levels. (With permission of [4].)

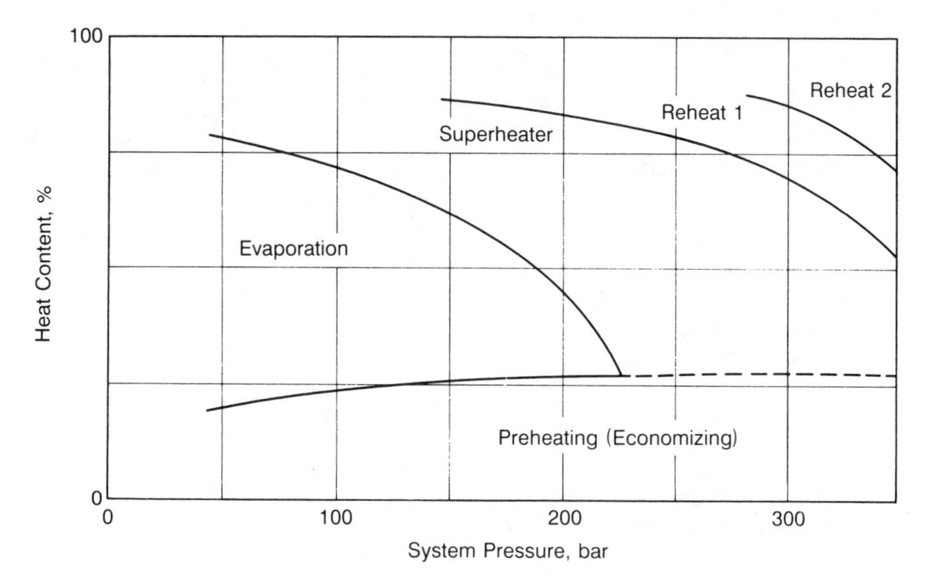

Fig. 7.17. Assignment of heat to different sections. (With permission of [4].)

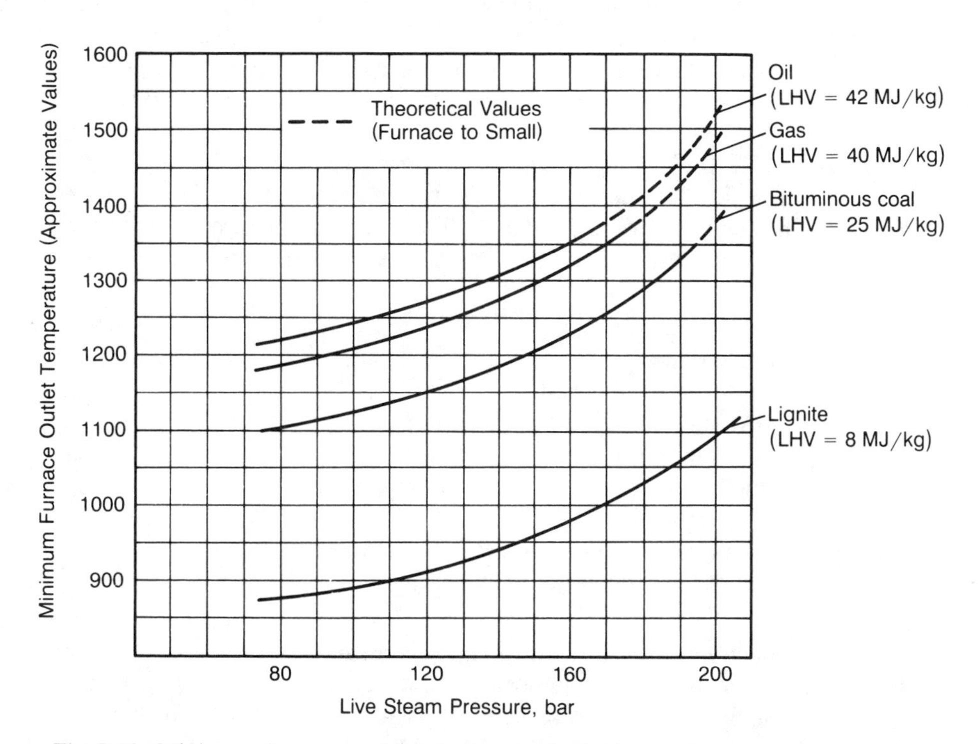

Fig. 7.18. Minimum furnace outlet temperature for boilers with evaporator recirculation (fixed evaporator endpoint) and different fuels (without wall superheater in the furnace and flue gas recirculation etc.). (With permission of [49].)

conducted. The results, as a function of pressure and fuel type are shown in Fig 7.18 and give at the same time the upper limit of the furnace outlet temperature for once-through steam generators.

The study was based on units with a single reheat system and the following parameters:

Live steam pressure	bar	200	160	120	80
Feed-water temperature	°C	240	220	195	165
Live steam temperature	°C	530	530	530	530
Reheat inlet pressure	bar	40	32	25	20
reheat inlet temperature	°C	366	347	325	295
Reheat outlet temperature	°C	530	530	530	530
High pressure part pressure drop	bar	20	16	12	8

A flue gas temperature to the stack of 150°C was assumed for brown coal, all other fuels being based on 130°C.

The following design criteria have been taken into account:

1. Air preheater must be as large as possible.
2. Economizer heating surface must be designed so that the temperature difference between the reheater inlet and flue gas does not exceed 100 K. When firing fuels such as brown coal, the economizer is to be designed for counterflow because of the low flue gas temperature.

For a specific fuel (brown coal, net calorific value 8.24 MJ/kg, ash = 12.7%, water = 51.6%), the effects of different main steam temperatures and air ratios and/or flue gas recirculation flow in the furnace on the furnace outlet temperature were studied and the results are shown in Fig. 7.19.

Given parameters:

Single reheat	
Feed-water temperature	200°C
Air temperature before air heater	90°C
Air temperature after air heater	255°C
Flue gas temperature to the stack	185°C
Air from air heater	89%
Temperature difference between reheater inlet and flue gas temperature	100 K

A low furnace outlet temperature is often considered indicative of a low slagging tendency. Even more important, however, is the burner belt release rate since the formation of slag deposits is to be prevented which would otherwise spread out and cause a fast increase of the furnace outlet temperature above design value. The use of radiant wall superheaters when firing

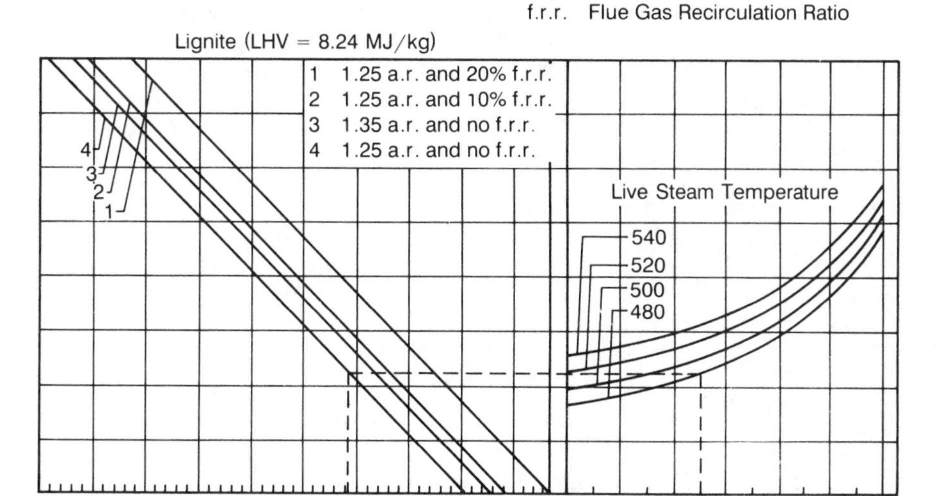

Fig. 7.19. Minimum furnace outlet temperature for lignite fired boilers (LHV = 8.24 MJ/kg) with evaporator recirculation (fixed evaporator end point), different air ratios, and different flue gas recirculation ratios without wall superheaters. (With permission of [49].)

coal prone to slagging in circulation boilers in an attempt to control the furnace outlet temperature is not recommended as this coal quality requires water blowers in the furnace. Radiant wall superheaters with material temperatures of above approximately 400°C, however, can withstand only a very limited number of cleaning cycles by water blowers. Therefore it is more reasonable either to use flue gas recirculation, which reduces the combustion temperature and slagging potential, or to reduce the pressure and/or main steam temperature.

Another study was conducted to find the limitations of the steam parameters for once-through steam generators given by the material used for the water walls and by the furnace outlet temperature [3]. The results are summarized in Fig. 7.20. High pressures, high temperatures, and double reheat, which result in high efficiency, compete with the demand for low furnace outlet temperatures to avoid fouling and slagging and are limited by the allowable material temperatures at the water wall outlet.

7.2.7 Differences in Heat Absorption and Flow Resistance in Individual Evaporator Tubes

Even under stable conditions (for instabilities, see Sections 6.4 and 6.5), temperature or at least enthalpy differences may occur in individual tubes

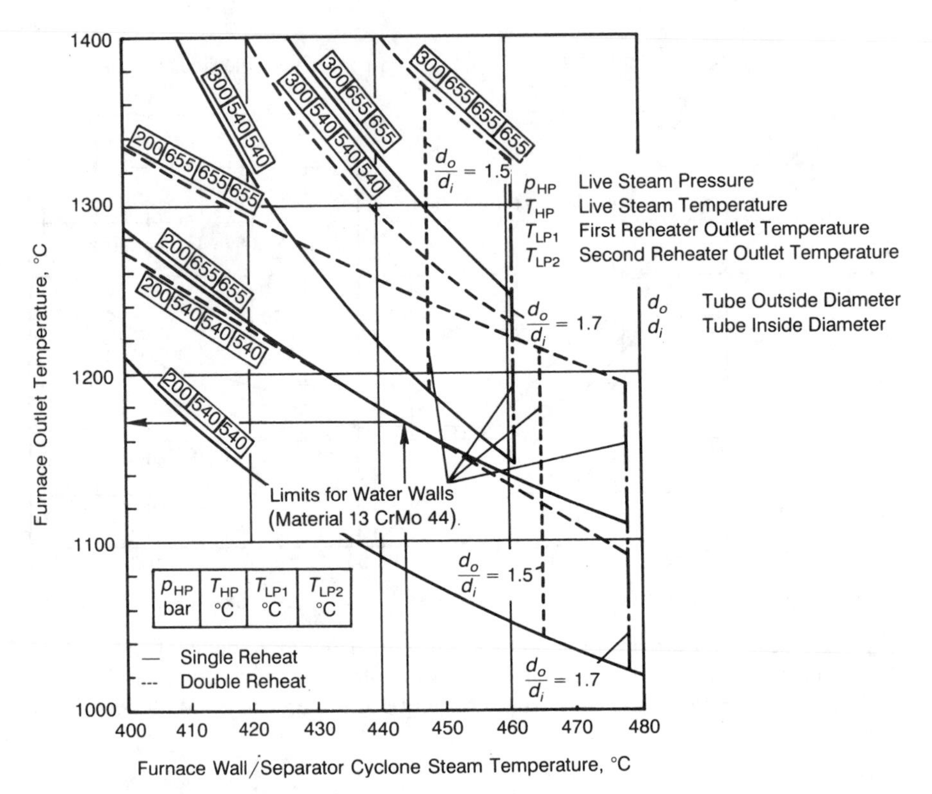

Fig. 7.20. Furnace outlet temperature and furnace wall–separator cyclone steam temperatures. (With permission of [3].)

[26, 27]. This phenomenon is based on the following:

1. Different heat absorption rates due to different tube lengths being exposed to heating as is inherent in design (e.g., bottom–hopper tubes, roof, burner openings, etc.), local slagging, or firing pattern (burners, burner levels, mills in operation, asymmetric fireside temperatures, etc., Fig. 7.21), which may be conditioned by load as is the case, for example, when operating the lower burner level (as is common practice for start-up from the cold to obtain low steam temperatures to roll the turbine), which can increase the heat pickup differences of the bottom–hopper tubes.
2. Different flow resistances (different tube lengths because of burner and other openings, hopper, etc.), admissible tube wall thickness tolerances (pressure loss inversely proportional to the fifth power of the inside diameter, welds, ripple roughness, etc.).

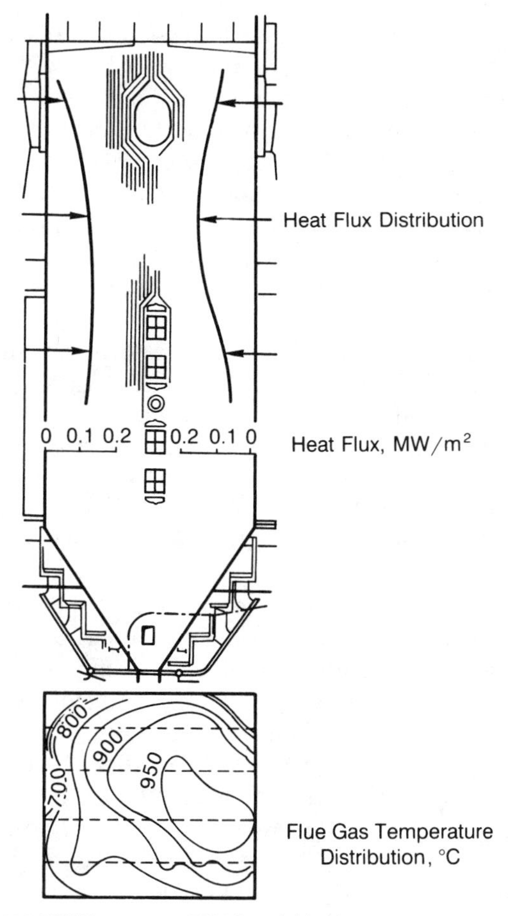

Fig. 7.21. Furnace readings. (With permission of [4].)

If required, the following counteractions are suggested:

1. Selection of a configuration that provides uniform heating surface for all tubes as far as possible (e.g., inclined hopper tubes instead of the combination of vertical tubes in the vertical hopper walls with hopper tubes parallel to the trajectory in the inclined hopper walls). An inclined furnace wall tube pattern (see Section 7.2.8) provides the best, though not complete, compensation for the differences in heat absorption.
2. Adjustment of the mass flow by orifices and/or throttle valves in individual tubes and/or tube banks.
3. Installation of wall blowers to prevent slag deposits.

Different heat pickup rates give rise to mass flow variations of individual tubes. Whether the mass flow in a tube with elevated heat absorption increases or decreases depends on the ratio of fluid friction loss and pressure loss due to the hydrostatic head and the density of the water–steam column, and consequently it also depends on the furnace level where the heat pickup difference occurs (lower or upper section).

As a general rule, fluid friction loss is dominant in once-through steam generators and the mass flow of an individual tube subject to increased heating will decrease; the outlet temperature of this tube being even higher than expected.

In natural-circulation systems, the mass flow of an individual tube or group of tubes increases with increasing heat absorption when the maximum heat absorption is located in the lower portion of the furnace, thus providing a certain self-compensating effect, whereas the effect of the maximum heat absorption being located in the upper furnace wall (slagging of the bottom part) can be to the contrary. Likewise, in once-through steam generators with superimposed circulation, the mass flow of the tube with elevated heat pickup rate will usually increase.

Differences in the flow resistance of individual tubes or groups of tubes entail variations of the mass flow and thus variations of the outlet enthalpy and steam quality or temperature. Specifically, different flow resistance in one tube can be compensated for by providing corrective resistances in other tubes at the same level which, however, is not always feasible. As a consequence, perfect compensation for all load conditions is not always possible.

The effects of differences in heat absorption and flow resistance are combined. Experience has shown that in once-through systems, their sum is equivalent to a heat pickup difference of at least approximately $\pm 5\%$. Natural-circulation, controlled-circulation, combined-circulation, and once-through steam generators with superimposed circulation (vertical tubing) have a somewhat higher percentage. The effects are as follows:

1. In natural-circulation, controlled-circulation, and once-through steam generators with superimposed circulation, the saturated steam–water flow through the respective tube, the enthalpy, and steam quality vary which, however, is negligible as no temperature differences arise. This is in contrast to the once-through steam generator where steam temperature differences develop, creating thermal stresses and possibly cracking of the tube wall (Fig. 7.22). Figure 7.23 shows the measured individual evaporator tube outlet temperatures of a bituminous coal-fired power station once-through boiler and the time histories of these temperatures during a feed-water shortage due to an insufficiently adapted feed-water control during a change of mills. To get the whole picture of the temperature distribution in the water walls, one would have to calculate from the individually measured evaporator tube outlet temperatures the temperatures along each tube. The resulting time-

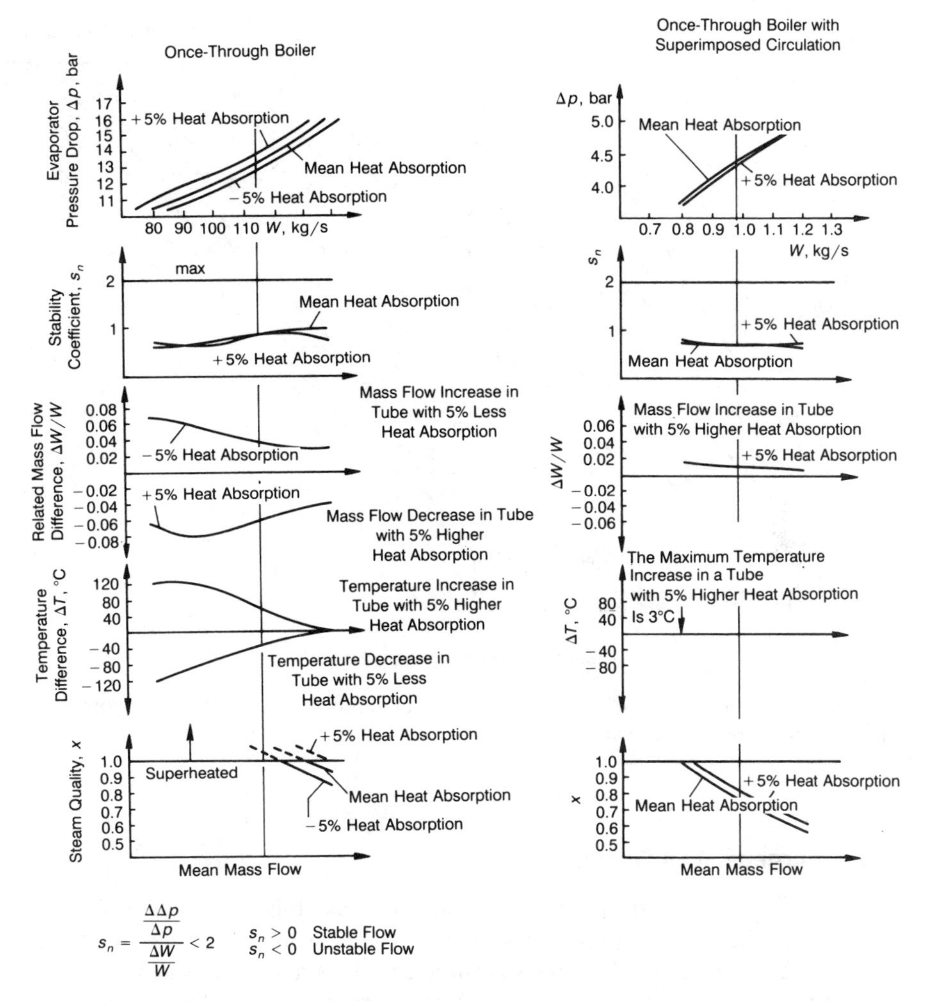

$$s_n = \frac{\frac{\Delta\Delta p}{\Delta p}}{\frac{\Delta W}{W}} < 2 \qquad \begin{matrix} s_n > 0 & \text{Stable Flow} \\ s_n < 0 & \text{Unstable Flow} \end{matrix}$$

Fig. 7.22. Comparison of once-through and once-through with superimposed circulation boiler concerning evaporator outlet parameter differences in tubes with different heat pickup. (With permission of [49].)

dependent thermal stresses are to be taken into consideration when designing the finned tube water wall (membrane wall).

Temperature differences in finned tube walls can be avoided by adding a final convective evaporator tube bank to the system [28].

2. Variations of the heat transfer coefficient and thus of the tube wall temperature (DNB or DO, particularly at points of elevated heat flux density) can cause tube failures.

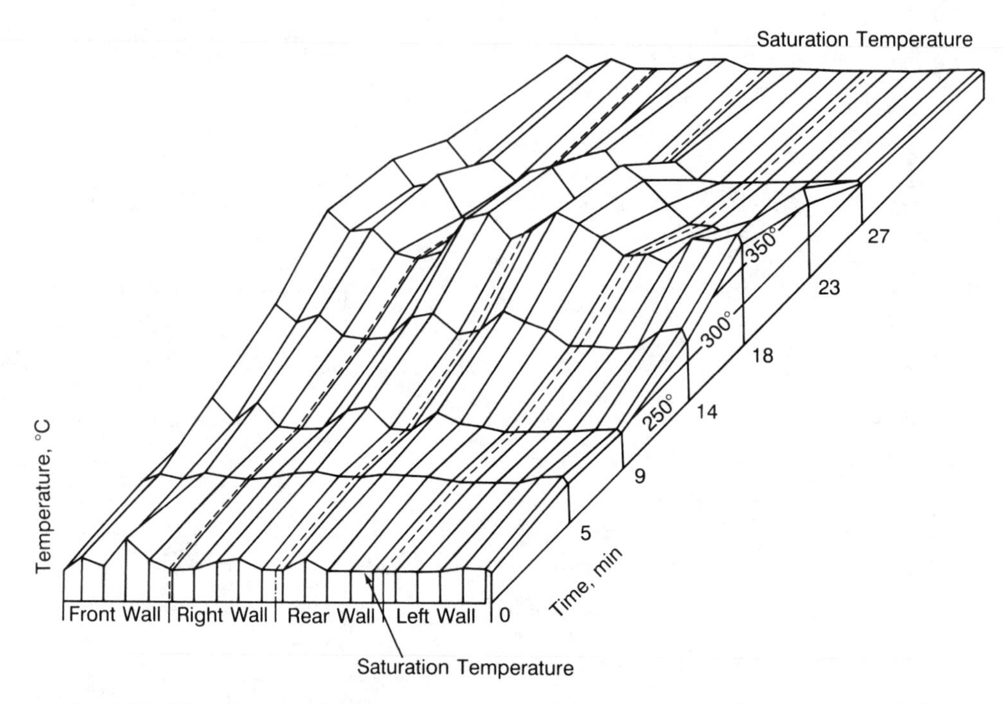

Fig. 7.23. Time history of evaporator tube outlet temperatures during a change of mills and unadapted feed-water control.

7.2.8 Furnace Wall Design

Boilers with recirculation in the evaporator can establish a sufficiently high mass flow in the evaporator such that the mass flux even in a large number of parallel tubes with large diameters is sufficient. This makes it possible to use vertical tubing for circulation boilers. Natural-circulation boiler water walls usually consist of 57 mm outside diameter vertical tubes with 75 mm pitching. For once-through steam generators with superimposed-circulation, tubes with smaller diameters (26.9 and 40 mm pitching as a minimum) in vertical arrangement are used (Fig. 7.24).

For once-through boilers [especially when a low load ($< 40\%$) in once-through operation is demanded], the feed-water flow is not sufficient to provide the necessary mass flux for a sufficient number of even small tubes in vertical arrangements. Therefore water walls with meandering, sectionalized risers or downcomers and inclined tubes were developed (Fig. 7.25). Today, mainly inclined tube water walls are used (Fig. 7.24). Inclined tubes for once-through boilers usually have diameters about 38 mm and 50 mm pitching.

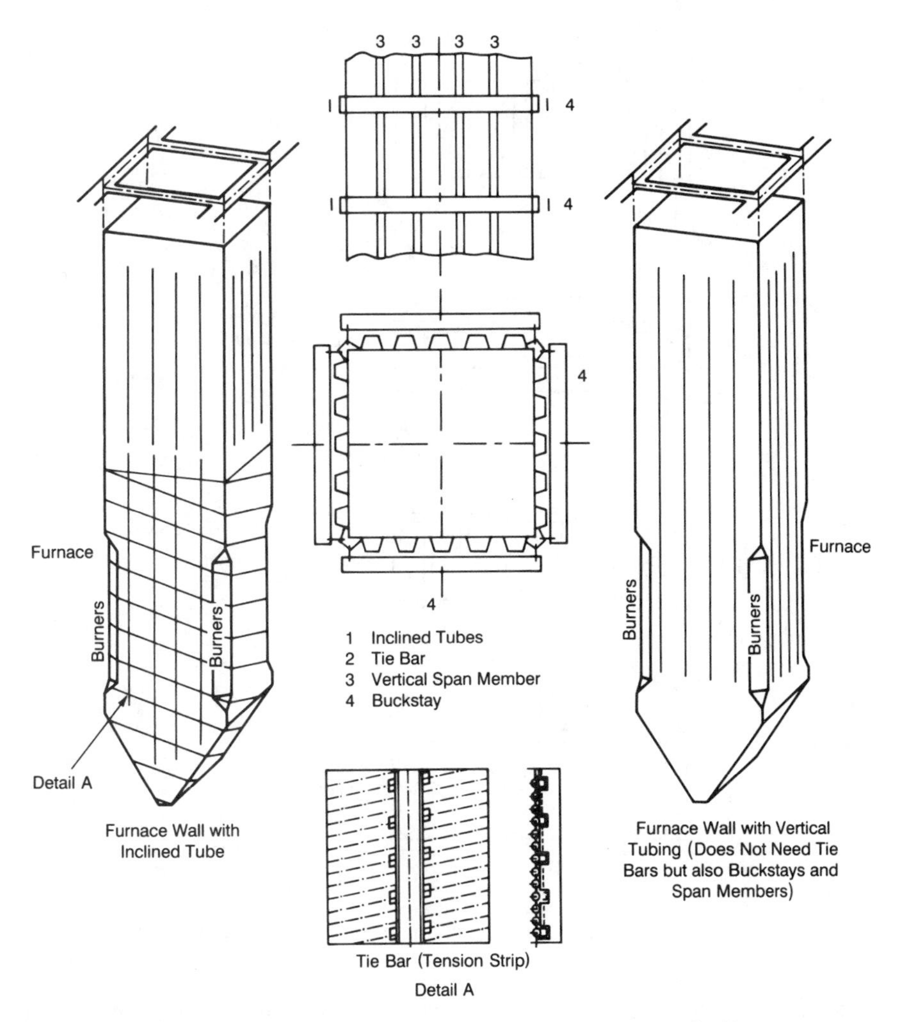

Fig. 7.24. Evaporator wall design. (With permission of [49].)

With vertical tubing, load due to self-weight and bottom ash is applied axially and is thus more favorable than load with an inclined tubing arrangement (radial loading) where tie bars (Fig. 7.24) are required. These are to be designed to avoid inadmissible thermal stresses because of temperature differences between the finned tube walls and tie bars in case of fast load changes and start-ups [29].

The load due to the flue gas pressure is taken up by buckstays and vertical span members (see Fig. 7.24). The space between buckstays and vertical span members is subject to high bending stresses in both the vertical and inclined

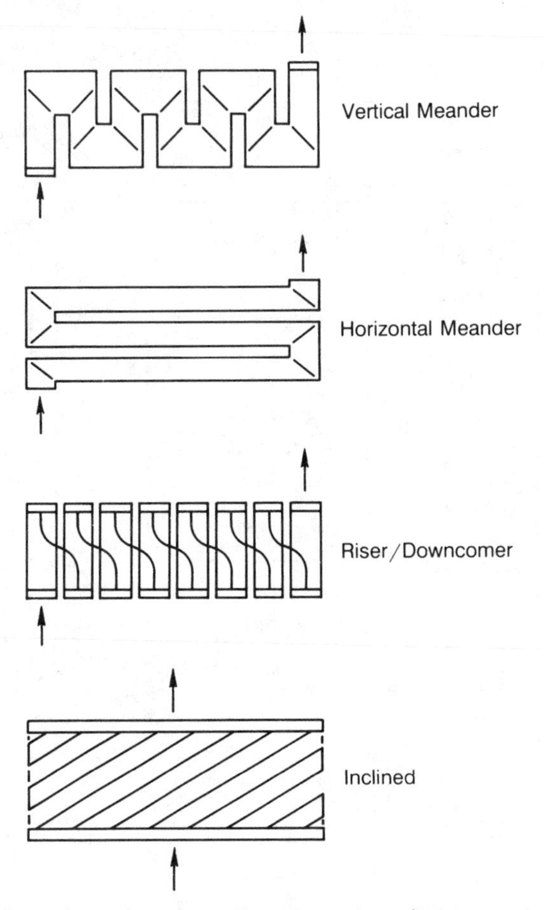

Fig. 7.25. Once-through water wall configurations. (With permission of [4].)

tube pattern. The forces from the buckstays are transmitted into the adjacent furnace tube walls which results in radial and axial loading for the vertical and inclined tubing, respectively.

Vertical furnace tubes are less prone to slagging than inclined tubes which offer a larger area for deposition.

7.2.9 Feed-Water Quality

According to the VGB* Directives for boiler feed water, boiler water, and steam [30], the required feed-water quality for water-tube boilers above 64

*VGB, Technische Vereinigung der Großkraftwerksbetreiber e.V., Klinkestraße 27-31, Postfach 10 39 32, D-4300 Essen, West Germany; Worldwide Boiler User Club (mainly utilities) founded in 1920 in Germany.

bar is basically the same for all systems, as for this pressure range demineralized feed water is used. Ripple formation in the evaporator is mentioned in Section 7.2.5. Erosion-corrosion problems are described in [18].

7.2.10 Disturbances

The effect of short-term feed-water pump failures is more detrimental in once-through systems than in once-through steam generators with superimposed-circulation, controlled-circulation, and natural-circulation boilers due to the fact that in these latter systems, the lacking feed water is substituted by the water in the drum or water separator, respectively.

In once-through systems, feed-water supply must be reinstated after a maximum period of 10 to 15 s; otherwise overheating of the evaporator tubes occurs. Once-through boilers with superimposed circulation allow a longer interruption; controlled-circulation and natural-circulation boilers allow even more according to the water stored.

In the case of high-pressure feed-water preheater failures (sudden drop of the feed-water temperature), the different behaviors of once-through steam generators (variable liquid–vapor phase transition point), once-through steam generators with superimposed circulation, and controlled- and natural-circulation steam generators (fixed liquid–vapor phase transition point) becomes particularly manifest. At constant heat absorption rates, the liquid–vapor phase transition point in the once-through system shifts toward the main steam outlet. As a consequence the feed-water flow is reduced to maintain the main steam temperature and spray water flows and the superheater temperatures decrease. In once-through systems with superimposed circulation and in controlled- and natural-circulation systems, the steam output decreases and the elevated temperatures of the superheaters are to be controlled by increasing the spray water flow rate.

For unallowable pressure increases, safety valves are provided at the superheater and reheater outlet. Safety valves at the separator (or drum) are (also for natural-circulation boilers) not useful because of superheater temperature excursions, if the fire is not tripped, when those safety valves open. Usually, the safety valve at the superheater outlet is replaced by a high-pressure bypass (see Section 7.4.1).

7.2.11 Storage Capacity, Load Changes, and Control

Of all systems, the smallest storage capacity [31] and thermal inertia are given in the once-through steam generator, the largest are given in the natural-circulation system.

Figure 7.26 shows the different behaviors at a ramp-type load increase from 40% to 100% within about 7 min under controlled sliding pressure operation in a bituminous-coal-fired unit.

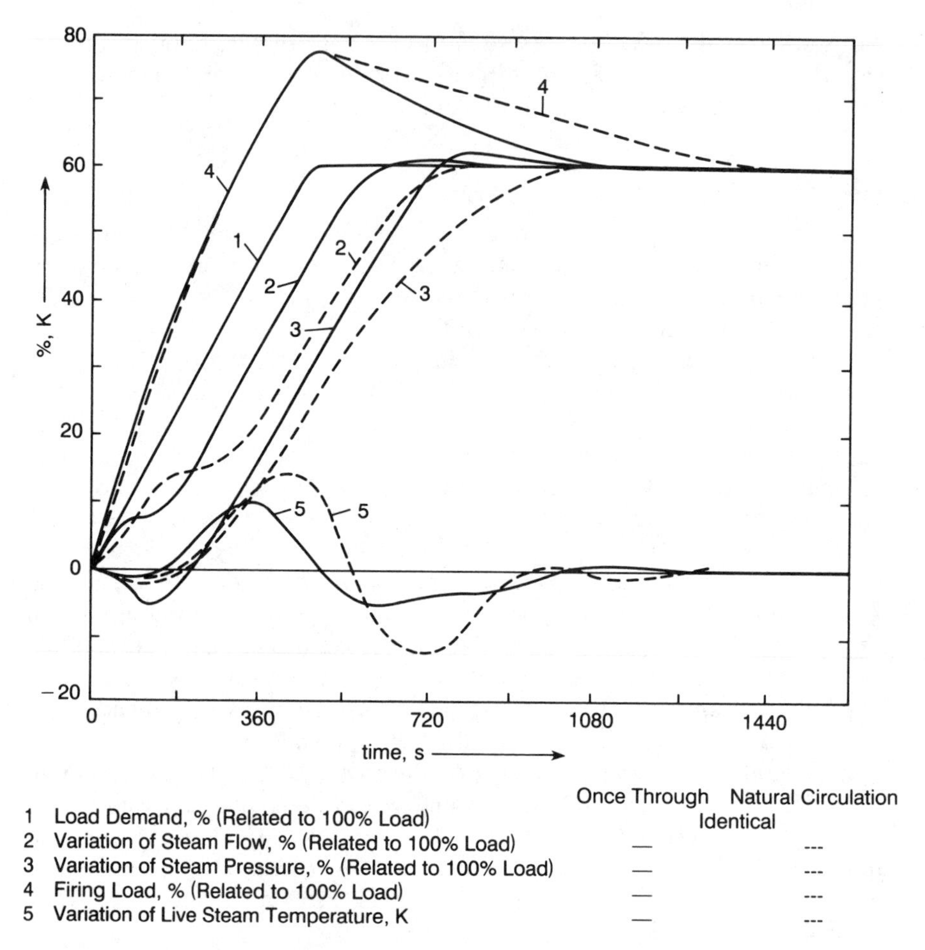

Fig. 7.26. Load change between 40% and 100%, coal-fired boiler, controlled sliding pressure operation. (With permission of [49].)

As expected, the once-through system shows the smallest main steam temperature variation and shortest time within which the actual load follows the load set value without delay; the natural-circulation system shows the largest temperature variations and longest time. The once-through steam generator with superimposed circulation has an intermediate behavior.

The advantage of the natural-circulation steam generator and the once-through system with superimposed circulation [that the time within which the actual load follows the set load value without delay is longer in natural-circulation and once-through with superimposed circulation steam generators than in the once-through steam generators] cannot be preserved in fast-

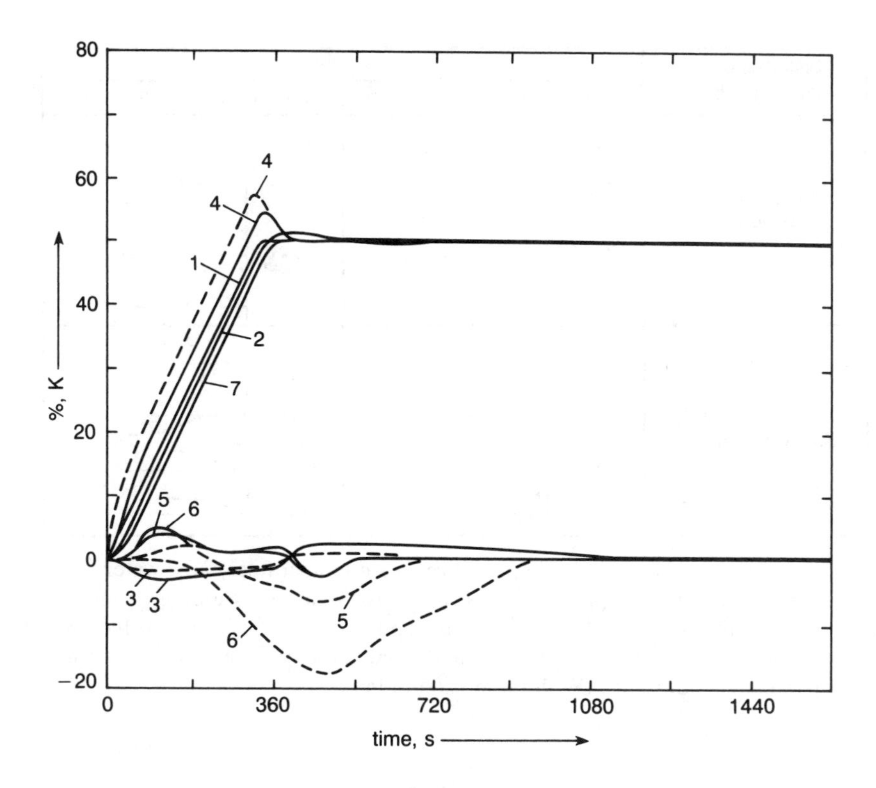

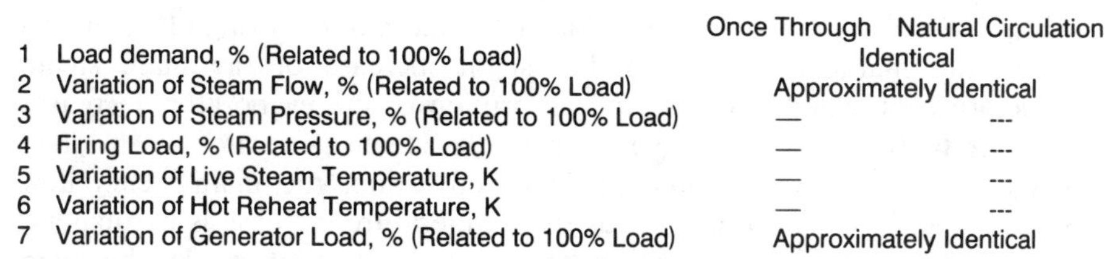

		Once Through	Natural Circulation
1	Load demand, % (Related to 100% Load)	Identical	
2	Variation of Steam Flow, % (Related to 100% Load)	Approximately Identical	
3	Variation of Steam Pressure, % (Related to 100% Load)	—	---
4	Firing Load, % (Related to 100% Load)	—	---
5	Variation of Live Steam Temperature, K	—	---
6	Variation of Hot Reheat Temperature, K	—	---
7	Variation of Generator Load, % (Related to 100% Load)	Approximately Identical	

Fig. 7.27. Load change between 50% and 100% oil–gas-fired boiler, fixed pressure operation. (With permission of [49].)

reacting firing systems (oil and gas) (Fig. 7.27). When using oil or gas in once-through boilers, the feed water can be used to obtain short-term load increases because the firing system is fast reacting (Fig. 7.27).

Figure 7.28 shows the different behavior in the case of varying firing rate and feed-water flow (nothing else changed). The behavior of natural-circulation and controlled-circulation steam generators and once-through steam generators with superimposed circulation is similar.

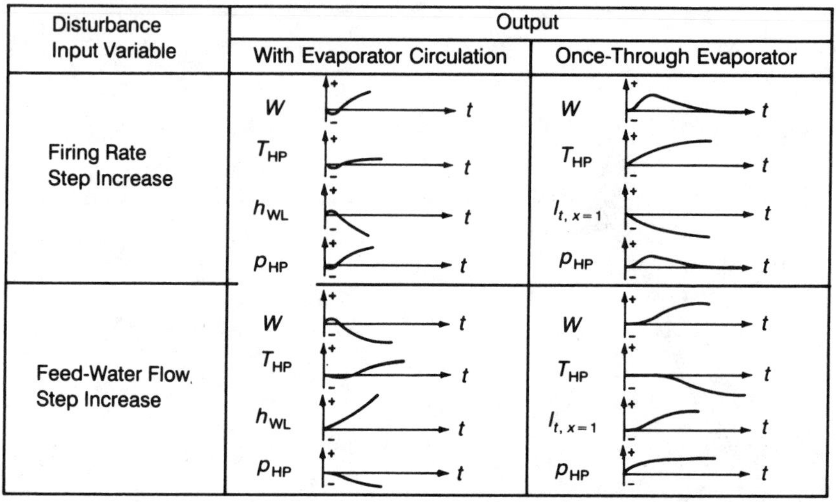

Disturbance Input Variable	Output	
	With Evaporator Circulation	Once-Through Evaporator
Firing Rate Step Increase	W	W
	T_{HP}	T_{HP}
	h_{WL}	$l_{t,x=1}$
	p_{HP}	p_{HP}
Feed-Water Flow Step Increase	W	W
	T_{HP}	T_{HP}
	h_{WL}	$l_{t,x=1}$
	p_{HP}	p_{HP}

Fig. 7.28. Behavior of live steam flow, W; live steam temperature, T_{HP}; water level in the storage vessel–water separator–drum. h_{WL}; live steam pressure, p_{HP}; and position of evaporator endpoint, $l_{t,x=1}$, due to a step increase in firing rate or feed-water flow. Comparison between once-through and circulation evaporator operation mode (other parameters constant). (With permission of [49].)

In the circulation mode of operation, an increase of the firing rate causes a rise in the main steam flow and a slight increase of the main steam temperature provided that the wall absorption of the evaporator decreases relatively to the total heat absorption of the steam generator as the load increases. In the once-through mode of operation, the main steam flow increase is only short term, whereas the temperature increase is permanent.

When increasing the feed-water flow in the circulation mode, the main steam flow decreases and the main steam temperature increases, whereas in the once-through mode of operation, the main steam flow increases and the main steam temperature decreases.

In the circulation mode of operation, return to the steady-state condition is impossible without feed-water control as a function of the water level. Theoretically, for the return to a steady-state condition without feed-water control, changeover to once-through operation would occur by either emptying or overfeeding the water separator or drum.

In the once-through mode of operation, return to a steady-state condition takes place automatically without correction but this new steady state may be not acceptable; for example, when reducing the feed-water flow, the reinstated steady-state condition can cause inadmissibly high material temperatures, when increasing the feed-water flow, two-phase flow in superheaters may cause problems.

The manipulated variables for controlling the main steam flow, main steam temperature, and water level are given in the following table:

Controlled Variable	Manipulated Variable	
	Once-through, combined circulation	Natural circulation, controlled circulation, once through with superimposed circulation
Main steam flow	Firing rate	Firing rate
Main steam	Feed water and spray water flow	Spray water flow
Water level	Not applicable	Feed-water flow

7.2.12 Unit Capacity, Dimensions, and Design

All steam generator systems can be designed as a single-pass, two-pass, multi-pass, box-type, or cube-type steam generator [32–48] of different size and output (Fig. 7.29). Even natural-circulation steam generators have been designed for ratings up to 900 MW.

Natural-circulation, controlled-circulation, combined-circulation steam generators, and once-through steam generators with superimposed circulation are particularly advantageous as regards furnace tube pattern (vertical tubing). This applies particularly to units up to about 150 MW as in once-through systems of such capacity also water walls surrounding the convection heating surfaces are either designed as horizontal meander or inclined tubing (unfavorable as regards wall penetration and pressure loss) or are designed as wall superheaters (causing temperature differences in the wall during start-up).

7.3 SPECIAL DESIGN CONSIDERATIONS

7.3.1 Water Wall Design

For the design of furnaces empirical correlations like cross section and volume heat release rate (Fig 7.30) are used. For a given total heat input to the furnace, the cross section (i.e., the length of a side l) and the volume (i.e., with the given cross section the height h) of the furnace can be calculated. Knowing the radiation heat fluxes to the furnace walls one can calculate, as explained in Section 7.2.5, the minimum mass flux W'' required to obtain acceptable wall temperatures. The pitching of the finned tubes is selected so that the temperature of the fin is not higher than the temperature in the tube

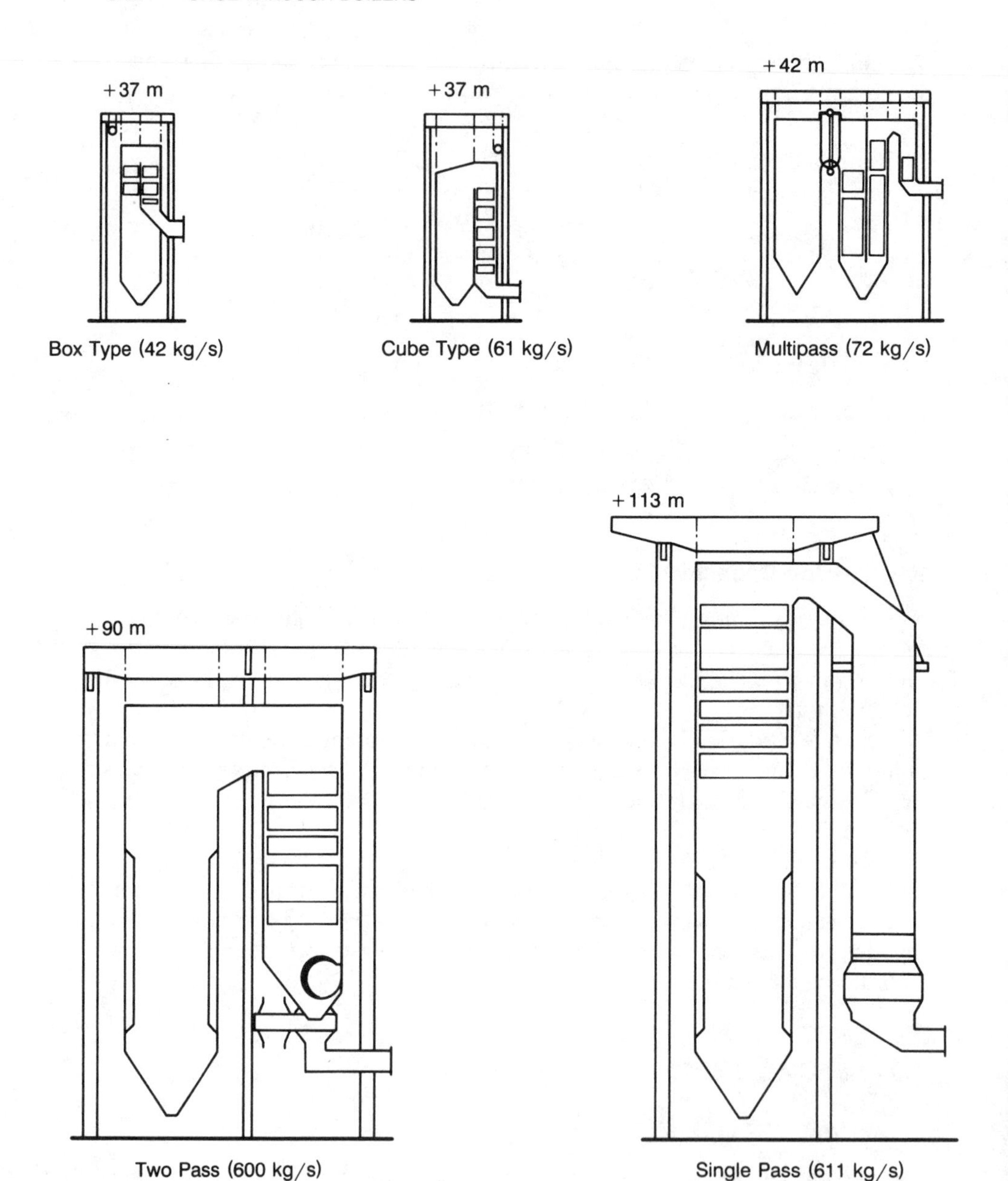

Fig. 7.29. Steam generator types. (With permission of [49].)

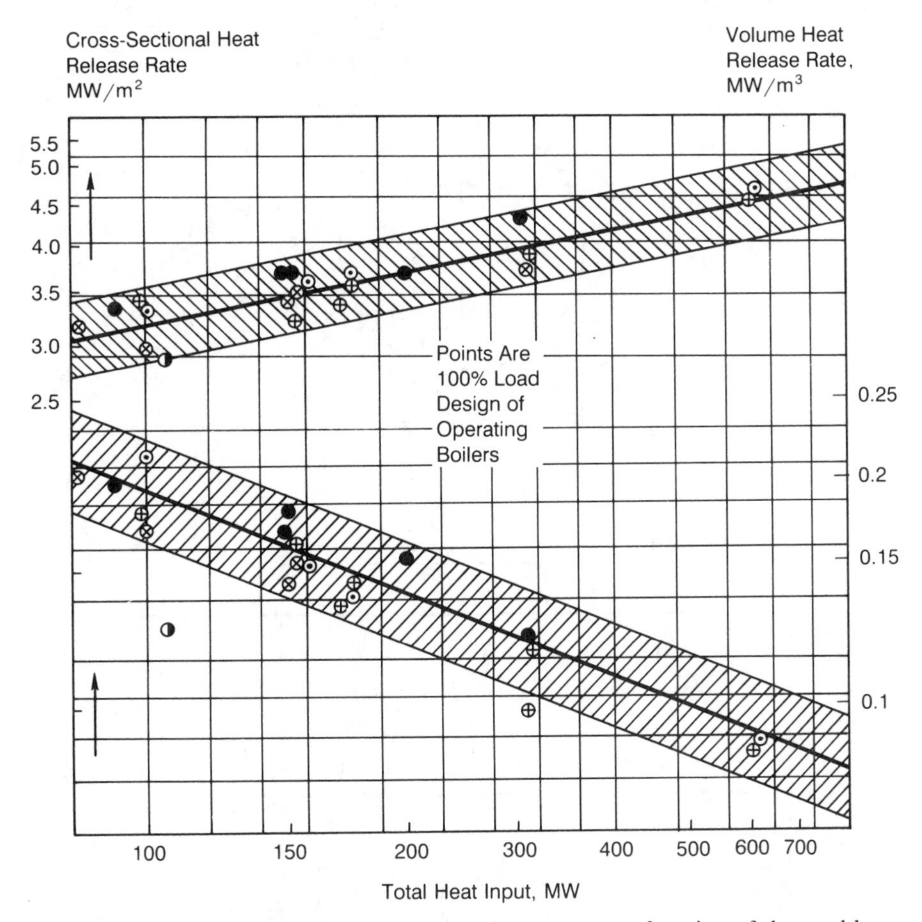

Fig. 7.30. Volume and cross-sectional heat release rate as a function of the total heat input of lignite furnaces. (Adapted from [50].)

wall; this applies for pitchings of k_p tube diameters. If one prefers a certain minimum once-through load, that is, a certain evaporator = feed-water flow W and a certain (inner) tube diameter d_i, one can calculate the angle of inclination to the horizontal according the following equation (Fig. 7.31):

$$\sin \alpha_h = \frac{pt}{pt_h} = \frac{k_p d_i}{\dfrac{4l}{n}} = \frac{k_p d_i W_{\min}}{4l \dfrac{d_i^2 \pi}{4} W''} = \frac{W k_p}{W'' \pi l d_i} \tag{7.27}$$

With the decreasing heat flux in the walls surrounding the convective tube banks, the mass flux can also be decreased and (for larger units) vertical wall tubing can be used, which makes the penetration of the convective tubes

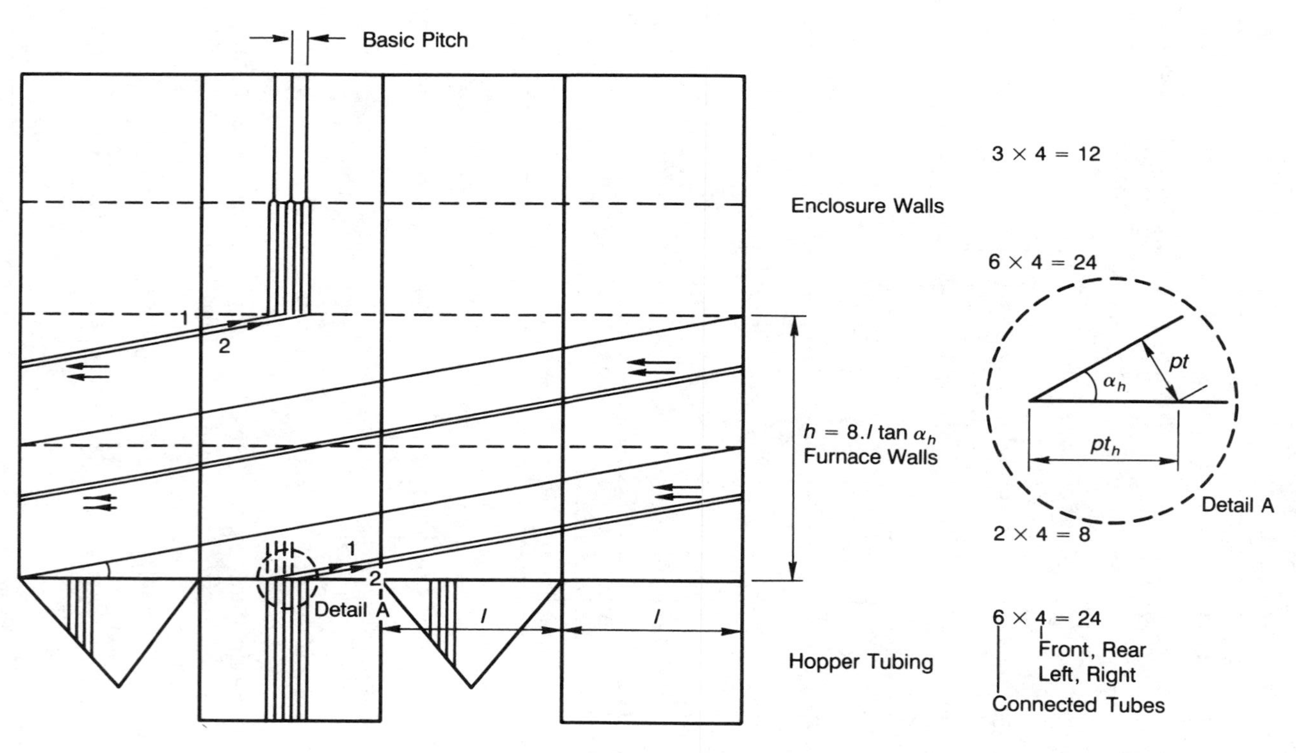

Fig. 7.31. Water wall design (example).

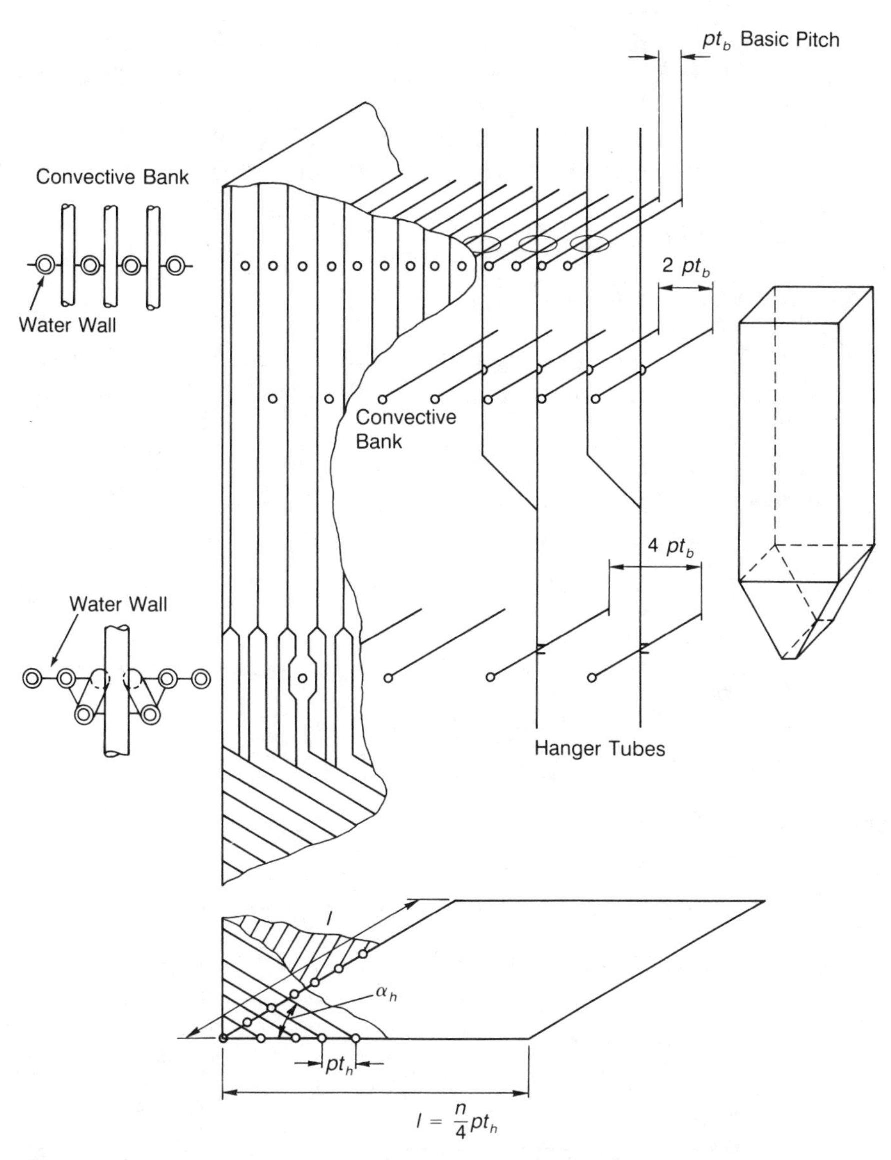

Fig. 7.32. Relation of water wall, convective tube banks, and hanger tube pitch.

through the circumferential walls simple and tie bars unnecessary.

For single-pass boilers the water wall pitches, convective tube banks pitches, and hanger tube pitches are interconnected (Fig 7.32). If one uses, for example, forged transition pieces from inclined to vertical tubing, which leads to a simple design (but also with intermediate headers to obtain a good meshing), one has to keep certain integer tube number ratios as shown in Fig. 7.31.

From Eq. (7.27), one can see that increasing unit capacity (approximately 700 MW for bituminous-coal-fired units), higher minimum once-through load (approximately 70%) [both increasing W in Eq. (7.27)], and decreasing mass flow density W'' by using rifled tubes lead to a vertical tubing also of the furnace, which is preferable because it is self-supporting. The limits are discussed in [51].

The furnace wall is subjected to the following loads [11, 29, 43]:

1. Water–steam pressure
2. Self-weight (including buckstays, burners etc.), deposits and ash in the hopper, and loads due to temperature differences between the water wall and tie bars
3. Flue gas pressure (therefore the wall needs buckstays and vertical span members) (Fig. 7.24)

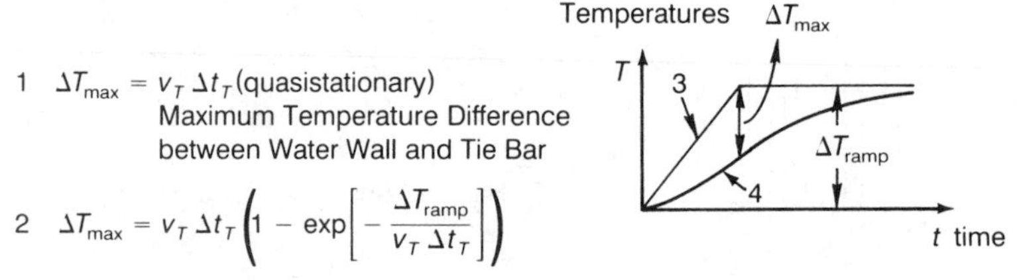

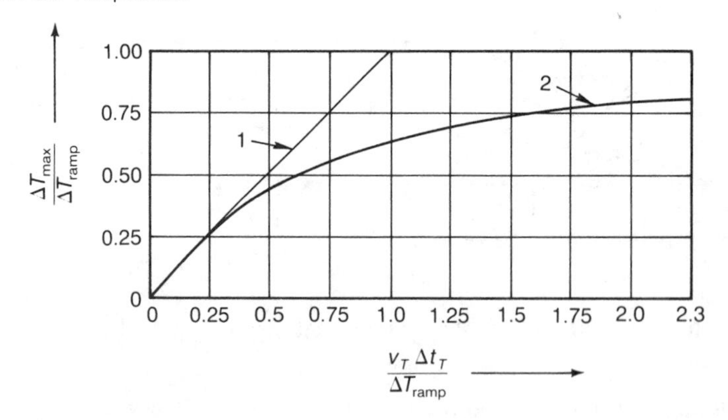

Fig. 7.33. Maximum temperature difference ΔT_{max} between furnace wall and ties for ramp-type temperature changes of the fluid.

4. Thermal stresses due to radiation (heat flux only on one side), due to temperature differences between the tubes (Fig. 7.23) and in the tube wall

If all other conditions are fixed, one can only choose the rate of temperature change in the evaporator (for sliding pressure operation connected to load because of the saturated temperature–pressure–load relation). The transfer function of the mean tie bar temperature related to the wall temperature is a first-order delay. This leads to a simple relation between the temperature change rate, delay time, and maximum temperature differences between the wall and tie bar for a ramp change of the wall temperature (Fig. 7.33).

As this temperature difference adds the rest to reach (according to cycle numbers) the chosen stress limits, one can calculate a diagram of allowable temperature–pressure–load (for sliding pressure operations) transients [11]. These transients should correspond to those of the water separator and live steam header.

7.3.2 Steam Preheating Equipment

Preheating a boiler with steam has several advantages [4]:

1. Very slow temperature transients can be realized, that is, very low life-time consumption of thick-walled parts due to low thermal stress.
2. The time during which the air heater is operated in a temperature range below the dew point of the flue gas is very short, the economizer is not operated below the dew point of the flue gas. Therefore corrosion and fouling are minimized.
3. The air temperature is higher and this improves the conditions for ignition and low firing rates.

Simple feeding of steam into a water-filled pipe will not work effectively: the steam necessarily has a higher pressure and will push the water column aside, creating a space full of steam like a piston. The inner metal surface of the pressure part will instanteneously heat up to saturation temperature corresponding to system pressure by the condensation of steam. Condensation will also take place at the boundaries of the steam plug. The preheating process is limited to the spot where steam is locally introduced.

The correct preheating process starts with the economizer and evaporator filled up with water and the circulation pump, establishing a certain water flow through the whole tube and pipe system and atmospheric pressure in the separator. By connecting the auxiliary steam source to the circuit, the introduced steam will mix with the circulating water, condense, and heat this up as long as the saturation temperature is not achieved. The water level has to be controlled to avoid overflooding of the superheater due to thermal expansions of the fluid in the loop. When saturation is reached, bubbles will be transferred in the water–steam mixture to the separator. From there the saturated steam will fill up the superheater tubes and pipes, displace the air, and condense at the colder inner surface. (Vacuum startup would be preferable, see Section 7.4.1.) Condensate has to be well drained to avoid blockage of individual tubes and to provide proper steam flow.

The pressure will increase in parallel in the superheater section filling with steam (continuous condensation—drainage) and in the circulating system filled with saturated water.

The spot for introducing steam has to be carefully selected: the static head in large boilers may be as high as a 100-m water column equivalent to 10 bar. To preheat such a system at the lowest point, say in the feed line, would require steam of accordingly higher pressure. Arranging it just upstream of the evaporator outlet (Fig. 7.34) requires consideration of only the small difference of static head and flow resistance, the latter being very small while circulating saturated water. But one must take into account that the mixing length available before entering the separator is restricted. Therefore a specially designed chamber should be foreseen (Fig. 7.35).

The mass flux of water is selected in the range of 1000 $kg/(m^2 \cdot s)$ and for design steam flow the pressure drop in the bores of the distribution pipe should be around critical, thus avoiding "banging" in the mixing vessel. A larger number of small-diameter bores provides a sufficient equalization effect and an optimum condensation–mix–heat-up effect.

To preheat a 600-MW single reheat, bituminous-coal-fired boiler from 60 to 180°C/10 bar, a steam line of about 20 bar is necessary and it requires 6 hr to achieve this with a maximum steam flow of 10 kg/s and a total steam consumption of 200 t.

The same system can be used to keep the steam generator in pressurized start-up conditions and avoid corrosion and fouling due to temperatures below the dew point of the flue gas.

Because the first 10-bar pressure increase from cold consumes 35% of the total time to reach full load, the start-up time is considerably reduced.

7.3.3 Water Separation

During low-load operation all boiler systems besides the combined-circulation boiler when operated at supercritical pressure need a water separation device to separate the steam from the water, which has to be recirculated through the evaporator, unless the once-through boiler evaporator is overfed [4]. Natural-circulation, controlled-circulation, and once-through boilers with superimposed circulation need such a device for all loads.

Usually natural-circulation and controlled-circulation boilers use a drum without (for lower pressure) or with drum internals (special steam dryers and/or a large number of small cyclones—for higher pressures—especially

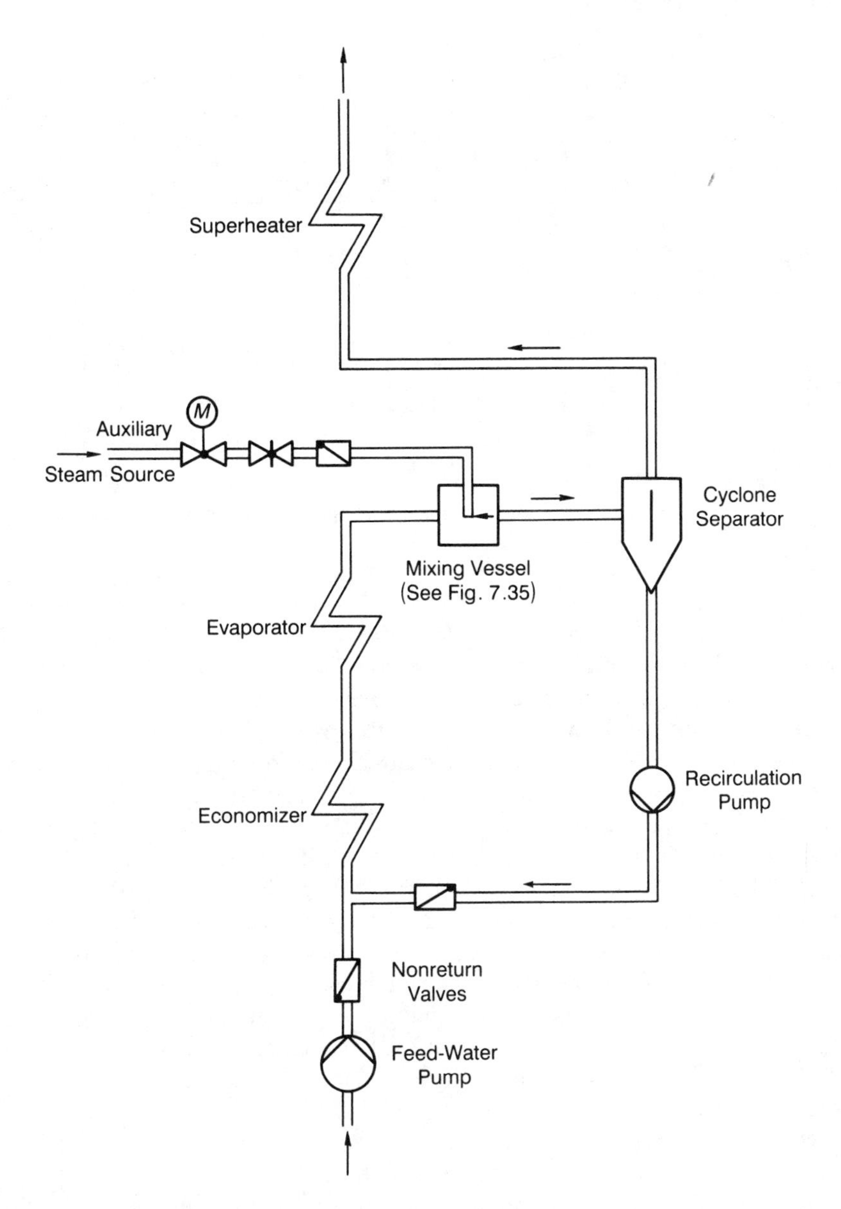

Fig. 7.34. Flow scheme for steam preheating. (With permission of [4].)

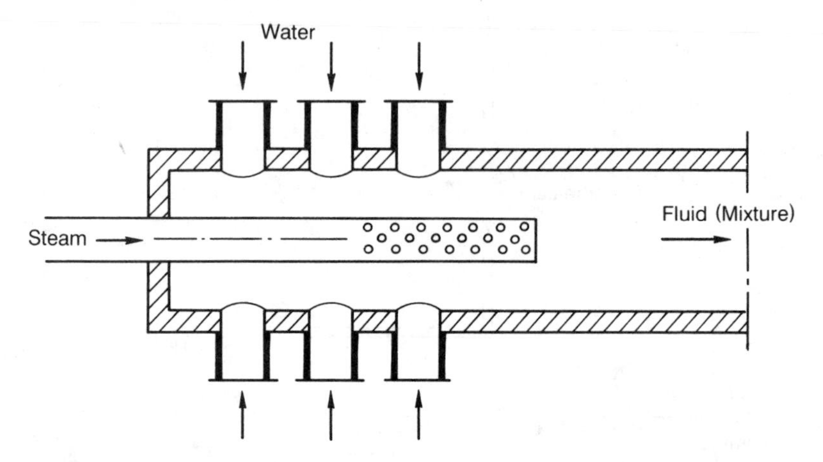

Fig. 7.35. Mixing vessel for steam preheating. (With permission of [4].)

at high pressures it is necessary to avoid carry-under of steam in drums of natural-circulation boilers to provide sufficient circulation [7]), whereas all others, that is, once-through boilers and once-through boilers with superimposed circulation, use one or a small number of cyclones.

One common vessel (cyclone and storage vessel combined) is arranged for once-through systems with superimposed circulation up to full load (Fig. 7.36). This design is very compact and allows a separation of water better than 95%. The mass flow versus load is nearly constant and the separation effect theoretically decreases with falling load. The fluid flow also changes and, in the case of natural pressure, the change in density overrules this effect and the separation results to be better so assisting the live stream temperature characteristic at partial load as the outlet condition increases from 95% steam quality nearly to saturated steam.

The mass flux related to the inner diameter may rank up to 800 $kg/(m^2 \cdot s)$. The inlet pipes should not exceed 40% of the vessel diameter. Downward inclination of these pipes improves the separation. The ratio between the total cross-sectional area of the inlet pipes, the mass flux in the vessel, and the steam pipe governs the total pressure drop. It is ideal to keep the inlet and outlet areas equal at a moderate mass flux.

The other design used is mainly applied for once-through systems with circulation only at low load or start-up (Fig. 7.36). The pressure drop at full load (no separation, superheated steam flow) is an important figure.

Multiseparation vessel design is applied to reduce the mass flux to 500 $kg/(m^2 \cdot s)$ and to limit the wall thickness along with the vessel diameter, otherwise a possibly limiting factor for start-up transients. The separation effect is better than 98%. The total flow resistance coefficient is 3 to 5 relative to the inlet area.

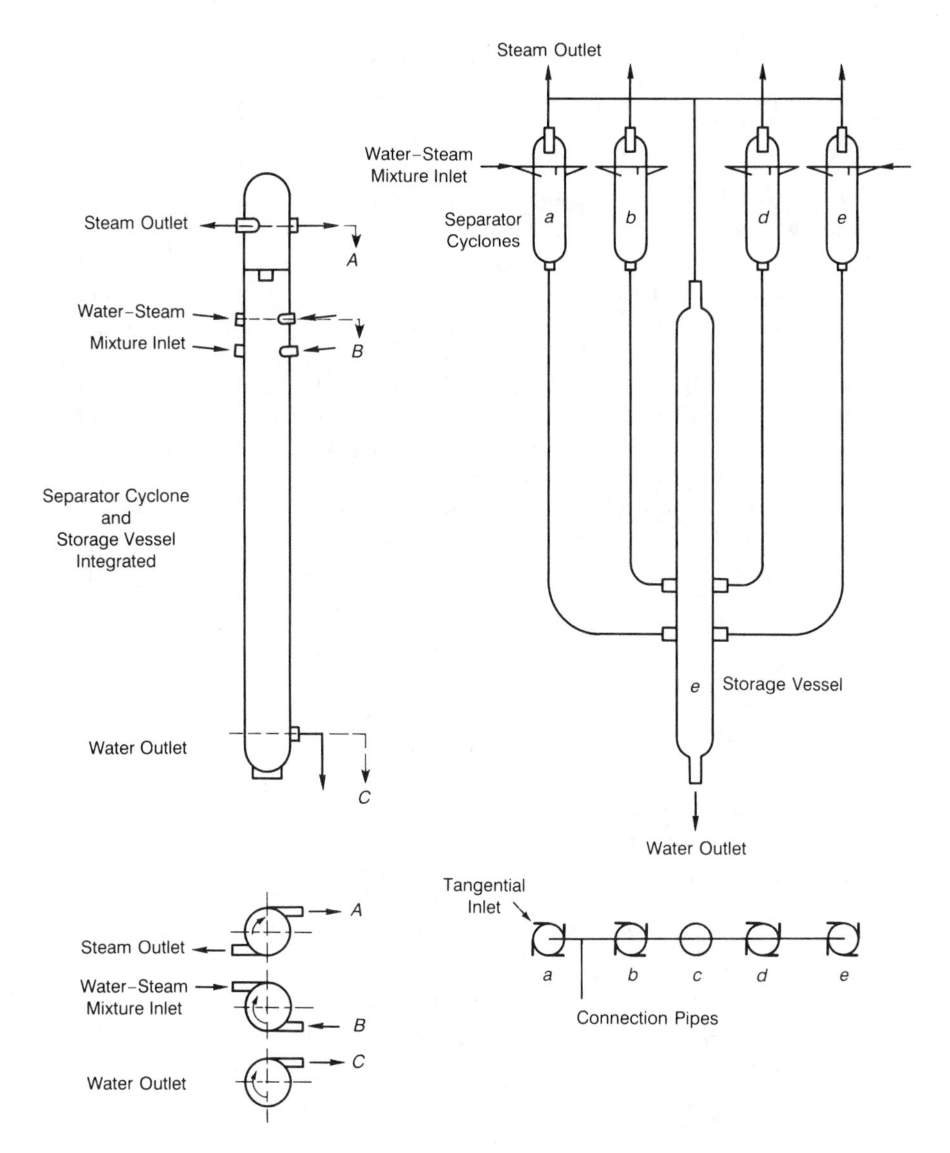

Fig. 7.36. Water–steam separator for once-through boilers. (Adapted from [4].)

Other design recommendations are similar, only a vortex breaker should be arranged upstream of the drain pipes to avoid annular flow of the water down to the water storage vessel, which is used for level control. A pressure balance pipe must be routed from the storage vessel to the steam lines to avoid water plugs in the drain pipes. The drain pipes have to be designed with a sufficient inner diameter.

7.4 START-UP SYSTEMS AND FEED-WATER CONTROL

7.4.1 Start-Up Systems

As already mentioned in Sections 7.2.4 and 7.2.5 below a certain load (once-through minimum load), the mass flux in the evaporator has to be kept constant by special means. There are four different methods in practical use for this purpose shown in Fig. 7.37 [4].

System A (wet superheater) is the classic once-through system and allows start-up only under a certain system pressure in the whole boiler. Evaporation will start after heat input mainly to the evaporator and the formation of saturated steam will create a water swell through the wet superheater and this must be dumped through the blowdown line. The water–steam mixture must be handled in most boiler sections including the superheater and only after blow-out of the water swell, the saturated steam can start to be superheated. The system is simple but the start-up time is long, heat losses are high, and the system cannot be used for larger boilers; in addition, other problems arise because of the different weight and problems with constant

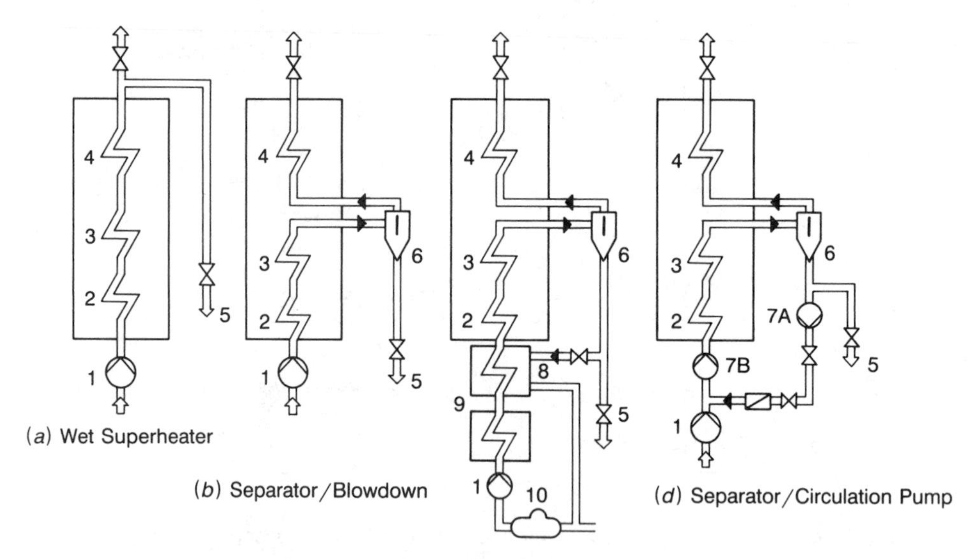

1	Feed Pump	6	Separator
2	Economizer	7	Circulation Pump (A—Bypass, B—Mainline)
3	Evaporator	8	Heat Exchanger
4	Superheater	9	High-Pressure Bled-Steam Feed-Water Preheaters
5	Blowdown	10	Feed-Water Tank

Fig. 7.37. Start-up system for once-through boilers. (Adapted from [4].)

hangers of the superheater connection piping initially filled with water and after start-up with steam.

The following systems are all capable of variable pressure mode of operation.

System B (separator/overfeeding): the boiler is started filled with water only up to the separator, leaving the superheater sections dry, fed with minimum evaporator flow, and on atmospheric pressure from cold. All saturated stream produced in the evaporator can immediately be superheated. The water flow through the blowdown line will be reduced with increasing steam production in the evaporator, so the heat losses are less compared with system A.

System C reduces the heat and water losses by recirculating most of the water via a heat exchanger and the feed-water tank by the feed-water pump. The heat exchanger is necessary to decrease the saturation pressure to the pressure level in the feed-water tank. Of course, one must consider that the thermal expansion from water to partial evaporation in the whole circuit from the feed line via economizer, evaporator, separator, and back via heat exchanger, feed water tank and pump to the feed line will also create a water swell which cannot be fully used in the start-up process and needs to be blown down. As a warning it should be mentioned that the control valve in the hot-water line to the feed-water tank must be treated as a safety stop valve to block the high-pressure boiler system from the low-pressure feed-water tank system. This system was chosen because of the supposed unavailability of the recirculation pumps.

Today, the most common system is the use of a recirculation pump (Fig. 7.37, system D), which reduces heat and water losses further than system C and has the best operating flexibility. The availability of the recirculation pump is no question. There are two possible arrangements of the recirculation pump:

1. The most common one is in the bypass (7A in Fig. 7.37, system D), pumping saturated water from the separator subcooled only by the hydrostatic head, which allows only slow pressure descreases (see Section 7.2.3). In addition, several precautions have to be taken:
   a. Minimum flow line to protect the pump switched on against evaporating the pump water content while the recirculation valve is closed
   b. Heating line to avoid cooling down in standby mode and thermal shocks when the pump is switched on
   c. Pressure difference control to trip the pump automatically in the case of an unallowable low pressure difference indicating evaporation followed by cavitation and imbalance (steam entrainment in the downcomer or steam formation due to an unallowable quick pressure drop)

2. These problems can be solved with the arrangement in the main feed line (7B in Fig. 7.37, system D) after the mixing of recirculated saturated water and feed water from the economizer, which gives sufficient subcooling for pressure decreases (provided the economizer outlet water is subcooled). The circulation pump works in this arrangement in series with the feed pump. But there are also disadvantages: trip of the feed-water pump or trip of bled-steam feed-water heaters during recirculation cause thermal shocks in the pump.

Heat losses through the blowdown line are also water losses. These may be losses due to vapor escaping from the venting of the blowdown tank or due to quality, as the water would have to be treated again, at least through the condensate polishing plant.

In parallel with evaporator protection, the other sections of a steam generator have to be discussed. To protect these from unallowable metal temperatures, the steam flow should be controlled in such a manner that sufficient heat transfer takes place during any start-up period.

To avoid the thermal shock during cold start due to sudden condensation when the steam (hopefully not with a higher pressure than 1 bar) displaces the air, a vacuum should be created in the superheater and reheater. This would be possible, when the high pressure (HP) and low pressure (LP) turbine bypasses are open (Fig. 7.38), all drains are routed to the condenser, no vent or connection to the atmosphere is open, and the condenser is at vacuum. But this requires all valves to be air tight in one direction and water tight in the other. Practical experience showed that such a large amount of maintenance on the valve gaskets was necessary that this method of start-up under effective steaming from a temperature level of about 60°C is no longer in use. It should be tried again, if modern valve technology allows this start-up method.

Well in use is the same installation but the LP bypass is closed and only the HP bypass is fully open, together with all superheaters, reheaters, pipework drains to the blowdown tank, and venting to the atmosphere. The economizer and evaporator have to be filled up to the separator water level. The next step is to adjust the minimum evaporator flow with the recirculation pump and to establish the operation of the feed pump via minimum flow line.

After light-off, the water in the recirculation line is heated and starts to evaporate, which causes (depending on the position of first evaporation) a smaller or higher water swell. The steam will entrain into the superheater and reheater. After heating up the tube metal by the inner condensation of saturated steam and the outer heat transfer from flue gas, the steam accumulates and the system pressure rises.

Reheater and superheater system pressures can be controlled by partially opening and closing the appropriate valves, at the same time allowing a certain flow and controlling the steam temperatures (warning: no flow—no steam temperature measurement!). A typical figure for preheating a cold

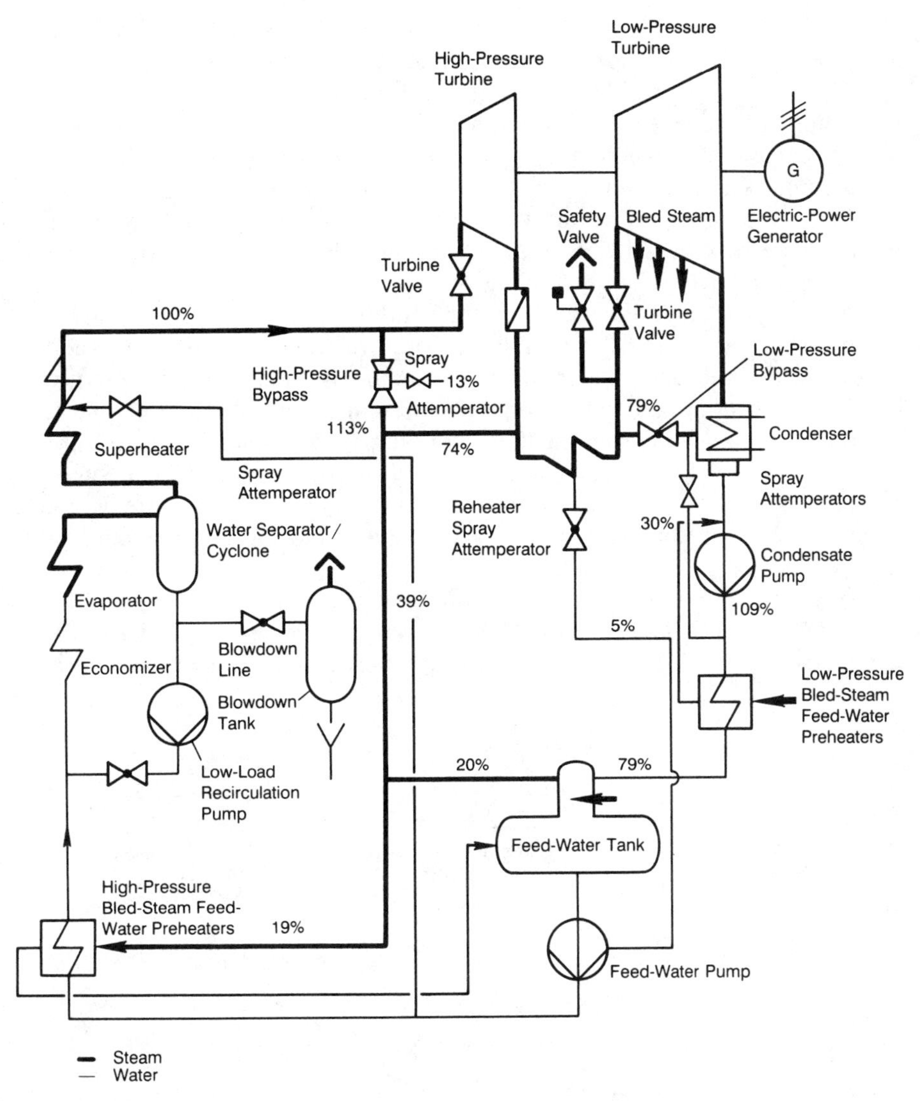

Fig. 7.38. Power plant start-up system mass flow parts at 100% boiler load and bypass of turbine. (With permission of [4].)

turbine is 50 K superheat, which may be achieved at a system pressure of about 40 bar with a corresponding saturation temperature of 250°C.

After rolling the turbine, the unit start-up procedure is continued by increasing in *parallel* the heat input, feed flow, system pressure, superheater, and reheater outlet temperatures. Temperatures have to be controlled to avoid rapid increase along with the risk of unallowable thermal stresses. Pressure follows according to the requirement of the steam flow to the turbine. The reheater in the beginning is controlled by the LP bypass until the turbine can accept the existing flow.

After a weekend shutdown (56 hr) the turbine is still rather hot and the boiler is only at 10 to 30 bar, the temperatures are down to 350°C both livesteam and reheater. The starting procedure is similar to cold start-up besides the venting and connection to atmosphere. However, the time to reach evaporation and superheat is reduced. Turbine metal temperatures with 450°C differ a significant amount from the originally available superheater–reheater temperatures. In this situation, the bypass systems can demonstrate their effectiveness to achieve acceptable turbine matching parameters.

Even more important is the HP and LP turbine bypass system for hot restarts (Fig. 7.39). This start-up is very common because of load requirements to accomplish two-shift operation or short turbine outages caused by

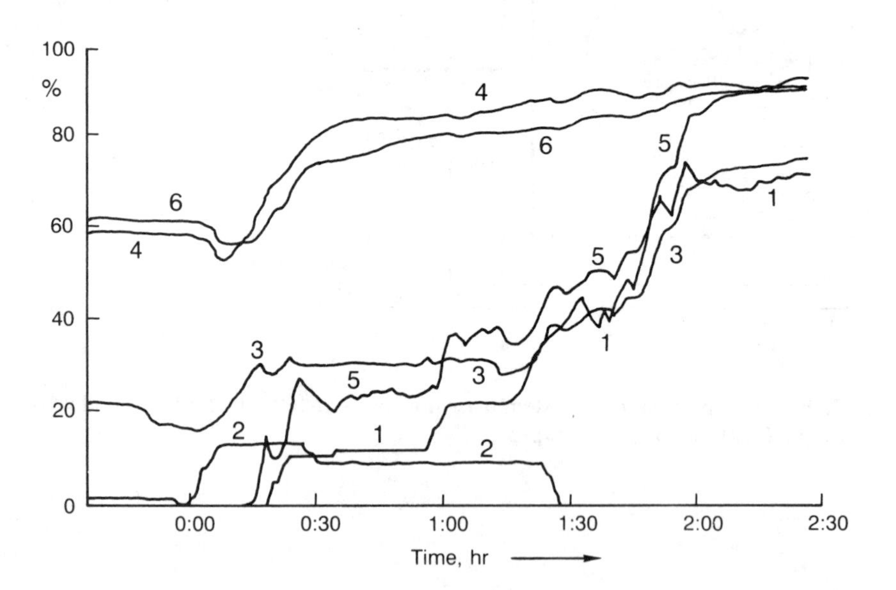

Fig. 7.39. Hot start after 7 hr "fire off". (With permission of [4].)

any protective device or component. The bypass systems have to be capable of full-load flow to avoid reheater safety valves opening. This may also be a requirement of noise protection as silencers cannot reasonably be designed to provide for allowable noise levels during the night. In densely populated areas blow-off is thus only allowed in emergency cases.

Such designed bypass systems allow a turbine trip without immediately tripping the boiler. The boiler can quickly be reduced in load say to 60% without risk to any equipment preparing the unit for restarting the turbine without delay and avoiding a start-up of the boiler if the turbine can be restarted within a reasonable time. For full-load bypass capacity, the nominal flow figures are given in Fig. 7.38.

7.4.2 Feed-Water Control

There are two different feed-water controls to be combined [22]:

1. The feed-water control during recirculation operation (start-up and low load)
2. The feed-water control during once-through operation

As already shown in Section 7.2.11 for recirculation operation, the feed-water flow is controlled according to the water level in the separator–storage vessel while for once-through operation, it is controlled according to the main steam temperature. The behavior of the boiler is also different (Fig. 7.28).

The differences explain why for many once-through boilers the change from recirculation operation to once-through operation (during start-up) and vice versa (during low-load operation and shutdown) is still a problem. The boilers are often not designed appropriately and the control systems are often not capable of managing this changeover without large disturbances. A simple method to avoid these problems is described in [22].

In the recirculation mode, the water level in the leveling vessel is controlled by means of a valve after the recirculation pump. The minimum evaporator flow is kept constant by means of the feed water. In once-through operation, the feed water is controlled as a function of the temperature or enthalpy at the evaporation outlet. If the set point for this temperature is given (cascade control beginning at the attemperators) such that the ratio of spray water and feed-water flow is constant, there will always be sufficient spray water flow; but the change from once-through operation to recirculation mode is rendered difficult, particularly in such cases where either the steam at the evaporator outlet is extremely superheated or the liquid–vapor phase transition point is located in the first superheating stage. Changeover from once-through to recirculation mode occurs automatically when the minimum evaporator flow is reached although the (leveling) storage vessel is either empty or full and the controlled variable "water level" required for this mode of operation is not available. Frequently, this entails either an excessive increase of the feed-water flow and a temperature decrease in superheaters if the discharge valve fails to open in time or an excessive reduction of the feed-water flow followed by the tripping of the recirculation pump and/or an evaporator flow below the minimum resulting in a unit trip. Conditions are aggravated by the fact that the discharge valve control in the once-through operation must be out of operation if the steam at the evaporator outlet entrains water. Otherwise, this water will be discharged and the controlled temperature or enthalpy will continue to rise, demanding more feed water and thus increasing the amount of water to be discharged.

Changeover from once-through to recirculation mode is possible without any disturbances provided that the temperature or enthalpy controller controls the feed-water flow such that within the transition stage the steam leaves the evaporator in a low superheated condition. In this case, however, the advantage of a constant ratio between feed-water and spray water flow under different fouling conditions is forfeited, at least in the lower end load range [22].

7.5 EXAMPLES AND OPERATING EXPERIENCES

Once-through boilers are used especially in large utility plants with high efficiency [33–48] but there are also some other special applications. In this chapter therefore three examples for utility boilers—a lignite-fired 600-MW unit, a bituminous-coal-fired 740-MW unit, both operated at subcritical pressure, and a bituminous-coal-fired 475-MW unit operated at supercritical pressure with double reheat—will be given. In addition—as an example for special purposes—a once-through steam generator for steam soak or steam drive in oil fields will be described.

7.5.1 Lignite-Fired 600-MW Once-Through Steam Generator

In 1971 the Rheinisch-Westfälische-Elektrizitätswerke (RWE) ordered two 600-MW units for the Neurath Power Station in West Germany [37]. The Neurath D and E units were commissioned in 1975 and 1976, respectively.

The fuel is Rheinische Braunkohle (lignite) with a lower heating value of 6.28 to 10.72 MJ/kg, an ash content of 1.8% to 20%, and a water content of 50% to 56.8%.

The design parameters are as follows:

Main steam flow	500 kg/s	Cold reheat pressure	34.8 bar
Live steam pressure	175 bar	Cold reheat temperature	311°C
Live steam temperature	530°C	Hot reheat pressure	32.7 bar
Feed-water temperature	235.4°C	Hot reheat temperature	530°C
Reheat stem flow	447 kg/s	Flue gas temperature before stack	140°C

Of particular interest is the simple single-pass design [29, 37], which is favorable with respect to:

1. Thermal expansion
2. Stress calculation
3. Flue gas flow (does not change the direction and is therefore less prone to erosion and allows for easier calculation of the heat transfer)
4. Drainage of the convective tube banks (horizontal)
5. Fouling (the upward decreasing temperature allows and keeping cross-sectional area and flue gas velocity constant asks for upward decreasing tube pitching and the latter allows pieces of fouling freely falling down)

Of further interest is the use of heat exchange between the high-pressure and reheat steam outside the flue gas flow (bifluxes) to avoid permanent reheat spray water flow and to increase the efficiency [37].

The mean availability is between 92% and 94% [37]. Every three years there is a revision.

7.5.2 Bituminous-Coal-Fired 740-MW Once-Through Steam Generator

The commissioning of the 740-MW unit Scholven F was completed in 1979 [43].

The fuel is bituminous coal from the Ruhr with a lower heating value of 25.7 to 30.6 MJ/kg; an ash content of 6% to 10%; a water content of 6% to 12%; volatile matter (with respect to combustible matter), 25% to 35%; ash softening temperature (in oxidizing atmosphere), 1100°C; ash half-sphere temperature (i.o.a), 1200°C; and ash fusion temperature, 1300°C.

The design parameters are as follows:

Main steam flow	611 kg/s	Cold reheat pressure	44.9 bar
Live steam pressure	201 bar	Cold reheat temperature	315°C
Live steam temperature	535°C	Hot reheat pressure	41.9 bar
Feed-water temperature	259°C	Hot reheat temperature	535°C
Reheat steam flow	558 kg/s	Flue gas temperature before stack	145°C

Figures 7.40 (cross section) and 7.41 (flow scheme) show that the design is even more simple, involving single pass and with a forged trifurcation piece for the transition from the inclined tubing of the furnace walls to the vertical tubing of the enclosure walls of the convective tube banks. There are also reheat spray water attemperators instead of heat exchangers because of the high investment cost of the latter and the small gain in efficiency because the pressure drop through the heat exchangers and the piping consumes at least partly the efficiency gain by avoiding permanent reheater spray water flow.

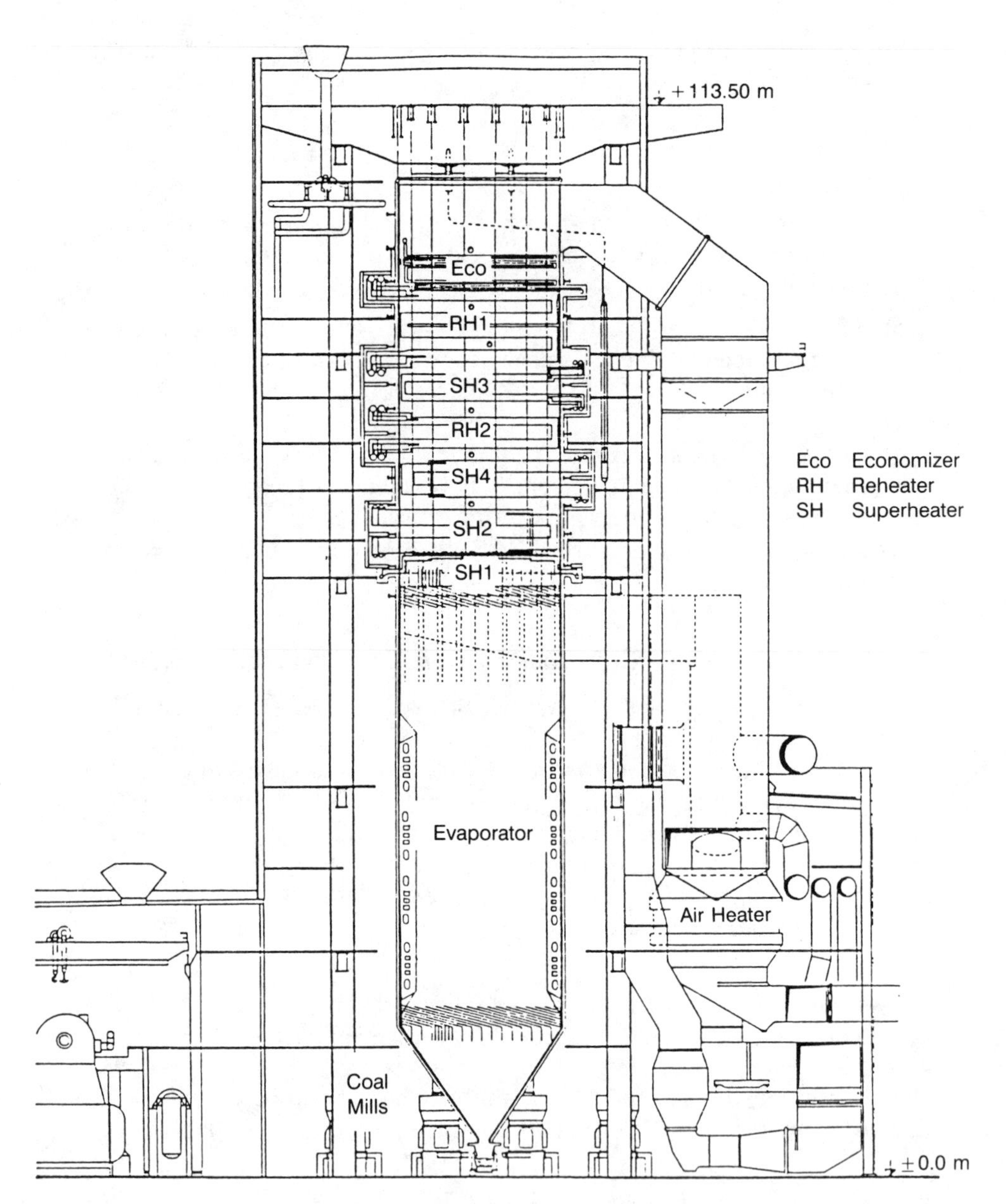

Fig. 7.40. Cross section of 740-MW bituminous-coal-fired once-through steam generator Scholven F.

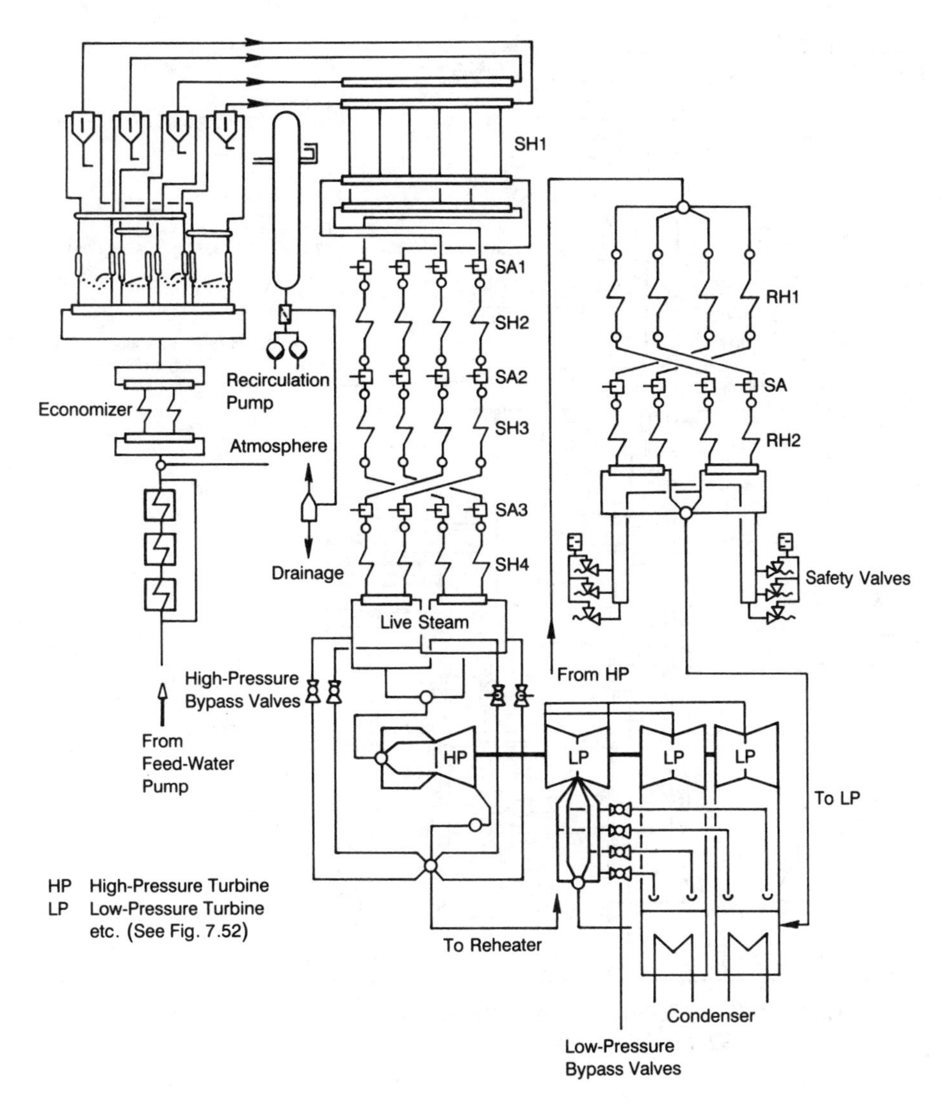

Fig. 7.41. Flow scheme of 740-MW bituminous-coal-fired once-through steam generator Scholven F.

The excellent agreement of the calculated design parameters and the measured values is documented in [37].

7.5.3 Power Boiler for Supercritical 475-MW Unit

This is a description of the single-pass boiler for unit 7 of the Mannheim Power Station in West Germany (Fig. 7.42) [4].

This utility is well known for the application of highly sophisticated technologies. The site-specific circumstances for fuel, power, and district heat

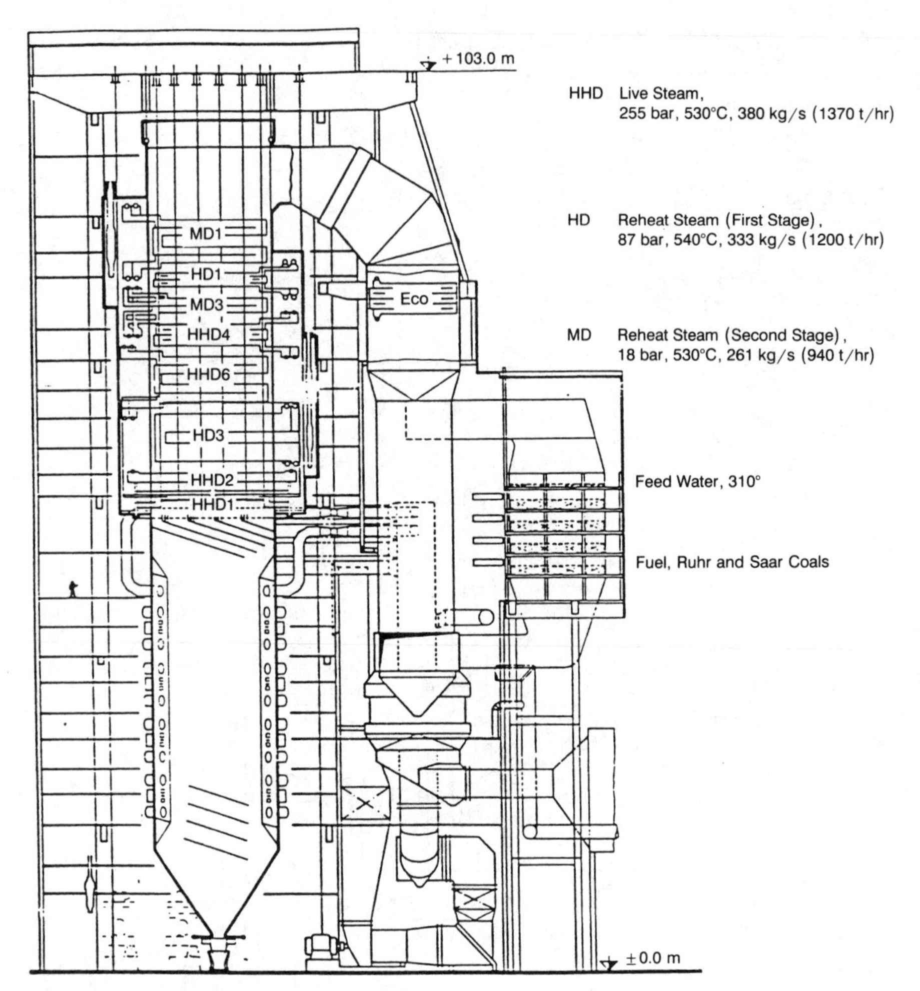

Fig. 7.42. 475-MW EVT-Sulzer supercritical once-through steam generator with double reheat and DeNox plant. (With permission of [4].)

required a design with double reheat and supercritical pressure under the aspects of overall cost optimization [33, 34, 42]. In addition, the first and second reheater attemperators are only in operation for rapid transients and the spray water flow will be reduced to 0 by using a double set of heat exchangers to maintain high outlet temperatures also under decreasing loads. The flow scheme is rather complex because of the necessity of these heat exchangers [4].

This boiler was the first to use the low-load circulation pump in series with the feed-water pump and control means for any system pressure. In other words, the very high pressure section (HHD) can be operated either in fixed *or* variable sliding pressure mode. The water–steam separation does not work above 210 bar effectively and above 225 bar in principle. Therefore only in the lower pressure range, the feed-water flow is controlled with feedback from the water level, whereas at higher pressures the feedback comes from the temperatures [38].

The furnace walls and also the furnace hopper are inclined. These walls, including the burner nozzles, are exposed to high radiation from the furnace. The enclosure walls for the convection banks with vertical tubes are connected with trifurcations to the inclined furnace wall tubes. The vertical tubes allow a simple design for convective superheater and reheater tube penetrations through these gas-tight welded water walls. The furnace surface and volume, including the burner belt area, are generously designed and have no limitation in the heat pickup in the "evaporator" ending with saturation. The upper part of the furnace walls is in fact a superheater.

The economizer is arranged in the "second" pass (Fig. 7.42) with the advantage of counterflow; that is, water flow upwards.

Normal operation above 35% load is supercritical due to the cross-compound unit interconnections. Only under very low load conditions also for the very high pressure turbine the system pressure is subcritical. But it has been proven that it does not matter for the boiler what system pressure is available, it can handle both ranges *and* the transition effectively [3].

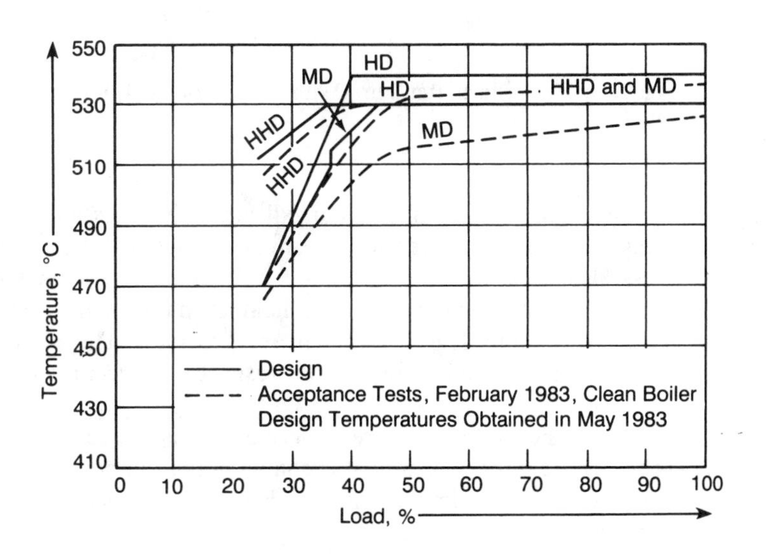

Fig. 7.43. Characteristic for superheaters and reheaters. (With permission of [4].)

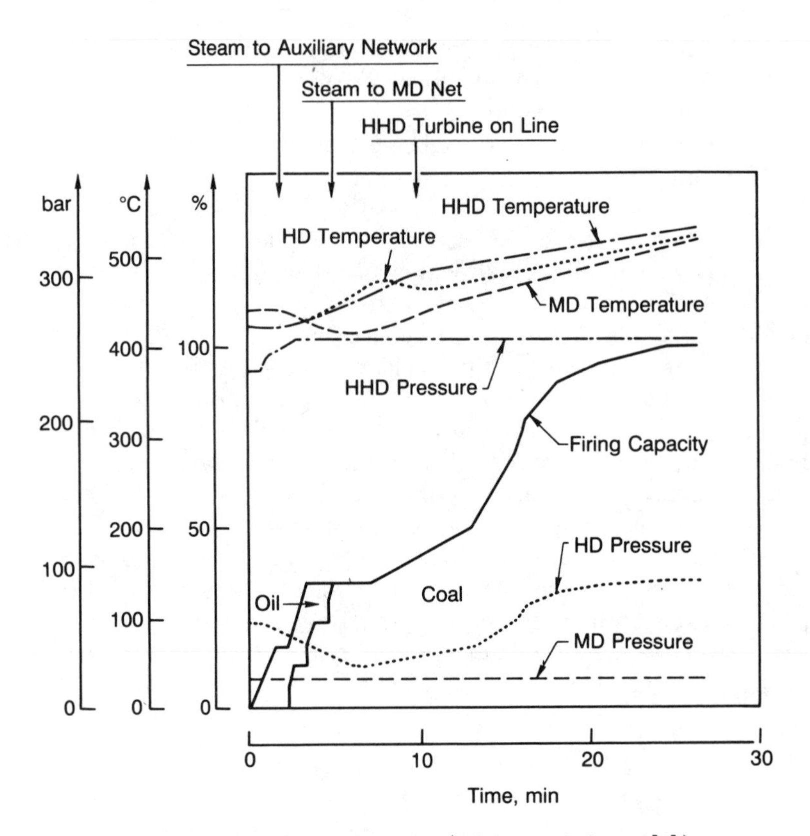

Fig. 7.44. Hot restart after 2 hr. (With permission of [4].)

Because of the design pressure of 270 bar, all larger headers and vessels have been increased in number to reduce the inner diameter and as a result the wall thickness also. This resulted in eight water separator cyclones with four-leg superheaters in parallel and for the heat exchangers from superheater to reheater again four legs in parallel are provided. The allowable temperature transients are excellent and do not restrict start-up time or load transients.

In Fig. 7.43 some operating data are described. This proves that with superheater–reheater steam-to-steam heat exchangers the temperature characteristics are excellent. The optimum efficiency is not quite reached because of the relative increase in the pressure drop.

Restart after a short outage is possible without delay and with a minimum heat loss. In Fig. 7.44 the diagram explains that already some minutes after light-off low-pressure steam is available in the reheater steam line. For synchronizing the very high pressure turbine only 10 min is necessary and full load is achieved within about half an hour.

This boiler has been in operation since 1983 with a total number of 36,000 operating hours and 255 starts (as of June 1987).

7.5.4 Steam Generator Unit for Steam Soak or Steam Drive in Oil Fields

The increasing cost of crude oil forced some companies to obtain more oil from the deposits than feasible with conventional techniques. After simply pressing water into an existing oil well the next step was to use hot water and/or steam [52].

As a natural behavior, the necessary operating pressure for a steam generator used in such an application may vary from 60 to 200 bar and the required steam quality from 50% to saturated steam. No other system than the once through can achieve this in a simple design, especially with the additional request of skid mounting and easy transport from well to well or field to field.

The feed-water treatment is only necessary in respect to decarbonizing and deaeration. Desalination is not required even for very high pressures. In order to keep the salt in the boiler water dissolved and thus to avoid deposits on the boiler tubes, the steam produced has a residual moisture of approximately 20%. The necessary moisture content over the total boiler operating range is controlled by the ratio between fuel and feed flow. If the quality of the steam is requested to be near saturation the separated water from the cyclone is routed to the blowdown vessel and after that through the heat exchanger for primary feed-water heating (Fig. 7.45) [32].

The combustion chamber enclosure is manufactured from tubes as a membrane wall. The tubes are welded together by means of longitudinal plate strips on a membrane wall multipass welding machine. Thus the combustion chamber cage is completely gas tight and self-supporting and also does not require any refractory material and a hot tube supporting system which are both likely to fail especially when firing high-sulfur fuel. In addition, any corrosion problems of an external plate casing due to penetration of acidic condensate from flue gas through the refractory are avoided. Only external insulation against heat loss from the tubes being on saturated steam temperature or lower is required. Such boilers are operated on forced draft and can use any liquid or gaseous fuel.

In the lower section, the extended combustion chamber tubes form the external casing for the convection evaporator and the flue gas deflection duct to the economizer. The multipass economizer is situated in an externally insulated plate casing which is bolted to the tubular casing and which also carries the stack (Fig. 7.46).

The burner opening is located in the front wall of the combustion chamber. All tube arrangements are such that the boiler is completely drainable.

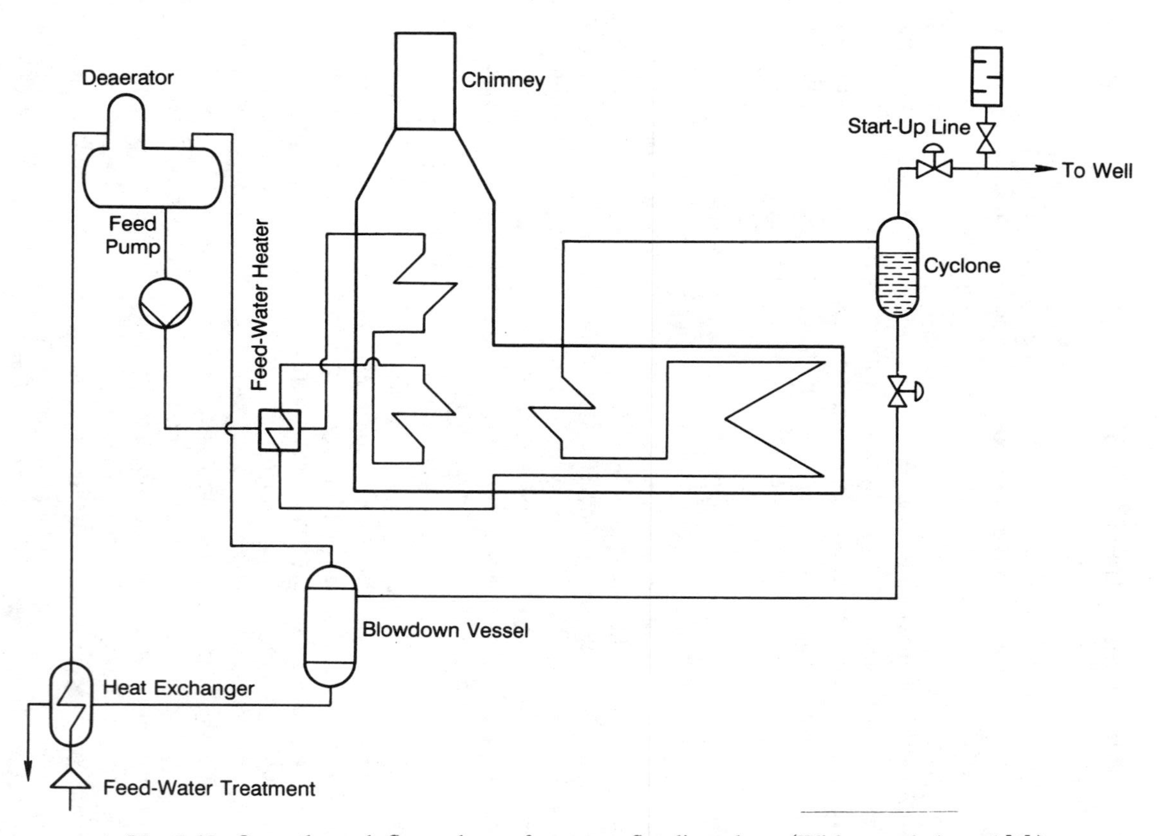

Fig. 7.45. Once-through flow scheme for steam flooding plant. (With permission of [4].)

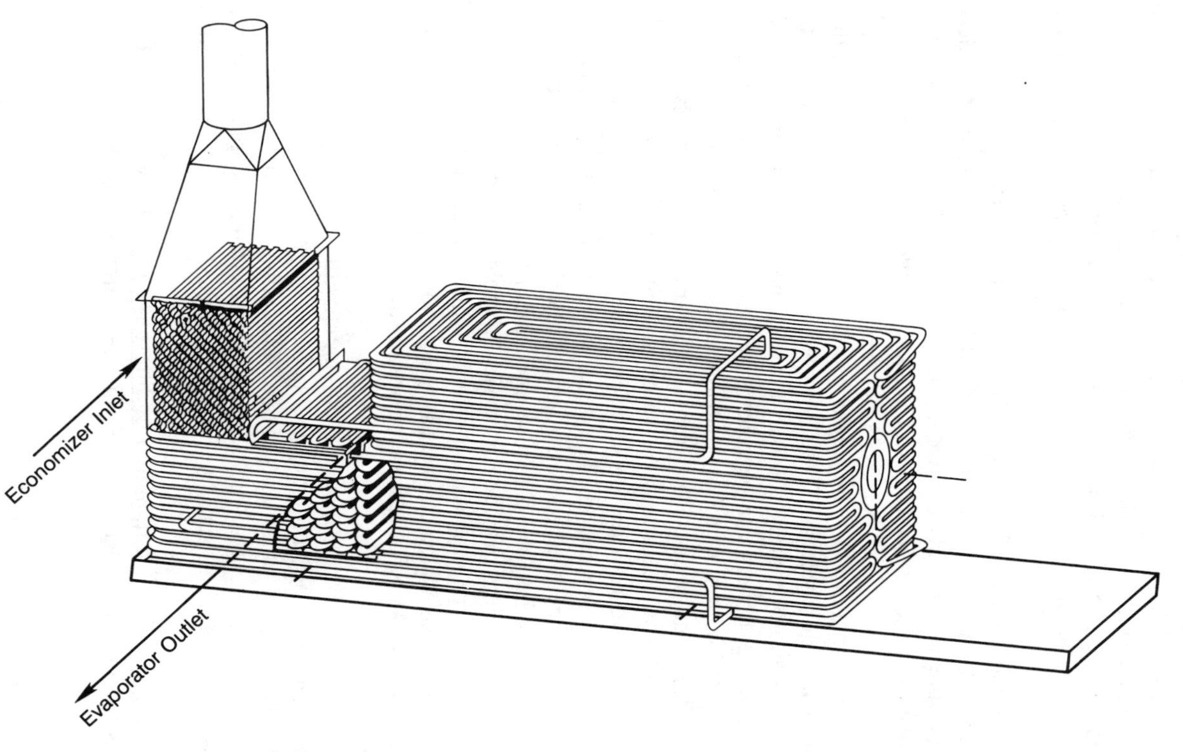

Fig. 7.46. Once-through boiler for steam flooding. (With permission of [4].)

The feed water from the pump passes through the economizer to the combustion chamber wall tubes. The water enters into the bottom, passes through the wall tubes, and exits through the roof tubes. The major portion of the combustion chamber wall tubes are still operating in the preheating zone. The steam content at the transition to the convection coils ranges at 100% load between 25% and 35%, therefore safely avoiding any depositing and hiding-out of solids in the combustion chamber tubes exposed to flame radiation.

The subsequent convection evaporator coil is a co-current flow arrangement so that the maximum steam content of 80% is in the area of moderate gas temperature (approximately 600°C) and low heat flux; these conditions again contribute to low sensitivity to feed-water deficiencies and operational safety (Fig. 7.46).

Up to approximately 30 t/hr steam production, the boilers are built as monotube evaporator systems. The tubes used in these boilers, designed for a pressure up to approximately 200 bar, are 88.9 mm OD. When higher unit rates are required, boilers can be constructed with several tubes in parallel. In order to avoid nonuniform flow distribution and unstable flow conditions, each tube will be equipped with a front-end distribution control system.

At high-sulfur content, the feed water is preheated in a water-supplied preheater in order to provide for sufficient tube surface temperature in the economizer to avoid corrosion. The feed water leaving the economizer is used as a heating medium; downstream of the feed-water preheater, the water is passed back into the combustion chamber tubes.

Field injection boilers provide for maximum mobility. Small units up to 10 t/hr steam outputs are supplied on wheels, while larger units normally are mounted on a chassis which can be easily equipped with wheel sets in order to move the boiler. Also flatbed trailer or railway transportation is a possible alternative.

Boilers are normally installed in containers with the local control panel inside the housing. The containerized boiler system can be equipped for operation under arctic conditions. Standby heating systems can be either electrically operated or fossil-fuel based; special circulation pumps provide for uniform heating of all sensitive pressure parts also in the depressurized state.

Large boiler systems are mounted in different containers which in the field can be easily connected together to form one operational unit.

7.6 SUMMARY

The once-through boiler design is presented starting with a short historical review. The once-through boiler system is compared with other systems, namely the natural-circulation, controlled-circulation, once-through with superimposed circulation, and combined-circulation systems, under the aspects of the applicable pressure range, the influence of pressure and temperature on plant efficiency, and the suitability of the systems for fixed and sliding pressure operation. Reference is made to mass flux in the evaporator (cooling of the evaporator tubes), arrangement of the recirculation pump, recirculation and feed-water pump power, furnace wall design, main steam temperature pattern under part-load operation and furnace slagging (feasible furnace outlet temperatures), required feed-water quality, dynamic behavior (start-up times, disturbances, load changes), preferred design practices, unit sizes, and advantages in specific applications.

Special attention is focused on the water wall design including DNB, DO, thermal stresses and temperature differences between individual furnace tubes, steam preheating and water separation devices, start-up systems, and feed-water control.

Finally three examples of large utility boilers and a special application for steam soak or steam drive in oil fields are described.

ACKNOWLEDGMENTS

I want to thank Mr. J. Merz who allowed me to use extensively his papers on boiler design [4], especially in Sections 7.3.2, 7.3.3, 7.4.1, 7.5.3, and 7.5.4, and all the other authors (see the references) whose published papers I used. Further I want to thank Mr. G. Bauer, Mr. P. Fritz, Mr. G. Groß, Mr. U. Krogmann, and Mr. C. T. Nguyen, who assisted me with the calculations. Last but not least I want to thank Mrs. Ch. Schröder, who typed the manuscript, for her patience with both my handwriting and the endless corrections; I also want to thank the several people who drafted the figures.

NOMENCLATURE

A_V	valve area, m^2
am	amortization $\text{am} = \dfrac{(1 + \text{per}/100)^{n_y}\,\text{per}/100}{(1 + \text{per}/100)^{n_y} - 1}$, 1/annum
C_{cap}	specific capital costs, \$/kW
C_{el}	electrical power costs, \$/kWh
C_{fuel}	fuel cost, \$/kWh
c_{FG}	specific heat at constant pressure of the flue gas, kJ/(kg · K)
$c_{1,2,3,4}$	constant, variable
d_i	tube inside diameter, m
FLH	full-load hours per year, hr/a
f_{av}	FLH/PFLH availability factor, dimensionless
f_{fc}	factor for fixed specific costs (e.g., part of maintenance and personnel costs independent of operation), dimensionless

f_{us}	PFLH/8760 usage factor, dimensionless
f_{var}	factor for variable specific costs (e.g., part of maintenance and personnel costs dependent on operation), dimensionless
g	gravitational acceleration, m/s^2
h	height, m
i	enthalpy, kJ/kg, J/kg
$i_{i,o,ad}$	enthalpy (inlet, outlet, adiabatic), J/kg
k_p	tube pitch diameter ratio in finned tube water walls, dimensionless
LHV	lower heating value, J/kg, MJ/kg
LRC	radiative and convective heat losses related to the maximum heat output, dimensionless
l	length of a side (depth, width), m
l_t	tube length, m
m_{FG}	specific flue gas mass, kg flue gas/kg fuel
n	number of parallel tubes, dimensionless
n_y	number of years in operation (deduction), a
PFLH	possible full-load hours per year, hr/a
P_T	turbine load, W
p, p_{cr}, p_1, p_2	pressure (cr, critical; 1, 2, numbering index), Pa, bar
Δp_b	pressure drop (*b*, boiler), bar
per	interest, %/a
pt	tube pitch, m
pt_h	horizontal tube pitch, m
$\dot{Q}$	heatflow, W, kW, MW (out, output; in, input; losses, sum of losses; FG, losses due to sensible heat in the flue gas; UF, losses due to unburned matter; RC, losses due to radiation and convection)
q''	heat flux (cr, critical), W/m^2
Re	Reynolds number, $\rho v d_i/\mu$, dimensionless
RR	evaporator water recirculation ratio, dimensionless
S	entropy kT/kgK
Sn	stability number, dimensionless
T	temperature (*s*, saturated; 1, 5, points shown in Fig. 7.11; TQ, mean value; inc, upper cycle temperature level)
ΔT_u	steam superheating, K
ΔT_{max}	maximum temperature difference between water wall and tie bar, K
ΔT_{ramp}	ramp temperature change, K
t	time, s
Δt_T	time constant of first order delay, s
v	velocity (opt, optimal), m/s
v''	specific volume of saturated steam, m^3/kg
v_T, v_{T1}, v_{T2}	rate of temperature change, K/s
W''	mass flux (mass velocity), $kg/(m^2 \cdot s)$

W	mass flow (evap, evaporator; feed, feed water; steam, live steam; min, minimum evaporator flow), kg/s
$\dot{x}$, $\dot{x}_{cr}$, x, x_{out}	mass steam quality (cr, critical; out, at the evaporator outlet), dimensionless
8760	hr/a
α	angle of inclination between tube and vertical downward direction, rad
α_h	angle of inclination between tube and horizontal direction, rad
Δ	difference (to steady state)
η	efficiency (cyc, cycle; i, turbine; SG, steam generator; F, firing), dimensionless
λ_f	tube friction coefficient, dimensionless
μ	dynamic viscosity, kg/(m · s)
ρ	density, kg/m^3

Subscripts

a	ambient
FB	fuel burned
FG	flue gas
FI	fuel input
FW	feed water
i	inlet
o	outlet, steady state
RH	reheater
SH	superheater
UF	unburned fuel
SW	spray water
HP	high pressure
LP	low pressure

REFERENCES

1. Fryling, G. R. (1966) *Combustion Engineering—A Reference Book on Fuel Burning and Steam Generation*, pp. 25-3 ff. Published by Combustion Engineering, Inc., New York. The Riverside Press, Cambridge, Mass.
2. Wiehn, H., et al. (1985) Trends and Lösungen im internationalen Dampferzeugerbau. *VGB-Kraftwerkstechnik* **65**(12) 1126–1132.
3. Leithner, R. (1984) Überkritische Dampferzeuger, Auslegungskriterien und Betriebserfahrungen. *Brennstoff-Wärme-Kraft* **36**(3) 71–82.
4. Merz, J. (1988) Design considerations for fossil-fired steam generators of the once-through system, Chapter 2; Selected fluid phenomena in water/steam, Chapter 6; Operating characteristics and experience with once through power

boilers, Section 7.1; Application of once through technique, Section 7.2. In *Two-Phase Flow Heat Exchangers*, S. Kakaç, A. E. Bergles, and E. O. Fernandes (eds.). Kluwer, Dordrecht.

5. Pich, R. (1979) Betrachtungen über den Einfluß der Stromerzeugungskosten auf die Entwicklung im Kraftwerksbau unter besonderer Berücksichtigung der Verfügbarkeit. *EVT-Register 35 / 1979*. Energie- und Verfahrenstechnik GmbH, Stuttgart.
6. Strauß, K. (1985/86) Kriterien für den Einsatz unterschiedlicher Dampferzeugersysteme bei Kraftwerks-Dampferzeugern. *Jahrbuch der Dampferzeugertechnik 1985 / 86*, pp. 332–342. Vulkan-Verlag, Essen.
7. Brockel, D., et al. (1985/86) Große Naturumlaufdampferzeuger. *Jahrbuch der Dampferzeuger-Technik 1985 / 86*, pp. 362–383. Vulkan-Verlag, Essen.
8. Gericke, B. (1978) Der natürliche Wasserumlauf in Abhitzedampferzeugern. *EVT-Bericht 46 / 79. Brennstoff-Wärme-Kraft* **30**(12) 459–468.
9. Lange, F. (1971) Kesselspeisepumpen—Zulaufsysteme bei gleitendem Entgaserdruck. *Brennstoff-Wärme-Kraft* **23**(7) 321–328.
10. *Technische Regeln für Dampfkessel* herausgegeben im Auftrage des Deutschen Dampfkesselausschusses von der Vereinigung der Technischen Überwachungsvereine e.V. Essen, Carl Heymann Verlag KG, Cologne.
11. Leithner, R. (1979) Dynamik im Großdampferzeugerbau. *Elektrizitätswirtschaft* **80**(8) 281–290. *EVT-Bericht 52 / 1980.*
12. Jens, W. H., and Lottes, P. A. (1951) Analysis of heat transfer, Burnout, Pressure drop and density data for high pressure water. USAEC Report ANL-4627.
13. Drescher, G., and Köhler, W. (1981) Die Ermittlung kritischer Siedezustände im gesamten Dampfgehaltsbereich für innendurchströmte Rohre. *Brennstoff-Wärme-Kraft* **33**(10) 416–422.
14. Doroshchuk, V. E., Levitan, L. L., and Lantsmann, F. P. (1975) Recommendations for calculating burnout in a round tube with uniform heat release. *Teploenergetika* **22**(12) 66–70.
15. Kon'kov, A. S. (1965) Experimental study of the conditions under which heat exchange deteriorates when a steam-water mixture flows in a heated tube. *Teploenergetika* **12**(12) 77.
16. Köhler, W. (1984) Einfluß des Benetzungszustandes der Heizfläche auf Wärmeübergang und Druckverlust in einem Verdampferrohr. Dissertation, Technical University of Munich.
17. Hein, D., Kastner, W., and Köhler, W. (1982) Einfluß der Rohrlage auf den Wärmeübergang in einem Verdampferrohr. *Brennstoff-Wärme-Kraft* **34** 489–493.
18. Heitmann, H.-G., and Kastner, W. (1982) Erosionskorrosion in Wasser-Dampfkreisläufen—Ursachen und Gegenmaßnahmen. *VGB Kraftwerkstechnik* **62**(3) 211–219.
19. Hein, D., and Wittchow, E. (1985/86) Verbesserung der Auslegung und des Betriebsverhaltens von Benson-Dampferzeugern. *Jahrbuch der Dampferzeuger-Technik 1985 / 86, pp. 342–361. Vulkan-Verlag, Essen.*
20. Teigen, B. C., and Peletz, L. J. (1981) Heat transfer data, rifled tubing. Paper presented at the CE–Sulzer Conference, August 1981. Combustion Engineering, Inc., Windsor, Conn.

21. Iwabuchi, M. et al. (1982) Heat transfer characteristics of rifled tube in near critical pressure region. *Proc. Seventh Int. Heat Transfer Conf., Munich*. Vol. 5, TF 21, *Heat Transfer*.

22. Läubli, F., Leithner, R., and Trautmann, G., (1984) Probleme bei der Speisewasserregelung von Zwangdurchlaufdampferzeugern und deren Lösung. *VGB-Kraftwerkstechnik* **64**(4) 279–291.

23. Hein, D., and Wittchow, E. (1980/81) Forschung und Entwicklung auf dem Gebiet des Benson Dampferzeugers. *Jahrbuch der Dampferzeugertechnik 1980 / 81*, pp. 218–230. Vulkan-Verlag, Essen.

24. Pfau, B. (1977) Riffelrauhigkeit im Wandbelag von Rohrleitungen vt. *Verfahrenstechnik* **11**(1).

25. Thom, J. R. S. (1964) Prediction of pressure drop during forced circulation boiling of water. *Int. J. Heat Mass Transfer* **7** 709–724.

26. Thelen, F. (1981) Strömungsstabilität in Verdampfern von Zwangdurchlaufdampferzeugern. *VGB-Mitteilungen* **61**(5) 357–367.

27. Ledinegg, M. (1966) *Dampferzeugung, Dampfkessel, Feuerungen einschließlich Atomreaktoren*, Vol. 2. Springer-Verlag.

28. Miszak, P. Zwangdurchlaufdampferzeugeranlage. Europäische Patentanmeldung, Veröffentlichungsnummer 0054601.

29. Leithner, R. (1980) Entwicklung großer Einzug-Zwangdurchlaufdampferzeuger. *Jahrbuch der Dampferzeugertechnik*, Vol. 4, pp. 230–245. *EVT-Bericht 57 / 1981.*

30. *VGB-Richtlinien* für Kesselspeisewasser, Kesselwasser und Dampf von Wasserrohrkesseln der Druckstufe ab 64 bar. Vereinigung der Großkraftwerksbetreiber, Essen.

31. Linzer, V. and Leithner, R. (1975) Einfaches Dampferzeugermodell (digitale Simulation). *Forstschritt-Berichte der VDI-Zeitschriften*, Vol. 6, No. 41. *EVT-Bericht 31 / 75.*

32. Mattern, J. and Merz, J. (1980) Dampferzeugung zur Bedampfung von Erdöl-Lagerstätten. *Jahrbuch der Dampferzeugungstechnik*, Vol. 4, 324–337.

33. Baumüller, F., and Richter, R. (1983) Sind Sammelschienenkraftwerke noch zeitgemäß? *VGB-Kraftwerkstechnik* **63**(5) 381–388.

34. Baumüller, F., Richter, R., and Strasser, P. (1985) Überkritischer 475-MW-Heizkraftwerksblock mit REA im Großkraftwerk Mannheim. *VGB Kraftwerkstechnik* **65**(3) 208–218.

35. Grünn, H., Seefeldt, K.-F., Waldmann, H., Reidick, H., and Schüler, U. (1973) Kessel und Feuerungen für 600-MW-Blöcke für Braunkohle. *VGB Kraftwerkstechnik* **53**(12) 772–791.

36. Komo, G. (1977) Errichtung und Betriebsergebnisse für den 600-MW-Braunkohlekessel des RWE. *Braunkohle*, October 1977, 403–412.

37. Vetter, H., and Leithner, R. (1980) Betriebserfahrungen mit den Dampferzeugern für Braunkohlefeuerung Neurath D and E. *Jahrbuch der Dampferzeugungstechnik 1980*, pp. 813–822. Vulkan-Verlag, Essen.

38. Leithner, R. (1981) Berechnung des Betriebsverhaltens überkritischer Dampferzeuger im Anfahr- und Umwälzbetrieb. *EVT-Bericht 54 / 81.* Energie- und Verfahrenstechnik GmbH, Stuttgart.

39. Bürkle, E., and Hackmaier, R. (1975) 740-MW-Block Kraftwerk Scholven, ein EVT-Sulzer-Dampferzeuger mit Steinkohlenfeuerung. *EVT-Register 29 / 1975*. Energie- und Verfahrenstechnik GmbH, Stuttgart.
40. Kübler, D. *Der steinkohlegefeuerte 740-MW-Block Scholven F*. Musteranlagen der Energiewirtschaft, Das Kraftwerk Scholven, Energiewirtschaft und Technik, Verlagsgesellschaft mbH, Federal Republic of Germany.
41. Kübler, D., and Eggers, H. J. (1978) 740-MW-Steinkohleblock Scholven F, Einflußnahme des Bestellers auf die Konstruktion an den Beispielen "Dampferzeuger" und "Rauchgasentschwefelung," *VGB Kraftwerkstechnik* **58**(12) 861–866.
42. Richter, R., Knisel, G., and Leithner, R. (1979) Überkritischer 475-MW-Zwangdurchlaufdampferzeuger für das Großkraftwerk Mannheim. *EVT Register 36 / 79*. Energie- und Verfahrenstechnik GmbH, Stuttgart.
43. Leithner, R., and Reidick, H. (1979) Auslegung und erste Betriebserfahrungen des 740-MW-steinkohlegefeuerten Blockes Scholven F. *EVT Register 36 / 1979*. Energie- und Verfahrenstechnik GmbH, Stuttgart.
44. Schlessing, J., and Strasser, P. (1985) Erfahrungen bei der Inbetriebnahme und beim Betrieb des überkritischen Dampferzeugers Kessel 18 im Großkraftwerk Mannheim AG. *VGB Kraftwerkstechnik* **65** 1000–1011.
45. Richter, R., and Strauß, K. (1985) Betriebserfahrungen mit dem kohlegefeuerten, überkritischen Dampferzeuger mit doppelter Zwischenüberhitzung des Großkraftwerks Mannheim AG. *EVT-Bericht 85 / 85*. Energie- und Verfahrenstechnik GmbH, Stuttgart.
46. Merz, J. (1983) The new steam generator no. 18 for power station unit 7 in the central power station Mannheim. Paper presented at the CE–Sulzer Conference. Combustion Engineering, Inc., Windsor, Conn.
47. Bieber, K.-H. (1979) Einsatz und Bereitschaft von Reservekraftwerken. *VGB Kraftwerkstechnik* **59**(7) 531–539.
48. Fischer, P., and Fröhlich, P. (1974) BAG Kraftwerk Pleinting II. *EVT Register 26 / 1974*. Energie- und Verfahrenstechnik GmbH, Stuttgart.
49. Leithner, R. (1983) Vergleich zwischen Zwangdurchlaufdampferzeuger, Zwangdurchlaufdampferzeuger mit Vollastumwälzung und Naturumlaufdampferzeuger. *VGB-Kraftwerkstechnik* **7** 553–568.
50. Geißler, Th. (1979) Feurungsanlagen für Braun- und Steinkohle. *EVT-Bericht 27 / 79*. Energie- und Verfahrenstechnik GmbH, Stuttgart.
51. Juzi, H., et al. (1984) Zwangdurchlaufkessel für Gleitdruckbetrieb mit vertikaler Brennkammerberohrung, *VGB Kraftwerkstechnik* **4** 292–302.
52. Schlemm, F. (1979) Planung und Betrieb von Dampfkesselanlagen bei Tertiär-Projekten. *Erdöl-Erdgas-Zeitschrift* **95** July 1979.
53. Linzer, W. (1984) *Lecture on Boiler Design*. Technical University Vienna, Austria.

APPENDIX 7.1: EXAMPLE FOR CALCULATING POWER GENERATION COSTS

From a bituminous-coal-fired power station the following information is available: interest rate, 8%; payout time (period of amortization), 20 years;

fuel costs, \$120/10^3 kg [LHV = 28.47 (MW · s)/kg]; overall plant efficiency, 37%; annual equivalent full-load operation hours, 5400 hr/a; the specific fixed costs are 6% of the specific plant costs; the sum of the specific variable costs is 5% of the specific fuel costs; and the specific plant costs:

$$C_{\text{cap}} = \frac{\text{plant costs}}{\text{installed electric capacity (kW)}}$$

$$C_{\text{cap}} = 1100\ \$/\text{kW} \qquad \text{from Table 7.3}$$

With the given information the costs for power generation in this power station can be calculated according to Eq. (7.1):

$$C_{\text{el}} = \frac{C_{\text{cap}} \text{am} f_{\text{fc}}}{8760 f_{\text{av}} f_{\text{us}}} + \frac{C_{\text{fuel}} f_{\text{var}}}{\eta_{\text{cyc}}}$$

The solution is as follows:
Specific fixed-costs factor:

$$f_{\text{fc}} = 1.06$$

Capital repayment factor:

$$\text{am} = \frac{(1+q)^{n_y} q}{(1+q)^{n_y} - 1}$$

where q is the interest rate per year and n_y is the payout time in years

$$\text{am} = \frac{1.08^{20} \times 0.08}{1.08^{20} - 1}$$

$$= 0.1019 \left[\frac{1}{a} \right]$$

Availability factor (operating time, etc., in equivalent full-load operating hours):

$$f_{\text{av}} = \frac{\text{annual operating time + annual standby time}}{\text{possible operating time per annum (8760hr/a)}}$$

Utilization factor:

$$f_{\text{us}} = \frac{\text{annual operating time}}{\text{annual opertaing time + annual standby time}}$$

TABLE 7.3 Characteristic Information for Different Power Plants

Power Plant	Specific Investment Costs, $/kW	Construction Period, a	Overall Plant Efficiency, %	Fuel	Suitable for	Cold Start-Up Period, hr
Hydroelectric power plant	1450–2020	3–5	($\approx$ 80)	"Sun"	Basic load or peak load	0.1–0.25
Nuclear power plant with pressurized or boiling-water reactor	1730–2310	6–8	30–40	Uranium	Basic load	30–50
Steam power plant						
With flue gas desulfurization	870–1270	4(–6)	37	Coal	Medium load or basic load	5–8
Without flue gas desulfurization	690–1040	3(–5)	40	Coal	Medium load or basic load	5–8
Gas turbine power plant	260–430	1–2	28–32	Natural gas Mineral oil	Peak load	0.25
Combined power plant	400–690	2–3	45–52	Natural gas Mineral oil	Medium load or basic load	0.5–2.5
Cogeneration (of heat and power)						
Steam power plant with desulfurization	920–1330	2–4	70–85	Coal Wood	Basic load	5–8
Combined power plant	420–720	2–3	70–85	Natural gas Mineral oil	Basic load	0.5–2.5
Gasification of coal with a combined power plant	1010–1450	3–4	38.5	Coal	Basic load	70

Source: Kraftwerke, *Charakteristische Daten verschiedener Kraftwerkstypen*. ABB Technik, 6/89, p. 21, Table 1.

The utilization factor and the availability factor can be solved easily together:

$$f_{av} f_{us} = \frac{5400}{8760} = 0.616$$

Specific variable costs:

$$f_{var} = 1.05$$

Overall plant efficiency:

$$\eta_{cyc} = 0.37$$

Specific fuel costs:

$$C_{fuel} = \frac{\text{fuel costs (\$/kg)}}{\text{lower heating value (kWh/kg)}}$$

$$= \frac{\$120}{10^3 \text{ kg}} \frac{1 \text{ kg}}{28.47 \times 10^3 \text{ (kW} \cdot \text{s)}} \frac{3600 \text{ s}}{1 \text{ hr}}$$

$$= 0.0152 \text{ \$/kWh}$$

The numerical calculation is

$$c_{el} = \frac{1100 \times 0.1019 \times 1.06}{8760 \times 0.616} + \frac{0.0152 \times 1.05}{0.37}$$

$$= 0.0220 + 0.0431$$

$$= 0.0651 \text{ \$/kWh}$$

APPENDIX 7.2: OPTIMAL DESIGN OF A RECIRCULATION PUMP SUCTION PIPE

The following information is available: the suction pipe (or downcomer) is vertical; the mass flow $W = 145$ kg/s of saturated (at the top, i.e., in the drum or storage vessel) water with a pressure of 1 bar; and from a water–steam table one can get the density for saturated water at 1 bar, $\rho = 958.41$ kg/m^3, and the dynamic viscosity, $\eta = 281.9 \times 10^{-6}$ kg/(m · s).

The allowable pressure decrease can be calculate according to Eq. (7.16):

$$\frac{dp}{dt} = v\rho \left(g + \frac{\lambda_f v^2}{2d_i} \right)$$

Using the continuity equation:

$$W = \rho v \frac{d_i^2 \pi}{4}$$

one gets

$$\frac{dp}{dt} = \frac{4W}{\pi d_i^2}\left(g - \frac{8\lambda_f W^2}{\rho^2 \pi^2 d_i^5}\right)$$

All variables are known besides of course dp/dt and d_i but also the tube friction coefficient λ_f. To simplify the calculation, we use an equation of Prandtl and von Kármán for turbulent flow and hydraulically smooth tubes:

$$\lambda_f = \frac{0.309}{[\log(Re/7)]^2}$$

with the Reynolds number

$$Re = \frac{4W}{d_i \pi \mu}$$

Using these equations, one gets the allowable pressure decreases as a function of the inner diameter (Fig. 7.47) and can see the optimal diameter.

The optimal diameter can also be calculated according to Eq. (7.19). Because of the dependence of the tube friction coefficient on the Reynolds number and therefore on the diameter, Eq. (7.19) can only be solved by

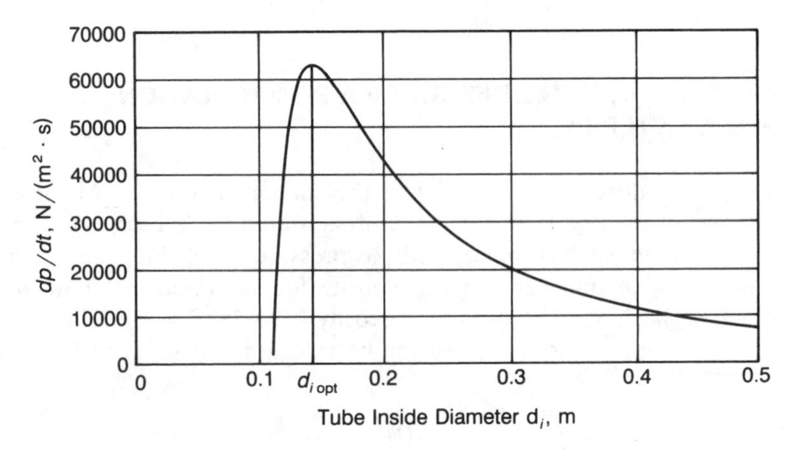

Fig. 7.47. Allowable pressure decrease.

iteration. The result of such a calculation is

$$d_i = 0.143$$

$$Re = \frac{4.145 \times 10^6}{0.143 \times \pi \times 281.9} = 4.57 \times 10^6$$

$$\lambda_f = \frac{0.309}{[\log(Re/7)]^2} = 0.00914$$

$$d_i = \sqrt[5]{\frac{28 \times 0.00914 \times 145^2}{958.41^2 \times 9.81 \times \pi^2}} = 0.143 \text{ m}$$

21-24

APPENDIX 7.3: STEAM GENERATOR ENERGY BALANCE

The efficiency of a boiler depends on the energy balance boundaries; the relation temperature, relation calorific value (gross or net = higher or lower heating value) of the fuel, and so on; and the load (steady state!).

It is therefore necessary to define the measurements of the efficiency exactly. This is done, for example, in the American Society of Mechanical Engineers (ASME) Power Test Code, Code for Acceptance tests on stationary steam generators of the power station type, British Standards Institution BSI 2885, and Deutsche Industrie Norm (German Industry Code) DIN 1942. The efficiency is defined as

$$\eta_{SG} = \frac{\dot{Q}_{out}}{\dot{Q}_{in}}$$

and with

$$\dot{Q}_{out} = \dot{Q}_{in} - \dot{Q}_{losses}$$

two further definitions are possible

$$\eta_{SG} = 1 - \frac{\dot{Q}_{losses}}{\dot{Q}_{in}}$$

$$= \frac{1}{1 + \dot{Q}_{losses}/\dot{Q}_{out}}$$

The heat output $\dot{Q}_{out}$ is defined as the sum of the products of the superheater and reheater steam flows with the respective enthalpy differences.

For example, the heat output of a boiler with a superheater SH and one reheater stage RH (using feed water FW as spray water SW in the superheater and reheater) is

$$\dot{Q}_{out} = W_{SHo}(i_{SHo} - i_{FW}) + W_{RHi}(i_{RHo} - i_{RHi}) + W_{RHSW}(i_{RHo} - i_{FW})$$

For example, the following data are given by the turbine manufacturer for a 600-MW (electrical) lignite-fired unit:

$$W_{SHo} = 517 \text{ kg/s}$$

$$\left.\begin{aligned} p_{SHo} &= 180 \text{ bar} \\ T_{SHo} &= 530°\text{C} \end{aligned}\right\} \text{ from steam–water table } i_{SHo} = 3359.0 \text{ kJ/kg}$$

$$\left.\begin{aligned} T_{FW} &= 240°C \\ p_{FW} &= p_{SHo} + \Delta p_{boiler} \\ &= 180 + 30 \text{ (estimated)} = 210 \text{ bar} \end{aligned}\right\} \quad i_{FW} = 1040.5 \text{ kJ/kg}$$

$$i_{SHo} - i_{FW} = 2318.5 \text{ kJ/kg}$$

$$W_{RHi} = 479 \text{ kg/s}$$

$$\left.\begin{aligned} p_{RHo} &= 31 \text{ bar} \\ T_{RHo} &= 530°\text{C} \end{aligned}\right\} \quad i_{RHo} = 3522.3 \text{ kJ/kg}$$

$$\left.\begin{aligned} p_{RHi} &= 33 \text{ bar (estimated)} \\ T_{RHi} &= 300°\text{C} \end{aligned}\right\} \quad i_{RHi} = 2985.5 \text{ kJ/kg}$$

$$i_{RHo} - i_{RHi} = 536.8 \text{ kJ/kg}$$

$$W_{RHSW} = 10 \text{ kg/s (usually 2\%)} \qquad i_{RHo} - i_{FW} = 2481.8 \text{ kJ/kg}$$

$$\begin{aligned} \dot{Q}_{out} &= 517 \times 2318.5 + 479 \times 536.8 + 10 \times 2481.8 \\ &= 1{,}480{,}000 \text{ kW} \end{aligned}$$

The heat input is (simplified)

$$\dot{Q}_{in} = W_{FI} \text{LHV}$$

To calculate the necessary fuel (lignite) flow, we need the efficiency or the losses and the calorific value of the fuel. We will use the net calorific value, that is, the lower heating value, LHV = 8000 kJ/kg. To simplify the calculation, we will only take into account the following losses (related to ambient conditions):

1. Losses due to the sensible heat of the flue gas $\dot{Q}_{\mathrm{FG}}$ (most important part)

$$\dot{Q}_{\mathrm{FG}} = W_{\mathrm{FG}} c_{\mathrm{FG}} (T_{\mathrm{FG}} - T_a) = W_{\mathrm{FB}} m_{\mathrm{FG}} c_{\mathrm{FG}} (T_{\mathrm{FG}} - T_a)$$

   With the ambient temperature $T_a = 25°\mathrm{C}$ and the flue gas temperature $T_{\mathrm{FG}} = 130°\mathrm{C}$ as given values and the values calculated for lignite of 8000 kJ/kg LHV and 25% excess air (e.g., according to F. Brandt: Brennstoffe and Verbrennungsrechnung FDBR, Fachbuchreihe Band 1, Fachverband Dampfkessel, Behälter, and Rohrleitungsbau e.V., Vulkan-Verlag, Essen): the specific flue gas mass $m_{\mathrm{FG}} = 4.78$ kg flue gas/kg fuel; the specific flue gas heat $c_{\mathrm{FG}} = 1.144$ kJ/kg flue gas K.

2. Losses due to unburned solid matter $\dot{Q}_{\mathrm{UF}}$ usually 1% of the heat input $\dot{Q}_{\mathrm{in}}$:

$$\dot{Q}_{\mathrm{UF}} = (W_{\mathrm{FI}} - W_{\mathrm{FB}})\mathrm{LHV}$$

$$\frac{\dot{Q}_{\mathrm{UF}}}{\dot{Q}_{\mathrm{in}}} = 1 - \frac{W_{\mathrm{FB}}}{W_{\mathrm{FI}}} = 1 - \eta_F = 0.01$$

$$\frac{W_{\mathrm{FB}}}{W_{\mathrm{FI}}} = \eta_F = 0.99$$

3. Losses due to radiation and convection to the environment $\dot{Q}_{\mathrm{RC}}$,

$$\dot{Q}_{\mathrm{RC}} = \mathrm{LRC}\,\dot{Q}_{\mathrm{outmax}}$$

   LRC from DIN 1942, LRC = 0.0035. For $\dot{Q}_{\mathrm{outmax}} \approx \dot{Q}_{\mathrm{in}}$:

$$\frac{\dot{Q}_{\mathrm{RC}}}{\dot{Q}_{\mathrm{in}}} \approx \mathrm{LRC}$$

Therefore the boiler efficiency is

$$\eta_{SG} = 1 - \frac{W_{FB} m_{FG} c_{FG} (T_{FG} - T_a) + \dot{Q}_{UF} + \dot{Q}_{RC}}{W_{FI} \text{LHV}}$$

$$= 1 - 0.99 \frac{4.78 \times 1.144(130 - 25)}{8000} - 0.01 - 0.0035$$

$$= 1 - 0.072 - 0.01 - 0.0035 = 0.9145$$

The fuel input therefore is

$$W_{FI} = \frac{\dot{Q}_{out}}{\eta_{SG} \text{LHV}} = \frac{1{,}480{,}000}{0.9145 \times 8000} = 202.3 \text{ kg/s} = 728 \text{ t/hr}$$

CHAPTER 8

THERMOHYDRAULIC DESIGN OF FOSSIL-FUEL-FIRED BOILER COMPONENTS

Z. H. LIN
Xi'an Jiaotong University
Xi'an, People's Republic of China

8.1 INTRODUCTION

A boiler is a device for generating steam for power, processing, and heating purposes, or for producing hot water for heating purposes and hot-water supplies. The former is called a steam boiler and the latter is called a hot-water boiler. Both boilers work on the same principle and a hot-water boiler is easier to design. In Chapter 6, the fundamentals and elements of fossil-fired boilers are presented. Chapter 7 discusses once-through boilers. In this chapter only the construction and design problems of steam boilers will be discussed.

8.1.1 Working Principle of a Steam Boiler

A boiler consists of two parts: a furnace in which combustion of fuel takes place, and a water–steam system through which feed water passes and is converted into steam by the absorption of heat produced by the combustion of fuel.

In boilers pulverized coal, fuel oil, or gas is burned in the furnaces through burners, while solid fuels are burned on stokers. To support combustion, it is necessary to supply a quantity of air and to remove the products of combustion by means of a draft caused by a chimney or draft fans.

Boilers, Evaporators and Condensers, Edited by Sadik Kakaç
ISBN 0-471-62170-6

In large boilers the incoming air is preheated in an air heater and the feed water (working fluid) is heated in an economizer by the discharged flue gases. This arrangement improves the boiler efficiency.

After leaving the economizer, the working fluid enters the furnace water wall tubes through a drum or distribution header and is heated and partially evaporated there. Then saturated steam is collected in a drum or a header. For common power plant boilers, saturated steam is further superheated to the required temperature in steam superheaters, while for reheat cycle power plant boilers, steam has to be reheated in reheaters.

The working principle of a steam boiler is shown in Fig. 8.1 (see also Fig. 6.15) which expresses the flow diagram of steam production in a steam-turbine power plant with a natural-circulation boiler fired with pulverized coal.

The boiler consists of two vertical shafts connected at the top by a horizontal gas duct. The left shaft serves as the boiler furnace. Water walls, formed by tubular panels, are arranged around the entire perimeter of the furnace chamber and are heated directly by the radiant heat of the flame.

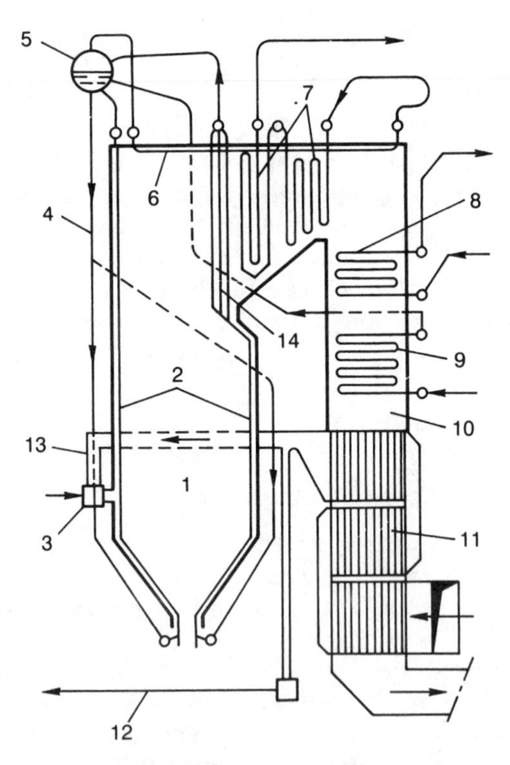

Fig. 8.1. Flow diagram of a natural-circulation boiler: 1—Furnace, 2—water walls, 3—burners, 4—downcomers, 5—drum, 6—radiant superheater, 7—convection superheaters, 8—reheaters, 9—economizer, 10—gas duct, 11—air heater, 12—primary air, 13—secondary air, 14—slag screens.

The reheater, economizer, and air heater are arranged in the right shaft while the superheater is located in the horizontal gas duct. These heating surfaces receive heat by convection and are called convective heating surfaces.

As shown in Fig. 8.1, there are three major flow systems in a boiler: the combustion products flow system, the steam–water flow system, and the air flow system.

The combustion products flow system: As pulverized coal is injected with air into the furnace, it is burned and forms the high-temperature combustion products (flue gases) which serve as a heat transfer agent on the heating surfaces. Flue gases give up part of their heat by radiation to the water walls and leave the furnace at a safe temperature (about 1000 to 1200°C, depending on the type of fuel) which will not cause slagging and fouling of the subsequent convective heating surfaces. After that, the flue gases pass through the external surfaces of the slag screen, superheaters, reheater, economizer, air heater, etc., successively and give up heat mainly by convection to these convective heating surfaces. Downstream of the air heater, the flue gases already have a rather low temperature (110 to 160°C) and are called the discharged flue gases. The discharged flue gases pass through a fly ash collector, as induced-draft fan, and are ejected through the chimney into the atmosphere.

The steam–water flow system: Feed water is passed through the feed-water pump into the economizer and is heated to a temperature below the saturation point. Water then flows into the drum and is distributed through the unheated downcomers and headers to the water walls (risers) in which it is heated and partially converted to steam. The difference in density between the steam–water mixture in the water walls and the water in the downcomers produces a natural circulation which causes the water to flow downwards from the drum into the downcomers and the steam–water mixture to flow upwards from the water walls into the drum. In the drum, steam is separated from the steam–water mixture discharged by the water walls. Saturated steam flows into the superheater and the remaining water is then recirculated together with the feed water to the water walls. In superheaters, saturated steam is heated to the required parameters and flows to the high-pressure turbine. In order to improve the power plant efficiency, part of the exhaust steam of the high-pressure turbine is returned to the reheater for reheating, and then flows to the inlet of the reheat turbine.

The air flow system: Cold air (30 to 60°C) is pressed by the forced-draft fan into the inlet of an air heater and flows across its tubes. The air is heated by flue gases flowing in the tubes to the desired hot-air temperature (200 to 400°C, depending on the kind of fuel) at the outlet of the air heater. With pulverized-coal combustion, the hot air is separated into two flows. The primary air is used for drying the fuel and transporting the fuel dust through the burners into the furnace. The secondary air is directed through the burners into the furnace. During burning, fuel leaves fly ash which is mostly carried off by the flue gas and is collected in a fly ash collector arranged upstream of the induced-draft fan. The collected ash is removed by means of ash-removal devices. Part of the ash falls onto the bottom of the boiler furnace and is removed continuously by the ash-handling system.

What has been described previously is the main working principle of a power plant boiler with natural circulation. In addition to this type of boiler, there are boilers with other kinds of circulation, the working principles of which differ from that described previously only in the steam–water system and which will be discussed in later sections.

8.1.2 Main Characteristics of Steam Boilers

The main characteristics of steam boilers are the rated steam-generating capacity and the superheated steam parameters.

The rated steam-generating capacity of a boiler expresses the highest load of the boiler in stable operation for long periods of time on special fuel and with the rated parameters of steam and feed water.

The superheated steam parameters are the pressure and temperature at the outlet header of the superheater. Steam boilers for pressures of 14 MPa or more are usually designed with a reheater. In this case, the inlet and exit pressure and the temperature of the reheated steam are also the main characteristics of the boiler.

8.2 TYPES OF BOILERS AND CONSTRUCTION OF BOILER COMPONENTS

8.2.1 Classification of Boilers

Steam boilers are built in a variety of sizes, shapes, and forms to fit conditions peculiar to the individual plant and to meet varying requirements. Generally speaking, steam boilers may be classified according to their uses, steam pressures, circulation methods, fuels, firing methods, methods of removing slag, and boiler layout forms, as is shown in Table 8.1.

The main components of a steam boiler, as is mentioned in Section 8.1, included the furnace, slag screen, superheater, reheater, economizer, air heater, and drum. Their construction and design problems will be described in the following section.

8.2.2 Construction and Design Problems of Furnaces

A furnace is the combustion chamber of a boiler in which fuel is burned efficiently; it usually consists of burning equipment, water-cooled tube enclosure surfaces (water-cooled walls), and refractory constructions.

TABLE 8.1 Classification of Steam Boilers

Number	Classification	Nomenclature	Brief Remarks
1	By uses	Utility boiler	To produce steam for electric power generation. Large capacity, high steam parameters, high boiler efficiency, completely water-cooled furnace with burners, when pressure is greater than or equal to 14 MPa usually with reheater.
		Industrial boiler	To produce steam for heating and process, etc. Smaller capacity, lower steam parameters, furnaces with burners, stokers or fluidized beds, no reheater.
		Marine boiler	As a source of motive power for ships. Compact general shape, lighter boiler weight, mostly fuel-oil fired, no reheater.
2	By steam–water circulation	Natural-circulation boiler	The circulation of the working fluid in the evaporating tubes is produced by the difference in density between the steam-water mixture in the risers and water in the downcomers. With one or two drums, can only operate at subcritical pressure.
		Forced multiple circulation boiler	The circulation of the working fluid in the evaporating tube is produced forcedly by means of a circulating pump included in the circulation circuit. With single drum or separators, can only operate at subcritical pressure.
		Once-through boiler	No drum, the working fluid forcedly passes through the evaporating tubes only under the action of the feed-water pump, can operate at subcritical and supercritical pressure.
		Combined-circulation boiler	There are a circulating pump, a back-pressure valve, and a mixer in the circuit. At starting the back-pressure valve is opened and the boiler operates as a forced multiple-circulation boiler, on attaining the specified load, the circulating pump is switched off, the back-pressure valve is closed automatically, and the boiler operates as a once-through boiler. It can operate at subcritical and supercritical pressure.
3	By pressure	Low- and middle-pressure boiler (< 10 MPa)	Used as industrial boilers, natural circulation, some with boiler bank, furnace with burners or with stockers, no reheater.
		High-pressure boiler (10–14 MPa)	Used as utility boilers, usually natural circulation, with reheater only when pressure is greater than or equal to 14 MPa.

TABLE 8.1 *(Continued)*

Number	Classification	Nomenclature	Brief Remarks
		Superhigh-pressure boiler (> 17 MPa)	Used as utility boilers; natural circulation or forced circulation depending on the engineering–economical approach, with reheater; the prevention of film boiling and high-temperature corrosion should be considered.
		Supercritical-pressure boiler (> 22.1 MPa)	Used as utility boilers; large capacity, once through or combined circulation, with reheater, the prevention of pseudo-film boiling and high-temperature corrosion should be considered.
4	By fuel or heat source	Solid-fuel-fired boiler	Coal is mainly used; the components of fuel and the characteristics of ash are important influential factors for boiler design.
		Fuel-oil-fired boiler	With higher flue gas velocity and smaller furnace volume.
		Gas-fired boiler	Natural gas or blast-furnace gas are mainly used; with higher flue gas velocity and smaller furnace volume.
		Waste-heat boiler	Utilizing waste heats from any industrial process as the heating source.
5	By firing method	Boiler with stoker	Mainly used as industrial boilers.
		Boiler with burners	Mainly used as utility boilers or large-capacity industrial boilers.
		Boiler with cyclone furnaces	Applicable to coals having low slag viscosity and low iron content; fuel is fired in a water-cooled cylinder, and the flame is whirled by either tangential coal dust–air jets from burners or tangential high-speed jets of secondary air (80–120 m/s); ash is removed from the furnace in liquid form.
		Boiler with fluidized bed	Solid-fuel particles (1–6 mm) are placed onto a grate and blown from beneath with an air flow at such a speed that the particles are lifted above the grate and are burned in suspending state; used as industrial boilers for burning low-grade solid fuels.
6	By method of removing slag in furnace	Boiler with dry ash furnace	Applicable to coals with high-ash fusion temperature; the ash removed from the hopper bottom of the furnace is solid and dry.
		Boiler with slag tap furnace	Liquid form slag flows to the wet bottom of the furnace (a pool of liquid slag) and tapped into a slag tank containing water.
7	By boiler layout form	Tower shape, inverted U shape, box shape, etc.	

Construction of Burning Equipment Depending on the fuels fired in the furnace and the boiler capacity, the following four firing methods are used in modern boilers: stoker firing, burner firing, cyclone-furnace firing, and fluidized bed firing.

The selection of the most suitable firing method for a boiler consists of balancing the investment, operating characteristics, efficiency, and type of fuel to give the most economical installation. For industrial boilers of a capacity less than 65 t/hr fired with coal, stoker firing is usually adopted; for utility boilers and larger industrial boilers fired with coal, pulverized-coal burner firing or cyclone-furnace firing is more economical; for industrial boilers fired with low-grade solid fuels, fluidized bed firing may be used, and for boilers fired with fuel oil or gas, burner firing is usually adopted.

Although the combustion principles of the four firing methods mentioned previously are different, their main burning components are stokers or grates (for stoker firing and fluidized bed firing) and burners (for burner firing and cyclone firing).

Mechanical stokers can be classified into four principal groups: spreader stokers, underfeed stokers, vibrating-grate stokers, and chain-grate or traveling-grate stokers (Fig. 8.2).

The spreader stoker (Fig. 8.2*a*) is used with boilers having a steam-generating capacity less than 65 t/hr, although in some cases, it may also be used with larger capacity boilers. The revolting rotor with blades of the spreader projects fuel into the furnace over the fire. The fine fuel particles are burned under suspension conditions, while heavier pieces fall on the grate and are burned there. Grates for the spreader stoker may be the stationary type, the dumping type, or the traveling-grate type. Since the traveling-grate type can discharge ash continuously, and its average burning rate is much higher than that of the former two, it is usually preferred for larger boilers. Ashes are removed at the front of the stoker; this fits the fuel distribution pattern and may provide more residence time on the grates for complete combustion of the fuel. Because the spreader stoker may response rapidly to load swings and burn a wide range of fuels, it is widely used in industrial boilers.

Figure 8.2*b* expresses the scheme of a side-ash discharge underfeed stoker. Coal is conveyed from the hopper to a central trough, called the retort, by a slow-speed feed screw and is continuously pushed out from the retort over the air-admitting grates. Coal is burned as it passes through the fuel bed. The incoming raw coal continuously forces the fuel bed to each side. By the time the coal reaches the side dumping grates, combustion is completed and the ash is discharged through the dumping grates into ash pits. The single-retort underfeed stoker is used with boilers of capacity less than 13 t/hr, while the multiple-retort type can be designed for boilers of larger capacity.

The vibrating-grate stoker may be air cooled or water cooled. The latter type is widely used due to the much better cooling effect of its grates, and its scheme is shown in Fig. 8.2*c*. Grates are mounted on a grid of water tubes

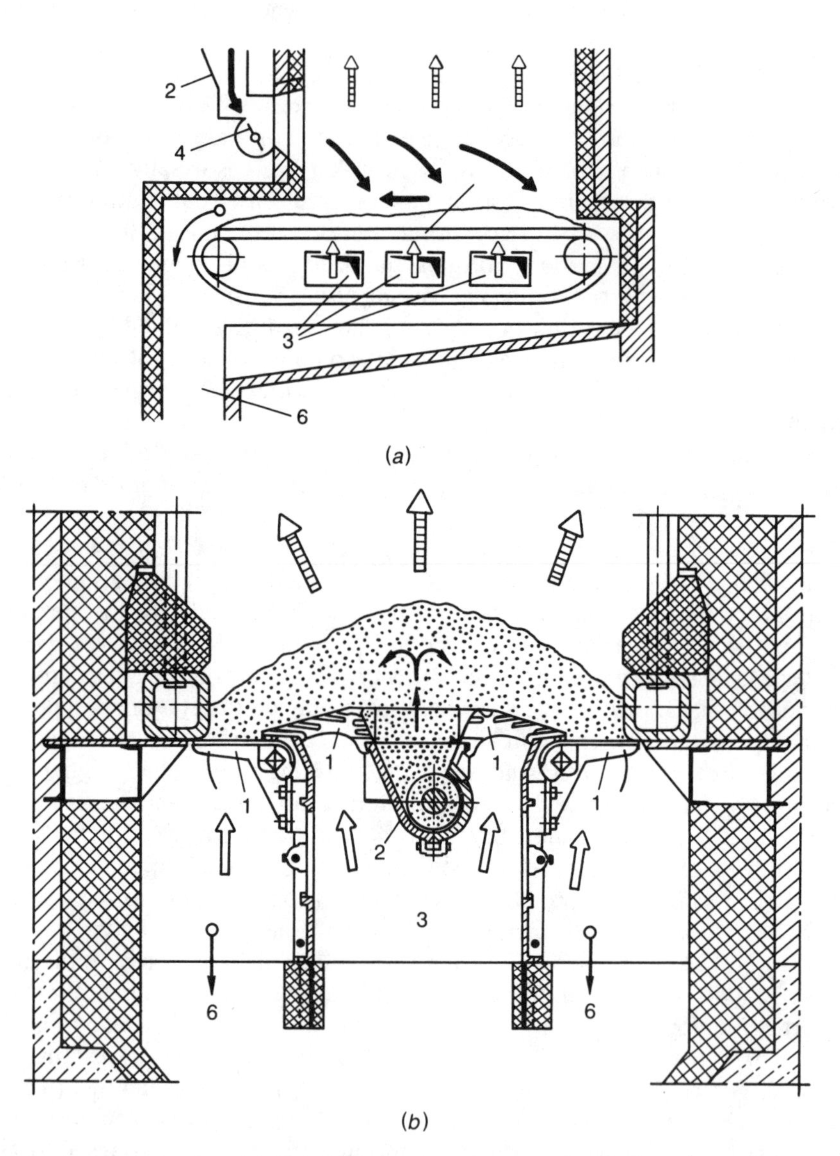

Fig. 8.2. Scheme of mechanical stokers: (*a*) spreader stoker, (*b*) underfeed stoker, (*c*) vibrating-grate stoker, (*d*) chain-grate or traveling-grate stoker. 1—stoker, 2—coal bunker, 3—air compartments, 4—spreader, 5—vibration generator, 6—ashpit.

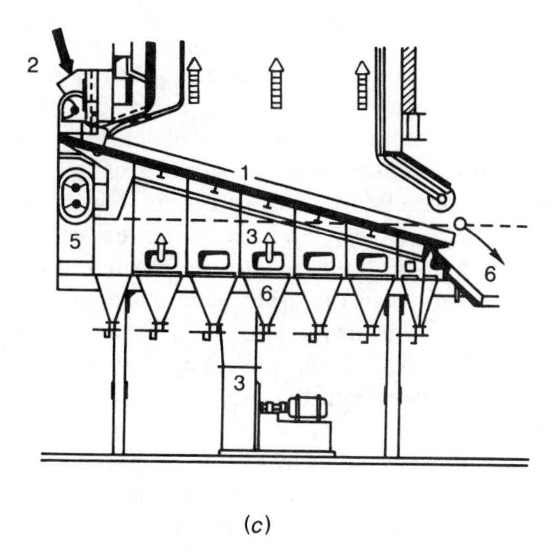

(c)

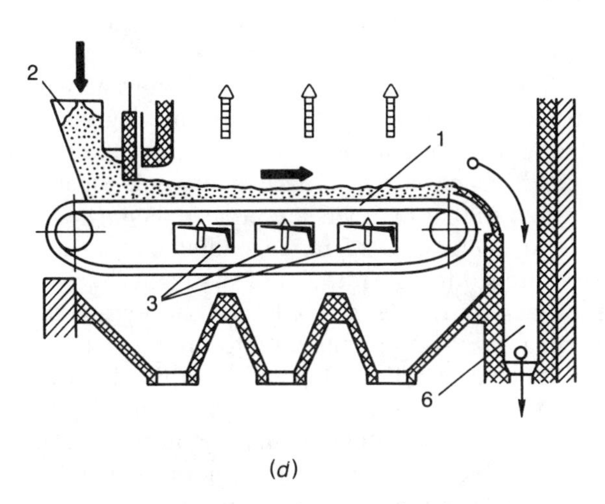

(d)

Fig. 8.2. *(Continued)*

equally spaced between the headers that are connected to the boiler circulating system. In new constructions, a water-cooled membrane with a tuyere on the membrane bars is used as the tube panel grate surface. The space beneath the stoker is divided into air compartments by means of flexible plates which support the entire structure and allow the grates to move freely during vibration. The vibration of the grates is caused by a vibration generator and the inclination of the grates conveys the coal from the feeding hopper onto the grate and moves the coal gradually to the rear of the stoker. It is widely used with boilers of capacity less than 65 t/hr.

The scheme of a chain-grate and traveling-grate stoker is shown in Fig. 8.2*d*. The chain-grate stoker employs an endless chain constructed by a series of links to form a grate surface to support the fuel bed. The grate surface of a traveling-grate stoker is comprised of a series of small grate bars mounted on carrier bars which are fastened to two or more endless driven chains. In either case, the chain travels over two sprockets, one at the front which is connected to a driving mechanism, and one at the rear of the furnace. Coal is fed by gravity from the coal hopper onto the moving grate surface and air enters through the openings in the grates. The entering coal on the grate is heated by radiation from the furnace gases and is burned as it moves along. Generally, these stokers use front and rear furnace arches to improve combustion by reflecting heat onto the fuel bed; they are used with boilers of steam capacity less than 40 t/hr.

There are two main types of pulverized-coal burners: the vortex burner and the straight-flow burner. Their schemes are shown in Fig. 8.3. In vortex burners, both a pulverized-coal–air mixture and secondary air, or secondary air alone is whirled by scrolls or vanes on the burners and enters the furnace with a whirling motion in the form of a conical flare. In straight-flow burners, the pulverized-coal–air mixture and secondary air are blown into it as parallel straight flows, and their intermixing in the furnace space is weak. For the sake of enhancing their intermixing and achieving efficient combustion, straight-flow burners are usually arranged in each of the four corners of the furnace. The cross section of the furnace is almost a square and the burner nozzles are directed so that their jets are projected along a line tangent to an imaginary circle (diameter 1 to 2.5 m, depending on the size of the furnace) at the furnace center to form a rotation of the flame.

A fuel oil burner consists of a nozzle and an air register; the former is used for atomizing, or dispersing the oil into the furnace as a fine mist, and the latter is used for air whirling. Depending on the method of atomization, oil burners may be classified as follows: pressure atomizing burners, steam or air atomizing burners, and rotary atomizing burners. Their schemes are shown in Fig. 8.4. In the pressure atomizing burners (Fig. 8.4*a*), atomization is completed by supplying fuel oil under high pressure (2.5 to 6.0 MPa) and oil is injected through a narrow hole of the nozzle into the furnace. In the steam or air atomizing burners (Fig. 8.4*b*), compressed air or steam is used as the medium for effecting atomization. Steam or air flows around a central oil tube; as the oil flows out from the tube, steam–oil or air–oil emulsion jets into the furnace. Oil is atomized through the rapid expansion of the steam or air. In the rotary atomizing burners (Fig. 8.4*c*), an atomizing cup is rotated at high speed (3000 to 10,000 r/min). The centrifugal force produced by this rotation causes the oil to leave the edge of the cup in an atomized fine spray. The nozzle of an oil burner is located in the center of the air register. Depending on the method of whirling air, the registers of various oil burners can be divided into three types: the scroll type, the tangential vane type, and the axial vane type. Their schemes are given in Fig. 8.5.

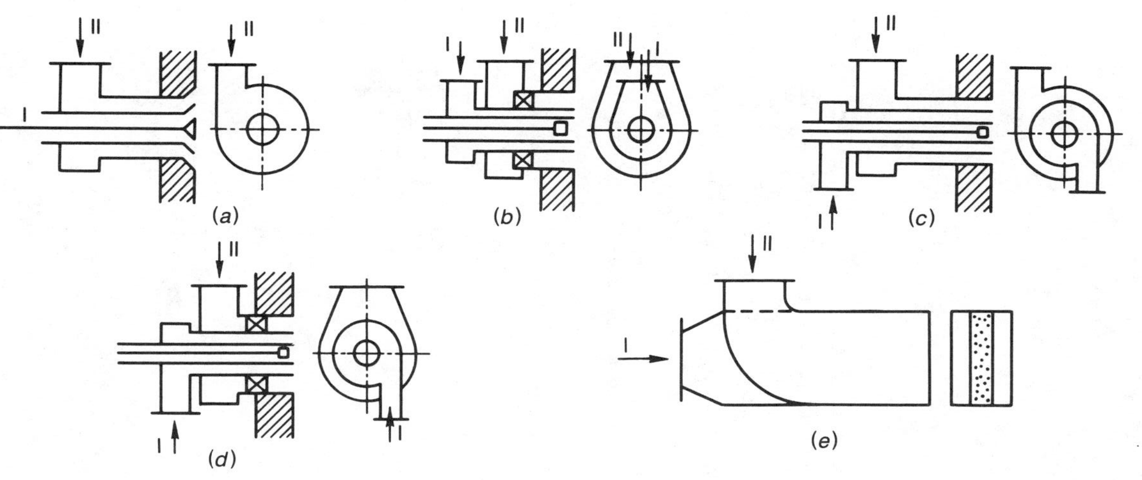

Fig. 8.3. Pulverized coal burners: (*a*) scroll type, (*b*) vane type, (*c*) two scroll type, (*d*) scroll-vane type, (*e*) straight-flow type. I—Primary air with coal dust, II—secondary air.

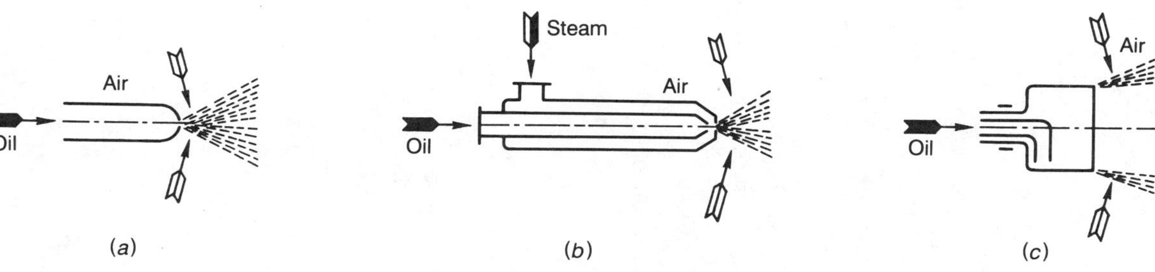

Fig. 8.4. Oil burners: (*a*) pressure atomizing type, (*b*) steam or air atomizing type, (*c*) rotary atomizing type.

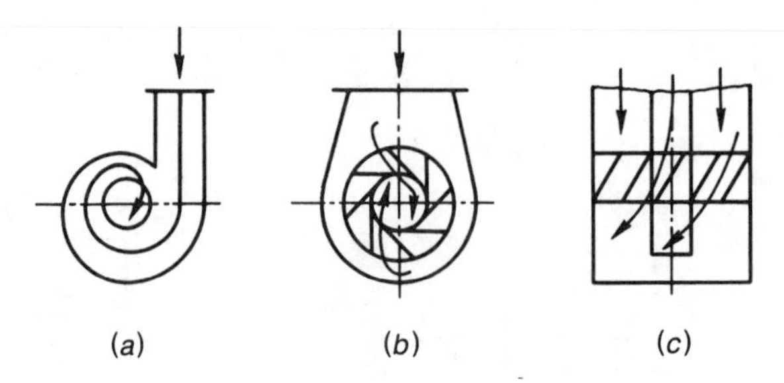

Fig. 8.5. Air registers: (*a*) scroll type, (*b*) tangential vane type, (*c*) axial vane type.

A gas burner consists of a gas element and an air register. Depending on the method of mixing the gas and air, gas burner elements may be classified as follows: external mixing type, internal mixing type, and partly internal mixing type (Fig. 8.6). The air registers of gas burners are similar to those shown in Fig. 8.5. The arrangements of burners in furnaces are shown in Fig. 8.7 (see also Fig. 6.16).

Construction of Water-Cooled Walls and Slag Screens The combustion space of a modern furnace is partially or completely surrounded by

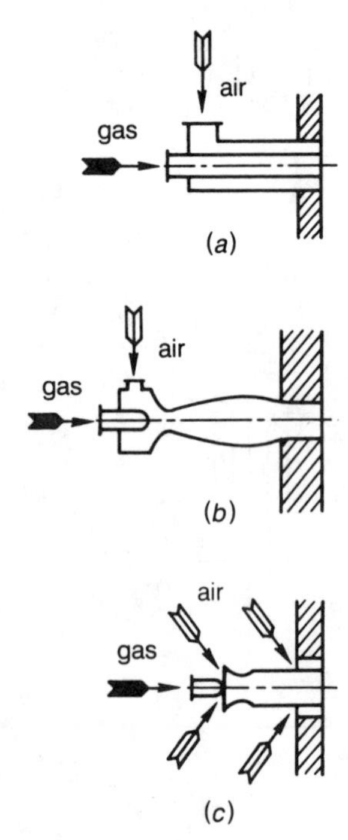

Fig. 8.6. Gas burners: (*a*) external mixing type, (*b*) internal mixing type, (*c*) partly internal mixing type.

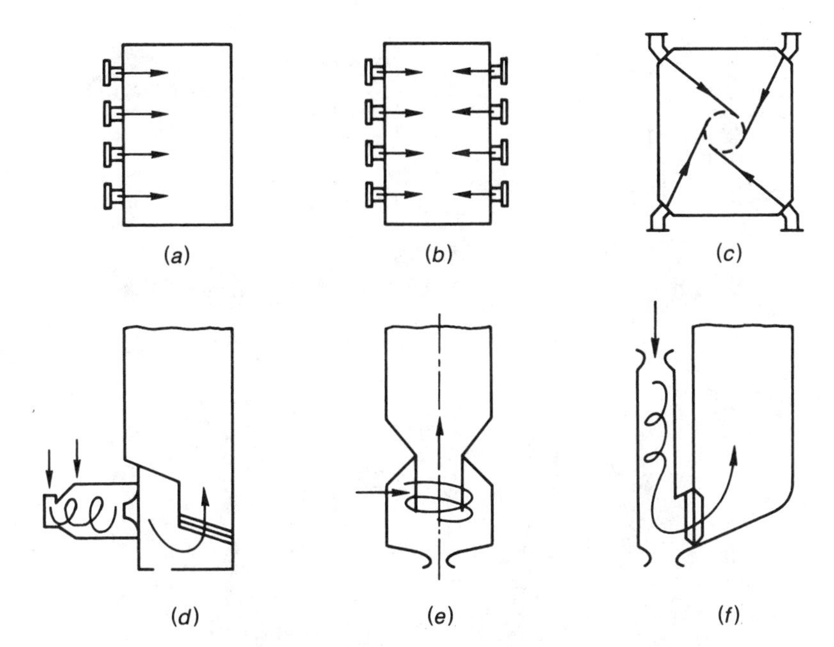

Fig. 8.7. Arrangement of burners in furnaces: (*a*) front type, (*b*) opposite type, (*c*) tangential type, (*d*)–(*f*) cyclone type.

water-cooled walls which absorb radiant heat. Generally, water-cooled walls may be divided into three groups: bare tube type, membrane type, and refractory-faced type (Fig. 8.8).

Bare tube water walls are widely employed in boilers with vacuum furnaces. Membrane water walls in which the tubes are welded together have an all-welded gas-tight structure and can be used both in boilers with vacuum furnaces and boilers with positive pressure furnaces (see also Fig. 6.8). Refractory-faced water walls are made of studded tubes coated with a refractory material on the studs. They are used in dry bottom furnaces burning low-volatile fuels to stabilize ignition in the burner region or in slag tap furnaces.

In natural-circulation boilers, water walls are usually arranged vertically, except in some special cases where tubes may be arranged at an incline. In once-through boilers and multiple forced-circulation boilers, water walls may be arranged vertically, horizontally, in an ascending–descending manner, or in other forms.

In some boilers, at the exit of the furnace, slag screens consisting of several rows of widely spaced tubes are arranged to prevent plugging with ash and slag. These tubes are formed by dividing the rear water walls of the furnace into several rows at the exit of the furnace (Fig. 8.1) and are usually arranged in staggered form.

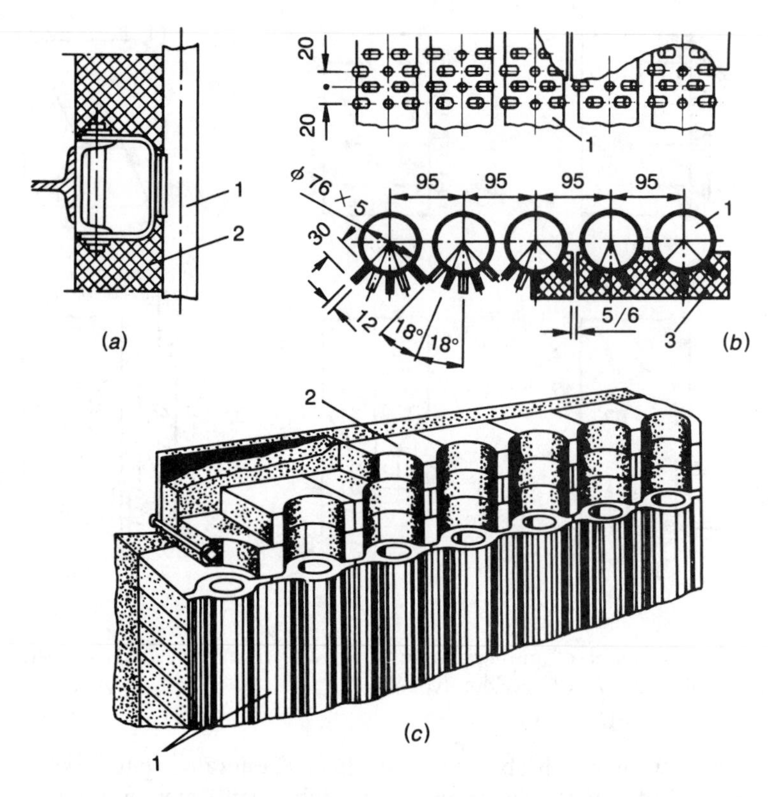

Fig. 8.8. Types of water walls: (*a*) bare tube type, (*b*) refractory-face type, (*c*) membrane type. 1—Tube, 2—Brick, 3—refractory face.

Design Problems of Furnaces The basic design requirements of a furnace are as follows:

1. Sufficient furnace volume for burning the fuel completely
2. Sufficient heating surface for cooling the combustion products to a safe temperature at the exit of the furnace
3. Proper arrangement of the burning equipment to avoid flame impingement on furnace walls
4. Maintaining similar heat fluxes of the heating surfaces
5. Reliability of the steam-generating circuit
6. Possibility of fuel flexibility

The determination of the cross-sectional area, the volume, and the linear dimensions of a furnace should satisfy the preceding requirements.

For furnaces with burners, the furnace cross-sectional area may be determined from the value of the heat release rate per unit furnace area q_F, kW/m^2, which expresses the ratio of the total heat released in the combustion zone of a furnace to its cross-sectional area A_F, m^2, and may be expressed as follows:

$$q_F = \frac{BH_1^w}{A_F} \tag{8.1}$$

where B is the fuel consumption, kg/s, and H_1^w is the lower heating value of the fuel, kJ/kg.

The highest allowable value of q_F is assigned depending on the kind of fuel and the arrangement and type of burners used, and is listed in Table 8.2. The values of q_F in Table 8.2 are suitable for boilers with a dry ash furnace; for a slag tap furnace the highest value of q_F is 5.2 MW/m^2.

For furnaces with stokers, the furnace cross-sectional area may be determined from the fuel burning rates of a stoker q_R:

$$q_R = \frac{BH_1^w}{R} \tag{8.2}$$

where R is the grate area, m^2.

The highest allowable value of q_R is also assigned based on the type of stoker and is listed in Table 8.3.

When BH_1^w is calculated and the rated value of q_F or q_R is selected from Table 8.2 or Table 8.3, the required furnace cross-sectional area can be determined from Eq. (8.1) or Eq. (8.2).

The width and depth of a furnace are determined by the arrangement of the burners, or are equal to the dimensions of the grate. The height of the furnace may be determined if the furnace volume is known.

The furnace volume can be obtained from the heat release rate per unit furnace volume q_V, kW/m^3, which is assigned depending on the kind of fuel and the firing method. q_V can be expressed as

$$q_V = \frac{BH_1^w}{V_F} \tag{8.3}$$

where V_F is the furnace volume, m^3.

TABLE 8.2 Statistical Value of the Heat Release Rate per Unit Area, q_F (for Dry Ash Furnace), MW/m^2 [1–3]

Steam Capacity, t/hr		130	220	400	670–950	1000–1600	1600
Tangential firing with corner burners	Lignite		2.1–2.56	2.9 –3.36	3.25–3.71	< 4.06	4.06–4.46
	Bituminous		2.32–2.67	2.78–4.06	3.71–4.64	< 6.38	< 6.38
	Anthracite		2.67–3.48	3.02–4.52	3.71–4.64	—	—
Front arrange burners or oppositely arranged coal burners			2.2–2.79	3.02–3.72	3.48–4.07	< 4.64	< 4.64
Oil or gas firing		< 4	4.07–4.77	4.19–5.23	5.23–6.16	< 6.38	< 6.38

TABLE 8.3 Statistical Fuel Burning Rate of Stokers q_R, MW/m^2 [1, 3]

	Chain Grate or Traveling Grate		Spreader Stoker		Fluidized Bed			
	Bituminous	Anthracite	Dumping Grate	Traveling Grate	Lignite	Other coal	Vibrating Grate	Underfeed Grate
q_R	0.58–1.10	0.58–0.80	1.10–1.30	1.30–1.80	4.60–7.00	2.10–2.90	0.82–1.2	1.20–1.70

The highest allowable value of q_V for complete combustion of fuel may be selected from Table 8.4, from which the minimal furnace volume can be determined. On the other hand, the furnace volume should satisfy the condition of cooling flue gases to the required furnace exit temperature T_{gFe}; the larger the furnace volume, the more the water wall may be arranged, that is, the lower the furnace exit temperature. When the boiler steam-generating capacity is greater than or equal to 400 t/hr, the furnace volume required for cooling gas temperature is larger than that required for complete combustion, and the former may be obtained from the heat transfer calculation of the furnace. Therefore the furnace volume may be determined by selecting a q_V, that is, from Eq. (8.3), only when the boiler capacity is less than 400 t/hr; otherwise, the furnace volume, that is, the height of the furnace, should be determined from the furnace heat transfer calculation results which will be discussed in Section 8.3.

Other geometrical dimensions for different furnaces may be selected from steam boiler handbooks [1] or steam boiler design standards [2].

During design of a boiler furnace, sufficient water-cooled walls should be arranged in a furnace so that the temperature of flue gases at the exit of the furnace is equal to or below a safe temperature to avoid slagging and fouling of the subsequent convective heating surface. Usually, this safe temperature is equal to 50°C below the ash initial deformation temperature or 150°C below the ash soften temperature. The required heating surfaces of water-cooled walls for reaching the safe exit temperature of a furnace may be determined by the heat transfer calculation of the furnace.

The tube outside diameters and the relative tube spacings, which are equal to the ratio of the riser centerline spacing, S, to the riser outside diameter, d, for water-cooled walls, are listed in Table 8.5.

Other geometric dimensions of a dry ash furnace with burners may be determined according to the values expressed in Fig. 8.9.

The vertical distance between the axis of the first row of burners and the midpoint of the furnace exit, h, denotes the height of the flame; it should be high enough for complete combustion; its minimum value is expressed in Table 8.6.

For large modern boilers, pendant superheaters are usually used. The vertical distance between the axis of the first row of burners and the lowest part of the pendant superheater, h', should be larger than 8 m; its statistical value is listed in Table 8.7.

The boundary surface of the calculated furnace volume is shown by the slanted lines in Fig. 8.9.

In some large-capacity boilers, for the sake of cooling the flue gas temperature to a safe value, a furnace dividing water-cooled wall is arranged which is located in the middle of the furnace and divides the furnace into two chambers. This water-cooled wall is usually made of low alloy steel with a tube outside diameter of 51 to 76 mm, and its relative tube spacing S/d is in the range of 1.0 to 1.2.

TABLE 8.4 Statistical Value of the Heat Release Rate per Unit Furnace Volume q_V, MW / m^3 [1, 3, 4]

Fuel	Dry Ash Furnace with burners	Chain grate and Traveling Grate	Vibrating Stoker	Underfeed Stoker	Spreader Dumping Grate	Stoker Traveling Grate	Fluidized Bed
Anthracite	0.12–0.15	0.25–0.35	0.25–0.3	0.2–0.25	0.23–0.25	0.3–0.4	1.7–2.1
Bituminous coal	0.14–0.20	0.25–0.35	0.25–0.3	0.2–0.25	0.23–0.25	0.3–0.4	1.7–2.1
Lignite	0.09–0.15	0.25–0.35	0.25–0.3	0.2–0.25	0.23–0.25	0.3–0.4	1.7–2.1
Oil	0.23–0.35	—	—	—	—	—	—
Natural gas	0.35	—	—	—	—	—	—

TABLE 8.5 Dimensions and Materials for Water Wall Tubes

Circulation form			Natural		Forced multiple	Once through	
Pressure	Low	Middle	High	Superhigh	Subcritical	Superhigh	Supercritical
Outside diameter (risers), mm	51–60	60	60	60–76	32–51	22–51	22–42
Outside diameter (downcomers), mm	51–106	108–133	159–426	≥ 426	325–426	—	—
S/d							
Bare tube	≤ 2.0	1.1–1.2	1.05–1.2	1.05–1.2	1.05–1.2	1.05–1.2	—
Membrane type	—	—	1.3–1.35	1.35–1.35	1.3–1.35	1.4–1.6	1.4–1.6
Material for risers	Low carbon steel	Low carbon steel	Low carbon steel	Low carbon or low alloy steel	Low carbon or low alloy steel	Low carbon or low alloy steel	Low alloy steel

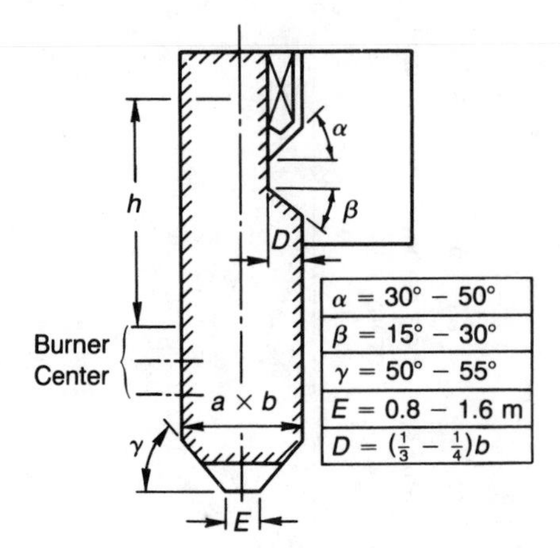

Fig. 8.9. Some geometrical dimensions of dry ash furnace with burners.

TABLE 8.6 Statistical Value of Minimum Flame Height, m

Steam capacity, t/hr	65–75	130	220	410	670	1000
Anthracite	8	11	13	17	18	—
Bituminous coal	7	9	12	14	17	18
Fuel oil	5	8	10	12	14	—

TABLE 8.7 Statistical Value of *h′*, m

Steam capacity, t/hr	670	670	1000	2000	2000
Fuel	Anthracite	Bituminous coal	Bituminous coal	Bituminous coal	Lignite
h', m	15	13	15–17	20	26–27

The reliability of the steam-generating circuit will be discussed in Section 8.5.

8.2.3 Construction and Design Problems of Superheaters and Reheaters

In order to improve the thermal efficiency of the power plant cycle and to reduce the vapor moisture content in the low-pressure stages of the turbine, superheated steam is always produced in a utility boiler and is then sent to the turbine.

The boiler element that superheats the saturated steam to the desired superheat steam temperature by absorbing heat from the flue gas is called the superheater.

A reheater is installed in the utility boiler of a reheat cycle to further increase the cycle efficiency.

Construction of Superheaters and Reheaters Superheaters and reheaters are similar in construction but differ in operating pressures (the operating pressure of the latter is about 3 to 6 MPa). Usually, they consist of unheated headers and a system of heated parallel tubes located in the path of flue gases. Steam flows inside the tubes and absorbs heat from the outside flue gases. Thus the steam temperature is raised and its volume is increased.

Due to the heat transfer modes, superheaters and reheaters may be classified as radiant type, convective type, and radiant-convective type (semiradiant type).

A radiant superheater or reheater is placed in the furnace as a part of the furnace enclosure and receives its heat mainly by radiation. It consists of parallel vertical tubes and headers as shown in Fig. 8.10. Its wall temperature is about 100°C over the inside steam temperature. To minimize tube failure, a high mass flow rate of saturated steam through it is necessary. Its tube outside diameter is in the range of 32 to 51 mm and the relative spacing of tubes is close to that of the water wall. This type of superheater or reheater is generally used in combination with other types of superheaters or reheaters.

A platen-type superheater or reheater is of the semiradiant type and is located in the upper part of the furnace, where it receives its heat by both radiation and convection. It consists of headers and parallel U-tube platens, each of which is made up of 15 to 30 U tubes with a relative spacing $S/d = 1.1$–1.2. The spacing between the platens is equal to 600 to 1000 mm, and its tube outside diameter is 32 to 42 mm. The tube wall temperature of this type of superheater is still high and may reach a temperature of 80°C higher than the inside steam temperature; therefore sufficient steam mass velocity is still required for cooling the tube material. This superheater is used in combination with convection superheaters or reheaters.

A convection superheater or reheater consists of parallel serpentine tubes and headers. It is placed in gas ducts, where it receives most of its heat by convection. Its tube outside diameter is 32 to 60 mm; the transverse relative spacing of tubes, $S_1/d = 2$–3, and the longitudinal relative spacing of tubes, $S_2/d = 1.6$–2.5. Usually, in-line tube banks predominate in convection superheaters or reheaters arranged in the high gas temperature region. This type of superheater or reheater can be used alone, as in some middle-capacity boilers, or in combination with another type of superheater or

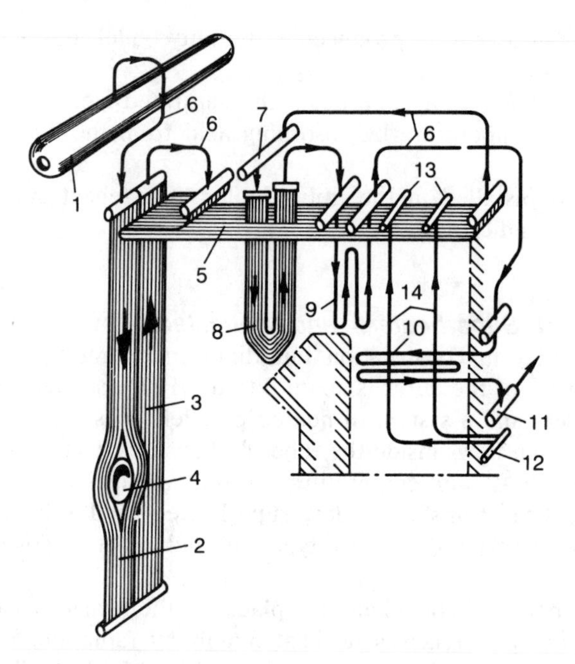

Fig. 8.10. Scheme of a superheater: 1—drum, 2, 3—downcomer and risers of radiant superheaters, 4—hole for placing a burner, 5—top radiant superheater, 6—connecting pipes, 7—attemperator, 8—platen-type superheater, 9, 10—convection superheater, 11—superheated steam outlet header, 12, 13—headers of pendant tubes, 14—pendant tubes.

reheater for large-capacity and high steam parameter boilers as shown in Fig. 8.10.

The flow system arrangement of a convection superheater or reheater may be counterflow, parallel flow, or combined flow (Fig. 8.11 or Fig. 6.10). The counterflow superheater or reheater may have the largest mean temperature difference, the smallest heating surface, and the highest tube wall temperature; the parallel-flow one may have the smallest mean temperature difference, the largest surface, and the lowest tube wall temperature; while the

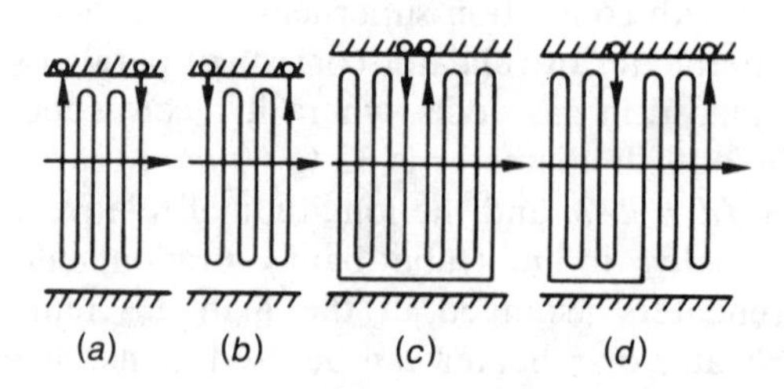

Fig. 8.11. Flow system arrangement of a convection superheater or a reheater: (*a*) counterflow, (*b*) parallel flow, (*c*) and (*d*) combined flow.

mean temperature difference, the amount of heating surface, and the tube wall temperature of the combined one, which is widely used, may stay in the middle level.

Design Problems of Superheaters or Reheaters The main design requirements of a superheater or a reheater are as follows:

1. The determination of an optimum flow system and the amount of heating surface required to give the rated steam temperature, which is usually given.
2. A system pressure drop less than 10% of the rated steam pressure for a superheater and less than 10% of the inlet reheat steam pressure for a reheater.
3. A reasonable steam and gas velocity for cooling the tube metal and reducing draft loss.
4. A uniform steam velocity and heat flux for each tube; the highest tube wall temperature has to be lower than the allowable tube metal temperature.
5. Enough spacing of the tubes to prevent accumulation of ash and slag.
6. The ability to regulate steam temperature within required limits.

The first requirement can be achieved by superheater or reheater performance calculations which will be discussed in Section 8.3. It may be necessary to compare several arrangements to obtain a design with optimum economic and operational characteristics.

The second, third, and fourth requirements may be solved by the correct selection of the steam mass velocity and gas velocity.

The steam velocity has to lower the tube temperature below the allowable tube metal temperature, but may not cause a system steam pressure drop over the permissible limit.

The recommended steam mass velocity is listed in Table 8.8. Depending on the economic and operational characteristics comparison, the flue gas velocity in a convection superheater for coal-fired boilers is equal to 10 to 14

TABLE 8.8 Steam Mass Velocity of Superheater or Reheater

Type	Mass Velocity, $kg/(m^2 \cdot s)$
Radiant type	1000–1500
Semiradiant type	800–1000
Convection type	
Middle pressure	250–400
High pressure	500–1000

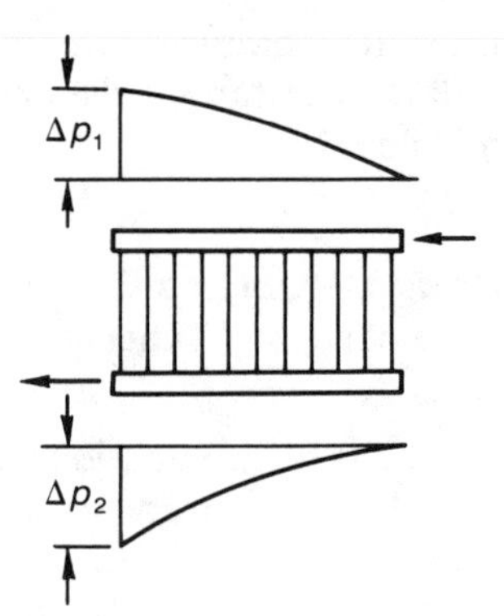

Fig. 8.12. Static pressure distribution in headers of a superheater (Z-shaped connection).

m/s; for oil- or gas-fired boilers, it may increase up to 20 m/s. The flue gas velocity in semiradiant types is usually about 5 to 6 m/s.

For the sake of maintaining a uniform steam flow rate in each tube and preventing overheating in the individual tubes of the parallel tube system, the effect of headers and their connection with the tube system of the heating surfaces should be considered. For different tube systems, the static pressure drop in each tube is different; that is, the steam flow rate in each tube is different. The steam flow rate increases with the increase of static pressure drop.

Figure 8.12 shows the static pressure distribution of the parallel tube system along the length of a header under the condition of a Z-shaped connection system; when steam is forced into the distributing header and is distributed among the coils, its axial velocity decreases and the static pressure increases toward the end of the header. In contrast, the static pressure toward the outlet of the discharge header decreases as shown in Fig. 8.12. The leftmost coil of the tube system operates at the highest pressure gradient, and the rightmost coil operates at the lowest pressure gradient, so the steam flow rate of the former will be greater than that of the latter.

In other circuits with a U-shaped connection or with concentrated supply and removal of steam through the middle part of the headers (Fig. 8.13), the header effect can be decreased and the steam flow rate may be distributed more evenly. The header effect can be decreased either by increasing the coil pressure drop or by decreasing the static pressure change in the headers.

In modern boilers, the superheater coil pressure drop is large, so the connecting system of headers has only a slight effect on the steam distribution among the coils. But in a reheater, where the resistance of the coil is relatively low while the resistance of the headers is high due to the high steam velocity, the effect of the header connection system can be substantial.

Due to the nonuniformity of the temperature and velocity fields of flue gases along the width and height of the furnace, heat absorption among parallel tubes is nonuniform; this may cause some of the superheater or reheater tubes to overheat. To avoid this, in large utility boilers, the connect-

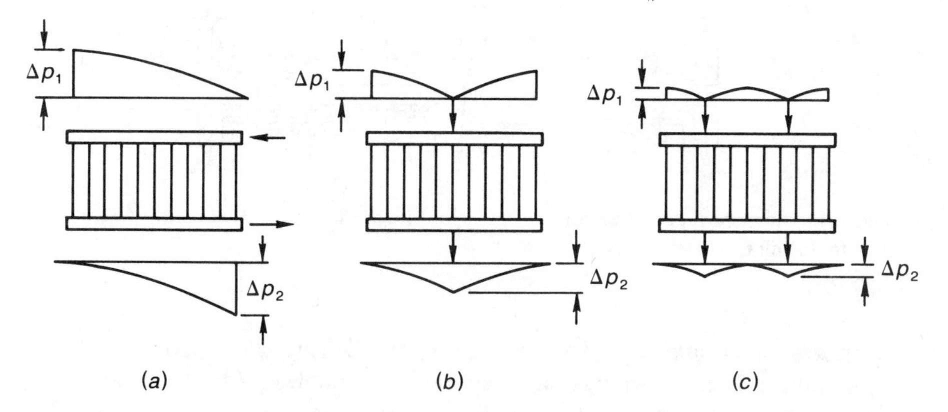

Fig. 8.13. U-shaped connection and connections of concentrated supply and removal of steam: (*a*) U shaped, (*b*) and (*c*) concentrated supply and removal of steam.

ing system of the superheaters or reheaters is usually divided into two or more stages, and mixing headers and crossover tubes are used for reducing the influence of nonuniform heat absorption among tubes as shown in Fig. 8.14.

The spacing of tubes for different superheaters and reheaters has already been mentioned. When the inlet gas temperature is close to 1000°C, it is necessary to locate several rows of convection superheater or reheater tubes with wide spacing ($S_1/d \geq 4.5$, $S_2/d \geq 3.5$) to prevent them from plugging with ash and slag.

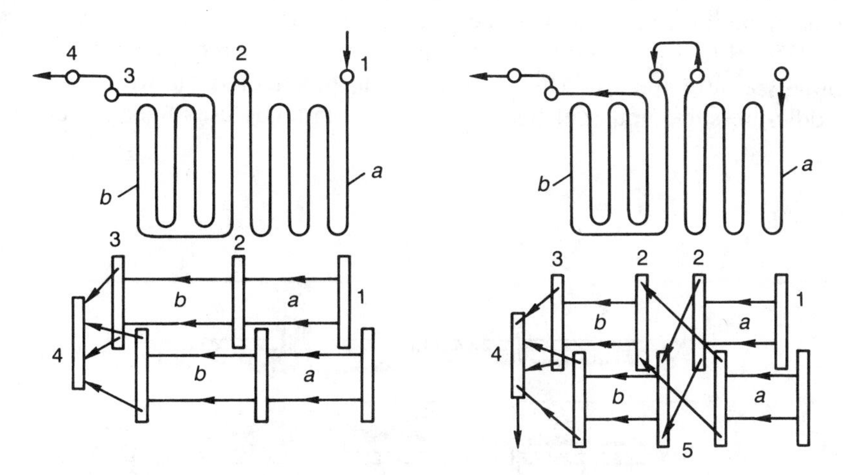

Fig. 8.14. Two kinds of superheater systems: (*a*) first stage, (*b*) second stage. 1—Inlet header, 2—mixing header, 3—exit header, 4—steam collector, 5—Cross-over pipes.

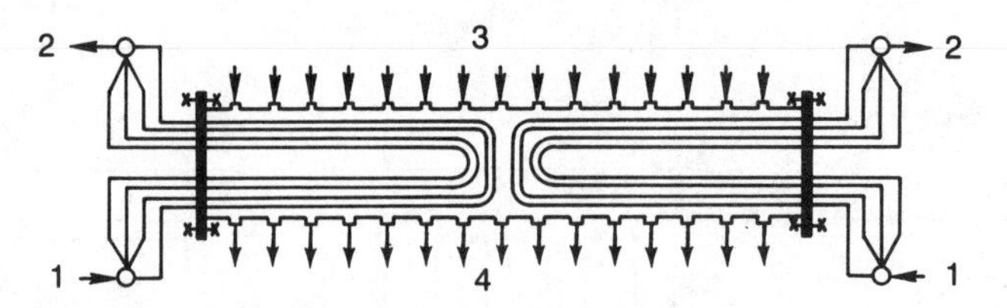

Fig. 8.15. Surface-type attemperator: 1—Cooling water inlet, 2—cooling water exit, 3—steam inlet, 4—steam exit.

Methods of Steam Temperature Control During operation, the steam temperature is affected by many operating variables. When the load increases, the quantity and temperature of the flue gases increase. In the convection superheater or reheater, steam temperatures increase with load, whereas in the radiant one steam temperatures decrease with load (see Fig. 6.12). A semiradiant superheater or a convection and a radiant superheater of proper proportions may maintain substantially constant steam temperatures over a certain range of load. In addition, the variations in the amount of excess air feed-water temperature, heating surface, cleanliness, etc., also affect the steam temperature.

A constant superheat temperature is desired to obtain maximum economy of the power plant and to avoid failures from overheating parts of the superheater, reheater, or turbine. As the permissible steam temperature tolerance for an operating modern utility boiler is only $\pm 5°C$, it is necessary to regulate the steam temperature within required limits by means of control.

There are two main methods of steam temperature control: steam control and gas control. Steam control is based on reducing the enthalpy of steam by transferring part of its heat to feed water through a surface-type attemperator (Fig. 8.15) or by injecting demineralized water into steam through a spray-type attemperator (Fig. 8.16). The former method is widely used in middle-pressure industrial boilers, while the latter is usually used in utility

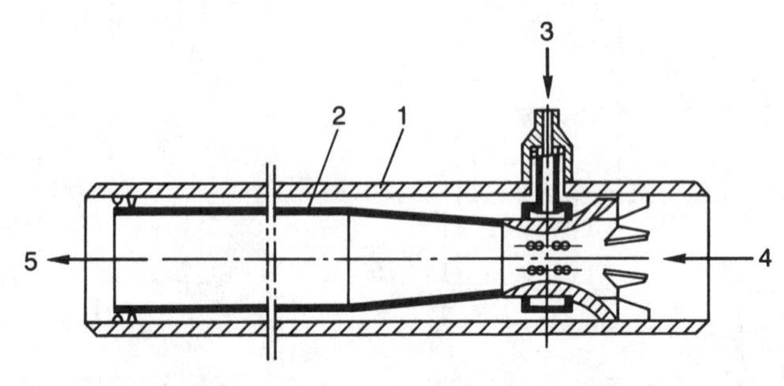

Fig. 8.16. Spray-type attemperator: 1—outer case, 2—Venturi tube, 3—water inlet, 4—steam inlet, 5—steam exit.

boilers. The attemperator may be located ahead of the superheater, between superheater stages, or at the superheater outlet. The first arrangement may protect the superheater from overheating, but it can cause uneven distribution of wet steam among tubes, which will cause excessive temperature variation problems in the tubes. Moreover, its control response is slow. The last arrangement cannot protect the superheater, although its control response is quick. Therefore most of the attemperators are installed between the superheater stages. Steam control methods are mainly used for controlling the superheater steam temperature.

Gas control is based on varying the heat absorption on the flue gas side of heating surfaces to maintain the steam temperature. Methods that belong to this type include gas recirculation, gas bypass, and tilting burners.

It is mainly employed for controlling the reheated steam temperature of reheat boilers and for controlling the superheated steam temperature of boilers without reheaters.

Gas recirculation control involves a method by which low temperature gas from the economizer outlet (250 to 350°C) is reintroduced into the hopper of the furnace by means of a recirculation fan and ducts. As the ratio of the recirculated gas to the total gas, r, increases, the change in heat absorption for convective heating surfaces increases due to the increase of gas mass velocity, while for radiant heating surfaces it decreases due to the decrease of furnace temperature. Thus the exit steam temperature of the convection superheater or reheater increases (Fig. 8.17). This method has a negligible effect on the furnace exit temperature, the total heat absorption, and the boiler efficiency. In the case of reintroducing the recirculated gas into the upper part of the furnace, the furnace exit gas temperature will decrease with the increase of recirculated gas. It is usually used as a method of gas tempering to avoid ash deposits on the convection superheater.

The principle of the gas bypass method is shown in Fig. 8.18. The gas duct is separated by gas-tight baffle walls into two parallel gas passes in which the superheater and the reheater are arranged separately. The proportion of gas flow over the superheater or the reheater may be varied by regulating dampers installed after the economizer (in the region of gas temperature lower than 500°C), and thus the steam temperature can be controlled.

A tilting burner control is usually used in tangentially fired furnaces equipped with vertically tilting burners. By directing the nozzles of the burners upward or downward, the main combustion zone and exit gas temperature of the furnace can be changed. Thus the regulation of the steam temperature can be accomplished by changes in the burner–nozzle position.

If in a boiler, the superheater is mainly of the radiant type and the reheater is of the convection type, the decrease of the boiler load will increase the superheated steam temperature and decrease the reheated steam temperature. In this case, for the sake of equalizing their temperatures, it is reasonable to transfer part of the heat from the superheated steam to the reheated steam through a steam–steam heat exchanger (Fig. 8.19). In

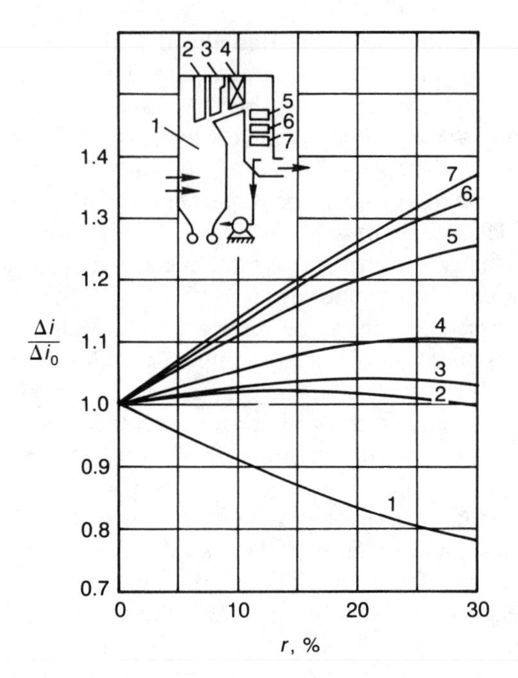

Fig. 8.17. Relative increase of working fluid enthalpy $\Delta i/\Delta i_o$ with the increase of r (Δi_o is the fluid enthalpy increase when $r = 0$): 1—furnace, 2, 3—platens, 4, 5—convection superheaters, 6—reheater, 7—economizer.

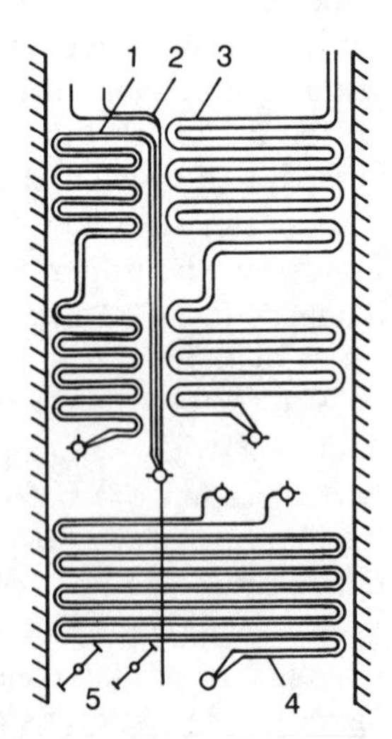

Fig. 8.18. Gas bypass control method: 1—superheater, 2—baffle wall, 3—reheater, 4—economizer, 5—regulating dampers.

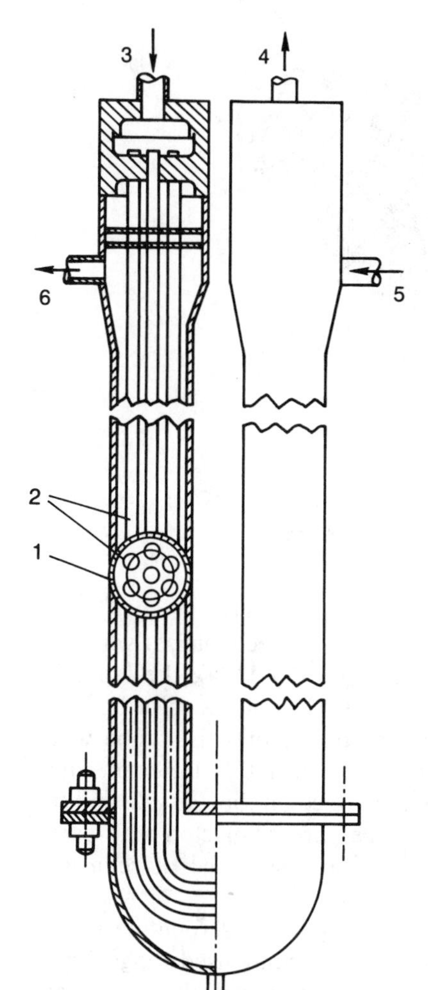

Fig. 8.19. Steam–steam heat exchanger: 1—outer case, 2—U tube, 3—superheated steam inlet, 4—superheated steam exit, 5—reheated steam inlet, 6—reheated steam exit.

this figure, superheated steam flows in the tubes and reheated steam flows in the header. The temperature is controlled by bypassing part of the reheated steam around the heat exchanger.

8.2.4 Construction and Design Problems of Economizers

An economizer is a heat exchanger located in the lower gas temperature region (450 to 600°C) designed to recover some of the heat from the discharged flue gas. It consists of a series of tubes through which feed water flows to the drum or to the inlet headers of furnace walls. Flue gases flow over the outside of the tubes.

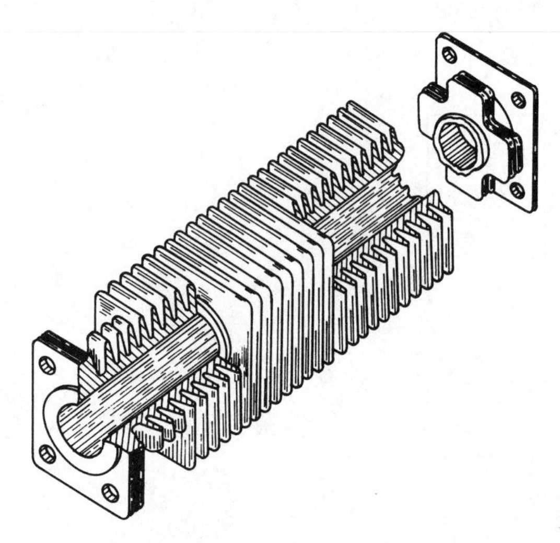

Fig. 8.20. Cast-iron economizer tube.

Construction of Economizers Economizers can be constructed of cast-iron tube or steel tube; the former is called the cast-iron economizer which is usually used in low-pressure industrial boilers ($p \leq 2.5$ MPa); the latter is called the steel tube economizer which may be used in all kinds of boilers.

Figure 8.20 shows the construction of a cast-iron economizer tube with outside fins (tube outside diameter 76×8 mm, fin size 150×150 mm). Fins are designed to give an extended heat transfer surface on the gas side and to minimize the effects of external corrosion. Tubes are connected with each other by return bends. Cast-iron material has the advantage of reducing internal and external corrosion.

A steel tube economizer consists of horizontal parallel serpentine tubes and headers (Fig. 8.21); tube outside diameters commonly used range from 25 to 38 mm with a wall thickness between 3 and 5 mm. Usually, the tubes are arranged in staggered form with a transverse relative spacing, $S_1/d =$ 2.5–3.0, depending on the gas velocity, and a longitudinal relative spacing, $S_2/d = 1.5$–2.0, depending on the permissive bend radius of the tube. In modern boilers, steel tube economizers with longitudinal fins or membrane-type economizers are also used for enhancing the gas-side heat transfer (Fig. 8.22). These economizers have economic advantages over bare tube surface ones, for example, lower initial cost and space required for installation.

Design Problems of Economizers The inlet working fluid of an economizer is treated and deaerated feed water, while the exit working fluid may be hot water or a steam–water mixture (the maximum steam quality is 20%). The former is called a nonsteaming economizer and the latter is called a

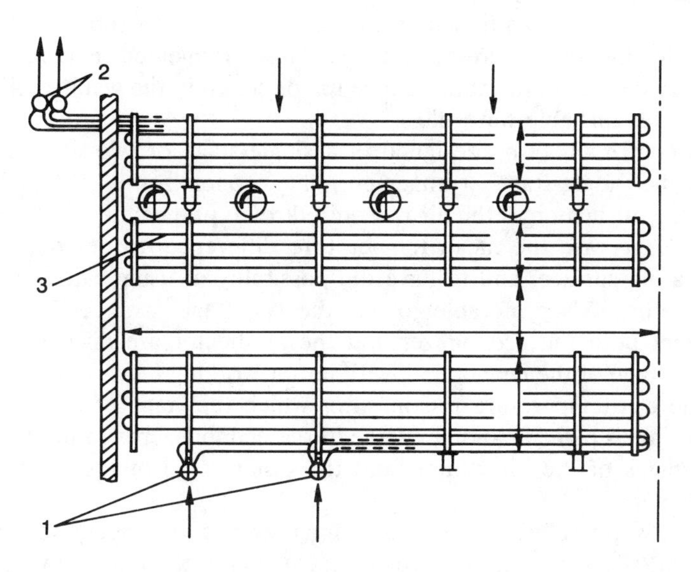

Fig. 8.21. Scheme of a steel tube economizer (showing half of it): 1—inlet header, 2—exit header, 3—surpentine tubes.

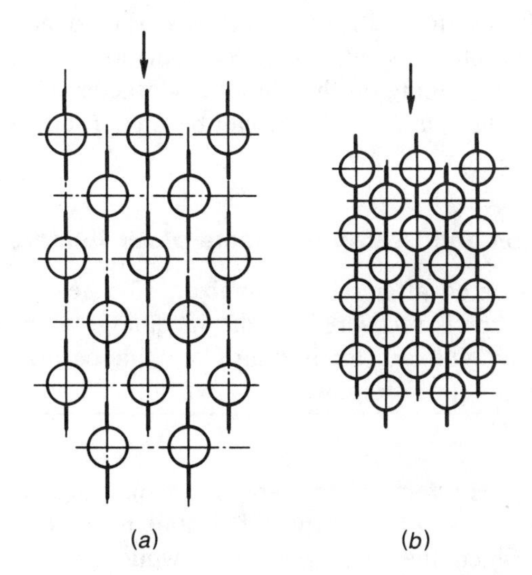

Fig. 8.22. New forms of economizers: (*a*) with longitudinal fins, (*b*) membrane type.

steaming economizer, which is usually used when the economizer surface costs less than other evaporating surfaces.

When arranging an economizer tube system, it is preferable for the water to enter at the bottom an flow up through the economizer tubes, and for the flue gases to flow down across the tubes. This arrangement may reduce the heating surface, allows the convenience gas or steam in the water to flow out, and eliminates unstable water flow.

An economizer may be designed in a single-stage or two-stage arrangement. In the single-stage arrangement, the economizer is always located ahead of the air heater. If the air is required to be preheated to 350 to 450°C in order to increase the mean temperature difference, to decrease the air heater heating surface, and to have the possibility of using carbon steel for the air heater, it is preferable to use the two-stage arrangement. In this arrangement both the economizer and the air heater are divided into two stages; the stage of the economizer with hotter working fluid is located in the higher flue gas temperature region, after which the stage of the air heater with hotter air is placed. Another stage of the economizer is located between the two stages of the air heater, and the other stage of the air heater is placed after it.

The feed-water velocity of a nonsteaming economizer usually ranges from 0.3 to 1.5 m/s, for preventing gases from staying on the inner tube wall and for keeping the pressure drop of the working fluid in the allowable region (5% of the boiler pressure for a high-pressure boiler and 8% of the boiler pressure for a middle-pressure boiler). For the steaming part of a steaming economizer, the feed-water velocity in the tube should not be lower than 1 m/s. The allowable flue gas velocity in economizers of coal-fired boilers is about 9 to 11 m/s, depending on the abrasive characteristics of ash. For oil- or gas-fired boilers, the flue gas velocity can be higher, but is limited by the increase of draft loss.

8.2.5 Construction and Design Problems of Air Heaters

The air heater is located after the economizer. The heat in the flue gases leaving the economizer is recovered by the incoming air in the air heater, thereby reducing the flue gas temperature and increasing the boiler efficiency. In addition, hot air may improve combustion conditions; this may also increase the efficiency.

Construction of Air Heaters There are two main types of air heaters, the tubular type and the regenerative type. The tubular type consists of a series of tubes through which the flue gases pass, while air passes around the outside of the tubes. A two-stage air heater unit is shown in Fig. 8.23.

Tubes in a tubular air heater can be arranged vertically, as is shown in Fig. 8.23, or horizontally. In the latter case, air passes through tubes while flue gases pass around the outside of the tubes.

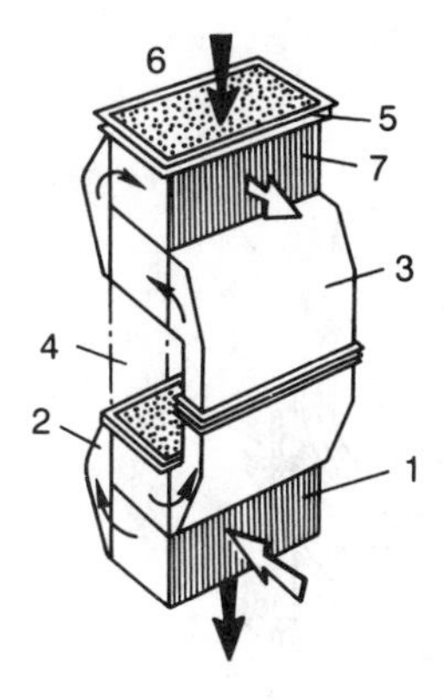

Fig. 8.23. Two-stages tubular air heater: 1—lower stage, 2, 3—connecting air duct, 4—space for placing economizer, 5—upper stage, 6—flue gas inlet, 7—hot-air exit.

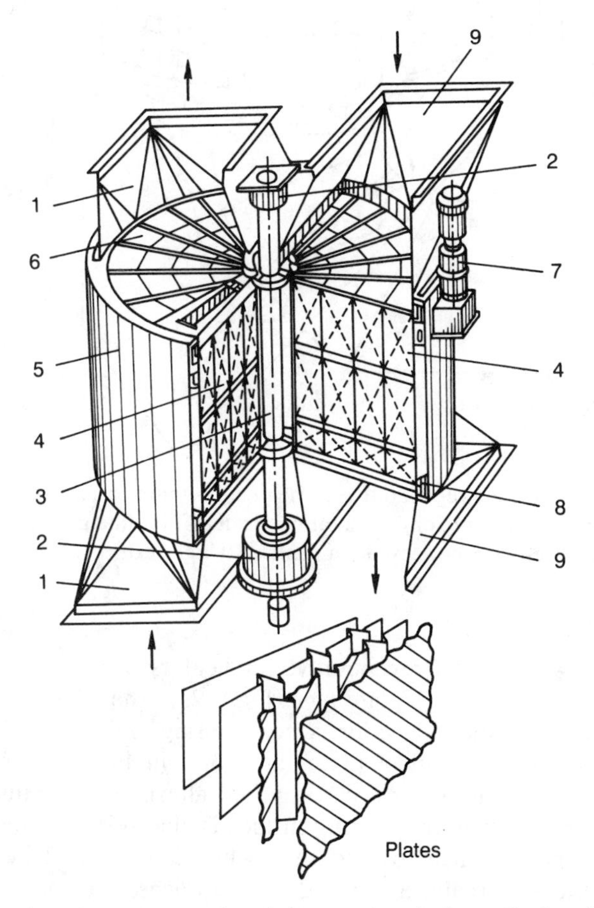

Fig. 8.24. Rotating-plate regenerative air heater: 1—air ducts, 2—bearings, 3—shaft, 4—plates, 5—outer case, 6—rotor, 7—motor, 8—sealings, 9—flue gas ducts.

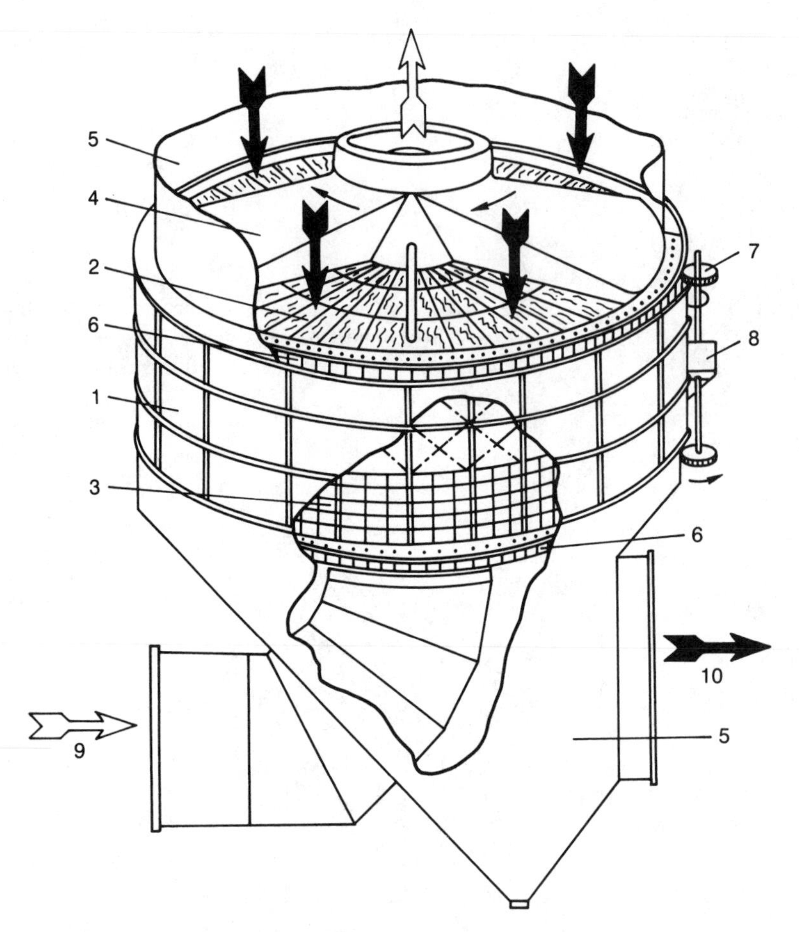

Fig. 8.25. Stationary-plate regenerative air heater: 1—outer case, 2—plates, 3—plates in the lower-temperature region, 4—rotating air ducts, 5—flue gas ducts, 6, 7—drive, 8—motor and drive-down devices, 9—air inlet, 10—gas exit.

There are two kinds of regenerative air heaters: the rotating-plate type, (Fig. 8.24) and the stationary-plate type (Fig. 8.25). The rotor of the rotating-plate air heater is mounted within a box housing and is installed with the heating surface in the form of plates as shown in Fig. 8.24. As the rotor rotates slowly, the heating surface is exposed alternatively to flue gases and to the entering air. When the heating surface is placed in the flue gas stream, the heating surface is heated, and then when it is rotated by mechanical devices into the air stream, the stored heat is released to the air flow. Thus the air stream is heated.

In the stationary-plate air heater, the heating plates are stationary, while cold-air hoods, both top and bottom, are rotated across the heating plates. Its

TABLE 8.9 Required Hot-Air Temperature, °C

Furnace Type	Fuel Type	Hot-Air Temperature
Dry ash furnace with burners	Bituminous	300–350
	Anthracite	350–400
	Lignite	350–400
Slag tap furnace and cyclone furnace		350–400
Oil- or gas-fired furnace		250–300
Furnace with stokers		< 200

heat transfer principles are the same as those of the rotating-plate regenerative air heater.

Design Problems of Air Heaters The size of an air heater depends on the required hot-air temperature for fuel combustion which is listed in Table 8.9.

The outside diameter of the tubes in a tubular air heater is usually 40 mm, and the tubes are arranged in a staggered pattern. The transverse relative spacing of the tubes, $S_1/d = 1.5–1.9$, while the longitudinal relative spacing, the $S_2/d = 1.0–1.2$. The tube length is usually less than 5 m. For solid fuel-fired boilers, the flue gas velocity of a vertical tubular air heater ranges from 10 to 16 m/s, and the air velocity is equal to half of the gas velocity. These velocity values should be vice versa for horizontal tubular air heaters. For oil- or gas-fired boilers, these velocities can be higher.

The rotating speed of a rotating-plate air heater ranges from 1.5 to 4 r/min; while the stationary plate ranges from 0.75 to 1.4 r/min. In these regenerative air heaters, air and gas velocities are nearly equal and range from 8 to 12 m/s.

When burning fuels contain sulfur, if the metal temperature of the heating surface falls below the dew point due to a low air-inlet or low gas-exit temperature, the corrosion of metal may occur. It may be prevented by preheating the air before it enters the air heater or by using corrosion-resistant materials and coatings in the low-temperature section of the heater.

Ash particulate residues from coal or oil firing are usually cleaned from the heating surface by means of a soot blower that uses steam or air as a cleaning medium. Sometimes for a tubular air heater with tube length greater than 6 m for for oil-fired boilers, ash-removal can be accomplished by circulating metal shot through the air heater tubes.

8.2.6 Construction and Design Problems of Steam Drums

All boilers operating under subcritical pressure, except for once-through types, are at least provided with a steam drum (See Fig. 6.13). In the steam

TABLE 8.10 Statistical Value of Inner Diameter, Wall Thickness, and Steel of Drums used in Natural-Circulation Boilers, mm

Pressure	Low	Middle	High	Superhigh
Inner diameter	800–1200	1400–1600	1600–1800	1600–1800
Thickness	16–25	32–46	60–100	80–100
Steel	Carbon steel	Carbon steel and alloy steel	Carbon steel and alloy steel	Alloy steel

drum saturated steam is separated from the steam–water mixture discharged from the risers in its steam space and in the separating equipment installed in it. The steam drum is also a water storage vessel which accomodates the changes in water level during load changes and internal water treatment; therefore the steam drum size must be large enough to house the separating equipment and to contain the required quantity of water. The length should be greater than the width and/or depth of the boiler, depending on the arrangement. The inner diameter of the steam drum depends on pressure, circulation form, steam-generating capacity, and separating equipment type. The statistical steam drum inner diameter, wall thickness, and steel used for natural-circulation boilers are listed in Table 8.10.

8.3 HEAT TRANSFER CALCULATIONS OF BOILER COMPONENTS

There are many methods used to calculate the heat transfer performance of a boiler. In the USSR the legal method for heat transfer calculation of a boiler is given in [2], which is also widely used in the People's Republic of China (PRC), notwithstanding that some of the empirical coefficients or data selected during calculation are different. Therefore the method mentioned in this section closely resembles the method of [2].

Before the heat transfer calculation of a boiler can be started, the boiler steam-generating capacity, steam pressure, steam temperature, and the feed-water temperature must be given by the customer.

8.3.1 Boiler Efficiency and Weight of Fuel Fired

The efficiency of a boiler, η_b, is defined as the ratio of the quantity of heat absorbed by the working fluid, H_1, to the available heat of fuel, H_{av}^w, kJ/kg, that is,

$$\eta_b = \frac{H_1}{H_{\mathrm{av}}^w} 100 \tag{8.4}$$

where H_1 and H_{av}^w are written for 1 kg of burned solid or liquid fuel. H_1 can be expressed as follows:

$$H_1 = \frac{1}{B}[W_{sh}(i_{ss} - i_{fw}) + W_{rh}(i_{rhe} - i_{rhi}) + W_{bw}(i_{sw} - i_{fw})] \quad (8.5)$$

where W_{sh}, W_{rh}, and W_{bw} are the flow rates of the superheated steam, the reheated steam, and the blow-off water from the boiler, respectively, kg/s; i_{ss}, i_{fw}, and i_{sw} are the enthalpies of the superheated steam, the feed water, and the saturated water, kJ/kg, respectively; and i_{rhe} and i_{rhi} are the enthalpies of the exit steam and the inlet steam of the reheater, kJ/kg, respectively. H_{av}^w is expressed as follows:

$$H_{av}^w = H_1^w + H_{ph} \quad (8.6)$$

where H_{ph} is the physical heat of the solid or liquid fuel before burning, which is equal to $C_F T_F$, where C_F and T_F are the specific heat of fuel and the fuel temperature, respectively.

H_{av}^w can be divided into two parts: the useful heat, H_1, and the lost heat, $\Sigma_{2-6} H_i$, and can be expressed as

$$H_{av}^w = H_1 + H_2 + H_3 + H_4 + H_5 + H_6 \quad (8.7)$$

Dividing Eq. (8.7) by H_{av}^w, the equation can be expressed as a percentage form:

$$100 = \eta_b + h_2 + h_3 + h_4 + h_5 + h_6 \quad (8.8)$$

where h_2 is the relative heat loss with waste gases, h_3 is the relative heat loss by incomplete combustion, h_4 is the relative heat loss with unburned carbon, h_5 is the relative heat loss by giving up heat to the environment, and h_6 is the heat with physical heat of slag.

If the sum of the relative heat losses of a boiler is known, the boiler efficiency can be obtained from Eq. (8.8):

$$h_2 = \frac{H_2}{H_{av}^w} 100 \quad (8.9)$$

where H_2 is the absolute heat loss and waste gases which is equal to

$$H_2 = \left(I_{wg} - \alpha_{wg} I_{ca}^0\right)(1 - 0.01 h_4) \quad (8.10)$$

where I_{wg} is the enthalpy of waste gases, kJ/kg; I_{ca}^0 is the enthalpy of the theoretically required volume of cold air, kJ/kg; α_{wg} is the excess air ratio in the waste gases, which represents the ratio of the actual air volume to the theoretically required air volume.

h_3 and h_4 depend on the fuel fired and the burning equipment used. For furnaces with burners, h_3 is equal to 0% for coal dust and 0.5% for oil or gas fuel; h_4 is equal to 2% for bituminous, 3% to 4% for anthracite, and 0% for oil or gas fuel. For furnaces with stokers, h_3 and h_4 can be selected from a steam boiler handbook or related references [1–3], generally $h_3 = 1\%–3\%$, $h_4 = 5\%–15\%$.

h_5 can be determined from Fig. 8.26. During heat transfer calculations, h_5 is assumed to be proportional to the heat absorbed by each heating surface of the boiler and is accounted for by a heat retention coefficient ϕ:

$$\phi = 1 - \frac{h_5}{\eta_b + h_5} \tag{8.11}$$

h_6 can be found by the following equation:

$$h_6 = \frac{a_{sl}(CT)_{sl} A^w}{H^w_{av}} \tag{8.12}$$

where a_{sl} is the fraction of the total ash removed as slag from the furnace; for boilers with stokers, $a_{sl} = 0.75–0.8$; for boilers with coal burners, $a_{sl} = 0.1$; $(CT)_{sl}$ is the enthalpy of slag kJ/kg; A^w is the ash mass content of fuel on a moist basis.

For dry ash furnaces with burners or furnaces with stokers $T_{sl} = 600°C$ and $(CT)_{sl} = 554$ kJ/kg.

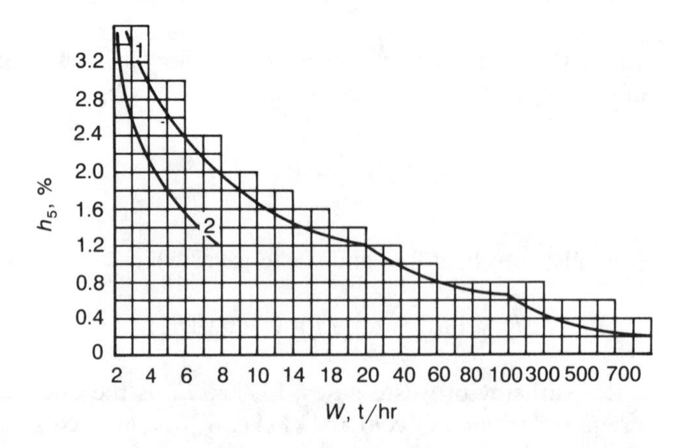

Fig. 8.26. Curves for determining h_5: 1—boiler with economizer and air heater, 2—boilers without economizer and air heater.

When η_b, $\Sigma_{2-6}h_i$, and H_{av}^w are determined, the fuel consumption of the boiler, B, can be obtained from Es. (8.4) and (8.5) as follows:

$$B = \frac{1}{H_{av}^w \eta_b}\left[W_{sh}(i_{ss} - i_{fw}) + W_{rh}(i_{rhe} - i_{rhi}) + W_{bw}(i_{sw} - i_{fw})\right] \quad (8.13)$$

B is used for calculating the coal pulverization system of the boiler. In order to find the actual volumes of the combustion products and the air flow rate for combustion, the rated fuel consumption, B_r, should be used. When there is a heat loss with unburned carbon during combustion, B_r can be determined as

$$B_r = B(1 - 0.01h_4) \quad (8.14)$$

8.3.2 Heat Transfer Calculation of Water-Cooled Furnace

The method for calculating the heat transfer in boiler furnaces in the USSR and the PRC is based on the semiempirical formula obtained by Gurvich [5]:

$$\frac{T_{gFe}}{T_{ga}} = \frac{Bo^{0.6}}{Bo^{0.6} + Ma_F^{0.6}} \quad (8.15)$$

where T_{gFe} is the gas temperature at the furnace outlet, K; T_{ga} is the adiabatic temperature of combustion, K; M is a coefficient relating to the pattern of the temperature field in the furnace; and Bo is the Boltzman which can be expressed as

$$Bo = \frac{\phi B_r \overline{VC}}{\sigma_0 \psi_{ef} A_w T_{ga}} \quad (8.16)$$

where $\overline{VC}$ is the average heat capacity of flue gases in the furnace in the temperature interval $(T_{ga} - T_{gFe})$ formed by 1 kg of burned fuel, kJ/(kg · K); σ_0 is the emissivity of the black body and is equal to 5.67×10^{-11} kW/(m^2 · K^4); ψ_{ef} is the average coefficient of thermal efficiency of the water walls; A_w is the surface area of the furnace walls; and ϕ is the heat retention coefficient. ψ_{ef} characterizes the fraction of heat absorbed by a water wall, and is expressed as

$$\psi_{ef} = x\zeta \quad (8.17)$$

where x is the angular coefficient of a water wall and can be determined from Fig. 8.27; ζ is the coefficient of fouling. For fuel oil $\zeta = 0.55$, for gas fuel $\zeta = 0.65$, for a furnace with stokers $\zeta = 0.60$, for a furnace with coal burners $\zeta = 0.35$–0.55; for tubes with refractory covers $\zeta = 0.1$–0.2.

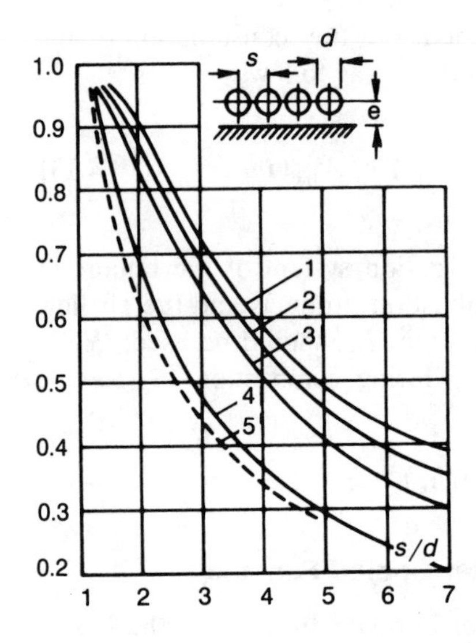

Fig. 8.27. Angular coefficient x: 1—$e \geq 1.4d$, 2—$e = 0.8d$, 3—$e = 0.5d$, 4—$e = 0$, 5—$e \geq 0.5d$, without considering the radiation of the furnace wall.

For water walls with varying ζ and x, the average value of ψ_{ef} is

$$\psi_{\text{ef}} = \frac{\Sigma x_i A_{wi} \zeta_i}{A_w} \tag{8.18}$$

where a_F is the coefficient of thermal radiation of the furnace and can be calculated from the following equation:

$$a_F = \frac{1}{1 + \left(\dfrac{1}{a_{\text{fl}}} - 1\right)\psi_{\text{ef}}} \tag{8.19}$$

where a_{fl} is the flame emissivity and can be calculated from Eqs. (8.20) and (8.26).

For solid fuels:

$$a_{\text{fl}} = 1 - e^{-k_F p S} \tag{8.20}$$

where k_F is the effective coefficient of absorption in the furnace; p is the pressure of flue gases in the furnace (for a common vacuum furnace, $p = 0.1$ MPa); and S is the effective thickness of the radiating layer, m, $S = 3.6V_F/A_w$, where V_F is the furnace volume, m^3.

$$k_F = k_g r + k_a \mu_a + k_c x_1 x_2 \tag{8.21}$$

where r is the total volume concentration of triatomic gases; k_g is the effective coefficient of absorption by triatomic gases; k_a is the effective coefficient of absorption by ash particles, $1/(\text{m} \cdot \text{MPa})$; μ_a is the dimensionless concentration of fly ash in the furnace, kg/kg; k_c is the effective coefficient of absorption of radiation by coke particles, $k_c = 1.0$; x_1 is a constant for a particular kind of fuel, for coal with low volatile matter $x_1 = 1.0$; for coal with high volatile matter $x_1 = 0.5$; and x_2 is a constant taking into account the influence of the burning equipment, for stokers $x_2 = 0.03$, for burners $x_2 = 0.1$.

$$r = r_{RO_2} + r_{H_2O} \tag{8.22}$$

where r_{RO_2} and r_{H_2O} are volume concentrations of $CO_2 + SO_2$ and H_2O, respectively.

$$k_g = 10\left[\frac{0.78 + 1.6r_{H_2O}}{(10pSr)^{1/2}} - 0.1\right]\left(1 - 0.37\frac{T_{gFe}}{1000}\right) \tag{8.23}$$

$$k_a = \frac{5990}{\left(T_{gFe}^2 d_a^2\right)^{1/3}} \tag{8.24}$$

where d_a is the average diameter of the ash particles, for pulverized coal $d_a = 13\text{–}16\ \mu\text{m}$; and T_{gFe} is the gas temperature at the furnace outlet, K.

$$\mu_a = \frac{A^w a_{fa}}{100 W_g} \tag{8.25}$$

where a_{fa} is the fraction of the total ash removed as fly ash from the furnace, $a_{fa} = 1 - a_{sl}$; and W_g is the mass of the flue gases per unit fuel burned, kg/kg.

For liquid and gas fuels,

$$a_{fl} = m a_{lum} + (1 - m) a_g \tag{8.26}$$

where a_{lum} is the emissivity of the luminous portion of the flame; a_g is the emissivity of the nonluminuous gaseous medium of the flame; and m is the fraction of the luminous portion of the flame, for natural-gas combustion $m = 0.1$, for fuel oil $m = 0.55$.

$$a_{lum} = 1 - e^{-k_{lum} pS} \tag{8.27}$$

where k_{lum} is the effective coefficient of absorption of the luminous portion, $1/(\text{m} \cdot \text{MPa})$.

$$k_{lum} = k_g r + k_s \tag{8.28}$$

where k_s is the effective coefficient of absorption by soot particles, 1/m(·MPa).

$$k_s = 0.3(2 - \alpha_{Fe})\left(1.6\frac{T_{gFe}}{1000} - 0.5\right)\frac{C^w}{H^w} \tag{8.29}$$

where C^w and H^w are the carbon and hydrogen mass content of fuel on a moist basis, and α_{Fe} is the excess air ratio at the furnace exit.

$$a_g = 1 - e^{-k_g rpS} \tag{8.30}$$

The coefficient M is expressed as

$$M = A - BX \tag{8.31}$$

where A and B are empirical coefficients depending on the kind of fuel used, for gas or fuel oil, $A = 0.54$ and $B = 0.2$; for coal burners, $A =$ 0.56–0.59 (a higher value is for high volatile matter) and $B = 0.5$; for furnaces with stokers, $A = 0.59$ and $B = 0.5$. X is the relative position of the highest temperature zone in the furnace; for a spreader stoker, $X = 0$; for furnaces with other stokers, $X = 0.14$; for furnaces with burners, X can be calculated from the following expression:

$$X = X_b + \Delta X \tag{8.32}$$

where X_b is the relative level of the burners which is equal to the ratio of the height of the burner center to that of the furnace exit center; both heights are counted from the lowest boundary of the furnace volume. ΔX is a corrective coefficient of X to account for the actual position of the flame core, for horizontally arranged burners $\Delta X = 0$, for tilting burners when the tilting angle is equal to $\pm 20°$, $\Delta X = \pm 0.1$, and for other types of burners $\Delta X = 0.05$–0.1.

The adiabatic temperature of combustion T_a, K, can be obtained as

$$T_a = \frac{H_u}{\overline{VC}} + 273 \tag{8.33}$$

where H_u is the useful heat release in the furnace, kJ/kg, and $\overline{VC}$ is the average specific heat of combustion products formed by 1 kg of fuel within the temperature interval 0–T_a, kJ/(kg · K).

Usually, for lignite and peat $T_a = 1973$–2123 K, for anthracite, fuel oil, and natural gas $T_a = 2123$–2373 K.

$$H_u = H_{av}^w \frac{100 - h_3 - h_4 - h_6}{100 - h_4} + H_a \tag{8.34}$$

where H_a is the heat introduced into the furnace by hot and cold air, kJ/kg,

$$H_a = (\alpha_{Fe} - \Delta\alpha_F)I^0_{\text{ha}} + \Delta\alpha_F I^0_{\text{ca}} \tag{8.35}$$

where $\Delta\alpha_F$ is the increment of the relative air ratio of the furnace considering the leakage of air from the surroundings into the furnace; for a vacuum furnace with burners, $\Delta\alpha_F = 0.05$, and for furnaces with stokers, $\Delta\alpha_F = 0.1$.

If the values of B_0, M, T_a, and a_F are determined, the gas temperature at the furnace outlet T_{gFe} can be obtained from Eq. (8.15).

The heat exchange in the boiler furnace can be obtained from the heat balance equation of the gas side H_r, kJ/kg:

$$H_r = \phi(H_u - I_{Fe}) = \phi(\overline{VC})(T_a - T_{gFe}) \tag{8.36}$$

Equation (8.15) can also be arranged as follows:

$$T_{gFe} = \frac{T_a}{M\left(\dfrac{5.67 \times 10^{-11}\psi_{\text{ef}} A_w a_F T_a^3}{\phi B_r \overline{VC}}\right)^{0.6} + 1} \tag{8.37}$$

To determine T_{gFe}, for the sake of performing the calculations to find k_g, k_s or k_a, we first have to assume this temperature and then correct it by making a comparison with the calculated value of T_{gFe}; if the discrepancy between them surpasses $\pm 100°\text{C}$, the calculation has to be repeated for a new assumed value of T_{gFe}.

The average heat flux of the furnace heating surfaces, q_{rF}, kW/m^2, may be calculated by

$$q_{rF} = \frac{B_r H_r}{A} \tag{8.38}$$

where A is the radiant heating surfaces of the furnace, m^2.

$$A = \frac{\Sigma x_i A_{wi}}{A_w} \tag{8.39}$$

where x_i is the angular coefficient of an individual water wall; and A_{wi} is the surface area of an individual furnace wall, m^2.

After calculation, we have to check whether T_{gFe} is below the safe temperature mentioned in Section 8.2.2. The value of the heat release rate per unit furnace cross-sectional area, q_F, and the value of the heat release rate per unit furnace volume, q_V should not exceed the recommended values listed in Tables 8.2 and 8.4 respectively.

8.3.3 Heat Transfer Calculation of Convection Heating Surfaces

The convection heating surfaces of a boiler are those heating surfaces arranged behind the furnace including platens, slag screen, convective superheater, convective reheater, economizer, and air heater, etc.

The principal equations of convective heat transfer are as follows.

The heat transfer equation is

$$H_t = \frac{U \Delta T A}{B_r} \tag{8.40}$$

where U is the overall heat transfer coefficient, kW/(m$^2 \cdot$ K); ΔT is the mean temperature difference, K; and H_t is the heat transfer quantity, kJ/kg.

The heat balance equation for the flue gas side is

$$H_b = \phi\left(I_i - I_e + \Delta\alpha\, I^0_{\text{ca}}\right) \tag{8.41}$$

where I_i and I_e are the enthalpies of the gases at the inlet and outlet of the heating surface, kJ/kg, respectively; $\Delta\alpha$ is the relative air in leakage in a gas duct; and I^0_{ca} is the cold-air enthalpy, kJ/kg.

The heat balance equation of heat absorption by the working fluid in heating surfaces just downstream of the furnace (such as platen, slag screen, etc.) is

$$H_b = \frac{W}{B_r}(i_e - i_i) - H_{rF} \tag{8.42a}$$

where W is the mass flow rate of the working fluid, kg/s; i_e and i_i are the enthalpies of the working fluid at the inlet and outlet of the heating surface, kJ/kg, respectively; H_{rF} is the radiant heat absorbed by the heating surface from the furnace, kJ/kg.

The heat balance equation of heat absorption by the working fluid in heating surfaces arranged in boiler flue ducts (such as convection superheater, reheater, economizer, and air heater, etc.) is

$$H_b = \frac{W}{B_r}(i_e - i_i) \tag{8.42b}$$

For an air heater, the heat absorbed by the air is

$$H_b = \left(\beta_a + 0.5\,\Delta\alpha_{\text{ah}}\right)\left(I^0_{\text{ha}} - I^0_{\text{ca}}\right) \tag{8.43}$$

where β_a is the excess air ratio at the outlet of the air heater, $\beta_a = \alpha_{Fe} - \Delta\alpha_F$; $\Delta\alpha_{\text{ah}}$ is the relative air leakage from the air heater; and I^0_{ha} and I^0_{ca} are

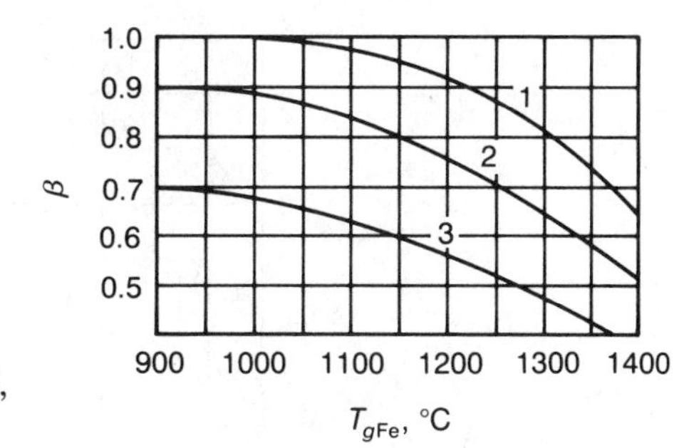

Fig. 8.28. Coefficient β: 1—coal, 2—fuel oil, 3—gas.

the enthalpies of the theoretical volume of hot air and cold air, kJ/kg, respectively.

The radiant heat absorbed by convection heating surfaces of platens from the furnace H_{rF} can be obtained as

$$H_{rF} = H_{ri} - H_{ro} \tag{8.44}$$

where H_{ri} and H_{ro} are the radiant heat flux at the inlet to the heating surface and the radiant heat flux at the outlet from the heating surface onto subsequent heating surfaces, kJ/kg.

$$H_{ri} = \beta \eta_h q_{rF} \frac{A_{Fe}}{B_r} \tag{8.45}$$

where β is the coefficient which takes into account heat exchange between the furnace and the calculated heating surfaces (Fig. 8.28); η_h is the coefficient of distribution of heat absorption along the furnace height (η_h for furnaces with front-type arrangement of burners is shown in Fig. 8.29); and A_{Fe} is the surface area of the furnace outlet, m^2.

$$H_{ro} = \frac{H_{ri}(1-a)x_e}{\beta} + H_{rpe} \tag{8.46}$$

where a is the emissivity of gases in the zone of the platens which is determined from Eq. (8.74); x_e is the angular coefficient of radiation from the inlet onto the outlet section of the calculated surfaces, for platens, it is determined by Eq. (8.47); and H_{rpe} is the radiation of gases from the platen zone onto the subsequent heating surfaces and can be determined by Eq. (8.48), kJ/kg.

$$x_e = \left[\left(\frac{b}{s_l}\right)^2 + 1\right]^{1/2} - \frac{b}{s_l} \tag{8.47}$$

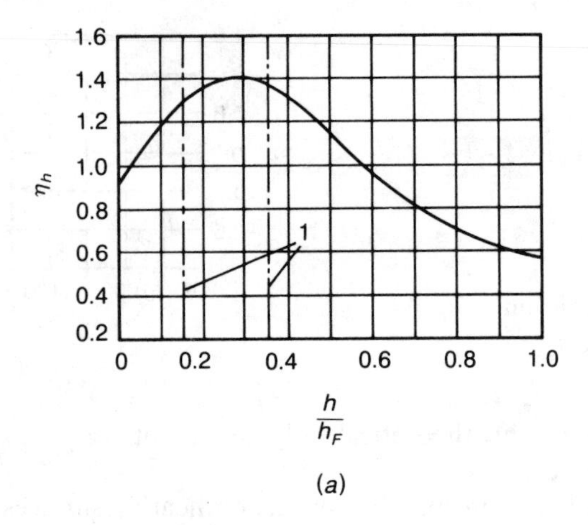

(a)

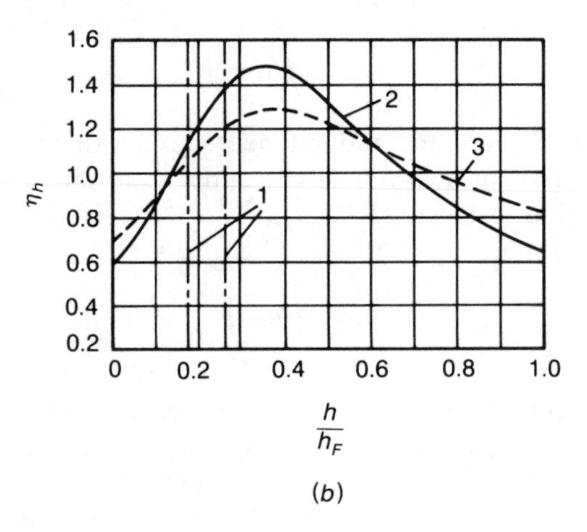

(b)

Fig. 8.29. Coefficient η_h (h_F is the height of the furnace): (*a*) for fuel-oil and gas furnace, (*b*) for coal-dust furnace. 1—Burner axis, 2—for anthracite and bituminous coal, 3—for lignite.

where s_l is the spacing between the platens, m, and b is the depth of the platens (in the direction of the flow), m.

$$H_{rpe} = \frac{5.67 \times 10^{-11} \times aA_{pe}\overline{T}^4\xi}{B_r} \tag{8.48}$$

where $\overline{T}$ is the average gas temperature of the platen, K; ξ is the coefficient which takes into account the influence of fuel type, for coal and fuel oil, $\xi = 0.5$, for natural gas, $\xi = 0.7$; and A_{pe} is the surface area of the platen outlet, m^2.

In the case of calculating slag screens,

$$H_{rF} = \frac{x_s \eta_h q_{rF} A_{Fe}}{B_r} \tag{8.49}$$

where x_s is the angular coefficient of the slag screens, $x_s = 1 - (1 - x)^n$; n is the number of the tube rows, and x is the angular coefficient of one row of tubes determined from Figure 8.27, curve 5.

Overall Heat Transfer Coefficient, U In the convection heating surfaces of a boiler, the heat transfer process consists of three parts: heat transfer from the flue gases to the tube wall, heat transfer through a multilayer wall (a metallic tube wall with deposits on its outside and inside surface), and heat transfer from the wall to the working fluid. The equation of the overall heat transfer coefficient U, $\mathrm{W/(m^2 \cdot K)}$, is as follows:

$$U = \frac{1}{\dfrac{1}{h_o} + \dfrac{\delta_{sl}}{\lambda_{sl}} + \dfrac{\delta_m}{\lambda_m} + \dfrac{\delta_{sc}}{\lambda_{sc}} + \dfrac{1}{h_i}} \tag{8.50}$$

where h_i and h_o are the coefficients of heat transfer from the gas to the wall and from the wall to the working fluid, $\mathrm{kW/(m^2 \cdot K)}$, respectively; δ_m, δ_{sl}, and δ_{sc} are the thickness of the tube wall, the external layer deposits, and the internal layer deposits, m, respectively; and λ_m, λ_{sl}, and λ_{sc} are the conductivities of the tube wall, the external layer deposits, and the internal layer deposits, $\mathrm{kW/(m \cdot K)}$, respectively.

As δ_m/λ_m is much less than $1/h_o$ and $1/h_i$, it can be neglected; δ_{sc}/λ_{sc} is nearly equal to 0 under normal operating conditions, therefore it is not considered during calculation. Let the coefficient of effectiveness ψ take into account the influence of δ_{sl}/λ_{sl} or let $\varepsilon = \delta_{sl}/\lambda_{sl}$; Eq. (8.50) can be expressed as follows:

$$U = \frac{\psi}{\dfrac{1}{h_o} + \dfrac{1}{h_i}} \tag{8.51}$$

or

$$U = \frac{1}{\dfrac{1}{h_o} + \varepsilon + \dfrac{1}{h_i}} \tag{8.52}$$

For slag screens, with an in-line arranged convection superheater, $\psi = 0.6$ when the boiler is fired with anthracite; $\psi = 0.65$, when the boiler is fired with bituminous coal; and $\psi = 0.6–0.65$ when the boiler is fired with fuel oil; for slag screens and superheaters, the smaller value is for high velocity, and for economizers, $\psi = 0.65–0.70$.

For a tubular air heater, the coefficient of utilization ξ is introduced into Eq. (8.50) which considers both the influence of deposits and the nonuniform sweeping of a heating surface by the gas flow. Equation (8.50) can then be expressed as follows:

$$U = \frac{\xi}{\dfrac{1}{h_o} + \dfrac{1}{h_i}} \tag{8.53}$$

where $\xi = 0.8–0.85$.

For regenerative air heaters

$$U = \frac{\xi c}{\dfrac{1}{x_g h_o} + \dfrac{1}{x_a h_i}} \tag{8.54}$$

where $\xi = 0.8–0.9$; c is a coefficient relating to rotating speed n; for $n =$ 0.5 r/min, $c = 0.85$; for $n = 1.0$ r/min, $c = 0.97$; for $n = 1.5$ r/min, $c = 1.0$; x_g and x_a are fractions of the heating surface of gas and air, respectively, usually $x_g = 0.5$ and $x_a = 0.333$.

For platens, direct radiation from the furnace increases the temperature of the deposits on the platen tubes and decreases the heat absorption from the gas flow sweeping these tubes, therefore a multiplier $(1 + H_{rF}/H_p)$ is introduced into Eq. (8.52) to take these circumstances into account, that is,

$$U = \frac{1}{\dfrac{1}{h_o} + \left(1 + \dfrac{H_{rF}}{H_p}\right)\left(\varepsilon + \dfrac{1}{h_i}\right)} \tag{8.55}$$

here H_p is the total heat absorption of the platens due to convective heat transfer and radiation of gases in the platen zone, kJ/kg; ε is the fouling coefficient, for fuel oil, $\varepsilon = 5.2\ (\mathrm{m^2 \cdot K})/\mathrm{kW}$, for gas fuel, $\varepsilon = 0$, for solid fuels, ε can be obtained from [1, 2].

Calculation of the Heat Transfer Coefficient from Gas to Wall, h_o The heat absorption by a convective heating surface from the gas flow is determined by the heat transfer coefficient h_o, kW/(m^2 · K), which takes into account both the radiant and convective heat transfer at the external side of the heating surface.

$$h_o = \xi(h_c + h_r) \tag{8.56}$$

where ξ is the coefficient of the nonuniform sweeping of the boiler component by gases, for cross-current flow $\xi = 1.0$, for most mixed-current flow $\xi = 0.95$; h_c is the convective heat transfer coefficient; and h_r is the radiant heat transfer coefficient of the space between the tubes.

Calculation of the Convective Heat Transfer Coefficient, h_c For flow across an in-line tube bundle (Fig. 8.30*a*) [2]:

$$h_c = c_s c_n \frac{\lambda}{d} Re^{0.65} Pr^{0.33} \tag{8.57}$$

where λ is the conductivity coefficient of a fluid, kW/(m · K); c_s is the corrective coefficient which takes into account the influence of the relative spacing of the tube bundle; and c_n is the corrective coefficient which takes into account the influence of the number of tube rows along the flow direction.

$$c_s = 0.2\left[1 + \left(2\frac{S_1}{d} - 3\right)\left(1 - \frac{S_2}{d}\right)^3\right]^{-2} \tag{8.58}$$

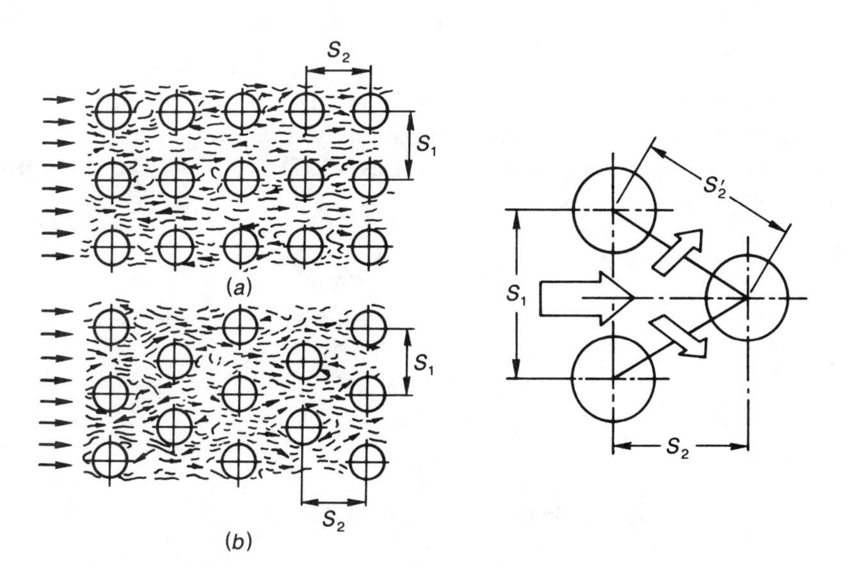

Fig. 8.30. Flow across tube bundle: (*a*) in-line tube bundle, (*b*) staggered tube bundle.

If $S_1/d \leq 1.5$ or $S_2/d \geq 2$, $c_s = 0.2$; when the number of rows,

$$\begin{aligned} n < 10 \qquad & c_n = 0.91 + 0.0125(n - 2) \\ n \geq 10 \qquad & c_n = 1.0 \end{aligned} \tag{8.59}$$

For flow across a staggered tube bundle (Fig. 8.30*b*):

$$h_c = c_s c_n \frac{\lambda}{d} Re^{0.6} Pr^{0.33} \tag{8.60}$$

Let $\phi = (S_1/d - 1)/(S_2'/d - 1)$; when

$$0.1 < \phi \leq 1.7 \qquad c_s = 0.34\phi^{0.1} \tag{8.61}$$

$$1.7 < \phi \leq 4.5 \text{ and } \frac{S_1}{d} < 3 \qquad c_s = 0.275\phi^{0.5} \tag{8.62}$$

$$1.7 < \phi \leq 4.5 \text{ and } \frac{S_1}{d} \geq 3 \qquad c_s = 0.34\phi^{0.1} \tag{8.63}$$

when the number of rows, $n < 10$, $S_1/d < 3.0$;

$$c_n = 3.12n^{0.05} - 2.5 \tag{8.64}$$

when $n < 10$, $S_1/d \geq 3.0$;

$$c_n = 4n^{0.02} - 3.2 \tag{8.65}$$

when $n \geq 10$, $c_n = 1.0$.

Equations (8.57) and (8.60) can be used in the range of $Re = 1.5 \times 10^3$–1.5×10^5; the boiler operating condition is always within this range.

For the simplification of calculation, Eqs. (8.57) and (8.60) can be rewritten as follows:

$$h_c = A_1 c_s c_n \frac{V^{0.65}}{d^{0.35}} \tag{8.66}$$

$$h_c = A_2 c_s c_n \frac{V^{0.6}}{d^{0.4}} \tag{8.67}$$

For flue gases (300°C $\leq T \leq$ 1000°C):

$$A_1 = 28.96(1 - 1.25 \times 10^{-4} T) \times 10^{-3}$$

$$A_2 = 16.98 \times 10^{-3}$$

For air (50°C $\leq T \leq$ 500°C):

$$A_1 = 29.77(1 - 5.28 \times 10^{-4}T) \times 10^{-3}$$

$$A_2 = 17.56(1 - 3.4 \times 10^{-4}T) \times 10^{-3}$$

For flow along the longitude direction of a tube bundle or in a tube:

$$h_c = 0.23\frac{\lambda}{d_e}Re^{0.8}Pr^{0.4}C_l \tag{8.68}$$

where d_e is the equivalent diameter, m, and C_l is a coefficient which takes into account the influence of the heating surface length and can be determined from Fig. 8.31.

The simplified equation for Eq. (8.68) is as follows:

$$\text{for air and flue gases} \qquad h_c = A\frac{V^{0.8}}{d^{0.2}} \tag{8.69}$$

$$\text{for superheated steam} \qquad h_c = B\frac{(\rho V)^{0.8}}{d^{0.2}} \tag{8.70}$$

when 50°C $\leq T$ (average temperature) $\leq$ 400°C, for air $A = 3.49(1 - 8.26 \times 10^{-4}T) \times 10^{-3}$; for flue gases $A = 3.7(1 - 8.26 \times 10^{-4}T) \times 10^{-3}$; for superheated steam when p = 4–4.4 MPa, T = 320–450°C, $B = 6.61 \times 10^{-3}$;

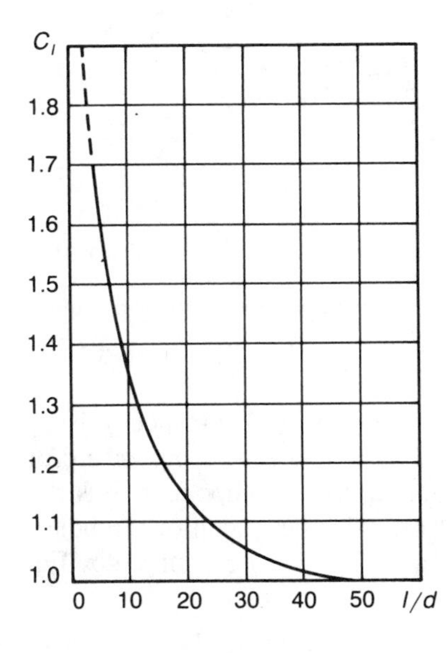

Fig. 8.31. Coefficient C_l.

$p = 10\text{–}11$ MPa, $T = 420\text{–}540°C$, $B = 7.5 \times 10^{-3}$; $p = 14\text{–}15$ MPa, $T = 460\text{–}550°C$, $B = 8.0 \times 10^{-3}$.

The preceding equations are applicable for $Re = 10^4\text{–}5 \times 10^5$, and all properties are evaluated at the average fluid temperature.

For regenerative air heaters, h_c for the gas side and the air side is

$$h_c = A \frac{\lambda}{d_e} Re^{0.8} Pr^{0.4} c_f C_l \tag{8.71}$$

where for common types of heating surfaces, $A = 0.027$, $d_e = 7.8$ mm, and C_l can be determined from Fig. 8.31; c_t is a coefficient which takes into account the influence of surface temperature; $c_t = 1.0$ for gas and $c_t = (T/T_w)^{0.5}$ for air, where T is the air temperature and T_w is the average wall temperature.

Calculation of the Radiant Heat Transfer Coefficient of the Space, h_r

The radiant heat transfer coefficient of the space h_r, kW/(m$^2 \cdot$ K) can be expressed as follows:

For solid fuel

$$h_r = 5.1 \times 10^{-11} a_g T_g^3 \left[\frac{1 - (T_{aw}/T_g)^4}{1 - (T_{aw}/T_g)} \right] \tag{8.72}$$

where a_g is the emissivity of gases; T_g and T_{aw} are the absolute average temperatures of the gases and the ashy tube wall, K.

For fuel oil or gas

$$h_r = 5.1 \times 10^{-11} a_g T_g^3 \left[\frac{1 - (T_{aw}/T_g)^{3.6}}{(1 - T_{aw}/T_g)} \right] \tag{8.73}$$

$$a_g = 1 - e^{-kpS} \tag{8.74}$$

where the gas pressure $p = 0.1$ MPa, and the effective coefficient of absorption, k, is equal to

$$k = k_g r + k_a \mu_a \tag{8.75}$$

where k_g and k_a can be determined by Eqs. (8.23) and (8.24) respectively; in these equations the gas temperature is equal to the average temperature of the inlet and exit temperatures of the calculated boiler component; r is the total volume concentration of the triatomic gases in the region of the boiler component; and μ_a can be determined by Eq. (8.25), S is determined by Eq. (8.78), m.

When using solid fuel or liquid fuel, for platens and convection superheaters

$$T_{aw} = T_f + \left(\varepsilon + \frac{1}{h_i}\right)\frac{B_r H}{A} \tag{8.76}$$

where T_f is the average temperature of the working fluid in the boiler component, K; H is the total quantity of heat absorbed in the component, kJ/kg; ε is the fouling coefficient; for burning solid fuel $\varepsilon = 4.3$, for burning liquid fuel $\varepsilon = 2.6$ $(m^2 \cdot K)/kW$; A is the heating surface, m^2; and h_i is the convective heat transfer coefficient from the wall to the working fluid, Eq. (8.70).

For the other components

$$T_{aw} = T_f + \Delta T \tag{8.77}$$

where for slag screens $\Delta T = 80$ K; for economizers with inlet-gas temperatures greater than 400°C, $\Delta T = 60$ K; with inlet-gas temperature less than 400°C, $\Delta T = 25$ K.

When using gas fuel, $\Delta T = 25$ K for all boiler components.

For air heaters, T_{aw} is equal to the average temperature of air and gas.

The effective thickness of the radiating layer, S, of tube bundles is

$$S = 0.9d\left(\frac{4s_1 s_2}{\pi d^2} - 1\right) \tag{8.78}$$

For platens

$$S = \frac{1.8}{\dfrac{1}{A} + \dfrac{1}{B} + \dfrac{1}{C}} \tag{8.79}$$

where A, B, and C are the height, depth, and width of the space between two platens, respectively.

For tubular air heaters, $S = 0.9d$.

When there is an empty room before the calculated component, for considering the radiation of this empty room, the radiant heat transfer coefficient should be calculated as follows:

$$h_r' = h_r\left[1 + A\left(\frac{T_R}{1000}\right)^{0.25}\left(\frac{L_R}{L_B}\right)^{0.07}\right] \tag{8.80}$$

where T_R is the temperature of gas in the room, K; L_R and L_B are the depths of the room and the tube bundle, m, respectively; and A is a constant, for fuel oil and gas $A = 0.3$; for bituminous coal and anthracite $A = 0.4$, and for lignite $A = 0.5$.

Calculation of the Mean Temperature Difference, ΔT The mean temperature difference ΔT depends on the flow system; for parallel-flow system (Fig. 8.11*b*):

$$\Delta T_p = \frac{(T_{gi} - T_{fi}) - (T_{ge} - T_{fe})}{\ln\left(\dfrac{T_{gi} - T_{fi}}{T_{ge} - T_{fe}}\right)} \tag{8.81}$$

where T_{gi}, T_{ge}, T_{fi}, and T_{fe} are the temperatures of the inlet gas, exit gas, inlet working fluid, and exit working fluid, respectively, K.

For a counterflow system (Fig. 8.11*a*):

$$\Delta T_c = \frac{(T_{gi} - T_{fe}) - (T_{ge} - T_{fi})}{\ln\left(\dfrac{T_{gi} - T_{fe}}{T_{ge} - T_{fi}}\right)} \tag{8.82}$$

TABLE 8.11 Determination of Corrective Coefficient, ψ

Flow System	Figure Used and Parameters
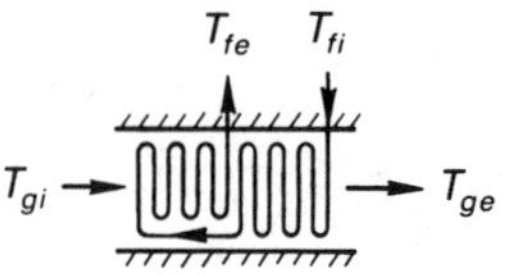 T_{fe} T_{fi} T_{gi} → T_{ge}	Fig. 8.32, $\tau_1 = T_{gi} - T_{ge}$, $\tau_2 = T_{fe} - T_{fi}$, $P = \tau_2/(T_{gi} - T_{fi})$, $R = \tau_1/\tau_2$, $C = A_p/A$
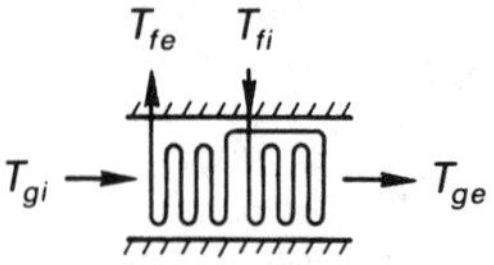 T_{fe} T_{fi} T_{gi} → T_{ge}	Fig. 8.32, $\tau_1 = T_{fe} - T_{fi}$, $\tau_2 = T_{gi} - T_{ge}$, $P = \tau_2/(T_{gi} - T_{fi})$, $R = \tau_1/\tau_2$, $C = A_p/A$
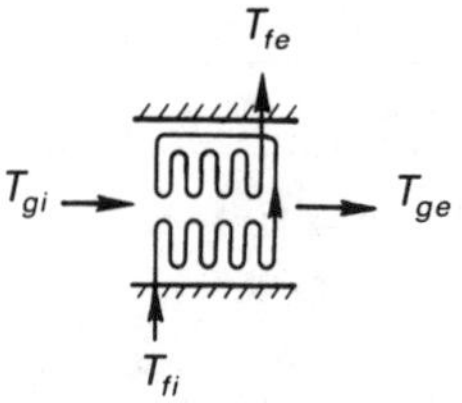 T_{fe} T_{gi} → T_{ge} T_{fi}	Fig. 8.33, curve 1; τ_L is the larger value between $(T_{gi} - T_{ge})$ and $(T_{fe} - T_{fi})$ while τ_s is the smaller value between them. P has the same meaning as that in Fig. 8.32
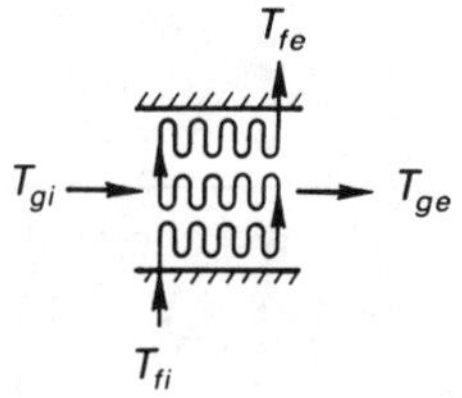 T_{fe} T_{gi} → T_{ge} T_{fi}	Fig. 8.33, curve 2

TABLE 8.11 *(Continued)*

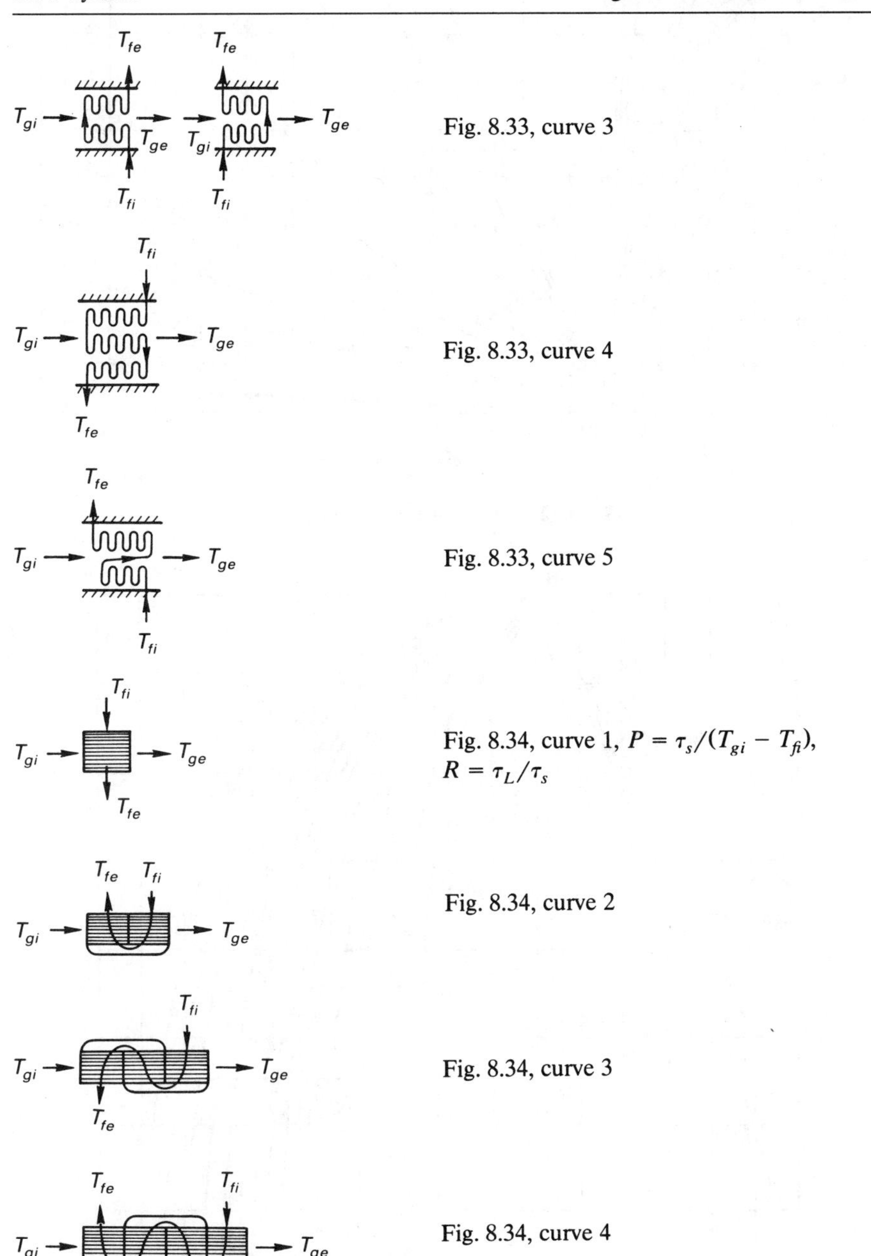

Flow System	Figure Used and Parameters
	Fig. 8.33, curve 3
	Fig. 8.33, curve 4
	Fig. 8.33, curve 5
	Fig. 8.34, curve 1, $P = \tau_s/(T_{gi} - T_{fi})$, $R = \tau_L/\tau_s$
	Fig. 8.34, curve 2
	Fig. 8.34, curve 3
	Fig. 8.34, curve 4

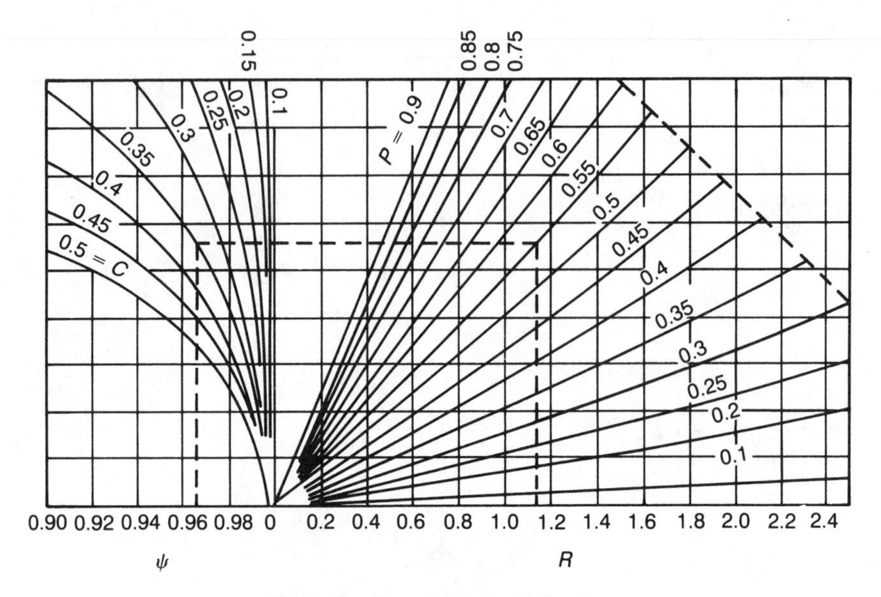

Fig. 8.32. Corrective coefficient ψ.

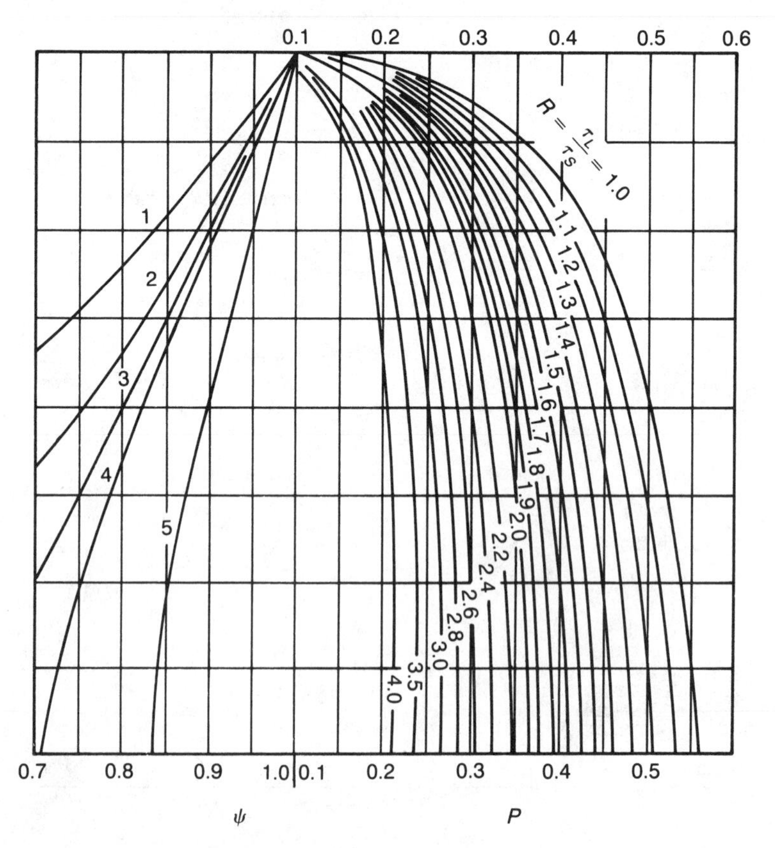

Fig. 8.33. Corrective coefficient ψ.

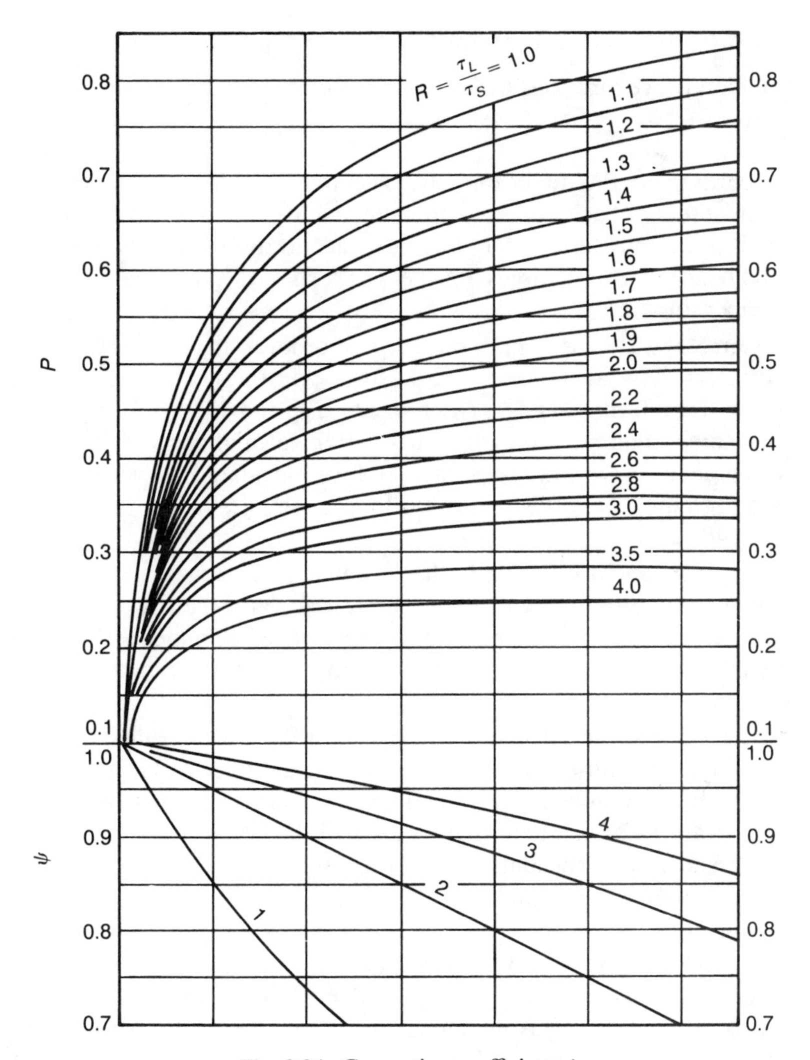

Fig. 8.34. Corrective coefficient ψ.

For the other kinds of flow systems listed in Table 8.11, ΔT can be determined as follows:

$$\Delta T = \psi \, \Delta T_c \tag{8.83}$$

where ψ is the corrective coefficient determined by Table 8.11 and Figs. 8.32 to 8.34. In Fig. 8.32, A_p is the parallel heating surface of the flow system, m^2, and A is the total heating surface of the flow system, m^2.

8.3.4 Procedure for Heat Transfer Calculation of a Boiler

The common procedure for heat transfer calculation of a boiler with burners is as follows:

1. According to the fuel contents, calculate the flue gas characteristic parameters of each boiler component, such as the actual volume of the flue gases and the air, the volume concentrations of $CO_2 + SO_2$ and H_2O, and the enthalpies of the gases and the air related to their temperature, etc.
2. Assume a waste gas temperature T_{wg} in order to determine h_2, η_b, and B; select a hot-air temperature T_{ha} to begin the heat transfer calculations of the furnace and its components. T_{wg} and T_{ha} will be checked by the corresponding values obtained after calculations are made. The discrepancy between the assumed and the calculated T_{wg} should be less than or equal to $\pm 10°C$, and that for T_{ha} should be less than or equal to $\pm 40°C$.
3. Select a q_F and a q_V value in the recommended range to determine the volume and other dimensions of the furnace and the convection ducts, and arrange the heating surface of every boiler component. Therefore, before the heat transfer calculation, the constructive parameters are known.
4. The heat transfer calculation begins from the furnace; during calculation, the trial-and-error method is used. In the furnace calculation, T_{gFe} has to be determined; for performing calculations we first assume this temperature and then check it by Eq. (8.37). If the discrepancy between the assumed and calculated T_{gFe} is less than or equal to $\pm 100°C$, then the calculation is complete.
5. In the heat transfer calculation of convection components, the trial-and-error method is also used. Usually, for a particular component the inlet-gas temperature (including enthalpy) and one of the working fluids are known. By assuming an exit-gas temperature, we may preset the heat quantity absorbed by this component, H_b, by employing the heat balance equation, Eq. (8.41), and then comparing the value obtained with the heat quantity, H_t, calculated from the heat transfer equation [Eq. (8.40)]. If the error $(H_b - H_t)100/H_b$ is less than or equal to $\pm 5\%$ for the slag screen, and less than or equal to $\pm 2\%$ for the other boiler components, then the heat transfer calculation of the individual convection component is complete.
6. After finishing the calculation of all the boiler components, the total heat quantity error ΔH should be checked as follows:

$$\Delta H = \frac{H_{av}^{w}\eta_b - (H_r + H_s + H_{ss} + H_R + H_E)\left(1 - \dfrac{h_4}{100}\right)}{H_{av}^{w}} 100\% \qquad (8.84)$$

where H_r, H_s, H_{ss}, H_R, and H_E are the heat absorbed in the furnace, slag screen, superheater, reheater, and economizer, respectively.

If $\Delta H \leq \pm 0.5\%$ and the discrepancies of T_{wg} and T_{ha} are allowable as mentioned previously, then the heat transfer calculation is complete. Otherwise, the calculation should begin again by assuming a new T_{wg} and a new T_{ha}.

The heat transfer calculations of a boiler can be substantially facilitated by using an electronic computer.

8.4 A NUMERICAL EXAMPLE OF THE HEAT TRANSFER CALCULATIONS OF BOILER COMPONENTS

To show the general procedure of the heat transfer calculation of a boiler, a middle-pressure boiler with a simplified arrangement of heating surfaces will serve as an example. The procedure and methods described are also applicable to the more complex arrangements of large modern boilers.

Main Parameters of the Example Unit The rated steam-generating capacity, $W = 130$ t/hr or 36.11 kg/s; the superheated steam pressure (absolute) at the exit of the superheater, $p_{ss} = 3.92$ MPa; the absolute drum pressure, $p_d = 4.41$ MPa; the superheated steam temperature, $T_{ss} = 450°\mathrm{C}$; the feed-water temperature, $T_{\mathrm{fw}} = 172°\mathrm{C}$; the feed-water inlet pressure, $p = 4.7$ MPa; the rate of water blow-off from the boiler drum, $W_{\mathrm{bw}} = 1.3$ t/hr (0.3611 kg/s); the waste gas temperature, $T_{\mathrm{wg}} = 160°\mathrm{C}$ (first selected and checked afterwards); the hot-air temperature, $T_{\mathrm{ha}} = 200°\mathrm{C}$; and the physical heat of oil before burning, $H_{\mathrm{ph}} = 266$ kJ/kg.

The contents of the fuel oil fired are as follows (moist basis, mass percentage) carbon $\mathrm{C}^w = 86.55$, hydrogen $\mathrm{H}^w = 12.68$, oxygen $\mathrm{O}^w = 0.03$, nitrogen $\mathrm{N}^w = 0.29$, sulfur $\mathrm{S}^w = 0.29$, moisture $W^w = 0$, and ash $A^w = 0.16$; the lower heating value of the fuel oil, $H_l^w = 41{,}242$ kJ/kg.

The example boiler consists of a furnace 1, slag screens 2, a superheater 3, an economizer 4, and an air heater 5 (Fig. 8.35), and the flue gases pass through these components, respectively.

Auxiliary Calculations The theoretical volume of air (at normal state) required for combustion of 1 kg of fuel oil is calculated as

$$V^0 = 0.0889(\mathrm{C}^w + 0.375\mathrm{S}^w) + 0.265\mathrm{H}^w - 0.0333\mathrm{O}^w = 11.063\ \mathrm{m}^3/\mathrm{kg}$$

The theoretical volume of RO_2 (at normal state) in the combustion products of 1 kg of fuel oil is

$$V_{\mathrm{RO}_2} = 1.866\frac{\mathrm{C}^w + 0.375\mathrm{S}^w}{100} = 1.617\ \mathrm{m}^3/\mathrm{kg}$$

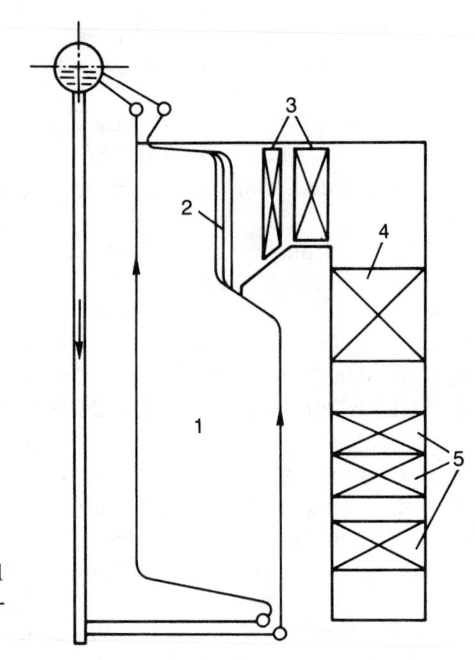

Fig. 8.35. Scheme of the example unit: 1 furnace, 2—slag screen, 3—superheater, 4—economizer, 5—air heater.

The theoretical volume of H_2O (at normal state) in the combustion products of 1 kg of fuel oil is

$$V_{H_2O}^0 = 0.111H^w + 0.0124W^w + 0.0161V^0 = 1.5855 \text{ m}^3/\text{kg}$$

The theoretical volume of N_2 (at normal state) in the combustion products of 1 kg of fuel oil is

$$V_{N_2}^0 = 0.79V^0 + 0.8\frac{N^w}{100} = 8.742 \text{ m}^3/\text{kg}$$

The actual volume of the combustion products of 1 kg of fuel oil in different boiler components are listed in Table 8.12. In this table, $\Delta\alpha$ is the increment of the air ratio, considering the leakage of air from the surroundings into the gas ducts.

The flue gas temperature and its corresponding enthalpy are listed in Table 8.13.

The available heat of fuel, H_{av}^w, Eq. (8.6):

$$H_{av}^w = 41{,}242 + 266 = 41{,}508 \text{ kJ/kg}$$

The relative heat loss, $h_3 = 0.5\%$. $h_4 = 0\%$ (from Section 8.3.1), h_2 can be

TABLE 8.12 Characteristics of Flue Gases in Different Boiler Components

Items and Equations	Unit	Furnace and Slag Screen	Super-heater	Economizer	Air Heater
Excess air ratio at exit of boiler component, α_e (selected)	—	1.1	1.16	1.18	1.21
Average excess air ratio, $\bar{\alpha}$	—	1.1	1.13	1.17	1.195
The air ratio increment, $\Delta\alpha$	—	0.05	0.06	0.02	0.03
Volume of excess air, $(\bar{\alpha} - 1)V^0$	m^3/kg	1.106	1.438	1.881	2.157
Actual volume of H_2O, $V_{H_2O} = V^0_{H_2O} + 0.0161(\bar{\alpha} - 1)V^0$	m^3/kg	1.603	1.609	1.616	1.620
Actual volume of flue gases, $V_g = V_{H_2O} + V^0_{N_2} + V_{RO_2} + (\bar{\alpha} - 1)V^0$	m^3/kg	13.068	13.406	13.856	14.136
$r_{RO_2} = V_{RO_2}/V_g$	—	0.1237	0.1206	0.1167	0.1144
$r_{H_2O} = V_{H_2O}/V_g$	—	0.1227	0.1200	0.1166	0.1146
$r/r_{RO_2} + r_{H_2O}$	—	0.2464	0.2406	0.2333	0.2290

obtained from Eq. (8.9):

$$h_2 = \frac{(3153 - 1.21 \times 440)}{41{,}508} 100 = 6.32\%$$

The relative heat loss, $h_5 = 0.7\%$ (Fig. 8.26), and h_6 for fuel oil can be neglected.

The boiler efficiency, $\eta_b = 100 - 6.31 - 0.5 - 0.7 = 92.48\%$. The weight of the fuel consumption, B, can be obtained from Eq. (8.13):

$$B = \frac{1}{41{,}508 \times 0.9248}[36.11(3332 - 730.2) + 0 + 0.3611(1116.3 - 730.2)]$$

$$= 2.451 \text{ kg}/s$$

The rated fuel consumption, $B_r = B$ [Eq. (8.14)]. The heat retention coefficient [Eq. (8.11)], $\phi = 1 - [0.7/(92.48 + 0.7)] = 0.9925$.

Construction Parameters of Boiler Components The furnace volume is obtained by selecting a heat release rate per unit furnace volume, q_V, from Table 8.4. For this example let $q_V = 263$ kW/m^3; therefore the furnace volume is calculated as $V_F = BH_1^w/q_V = 2.451 \times 41{,}242/263 = 384$ m^3.

Select $q_F = 3260$ kW/m^2 (Table 8.2); the furnace cross-sectional area becomes $A_F = BH_1^w/q_F = 2.451 \times 41{,}242/3260 = 31$ m^2. Assume the

TABLE 8.13 Flue Gas Temperature and Its Corresponding Enthalpy

Temperature °C	$V_{RO_2} = 1.617$, m^3/kg		$V^0_{NO_2} = 8.742$, m^3/kg		$V^0_{H_2O} = 1.5855$, m^3/kg	
	$C_{CO_2}T$, kJ/m^3	$V_{RO_2}C_{CO_2}T$, kJ/kg	$C_{N_2}T$, kJ/m^3	$V^0_{N_2}C_{N_2}T$, kJ/kg	$C_{H_2O}T$, kJ/m^3	$V^0_{H_2O}C_{H_2O}T$, kJ/kg
100	170.1	274.9	130	1,136.5	151	239.4
200	357.6	578.2	260	2,272.9	305	483.6
300	559.0	903.9	392	3,426.7	463	734.1
400	772.0	1,248.3	527	4,607.0	626	992.5
500	996.5	1,611.3	664	5,804.7	795	1,260.5
600	1,222.6	1,976.9	804	7,028.6	969	1,536.4
700	1,461.2	2,362.8	948	8,287.4	1,149	1,821.7
800	1,704.1	2,755.5	1,094	9,563.7	1,334	2,115.1
900	1,951.1	3,154.9	1,242	10,857.6	1,526	2,419.5
1,000	2,202.4	3,561.3	1,392	12,168.9	1,723	2,731.8
1,100	2,457.8	3,974.3	1,544	13,497.7	1,925	3,052.1
1,200	2,717.4	4,394.0	1,697	14,835.2	2,132	3,380.3
1,800	4,304.2	6,959.9	2,644	23,113.9	3,458	5,482.6
1,900	4,572.2	7,393.3	2,804	24,512.6	3,690	5,850.5
2,000	4,844.4	7,833.4	2.965	25,920.0	3,926	6,224.7
2,100	5,116.5	8,273.4	3,128	27,345.0	4,162	6,598.9

Temperature °C	$I_g^0 = V_{RO_2}C_{CO_2}T + V_{N_2}^0 C_{N_2}T + V_{H_2O}^0 C_{H_2O}T$, kJ/kg	$V^0 = 11.063$, m^3/kg		$I_g = I_g^0 + (\alpha - 1)I_a^0$			
		C_aT, kJ/m^3	$I_a^0 = V^0C_aT$, kJ/kg	$\alpha_{Fe} = 1.1$	$\alpha_{Se} = 1/16$	$\alpha_{Fe} = 1.18$	$\alpha_{Ae} = 1.21$
100	1,650.8	132	1,463.7				1,955
200	3,334.7	266	2,945.9			3,865	3,953
300	5,064.7	403	4,458.4			5,866	6,000
400	6,847.8	542	599.62			7,926	8,106
500	8,676.5	684	7,567.1	9,433	9,886	10,040	
600	10,541.9	830	9,182.3	11,456	12,008	12,193	
700	12,471.9	978	10,819.6	13,537	14,190		
800	14,434.3	1,129	12,490.1	15,676	16,426		
900	16,432.0	1,282	14,182.8	17,862			
1,000	18,462.0	1,435	15,875.4	20,073			
1,100	20,524.1	1,595	17,645.5	22,300			
1,200	22,609.5	1,753	19,393.4	24,540			
1,800	35,556.4	2,732	30,224.1	38,558			
1,900	37,756.4	2,899	32,071.6	40,970			
2,000	39,978.1	3,066	33,919.2	43,365			
2,100	42,217.3	3,232	35,755.6	45,789			

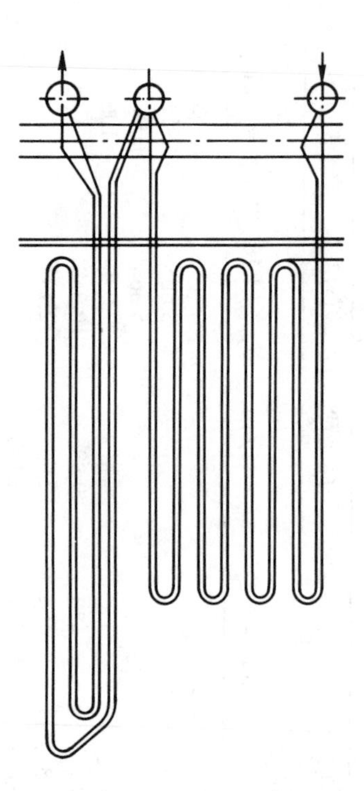

Fig. 8.36. The scheme of the superheater.

furnace width to be $a = 6.2$ m, then the furnace depth, $b = A_F/a = 31/6.2 = 5$ m.

The outside diameter of the water wall tubes, $d = 60$ mm with a tube wall thickness of 3 mm; the distance from the centerline of the riser to the furnace wall, $e = 0$; and the spacing of the water wall tubes, $s = 64$ mm. The total number of risers, $n = 348$; and the total surface area of the furnace walls, $A_w = 362.84\ m^2$ (including the bottom surface area of the furnace 31.5 m^2 which is covered with refractory brick, and the exit surface area of the furnace 29.08 m^2).

The staggered arranged slag screens are formed by the rear water wall tubes. Three rows of 60 mm outside diameter tubes are spaced on 250 mm centers; 24 tubes per row are spaced on 256 mm centers; the total heating surface, $A = 73.52\ m^2$; and the flow area for flue gases, $A_g = 21.73\ m^2$. $S_1/d = 256/60 = 4.27$, $S_2/d = 250/60 = 4.17$, $x_s = 0.685$ [Eq. (8.49) and Fig. 8.27].

The superheater (Fig. 8.36) consists of 24 rows of tubes with an outside diameter, $d = 38$ mm (3.5 mm thickness). The tubes are arranged in the in-line form. $S_1 = S_2 = 84$ mm; the total number of serpentine tubes, $n = 146$; and each row consists of 73 tubes. The total heating surface, $A = 641\ m^2$, and the parallel-flow heating surface area, $A_p = 68\ m^2$. The flow area for flue gases, $A_g = 13.13\ m^2$; and the flow area for steam, $A_{ss} = 0.11\ m^2$. The depth of the empty room before the superheater, $L_R = 0.65$ m, while the depth of the superheater tube bundle, $L_b = 1.725$ m.

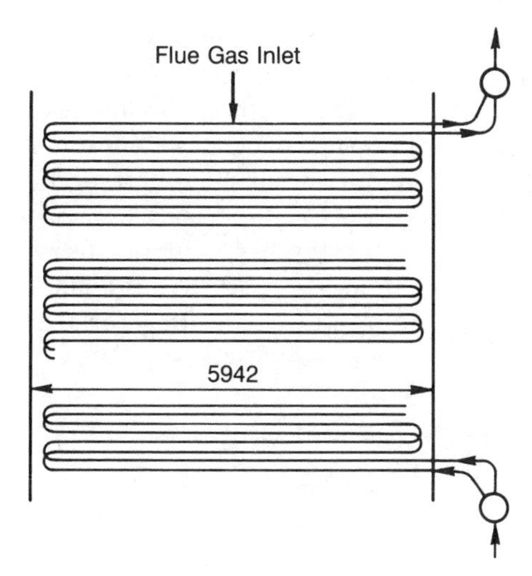

Fig. 8.37. Scheme of the economizer.

The economizer (Fig. 8.37) consists of 56 rows of 32 mm outside diameter tubes (3 mm thick). Tubes are arranged in the staggered form, with $S_1/d = 45/32 = 1.4$, $S_2/d = 75/32 = 2.34$, and $S_2'/d = 2.73$. The total number of serpentine tubes, $n = 73$, and 36 or 37 tubes per row are placed horizontally. The total heating surface area, $A = 1215$ m^2; the flow area of flue gases, $A_g = 10.206$ m^2; and the flow area for water, $A_w = 0.0387$ m^2. The height of the empty room before the economizer, $L_R = 3.8$ m; and the height of the economizer tube bundle, $L_B = 2.43$ m. The depth of the flue gas duct, $b = 2.86$ m; and the width of the flue gas duct, $a = 5.942$ m.

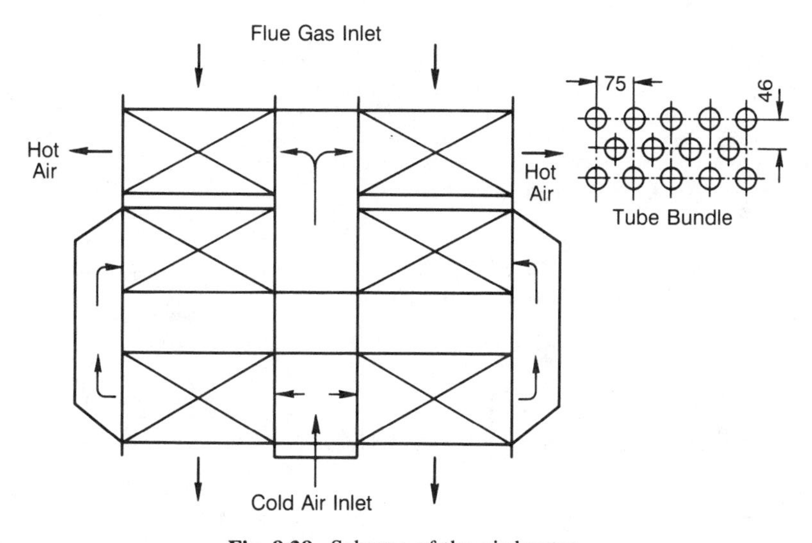

Fig. 8.38. Scheme of the air heater.

The tubular air heater is arranged horizontally, and the air passes through the tubes while flue gases pass around the outside of the tubes. The flow systems of the air and gases is shown in Fig. 8.38. The width of the flue gas duct, $a = 5.942$ m; and the depth of the flue gas duct, $b = 3.173$ m. Ninety-six rows of 40 mm outside diameter tubes (1.5 mm thick) are arranged in staggered form with $S_1/d = 75/40 = 1.88$, $S_2/d = 46/40 = 1.15$, and $S_2'/d = 88/40 = 2.2$. Each row consists of 40 or 41 tubes, and the total number of tubes, $n = 7776$. The flow area of the air, $A_a = 2.785$ m^2, while the flow area of the flue gases, $A_g = 7.665$ m^2. The total heating surface area, $A = 2382$ m^2.

Heat Transfer Calculation of the Furnace The heat introduced into the furnace by the hot air and the cold air is [Eq. (8.35)]:

$$H_a = (1.1 - 0.05)2945.9 + 0.05 \times 440 = 3115 \text{ kJ/kg}$$

where the enthalpies of air are obtained from Table 8.13, and α_{Fe} and $\Delta\alpha_F$ are obtained from Table 8.12.

The useful heat released in the furnace [Eq. (8.34)]:

$$H_u = 41{,}508\frac{100 - 0.5}{100} + 3115 = 44{,}416 \text{ kJ/kg}$$

The adiabatic temperature of combustion, T_a, according to H_u and $\alpha_{Fe} = 1.1$, can be found from Table 8.13; $T_a = 2043°\text{C}$.

Assume the flue gas temperature at the furnace outlet is $T_{gFe} = 1100°\text{C}$, its corresponding enthalpy $I_{Fe} = 22{,}300$ kJ/kg (obtained from Table 8.13); T_{gFe} will be checked afterwards.

The effective thickness of the radiating layer:

$$S = \frac{3.6V_F}{A_w} = \frac{3.6 \times 384}{362.84} = 3.81 \text{ m}$$

The effective coefficients of absorption by triatomic gases, k_g, and by soot particles, k_s [Eqs. (8.23) and (8.29)]:

$$k_g = 10\left[\frac{0.78 + 1.6 \times 0.1227}{(10 \times 0.1 \times 3.81 \times 0.2644)^{0.5}} - 0.1\right]\left(1 - 0.37 \times \frac{1373}{1000}\right)$$

$$= 4.444 \text{ } 1/(\text{m} \cdot \text{MPa})$$

$$k_s = 0.3(2 - 1.1)\left(1.6 \times \frac{1373}{1000} - 0.5\right)\frac{86 \times 55}{12.68}$$

$$= 3.127 \text{ } 1/(\text{m} \cdot \text{MPa})$$

The flame emissivity, a_{fl} [Eqs. (8.26), (8.27), and (8.30)]:

$$a_{fl} = 0.55[1 - e^{-(4.444\times 0.2464+3.127)\times 0.1\times 3.81}] + (1 - 0.55) \times [1 - e^{-4.444\times 0.2464\times 0.1\times 3.81}] = 0.5957$$

The bottom surface area of the furnace (31.5 m^2) is covered with refractory brick; its angular coefficient, $x = 1.0$ and its coefficient of fouling, $\zeta = 0.1$; for the exit surface area of the furnace (29.08 m^2), $x = 1.0$ and $\zeta = 0.55$; for the other water walls (302.26 m^2), $\zeta = 0.55$ (Table 8.10) and $x = 0.99$ (Fig. 8.25): therefore the average value of the coefficient ψ_{ef} is equal to [Eq. (8.18)]:

$$\psi_{ef} = \frac{0.99 \times 302.26 \times 0.55 + (31.5 \times 0.1 + 29.08 \times 0.55) \times 1.0}{362.84} = 0.5063$$

The coefficient of thermal radiation of the furnace, a_F, is calculated from Eq. (8.19) as

$$a_F = \frac{1}{1 + \left(\dfrac{1}{0.5957} - 1\right)^{0.5063}} = 0.7443$$

The average heat capacity of the gases in the temperature interval of T_a–T_{gFe} can be obtained from Eq. (8.36):

$$\overline{VC} = \frac{H_u - I_{Fe}}{T_a - T_{gFe}} = \frac{44{,}416 - 22{,}300}{2043 - 1100} = 23.45 \text{ kJ/(kg} \cdot {}^\circ\text{C)}$$

Burners are arranged horizontally, and the average relative level, $X_b = 0.239$. The coefficient M, can be obtained [from Eq. (8.31)]:

$$M = 0.54 - 0.2 \times 0.239 = 0.4922$$

The calculated flue gas temperature at the furnace outlet [Eq. (8.37)]:

$$T_{gFe} = \frac{2316}{0.4922\left(\dfrac{5.67 \times 10^{-11} \times 0.5063 \times 362.84 \times 0.7443 \times 2316^3}{0.9925 \times 2.451 \times 23.45}\right)^{0.6} + 1} - 273$$

$$= 1110^\circ\text{C}$$

The discrepancy between the calculated T_{gFe} and the assumed T_{gFe} is 10°C; it is smaller than the allowable discrepancy of ±100°C. Therefore, we consider T_{gFe} to be equal to 1110°C, and need not calculate it again; the corresponding flue gas enthalpy can be found from Table 8.13 as $I_{Fe} = 22{,}524$ kJ/kg.

The quantity of heat transferred in the furnace [Eq. (8.36)]:

$$H_r = 0.9925(44{,}416 - 22{,}524) = 21{,}728 \text{ kJ/kg}$$

The average heat flux of the furnace heating surfaces [Eq. (8.38)]:

$$q_{rF} = \frac{2.451 \times 21{,}728}{359.82} = 148 \text{ kW/m}^2$$

The radiant heat absorbed by the slag screens from the furnace [Eq. (8.49)]:

$$H_{rF} = \frac{0.685 \times 0.72 \times 148 \times 29.08}{2.451} = 866 \text{ kJ/kg}$$

In the preceding equation, the ratio of the burner axis height to the furnace outlet center height is equal to 0.8; using this value and Fig. 8.29, we may obtain $\eta_h = 0.72$.

The radiant heat absorbed by the superheater

$$H'_{rF} = \frac{(1 - x_s)\eta_h q_{rF} A_{Fe}}{B_r} = \frac{(1 - 0.685)0.72 \times 148 \times 29.08}{2.451} = 398.3 \text{ kJ/kg}$$

The total radiant heat absorbed by the water walls

$$H_{\text{ww}} = H_r - (H_{rF} + H'_{rF}) = 21{,}728 - (866 + 398.3) = 20{,}463.7 \text{ kJ/kg}$$

Heat Transfer Calculations of Slag Screens The inlet flue gas temperature is $T_{gSi} = T_{gFe} = 1110$°C and $I_{gSi} = 22{,}524$ kJ/kg. Assume that the exit flue gas temperature, $T_{gSe} = 1040$°C; its corresponding enthalpy is $I_{gSe} = 20{,}964$ kJ/kg. In the slag screen duct $\Delta\alpha = 0$.

According to the heat balance equation [Eq. (8.41)]:

$$H_b = 0.9925(22{,}524 - 20{,}964) = 1560 \text{ kJ/kg}$$

The average flue gas velocity, V:

$$V = \frac{B_r V_g}{A_g}\left(1 + \frac{T}{273}\right) = \frac{2.451 \times 13.068}{21.73}\left(1 + \frac{1075}{273}\right) = 7.28 \text{ m/s}$$

The coefficient $\phi = (4.27 - 1)/(5.863 - 1) = 0.659$; since $0.1 < \phi \leq 1.7$, this value can be used with Eq. (8.61) to determine c_s:

$$c_s = 0.34 \times 0.659^{0.1} = 0.326$$

Because $S_1/d = 4.27 \geq 3.0$, and $n < 10$, therefore Eq. (8.65) is used to calculate c_n:

$$c_n = 4.3^{0.02} - 3.2 = 0.889$$

The calculation of the convective heat transfer coefficient, h_c [Eq. (8.67)]:

$$h_c = 16.98 \times 10^{-3} \times 0.326 \times 0.889 \times \frac{7.28^{0.6}}{0.06^{0.4}} = 0.0499\ \text{kW}/(\text{m}^2 \cdot \text{K})$$

The effective thickness of the radiating layer, S [Eq. (8.78)]:

$$S = 0.9 \times 0.06\left(\frac{4 \times 0.256 \times 0.250}{\pi \times 0.06^2} - 1\right) = 1.17\ \text{m}$$

The effective coefficient of absorption, k [Eq. (8.75)], as $\mu_a = 0$,

$$k = k_g r = 10\left[\frac{0.78 + 1.6 \times 0.1227}{(10 \times 0.1 \times 1.17 \times 0.2464)^{0.5}} - 0.1\right]$$

$$\times\left(1 - 0.37\frac{1040 + 273}{1000}\right)0.2464$$

$$= 2.124\ 1/(\text{m} \cdot \text{MPa})$$

The emissivity of the gases [Eq. (8.74)]:

$$a_g = 1 - e^{-2.124 \times 0.1 \times 1.17} = 0.22$$

The absolute temperature of the ashy tube wall, since the working fluid is steam–water mixture at 4.41 MPa and its saturated temperature is 256.2°C, can be found using Eq. (8.77), $T_{aw} = 256.2 + 80 + 273 = 609.2$ K.

The radiant heat transfer coefficient of the space, h_r, can be determined by Eq. (8.73):

$$h_r = 5.1 \times 10^{-11} \times 0.22 \times 1348^3\left[\frac{1 - \left(\frac{609.2}{1348}\right)^{3.6}}{1 - \left(\frac{609.2}{1348}\right)}\right]$$

$$= 0.04729\ \text{kW}/(\text{m}^2 \cdot \text{K})$$

The heat transfer coefficient from the gas to the tube wall, h_o [Eq. (8.56)]:

$$h_o = 1.0(49.9 + 47.29) = 97.19\ \text{W}/(\text{m}^2 \cdot \text{K})$$

Since the heat transfer coefficient from the wall to the working fluid, h_i, is very large, $1/h_i$ can be neglected; the overall heat transfer coefficient, U, can be obtained from Eq. (8.51), where ψ is selected as $\psi = 0.63$, $U = \psi h_o = 0.63 \times 97.19 = 61.23\ \text{W}/(\text{m}^2 \cdot °\text{C})$.

The mean temperature difference, ΔT [Eq. (8.81) or Eq. (8.82)]:

$$\Delta T = \frac{(1110 - 256.2) - (1040 - 256.2)}{\ln\left(\dfrac{1110 - 256.2}{1040 - 256.2}\right)} = 818.3°\text{C}$$

The quantity of heat transfer calculated by using the heat transfer equation [Eq. (8.40)]:

$$H_t = \frac{61.23 \times 818.3 \times 73.52}{2.451 \times 10^3} = 1502.93\ \text{kJ/kg}$$

Since

$$\frac{H_b - H_t}{H_b} = \frac{1560 - 1502.93}{1560} \times 100 = 3.65\% < \pm 5\%$$

this calculation is acceptable, and the total convective heat absorbed by the working fluid in the slag screen is $H_s = 1560$ kJ/kg.

Heat Transfer Calculations of the Superheater From the preceding calculations and the given data, we know the flue gas parameters at the inlet are $T_{gSSi} = 1040°\text{C}$ and $I_{gSSi} = 20{,}964$ kJ/kg; the working fluid parameters at the inlet are $T_{SSi} = 256.23°\text{C}$ and $I_{SSi} = 2798.6$ kJ/kg (saturated temperature and enthalpy at 4.41 MPa); and the working fluid parameters at the outlet are $T_{SSe} = 450°\text{C}$ and $I_{SSe} = 3322$ kJ/kg.

The quantity of convective heat transfer that must be absorbed by the steam to satisfy the heat balance equation on the steam side, Eq. (8.42a):

$$H_b = \frac{36.11}{2.451}(3322 - 2798.6) - 398.3 = 7460.3\ \text{kJ/kg}$$

The flue gas parameters at the outlet of the superheater can be obtained from Eq. (8.41):

$$7460.3 = 0.9925(20{,}964 - I_{gSSe} + 0.06 \times 440)\ \text{kJ/kg}$$

Therefore $I_{gSSe} = 13{,}473.7$ kJ/kg and its corresponding temperature is $T_{gSSe} = 697°\text{C}$ (from Table 8.13).

The mean temperature difference for counterflow [Eq. (8.82)]:

$$\Delta T_c = \frac{(1040 - 450) - (697 - 256.23)}{\ln\left(\dfrac{1040 - 450}{697 - 256.23}\right)} = 511.94°\text{C}$$

Since $\tau_1 = 1040 - 697 = 343°\text{C}$, $\tau_2 = 450 - 256.23 = 193.77°\text{C}$, $P = 193.77/(1040 - 256.23) = 0.2472$, $R = 343/193.77 = 1.77$, and $A = 68/641 = 0.106$, from Fig. 8.32 we may obtain $\psi = 0.998$. Therefore the actual mean temperature difference [Eq. (8.83)] $\Delta T = \psi\,\Delta T_c = 0.998 \times 511.94 = 509.4°\text{C}$. The average specific volume of steam, $\overline{V} = 0.06469\ \text{m}^3/\text{kg}$ (for $p = 4.16$ MPa and $T_{SS} = 353°\text{C}$). The average steam velocity in the superheater:

$$V = \frac{36.11 \times 0.06469}{0.11} = 21.24\ \text{m/s}$$

The heat transfer coefficient from the wall to the steam, h_i [Eq. (8.70)]:

$$h_i = 6.61 \times 10^{-3}\,\frac{(21.24 \times 15.458)^{0.8}}{0.031^{0.2}} = 1.365\ \text{kW}/(\text{m}^2 \cdot °\text{C})$$

The average gas temperature, $\overline{T}_g = (1040 + 697)/2 = 868.5°\text{C}$. The average flue gas velocity:

$$V = \frac{2.451 \times 13.406}{13.13}\left(1 + \frac{868.5}{273}\right) = 10.46\ \text{m/s}$$

Because $n > 10$ and $S_2/d = 2.21 > 2$, $c_n = 1.0$ and $c_s = 0.2$ [Eqs. (8.58) and (8.59)].

The convective heat transfer coefficient, h_c [Eq. (8.66)]:

$$h_c = 28.96(1 - 1.25 \times 10^{-4} \times 868.5) \times 10^{-3} \times 0.2 \times 1.0 \times \frac{10.46^{0.65}}{0.038^{0.35}}$$

$$= 0.07467\ \text{kW}/(\text{m}^2 \cdot °\text{C})$$

The absolute temperature of the ashy tube wall [Eq. (8.76)]:

$$T_{\text{aw}} = \frac{450 + 256.23}{2} + \left(2.6 + \frac{1}{1.365}\right)\frac{2.451(7460.3 + 398.3)}{641} + 273$$

$$= 726.3\ \text{K}$$

The effective thickness of the radiating layer, S [Eq. (8.78)]:

$$S = 0.9 \times 0.038\left(\frac{4 \times 0.084 \times 0.084}{\pi \times 0.038^2} - 1\right) = 0.1787 \text{ m}$$

The effective coefficient of absorption, k [Eq. (8.75)]:

$$k = k_g r = 10\left[\frac{0.78 + 1.6 \times 0.12}{(10 \times 0.1 \times 0.1787 \times 0.2406)^{0.5}} - 0.1\right]$$

$$\times\left(1 - 0.37\frac{697 + 273}{1000}\right)0.2406$$

$$= 7.09 \text{ 1/(m} \cdot \text{MPa)}$$

The emissivity of the gases [Eq. (8.74)]:

$$a_g = 1 - e^{-7.09 \times 0.1 \times 0.1787} = 0.119$$

The radiant heat transfer coefficient of the space, h_r [Eq. (8.73)]:

$$h_r = 5.1 \times 10^{-11} \times 0.119 \times 1141 \times 5^3\left[\frac{1 - \left(\frac{726.3}{1141.5}\right)^{3.6}}{1 - \left(\frac{726.3}{1141.5}\right)}\right]$$

$$= 0.01986 \text{ kW/(m}^2 \cdot \text{K)}$$

Considering the empty room before the superheater, the corrected radiant heat transfer coefficient will be [Eq. (8.80)]:

$$h_r' = 0.01986\left[1 + 0.3\left(\frac{1040 + 273}{1000}\right)^{0.25}\left(\frac{0.65}{1.725}\right)^{0.07}\right] = 0.0258 \text{ kW/(m}^2 \cdot \text{K)}$$

The heat transfer coefficient from the gas to the tube wall, h_o [Eq. (8.56)]:

$$h_o = 1.0(74.67 + 25.80) = 100.47 \text{ W/(m}^2 \cdot \text{K)}$$

The overall heat transfer coefficient, U [Eq. (8.51)]:

$$U = \frac{0.6}{\dfrac{1}{100.47} + \dfrac{1}{1365}} = 56.18 \text{ W/(m}^2 \cdot \text{K)} = 0.05618 \text{ kW/(m}^2 \cdot \text{K)}$$

The quantity of heat transfer calculated from the heat transfer equation [Eq. (8.40)]:

$$H_t = \frac{0.05618 \times 509.4 \times 641}{2.451} = 7484.3 \text{ kJ/kg}$$

Since

$$\frac{H_b - H_t}{H_b} = \frac{7460.3 - 7484.3}{7460.3} 100 = -0.32\% < \pm 2\%$$

this calculation is acceptable, and the total convective heat absorbed in the superheater, $H_{SS} = 7460.3$ kJ/kg.

Heat Transfer Calculations of the Economizer From the calculations for the superheater and the given data, we know that the flue gas parameters at the inlet are $T_{gEi} = 697$°C and $I_{gEi} = 13{,}473.7$ kJ/kg; the working fluid parameters at the inlet are $T_{Ei} = T_{\text{fw}} = 172$°C and $I_{\text{fw}} = 730.2$ kJ/kg.

The total quantity of heat that must be absorbed by the working fluid, H_1 [Eq. (8.5)]:

$$H_1 = \frac{1}{2.451}[36.11(3332 - 730.2) + 3.611(1116.3 - 730.2)]$$

$$= 38{,}389 \text{ kJ/kg}$$

The quantity of heat that must be absorbed by the working fluid in the economizer, H_E, can be obtained from the following heat balance equation:

$$H_1 = H_r + H_S + H_{SS} + H_E$$

or

$$H_E = H_1 - H_r - H_S + H_{SS} = 38{,}389 - 21{,}782 - 1560 - 7460.3$$

$$= 7586.7 \text{ kJ/kg}$$

The flue gas enthalpy, I_{gEe}, and temperature, T_{gEe}, can be determined from Eq. (8.41):

$$7586.7 = 0.9925(13{,}473.7 - I_{gEe} + 0.02 \times 440) \text{ kJ/kg}$$

Therefore $I_{gEe} = 5838.47$ kJ/kg and $T_{gEe} = 298$°C (from Table 8.13 with I_{gEe} and $\alpha_{Ee} = 1.18$).

The enthalpy of the working fluid at the exit of the economizer

$$i_{Ee} = i_{\text{fw}} + \frac{H_E B_r}{(W_{\text{sh}} + W_{\text{bw}})} = 730.2 + \frac{7586 \times 2.451}{(36.11 + 3.611)} = 1240 \text{ kJ/kg}$$

The exit pressure is 4.41 MPa; at this pressure, the enthalpy of saturated water is $i_{sw} = 1116.6$ kJ/kg and the latent heat of evaporation $i_{lg} = 1681.9$ kJ/kg; therefore the steam quality at the exit is

$$x = \frac{i_{Ee} - i_{sw}}{i_{lg}} 100 = \frac{1240 - 1116.6}{1681.9} 100 = 7.33\%$$

The exit temperature of the working fluid is equal to the saturated temperature, $T_{Ee} = 256.23$°C.

The average velocity of water

$$V = \frac{(36.11 + 3.611) \times 0.00117}{0.0387} = 1.1 \text{ m/s}$$

The flow system is counterflow; therefore the mean temperature difference [Eq. (8.82)] is

$$\Delta T_c = \frac{(697 - 256.23) - (298 - 172)}{\ln\left(\dfrac{697 - 256.23}{298 - 172}\right)} = 251.3\text{°C}$$

The average gas temperature,

$$\bar{T}_g = (697 + 298)/2 = 497.5\text{°C}$$

The average flue gas velocity

$$V = \frac{2.451 \times 13.856}{10.206}\left(1 + \frac{497.5}{273}\right) = 9.93 \text{ m/s}$$

Because $n > 10$, from Eq. (8.65), $c_n = 1.0$; $\phi = (2.34 - 1)/(2.73 - 1) = 0.775$, from Eq. (8.61), $c_s = 0.34 \times \phi^{0.1} = 0.34 \times 0.775^{0.1} = 0.33$.

The absolute temperature of the ashy tube wall, $T_{aw} = (172 + 265.23)/2 + 60 + 273 = 547.2$ K.

The convective heat transfer coefficient, h_c [Eq. (8.67)]:

$$h_c = 16.98 \times 10^{-3} \times 0.33 \times 1.0 \times \frac{9.39^{0.6}}{0.032^{0.4}} = 0.085 \text{ kW/(m}^2 \cdot \text{°C)}$$

The effective thickness of the radiating layer, S [Eq. (8.78)]:

$$S = 0.9 \times 0.032\left(\frac{4 \times 0.045 \times 0.075}{\pi \times 0.032^2} - 1\right) = 0.0921 \text{ m}$$

The effective coefficient of absorption, k [Eq. (8.75)]:

$$k = 10\left[\frac{0.78 + 1.6 \times 0.1166}{(10 \times 0.1 \times 0.092 \times 0.2333)^{0.5}} - 0.1\right]\left(1 - 0.37\frac{298 + 273}{1000}\right)0.2333$$

$$= 11.94\ 1/(\text{m} \cdot \text{MPa})$$

The emissivity of the gases [Eq. (8.74)]:

$$a_g = 1 - e^{-11.94 \times 0.1 \times 0.0921} = 0.104$$

The radiant heat transfer coefficient of the space, h_r [Eq. (8.73)]:

$$h_r = 5.1 \times 10^{-11} \times 0.104 \times 770.5^3\left[\frac{1 - \left(\dfrac{547.2}{770.5}\right)^{3.6}}{1 - \left(\dfrac{547.2}{770.5}\right)}\right]$$

$$= 0.0059\ \text{kW}/(\text{m}^2 \cdot \text{K})$$

Considering the influence of the empty room before the economizer, the corrected radiant heat transfer coefficient [Eq. (8.80)]:

$$h_r' = 0.0059\left[1 + 0.3\left(\frac{697 + 273}{1000}\right)^{0.25}\left(\frac{3.8}{2.43}\right)^{0.07}\right] = 0.0077\ \text{kW}/(\text{m}^2 \cdot \text{K})$$

The heat transfer coefficient from the gas to the tube wall, h_o [Eq. (8.56)]:

$$h_o = 1.0(85 + 7.7) = 92.7\ \text{W}/(\text{m}^2 \cdot \text{K})$$

Because the heat transfer coefficient from the wall to the working fluid, h_i, is very large, $1/h_i$ can be neglected in Eq. (8.51), and the overall heat transfer coefficient, $U = \psi h_o = 0.65 \times 92.7 = 60.25\ \text{W}/(\text{m} \cdot \text{K}) = 0.06025\ \text{kW}/\text{m}^2 \cdot \text{K})$.

The quantity of the heat transfer calculated from the heat transfer equation [Eq. (8.40)]:

$$H_t = \frac{0.06025 \times 251.3 \times 1215}{2.451} = 7505.6\ \text{kJ}/\text{kg}$$

Since

$$\frac{H_b - H_t}{H_b} = \frac{7686.7 - 7505.6}{7586.7}100 = 1.07\% < \pm 2\%$$

this calculation is acceptable and the total heat absorption by the economizer, $H_E = 7586.7$ kJ/kg (obtained from the heat balance equation).

Heat Transfer Calculations of the Air Heater From the preceding calculations and given data, we know the flue gas parameters at the inlet of the air heater, $T_{gAi} = 298°C$ and $I_{Ai} = 5838.47$ kJ/kg; the inlet cold-air temperature, $T_{ca} = 30°C$ and the exit hot-air temperature, $T_{ha} = 200°C$; their enthalpies are $I_{ca} = 440$ kJ/kg and $I_{ha} = 2945.9$ kJ/kg, respectively.

According to the heat balance equation at the air side, Eq. (8.43), the quantity of heat transfer needed to be absorbed by the air in the air heater is

$$H_b = (1.1 - 0.05 + 0.5 \times 0.03)(2945.9 - 440) = 2668.8 \text{ kJ/kg}$$

The flue gas enthalpy at the exit of the air heater I_{Ae} can be obtained from Eq. (8.41), where the enthalpy of the air leaked into the flue gases is equal to $(I_{ha} + I_{ca})/2$:

$$H_b = 0.9925\left[5838.47 - I_{Ae} + 0.03\left(\frac{2945.9 + 440}{2}\right)\right] = 3200.3 \text{ kJ/kg}$$

The corresponding gas temperature, $T_{gAe} = 163°C$ (Table 8.13). The average temperature of the flue gases, $\overline{T} = (298 + 163)/2 = 230.5°C$. The average flue gas velocity

$$V = \frac{2.451 \times 14.136}{7.665}\left(1 + \frac{230.5}{273}\right) = 8.34 \text{ m/s}$$

Because $n > 10$, $c_n = 1.0$; since $\phi = (1.88 - 1)/(2.2 - 1) = 0.77$, $c_s = 0.34\phi^{0.1} = 0.33$.

The convective heat transfer coefficient, h_c [Eq. (8.67)]:

$$h_c = 16.98 \times 10^{-3} \times 0.33 \times 1.0 \times \frac{8.34^{0.6}}{0.040^{0.4}} = 0.07249 \text{ kW/(m}^2 \cdot \text{K)}$$

The average air temperature, $\overline{T} = (30 + 200)/2 = 115°C$. The average air velocity

$$V = \left(\alpha_{Fe} - \Delta\alpha_F + \frac{\Delta\alpha_{ah}}{2}\right)\frac{B_r V^0}{A_0}\left(\frac{\overline{T} + 273}{273}\right)$$

$$= \left(1.1 - 0.05 + \frac{0.03}{2}\right)\frac{2.451 \times 11 \times 0.63}{2.785}\left(\frac{115 + 273}{273}\right) = 14.75 \text{ m/s}$$

The heat transfer coefficient from the tube wall to the air, h_i [Eq. (8.69)]:

$$h_i = 3.49(1 - 8.26 \times 10^{-4} \times 115) \times 10^{-3} \frac{14.75^{0.8}}{0.037^{0.2}} = 0.0533 \text{ kW/(m}^2 \cdot \text{K)}$$

The overall transfer coefficient, U [Eq. (8.53)], for $\xi = 0.8$:

$$U = \frac{0.8}{\dfrac{1}{72.49} + \dfrac{1}{53.3}} = 24.57 \text{ W/(m}^2 \cdot \text{K)} = 0.02457 \text{ kW/(m}^2 \cdot \text{K)}$$

According to the flow system of the air heater, from Table 8.11, the following parameters can be obtained: $\tau_1 = 200 - 30 = 170°\text{C}$, $\tau_2 = 298 - 163 = 135°\text{C}$, $P = 136/(298 - 30) = 0.5132$, and $R = 170/135 = 1.26$; from Fig. 8.34, curve 3, we obtain $\psi = 0.972$.

The mean temperature difference [Eq. (8.83)]:

$$\Delta T = 0.972 \frac{(163 - 30) - (298 - 200)}{\ln\left(\dfrac{163 - 30}{298 - 200}\right)} = 110.3°\text{C}$$

The quantity of heat transfer calculated from the heat transfer equation [Eq. (8.40)]:

$$H_t = \frac{0.02457 \times 110.3 \times 2382}{2.451 \times 10^3} = 2634 \text{ kJ/kg}$$

Since

$$\frac{H_b - H_t}{H_b} = \frac{2668.8 - 2634}{2668.8} 100 = 1.3\% < \pm 2\%$$

this calculation is acceptable, and the total heat absorbed by the air heater, $H_A = 2668.8$ kJ/kg. The error of the total heat balance calculation

$$\Delta H = \frac{H_{av}^w \eta_b - (H_r + H_s + H_{ss} + H_E)\left(1 - \dfrac{h_4}{100}\right)}{H_{av}^w} 100$$

$$= \frac{41{,}508 \times 0.9248 - (21{,}728 + 1560 + 7460.3 + 7586.7)\left(1 - \dfrac{0}{100}\right)}{41{,}508} 100$$

$$= 0.124\% < \pm 0.5\%$$

TABLE 8.14 Results of Heat Transfer Calculation of the Example Unit

Number	Nomenclature	Furnace	Slag Screen	Superheater	Economizer	Air Heater
1	Heating surface, m^2	359.82	73.52	641	1,215	2,382
2	Flue gas temperature at inlet, °C	2,043	1,110	1,040	697	298
3	Flue gas temperature at exit, °C	1,110	1,040	697	298	163
4	Working fluid temperature at inlet, °C	256.23	256.23	256.23	172	30
5	Working fluid temperature at exit, °C	256.23	256.23	450	256.23	200
6	The average flue gas velocity, m/s	—	7.28	10.46	9.39	8.34
7	The average working fluid velocity, m/s	—	—	21.24	1.1	14.75
8	The mean temperature difference, °C	—	818.3	509.4	251.3	110.3
9	The overall heat transfer coefficient, $W/(m^2 \cdot K)$	—	61.23	55.95	60.25	24.57
10	The quantity of heat transfer kJ/kg	21,728	1,560	746.03	7,586.7	2,668.8

Before the calculation we assumed $T_{wg} = 160°C$, after the calculation we obtain $T_{wg} = 163°C$. The discrepancy between these values is $3°C < \pm 10°C$; therefore the assumed hot-air temperature is equal to the calculated one, and the ΔH is less than $\pm 0.5\%$; therefore the whole calculation is complete. Only the relative heat loss with waste gases, h_2, the boiler efficiency, and the fuel consumption should be corrected by using $T_{wg} = 163°C$.

After correction, $h_2 = 6.43\%$, $\eta_b = 92.37\%$, and $B = 2.45$ kg/s.

The results of the heat transfer calculation are listed in Table 8.14.

8.5 STEAM–WATER SYSTEMS OF BOILERS AND CIRCULATION CALCULATIONS

The flow of water, steam, or steam–water mixture within the steam boiler is called circulation. To remove heat from the boiler heating surfaces, it is necessary that the proper circulation be provided throughout the boiler circuits. Depending on the types of circulation, boilers may be divided into three kinds: natural-circulation boilers, controlled-circulation boilers, and once-through boilers. One of the important parameters for boiler circulation is the circulation ratio, K, which is equal to the ratio by weight of the water fed to the heated tubes, W_w, to the steam actually generated, W_s, $K = W_w/W_s$.

TABLE 8.15 Recommended Values of *K* for Natural-Circulation Boilers

Pressure, MPa	Steam Generation Capacity, t/hr	K
17–19	$\geq$ 800	4–6
14–16	185–670	5–8
10–12	160–420	8–15
2–3	35–240	15–25
$\leq$ 1.5	20–200	45–65
	$\leq$ 15	100–200

For the aforementioned three kinds of boilers, the values of K are quite different. K values for natural-circulation boilers are listed in Table 8.15; for controlled-circulation boilers, the values of K are between 3 and 10, while for once-through boilers, $K = 1.0$.

8.5.1 Steam – Water System of Natural-Circulation Boiler and Design Problems

Natural-circulation boilers usually have a steam–water system as shown in Fig. 8.39. The simplest form of this system consists of a drum, headers, risers, and downcomers. Risers are arranged in the furnace and when heated, the water in the risers evaporates, decreases in density, and tends to rise; downcomers are placed outside the furnace and are unheated. Cooler and heavier water in them flows downwards. This makes a circulation in the

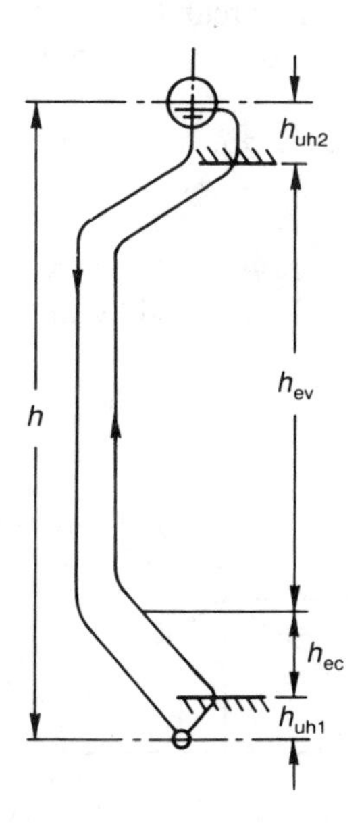

Fig. 8.39. Circulation circuit of a natural-circulation boiler.

circuit. For a steady flow, the following equation can be established for the circuit:

$$\rho_1 gh - \Delta p_d = \sum \rho_i h_i g + \Delta p_r + \Delta p_{se} \tag{8.85}$$

where ρ_i and ρ_1 are the average densities of the water or the steam–water mixture in the risers and water in the downcomers, kg/m^3; Δp_d, Δp_r, and Δp_{se} are the hydraulic resistances of the downcomers, risers, and steam–water separators in the drum, Pa.

If the left-hand side terms of Eq. (8.85) are set equal to Y_d, which expresses the total pressure difference of the downcomer, and the right-hand side terms are set equal to Y_r, which expresses the total pressure difference of the riser, then at the working point of a circuit with a steady flow, $Y_d = Y_r$.

The aim of the circulation calculation of a boiler steam–water system is to determine the flow rates in the risers and to check the reliability of the flow for the safe operation of the boiler circuit.

In Eq. (8.85), Y_d and Y_r both depend on the mass flow rate in the circuit (circulation flow rate) W, kg/s, or depend on the inlet water velocity of the risers (circulation velocity) V_0, m/s. $W = \rho_1 V_0 A_r$ (where A_r is the flow area of the risers, m^2). With an increase in W or V_0, Δp_d increases; that is, Y_d decrease while Y_r increases.

For a simple circuit (all risers have the same geometrical characteristics), the circulation calculation can be solved graph-analytically as follows [7]: first take three values of V_0 from which one may obtain three corresponding circulation mass flow rates, W, for establishing curves $Y_d = f(w)$ and $Y_r = f(w)$; the intersection of the two curves determines the working point A of the circulation circuit (Fig. 8.40). The actual quantity of the circulation flow rate, W, or the circulation velocity, V_0, can be obtained from the working point A as shown in the figure. For establishing curves $Y_d = f(w)$ and $Y_r = f(w)$, Δp_d, Δp_r, and the steam–water mixture density, ρ_m, have to be determined.

Calculation of the Hydraulic Resistance of the Downcomers, Δp_d As the fluid flowing in the downcomers is water, Δp_d can be determined by the

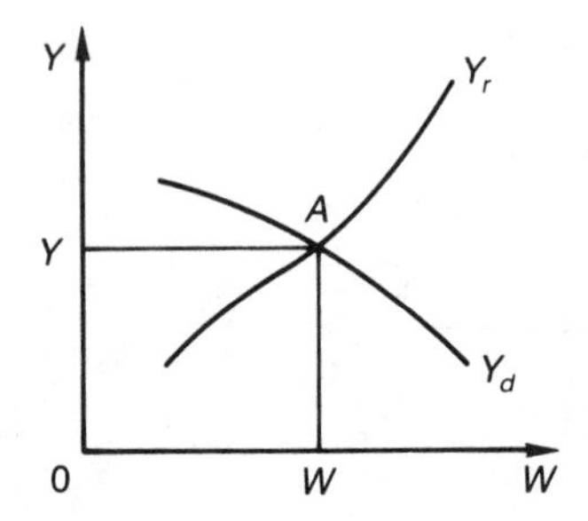

Fig. 8.40. Circulation characteristic curves of a simple circuit.

following equation [1, 7]:

$$\Delta p_d = \left(\lambda \frac{L}{d} + \sum \xi_M\right) \frac{V^2}{2} \rho_1 \tag{8.86}$$

where λ and ξ_M are the frictional coefficient and minor loss coefficients, L and d are the length and inner diameter of the calculated tube section, and V is the water velocity in the downcomers.

$$\lambda = \frac{1}{4\left[\log\left(3.7 \frac{d}{\Delta}\right)\right]^2} \tag{8.87}$$

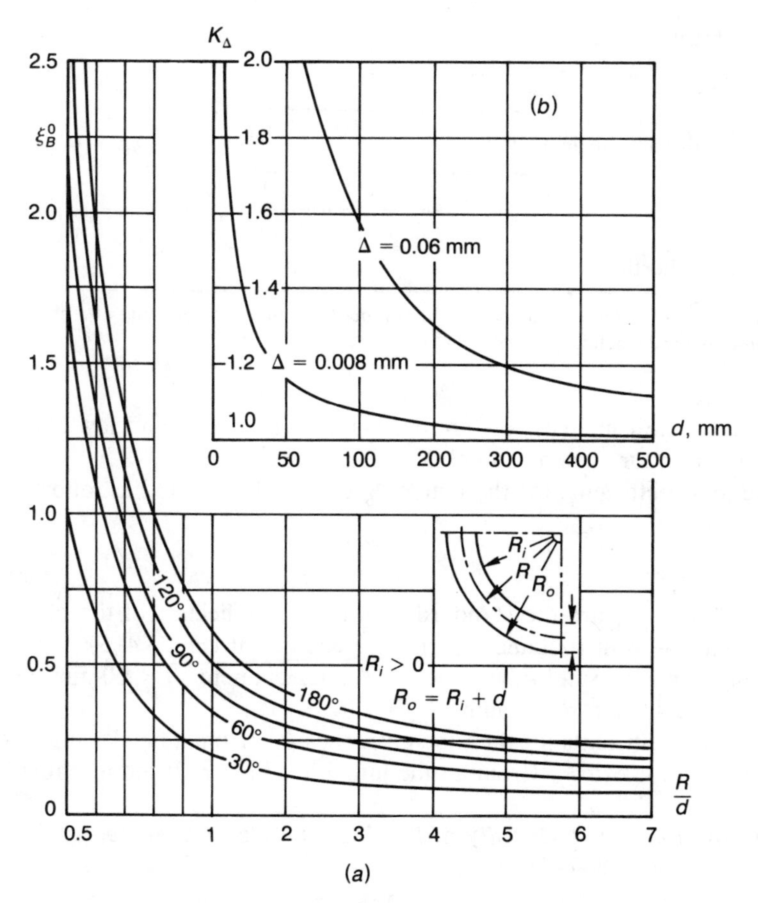

Fig. 8.41. Loss coefficient of bend ξ_B^0 and coefficient of roughness K_Δ: (*a*) ξ_B^0, (*b*) K_Δ.

TABLE 8.16 ξ_{en} and ξ_{ex} of Single-Phase Flow [1, 7][a]

Flow	Form	$d/d_h \leq 0.1$	$d/d_h > 0.1$
$n \leq 30$		$\xi_{en} = 0.5$	$\xi_{en} = 0.7$
$n > 30$		$\xi_{en} = 0.6$	$\xi_{en} = 0.8$
		$\xi_{en} = 0.4$	$\xi_{en} = 0.4$
		$\xi_{en} = 0.5$	$\xi_{en} = 0.5$
From drum into a tube		$\xi_{en} = 0.5$	$\xi_{en} = 0.5$
Flow into a drum		$\xi_{ex} = 1.0$	
Flow into a distributing header		$\xi_{ex} = 0.7$	
		$\xi_{ex} = 1.1$	
Flow into a collective header		$\xi_{ex} = 1.1$	

[a] In the table n is the ratio of number of outlet tubes to number of inlet tubes of a header; d_h is the header inner diameter.

where Δ is the roughness of tube wall: for carbon steel and low alloy steel $\Delta = 0.06$ mm, for austenite steel $\Delta = 0.008$ mm.

The loss coefficients for the bends ξ_B can be determined as follows:

$$\xi_B = \xi_B^0 K_\Delta \tag{8.88}$$

where ξ_B^0 and K_Δ are the standard minor loss coefficients of the bend (Fig. 8.41a) and the wall roughness corrective coefficient (Fig. 8.41b); for carbon steel and low alloy steel tubes when the inner diameter $d < 60$ mm, and for austenite steel when $d < 8$ mm, $K_\Delta = 1.0$.

The loss coefficients for the tube entrances ξ_{en} (from a header into a tube) and for the tube exit ξ_{ex} (from a tube into a header) are listed in Table 8.16.

Determination of the Density of the Steam – Water Mixture, ρ_m ρ_m can be expressed as follows [1, 7]:

$$\rho_m = \alpha \rho_g + (1 - \alpha)\rho_1 \tag{8.89}$$

where ρ_g and ρ_1 are the saturated densities of steam and water, kg/m^3; α is the steam void fraction.

$$\alpha = \frac{1}{1 + S\left(\dfrac{1-\beta}{\beta}\right)} \tag{8.90}$$

where S is the slip ratio and can be determined as follows [7, 8]:

$$S = 1 + \frac{0.4 + \beta^2}{V_0^{0.5}} \cdot \left(1 - \frac{p}{22.1}\right) \tag{8.91}$$

where P is the absolute pressure, MPa, and β is the volumetric quality

$$\beta = \frac{1}{1 + \left(\dfrac{1}{x} - 1\right)\dfrac{\rho_g}{\rho_1}} \tag{8.92}$$

where x is the steam mass quality of the mixture.

***Calculation of the Hydraulic Resistance of the Risers, Δp_r* [1, 7]** The calculation of Δp_r is a complicated problem. For a boiler with nonsteaming economizers, water entering the drum is below the saturated state, so water in the downcomers and at the inlet of the riser is subcooled. The subcooling of water is

$$\Delta i_{sub} = \frac{i_{sw} - i_{Ee}}{K} \text{ kJ/kg} \tag{8.93}$$

where i_{Ee} is the exit water temperature of the economizer, kJ/kg; K is the circulation ratio, which is the first selected according to Table 8.15 and should be checked after the entire calculation is completed, [if $(\Delta i'_{sub} - \Delta i_{sub})100/\Delta i'_{sub} \leq 50\%$, the assumption of K is correct, where $\Delta i'_{sub}$ is the calculated value]; and i_{sw} is the enthalpy of saturated water, kJ/kg.

For steaming economizers, $\Delta i_{sub} = 0$.

When the subcooled water flows into the risers from the downcomers, it first has to be heated to the boiling point in the water section, or so-called economizer section of the riser. The height of the economizer section, h_{ec}, is shown in Fig. 8.39, and can be determined as follows:

$$h_{ec} = \frac{\Delta i_{sub} + \dfrac{\Delta i}{\Delta p}\rho_1 g \times 10^{-6}\left(h - \dfrac{\Delta p_d}{\rho_1 g}\right)}{\dfrac{H_1}{h_1 W} + \dfrac{\Delta i}{\Delta p}\rho_1 g \times 10^{-6}} \tag{8.94}$$

where $\Delta i/\Delta p$ is the change in water enthalpy per unit pressure, kJ/kg · MPa; H_1 is the heat absorption of the first section of the risers, kJ/kg; W is the total flow rate of circulating water, kg/s; and h_1 is the heated height of the first section of risers, m. For boilers with steaming economizers, h_{ec} can be considered as 0.

The hydraulic resistance of the economizer portion of the riser can be obtained by using Eq. (8.86).

The evaporating portion of the riser is above the boiling point; the quantity of steam generated in the first section of the riser is equal to

$$W_{gl} = \frac{H_1 - W(\Delta i_{\text{sub}})}{i_{lg}} \tag{8.95}$$

The hydraulic resistance of the evaporating portion of the risers, Δp_{ev}, when the risers are heated uniformly along the tube length is

$$\Delta p_{\text{ev}} = \left(\sum \xi'_M + \psi\lambda \frac{L_{\text{ev}}}{d}\right)\frac{V_0^2}{2}\rho_1\left[1 + \frac{x_e}{2}\left(\frac{\rho_1}{\rho_g} - 1\right)\right] + \xi'_{\text{ex}}\frac{V_0^2}{2}\rho_1\left[1 + x_e\left(\frac{\rho_1}{\rho_g} - 1\right)\right] \tag{8.96}$$

where ξ'_M is the coefficient of minor losses for the steam–water mixture; L_{ev} is the length of the evaporating portion of the riser, m; x_e is the exit steam quality of the riser; ξ'_{ex} is the loss coefficient of the tube exit, for the steam–water mixture, it is equal to 1.2; and ψ is the two-phase frictional corrective coefficient, when $\rho V = 1000$ kg/m^2 · s, $\psi = 1.0$, in other cases, ψ can be obtained from Eqs. (8.97) and (8.98) [7, 8]; when $\rho V < 1000$ kg/(m^2 · s):

$$\psi = 1 + \frac{\bar{x}(1 - \bar{x})\left(\dfrac{1000}{\rho V} - 1\right)\dfrac{\rho_1}{\rho_g}}{1 + \bar{x}\left(\dfrac{\rho_1}{\rho_g} - 1\right)} \tag{8.97}$$

when $\rho V > 1000$ kg/(m^2 · s):

$$\psi = 1 + \frac{\bar{x}(1 - \bar{x})\left(\dfrac{1000}{\rho V} - 1\right)\dfrac{\rho_1}{\rho_g}}{1 + (1 - \bar{x})\left(\dfrac{\rho_1}{\rho_g} - 1\right)} \tag{8.98}$$

TABLE 8.17 Loss Coefficients of Tube Entrances ξ'_{en} for Steam–Water Mixture [6, 7]

The Form of Outletting Tube	The Relative Height of Tube				h/d			
	10	20	50	≥ 80	10	20	50	≥ 80
	$p \leq 6$ MPa				$p > 6$ MPa			
h	0.3	0.5	0.8	1.0	0.6	0.9	1.1	1.2
h	0.5	1.1	1.7	2.2	1.0	1.2	1.4	1.5

The loss coefficients for the tube entrances, ξ'_{en}, for the steam–water mixture can be obtained from Table 8.17; for horizontal outlet tubes, ξ'_{en} is equal to that for single-phase flow.

The loss coefficients of bends for the steam–water mixture, ξ'_B, can be determined as follows:

For horizontal bends $\xi'_B = \xi_B$; for bends with inclined upward outlets and an inclined angle less than 15°, $\xi'_B = 2\xi_B$; for bends with a vertical or inclined upward outlet bend angle greater than 90°, and an inclined angle greater than 15°, $\xi'_B = 2\xi_B$; for bends with a vertical upward or downward outlet or with an upward inclined outlet, an inclined angle greater than 15°, and a bend angle less than 90°, $\xi'_B = 4\xi_B$; for a vertical U-type bend, $\xi'_B = 3.64\xi_B$; for a vertical inverted U-type bend, $\xi'_B = 2.19\xi_B$, where ξ_B is the loss coefficient of the bend for single-phase flow.

The total hydraulic resistance of the riser (Fig. 8.39), Δp_r, can be calculated by

$$\Delta p_r = \Delta p_{uh1} + \Delta p_{ec} + \Delta p_{ev} + \Delta p_{uh2} \tag{8.99}$$

For complex circulation, we may use a similar method to solve the circulation calculation problems. Figure 8.42 shows a complex circulation circuit with common downcomers which supply water to two parallel connected riser sections. For solving this circulation problem, first take three different values of V_0, from which three values of W can be obtained to

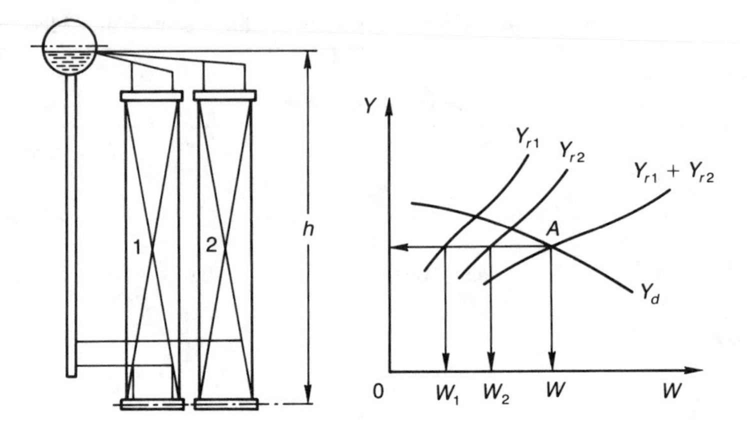

Fig. 8.42. Circulation characteristic curves of a complex-circulation circuit.

establish the curves $Y_{r1} = f(w)$ and $Y_{r2} = f(w)$ for the two riser sections and $Y_d = f(w)$ for the downcomers as shown in Fig. 8.42. Since both riser sections operate in parallel at the same pressure difference, their circulation characteristics are summed up by adding together the water flow rates, W, at the same value of Y_r (by the abscissas) for Y_{r1} and Y_{r2} to obtain the total circulation characteristic curve, $Y_r = Y_{r1} + Y_{r2} = f(w)$. After establishing the curve of $Y_d = f(w)$, the intersection of curves $Y_d = f(w)$ and $Y_r = f(w)$ gives the working point A of the circuit, from which the total circulation flow rate of the circuit, W, can be determined. The water flow rates for the two riser sections are found by drawing a horizontal line through the working point up to the intersection with the $Y_{r1} = f(w)$ and $Y_{r2} = f(w)$ curves. As shown in Fig. 8.42, W_1 is the water flow rate for one riser section and W_2 is that for another section.

When the water flow rate and the quantities of steam generated in each riser section are known, the circulation ratio K for each section and for the whole circuit can be determined.

Checking Circulation Reliability In a circulation circuit many risers are connected in parallel, but may be heated nonuniformly. In tubes with low heat flux, less steam is generated.

When risers are connected to the water space of the drum, the phenomenon of circulation stagnation may occur. With circulation stagnation, water in the riser moves very slowly upwards or downwards and steam bubbles move upwards through the column of water in the tube. When steam accumulates on some part of the heated tube (such as the bends), the tube metal may be overheated.

When risers are connected to the steam space of the drum, instead of circulation stagnation, a free water level may form in a low heat flux riser.

Circulation circuits have to be checked for circulation reliabilities for the least heated tubes by using the following reliability criteria.

The reliability criterion for escaping from circulation stagnation is

$$\frac{Y_r}{\Delta p_s} \geq 1.05 \tag{8.100}$$

where Y_r is the total pressure difference of the calculated riser section (the resistances of the outlet pipe and separator are not included); and Δp_s is the pressure difference of the least heated riser under the state of circulation stagnation, Pa.

$$\Delta p_s = \rho_1 g(h_{\text{uh1}} + h_{\text{ec}}) + \left[(1 - \alpha_s)\rho_1 + \alpha_s \rho_g\right] g h_{\text{ev}} + \left[(1 - \alpha_s')\rho_1 + \alpha_s' \rho_g\right] g h_{\text{uh2}} \tag{8.101}$$

where h_{uh1}, h_{ec}, h_{ev}, and h_{uh2} are the heights if the riser as shown in Fig. 8.39; α_s and α_s' are the steam void fractions of the heated tube and the unheated tube, respectively, under the state of circulation stagnation.

$$\alpha_s = \frac{V_0''}{V_0''A + B} \tag{8.102}$$

where V_0'' is the average superficial steam velocity in the heated portion of the least heated riser, m/s; and A and B are the coefficients listed in Table 8.18.

$$\alpha_s' = \frac{V_0''}{0.95V_0'' + B} \tag{8.103}$$

TABLE 8.18 Coefficients *A* and *B* for Eqs. (8.102) and (8.103) [7]

Pressure $p \times 1.02$ MPa	A	B
1	0.965	0.0661
2	0.984	0.612
3	0.992	0.544
4	0.999	0.476
6	1.019	0.385
8	1.071	0.306
10	1.086	0.246
12	1.113	0.180
14	1.135	0.127
16	1.182	0.095
18	1.217	0.091
20	1.290	0.082

where V_0'' is the superficial steam velocity in the unheated portion of the least heated riser, m/s; B can be determined from Table 8.18; after calculation, if $\alpha_s' > 1.0$, take $\alpha_s' = 1.0$.

The reliability criterion for avoiding the free water level in a riser connected to the steam space of the drum is

$$\frac{Y_r}{\Delta p_s + \Delta p_{\text{wl}}} \geq 1.05 \tag{8.104}$$

where Δp_{wl} is the pressure loss due to raising the steam–water mixture above the water level in the drum, Pa; and may be calculated by

$$\Delta p_{\text{wl}} = h_{\text{wl}}(1 - \alpha_e)(\rho_1 - \rho_g)g \tag{8.105}$$

where h_{wl} is the distance from the highest point of the riser to the water level in the drum, m; and α_e is the steam void fraction at the exit of the riser.

The circulation ratio, K, should also be checked. When its value is within the recommended value listed in Table 8.15, the circulation is considered to be reliable.

For boilers of $p = 17$–19 MPa or $p = 14$–6 MPa and $K \leq 4$, the heat transfer crisis phenomenon must be checked. The threshold of the heat transfer crisis depends on the steam mass quality, heat flux, mass velocity, pressure, tube diameter, flow direction of fluid, and internal surface conditions of the tubes. Many correlations have been presented for the estimation of the heat transfer crisis threshold conditions, and these have been introduced in the foregoing chapters. In the PRC and the USSR, the method recommended by [6] is used. This method takes the critical steam quality, x_c, to express the margin of the threshold of the heat transfer crisis; the actual steam quality, x, in the risers should be below x_c which can be determined by the various figures and equations listed in [6, 7].

Design Problems of the Steam – Water System of Natural-Circulation Boilers Disturbances in circulation conditions mainly occur because of nonuniform heating across the width of a steam–water system. For the sake of ensuring the reliability circulation, the following design requirements are recommended.

1. Water walls should be sectionalized; that is, a group of risers which are heated similarly and have close geometrical shape are combined into an independent section with independent downcomers.
2. Tube diameters of the risers and the downcomers should be selected according to Table 8.5.
3. For decreasing hydraulic resistances, it is better for the riser to be connected directly to the drum without any upper headers or outlet

pipes. If outlet pipes are needed, according to construction requirements, the ratio of the total cross-sectional area of the outlet pipes, A_o, to that of the risers, A_r, is recommended as follows: when the drum pressure, p = 4–6 MPa, A_o/A_r = 0.35–0.45; p = 10–12 MPa, A_o/A_r = 0.4–0.5; p = 14–16 MPa, A_o/ A_r = 0.5–0.7; p = 17–19 MPa, A_o/A_r = 0.6–0.8.

4. The ratio of the total cross-sectional area of the downcomers, A_d, to that of the risers, A_r, can be selected as follows: for downcomers with inner diameter, d = 80–140 mm: when the drum pressure, p = 4–6 MPa, A_d/A_r = 0.2–0.3; p = 10–12 MPa, A_d/A_r = 0.35–0.45; p = 14–6 MPa, A_d/A_r = 0.5–0.6; p = 17–19 MPa, A_d/A_r = 0.6–0.7; for downcomers with d = 180–550 mm, the value of A_d/A_r can be 0.1 less than the previously recommended value.
5. For the sake of avoiding evaporation at the inlet of the downcomers, the inlet water velocity of the downcomers, V_d, should not exceed the following values: when the drum pressure, p = 4–6 MPa, $V_d \leq 3$ m/s; p = 10–16 MPa, $V_d \leq 3.5$ m/s; p = 17–19 MPa, $V_d \leq 4$ m/s.
6. For the sake of avoiding stratified flow, the inclined angle from the horizontal at any section of the heated risers should not be less than 15°.

8.5.2 Steam-Water System of Controlled-Circulation Boilers and Design Problems

As the density difference between the water and steam decreases with the increase of pressure, and the pressure exceeds 17.5 Mpa, the reliability of a natural-circulation boiler greatly decreases. Thus forced-circulation systems are adopted at superhigh pressures. Controlled-circulation system and low circulation ratio systems are two kinds of forced-circulation systems.

The existance of one or more circulating pumps in the steam–water system is a feature of controlled-circulation boilers (or forced multiple circulation boilers).

In this kind of system, the working fluid is moved forcefully with the required velocities. This makes it possible to arrange the evaporating tubes in any form, and the steam–water mixture may flow not only vertically upwards, but also horizontally or even downwards. Also, tubes with a small inner diameter can be used; this may decrease the tube thickness and thus the tube weight.

These two circulation systems are shown in Fig. 8.43 (see also Fig. 6.14). The circulation ratio for the controlled-circulation boiler is 3 to 10, while for the low circulation ratio boiler it is 1.2 to 2.0.

The purpose of the circulation calculations for these forced-circulation boilers is to ensure the reliability of the evaporating tubes and the circulating pump. The circulation can also be solved graphically. In a forced-circulation

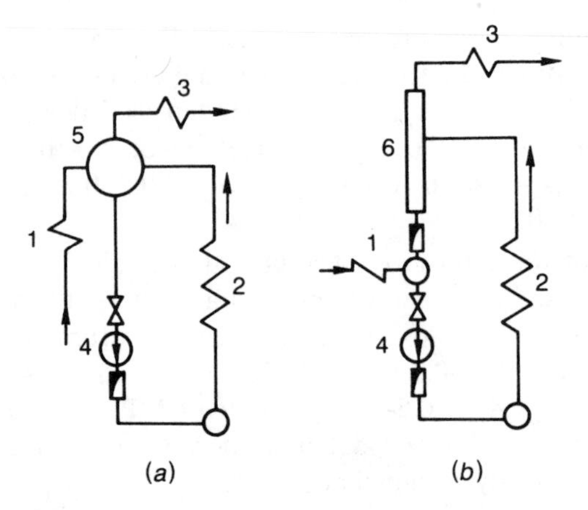

Fig. 8.43. Two kinds of controlled-circulation system: (*a*) multiple circulation boiler system, (*b*) low circulation ratio boiler system. 1—economizer, 2—water walls, 3—superheater, 4—circulating pump, 5—drum, 6—separator.

system, the hydraulic resistance of the whole circuit, Δp_c (including the resistances of the risers and downcomers), is overcome by the circulating pump head and the natural-circulation head; Δp_p; Δp_c and Δp_p both depend on the flow rate, W, in the circuit. For a simple circuit as shown in Fig. 8.44, if we take three values of W and establish curves $\Delta p_c = f(w)$ and $\Delta p_p = f(w)$, then the intersection of the two curves determines the working point A of the circuit.

For a complex circuit as shown in Fig. 8.45, the circulation calculations can be solved as follows. First take three flow rate values of W and determine the pressure drop Δp_r of the riser for the two riser sections and establish $\Delta p_{r1} = f(w)$ and $\Delta p_{r2} = f(w)$. Add the water flow rates of the aforementioned two curves together at the same value of Δp_r to form the curve $\Sigma \Delta p_r = f(w)$. Then calculate the pressure drop of the downcomers, Δp_d, and establish the $\Delta p_d = f(w)$ curve. Add the pressure drops of curves $\Delta p_r = f(w)$ and $\Delta p_d = f(w)$ together at the same value of W to form the curve $\Delta p_c = f(w)$, where Δp_c is the total pressure losses of the circuit. Draw the hydraulic characteristic curve $\Delta p_p = f(w)$. The intersection of $\Delta p_p =$

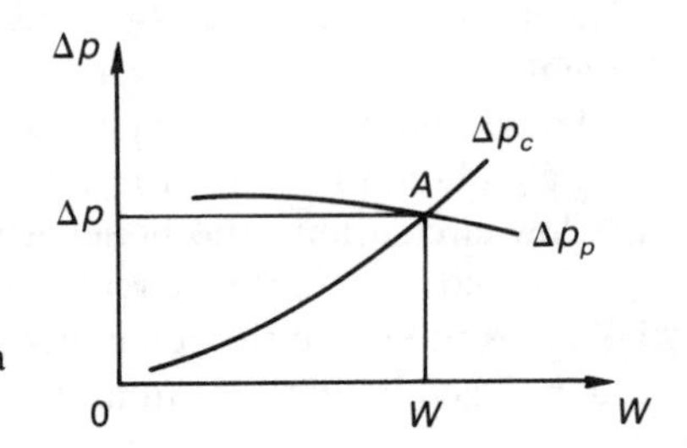

Fig. 8.44. Circulation characteristic curves of a simple circuit of a forced circulation boiler.

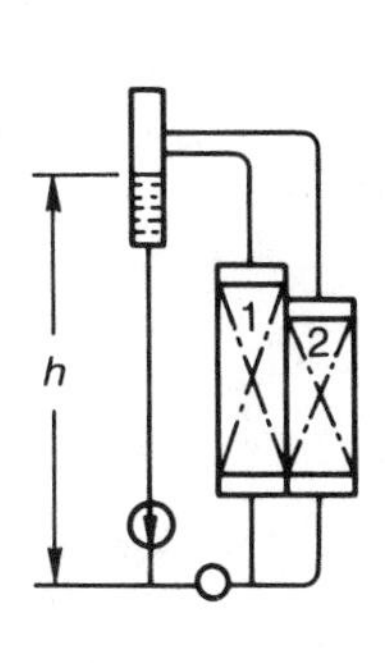

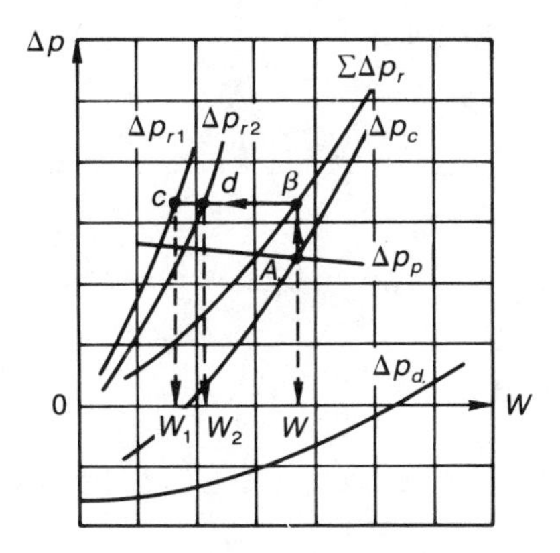

Fig. 8.45. Circulation characteristic curves of a complex circuit of a forced circulation boiler.

$f(w)$ and $\Delta p_c = f(w)$ determines the working point A, and thus the total flow rate, W, in the circuit. Make a vertical line to meet the curve $\Sigma\Delta p_r = f(w)$ at point B and from B draw a horizontal line to intersect curves $\Delta p_{r1} = f(w)$ and $\Delta p_{r2} = f(w)$ at c and d. The flow rate, W_1, in riser section 1 and W_2 in riser section 2 can be obtained by making vertical lines from point c and d to the axis of the abscissa as shown in Fig. 8.45.

If the distribution of flow rates among the riser sections is not reasonable, orifices may be installed at the inlets of the risers to regulate the distribution of flow rates among the riser sections.

Calculation of the Circuit Pressure Losses, Δp_c

$$\Delta p_c = \Delta p_r + \Delta p_d \tag{8.106}$$

$$\Delta p_r = \Delta p_f + \Delta p_M + \Delta p_e \tag{8.107}$$

where Δp_f, Δp_M, and Δp_e are the pressure drops due to frictional resistance, minor losses, and elevation, Pa, respectively; the former two can be obtained by using the same methods applied for the natural-circulation system, and Δp_e can be calculated as follows:

$$\Delta p_e = \rho_1 g(h_{\text{uh1}} + h_{\text{ec}}) + h_{\text{ev}} \rho_m g \tag{8.108}$$

where h_{uh1}, h_{ec}, and h_{ev} are the heights of the initial, unheated portion of a riser just beyond the inlet, the economizer portion, and the evaporating portion of a riser, m, respectively; and ρ_m is the mixture density, kg/m^3. ρ_m can be calculated by using Eq. (8.89).

$$\Delta p_d = \Delta p_f + \Delta p_M + \Delta p_e \tag{8.109}$$

TABLE 8.19 Coefficient $\tau = (\delta / d_o)$

δ/d_o	0	0.2	0.4	0.6	0.8	1.0	1.2	1.6	2.0	2.4
τ	1.35	1.22	1.10	0.84	0.42	0.24	0.16	0.07	0.02	0

where Δp_f and Δp_M can be determined by using Eq. (8.96).

$$\Delta p_e = -h\rho_1 g \tag{8.110}$$

$$h_{ec} = \frac{\Delta i_{sub} + \dfrac{\Delta i}{\Delta p}\rho_1 g\left(h - \dfrac{\Delta p_o}{\rho_1 g} + \dfrac{\Delta p_p}{\rho_1 g}\right) \times 10^{-6}}{\dfrac{H_1}{h_1 W} + \dfrac{\Delta i}{\Delta p}\rho_1 g \times 10^{-6}} \tag{8.111}$$

where Δp_o is the pressure drop of an orifice installed at the inlet of a riser, Pa; Δp_p is the pressure head of the circulating pump, Pa, for calculation, it can be selected to be equal to 3×10^5 Pa; and the definitions of the other symbols are the same as those in Eq. (8.94).

$$\Delta p_o = \left\{0.5 + \left[1 - \left(\frac{d_o}{d}\right)^2\right]^2 + \tau\left[1 - \left(\frac{d_o}{d}\right)^2\right]\right\}\left(\frac{d}{d_o}\right)^4 \frac{V_0^2}{2}\rho_1 \tag{8.112}$$

where d_o and d are the orifice diameter and the tube inner diameter, m, respectively. V_0 is the inlet water velocity of the tube, m/s; and τ is a coefficient depending on the ratio of the thickness, δ, to the orifice diameter, d_o, and can be determined from Table 8.19.

Checking Circulation Reliability To calculate the reliability criterion for avoiding circulation stagnation in the least heated tube, Eq. (8.100) may be used.

To calculate the reliability criterion for avoiding heat transfer crisis, the methods described for natural-circulation boilers [6, 7] may be used.

The check for safety of the water supply to the circulating pumps is as follows:

$$\rho_1 gh \geq 1.1\Delta p_{suc} + \Delta p_d \tag{8.113}$$

where h is the height of the downcomer; m, Δp_d is the pressure drop in the downcomer tubes, Pa; and Δp_{suc} is the pump positive suction head, specified by the pump manufacturer, Pa.

The absence of flow rate pulsation in steam–water systems can be determined empirically as follows [7, 10, 11]:

$$\frac{\Delta p_o + \Delta p_{ec}}{\Delta p_{ev}} > a \tag{8.114}$$

where Δp_o, Δp_{ec}, and Δp_{ev} are the pressure drops of the orifices at the inlet, the water portion, and evaporating portion of a riser, Pa, respectively; a is a constant depending on the pressure, p, MPa, and the mass velocity of the working fluid, ρV, kg/(m$^2 \cdot$ s), when $p = 4.0$, $a = 0.8$, when $p = 6.0$, $a = 0.65$, when $p = 8.0$, $a = 0.52$, when $p = 10.0$, $a = 0.37$, when $p = 12.0$, $a = 0.16$, when $p = 14.0$, $a = 0.10$; when $\rho V = 500$, $a = 1.0$, when $\rho V = 750$, $a = 0.5$, when $\rho V = 1000$, $a = 0.27$, when $\rho V = 1300$, $a = 0.1$, when $\rho V = 1500$, $a = 0.03$; the value of a in Eq. (8.114) is equal to the larger value between values of a determined by the mass velocity condition and the pressure condition.

For horizontal tubes, a method obtained by Habenski et al. [9] may be used to determine the absence of flow rate pulsation. The required critical mass velocity for avoiding flow rate pulsation, $(\rho V)_c$, is

$$(\rho V)_c = 4.62 \times 10^{-9} (\rho V)_c^* k_p \frac{qL}{d} \tag{8.115}$$

where $(\rho V)_c^*$ is the critical mass velocity obtained under a standard pressure ($p = 9.8$ MPa), kg/(m$^2 \cdot$ s); k_p is the pressure factor; q is the heat flux of the tube, W/m^2, L is the length of the tube, m; and d is the tube inner diameter, m.

For vertical tubes, $(\rho V)_c$ can be predicted as $(\rho V)_c$ of a horizontal tube times a coefficient, c. The values of $(\rho V)_c^*$, k_p and c may be obtained from the figures listed in [6, 7, 10].

The absence of multivalueness of the hydraulic characteristic curve for horizontal and vertical heated tubes can be checked by the inlet subcooling of Δi_{sub}, kJ/kg.

$$\Delta i_{sub} \le \left(1 + \frac{\xi_o}{z}\right) \frac{7.46 i_{lg}}{c\left(\dfrac{\rho_l}{\rho_g} - 1\right)} \tag{8.116}$$

where ξ_o and z are the pressure loss coefficient of the inlet orifice and the total resistance coefficient (ξ_o is not included) of a tube, respectively; i_{lg} is the latent heat of evaporation, kJ/kg; c is a coefficient of pressure, when $p < 10$ MPa, $c = 2$, when $p > 14$ MPa, $c = 3$, and when $p = 10$–14 MPa, $c = 0.25p$–0.5.

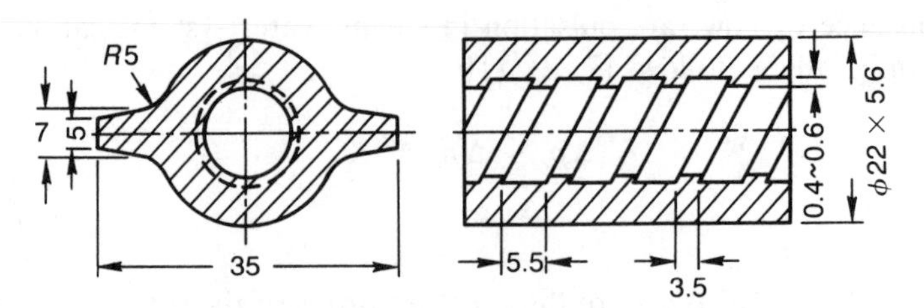

Fig. 8.46. Four-lead spiral ribbed tube.

Design Problems of Controlled-Circulation Boilers Generally speaking, the circulation ratio, K, should be greater than 3; for small-capacity boilers, $K = 6–10$, while for low-circulation boilers $K = 1.2–2.0$.

The mean velocity of the working fluid should ensure the cooling effect of the heated tube metal. In the downward flow tubes, for the sake of carrying away steam from the tube inner wall, the circulation velocity should be greater than 0.5 m/s.

In order to avoid multivalueness of the hydraulic characteristic curve, the mass velocity of low-circulation ratio boilers should maintain $\rho V \geq 850–1000$ kg/(m$^2 \cdot$ s) for coal-fired boilers and $\rho V \geq 1200–1350$ kg/(m$^2 \cdot$ s) for fuel-oil-fired boilers.

In order to increase the critical steam quality, x_c, tubes with ribs, multi-lead ribs, or twisters are used (see Fig. 6.41). In PRC, the four-lead spiral ribbed tubes shown in Fig. 8.46 are used; x_c for these ribbed tubes can be obtained as follows:

$$x_c = 11.16(59.29 - 2.25p)q^{-0.6}(\rho V)^{0.33} \tag{8.117}$$

where p is the pressure, MPa; q is the inner wall, heat flux, W/m^2; V is the mass velocity, kg/(m$^2 \cdot$ s). If the calculated x_c is greater than 1.0, it means that a heat transfer crisis is impossible in the calculated riser.

The ratio of the total flow area of the outlet tubes to that of the risers is generally equal to 0.4 to 0.6.

8.5.3 Steam – Water System of Once-Through Boilers

Another kind of forced-circulation boiler is a once-through boiler. The steam generation rate is numerically equal to the water content supplied; thus its circulation ratio, $K = 1.0$. The working principle and design problems of once-through boilers have been discussed in Chapter 7 and need not be discussed here.

For once-through boilers, the heat transfer crisis phenomenon, flow stagnation, flow rate pulsation, and multivalueness of the hydraulic characteristics of tube systems have to be checked. The same methods introduced in the preceding sections may be used to check them.

8.6 A NUMERICAL EXAMPLE OF BOILER CIRCULATION CALCULATIONS

The example boiler is the same boiler that is used in Section 8.4 as a numerical example of the heat transfer calculation of boiler components. Due to the limitation of space in this chapter, we take the circulation of a side water wall circuit of this natural-circulation boiler as a calculation example; its scheme is shown in Fig. 8.47. The circuit consists of a drum, downcomer, middle riser section, rear riser section, upper and lower headers, mixture outlet tubes, and water supply tubes.

The height from the drum level to the axis of the lower headers, h = 18.05 m, the downcomer length, L = 17.25 m, with an inner diameter, d = 299 mm (flow area, $A_d = 0.07\ \text{m}^2$). The ratio of the downcomer area to that of the risers, $A_d/A_r = 0.588$. The roughness, $\Delta = 0.08$ mm, while the entrance loss coefficient, $\xi_{en} = 0.5$.

The inner diameter of the water supply tube, d = 125 mm; there are two tubes for each riser section, so the water flow area, $A_{ws} = 0.024\ \text{m}^2$, the ratio of A_{ws} to the total flow area of a section of risers, A_r, $A_{ws}/A_r = 0.024/0.0595 = 0.403$. The length of the water supply tube for the middle riser section is 4.97 m, for the rear section it is 6.73 m. Tube roughness, $\Delta = 0.08$ mm. The total minor losses, $\Sigma\xi_M = 2.4$.

The mixture outlet tubes have the same inner diameter and number as the water supply tubes, so their flow area ratio, $A_o/A_r = 0.024/0.059 = 0.403$. The length of the tube for the middle riser section is 7.99 m, for the rear section it is 6.47 m. The total minor losses of the tube for the middle riser

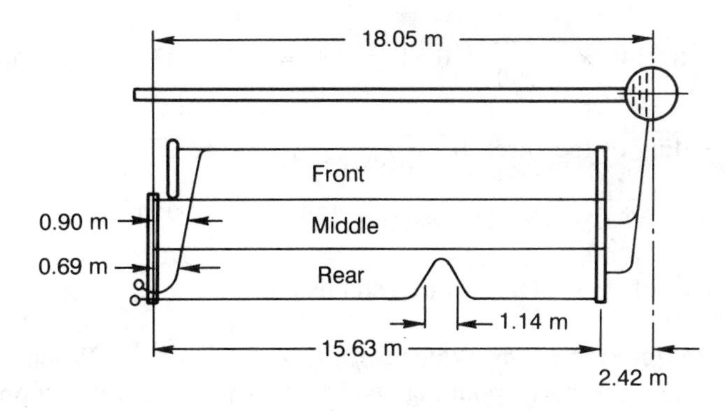

Fig. 8.47. Scheme of the example unit.

section, $\Sigma\xi_M = 3.3$, and that for the rear section, $\Sigma\xi_M = 2.5$; the tube roughness, $\Delta = 0.08$ mm.

Each riser section has 26 tubes with an inner diameter of 54 mm. The working fluid flow area of each section, $A_r = 0.059\ \text{m}^2$. The height between the upper and lower headers is equal to 15.63 m, and the height of the mixture outlet tube is 2.42 m. The height of the unheated portion of the middle riser section is 0.9 m, and that of the rear section is 0.69 m. The average heated height of the middle section is 14.73 m, while for the rear one it is 13.8 m (the remaining height of the rear section is 14.94 m, including 1.14 m that is unheated on account of the covering of the rear water wall). The tube roughness, $\Delta = 0.08$ mm. The tube entrance loss coefficient, $\xi_{\text{en}} = 1.2$.

From the numerical example of the boiler heat transfer calculation of this chapter, we know that the total radial heat absorbed by the water walls, $H_{\text{ww}} = 20{,}463.7$ kJ/kg. As the effective heating surface, $A_e = \zeta A$, where ζ is the coefficient of fouling, the average heat flux of the water walls, q_{ww}, is

$$q_{\text{ww}} = \frac{B_r H_{\text{ww}}}{\Sigma \zeta_i x_i A_{wi}} = \frac{2.451 \times 20{,}463.7}{0.55 \times 0.99 \times 302.26 + 0.1 \times 1.0 \times 31.5}$$

$$= 299.4\ \text{kW/m}^2$$

Calculation of Y_d Since the drum pressure, $p = 4.41$ MPa, the saturated water density, $\rho_1 = 790\ \text{kg/m}^3$, and $h = 18.05$ m, then $\rho_1 gh = 790 \times 9.8 \times 18.05 = 139{,}743$ Pa.

Taking three circulating velocities to be $V_0 = 0.5$, 1.0, and 1.5 m/s, the respective mass flow rate of the working fluid in one riser section, $W = 23.5$, 47, and 70.5 kg/s ($W = \rho_1 A_r V_0$). The water velocity in the downcomer V_d will be $V_d = V_o A_r / A_d = 0.85$, 1.7, and 2.55 m/s, respectively.

As $\lambda = 1/\{4[\log(3.7 \times 299/0.08)]^2\} = 0.0146$ [Eq. (8.87)], $\xi_{\text{en}} = 0.5$ (Table 8.16); therefore, for the three V_d values, Δp_d is equal to [Eq. (8.86)]:

$$\Delta p_d = \left(0.0146 \times \frac{17.25}{0.299} + 0.5\right) \frac{V_d^2}{2} 790 = 383,\ 1533,\ \text{and}\ 3448\ \text{m/s}$$

Corresponding to the three W or V_0, Y_d will be equal to

$$Y_d = \rho_1 gh - \Delta p_d = 139{,}743 - \Delta p_d = 139{,}360,\ 138{,}210,\ \text{and}\ 136{,}295\ \text{Pa}$$

Thus the curve of $Y_d = f(w)$ can be established.

Hydraulic Resistance of the Water Supply Tubes, Δp_{ws} When $V_0 = 0.5$, 1.0, and 1.5 m/s, the corresponding water velocity in the water supply tube, $V_{\text{ws}} = V_0 A_r / A_{\text{ws}} = V_0/0.403 = 1.24$, 2.48, and 3.72 m/s. The hydraulic resistance of the water supply tube for the middle riser section

$$\Delta p_{ws} = \left(\lambda \frac{L}{d} + \sum \xi_M\right) \frac{V_{ws}^2}{2} \rho_1 = \left(0.01766 \frac{4.97}{0.125} + 2.4\right) \frac{V_{ws}^2}{2} 790$$

$$= 1884, 7526.4, \text{ and } 16934.4 \text{ Pa}$$

where $\lambda = 1/\{4[\log(3.7 \times 125/0.08)]^2\} = 0.01766$, $L = 4.97$ m. The hydraulic resistance of the water supply tube for the rear riser section

$$\Delta p_{ws} = \left(0.01766 \frac{6.73}{0.125} + 2.4\right) \frac{V_{ws}^2}{2} 790 = 2038.4, 8153.6, \text{ and } 18{,}326 \text{ Pa}$$

Hydraulic Resistance of the Riser, Δp_r Δp_r can be calculated according to Eq. (8.99). In this equation, since the economizer is a steaming economizer, Δp_{ec} does not exist

$$\Delta p_r = \Delta p_{uh1} + \Delta p_{ev}$$

For $V_0 = 0.5$, 1.0, and 1.5 m/s or $W = 23.5$, 47, and 70.5 kg/s, the resistance of the unheated portion of the riser, Δp_{uh1}, can be obtained as follows: for the middle riser section

$$\Delta p_{uh1} = \left(\lambda \frac{L}{d} + \sum \xi_M\right) \frac{V_0^2}{2} \rho_1 = \left(0.0216 \frac{0.9}{0.054} + 0.9\right) \frac{V_0^2}{2} 790$$

$$= 124.43, 497.72, \text{ and } 1119.87 \text{ Pa}$$

where $\lambda = 0.0216$ [Eq. (8.87)], $L = 0.9$ m, and $\Sigma \xi_M = 0.9$.

For the rear riser section, by using the same equation,

$$\Delta p_{uh1} = \left(0.0216 \frac{0.69}{0.054} + 0.9\right) \frac{V_0^2}{2} 790 = 113.4, 453.6, \text{ and } 1020.6 \text{ Pa}$$

The radiant heat absorbed by the working fluid in the middle riser section

$$H_r^m = q_{ww}\left(\sum \zeta_i x_i A_w\right)_m = 299.4 \times 0.55 \times 0.99 \times 14.73 \times 1.664$$

$$= 3995.82 \text{ kW}$$

where the height of the section is 14.73 m, the width of the section is 1.664 m.

The radiant heat absorbed by the working fluid in the rear riser section

$$H_r^r = q_{ww}\left(\sum \zeta_i x_i A_w\right)_r = 299.4 \times 0.55 \times 0.99 \times 13.8 \times 1.664$$

$$= 3743.54 \text{ kW}$$

The quantity of steam generated in the middle riser section (the latent heat of evaporation $i_{lg} = 1680.7$ kJ/kg):

$$W_s = \frac{H_r^m}{i_{lg}} = \frac{3995.82}{1680.7} = 2.377 \text{ kg/s}$$

The quantity of steam generated in the rear riser section

$$W_s = \frac{H_r^r}{i_{lg}} = \frac{3743.54}{1680.7} = 2.277 \text{ kg/s}$$

Steam quality at the exit of the risers, $x_e = W_s/W$; for the middle riser section when $W = 23.5$, 47, and 70.5 kg/s, $x_e = 0.101$, 0.050, and 0.0337, respectively; for the rear section $x_e = 0.0948$, 0.047, and 0.0316.

Mass velocity, $\rho V = \rho_1 V_0$; for the middle riser section and the rear riser section; when $V_0 = 0.5$, 1.0, and 1.5 m/s, the corresponding $\rho V = 395$, 790, and 1185 kg/(m$^2 \cdot$ s).

The two-phase frictional corrective coefficient, ψ, can be determined by Eqs. (8.97) and (8.98), depending on the values of ρV.

When pressure $p = 4.41$ MPa, $\rho_1/\rho_g = 790/22.3 = 35.43$. Substituting this ratio and $x = x_e/2$ into Eqs. (8.97) and (8.98), we may obtain, when $\rho V = 395$, 790, and 1185 kg/(m$^2 \cdot$ s), for the middle riser section, $\psi = 1.95$, 1.123, and 0.997, respectively, and for the rear riser section, $\psi = 1.93$, 1.12, and 0.9975, respectively.

Δp_{ev} can be determined by Eq. (8.96). For the middle riser section, by substituting the values of ψ, V_0, and x_e into the equation, we may obtain

$$\begin{aligned}\Delta p_{ev} &= \left(\psi \times 0.0216 \times \frac{14.73}{0.054}\right)\frac{V_0^2}{2}790\left[1 + \frac{x_e}{2}\left(\frac{790}{22.3} - 1\right)\right] \\ &\quad + 1.2 \times \frac{V_0}{2}790\left[1 + x_e\left(\frac{790}{22.3} - 1\right)\right] \\ &= 3638, 6152.4, \text{ and } 10{,}553 \text{ Pa}\end{aligned}$$

where $\lambda = 0.0216$, $L = 14.73$, $\xi'_{ex} = 1.2$.

For the rear riser section $L = 14.94$ m, and taking $V_0 = 0.5$, 1.0, and 1.5 m/s,

$$\begin{aligned}\Delta p_{ev} &= \left(\psi \times 0.0216 \times \frac{14.94}{0.054}\right)\frac{V_0^2}{2}790\left[1 + \frac{x_e}{2}\left(\frac{790}{22.3} - 1\right)\right] \\ &\quad + 1.2 \times \frac{V_0^2}{2}790\left[1 + x_e\left(\frac{790}{22.3} - 1\right)\right] \\ &= 3503, 6023.5, \text{ and } 10{,}406.3 \text{ Pa}\end{aligned}$$

So when $W = 23.5$, 47, and 70.5 kg/s, the hydraulic resistance of the middle riser section is equal to

$$\Delta p_r = \Delta p_{\text{uh1}} + \Delta p_{\text{ev}} = 3762.43,\ 6650.12,\ \text{and}\ 11672.87\ \text{Pa}$$

and for the rear riser section

$$\Delta p_r = 3616.4,\ 6477.1,\ \text{and}\ 11{,}426.9\ \text{Pa}$$

Hydraulic Resistance of the Mixture Outlet Tube, Δp_o When $V_0 = 0.5$, 1.0, and 1.5 m/s, the corresponding circulating velocities in the outlet tube are equal to $V_0 A_r/A = V_0/0.403 = 1.24$, 2.48, and 3.72 m/s. The corresponding mass velocities are $\rho V = 979.6$, 1959.2, and 2938.8 kg/(m$^2 \cdot$ s).

The hydraulic resistance of the unheated mixture outlet tube, Δp_o, can also be calculated by Eq. (8.96), but in that equation x_e should be used instead of $x_e/2$. For the outlet tube of the middle riser section, $\Sigma \xi_M = 3.3$, $L = 7.99$ m; for the rear riser section, $\Sigma \xi_M = 2.5$ and $L = 6.47$ m. The tube roughness, $\Delta = 0.08$ mm and $\lambda = 0.01766$.

The two-phase frictional corrective coefficient, ψ, can be determined by Eqs. (8.97) and (8.98), depending on the values of ρV. After calculation $\psi = 1.015$, 0.976, and 0.978.

For the outlet tube of the middle riser section:

$$\Delta p_o = \left(\psi \times 0.01766 \times \frac{7.99}{0.125} + 3.3\right)\frac{V_0^2}{2}790\left[1 + x_e\left(\frac{790}{22.3} - 1\right)\right]$$

$$= 12{,}088.5,\ 29{,}100,\ \text{and}\ 51{,}959\ \text{Pa}$$

For the outlet tube of the rear riser section:

$$\Delta p_o = \left(\psi \times 0.01766 \times \frac{6.47}{0.125} + 2.5\right)\frac{V_0^2}{2}790\left[1 + x_e\left(\frac{790}{22.3} - 1\right)\right]$$

$$= 8876.3,\ 21{,}573.8,\ \text{and}\ 38{,}744\ \text{Pa}$$

Calculation of $\Sigma \rho_i h_i g$ of the Riser

$$\sum \rho_i h_i g = \rho_1 h_{\text{uh1}} g + \rho_m h_{\text{ev}} g + \rho'_m h_{\text{uh2}} g$$

where ρ_m and ρ'_m are the mixture densities in the heated evaporating portion of the riser and in the mixture outlet tube, kg/m^3, respectively; and h_{uh2} is the height of the mixture outlet tube, m.

The calculation procedures for the middle riser section and the rear riser section are listed in Table 8.20.

TABLE 8.20 Calculation of $\Sigma\rho_i h_i g$ of the Riser

Nomenclature and Equation	Unit	Middle Riser Section			Rear Riser Section		
Circulating velocity, V_0 (selected)	m/s	0.5	1.0	1.5	0.5	1.0	1.5
Circulating flow rate, W	kg/s	23.5	47	70.5	23.5	47	70.5
h_{uh1} (given)	m		0.9			0.69	
$\rho_1 h_{uh1} g = 790 h_{uh1} \cdot 9.8$	Pa		6,967.8			5,341.98	
h_{ev} (given)	m		14.73			14.94	
Exit quality, x_e	—	0.101	0.050	0.0337	0.0948	0.047	0.0316
Average quality, $\bar{x}$	—	0.0505	0.025	0.0169	0.0474	0.0235	0.0158
Average volumetrical quality, $\bar{\beta}$ [Eq. (8.92)] (use $\bar{x}$)	—	0.653	0.476	0.378	0.638	0.46	0.363
Slip ratio, S [Eq. (8.91)]	—	1.935	1.5	1.354	1.913	1.489	1.347
Average void fraction, α [Eq. (8.90)]	—	0.493	0.377	0.31	0.4795	0.364	0.297
Average mixture density							
ρ_m [Eq. (8.89)]	kg/m^3	411.52	500.57	552	421.89	510.56	562
$\rho_m h_{ev} g$	Pa	59,404.6	72,259.3	79,683.4	61,769.8	74,752.1	82,283.5
h_{uh2} (given)	m		2.42			2.42	
Average quality, $\bar{x}$	—	0.101	0.050	0.0337	0.0948	0.047	0.0316
$\bar{\beta}$ [Eq. (8.92)]	—	0.799	0.65	0.5526	0.788	0.636	0.536
S [Eq. (8.91)]	—	1.745	1.418	1.292	1.733	1.409	1.233
α [Eq. (8.90)]	—	0.695	0.567	0.489	0.682	0.5536	0.484
ρ'_m [Eq. (8.89)]	kg/m^3	256.44	354.71	414.6	266.43	365	418.43
$\rho'_m h_{uh2} g$	Pa	6,082	8,412.4	9,832.7	6,318.7	8,656.3	9,925.3
$\Sigma\rho_i h_i g$	Pa	72,454.4	87,639.5	96,483.9	73,430.4	88,750.4	97,549

Hydraulic Resistance of the Steam – Water Separator, Δp_{se} The mixture density, ρ_m, in the steam–water separator is equal to that in the mixture outlet tubes, ρ'_m (listed in Table 8.20).

The flow area of the separator for each riser section is equal to $A_{se} = 0.0342\ m^2$; the area ratio of A_{se} to the flow area of a riser section, A_r, is equal to $A_{se}/A_r = 0.0342/0.0595 = 0.577$. The velocity of the mixture in the separator, $V_{se} = \rho_1 V_0 A_r/(\rho'_m A_{se}) = 790\ V_0/(0.577\ \rho'_m)$, where V_0 is the circulating velocity of the riser, m/s. For the middle riser section, when $V_0 = 0.5$, 1.0, and 1.5 m/s, $V_{se} = 2.68$, 3.88, and 4.976 m/s; for the rear riser section, when $V_0 = 0.5$, 1.0, and 1.5 m/s, $V_{se} = 2.58$, 3.76, and 4.92 m/s.

The hydraulic resistance of the separator, Δp_{se}, can be calculated as follows:

$$\Delta p_{se} = \xi_{se}\frac{V_{se}^2}{2}\rho'_m = 4.5\frac{V_{se}^2}{2}\rho'_m$$

For the middle riser section, when $V_0 = 0.5$, 1.0, and 1.5 m/s, $\Delta p_{se} = 4144$, 12,015, and 23,098 Pa; for the rear riser section, when $V_0 = 0.5$, 1.0, and 1.5 m/s, $\Delta p_{se} = 3990$, 11,610, and 22,789 Pa.

Calculation of Y_r and Y_r'

$$Y_r = \sum \rho_i h_i g + \Delta p_r + \Delta p_o + \Delta p_{se} + \Delta p_{ws}$$

$$Y_r' = \sum \rho_i h_i g + \Delta p_r$$

Let the subscripts m and r denote the middle section and the rear section, respectively. For the middle riser section, when $V_0 = 0.5$, 1.0, and 1.5 m/s or $W = 84.5$, 169, and 253.5 t/hr, Y_{rm} is equal to 94,333, 142,920, and 200,189 Pa, $Y'_{rm} = 76{,}216.63$, 94,289.6, and 108,156.7 Pa. Thus the curves of $Y_{rm} = f(w)$ and $Y'_{rm} = f(W)$ can be established.

For the rear riser section, when $V_0 = 0.5$, 1.0, and 1.5 m/s or $W = 84.5$, 169, and 253.5 t/hr, Y_{rr} is equal to 91,952, 136,563, and 188,835 Pa and $Y'_{rr} = 77{,}046.8$, 95,227.5, and 108,975.9 Pa. The curves of $Y_{rr} = f(W)$ and $Y'_{rr} = (W)$ can also be established.

In Fig. 8.48 the curves of $Y_d = f(W)$, $Y_{rm} = f(W)$, $Y_{rr} = f(W)$, $Y'_{rm} = f(W)$, and $Y'_{rr} = f(W)$ are drawn.

The total circulation characteristic curve, $Y_r = Y_{r1} + Y_{r2} = f(W)$.

The intersection of the curves $Y_d = f(W)$ and $Y_r = f(W)$ gives the working point, A, of the circuit, from which we obtain the total mass flow rate, $W = 325$ t/hr; the mass flow rate in the middle riser section, $W_m = 160$ t/hr; and that for the rear riser section, $W_r = 165$ t/hr. $Y'_{rm} = 9.1 \times 10^4$ Pa, $Y'_{rr} = 9.4 \times 10^4$ Pa.

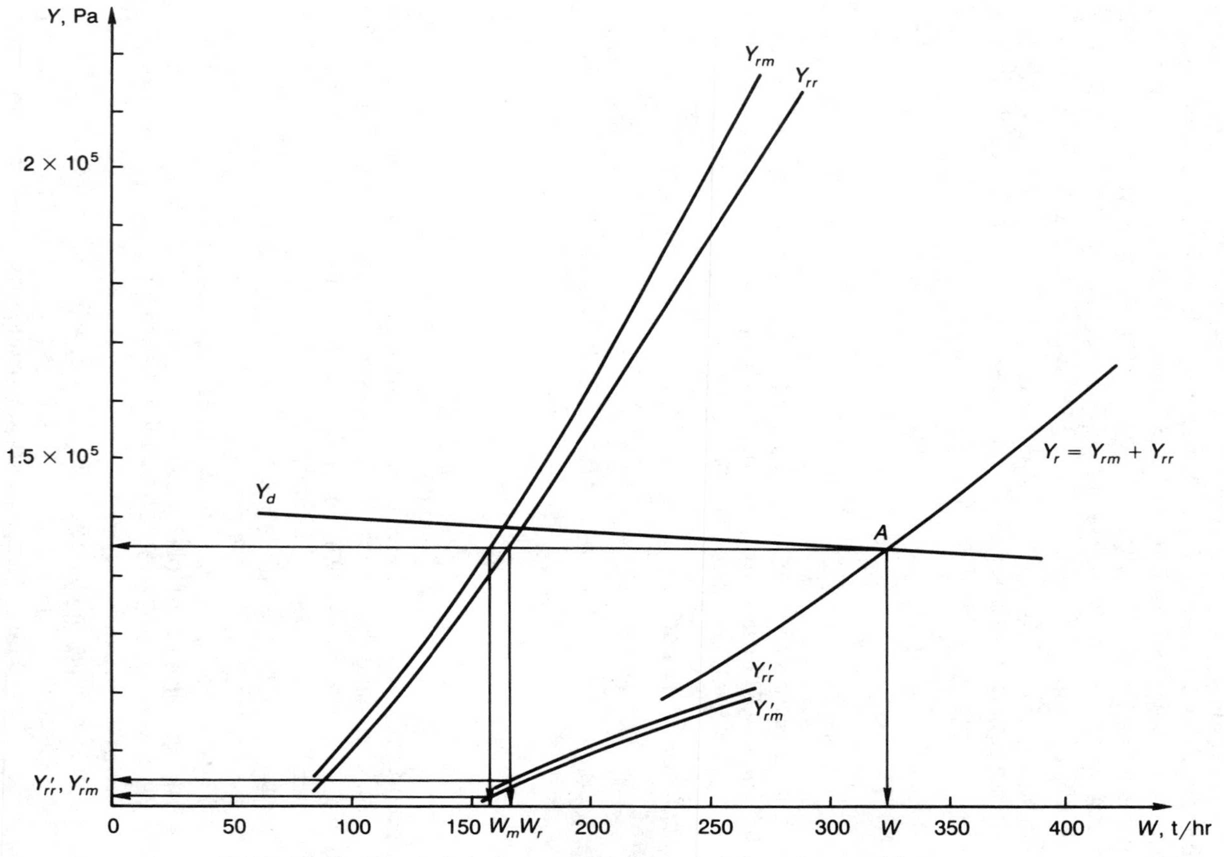

Fig. 8.48. Circulation characteristic curves of the example unit.

Checking Reliability The circulation ratio in the middle riser section, $K = W_m/W_s = 160 \times 10^3/(3600 \times 2.377) = 18.7$; in the rear riser section, $K = 165 \times 10^3/(3600 \times 2.227) = 20.58$; both are in the recommended range of Table 8.15.

The check for stagnation should be conducted for the least heated tube of the riser section. The coefficient of distribution of the heat absorption along the width of the middle riser section, $\eta_{wm} = 0.8$; and that for the rear riser section, $\eta_{wr} = 0.6$.

The superficial steam velocity at the exit of the middle riser section is equal to $W_s/(A_r\rho_g) = 2.377/(0.0595 \times 22.3) = 1.79$ m/s, and for the rear section it is equal to 1.678 m/s.

The average superficial steam velocity of the middle riser, $V_m'' = 1.79/2 = 0.895$ m/s, and for the rear riser section, $V_r'' = 1.678/2 = 0.839$ m/s.

The average superficial steam velocity of the least heated tube of the middle riser section, $V_0'' = \eta_{wm}V_m'' = 0.8 \times 0.895 = 0.716$ m/s, and for the rear riser section, $V_0'' = \eta_{wr}V_r'' = 0.6 \times 0.839 = 0.503$ m/s.

The void fraction, α_s can be obtained from Eq. (8.102). For the middle riser section, $\alpha_s = 0.716/(0.716 \times 1.004 + 0.4532) = 0.611$; for the rear riser section, $\alpha_s = 0.503/(0.503 \times 1.004 + 0.4532) = 0.525$.

The pressure difference of stagnation of the least heated riser, Δp_s, can be obtained from Eq. (8.101).

For the middle riser section

$$\Delta p_{sm} = 790 \times 98 \times 0.9[(1 - 0.611)790 + 0.611 \times 22.3] \times 9.8 \times 14.73$$

$$= 53{,}296 \text{ Pa}$$

For the rear riser section

$$\Delta p_{sr} = 790 \times 9.8 \times 0.69 + [(1 - 0.525)790 + 0.525 \times 22.3] \times 9.8 \times 14.94$$

$$= 56{,}655 \text{ Pa}$$

The reliability criterion [Eq. (8.100)] for the middle riser section, $Y_{rm}'/\Delta p_{sm} = 9.1 \times 10^4/53{,}296 = 1.7 > 1.05$; for the rear riser section, $Y_{rr}'/\Delta p_{sr} = 9.4 \times 10^4/56{,}655 = 1.66 > 1.05$; therefore circulation stagnation will not occur.

Since the working pressure, $p < 14$ MPa, and circulation ratio, $K > 4$, the heat transfer crisis need not be checked.

The velocity in the downcomer, $V_d = W/(A_d\rho_1) = 325 \times 10^3/(3600 \times 0.07 \times 790) = 1.63$ m/s < 3 m/s; therefore evaporation will not occur at the inlet of the downcomer.

The calculation shows the calculated circuit is reliable.

NOMENCLATURE

A	surface area; cross-sectional area; flow area, m^2
a	emissivity; fraction of ash
a	width, m
a_F	the coefficient of thermal radiation of furnace
B	fuel consumption of the boiler, kg/s
B_r	the rated fuel consumption of the boiler, kg/s
Bo	the Boltzmann number
b	depth, m
C	specific heat, kJ/(kg · K)
c	coefficient
d	tube diameter, m
g	gravitational acceleration, m/s^2
H	heat, kJ/kg
H_1^w	lower heating value of fuel, kJ/kg
H_{av}^w	available heat of fuel, kJ/kg
h	height, m
h	relative heat losses defined by Eq. (8.8)
h	heat transfer coefficient, $kW/(m^2 \cdot K)$
I	enthalpy of flue gases or air, kJ/kg
i	enthalpy of working fluid, kJ/kg
i_{lg}	latent heat of evaporation, kJ/kg
K	circulation rate
k	effective coefficient of absorption, 1/(m · MPa)
L	length or depth, m
M	coefficient relating to temperature field pattern in the furnace, defined by Eq. (8.31)
m	fraction of the luminous portion of the flame
n	rotation speed, r/min
n	number of tubes
Pr	Prandtl number, $\mu c_p/k$
p	pressure, MPa
q	heat flux, kW/m^2
q_F	heat release rate per unit cross-sectional area of the furnace, kW/m^2
q_R	fuel burning rate per unit volume, kW/m^2
q_v	heat release rate per unit volume, kW/m^3
R	grate area, m^2
Re	Reynolds number, $\rho V d/\mu$
r	volume concentration of triatomic gases; gas recirculation ratio
S	effective thickness of the radiating layer; spacing of tubes, m
S	slip ratio
T	temperature, K, °C
U	overall heat transfer coefficient, $kW/(m^2 \cdot K)$
V	volume, m^3
V	velocity, m/s

v specific volume, m^3/kg
W mass flow rate, kg/s
X the relative position of the highest temperature zone in the furnace defined by Eq. (8.32)
x angular coefficient; steam quality; fraction of heating surface of gas or air in a regenerative air heater; constant defined by Eq. (8.21)
Y the total pressure difference of the riser or the downcomer portion of a circuit defined by Eq. (8.85), Pa

Greek Symbols

α void fraction; excess air ratio
β volumetric steam quality; excess air ratio of air heater; coefficient of the influence of mutual heat exchange defined by Eq. (8.45)
δ thickness, m
Δ roughness of tube wall, m
ε fouling coefficient defined by Eq. (8.76)
η_b efficiency of boiler
η_h coefficient of distribution of heat absorption along the furnace height
ζ fouling coefficient of water wall tubes
λ frictional coefficient
λ thermal conductivity, $kW/(m \cdot K)$
μ_a dimensionless concentration of fly ash defined by Eq. (8.25)
ξ coefficient of utilization; coefficient considering the influence of fuel
ξ_M coefficient of minor losses
ρ density of fluid, kg/m^3
σ_0 emissivity of the black body, $kW/(m^2 \cdot K^4)$
τ coefficient of orifice thickness defined by Eq. (8.112)
π transcendental irrational number = 3.14159...
ϕ heat retention coefficient defined by Eq. (8.11); coefficient relating to tube spacing defined by Eq. (8.61)
ψ connective coefficient of mean temperature difference; coefficient of effectiveness defined by Eq. (8.51); two-phase frictional corrective coefficient
ψ_{ef} average coefficient of thermal efficiency of water walls

Subscripts

A air heater
a adiabatic; air; ash
aw ashy tube wall
B bend, tube bundle
b heat balance condition
bw blow-off water
c critical; circuit; counterflow system; coke particles
ca cold air
d downcomer

e	elevation; exit condition; equivalent
E	economizer
ec	economizer portion
en	entrance
ev	evaporating portion
ex	exit
f	frictional; working fluid
F	furnace; fuel
fa	fly ash
fl	flame
fw	feed water
g	flue gas; vapor
h	along height direction
ha	hot air
i	inlet condition; inside of a tube
l	length; liquid
o	outlet; orifice; outside of a tube
P	pump
p	parallel-flow system, platen
hp	physical heat
R	grate area; reheater; empty room
r	riser; radiation
rh	reheater
S	slag screen
s	steam; stagnation; soot particles
SS	superheater
sub	subcooling
t	heat transfer condition; temperature
u	useful
uh	unheated
V	volume
w	water; wall
wg	waste gases
wl	water level
ws	water supply
ww	water wall
–	mean value
0	theoretical

REFERENCES

1. Lin, Z. H., and Zhang, Y. Z. (1988) *Handbook of Boilers*. Mechanical Industry Publishing House, Peking.
2. Kuznetsov, N. V., and Mitor, V. V. (eds.) (1973) *Heat Calculations of Boiler Plants (Standard Method)*. Energiya Publishing House, Moscow.

3. *Mechanical Engineering Handbook*, Vol. 71, compiled by the Shanghai Boiler Institute. Mechanical Industry Publishing House, Peking.
4. He, B. A., Zhao, Z. H., and Qin, Y. K. (1987) *The Design and Operation of Pulverized Coal Burners*. Mechanical Industry Publishing House, Peking.
5. Blokh, A. G. (1984) *Heat Transfer in Steam Boiler Furnaces*. Energiatom Publishing House, Leningrad. Its English translation was published by Hemisphere, New York, 1988.
6. Lokshin, V. A., Peterson, D. F., and Schwarz, A. L. (eds.) (1978) *Standard Methods of Hydraulic Design for Power Boilers*. Energia Publishing House, Moscow. Its English translation was published by Hemisphere, New York, 1988.
7. *Hydraulic Calculations of Utility Boilers*. JB/Z 201-83, approved by the Ministry of Mechanical Industry of China, 1983.
8. Lin, Z. H. (1978) *The Calculation of Void Fraction and Frictional Resistance of Steam–Water Two-Phase Flow* (A Special Report for Making the Method of Hydraulic Calculation of Utility Boilers). Xi'an Jiaotong University Scientific Report 78-035.
9. Habenski, V. B., Baldina, O. M., and Kalinin, R. I. (1973) *Achievements in the Studying Region of Two-Phase Heat Transfer and Hydraulics in Elements of Power Devices*. Nauka, Moscow.
10. Lin, Z. H. (1983) Soviet and Chinese research works on vapor–liquid two-phase flows. In *Thermal Science 16*, T. N. Veziroglu (ed.), Vol. 2. Hemisphere, New York.
11. *Hydraulic Calculations for Power Boilers* (Standard Method), compiled by the Soviet Thermal Engineering Institute and Central Boiler and Turbine Institute. Energiya Publishing House, Moscow, 1966.

CHAPTER 9

NUCLEAR STEAM GENERATORS AND WASTE HEAT BOILERS

J. G. COLLIER

Nuclear Electric plc
Barnett Way
Barnwood
Gloucester GL4 7RS, United Kingdom

9.1 ABSTRACT

This chapter describes the principal types of modern unfired steam raiser, concentrating on land-based units in service in the power and process chemical industries, the methods used for the thermal design of the various types, and the common problems encountered during the operation of steam-raising equipment.

9.2 INTRODUCTION

The idea of boiling water to create steam seems to have originated with the Greeks and Romans who used boilers in their households. One of the earliest recorded boilers operating on the water tube principle supplied steam to Hero's engine, a hollow sphere mounted on hollow trunnions which permitted steam to pass into the sphere. The steam exhausted through two offset nozzles that caused the sphere to revolve, thus providing the world's first steam turbine. That was in 130 A.D. For the next 1600 years boilers seemed to be little used until around 1700 when the first commercial steam engines were produced by Savery in 1968 and Newcomen in 1705. This was the start of the Industrial Revolution and since then water has been evaporated into steam to meet every need of transportation and industry.

Basically, a boiler, alternatively known as a steam "raiser" or steam "generator," consists of a means of containing a volume of water within a

Boilers, Evaporators and Condensers, Edited by Sadik Kakaç
ISBN 0-471-62170-6

tank or tubes and a method of heating either by combustion of a fuel or by use of a hot fluid, gas, or liquid. When a boiler is heated by the direct combustion of a fuel, it is referred to as a "fired" boiler; when it is heated indirectly by a gas or a liquid from an industrial process, it is called an "unfired" or "waste heat" boiler. This chapter concentrates on the latter type.

More recently nuclear power, through the fission of the uranium atom, has provided a further means of raising steam for electricity production. Although steam can be raised directly in the core of nuclear reactor, this subject is beyond the scope of this book. However, the various designs of steam generators heated indirectly by the nuclear reactor coolant will be discussed since they are a particular form of "waste heat boiler."

In the early boilers the pressure at which the steam was raised was low, typically only 1 bar above atmospheric pressure. Watt appreciated the advantages to be gained by higher pressures in terms of thermal efficiency and around 1770 attempted to construct a boiler operating at 4 bar pressure. Nevertheless up until about 100 years ago pressures remained low. Despite this, boiler explosions in locomotives, ships, and on land were all too frequent. In the period 1816–1848, at least 233 steamboats used on U.S. waterways exploded resulting in the deaths of approximately 2560 persons plus 2100 injuries.

These tragedies prompted the enactment of the first sets of steam boiler construction rules from which the modern pressure vessel codes like ASME III and VIII and BS5500 are derived. Slowly boiler plants become more reliable and steam pressures were raised. By 1900 pressures of 20 bar were common and there followed a steady increase in both the pressure and size of land-based boilers up to the 1960s where, for electricity production, high subcritical and supercritical pressures were standard and unit sizes for fossil-fired plant were in the 2000 to 3000 MW(t) range.

9.3 THE PRINCIPAL TYPES OF BOILER

9.3.1 Nuclear Power Plants Nuclear power plants [1] in common use around the world can be classified into four main types.

Light-Water Reactor (LWR) Two types of light-water reactor are in operation, namely:

1. The pressurized-water reactor (PWR) where the reactor core is cooled by ordinary (light) water at a pressure around 160 bar. Hot water from the reactor pressure vessel is transferred via two or more coolant loops each containing a steam generator.
2. The boiling-water reactor (BWR) where boiling occurs directly within the core of the reactor and the water–steam mixture passes to separa-

tors where the steam is dried before going directly to the turbine. The separated water is returned to the core with the aid of recirculating pumps.

Heavy-Water Reactor (HWR) In this case the nuclear fuel is contained inside individual pressure tanks within the reactor core and is cooled by heavy water at a pressure of around 100 bar. The heat generated in the reactor is transferred to steam generators (of a similar design to those used in pressurized-water reactors) via a "figure of eight" coolant circuit.

Gas-Cooled Reactor (GCR) In this case the nuclear fuel, contained in channels within a graphite moderator, is cooled by either carbon dioxide or helium at a pressure up to 60 bar. The primary coolant system for a modern gas-cooled reactor is usually contained within a prestressed concrete pressure vessel with the steam generators housed either within the vessel or in cavities within the wall of the vessel.

Liquid-Metal-Cooled Reactor (LMFBR) This reactor, which operates with high-energy ("fast") neutrons and utilizes uranium fuel significantly more efficiently that other nuclear reactor types, is cooled by low-pressure liquid sodium. The reactor may be of the "loop" or "pool" design. In the former case, hot sodium from the reactor, is passed via pipework to an intermediate heat exchanger where the heat is given up to a secondary sodium circuit which in turns heats the steam generator. In the "pool" design all the primary circuit components including the core, the pumps, and the intermediate heat exchanger are immersed in a large pool of liquid sodium.

Steam Generators for Water-Cooled Reactors Most steam generators used with PWRs and HWRs consist of a vertically mounted shell containing a bundle of tubes in the form of an inverted U (Fig. 9.1). The shell consists of two separate sections; an evaporator section containing the tube bundle and the larger-diameter stream drum section where the steam is separated and dried.

The hot high-pressure water from the reactor core flows into the channel head at the base of the unit, through the inside of the inverted U-tube bundle containing some 78 km of tubing and back to the channel head. A partition plate divides the channel head into inlet and outlet sections. The channel head is fabricated from ferritic steel and clad internally with stainless steel. The tubes are usually fabricated from Inconel 600 or Incoloy 800 and are mounted on a thick ferritic steel tube plate also clad on the primary side with Inconel. The tubes are rolled into the tube plate, welded to the primary side cladding, and supported at intervals by tube support plates. Feed water enters the steam generator in the upper shell and mixes with water separated from the steam by the swirl vane separators. This water flows down the

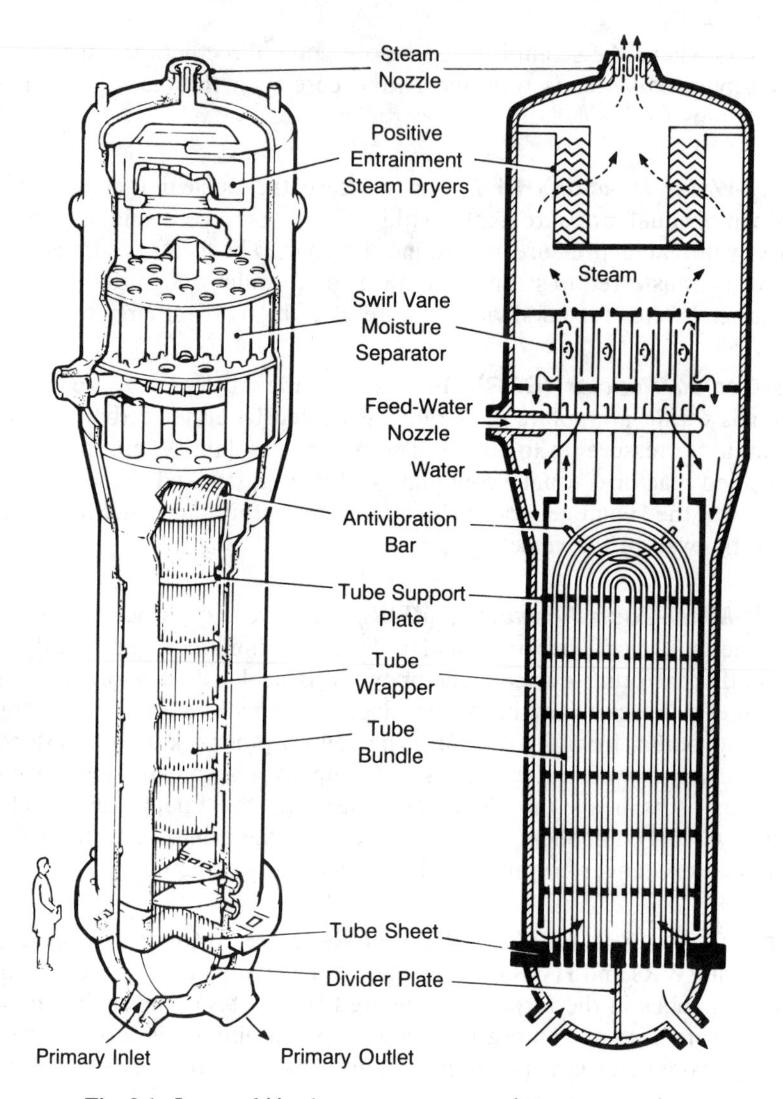

Fig. 9.1. Inverted U-tube steam generator (Westinghouse).

annulus between the steam generator shell and a baffle surrounding the tube bundle. When the water reaches the tube plate, it flows radially across the upper surface of the tube plate into the tube nest. Boiling occurs on the outside surfaces of the tubes within the bundle and the steam–water mixture passes upwards into the swirl vane separators. Natural circulation is induced as a result of the density difference within the bundle and the annular downcomer. The steam from the separators passes through impingement-type driers and exits from the top of the shell. Difficulties have been experienced with some designs of inverted U-tube steam generators in maintaining the integrity of the boundary between the primary and secondary (shell) side. A detailed review of these difficulties has been given by Green [2].

Whilst the majority of PWR and HWR power plants are equipped with steam generators of the vertical shell, inverted U-tube recirculating design, other steam generator designs are employed in some PWR plants.

In the original Shippingport PWR plant [3], the steam generator equipment consisted of four units each comprising a heat exchanger, a steam drum, and connecting piping. Two different types of heat exchanger, a horizontal U-tube design supplied by Babcock and Wilcox and a horizontal straight-tube design supplied by Foster Wheeler, were installed to evaluate the relative performance of the two designs (Fig. 9.2). Each heat exchanger was of the shell-and-tube type. Primary coolant flowed through the tubes and steam was generated on the shell side. The steam–water mixture passed up the risers to the steam drum where standard separators and driers were used to separate the water from the steam. The water returned to the lower heat exchanger via the downcomers. The Babcock and Wilcox design was rated at 75 MW(t) and contained 921 × 19 mm stainless steel tubes 15 m long. The Foster Wheeler design was also rated at 75 MW(t) and contained 2096 × 12.7-mm stainless steel tubes 9.5 m long.

Horizontal natural-circulation steam generators are also widely used in PWRs constructed in the USSR [4]. The units for the 440-MW(e) plant consist of a horizontal shell 11.5 m long and 3 m in diameter (Fig. 9.3). Vertical tubular headers located half way along the shell act as the inlet and outlet for the primary coolant. Horizontal bundles of 5536 U tubes mounted on these headers provide the heat transfer surface. These particular units are rated at 250 MW(t), but units of 800 MW(t) have been manufactured for the 1000-MW(e) plants.

In the United States, one PWR supplier, Babcock and Wilcox, has equipped its reactors with a vertical shell, straight-tube once-through steam generator (Fig. 9.4). The primary coolant enters the header at the top of the unit and flows down through the tubes to exit at the base. On the secondary side, the feed water is boiled in the interspace between the tubes, totally evaporated, and slightly superheated (by 30°C). The positioning of the feed nozzles and steam outlet on the shell and the use of some of the steam to preheat the feed water in the annulus around the tube bundle overcome the problem of differential thermal expansion of the tubes and shell. A feature of the once-through steam generator, which was significant in the accident at Three Mile Island in 1979, is the reduced water inventory in the unit compared with the recirculating design which, in turn, leads to a shorter time before the unit "dries out" in the event of a loss of feed water.

Even with the vertical shell, inverted U-tube recirculating units, there are significant differences between vendors in respect to design details such as thermohydraulic parameters, methods of construction, tube supports, and materials which profoundly influence their performance.

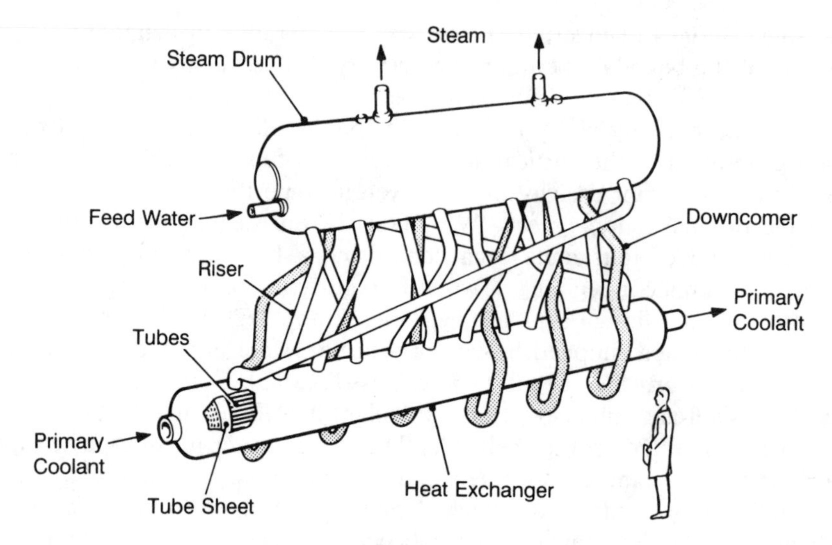

Foster Wheeler Steam Generator

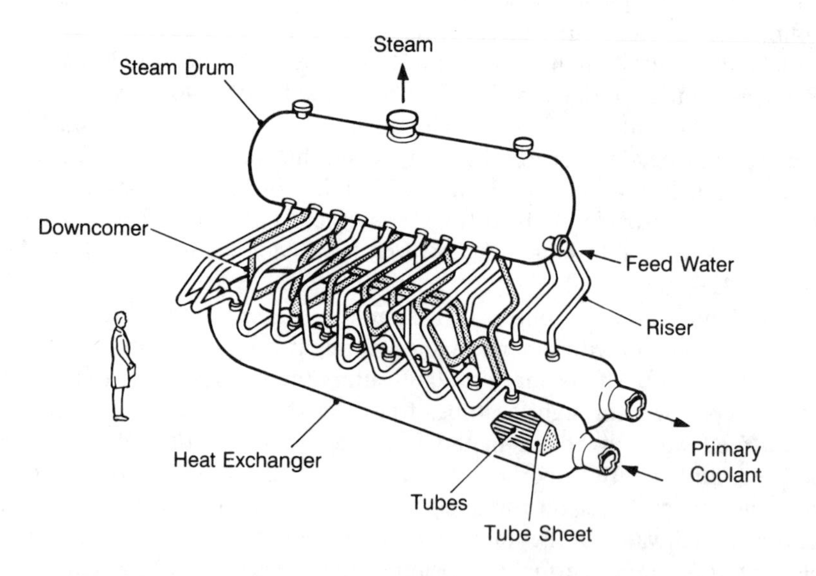

Babcock and Wilcox Steam Generator

Fig. 9.2. Two different designs of steam generator used at Shippingport (Babcock and Wilcox and Foster Wheeler).

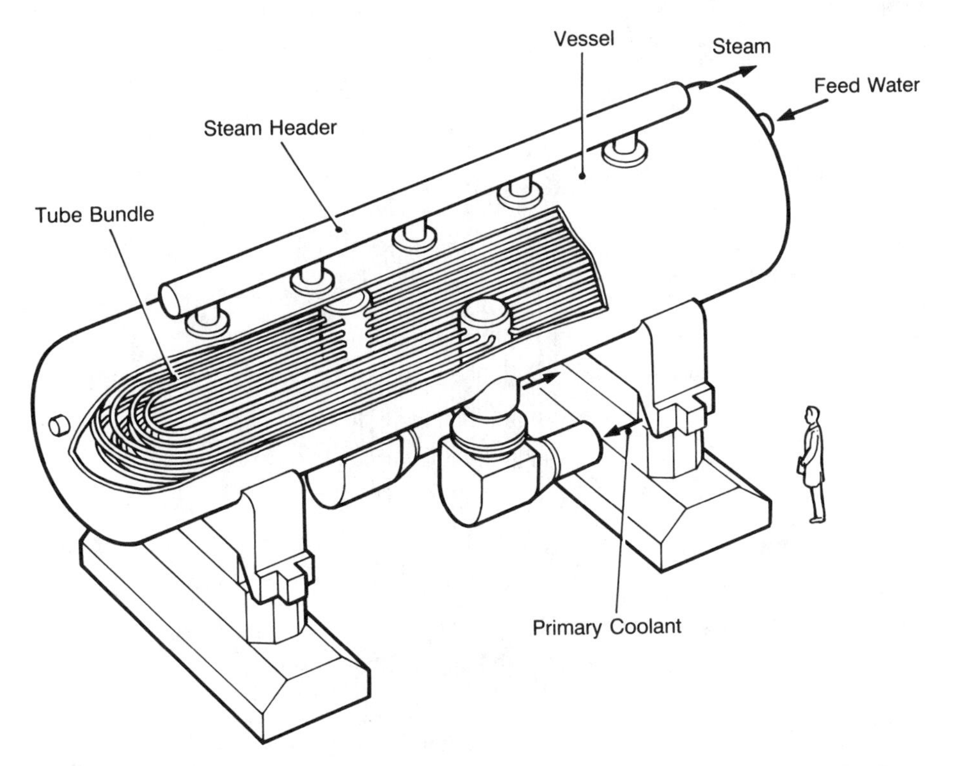

Fig. 9.3. Horizontal shell natural-circulation steam generator designs used in PWRs constructed in the USSR.

Some units are equipped with a feed-water preheating section or economizer located just above the tube plate on the cold leg side of the U tubes. The feed water enters the preheater and is heated almost to saturation temperature by countercurrent heat transfer from the reactor coolant within the tubes.

In the design of the unit offered by Foster Wheeler [5] (Fig. 9.5), the massive thick tube plate and channel head is dispensed with and is replaced by two cylindrical horizontal headers upon which the tube bundle is mounted directly. The primary reactor coolant passes through a vertical penetration in the steam generator shell to this horizontal header feeding the tube bank and exits by way of a similar header and shell penetration. The advantages claimed for this design include the avoidance of sludge deposition on the tube plate and the elimination of tube-to-tube plate crevices.

Tube support designs [6] are particularly important because of the consequences of corrosion of the tube support material. Since the corrosion products of carbon steel occupy about twice the volume of the original metal, it is possible, with some tube support designs, for the corrosion to dent the tubes and to distort the support plate itself. Figure 9.6 shows a variety of tube

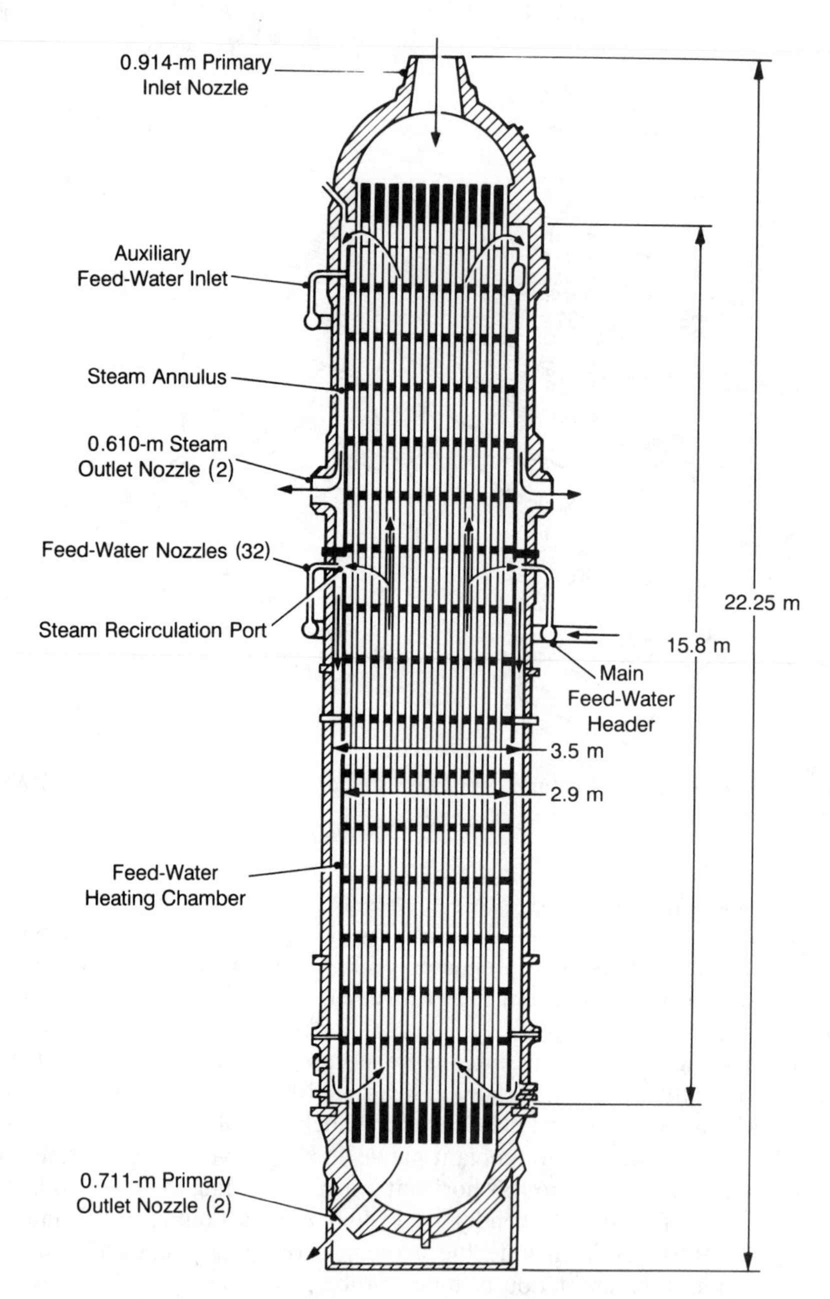

Fig. 9.4. Once-through steam generator supplied by Babcock and Wilcox.

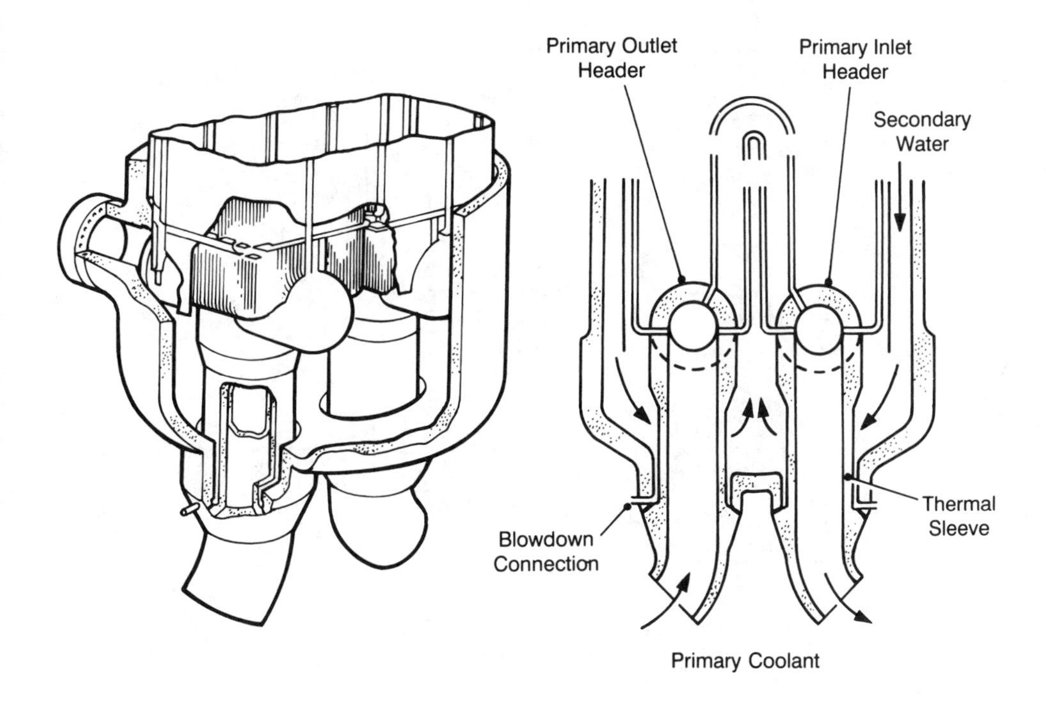

Fig. 9.5. Alternative design of inverted U-tube steam generator avoiding thick tube plate (Foster Wheeler).

support arrangements used in the steam generator units offered by the various PWR suppliers.

Steam Generators for Gas-Cooled Reactors Whilst early gas-cooled reactors (Magnox) built in the United Kingdom employed recirculating boilers to which the reactor coolant gas was passed via large-diameter ducts, steam generators used on modern gas-cooled reactors are of the once-through type–that is, the water is evaporated to steam and superheated in one single pass. This design associated with the integral circuit contained within the prestressed concrete pressure vessel gives a simple compact arrangement avoiding the need for steam drums and the associated penetrations in the pressure vessel wall. The steam generator not only has economizer, evaporator, and superheater sections but also a reheater section to reheat the steam between the high pressure and intermediate cylinders of the turbine.

Examples of two different designs of steam generator are given below. In the case of the nuclear stations at Hinkley Point B, Hunterston B, Heysham II, and Torness each 660-MW(e) reactor has four boilers each made up of three modules. The boilers are installed in the annulus between the reactor

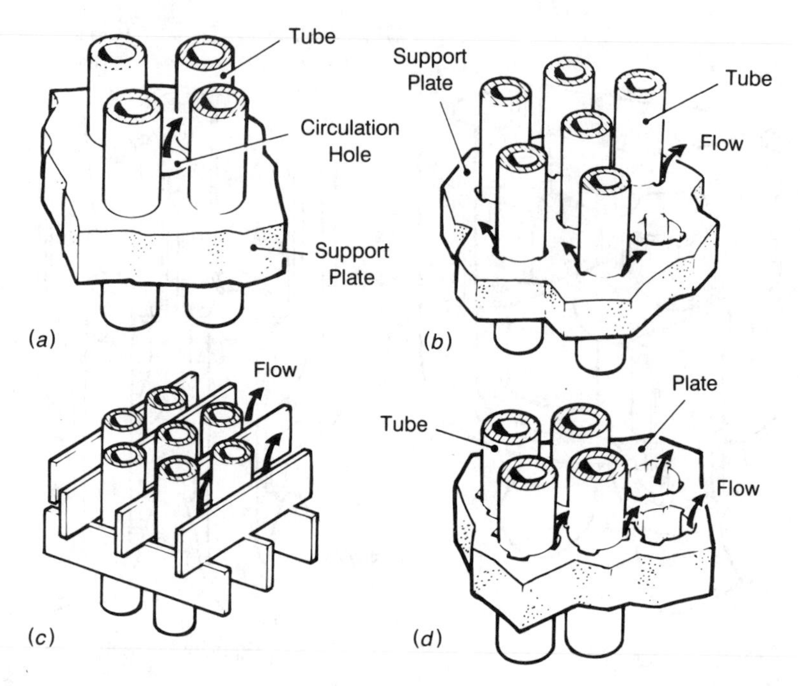

Fig. 9.6. Tube support plate designs (*a*) Westinghouse drilled hole, (*b*) Babcock and Wilcox trifoil plate for OTSG units, (*c*) KWU egg crate, and (*d*) Westinghouse Quatrafoil for model F units.

core and pressure vessel wall. Each of the 12 modules is 16 m high, weighs 120 tons, and contains 50 km of boiler tubing. The units (Fig. 9.7) are built up from 44 serpentine tube platens placed side by side and supported in a rectangular steel casing open at the top and bottom. The feed and steam connectors are taken out through penetrations in the side wall of the pressure vessel. The hot gas from the reactor flows downwards first through the reheater section and then through the superheater, evaporator, and economizer. The construction materials are austenitic steel (18% Cr, 12% Ni) above 520°C, 9% chrome steel between 520°C and 350°C for the evaporator section, and ferritic steel (1% Cr, 0.5% Mo) below 350°C. A transition section of Inconel 600 is placed between the 9% Cr and austenitic sections. This transition is so located that the local gas temperature does not exceed 550°C and the superheat in the steam at this point is not less than 70°C to avoid the risk of stress corrosion.

An alternative design of steam generator is employed at the Heysham A and Hartlepool reactors. In this case the boilers and gas circulators are unitized with the boilers arranged in eight vertical cavities or pods within the pressure vessel wall. Adjacent pairs of boiler units are connected via external circuits to form four boilers from an operational viewpoint. The helical boiler

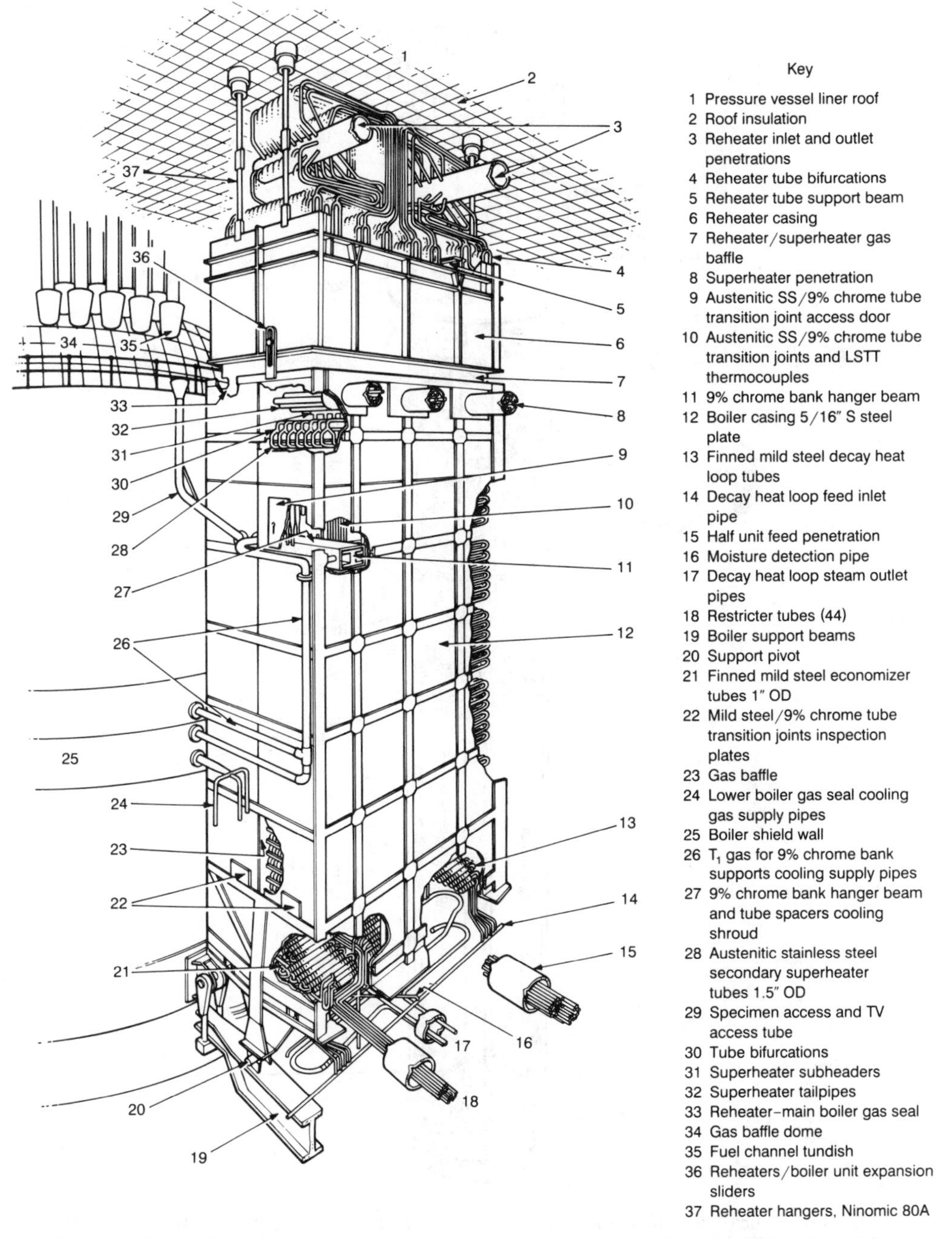

Fig. 9.7. Once-through steam generator for advanced gas-cooled reactor (Heysham II).

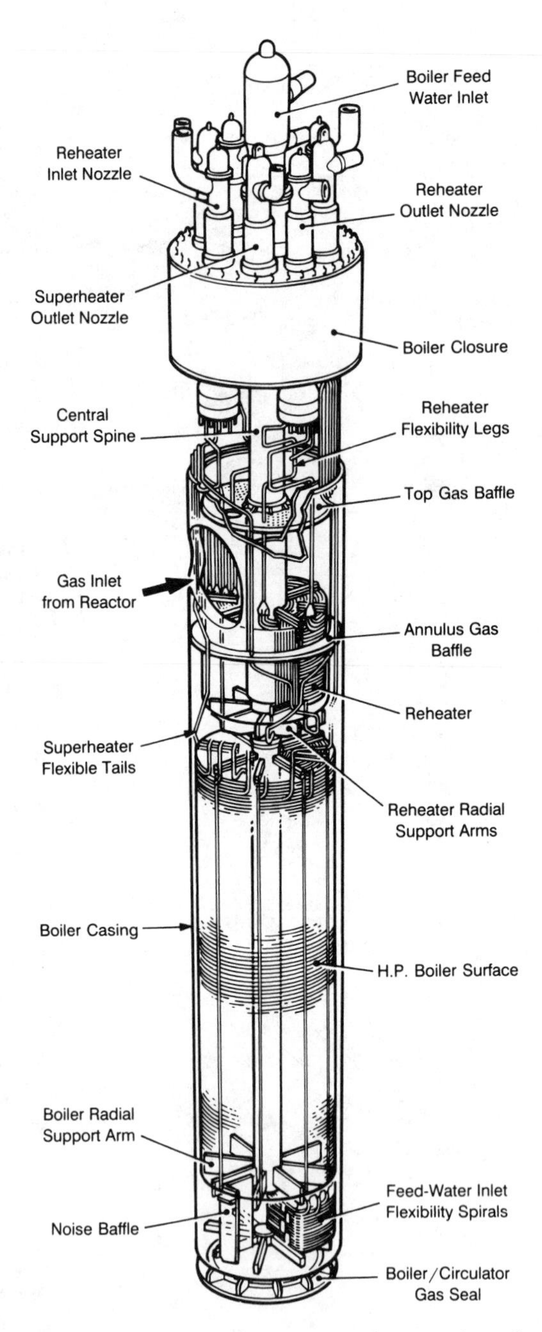

Fig. 9.8. Helical design of once-through steam generators for advanced gas-cooled reactor (Heysham I/Hartlepool).

TABLE 9.1 Main Design Parameter for Steam Generators of Gas-Cooled Reactors and Liquid-Metal-Cooled Reactors

	MAGNOX		AGR		HTR	LMFBR	
	Trawsfynydd	Oldbury	Hinkley B	Hartlepool	—	PFR	—
Type of boiler	Drum assisted recirculation	Once through	Once through	Once through	Once through	Forced circulation	Once through or forced circulation
Steam pressure, bar	65 20.6	96 52	165	165	165	165	165
Tube geometry	Serpentine (finned)	Serpentine (finned)	Serpentine (finned)	Helical (finned)	Helical	U tube	U-tube helical
Tube material							
Evaporator	Carbon steel	Carbon steel	Mild steel followed by 9Cr–1Mo	Mild steel followed by 9Cr–1Mo	As for AGR	2.25Cr–1Mo stabilized	—
Superheater primary	Carbon steel	Carbon steel	9Cr–1Mo	9Cr–1Mo	As for AGR	Austenitic stainless steel	Incoloy
Secondary	Carbon steel	Carbon steel	Austenitic stainless	Austenitic stainless	As for AGR	—	—
Tube dimensions							
Evaporator							
OD, mm	56	38	28.6	18 (fin base)	20–30	25	15–25
Thickness, mm	7	5	3.25	3	3–5	2.3	2–4
Superheater							
OD, mm	—	38	38	20	25–35	16	15–25
Thickness, mm	—	5	4.1	3	2.5	2	2–4
Heat flux, kW/m^2							
Evaporator							
Inlet	25	54 19	50	140	160	630	
Outlet	63	190 110	137	320	380	110	700 150
Peak	—	— —	137	320	380	700	110 700
Mass velocity, $kg/m^2 \cdot s$							
Evaporator at full load	575	410 540	1200	1650	2030	2580	3000

(Fig. 9.8) has the feed pipework running down a central core which is surrounded by a helically wound boiler tubing giving a long cylindrical shape to fit into the circular cavities. All the feed, superheater, and reheater connections are brought out through the top of the boiler. Thus the units could be completely works-fabricated ready for insertion into the pressure vessel.

Technical details and main design parameters of these gas-cooled reactor units are given in Table 9.1.

Steam Generators for Liquid-Metal Reactors A range of boiler designs have been employed with sodium-cooled reactors—both recirculating and once-through. The design of such units is primarily determined by the need to prevent or at least minimize the consequence of a sodium–water chemical reaction in the case of a leak in one of the boiler tubes.

Figure 9.9 shows an isometric view of the U-tube design of the evaporator used on the UK prototype fast reactor (PFR). A total of nine separate steam

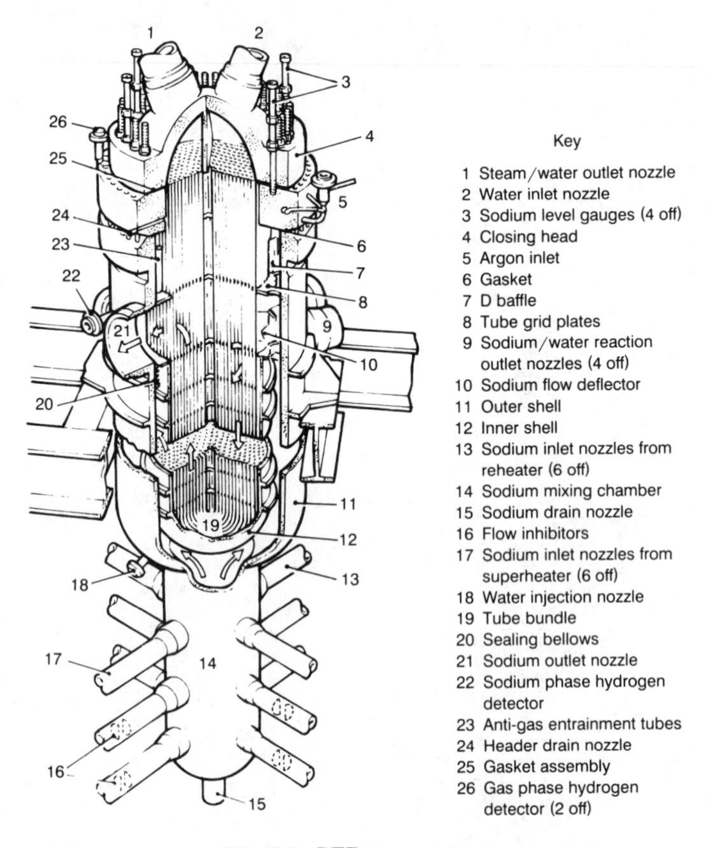

Fig. 9.9. PFR evaporators.

generator units are installed on this reactor—each consisting of a U-tube bundle installed within its own shell. The evaporator employs parallel flow whilst the superheater and reheater units operate with counterflow.

An alternative design of once-through steam generator is used on the French 1200-MW(e) Super Phenix fast reactor. Four secondary circuits each contain a single 750-MW(t) steam generator unit. Each unit comprises a vertical cylindrical shell enclosing a helically coiled tube bundle (Fig. 9.10).

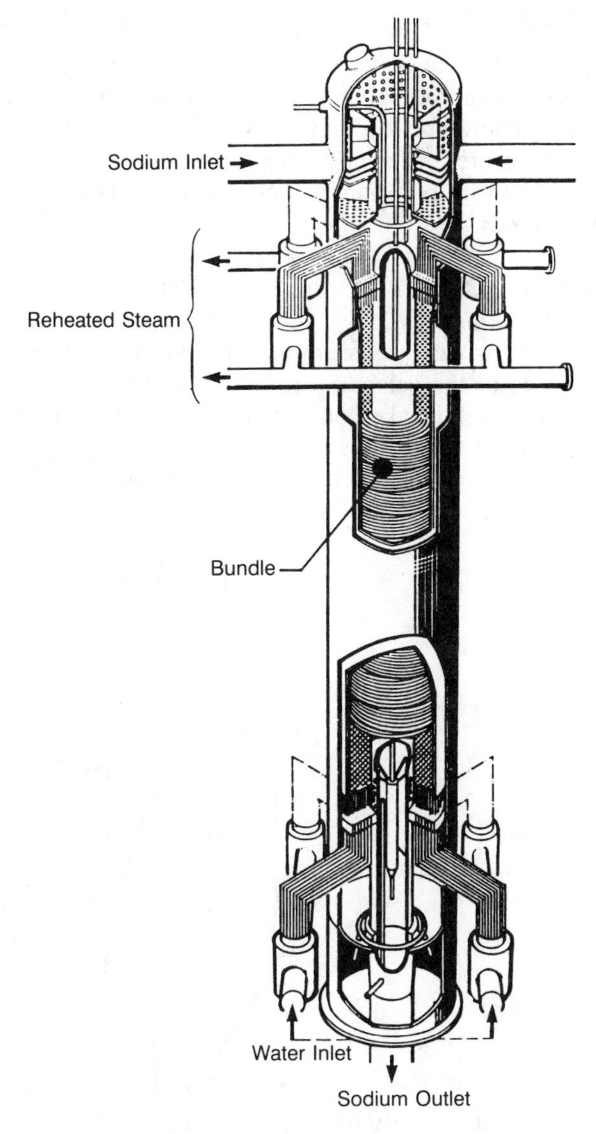

Fig. 9.10. SuperPhenix steam generator.

Sodium enters at the top of the unit via two opposed inlet nozzles and flows through the tube bundle to exit through a single outlet nozzle at the base of the unit. Feed water enters through thermally sleeved nozzles at the base of the units and water flows countercurrent to the sodium being first evaporated and then superheated to 490°C in a single pass before the steam exits the outer shell again via thermal sleeves. The 357 tubes per unit are fabricated from Incoloy 800 and 25 mm ID/2.5 mm wall thickness 91.5 m long.

9.3.2 Waste Heat Boilers

A wide variety of waste heat boiler designs have been constructed to recover heat from various chemical processes. Such boilers may have to accept process gases at temperatures up to 1000 to 1200°C and pressures up to 30 bar. As a result severe problems often arise due to local overheating sometimes associated with both fouling and corrosion. Typical of the current designs are the following.

Vertical Calandria Units In this fire-tube boiler design, a vertical calandria (Fig. 9.11) is formed by tubes located between two horizontal tube plates. The hot process gas passes upwards inside the tubes whilst water is fed to the shell side of the calandria at its base. The steam–water mixture generated within the shell is taken off at the top to a steam drum to separate the steam. The bottom tube sheet is exposed to the hot process gas, and cooling by the water on the shell side may become inadequate if the

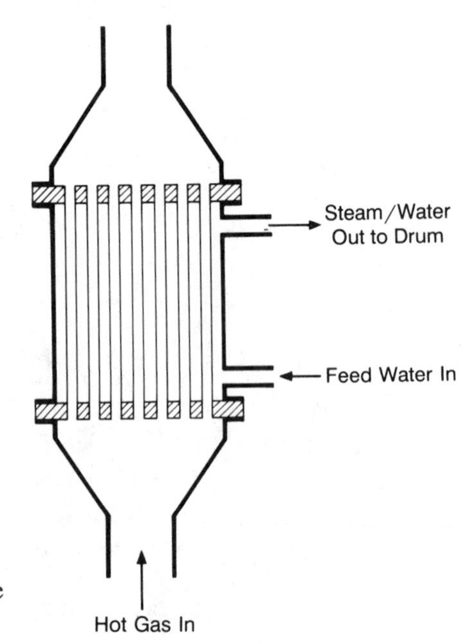

Fig. 9.11. Vertical calandria type waste heat boiler.

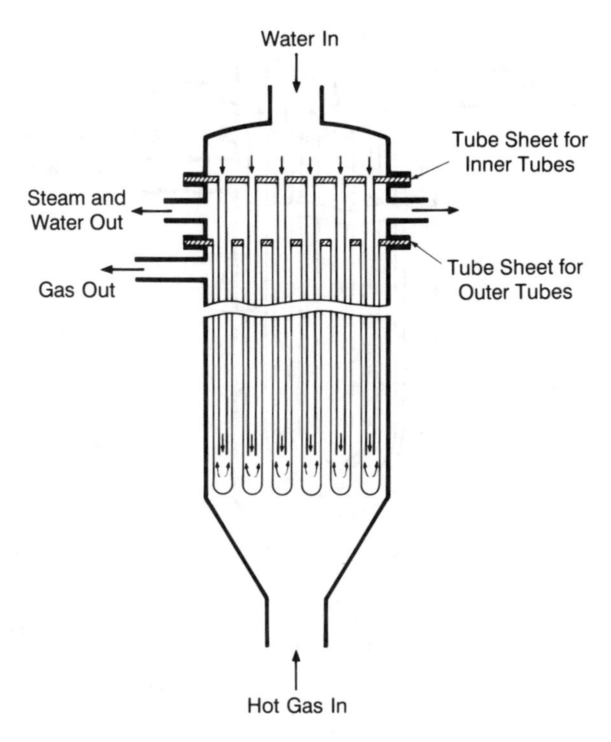

Fig. 9.12. Vertical bayonet-tube waste heat boiler.

circulation is impeded by the large volume of steam generated in this region. The use of ferrules at the entrance to the tubes can alleviate the problem, but it is difficult to find materials that will withstand the arduous conditions.

Bayonet-Tube Units A common waste heat boiler design, particularly in ammonia plants, is the bayonet-tube design (Fig. 9.12). In this design the problems of the lower tube plate are eliminated by having a reentrant design of water tube boiler. The water enters the unit from above, into an upper plenum, and thence to a series of downcomer tubes. At the bottom of the downcomer tube the flow reverses, and the steam–water mixture then flows upwards in the annulus between the inner downcomer tube and an outer tube which is in contact with the gas. The hot gas enters at the bottom of the vessel, flows across baffles, and exits at the top of the unit. These units have in the past suffered from problems of overheating at the point where the flow reverses at the base of the bayonet tube, from erosion, and from a rather complex tube sheet design.

Vertical U-tube Units In this arrangement the bayonet tubes are replaced by U tubes (Fig. 9.13); otherwise, the arrangement is similar to the bayonet-

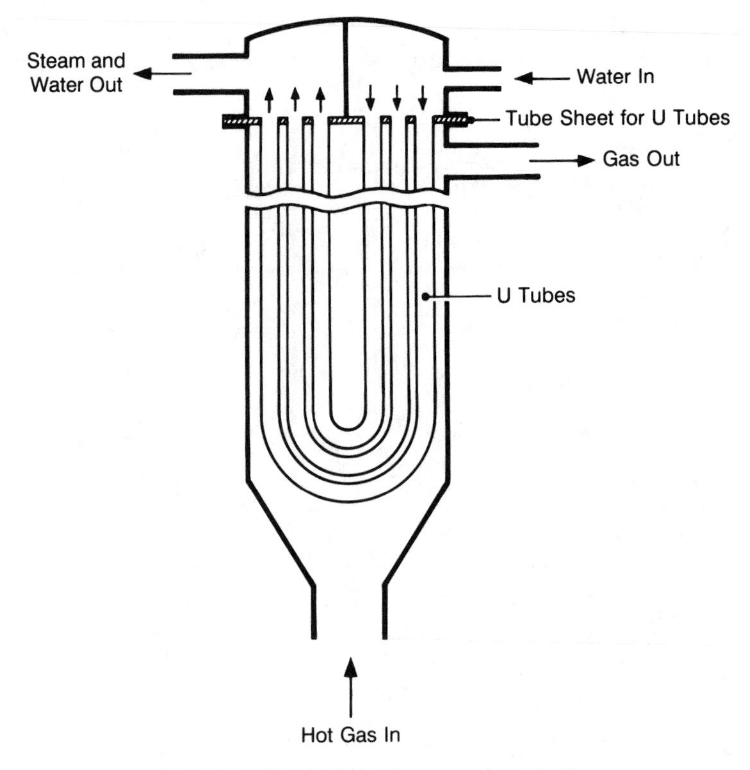

Fig. 9.13. Vertical U-tube waste heat boiler.

tube design. Forced circulation is preferred with this arrangement as natural circulation may be slow starting or may even occur in the wrong direction. The sodium-heated evaporator shown in Fig. 9.9 is an example of such a U-tube waste heat boiler design.

Horizontal U-Tube Units In this design a horizontal bundle of U tubes is mounted in the form of a shell-and-tube heat exchanger (Fig. 9.14). Water is fed to the lower limb of each U tube and the steam–water mixture generated within the unit passes out to a steam drum from the upper limb. Dryout heat fluxes are much lower in horizontal tubes compared with vertical tubes, and the situation is made more complex by the presence of the U bend. Flow separation effects can occur downstream of the bend inducing premature dryout and overheating. The problem may be alleviated by the insertion of twisted tapes into the tubes, but this increases the pressure drop. The problems of U-tube waste heat boilers have been discussed by Robertson [7].

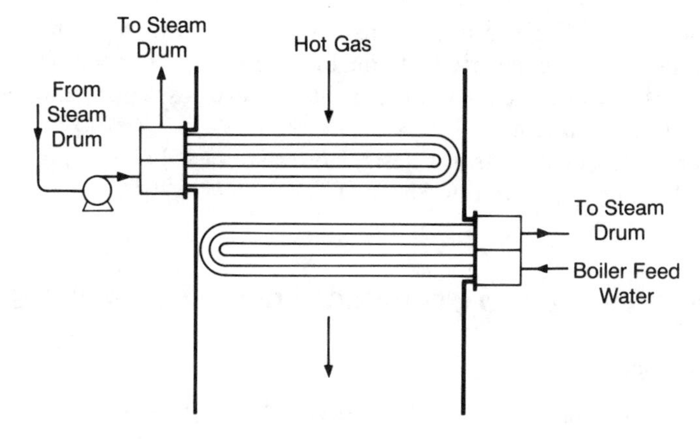

Fig. 9.14. Horizontal U-tube waste heat boiler.

Horizontal Crossflow Units This design of a fire-tube unit involves the hot process gases passing through a series of horizontal smoke tubes mounted between two vertically oriented tube plates (Fig. 9.15). The steam–water mixture generated on the shell side is taken off through a series of riser pipes to a steam drum and the separated water is returned to the shell via

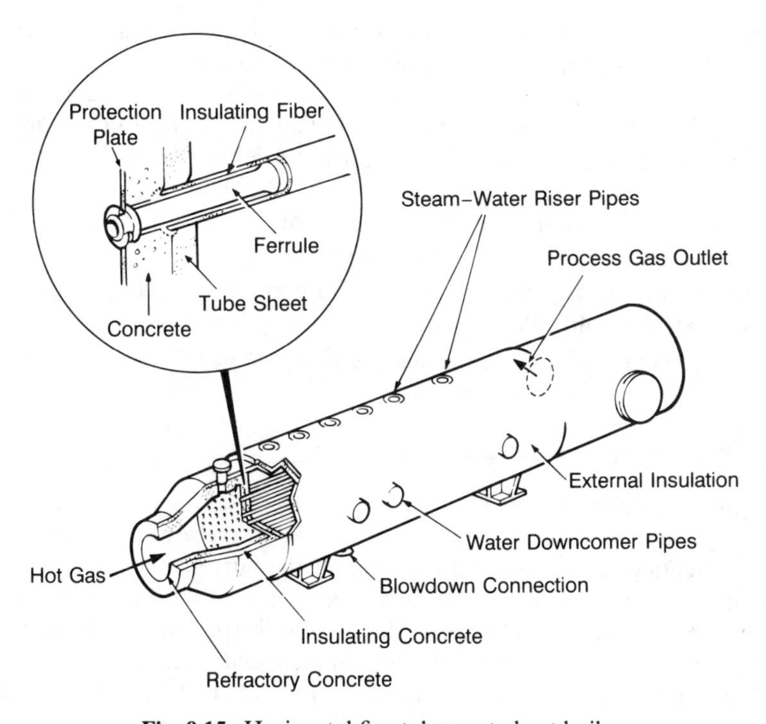

Fig. 9.15. Horizontal fire-tube waste heat boiler.

downcomer pipes. The gas inlet is insulated using refractory and insulating concrete. Ferrules are inserted at the tube entrance, but with the vertical tube plate less fouling occurs and the inside surface is swept by the recirculating steam–water mixture.

Further details of the various designs of waste heat boiler units have been given by Hinchley [8, 9] and by Smith [10].

9.4 THE THERMAL AND MECHANICAL DESIGN OF BOILERS

9.4.1 General

The design basis for waste heat boilers and for nuclear power plant boilers are similar. The basic steps are:

1. Selection of the type of waste heat boiler based on previous experience.
2. Preparation of a rough design to establish the basic dimensions of the unit. The surface area of the boiler can be estimated from

$$A = \frac{Q}{U \Delta T}$$

   where Q is the total heat load or duty, U is the overall heat transfer coefficient, and ΔT is the log-mean temperature difference between the primary side and the evaporating side. Guidance on values of the primary-side heat transfer coefficients have been given in a number of sources including Smith [10]. For hot water a value of 4000 $W/(m^2 \cdot K)$ might be used; whilst for hot gas 100 to 200 $W/(m^2 \cdot K)$, would be acceptable depending on the gas pressure. For liquid sodium 10,000 to 20,000 $W/(m^2 \cdot K)$ would be appropriate.
3. Estimation of the number, length, and diameter of tubes based on the appropriate fluid velocity and pressure drop limitations.
4. Refinement of the design using more accurate analytical procedures.

9.4.2 Primary Side (Unfired Boiler) Design

Certainly for nuclear power plant steam generators, it is often necessary to know in some considerable detail the flow and temperature patterns on the primary (heating) shell side. An example of the current techniques used to assess these flow–temperature fields has been given by Webster [11] for the PFR evaporator described earlier. The shell-side sodium flow is very complex, being convoluted and strongly three dimensional. The simulation model makes use of the "porous medium model" approach by Sha [12] in which the transport equations are solved numerically in conjunction with anisotropic continuum properties of porosity, permeability, and resistance drag. These properties reflect the detailed tube bundle and tube support geometries. The model grid involved 47,000 cells, each 68 mm square and 100 mm high. The results were provided in the form of velocity vectors and temperature contours. Similar three-dimensional models are available for steam generators installed in gas-cooled and water-cooled reactors. The porous medium approximation is a very powerful method for making detailed analyses of complex plant components such as U tubes, helical boilers, and the like. An example of the application of a two-dimensional computer code BESANT to predict the detailed flow and temperature distribution on the primary side of a sodium-heated once-through steam generator has been given by Brown [13].

9.4.3 Water-Side (Evaporator) Design

Boiling within Tubes The physical processes that occur when water is boiled within a vertical heated tube are complex, involving as they do the two-phase flow of steam and water in a variety of flow patterns and associated with these flow patterns many different heat transfer regimes. These processes have been described in detail by Collier [14] and it is not necessary to review them here. Basically, the boiler designer is concerned with just two questions:

1. What is the circulation rate within the boiler, is it sufficient, and is it stable for all likely operating conditions of the boiler?
2. Is the circulation and the heat flux distribution in all parts of the boiler sufficiently well matched to prevent the tube material overheating?

Circulation In a natural-circulation boiler the driving force to induce water circulation is the difference in density between the water in the downcomer and the steam–water mixture in the waste heat boiler and riser tubes. This circulation is only induced once boiling starts in the steam generator and increases as the volume of steam in the riser section increases and the density falls. Since the pressure drop around the complete circuit must be 0, the circulation is set by the condition that the difference in head between the riser and downcomer balances the frictional and momentum losses in the tubes, bends, headers, and separators.

Clearly, as the steam pressure rises toward the critical pressure, the difference in density between steam and water reduces with a consequent reduction in the driving force for natural circulation. Thus at high subcritical pressures, pumps are often installed at the base of the downcomers to assist circulation.

The thermohydraulic conditions within the boiler are usually incorporated into a computer-based model [15, 16]. The hydraulic circuit is modelled in a series of parallel–series pathways with nodes representing the steam drums, headers, and so on, connected by pathways representing a group of geometrically similar tubes exposed to similar heating conditions. Further subdivision is possible to distinguish between say tubes near the shell wall and those in the center of the bundle. Apart from geometrical data other input data required include the pressure and feed-water temperature in the drum. An essential element of the calculation is the evaluation of the pressure loss along the steam generator tube in which boiling occurs. This involves integrating a local pressure gradient along the tube length. This local pressure gradient is made up of three separate components: a frictional term, an accelerational term, and a static head term.

$$-\left(\frac{dP}{dz}\right)_{\mathrm{TP}} = -\left(\frac{dP}{dz}F\right) - \left(\frac{dP}{dz}A\right) - \left(\frac{dP}{dz}z\right) \tag{9.1}$$

Using the separated flow model, in which the steam and water phases are assumed to flow at different mean velocities each occupying a fraction of the total flow cross section, these three components are, respectively, given by

$$-\left(\frac{dP}{dz}F\right) = -\left(\frac{dP}{dz}F\right)_{\mathrm{LO}} \phi_{\mathrm{LO}}^2 \tag{9.2}$$

where ϕ_{LO}^2 is known as the two-phase frictional multiplier and

$$-\left(\frac{dP}{dz}F\right)_{\mathrm{LO}}$$

is the frictional pressure gradient calculated from the Fanning equation for the total flow (water and steam) assumed to be water:

$$-\left(\frac{dP}{dz}A\right) = G^2 \frac{d}{dz}\left(\frac{x^2 v_G}{\alpha} + \frac{(1-x)^2 v_L}{(1-\alpha)}\right) \tag{9.3}$$

where G is the total mass velocity, x is the steam quality, α is the void fraction (fraction of cross section occupied by steam), and v_G and v_L are, respectively, the steam and water specific volumes;

$$-\left(\frac{dP}{dz}z\right) = g \sin\theta \left(\frac{\alpha}{v_G} + \frac{(1-\alpha)}{v_L}\right) \tag{9.4}$$

where g is the acceleration due to gravity, and θ is the angle of inclination of the tube to the horizontal.

To evaluate the local pressure gradient, expressions are required for the functions ϕ_{LO}^2 and α. A very large number of correlations have been proposed for these functions, and a summary of the better known correlations for ϕ_{LO}^2 is given in Table 9.2. A number of workers have carried out

TABLE 9.2 Two-Phase Pressure Drop Correlations for Steam–Water Mixtures

	Correlation
Homogeneous Collier [14]	$\phi_{LO}^2 = \left[1 + x\left(\frac{\rho_L - \rho_G}{\rho_G}\right)\right]\left[1 + x\left(\frac{\mu_L - \mu_G}{\mu_G}\right)\right]^{-1/4}$
Baroczy [17]	$\phi_{LO}^2 = \Omega\phi_{LO}^2[G = 1356\ \text{kg}/(\text{m}^2 \cdot \text{s})]$ where $\phi_{LO}^2(G = 1356) = f\left(\left[(\mu_L/\mu_G)^{0.2}(\rho_G/\rho_L)\right], x\right)$, $\Omega = f\left(\left[(\mu_L/\mu_G)^{0.2}(\rho_G/\rho_L)\right], x\right)$
Chisholm [18]	$\phi_{LO}^2 = 1 + (\Gamma^2 - 1)[Bx^{(2-n)/2}(1 - x)^{(2-n)/2} + x^{2-n}]$ where $\Gamma = [(dP/dz)_{GO}/(dP/dz)_{LO}]^{1/2}$, $B = f(G, \Gamma)$
CISE Lombardi [19]	$\left(\frac{dP}{dz}\right)_{TP} = \left[\frac{KG^n\bar{v}^{0.86}\sigma^{0.4}}{D^{1.2}}\right]$ where $\bar{v} = [x/\rho_G + (1 - x)/\rho_L]$, $K = f$ (geometry), $n = f$ (geometry)
Martinelli and Nelson [20]	$\phi_{LO}^2 = fn(P, x)$
Smith and Macbeth [21]	$\phi_{LO}^2 = \left[\left\{\left(e(1 - x) + \left(\frac{\rho_L}{\rho_G}\right)x\right)(e(1 - x) + x)\right\}^{1/2} + (1 - e)(1 - x)\right]^2$ where $e = 0.4$

systematic comparisons between these various correlations and data banks containing large numbers of experimental pressure drop measurements for steam–water mixtures. A summary of these various comparisons is given in Table 9.3. Although Idsinga [24] concluded from his study that the homogeneous model was best when compared with his database, other studies agree that the most accurate correlations for ϕ_{LO}^2 are those of Baroczy [17], Chisholm [18], and CISE [19] but that in each case the standard deviation of errors about the mean is 30% to 35%. Perhaps this agreement is not so surprising given the considerable overlap in the database for steam–water pressure drops used by the various studies. More recently, Friedel [26] has published what is probably regarded as the most accurate generally available correlation for ϕ_{LO}^2.

To evaluate the changes in momentum (or kinetic energy) and also the mean density of a two-phase flow, it is necessary to be able to establish the local void fraction or fraction of the flow cross section occupied by steam. Once again a large number of correlations have been proposed for the evaluation of void fraction, α. Sometimes the correlation is expressed in terms of the slip velocity ratio, S, which is defined as the mean velocity of the steam phase divided by the mean velocity of the water phase and related to the void fraction by the identity:

$$S = \left(\frac{x}{1-x}\right)\left(\frac{\rho_L}{\rho_G}\right)\left(\frac{1-\alpha}{\alpha}\right) \tag{9.5}$$

A summary of the better known correlations for S or α is given in Table 9.4. Similarly, systematic comparisons have been carried out between the various void fraction correlations and data banks containing large numbers of experimental measures of either void fraction or fluid density measurements for steam–water mixtures. A summary of these comparisons is given in Table 9.5. It can be seen that the various studies agree that the most accurate void fraction correlations are those of Smith [29], CISE [30], and Chisholm [31], with the latter having the added advantage of great simplicity. Again, the standard deviation of error on the mean density is about 20% to 30%.

Having established a means of calculating the two-phase pressure loss in the steam generator tubes, the computer-based model is used to establish a set of flows such that the net pressure loss around the circuit is 0. The output from the model gives the flow, pressure, and steam quality for each pathway.

Dryout Flowing high-pressure water or steam–water flows provide good heat transfer and can remove relatively high heat flux levels with only small increases of tube wall temperature over the saturation temperature. However, if the heat flux is increased above a critical level, than a disproportionate increase in tube wall temperature occurs often resulting in damage to the boiler tube. This critical level of heating is usually referred to as the "critical" heat flux or "dry out" heat flux. It is a complex function of the tube diameter and the inclination, steam pressure, steam quality, and flow rate, as well as a

TABLE 9.3 Two-Phase Pressure Drop Data for Steam–Water Mixtures

Data Bank	ESDU [22]			Friedel [23]			Idsinga [24]			Ward [25]		
Flow Direction	Upflow, downflow, and horizontal			Upflow only			Upflow and horizontal			Upflow and horizontal		
Correlation	n	e	σ	n	e	σ	n	e	σ	n	e	σ
Homogeneous												
Collier [14]	1709	−13.0	34.2	2705	−19.9	42.0	2238	−26.0	22.8	4313	−23.1	34.6
Baroczy [17]	1447	4.2	30.5	2705	−11.6	36.7	2238	−8.8	29.7	4313	−2.2	30.8
Chisholm [18]	1536	19.0	36.0	2705	−3.8	36.0	2238	0.5	40.5	4313	13.9	34.4
CISE Lombardi [19]	—	—	—	2705	16.3	28.0	2225	22.6	28.9	—	—	—
Martinelli and Nelson [20]	1422	16.3	36.6	—	—	—	2238	47.8	43.7	—	—	—
Smith and Macbeth [21]	—	—	—	—	—	—	—	—	—	4313	−16.6	24.1

n = number of data points analyzed; e = mean error, % = $(\Delta p_{cal} - \Delta p_{exp}) \times 100/\Delta p_{exp}$; σ = standard deviation of errors about the mean, %.

TABLE 9.4 Void Fraction Correlations for Steam–Water Mixtures

	Correlation
Lockhart and Martinelli [27]	$\alpha = f(X)$ where $X = \left[\left(\frac{dP}{dz}\right)_L \Big/ \left(\frac{dP}{dz}\right)_G\right]^{1/2}$
Hughmark [28]	$\alpha = K$ where $K = f[\text{Re}, \text{Fr}, (1-\beta)]$
Smith [29]	$S = e + (1 - e)$ where $e = 0.4\sqrt{\frac{\rho_L/\rho_G + e(1/x - 1)}{1 + e(1/x - 1)}}$
CISE Premoli [30]	$S = f(G, D, \rho_L, \rho_G, \mu_L, \sigma, \beta) + 1$
Chisholm [31]	$S = \left[x\left(\frac{\rho_L}{\rho_G}\right) + (1 - x)\right]^{1/2}$
Thom [32]	$S = f\left(\frac{\rho_L}{\rho_G}\right)$
Bankoff and Jones [33]	$S = \left[\frac{1-\alpha}{A - \alpha + (1-A)\alpha^B}\right]$ where $A, B = f(p)$
Bryce [34]	$S = \left[\frac{1-\alpha}{A - \alpha + (1-A)\alpha^B}\right]$ where $A = f(P, G, X, \rho_G, \rho_L)$, $B = f(P, \rho_G, \rho_L)$

number of secondary variables. Boiler designers usually generate critical heat flux data experimentally simulating, as closely as possible, the expected thermal conditions within the steam generator. One experimental facility is that operated by the CEGB at its Marchwood Engineering Laboratories [36]. It has the capacity to test full-size tubes (32 to 52 mm ID) under realistic pressure, flow, steam quality, tube inclination, and heating. Typical results are shown in Figure 9.16.

The critical heat flux:

1. Decreases with increasing steam quality
2. Decreases with increasing pressure for a given steam quality and mass flux
3. Decreases at low flow rate and then increases again at high flow rate for a given steam quality and pressure

TABLE 9.5 Void Fraction Data for Steam–Water Mixtures

	Analysis of Mean Density						Analysis of Slip Ratio					
Data Bank	Freidel [23]			Bryce [34]			ESDU [35]			Bryce [34]		
Correlation	n	e	σ	n	e	σ	n	e	σ	n	e	σ
Lockhart and Martinelli [27]	—	—	—	—	—	—	598	−57.6	50.3	—	—	—
Hughmark [28]	484	−10.8	33.0	—	—	—	598	−9.1	29.2	—	—	—
Smith [29]	484	0.5	26.8	639	8.6	31.5	—	—	—	639	18.0	77.8
CISE Premoli [30]	484	9.3	35.0	639	−1.4	22.7	598	−23.7	27.2	639	−1.2	68.6
Chisholm [31]	484	−0.4	26.0	—	—	—	598	−14.5	30.8	—	—	—
Thom [32]	484	7.4	36.5	639	43.3	61.7	—	—	—	639	132.6	200.0
Bankoff and Jones [33]	—	—	—	639	9.32	31.6	—	—	—	639	34.5	137.6
Bryce [34]	—	—	—	639	0.1	20.7	—	—	—	639	6.1	86.9

n = number of data points analyzed; e = mean error, % = (cal − exp) × 100/exp; σ = standard deviation of error about the mean, %.

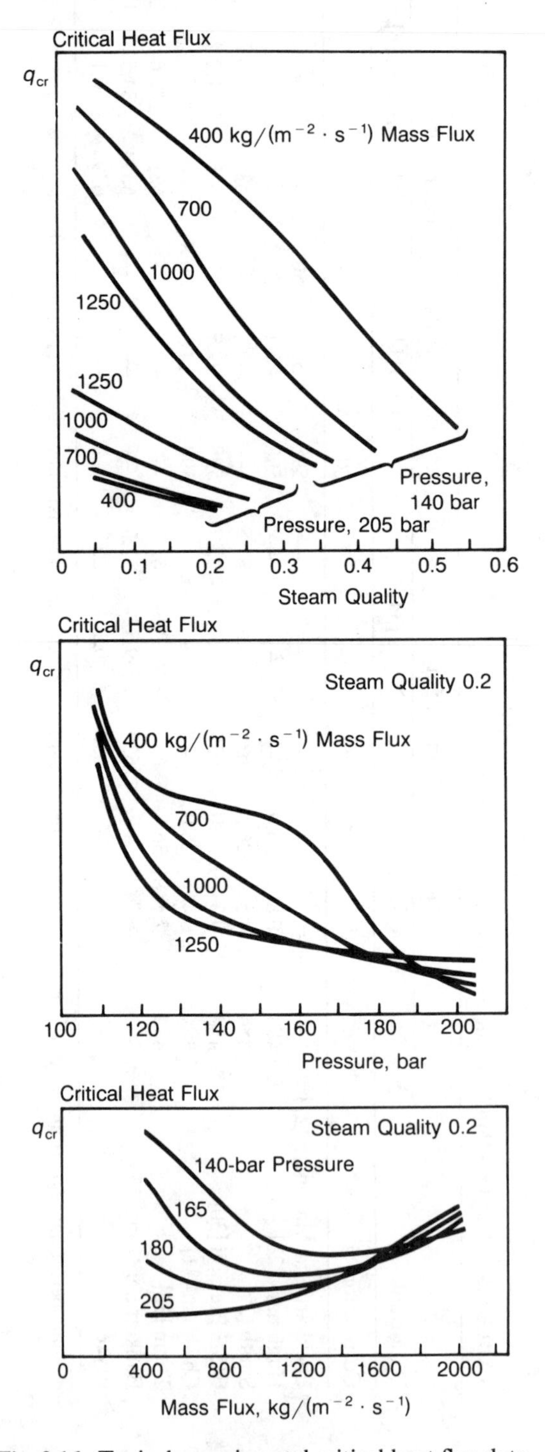

Fig. 9.16. Typical experimental critical heat flux data.

There are also effects due to nonuniform versus uniform heating and between vertical and horizontal tubes.

Dryout in a horizontal tube differs from that in a vertical tube in two ways:

1. Stratification of the flow can occur at low velocities for both low-quality and subcooled conditions. Such conditions can lead to overheating of steam boiler tubes at quite modest heat fluxes.
2. Dryout of the tube at high vapor qualities occurs over a relatively long tube length, starting at the top of the tube where the film thickness and flow rate are lowest and ending up with the final evaporation of the rivulet running along the bottom of the tube. Under these conditions the vapor flow in the upper part of the tube may become superheated before dryout occurs at the base of the tube.

Styrikovich and Miropolskii [37] reported the effects of stratification of a high-pressure steam–water mixture in a horizontal pipe. These caused wide temperature differences between the top and bottom of the boiler tube. Experiments were carried out on a single 7.5 m, 56 mm ID tube at pressures between 10 and 220 bar with heat fluxes in the range 22 to 135 kW/m^2 and inlet velocities between 0.24 and 1 m/s. It was found that there was a critical two-phase velocity, j, below which stratification occurred and above which it did not. Using an alcohol–water analog for steam–water flow, Gardner and Kubie [38] established an expression for the critical velocity, j.

$$j^{1.8} = 30.43 \frac{D^{0.2}\{\sigma g(\rho_L + \rho_G)\}^{0.5}}{\rho_L^{0.8}\mu_L^{0.2}} \tag{9.6}$$

Table 9.6 gives the values of j calculated from their equation for steam–water flows over a range of pressures and tube diameters. Excellent agreement was seen with the data of Styrikovich and Miropolskii and this approach is recommended as giving the minimum single- or two-phase velocity below which stratification will occur in a horizontal tube.

TABLE 9.6 Values of j, m / s, to Prevent Stratification for Steam–Water Flow in Horizontal Tubes

D, mm \ p, bar	33.5	64.2	112.9	146	165	187
20	2.71	2.47	2.11	1.82	1.63	1.37
40	2.92	2.67	2.28	1.97	1.77	1.48
60	3.06	2.80	2.38	2.06	1.85	1.55

Stratification may also occur in vertical bends and in helical coils. For the case of a vertical bend, Bailey [39] showed that the gravitational force maintaining stratified flow is supplemented by centrifugal forces.

Boiling on the Shell Side Boiling on the outside of a bundle of horizontal tubes occurs in shell boilers and in Soviet-designed nuclear steam generators. Although a large amount of research has been carried out with respect to the boiling of water at various pressures from submerged horizontal isolated tubes, considerably less information is available for tube arrays. What experimental evidence there is suggests an enhancement of heat transfer coefficients, especially for those locations within the tube bundle bathed by steam from tubes lower down in the bundle. One study by Cornwell et al. [40, 41] indicates that the heat transfer coefficient increases from the base to the top of the tube bundle by a factor of 3 to 6 as the steam quality rises. Those coefficients at the base of the bundle correspond to those expected for pool boiling on isolated tubes.

For a tube bundle the critical heat flux level is reduced compared with that for an isolated horizontal tube [42]. Various mechanisms are possible whereby a critical or limited heat flux (or bundle power) is reached within the tube bundle. These depend on the extent to which the bundle geometry permits or does not permit recirculation of the water to the base of the tube bundle. They may be briefly described as follows:

1. *Pool boiling critical heat flux*. This mechanism will occur in small bundles with widely spaced tubes. In this case the flow passages within the bundle are essentially filled with liquid and the limiting process is the same as that for an isolated single tube.
2. *Zero circulation and flooding limited*. This mechanism occurs when there is no net circulation through the tube bundle. Liquid entering the bundle can do so only from above and is hindered in doing so by the vapor being released from the bundle. A flooding condition is reached where the vapor release is such as to prevent sufficient downflow of the liquid to reach all the heating surfaces within the bundle. This mechanism is more likely to occur in large bundles with closely spaced tubes. Downflow can occur in some lanes and with upflow in others, giving an apparent internal circulation.
3. *Circulation and flooding limited*. This mechanism is similar except that some small circulation into the bundle from the sides and base occurs. This low inflow of liquid is evaporated within the bundle and the resulting vapor passes upward to join that generated from the liquid passing into the bundle from above. Again, the limiting condition is reached when this total vapor flow is such as to prevent a sufficient liquid inflow from above. This mechanism occurs in large bundles with closely spaced tubes.

4. *Circulation limited*. As the circulation through the bundle increases, a condition is reached where the vapor flow produced from this circulation alone is sufficient to prevent any liquid entering the bundle from above. The limiting condition is then complete evaporation of the liquid feed, starving the upper tubes in the bundle of liquid. Such a condition might occur in narrow but tall tube bundles.

In practice it is difficult to distinguish among some of these mechanisms in actual tube bundle geometries.

5. *Entrainment limited*. As the circulation through the bundle is further increased, a substantial amount of liquid may be entrained as droplets within the vapor flow and thus dryout will occur on the upper tubes on the bundle at vapor qualities considerably less than 100%.

A simple analytical expression can be derived for the condition where the bundle is limited by zero circulation on the assumption that for a bundle of circular cross section located within a blind-ended rectangular channel, flooding will occur first at the horizontal diameter of the bundle.

This expression for the critical heat flux in a tube bundle can be considered the product of two terms: a bundle geometry-characterizing parameter and a liquid physical property group. As might be expected q_{cr} increases as the bundle size decreases (D_b/N increases) and as the pitch, p, to tube diameter, D, ratio increases. Should q_{cr} predicted from this equation exceed the value calculated for an isolated single tube, then the latter value should be used. The situation is then that corresponding to pool boiling described previously.

Palen and Small [43] has given a correlation for the critical heat flux within a tube bundle under conditions where the liquid circulation is limited. It was arrived at by modifying the isolated single-tube value. The critical heat flux is again expressed as the product of a dimensionless tube density factor ϕ and a dimensional physical property factor φ.

$$q_{cr} = K(\phi)(\varphi) \tag{9.7}$$

The variation of the bundle critical heat flux as a function of these two parameters is shown in Fig. 9.17. Available experimental evidence suggests that the predictions of both are conservative, and, given the onset of steam blanketing within a bundle, Schuller and Cornwell [40] recommend that, provided the steam quality from the tube bundle does not exceed 15%, the critical heat flux is unlikely to be less than one-third of the isolated horizontal tube value.

Computer codes (cf. THIRST [44]) based on the porous media approach [11] are available to compute the three-dimensional steady-state thermohydraulic characteristics of boiling on the shell side of vertical inverted U-tube steam generators for light-water or heavy-water cooled reactors.

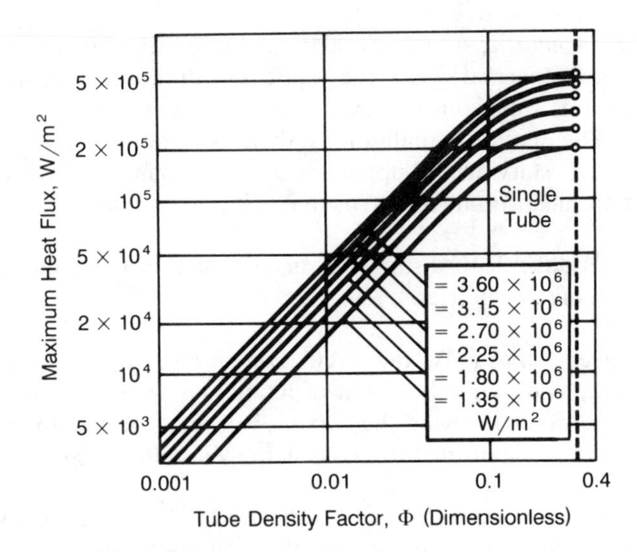

Fig. 9.17. Maximum heat flux in horizontal tube bundles (Palen and Small).

Use of Enhanced Surfaces [45] Although both high-fin and low-fin surfaces are used on the gas side of nuclear steam generators and waste heat boilers, enhancement devices are not on the water–steam side except in special circumstances. Twisted tape inserts have been used to increase the critical heat flux in horizontal U-tube waste heat boilers [7] in cases where stratification has occurred due to low flow.

In general, nuclear steam generators and waste heat boilers operate at lower pressures and lower peak heat fluxes than fossil-fired central station boilers. The increase in critical heat flux with decreasing pressure is such that even for horizontal tubes there is usually no requirement for enhancement devices.

9.4.4 An Example: PWR Inverted U-Tube Recirculating Steam Generator

The following is an example of the general design method outlined in Section 9.4.1. Steam generators are required for each loop of a four-loop PWR with the following characteristics:

Steam Pressure	69 bar
Steam flow rate	470 kg/s
Steam temperature	285°C
Feed-water temperature	277°C
Primary reactor coolant	
Inlet	325°C
Outlet	293°C
Thermal rating	850 MW(t)

The design is to be an inverted U-tube recirculating type (Fig. 9.1) with a single shell pass and two tube passes. The temperature distribution (Fig. 9.18) is as follows.

Because the shell-side steam temperature is essentially constant, no correction factor is needed for the LMTD:

$$\text{LMTD} = \frac{(40 - 8)}{\ln(40/8)} = 19.9°\text{C} \approx 20°\text{C}$$

An appropriate value of the overall heat transfer coefficient for this design is 7500 $W/(m^2 \cdot K)$.

$$A = \frac{Q}{U\Delta T} = \frac{850 \times 10^6}{7.5 \times 10^3 \times 20} = 5666 \text{ m}^2$$

Studies have shown that the optimum design of vertical natural-circulation steam generators, corresponding to the minimum cost (capital and operating), will be achieved with the smallest tubes consistent with fouling, vibration, and inspection considerations. An appropriate minimum tube diameter would be 17.5 mm OD × 1 mm tube thickness Inconel 690. To fit into the steam generator shell, an appropriate overall length of the average U tube would be 18 m. The number of tubes and the diameter of the shell can now be

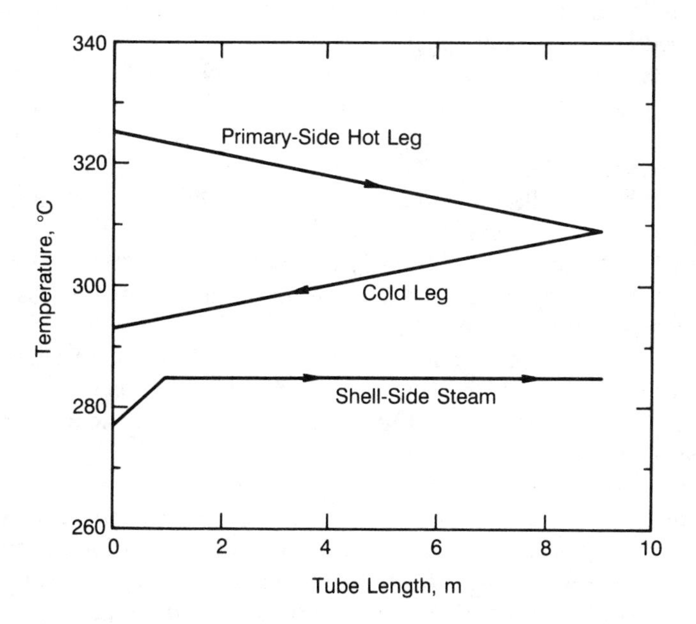

Fig. 9.18. Temperature distribution in PWR U-tube recirculating steam generator.

estimated:

$$N = \frac{A}{\pi DL}$$

$$= \frac{5666}{\pi} \times 17.5 \times 10^{-3} \times 18$$

$$= 5725 \text{ tubes}$$

These tubes can be fitted into a 3.5-m vertical shell.

The next stage is to fill out the basic mechanical design dimensions. Details of the wrapper around the tube bundle to form the downcomer; the tube sheet; the number, type, and material of the tube supports and the antivibration supports in the U-tube region need to be established. The basic "first cut" dimensions can then be used as initial input to one of the proprietary computer-based steam generator thermohydraulics codes such as THIRST [44]. These treat the shell side as a porous medium (the presence of tubes, baffles, and other objects is described in terms of a spatially variable porosity) in which the basic conservation equations of continuity, axial, radial, and circumferential momentum, and energy are solved. These codes, however, do require specification of various fluid flow and heat transfer correlations to effect closure of the transport equations. Since these correlations are derived from experiments in simple geometries such as tubes, annuli, or simple bundles, there is likely to be considerable uncertainty associated with their application in a steam generator code.

The output from the code provides both contours of steam quality and mass velocity throughout the steam generator shell, as well as spatial variations of tube wall temperature and heat flux. Important design parameters which can be established include the recirculation ratio (the ratio of the total flow through the bundle to the flow rate of steam out of the steam nozzle), the heat duty, the maximum local quality, and the maximum quality at the tube sheet (as a result of heat transfer and boiling in the hot leg region), as well as the performance of the unit at part-load conditions and under changes of plant state. Although recirculation ratios as low as 3 or 4 can be used with careful design, higher values (6 to 7) allow increased velocities across the tube sheet, which reduce debris buildup, and lower steam qualities, which reduce the probability of local dryout and concentration of aggressive chemicals near tube supports.

The effects of various alternative empirical correlations for pressure drop, void fraction, and heat transfer on the predictions of the THIRST code have been reported by Carlucci and Sutherland [46]. The various correlations investigated are shown in Table 9.7, and the results for a typical inverted U-tube steam generator [working pressure 60 bar, thermal rating 632 MW(t)]

TABLE 9.7 Correlations Investigated in THIRST Code Study (Taken from [46])

Parameter	Name of Source of Correlation	
	Reference	Variant
Pressure drop		
Single-phase parallel-flow friction factor	Rehme [47]	Miller [48]
Single-phase crossflow friction factor	Zukauskas [49]	Grimison [50]
Two-phase friction multiplier	Homogeneous	Baroczy-Chisholm [18]
		Martinelli and Nelson [20]
		Martinelli and Nelson, Jones [51]
		Thom [32]
Void fraction	Homogeneous	Thom [32] and Smith [29]
		Chisholm [31]
		Armand and Treshchev [52]
Secondary-side heat transfer		
Single-phase parallel flow	Groeneveld [53]	Inayatov [54]
Single-phase crossflow	Zukauskas [49]	Grimison [50]
Two-phase heat transfer	Chen [55]	Shrock and Grossman [14]
		Thom [56] and Jens-Lottes [14]
		Rohsenow [14]

are shown in Table 9.8. It can be concluded that:

1. The choice of single-phase tube bundle friction factor or heat transfer coefficient correlation does not significantly influence the magnitude of the global or local parameters.
2. The choice of two-phase void fraction or friction multiplier correlation does markedly affect the code predictions. Compared with the prediction made using the homogeneous model, circulation ratios, local velocities, and local qualities are considerably reduced. The choice of correlation for the two-phase friction multiplier is of greater importance than that for void fraction (see Section 9.4.3). The choice of two-phase heat transfer correlation is not particularly important in relation to global design parameters such as the recirculation ratio or heat duty. However, the various correlations do have a marked effect on the predicted wall temperature and heat flux distributions.

Very little experimental evidence is available to check the choice of the most appropriate correlation in the specific geometry of the steam generator. The only reported data are from CISE [57, 58]. These tests were carried out in a test section consisting of nine tubes arranged vertically on a square pitch.

TABLE 9.8 Variation of Key Output Parameters as Percentage Deviation from Reference Case

	Recirculation Ratio, %	Heat Duty MW(t), %	Maximum Quality, %	Maximum Tube Sheet Quality, %
Reference Value	5.97	632	25.3	3.0
Variant correlation				
Single-phase pressure drop				
Miller	−3.4	−0.2	+4.8	+5.0
Grimison	−3.4	−0.2	+5.7	+18.2
Two-phase friction pressure drop				
Baroczy and Chisholm	−18.1	−0.2	+23.3	+29.9
Martinelli and Nelson	−35.2	−0.3	+56.7	+76.8
Martinelli, Nelson, and Jones	−33.7	−0.3	−55.3	+71.2
Thom	−4.8	0.0	+4.0	+7.3
Void fraction				
Smith	−5.9	0.0	+5.8	+6.6
Chisholm	−5.5	0.0	+6.2	+6.3
Thom	−6.9	0.0	+5.0	+8.3
Armand and Treshchev	−2.2	0.0	+2.8	+2.6
Single-phase heat transfer				
Inayatov	−0.3	+1.1	+1.2	+0.3
Grimison	0.0	−0.3	+0.8	−1.3
Two-phase heat transfer				
Thom	−1.2	−1.7	−2.6	+7.9
Jens-Lottes	−0.5	−0.9	−2.1	+4.6
Rohsenow	−1.3	+1.7	+2.8	+11.9
Shrock and Grossman	+0.2	−1.1	−0.7	−5.6

The tube material was Inconel 600; the tube outside diameter was 19.05 mm; the wall thickness was 1.09 mm; the pitch was 27 mm; and the test section was 8.22 m long. It was heated by passing water through the tubes at 155 bar, a mass flux of 4250 kg/(m^2 · s), and inlet temperatures of 325°C for hot leg tests and 302°C for cold leg tests. The shell side was simulated by a vertical upflow of water at pressures between 35 and 80 bar, mass velocities of 180 to 1000 kg/(m^2 · s), and exit qualities of 5% to 80%. The experimental program investigated void fraction pressure drop, heat transfer coefficients, and dryout on the secondary side. Experimental measurements of the heat transfer coefficient were compared with the predictions of a number of standard correlations. All the correlations underpredict the observed heat transfer coefficient; the best agreement was obtained by Jens-Lottes and Chen. Even better agreement was obtained by setting the nucleate boiling suppression factor, S, in the latter correlation to unity. It was suggested that at low mass velocities, the thickness of the water film on the tubes will be higher than in simple tubular geometries, leading to less suppression of nucleate boiling.

9.5 COMMON PROBLEMS IN THE OPERATION OF BOILERS

It is perhaps appropriate to indicate the order of magnitude of the costs stemming from operational problems with boilers. Large companies are often reluctant to publicize the details of specific plant failures and consequent production losses for obvious reasons. However, some details have emerged. On fossil-fired central station boilers operated by the CEGB, these losses amount to around £23 million per year, whilst for nuclear plants the equivalent figure is around £0.2 million per year. The loss of electricity generation attributable to steam generator unreliability for U.S. pressurized water reactors averages about 3% or 10 to 11 days per year. Each individual outage lasts an average of 20 days, equivalent to an economic loss of £4 million. A study of 27 ammonia plants carried out in 1973 showed that waste heat boiler problems were the second most frequent cause of shutdowns. On average, one failure could be expected every 3 years and this failure caused 5 days of lost output. Thus the average annual loss of income due to boiler failures on ammonia plants was then around £100,000 (considerably higher at today's prices). For some plants the failures were more serious, causing the plants to be shut down for 50 to 100 days.

9.5.1 Causes of Steam Generator Problems

Some of the causes of boiler problems in service are [59]:

1. Fouling
2. Maldistribution of flow
3. Corrosion
4. Erosion
5. Vibration
6. Thermal fatigue
7. Corrosion fatigue
8. Maloperation
9. Water hammer

Sometimes two or more of these mechanisms can act simultaneously. For example, fouling in steam-generating systems is invariably linked with corrosion and erosion. These are topics to which the boiler designer, fabricator, and operator must give special attention. Of the factors listed previously, the first two can result in a degraded thermal performance from the unit, whilst the remainder if not checked in a greater or lesser loss of integrity of the unit.

Fouling Scaling in steam generators can cause, in extreme circumstances,

1. The bursting of tubes in water tube boilers
2. The cracking of tube ends and ligaments in tube plates in waste heat boilers

Usually the water quality in both process chemical waste heat boilers and central power station nuclear units is tightly controlled so that fouling is not a problem and indeed control over water quality and purity is the only satisfactory way of preventing scaling.

In PWR steam generators water-side corrosion products and other solids carried into the units tend to accumulate on the upper side of the tube sheet. Tens to hundreds of kilograms of magnetite have been removed from such units by sludge lancing. If such material is allowed to remain, high concentrations of aggressive salts can build up and induce tube thinning and other types of corrosion.

Maldistribution of Flow A significant cause of steam generators failing to meet their design rating is that the fluid flows do not behave on the idealized one-dimensional basis assumed. For example, it is unlikely that all the many tubes in the steam generator will see equal flows, while stagnant regions and leakages past baffles and around the tube bundles reduce the effectiveness of the shell-side flow.

A particularly striking example of how flow and temperature maldistribution can limit steam generator performance has been given by Collier and Whitmarsh-Everiss [60]. In this case extreme sensitivity to minor changes in the physical geometry of the tube bundle, coupled with the choice of an operating point on the boiler characteristic where the performance of the unit was very sensitive to primary-side flow and temperature distributions, limited the boiler's capacity to about 75% of design. The lessons learned from this study included:

1. The need for uniformity of the shell-side inlet temperature.
2. The choice of an operating point insensitive to shell-side flow–temperature maldistribution; this can be achieved by greater levels of orificing on the tube (water) side to reduce the steady-state gain.
3. The need to study the various plant operating states which can introduce asymmetries into the boundary conditions for the boiler.
4. The importance of not having an overrigid specification of materials at the design stage with their attendant limitations on operating tempera-

ture, especially for plants required to operate over a wide range of operating states.

A particularly serious problem is the distribution of two-phase gas–liquid flows. Basically, the gas and the liquid have a tendency to separate at any obstruction, bend, or the like. This makes prediction of the behavior of a heat exchanger involving such flows very difficult and whenever possible this problem should be avoided at the design stage. Stratification of a flow in a horizontal channel due to gravitational forces can also be considered as a form of maldistribution. A number of cases have been reported where waste heat boilers have failed due to dryout and overheating of horizontal boiler tubes.

Corrosion *On-load corrosion* is a particularly rapid attack of boiler tubes in zones of steam generators where steam is raised. In particular, pioneering work carried out by Masterson, Castle, and Mann [61] elucidated three mechanisms whereby salts normally in the ppm range in boiler feed water may be concentrated by factors of 10^3 or more.

1. Dryout–either complete as in a once-through boiler or partial as may occur due to maldistribution or stratification in horizontal boiler tubes. A number of examples have been reported were acid chloride attack has resulted in the deep gouging "tramline" corrosion of such tubes at the position of the water–steam interface.
2. Crevices—a particularly severe example of this type of attack has been in the steam generators of pressurized-water reactors (PWRs). In certain designs of the inverted U-tube boilers, the tubes were supported by carbon steel plates drilled to permit passage of the tube with little clearance (Fig. 9.6). Boiling occurs on the shell side of these units. A crevice is formed between the tube support plate and the tube. Severe corrosion of the carbon steel support plate has occurred resulting from the concentration of acid chloride in the crevice. Since the corrosion product occupies approximately twice the volume of the metal consumed, this expansion crushes the tubes and distorts the support plate. One other consequence of the physical distortion caused by this "denting" has been the increased strain at the apex of the U tube. The tubes have distorted into an oval cross section and some stress corrosion cracking has occurred in this region. In this specific example the problem is so serious that a redesign of the steam generator has been undertaken. The crevice has been removed by replacing the drilled support plate by a broached "quatre-foil" design (Fig. 9.6). The material of the support plate has been changed from a mild steel to a 12% Cr steel and full-flow condensate polishing and high-integrity condensor designs have been recommended.

3. Porous deposits—when steam-generating surfaces become fouled, water is drawn into the porous deposit by a "wicking" effect, whilst steam is released into "tunnels" in the deposit. Very high concentrations of aggressive salts occur within deposits leading to the type of pitting corrosion referred to previously. Instances have been reported where massive formations of magnetite deposits have resulted in the failure of unheated pipework in boilers as well as heated tubes.

Stress corrosion cracking. For stress corrosion cracking to occur, three factors must be present together. First, the levels of stress in the component must be at or close to the yield stress of the material concerned. This is often the case where the residual stresses due to welding, say at the tube-to-tube plate joint, have not been relieved by heat treatment. Second, an aggressive agent such as chloride, caustic, or nitrate ion needs to be present in the fluid. Third, the material needs to be in sensitized condition. Austenitic stainless steels, particularly close to welds, are often in such a condition, and care must be exercised when such materials are used in superheaters or reheaters to prevent carry-over of water droplets which may contain chlorides. Some alloys are resistant to stress corrosion cracking, particularly those with high nickel contents. Stress relieving of welds may prove effective and of course the elimination of the aggressive agent in the water by condensate polishing will be beneficial.

Erosion Accelerated metal wastage may occur in regions of high velocity particularly for metals that rely for their protection on the formation of a protective surface film. This protective film is eroded by cavitation, exposing the bare metal to chemical or electrochemical attack. The effect of velocity is usually important. One example of a significant outage involving the loss of approximately 1000 GWh of electrical power production concerned the erosion of the boiler feed regulating orifices for the Hinkley B Advanced Gas-Cooled Reactor. These boiler orifices are fitted at the entry to each boiler inlet tube to provide flow regulation and overall hydrodynamic stability. It was found that the screwed carriers holding the orifice plates were leaking along the screw threads on about half the tubes. The leakage flow increased rapidly due to erosion to the extent that some orifice plates were completely bypassed and many tube ends were completely eroded away. All the tube ends were eventually cut off and replaced with new tube ends of more resistant material. A similar occurrence has been reported for the steam generators of the French liquid-sodium-cooled Phenix reactor. Wastage has also been observed in the low-temperature sections of some serpentine boilers in the region of the 180° return bends. Here again, erosion of material at temperatures below that at which the protective magnetite film is formed is suspected.

Prevention or reduction of erosion damage involves the use of more erosion-resistant materials either locally or overall (e.g., Inconel), improvements in design to limit as far as possible local regions of high velocity, and sometimes controlled additions of oxygen to promote the formation of a protective magnetite film.

Vibration A major cause of failure in steam generators is tube vibration. Tube vibration may result in failure by mechanical fatigue, by fretting corrosion of the tubes at the tube sheet, or, most likely, by impact and rubbing of the tubes with the baffles or with one another at the midspan. Although there are many mechanisms that can induce vibration, the mechanisms of most concern relates to flow-induced vibration. Basically, there are three such mechanisms.

1. Fluid–elastic instability which occurs for the core of a tube bundle exposed in total or in part to a crossflow. Above a critical velocity a coupled orbital whirling motion occurs for a number of tubes in a given row.
2. Turbulent buffeting can induce random excitation forces which increase as the square of the velocity and excite the tube natural frequency and induce damage if the damping is low.
3. Periodic vortex shedding which may occur from tubes at the edge of the tube bundle, particularly the last row of tubes.

Various analytical methods [62] are available for checking for such mechanisms, but these methods require a detailed knowledge of the flow patterns and local velocity distributions in the unit. In the case of waste heat boilers heated by hot-gas, acoustic oscillations may be set up. The various flow-induced excitation mechanisms can be amplified when the frequency of this excitation coincides with the acoustic frequency of a standing wave across the diameter of the shell. Such acoustic oscillations can produce intense noise but are seldom damaging. They can be readily overcome by detuning the system by inserting suitable transverse acoustic baffles.

Before leaving vibration it should be noted that failures often occur due to excitation of foreign objects within the steam generator. Such objects left after a maintenance period will bear on tubes, causing fretting damage and ultimately penetration.

Thermal Fatigue Thermal fatigue results from the presence of alternating thermal stresses which, in turn, arise as a result of changes in temperature. These temperature changes could be as a result of random changes in temperature at normal operating conditions, as a result of starting up or shutting down a steam generator (which might occur hundreds of times during its lifetime), or as a result of the turbulent mixing of a hot and cold stream (e.g., near the feed point of a shell boiler). In the case of the steam generators for water-cooled nuclear reactors, thermal fatigue of the feed-water nozzles has occurred in 16 instances. The problem seems to arise when unheated auxiliary feed water is supplied to the steam generator. At these low flows (just a few percent of normal feed-water flows), temperature differences occur due to stratification in the pipe and these induce high local stresses in the areas where cracking has been observed. The mixing of hot and cold water, as well as these temperature changes, are prime factors in inducing and propagating cracks by a thermal fatigue mechanism.

Corrosion Fatigue The fatigue process may be considerably enhanced when the induced defect is exposed to a corrosive fluid. For example, crack extension rates may be increased by a factor of 10 when ferritic steels are exposed to high-temperature water.

Extensive cracking has been found in the main girth welds on the shells of all four steam generators of a U.S. PWR. The girth weld in question is located just below the feed-water distribution ring manifold in the normal operating water level where it may be subjected to thermal cycling. In addition, problems had been experienced with poor secondary-side water chemistry. This, together with the fact that the steam generators underwent numerous weld repairs during manufacture, points to corrosion fatigue as the likely cause of the cracking.

Maloperation Included under this heading are a wide variety of concerns not all directly related to the operation of the unit. Perhaps the most important is for the operator to check regularly that those protection devices such as safety relief valves and interlocks are in correct working order.

Another area of general concern relates to the need to assess very carefully all the consequences of a modification made to the equipment or to the operation of the unit. The importance of the operator keeping within the technical operating specifications in relation to rates of heat-up and cool-down and with respect to water purity and quality cannot be overstressed.

Damage can occur to steam generators during storage on site while awaiting installation. One example of such damage relates to the steam generator units for the prototype fast reactor at Dounreay. After being stored in its shell on site for several months, one of the tube bundles was found to have several tubes contaminated with water. Extensive inspection and chemical cleaning were necessary. Corrosion pitting was observed and as a result some 38 tubes (out of several hundred) were explosively plugged. Therefore it is important that specifications call for adequate protection of both internal and external surfaces during storage and for regular inspection during this period.

Finally, a careful check on installation and maintenance work is advisable.

Water Hammer The problem of "water hammer" relates to the generation of damaging pressure pulses due to condensation of steam onto cold feed water or condensate. Typically, most occurrences have taken place during a boiler warm-through or on trying to restart feed-water flow following an operational transient. Under these circumstances cold water may be introduced into a long horizontal pipe filled initially with steam. A water slug may form to trap the steam void and condensation of this steam void may create a large pressure difference across the water slug so that the latter will accelerate rapidly, collapsing the steam void. The resultant pressure wave, if severe enough, can seriously damage the piping and the boiler.

The problem can be overcome by eliminating horizontal pipe runs in which feed water or condensate may be deliberately or accidentally introduced or alternatively ensuring such pipework is always water filled.

9.5.2 Worked Solutions

Example 9.1. Estimate the frictional pressure gradient in a 50.8-mm bore evaporating tube for the following conditions:

Fluid	Steam–water
Pressure	180 bar
Inlet mass flow of saturated water	2.14 kg/s
Outlet steam quality	18.25%

Solution: At this high subcritical pressure the homogeneous model will give satisfactory results. The estimated value can however be checked against other methods:

1. Physical properties:

steam–water at 180 bar

$$v_L = 1.84 \times 10^{-3}\ \text{m}^3/\text{kg}$$

$$v_G = 7.50 \times 10^{-3}\ \text{m}^3/\text{kg}$$

$$\mu_L = 6.44 \times 10^{-5}\ (\text{N} \cdot \text{s})/\text{m}^2$$

$$\mu_G = 2.57 \times 10^{-5}\ (\text{N} \cdot \text{s})/\text{m}^2$$

2. Assumptions:
   (a) The overall change in absolute pressure is small, the generation of vapor due to flashing is small, there are no compressibility effects, and the physical property changes can be neglected.
   (b) The evaporating tube is uniformly heated; the steam quality varies linearly with the tube length.
3. General quantities:

$$\text{mass velocity } (G) = \frac{W}{A} = \frac{2.14 \times 4}{\pi \times 0.0508^2} = 1056\ \text{kg/m}^2 \cdot \text{s}$$

4. Pressure gradient:

$$-\left(\frac{dp}{dz}F\right) = \frac{1}{x}\int_0^x \frac{2f_{TP}G^2}{D} v_L\left[1 + x\left(\frac{v_G - v_L}{v_L}\right)\right] dx$$

$$= \frac{2G^2 v_L}{Dx}\int_0^x f_{TP}\left[1 + x\left(\frac{v_G - v_L}{v_L}\right)\right] dx$$

if f_{TP} is a constant (i.e., $f_{TP} = f_{LO}$) the preceding equation reduces to

$$-\left(\frac{dp}{dz}F\right) = \frac{2G^2 f_{TP}}{D} v_L\left[1 + \frac{x}{2}\left(\frac{v_G - v_L}{v_L}\right)\right]$$

For $f_{TP} = f_{LO}$, $\bar{\mu} = \mu_L$,

$$\therefore Re = \frac{GD}{\bar{\mu}} = \frac{1056 \times 0.0508}{6.44 \times 10^{-5}} = 8.33 \times 10^5$$

$$\therefore f_{TP} = 0.00285 \qquad \text{(Moody)}$$

$$\therefore -\left(\frac{dp}{dz}F\right) = \frac{2 \times 1056^2 \times 0.00285 \times 1.84 \times 10^{-3}}{0.0508}$$

$$\times\left[1 + \frac{0.1825}{2}\left(\frac{7.50 \times 10^{-3} - 1.84 \times 10^{-3}}{1.84 \times 10^{-3}}\right)\right]$$

$$= 295\ \text{N}/(\text{m}^2 \cdot \text{m})$$

If we use other methods we get: Matrinelli–Nelson method, 346 $\text{N}/(\text{m}^2 \cdot \text{m})$; Baroczy method, 378 $\text{N}/(\text{m}^2 \cdot \text{m})$.

Example 9.2. The steam generators employed on PWRs constructed in the USSR consist of a horizontal shell within which fits a horizontal tube bundle (see Fig. 9.3). For the 1000-MW(e) design there are four such units each with a shell 14 m long and 4.0 m ID. Each shell contains 15,648 tubes 12 mm ID × 1.2 mm wall thickness with an average tube length of 8.5 m. The total heat transfer surface is 5040 m^2. The operating pressure is 60 bar. Under these circumstances will dryout occur within the tube bundle?

Solution: One method of estimating the critical heat flux for boiling on the shell side is that of Palen and Small:

$$\ddot{q}_{cr} = K(\Phi)(\varphi)$$

where K is an empirical constant (for $0.06 < \Phi$), $K = 1.23$. Φ is a dimensionless tube density factor

$$\left(\simeq \frac{D_b L}{A} \right)$$

φ is a dimensional physical property factor

$$\varphi = \Delta h_v \rho_G^{1/2} [\sigma g(\rho_L - \rho_G)]^{1/4}$$

At 60 bar for water

$$\Delta h_v = 1211 \text{ kJ/kg}$$

$$\rho_G = 30.39 \text{ kg/m}^3$$

$$\rho_L = 759 \text{ kg/m}^3$$

$$\sigma = 0.020 \text{ N/m}$$

$$\varphi = 1211 \times 5.512 \times [0.020 \times 9.807 \times 729]^{0.25}$$

$$= 23{,}025 \text{ kW/m}^2$$

$$\Phi = \frac{4 \times 14}{5040} = 0.0111$$

Therefore

$$\ddot{q}_{cr} = 1.23 \times 23{,}025 \times 0.0111$$

$$= 314.4 \text{ kW/m}^2$$

Now the power from one steam generator is 750 MW(t), that is, 750,000 kW. The surface area is 5040 m^2. Therefore the average heat flux is 148.8 kW/m^2.

Thus there is approximately a factor of 2 between the bundle average heat flux and that predicted to cause dryout. This should be sufficient to accommodate nonuniformities and uncertainties in the analysis.

Example 9.3. A waste heat boiler is being designed in which the tubes are formed into a serpentine composed of horizontal 56 mm ID tubes connected by a 180° return bend. The unit is to be operated with an exit steam quality of 25% and at a pressure of 55 bar. The initial choice of mass velocity is 600 $kg/(m^2 \cdot s)$. What will, be the consequences of prolonged operation under these conditions? What is the minimum value of mass velocity required to prevent stratification and overheating of the upper tube surface?

Solution: In order to estimate whether stratification could occur at the conditions given, it is proposed to use Eq. (9.6). At a pressure of 55 bar,

$$T_{SAT} = 270°C \qquad \rho_L = 768\ \text{kg/m}^2$$

$$\rho_G = 28.09\ \text{kg/m}^3$$

$$\sigma = 0.021\ \text{N/m}$$

$$\mu_L = 1.01 \times 10^{-4} (\text{N} \cdot \text{s})/\text{m}^2$$

Thus, for a mass velocity of 600 kg/(m^2 · s) and a steam quality between 0 and 25%, we have

$$j_L = \frac{G(1-x)}{\rho_L} \qquad \begin{array}{ll} \text{if } x = 0 & j_L = 0.78\ \text{m/s} \\ \text{if } x = 0.25 & j_L = 0.586\ \text{m/s} \end{array}$$

$$j_G = \frac{Gx}{\rho_G} \qquad \begin{array}{ll} \text{if } x = 0 & j_G = 0\ \text{m/s} \\ \text{if } x = 0.25 & j_G = 5.34\ \text{m/s} \end{array}$$

$$j = j_L + j_G \qquad \begin{array}{ll} \text{if } x = 0 & j = 0.78\ \text{m/s} \\ \text{if } x = 0.25 & j = 5.93\ \text{m/s} \end{array}$$

The critical velocity below which stratification occurs is given by

$$j^{1.8} = 30.43 \frac{D^{0.2}\{\sigma g(\rho_L - \rho_G)\}^{0.5}}{\rho_L^{0.8} \mu_L^{0.2}} \tag{9.6}$$

$$j^{1.8} = 30.43 \frac{(56 \times 10^{-3})^{0.2} [0.021 \times 9.807 \times 740]^{0.5}}{768^{0.8} \times (1.01 \times 10^{-4})^{0.2}} = 6.53$$

$$j = 2.83\ \text{m/s}$$

Therefore, for areas of the horizontal tube where j is below 2.83 m/s, stratification, overheating, and on-load corrosion are possible. This corresponds to parts of the tube where the steam quality is below 10%.

One remedy would be to increase the mass velocity so that j at all locations is above the minimum value determined previously. Therefore the mass velocity must be raised above (2.83 × 768) = 2173 kg/(m^2 · s).

Example 9.4. A waste heat boiler consists of 78 vertical tubes 44.5 mm OD, 35.6 mm ID, each 10 m long. Saturated water taken from a steam drum is fed to the base of the tubes at a pressure of 132.5 bar (T_{SAT} = 332°C). The water mass velocity in the tubes is 912 kg/(m^2 · s). The tubes are heated by a gas stream on the shell side passing co-current with the evaporating water. Decide whether a critical heat flux (dryout) condition will occur on the water side for these conditions. Use Bowring's correlation together with the "overall power" hypothesis, that is, that the tube power up to dryout is independent of the heat flux profile.

Solution:

	Inlet $z = 0$	$z = 4$ m	Outlet $z = 10$ m
Local overall value of heat transfer coefficient, U, kW/(m^2 · °C) (referred to gas-side area)	0.960	0.875	0.850
Local gas temperature, °C	957	571	390
Temperature driving force, °C	625	239	58
Local heat flux, kW/m^2 (referred to gas side)	600	209	49
Local heat flux, kW/m^2 (referred to water side)	750	261	61.6

The variation of the mass fraction steam $x(z)$ can be calculated from

$$\Delta x = \frac{4\ddot{q}(z)}{DG\,\Delta h_v}\,\Delta z$$

where $D = 0.0356$ m, $\Delta h_v = 1100$ kJ/kg, and $G = 912$ kg/(m^2 · s). The variation of the inside heat flux and quality are given in the following table:

z, m	0	2	4	6	8	10
$\ddot{q}(z)$, kW/m^2	750	397	261	170	110.4	61.6
$x(z)$	0	0.128	0.188	0.236	0.266	0.286

The average heat flux for this profile $\ddot{q} = 225.3$ kW/m^2:

$$A' = 3215$$

$$C' = 1.6209$$

Therefore

$$\ddot{q}_{\text{cr}} = \frac{3215}{1.6209 + 10} = 276.65 \text{ kW/m}^2$$

So the average heat flux at the critical condition for a tube 10 m long for a saturated inlet flow is this value. It will be seen that the margin to a critical condition is quite small: (276.65/255.3) 1.08 or 8%. The uncertainty in the values of A' and C' derived from the correlation and the inaccuracy of the "overall power" hypothesis is considerably greater than this small margin. One alleviating factor is that for this particular heat flux profile (i.e., a sharply decreasing heat flux with length) it has been found that the power that can be extracted from a long tube is slightly greater (up to 10%) than for a uniform profile.

9.6 CONCLUSIONS

Experience shows that the reliability of individual boilers tends to improve over their early operating life, as weaknesses due to manufacture are weeded out. If the necessary changes are cost effective, similar weaknesses due to design or operating strategy can also be reduced. Experience also shows that, when changes between successive units are gradual, analysis of past operating history points to steps that can be taken to improve future plants. Improved fabrication methods and increased levels of preservice inspection can avoid or detect defects before these are translated into costly in-service failures.

ACKNOWLEDGMENT

I would particularly like to acknowledge the help of Roger Pearce in preparing this manuscript for publication.

NOMENCLATURE

A	function in Bankoff-Jones correlation (Table 9.4); function in Bryce correlation (Table 9.4)
B	function in Chrisholm correlation (Table 9.2); function in Bryce correlation (Table 9.2)
D	tube diameter, m
D_b	bundle diameter, m
e	fraction of liquid entrained
Fr	Froude number, $G^2/\rho^2 gD$
g	acceleration due to gravity, m/s^2
G	mass velocity, $kg/(m^2 \cdot s)$
j	two-phase velocity, m/s
K	function in CISE correlation (Table 9.2); function in Hughmark correlation (Table 9.4); function described by Eq. (9.7)
n	function in CISE correlation (Table 9.2)
N	number of tubes
p	tube pitch, m
P	static pressure, N/m^2
q''_{cr}	critical heat flux, W/m^2
Re	Reynolds number, GD/μ

S	slip ratio
v	specific volume, m^3/kg
x	mass vapor quality
X	Matrinelli parameter (Table 9.4)
z	axial coordinate, m
(dP/dz)	pressure gradient, $N/(m^2 \cdot m)$

Greek Symbols

α	void fraction
β	volumetric quality
Γ	function in Chrisholm correlation (Table 9.2)
θ	angle to horizontal plane, °
μ	viscosity, $(N \cdot s)/m^2$
ρ	density, kg/m^3
σ	surface tension, N/m
ϕ^2_{LO}	two-phase frictional multiplier based on pressure gradient for total flow assumed liquid
ϕ	dimensionless tube density factor [Eq. (9.6)]
Φ	tube density factor
φ	dimensional physical property factor [Eq. (9.6)], W/m^2
Ω	function in Baroczy correlation (Table 9.2)

Subscripts

A	accelerational
b	bundle
cr	critical
F	frictional
G	gas or vapor
GO	assuming total flow to be gas or vapor
L	liquid
LO	assuming total flow to be liquid
TP	two phase
z	static head

Superscripts

average based on homogeneous model

REFERENCES

1. Collier, J. G. (1981) *The Design of Boilers, in Heat Exchangers: Thermal Hydraulic Fundamentals and Design*, S. Kakaç, A. E. Bergles, and F. Mayinger (eds.), pp. 619–646. Hemisphere, New York.

2. Green S. J. (1988) Thermal, hydraulic and corrosion aspects of PWR steam generator problems. *Heat Transfer Eng.* **9**(1) 19–68.
3. *The Shippingport Pressurized Water Reactor* (1958) Addison-Wesley, Reading, Mass.
4. Styrikovich, M. (1978) The role of two-phase flows in nuclear power plants. *Int. Sem. Momentum, Heat and Mass Transfer in Two-Phase Energy and Chemical Systems*, Dubrovnik, Yugoslavia. ICHMT.
5. Davis, R. J., and Hirst, B. (1979) Twin header bore welded steam generator for pressurised water reactors. *Nucl. Energy* **18**(2) 133–140.
6. Garnsey, R. (1979) Corrosion of PWR steam generators. *Nucl. Energy* **18**(2) 117–132.
7. Robertson, J. M. (1973) Dryout in horizontal hair-pin waste heat boiler tubes. *AIChE Symp. Ser.* **69**(131) 55.
8. Hinchley, P. (1977) The engineering of reliability into waste heat boiler systems. *Proc. Inst. Mech. Eng.* **193**(8).
9. Hinchley, P. (1977) Waste heat boilers: problems and solutions. *Chem. Eng. Prog.* **73** 90; *Chem. Ing. Tech.* **49** 553.
10. Smith, R. A. (1986) *Vaporisers—Selection, Design and Operation*. Longmans/ Wiley, New York.
11. Webster, R. A. (1984) Three-dimensional thermal-hydraulic analysis of the PFR steam generator. *First UK Nat. Heat Transfer Conf.* pp. 241–258, Leeds.
12. Sha, W. T. (1978) A new approach for rod bundle thermal hydraulic analysis. *Proc. Int. Meeting on Nuclear Power Reactor Safety*, Brussels.
13. Brown, G. A. (1985) Thermal-hydraulic analysis for the CDFR steam generator unit (a once through boiler) using the Besant computer code. *Proc. Conf. Boiler Dynamics and Control in Nuclear Power Stations*, pp. 45–53. British Nuclear Energy Society.
14. Collier, J. G. (1981) *Convective Boiling and Condensation*, 2nd ed. McGraw-Hill, New York.
15. Scruton, B., Gibb, J., and Chojnowski, B. (1985) Conventional power station boilers: assessment of limiting thermal conditions for furnace wall tubes. *CEGB Research*, pp. 3–11.
16. Scruton, B., and Chojnowski, B. (1980) The assessment of critical heat flux margins for furnace wall tubes. *Forschung in der Kraftwerkstechnik*, pp. 160–165.
17. Baroczy, C. J. (1966) A systematic correlation of two-phase pressure drop. *Chem. Eng. Prog. Symp.* **62**(64) 323.
18. Chisholm, D. (1973) Pressure gradients due to friction during flow of evaporating two-phase mixtures in smooth tubes and channels. *Int. J. Heat. Mass Transfer* **16** 347–358.
19. Lombardi, C., and Peddrochi, E. (1972) A pressure drop correlation in two-phase flow. *Energia Nucleare* **19**(2).
20. Martinelli, R. C., and Nelson, D. B. (1948) Prediction of pressure drop during forced circulation boiling of water. *Trans. ASME* **70** 695.
21.. Macbeth, R. V. Private communication quoted in Brittain, I., and Fayers, F. J. (1976) A review of UK developments in thermal-hydraulic methods for loss of

coolant accidents. Paper presented at CSNI Meeting on Transient Two-Phase Flow, Toronto.

22. ESDU (1976) The frictional component of pressure gradient for two-phase gas or vapour/liquid flow through straight pipes. Engineering Sciences Data Unit (ESDU), London.
23. Friedel, L. (1977) Mean void fraction and friction pressure drop: comparison of some correlations with experimental data. European Two-Phase Flow Group Meeting, Grenoble, Paper A7.
24. Idsinga, W. (1975) An assessment of two-phase pressure drop correlations for steam–water systems. M.Sc. thesis, MIT.
25. Ward, J. A. (1975) Private communication (HTFS).
26. Freidel, L. (1979) Improved friction pressure drop correlations for horizontal and vertical two-phase flow. European Two-Phase Flow Group Meeting, Ispra, Italy.
27. Lockhart, R. W., and Martinelli, R. C. (1949) Proposed correlation of data for isothermal two-phase, two component flow in pipes. *Chem. Eng. Proc.* **58**(4) 62–65.
28. Hughmark, G. A. (1962) Hold-up in gas–liquid flow. *Chem. Eng. Proc.* **58**(4) 62–65.
29. Smith, S. L. (1969) Void fractions in two-phase flow. A correlation based on an equal velocity head model. *Proc. Inst. Mech. Eng.* **184**(36) 647–664.
30. Premoli, A., Di Francesco, D., and Prima, A. (1970) An empirical correlation for evaluating two-phase mixture density under adiabatic conditions. European Two-Phase Flow Group Meeting, Paper B9, Milan.
31. Chisholm, D. (1973) Research note: void fraction during two-phase flow. *J. Mech. Eng. Sci.* **15**(3) 235–236.
32. Thom, J. R. S. (1964) Prediction of pressure drop during forced circulation boiling of water. *Int. J. Heat Mass Transfer* **7** 709–724.
33. Jones, A. B. (1961) Hydrodynamic stability of a boiling channel. KAPL-2170.
34. Bryce, W. M. (1977) A new flow-dependent slip correlation which gives hyperbolic steam–water mixture flow equations. AEEW-R1099.
35. ESDU (1977) The gravitational component of pressure gradient for two-phase gas or vapour/liquid flow through straight pipes. Engineering Sciences Data Unit (ESDU), London.
36. Humphries, P., Derrett, G. F., and Scruton, B. (1984) Critical heat flux characteristics for vertical steam generating tubes with circumferentially non-uniform heating. *First UK Nat. Heat Transfer Conf.*, pp. 817–828, Leeds.
37. Styrikovich, M. A., and Miropolskii, Z. L. (1950) *Dokl. Akad. Nauk SSSR* **71**(2).
38. Gardner, G. C., and Kubie, J. (1976) Flow of two liquids in sloping tubes: an analogue of high pressure steam and water. *Int. J. Multiphase Flow* **2** 435–451.
39. Bailey, N. A. (1977) Dryout in the bend of a vertical U-tube evaporator. Personal communication.
40. Schuller, R. B., and Cornwell, K. (1984) Dryout on the shell side of tube bundles. *First Nat. Heat Transfer Conf.*, pp. 795–804, Leeds.
41. Leong, L. S., and Cornwell, K. (1979) Heat transfer coefficients in a reboiler tube bundle. *The Chemical Engineer* **343** 219–221.

42. Collier, J. G. (1983) Boiling outside tubes and tube bundles. *Heat Exchanger Design Handbook*, Section 2.7.5. Hemisphere, New York.

43. Palen, J. W., and Small, W. M. (1964) A new way to design kettle and internal reboilers. *Hydrocarbon Proc.* **43**(11) 199–208.

44. Carver, M. B., Carlucci, L. N., and Inch, W. W. R. (1981) *Thermal-Hydraulics in Recirculating Steam Generators: THIRST Code User Manual.* Atomic Energy of Canada, Report AECL-7254.

45. Smith, R. A. (1987) Private communication.

46. Carlucci, L. N., and Sutherland, D. (1981) The effects of various empirical correlations on the predictions of a steam generator thermal-hydraulics code. Paper presented at the Winter Annual Meeting of the ASME, November 15–20, Washington, D.C. (81-WA/NE-5).

47. Rehme, K. (1973) Pressure drop performance of rod bundles arranged in hexagonal arrangements. *Int. J. Heat Mass Transfer* **15** 2499–2517.

48. Tong, L. S. (1968) Pressure drop performance of a rod bundle. *ASME Symp. Heat Transfer in Rod Bundles*, New York.

49. Zukauskas, A. (1972) Heat transfer from tubes in crossflow. *Adv. in Heat Transfer* **8** 93–160.

50. Grimison, E. D. (1937) Correlation and utilization of new data on flow resistance and heat transfer for cross flow of gases over tube banks. *Trans. ASME* **59** 583–594.

51. Lahey, R. T., and Moody, F. J. (1977) *The Thermal-Hydraulics of a Boiling Water Nuclear Reactor*, p. 230. American Nuclear Society.

52. Armand, A. A., and Treshchev, G. G. (1947) Investigation during the movement of steam–water mixtures in a heated boiler pipe at high pressures. AERE Transl. 816, 1959.

53. Groeneveld, D. C. (1973) Forced convective heat transfer to superheated steam in rod bundles. AECL-4450.

54. Inayatov, A. Y. A. Correlation of data on heat transfer: flow parallel to tube bundles at relative tube pitches of $1.1 < p/d < 1.6$. *Heat Transfer–Soviet Research* **7**(3) 84–88.

55. Chen, J. C. Correlation for boiling heat transfer to saturated fluids in convective flow. *I & E C Process Design Development* **5**(3) 322–329.

56. Thom, J. R. S., et al. (1965–1966) Boiling in sub-cooled water during flow up heated tubes or annuli. *Proc. Inst. Mech. Eng.* **180** 226–246.

57. Cattadori, G., Masini, G., and Mazzocchi, L. (1983) Experimental tests on U-tube steam generator thermal hydraulics. European Two-Phase Flow Group Meeting, Zurich.

58. Cattadori, G., Masini, G., and Mazzocchi, L. (1984) Steady state tests on U-tube steam generator thermal hydraulics. European Two-Phase Flow Group Meeting, Rome.

59. Collier, J. G. (1983) Reliability problems of heat transfer equipment. Paper presented at Reliability 83, The Fourth National (UK) Reliability Conference, July, NEC, Birmingham.

60. Collier, J. G., and Whitmarsh-Everiss, M. J. (1985) Opening address: the orificing of once-through boilers for gas-cooled reactors. *Proc. Conf. Boiler Dynamics and Control in Nuclear Power Stations*, pp. 1–11. British Nuclear Energy Society.
61. Masterson, H. G., Castle, J. E., and Mann, G. M. W. (1969) Waterside corrosion of power station boiler tubes. *Chem. and Ind.* 1261–1266.
62. Carlucci, L. N., Campagna, A. O., and Pettigrew, M. J. (1985) Thermal-hydraulic and vibration analysis of a nuclear recirculating steam generator. *Proc. Conf. Boiler Dynamics and Control in Nuclear Power Stations*, pp. 175–180. British Nuclear Energy Society.

CHAPTER 10

HEAT TRANSFER IN CONDENSATION

P. J. MARTO

Naval Postgraduate School
Monterey, California 93943

10.1 INTRODUCTION

Condensation heat transfer occurs in numerous engineering applications. It may occur homogeneously as a fog or cloud of microscopic droplets when a vapor is cooled below its saturation temperature, or when a vapor–gas mixture is cooled below its dew point. It may also occur when vapor comes in direct contact with a subcooled liquid, such as introducing a spray of liquid drops, jets, or sheets into the vapor space, or injecting vapor into a pool of subcooled liquid.

The most common type of condensation involved in heat exchangers is surface condensation where a cooled wall, at a temperature less than the local saturation temperature of the vapor, is placed in contact with the vapor. In this situation, the vapor molecules that strike the cold surface may stick to it and condense into liquid. The resulting liquid (i.e., condensate) will accumulate in one of two ways. If the liquid "wets" the cold surface, the condensate will form a continuous film, and this mode of condensation is referred to as filmwise condensation. If the liquid does not "wet" the cold surface, it will form into numerous microscopic droplets. This mode of condensation is referred to as dropwise condensation and results in much larger heat transfer coefficients than during filmwise conditions. Since long-term dropwise condensation conditions are very difficult to sustain, all surface condensers today are designed to operate in the filmwise mode.

Boilers, Evaporators and Condensers, Edited by Sadik Kakaç
ISBN 0-471-62170-6

During condensation heat transfer, thermal resistances exist in the condensate, in the vapor, and across the liquid–vapor interface. The resistance in the condensate is due primarily to conduction of heat across the condensate film. In the vapor, the resistance depends on whether the vapor is pure or is mixed with other vapors or with noncondensable gases. The thermal resistance at the liquid–vapor interface is due to the nonequilibrium mass flux of molecules toward and away from the interface. An approximate interfacial heat transfer coefficient may be written as [1]:

$$h_i = \left(\frac{2\sigma}{2-\sigma}\right)\left(\frac{M}{2\pi RT}\right)^{1/2}\left(\frac{i_{lg}^2 PM}{RT^2}\right) \tag{10.1}$$

where σ is the condensation coefficient (i.e., the fraction of vapor molecules striking the condensate surface that actually stick and condense on the surface). In recent years, the value of σ has been measured to be near 1, so the first factor in Eq. (10.1) is not far from 2. The interfacial thermal resistance is normally very small for ordinary fluids and can be neglected without too much consequence. However, it is of crucial importance for dropwise condensation [2] and condensation of liquid metals, particularly at low pressures [3].

This chapter stresses condensation of a pure vapor where the interfacial resistance is assumed to be 0. Condensation with noncondensable gases and condensation of vapor mixtures are briefly described. Emphasis is placed on providing information that is suitable for designing condenser equipment. For additional information on condensation heat transfer phenomena, the reader is referred to Rohsenow [4], Butterworth [5], Marto [6], and Rose [7].

10.2 FILM CONDENSATION ON A SINGLE HORIZONTAL TUBE

10.2.1 Natural Convection

Nusselt [8] treated the case of laminar film condensation of a quiescent vapor on an isothermal horizontal tube as depicted in Fig. 10.1. In this situation, the motion of the condensate is determined by a balance of gravitational and viscous forces. Nusselt's analysis yields a solution for the local film thickness δ as a function of circumferential angle ϕ which is given by

$$X = \frac{\delta(\phi)^4}{\left\{\dfrac{3\mu_l k_l \Delta TD}{2\rho_l(\rho_l - \rho_g) g i_{lg}}\right\}} = \frac{\frac{4}{3}}{\sin^{4/3}\phi}\int_0^{\phi}\sin^{1/3}\phi \, d\phi \tag{10.2}$$

The function X is tabulated very accurately by Abramowitz [9]. In the Nusselt analysis, convection terms in the energy equation are neglected, so

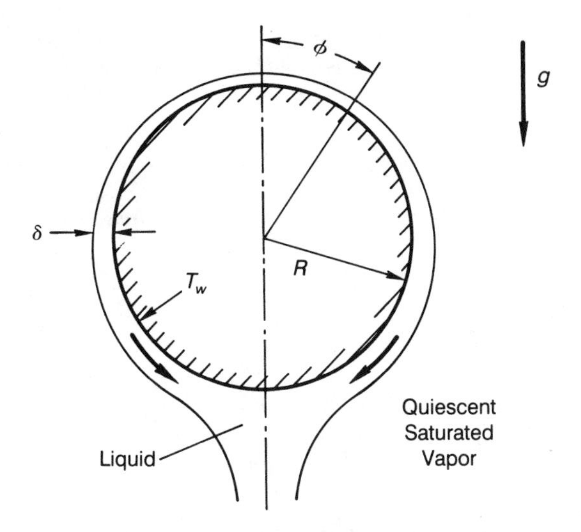

Fig. 10.1. Film condensation profile on a horizontal tube.

the local heat transfer coefficient around the tube can be written simply as

$$h(\phi) = \frac{k_l}{\delta(\phi)} \tag{10.3}$$

Clearly, at the top of the tube, where the film thickness is a minimum, the heat transfer coefficient is a maximum. Conversely, at the bottom of the tube, the heat transfer coefficient falls to 0 as the film thickness increases toward ∞. The Nusselt theory yields the following average heat transfer coefficient:

$$\frac{h_m D}{k_l} = 0.728 \left\{ \frac{\rho_l(\rho_l - \rho_g) g i_{lg} D^3}{\mu_l (T_{\text{sat}} - T_w) k_l} \right\}^{1/4} \tag{10.4}$$

For ordinary liquids, Eq. (10.4) generally underpredicts the experimental data by about 10% to 20%. A convenient alternative form for the average coefficient in terms of the film Reynolds number Re_Γ is given by

$$\frac{h_m}{k_l} \left\{ \frac{\mu_l^2}{\rho_l(\rho_l - \rho_g) g} \right\}^{1/3} = 1.51 Re_\Gamma^{-1/3} \tag{10.5}$$

where Re_Γ equals $4\Gamma/\mu_l$ and Γ is the liquid film flow rate per unit length.

Example 10.1. Quiescent refrigerant-22 vapor at a saturation temperature of 47°C is condensing on a horizontal smooth copper tube whose outside wall temperature is maintained constant at 40°C. The outside diameter of the tube is 19 mm. Calculate the average condensation heat transfer coefficient on this tube.

Solution: The average heat transfer coefficient can be calculated using the Nusselt expression, Eq. (10.4).

The thermophysical properties of R-22 at 47°C are as follows:

$$\rho_l = 1099.4 \text{ kg/m}^3$$

$$\rho_g = 79.1 \text{ kg/m}^3$$

$$k_l = 0.077 \text{ W/(m} \cdot \text{K)}$$

$$\mu_l = 0.176 \times 10^{-4} \text{ (N} \cdot \text{s)/m}^2 \text{ [kg/(m} \cdot \text{s)]}$$

$$i_{lg} = 158.1 \text{ kJ/kg} = 1.58 \times 10^5 \text{ J/kg}$$

Upon substitution into Eq. (10.4), we get

$$h_m = 0.728 \frac{(0.077)}{(0.019)} \left\{ \frac{(1099.4)(1099.4 - 79.1)(9.81)(1.58 \times 10^5)(0.019)^3}{(0.176 \times 10^{-4})(47 - 40)(0.077)} \right\}^{1/4}$$

$$= 3124 \text{ W/(m}^2 \cdot \text{K)}$$

10.2.2 Forced Convection

When the vapor surrounding a horizontal tube is moving at a high velocity, the analysis for film condensation is affected in two important ways: (1) the surface shear stress between the vapor and the condensate must be included, and (2) the effect of vapor separation must be accurately treated. Rose [7] has recently provided an excellent review of forced convection condensation.

The early analytical investigations of this problem were extensions of Nusselt's analysis to include the interfacial shear boundary condition at the edge of the condensate film. Shekriladze and Gomelauri [10] realized that the mass flow across a condensing interface is very important. They therefore assumed that the primary contribution to the surface shear stress was due to the change in momentum across the interface. Their simplified solution for an isothermal cylinder without separation and with no body forces is

$$Nu_m = \frac{h_m D}{k_l} = 0.9 \widetilde{Re}^{1/2} \tag{10.6}$$

where $\widetilde{Re}$ is defined as a two-phase Reynolds number involving the vapor velocity and condensate properties $u_g D/\nu_l$. When both gravity and velocity are included, they recommended the relationship

$$\frac{Nu_m}{\widetilde{Re}^{1/2}} = 0.64\{1 + (1 + 1.69F)^{1/2}\}^{1/2} \tag{10.7}$$

where

$$F = \frac{gD\mu_l i_{lg}}{u_g^2 k_l \, \Delta T} \tag{10.8}$$

Equation (10.7) neglects vapor separation, which occurs somewhere between 82 and 180° from the stagnation point of the cylinder. After the separation point, the condensate film rapidly thickens and, as a result, heat transfer is deteriorated. A conservative approach suggested by Shekriladze and Gomelauri [10] is to assume that there is no heat transferred beyond the separation point. If the minimum separation angle of 82° is then chosen, the most conservative equation results, and the heat transfer decreases by approximately 35%. Therefore Eq. (10.6) reduces to

$$Nu_m = 0.59\widetilde{Re}^{1/2} \tag{10.9}$$

An interpolation formula based on this conservative approach, which satisfies the extremes of gravity-controlled and shear-controlled condensation, was proposed by Butterworth [11]:

$$\frac{Nu_m}{\widetilde{Re}^{1/2}} = 0.416\{1 + (1 + 9.47F)^{1/2}\}^{1/2} \tag{10.10}$$

A variety of more complex analytical models exist in the literature and are thoroughly discussed by Rose [7]. Vapor boundary layer effects, especially separation, and the effect of the pressure gradient around the lower part of the tube provide significant difficulties in arriving at an accurate analytical solution. As a result, approximate, conservative expressions are used. Rose [7] compared the experimental data of a number of investigators and discovered a considerable spread especially at high vapor velocities [i.e., low values of F as defined in Eq. (10.8)]. Figure 10.2 compares the predictions of Eqs. (10.9) and (10.10), along with the Nusselt equation, Eq. (10.4), to the data provided in Rose [7]. In general, Eq. (10.10) is conservative and can be used with reasonable confidence.

Example 10.2. Suppose that the refrigerant-22 vapor in Example 10.1 were moving downward over the tube at a velocity of 5 m/s instead of being quiescent as originally stated. Calculate the average heat transfer coefficient in this situation.

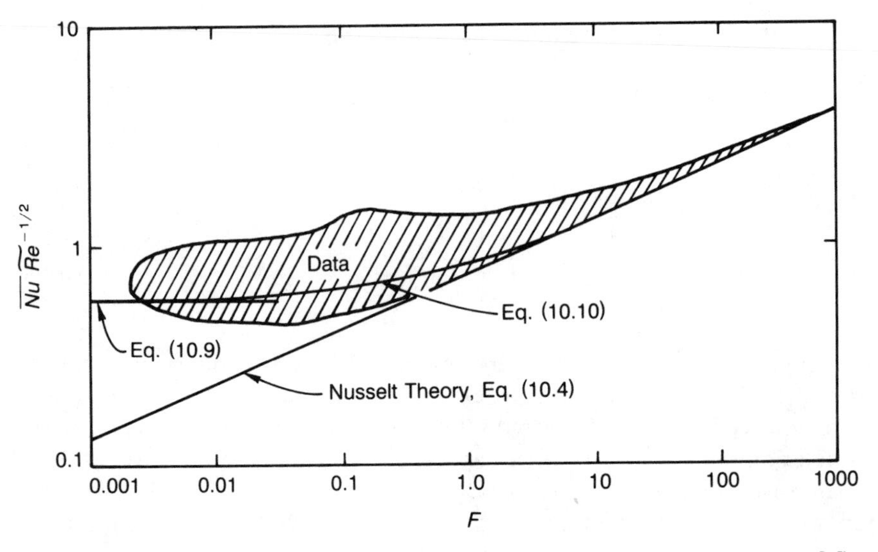

Fig. 10.2. Condensation in downflow over horizontal tubes (adapted from Rose [7]).

Solution: For downward-moving vapor over a horizontal tube, the average heat transfer coefficient can be calculated using Eq. (10.10):

$$\frac{h_m D}{k_l} = 0.416\{1 + (1 + 9.47F)^{1/2}\}^{1/2}\widetilde{Re}^{1/2}$$

From Eq. (10.8):

$$F = \frac{gD\mu_l i_{lg}}{u_g^2 k_l\,\Delta T}$$

Upon substitution of the R-22 properties as listed in Example 10.1,

$$F = \frac{(9.81)(0.019)(0.176 \times 10^{-4})(1.58 \times 10^5)}{(5)^2(0.077)(47 - 40)}$$

$$= 0.0385$$

The two-phase Reynolds number is

$$\widetilde{Re} = \frac{\rho_l u_g D}{\mu_l}$$

$$= \frac{(1099.4)(5)(0.019)}{(0.176 \times 10^{-4})}$$

$$= 5.93 \times 10^6$$

When these values for F and $\widetilde{Re}$ are substituted into Eq. (10.10), we get

$$h_m = 0.416 \frac{(0.077)}{(0.019)} \{1 + (1 + (9.47)(0.0385))^{1/2}\}^{1/2} (5.93 \times 10^6)^{1/2}$$

$$= 6047 \text{ W/m}^2 \cdot \text{K}$$

which represents a 94% increase over the quiescent vapor case of Example 10.1.

10.3 FILM CONDENSATION IN TUBE BUNDLES

During film condensation in tube bundles, the conditions are much different than for a single tube. The presence of neighboring tubes creates several added complexities as depicted schematically in Fig. 10.3. In the idealized case, Fig. 10.3*a*, the condensate from a given tube is assumed to drain by gravity to the lower tubes in a continuous, laminar sheet. In reality, depending on the spacing-to-diameter ratio of the tubes and depending on whether they are arranged in a staggered or in-line configuration, the condensate from one tube may not fall on the tube directly below it but instead may flow sideways, Fig. 10.3*b*. Also, it is well known experimentally that condensate does not drain from a horizontal tube in a continuous sheet but in discrete droplets along the tube axis. When these droplets strike the lower tube, considerable splashing can occur, Fig. 10.3*c*, causing ripples and turbulence in the condensate film. Perhaps most important of all, large vapor velocities can create significant shear forces on the condensate, stripping it away, independent of gravity, Fig. 10.3*d*.

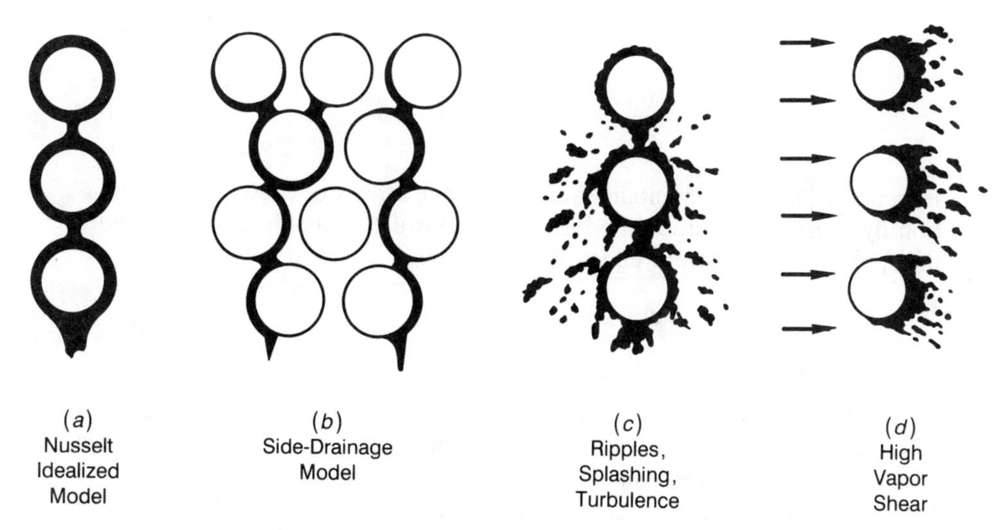

Fig. 10.3. Schematic representation of condensate flow (from Marto [21]).

10.3.1 Effect of Condensate Inundation

In the absence of vapor velocity, as condensate flows by gravity onto lower tubes in a bundle, the condensate thickness around the lower tubes should increase, and the condensation heat transfer coefficient should therefore decrease.

Nusselt [8] extended his analysis to include film condensation on a vertical in-line column of horizontal tubes. He assumed that all the condensate from a given tube drains as a continuous laminar sheet directly onto the top of the tube below it. With this assumption, together with the assumption that the temperature difference across the condensate film $(T_{sat} - T_w)$ remains the same for all the tubes, he showed that the average coefficient for a vertical column of N tubes, compared to the coefficient for the first tube (i.e., the top tube in the row), is

$$\frac{h_{m,N}}{h_1} = N^{-1/4} \tag{10.11}$$

In Eq. (10.11), h_1 is calculated using Eq. (10.4). In terms of the local coefficient for the Nth tube, the Nusselt theory gives

$$\frac{h_N}{h_1} = N^{3/4} - (N-1)^{3/4} \tag{10.12}$$

Kern [12] proposed a less conservative relationship

$$\frac{h_{m,N}}{h_1} = N^{-1/6} \tag{10.13}$$

or in terms of the local value,

$$\frac{h_N}{h_1} = N^{5/6} - (N-1)^{5/6} \tag{10.14}$$

Eissenberg [13] experimentally investigated the effects of condensate inundation by using a staggered tube bundle. He postulated a side-drainage model that predicts a less severe effect of inundation

$$\frac{h_{m,N}}{h_1} = 0.60 + 0.42N^{-1/4} \tag{10.15}$$

Numerous experimental measurements have been made in studying the effect of condensate inundation. The data, however, are very scattered. As a result, it is not too surprising that there is no successful theoretical model today that can predict accurately the effect of condensate inundation on condensation performance for a variety of operating conditions. For design

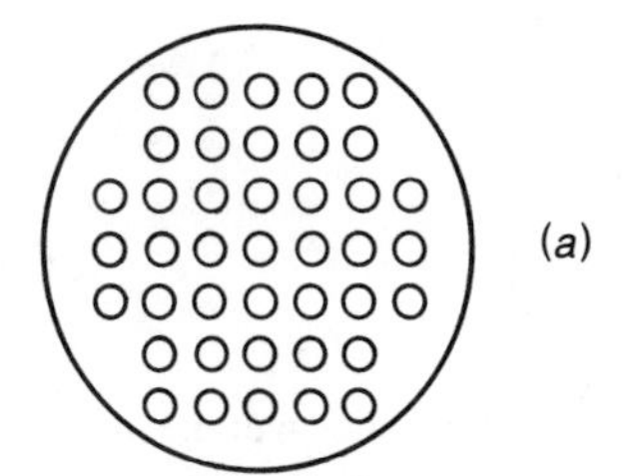

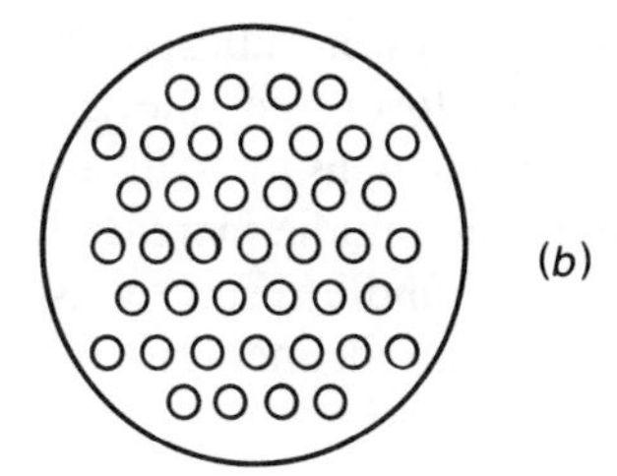

Fig. 10.4. Tube bundle layout: (*a*) square, in-line arrangement, (*b*) triangular, staggered arrangement.

purposes, the Kern expressions [either Eq. (10.13) or (10.14)] are conservative, and have been recommended by Butterworth [11].

Example 10.3. Suppose that the refrigerant-22 in Example 10.1 is condensing under quiescent conditions on the shell side of a bundle of 41 tubes. The bundle can be configured in a square, in-line arrangement or in a triangular, staggered arrangement as shown in Fig. 10.4. Find the average shell-side coefficient for each of the configurations.

Solution: To find the average heat transfer coefficient for the bundle, we correct the Nusselt expression for a single tube, Eq. (10.4), using the Kern relationship, Eq. (10.13).

1. *Square, in-line arrangement, Fig. 10.4(a).* With this configuration, there are five columns of seven tubes each and two columns of three tubes each. This arrangement would be equivalent to approximately seven columns of six tubes each. Therefore $N \simeq 6$. From Eqs. (10.4) and (10.13),

$$h_{m,6} = h_1(6)^{-1/6} = 3124(0.7418)$$

$$= 2317 \text{ W/m}^2 \cdot \text{K}$$

2. *Triangular, staggered arrangement, Fig. 10.4(b).* With this configuration, assuming that the condensate falls straight down and not sideways, there are seven columns of three tubes, four columns of four tubes, and two columns of two tubes. This arrangement would be equivalent to approximately thirteen columns of three tubes each. Therefore $N \simeq 3$. Once again, from

Eqs. (10.4) and (10.13),

$$h_{m,3} = 3124(3)^{-1/6} = 3123(0.8327)$$

$$= 2601 \mathrm{W/m^2 \cdot K}$$

Therefore, by using the staggered arrangement of tubes, the average heat transfer coefficient is 12% larger than the in-line arrangement.

One way of preventing inundation of condensate on lower tubes is to incline the tube bundle with respect to the horizontal. Shklover and Buevich [14] conducted an experimental investigation of steam condensation in an inclined bundle of tubes. They found that inclination of the bundle increased the average heat transfer coefficient over the horizontal bundle result by as much as 25%. These favorable results led to the design of an inclined-bundle condenser with an inclination angle of 5°.

10.3.2 Effect of Vapor Shear

In tube bundles, the influence of vapor shear has been measured by Nobbs and Mayhew [15], Kutateladze et al. [16], Fujii et al. [17, 18], and Cavallini et al. [19]. Fujii et al. [18] found that there was little difference between the downward-flow and horizontal-flow data obtained, but the upward-flow data were as much as 50% lower in the range $0.1 < F < 0.5$. They arrived at the following empirical expression which correlated the downward-flow and horizontal-flow data reasonably well

$$\frac{Nu_m}{\widetilde{Re}^{1/2}} = 0.96F^{1/5} \tag{10.16}$$

for $0.03 < F < 600$. Cavallini et al. [19] compared their data with the prediction of Shekriladze and Gomelauri [10] [Eq. (10.7)], and found the prediction to be conservative.

In a tube bundle, it is not clear which local vapor velocity should be used to calculate vapor shear effects. Butterworth [11] points out that the use of the maximum cross-sectional area would give a conservative prediction. Nobbs and Mayhew [15] have used the mean local velocity through the bundle. They calculate this velocity based on a mean flow width given by

$$w = \frac{p_l p_t - \pi D^2/4}{p_l} \tag{10.17}$$

where p_l and p_t are the tube pitches (i.e., centerline-to-centerline distance) in the longitudinal and transverse directions, respectively.

10.3.3 Combined Effects of Inundation and Vapor Shear

Initially, the effects of inundation and vapor shear were treated separately. The combined average heat transfer coefficient for condensation in a tube bundle was simply written as

$$h_{m,N} = h_1 C_N C_{ug} \tag{10.18}$$

where h_1 represents the average coefficient for a single tube from Nusselt theory [Eq. (10.4)] and C_N and C_{ug} are correction factors to account for inundation and vapor shear, respectively.

However, in a tube bundle, a strong interaction exists between vapor shear and condensate inundation, and local heat transfer coefficients are very difficult to predict. Butterworth [11] proposed a relationship for the local heat transfer coefficient in the Nth tube row which separates out the effects of vapor shear and condensate inundation. A slightly modified form of his equation is

$$h_N = \left[\tfrac{1}{2}h_{\text{sh}}^2 + \left(\tfrac{1}{4}h_{\text{sh}}^4 + h_1^4\right)^{1/2}\right]^{1/2}\left[N^{5/6} - (N-1)^{5/6}\right] \tag{10.19}$$

where h_{sh} is obtained using Eq. (10.9) and h_1 is obtained using Eq. (10.4).

McNaught [20] suggested that shell-side condensation may be treated as two-phase forced convection. He therefore proposed the following relationship for the local coefficient for the Nth row:

$$h_N = \left(h_{\text{sh}}^2 + h_G^2\right)^{1/2} \tag{10.20}$$

where h_G is given by Eq. (10.14); that is,

$$h_G = h_1\left(N^{5/6} - (N-1)^{5/6}\right)$$

and h_{sh} is given by

$$h_{\text{sh}} = 1.26\left[\frac{1}{X_{tt}}\right]^{0.78} h_l \tag{10.21}$$

In Eq. (10.21), X_{tt} is the Lockhart–Martinelli parameter, defined as

$$X_{tt} = \left(\frac{1-x}{x}\right)^{0.9}\left(\frac{\rho_g}{\rho_l}\right)^{0.5}\left(\frac{\mu_l}{\mu_g}\right)^{0.1} \tag{10.22}$$

and h_l is the liquid-phase forced convection heat transfer coefficient across a bank of tubes. This is generally expressed as

$$h_l = C\frac{k_l}{D}Re_l^m Pr_l^n \tag{10.23}$$

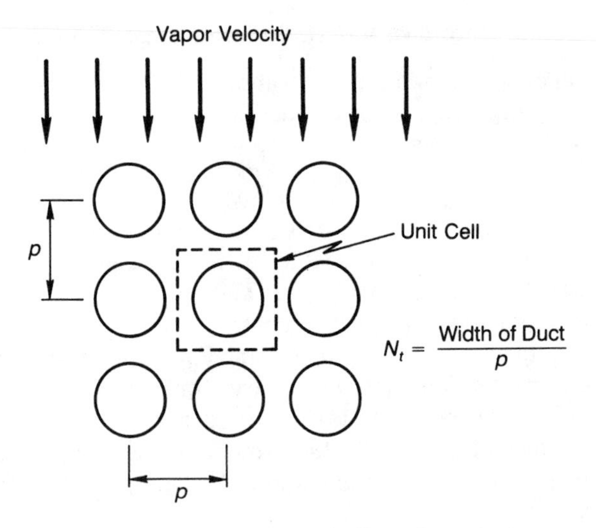

Fig. 10.5. Schematic of square, in-line tube arrangement.

where C, m, and n depend on the flow conditions through the tube bank. The numerical values in Eq. (10.21) were obtained for steam condensing in a bank of in-line or staggered tubes under the following conditions [15]:

$$\frac{p}{D} = 1.25$$

$$10 \leq G \leq 70 \text{ kg/m}^2 \cdot \text{s}$$

$$0.025 \leq x \leq 0.8$$

$$0.008 \leq X_{tt} \leq 0.8$$

The correlation includes the effect of condensate inundation. McNaught [20] found that Eqs. (10.20) and (10.21) correlated 90% of the steam data to within ±25%. Care must be taken to avoid using this correlation when the operating conditions fall outside of the ranges indicated previously.

Example 10.4. Steam at a saturation temperature of 100°C is condensing in a bundle of 320 tubes within a 0.56-m wide duct. The tubes are 25 mm in diameter and are 4 m long. They are arranged in a square, in-line pitch ($p = 35.0$ mm) as shown in Fig. 10.5. The bundle is made up of 20 rows of tubes with 16 tubes in each row. The tube wall temperature in each row is kept constant at 93°C. The steam flows downward in the bundle and at the sixth row of tubes, the local mass flow rate of vapor is

$$\dot{m}_g = 14.0 \text{ kg/s}$$

Find the local heat transfer coefficient for this sixth row of tubes using the method of Butterworth [11].

Solution: The thermophysical properties of steam at 100°C are as follows:

$$\rho_l = 957.9 \text{ kg/m}^3$$

$$\rho_g = 0.598 \text{ kg/m}^3$$

$$k_l = 0.681 \text{ W/(m} \cdot \text{k)}$$

$$\mu_l = 2.79 \times 10^{-4} \text{ kg/(m} \cdot \text{s)}$$

$$\mu_g = 1.2 \times 10^{-5} \text{ kg/(m} \cdot \text{s)}$$

$$i_{lg} = 2.257 \times 10^3 \text{ kJ/kg}$$

$$c_{pl} = 4.219 \text{ kJ/(kg} \cdot \text{K)}$$

The local steam velocity can be calculated as follows:

$$u_g = \frac{\dot{m}_g}{\rho_g A_m}$$

where A_m is the mean flow area. This mean flow area can be written in terms of the number of units cells N_t and the mean width per cell w:

$$A_m = wN_t L$$

where w is given by Eq. (10.17):

$$w = \frac{p_l p_t - \pi D^2/4}{p_l}$$

In this example, $p_l = p_t = p = 0.035$ m. Therefore

$$w = \frac{(0.035)^2 - \pi(0.025)^2/4}{(0.035)}$$

$$= 0.021 \text{ m}$$

The mean flow area is then

$$A_m = (0.021)(16)(4)$$

$$= 1.344 \text{ m}^2$$

The local steam velocity is

$$u_g = \frac{(14.0)}{(0.598)(1.344)}$$

$$= 17.4 \text{ m/s}$$

Using the method of Butterworth [11], the local heat transfer coefficient is given by Eq. (10.19):

$$h_N = \left[\tfrac{1}{2}h_{sh}^2 + \left(\tfrac{1}{4}h_{sh}^4 + h_1^4\right)^{1/2}\right]^{1/2}\left[N^{5/6} - (N-1)^{5/6}\right] \quad (10.19)$$

where

$$h_{sh} = 0.59\frac{k_l}{D}\widetilde{Re}^{1/2} \quad (10.9)$$

and

$$h_1 = 0.728\frac{k_l}{D}\left\{\frac{\rho_l(\rho_l - \rho_g)g i_{lg} D^3}{\mu_l \Delta T k_l}\right\}^{1/4} \quad (10.4)$$

The two-phase Reynolds number is

$$\widetilde{Re} = \left(\frac{\rho_l u_g D}{\mu_l}\right)$$

$$= \frac{(957.9)(17.4)(0.025)}{(2.79 \times 10^{-4})}$$

$$= 1.49 \times 10^6$$

Therefore

$$h_{sh} = 0.59\frac{(0.681)}{(0.025)}(1.49 \times 10^6)^{1/2}$$

$$= 19{,}618\ \text{W}/(\text{m}^2 \cdot \text{K})$$

From Eq. (10.4),

$$h_1 = 0.728\frac{(0.681)}{(0.025)}\left\{\frac{(957.9)(957.9 - 0.598)(9.81)(2.257 \times 10^6)(0.025)^3}{(2.79 \times 10^{-4})(100 - 93)(0.681)}\right\}^{/4}$$

$$= 13{,}859\ \text{W}/(\text{m}^2 \cdot \text{K})$$

Therefore, upon substitution into Eq. (10.19), we get

$$h_6 = \left[\tfrac{1}{2}(19{,}618)^2 + \left[\tfrac{1}{4}(19{,}618)^4 + (13{,}859)^4\right]^{1/2}\right]^{1/2}[6^{5/6} - 5^{5/6}]$$

$$h_6 = 13{,}519\ \text{W}/(\text{m}^2 \cdot \text{K})$$

10.3.4 Computer Modeling

Computer modeling must be used to predict accurately the average shell-side heat transfer coefficient in a tube bundle. Marto [21] has reviewed the evolvement of computer methods to predict shell-side condensation. The vapor flow must be followed throughout the bundle and within the vapor flow lanes. Knowing the local vapor velocity and the local vapor pressure and temperature, as well as the distribution of condensate from other tubes (and the local concentration of any noncondensable gases), it is possible to predict a local coefficient in the bundle which can then be integrated to arrive at the overall bundle performance. Early efforts to model the thermal performance of condensers were limited essentially to one-dimensional routines in the plane perpendicular to the tubes [22–25]. Today, more sophisticated two-dimensional models exist which have been utilized successfully to study the performance of complex tube bundle geometries [26–31].

10.4 FILM CONDENSATION INSIDE TUBES

10.4.1 Flow Patterns

During film condensation inside tubes, a variety of flow patterns can exist as the flow passes from the tube inlet (with a quality near 1.0) to the exit (with a quality near or below 0.0). This is illustrated in Fig. 10.6 for a low and high mass flux situation. The flow can pass through mist, annular, wavy, and slug patterns. Eventually, depending on the heat flux, all the vapor can condense, resulting in single-phase, subcooled liquid at the tube exit. Of course, these different flow patterns can alter the heat transfer considerably, so that local calculations must be made along the length of the tube.

The previously mentioned flow patterns have been studied by numerous investigators in recent years [32–37]. The transition from one flow pattern to another must be predicted in order to make the necessary heat transfer calculations. Breber et al. [36] have proposed a simple method of predicting flow pattern transitions that depend on the dimensionless mass velocity j_g^*, defined as

$$j_g^* = \frac{xG}{\left[gD\rho_g(\rho_l - \rho_g)\right]^{1/2}} \tag{10.24}$$

and the Lockhart–Martinelli parameter X_{tt}, Eq. (10.22). Their flow pattern criteria are

$$j_g^* > 1.5 \quad X_{tt} < 1.0: \quad \text{mist and annular} \tag{10.25a}$$

$$j_g^* < 0.5 \quad X_{tt} < 1.0: \quad \text{wavy and stratified} \tag{10.25b}$$

$$j_g^* < 0.5 \quad X_{tt} > 1.5: \quad \text{slug} \tag{10.25c}$$

$$j_g^* > 1.5 \quad X_{tt} > 1.5: \quad \text{bubble} \tag{10.25d}$$

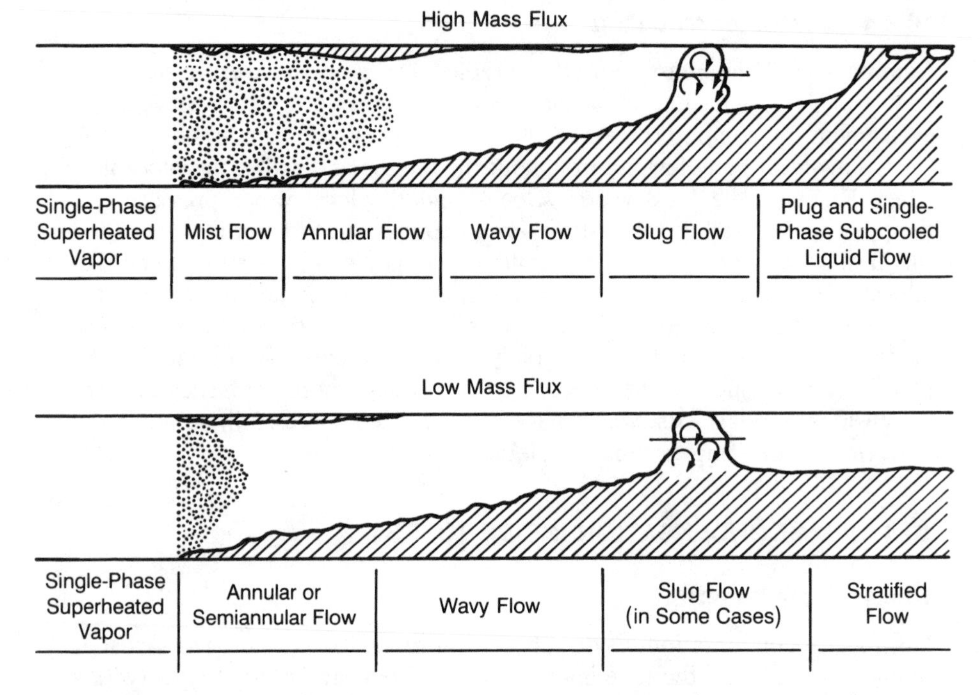

Fig. 10.6. Flow pattern development during horizontal in-tube condensation (from the *Encyclopedia of Fluid Mechanics*, volume 3, by Nicholas P. Cheremisinoff. Copyright ©1986 by Gulf Publishing Company Houston, TX. Used with permission. All rights reserved.)

From these criteria, it is clear that transition bands separate each of the flow patterns. Recently, Rahman et al. [34] obtained new data for steam condensing inside horizontal tubes. They compared their data to the various flow pattern classification methods and found good agreement with the method of Breber et al. [36]. During horizontal in-tube condensation, the transition from annular to stratified flow is most important. Equations (10.25a) and (10.25b) can therefore be used for design purposes. Soliman [35] has recently provided a Froude number criterion for both the spray–annular transition and the annular–wavy transition. Soliman [38] has also proposed a Weber number criterion for the mist–annular transition.

10.4.2 Condensation in Horizontal Tubes

During condensation inside horizontal tubes, different heat transfer models are used, depending on whether vapor shear or gravitational forces are more

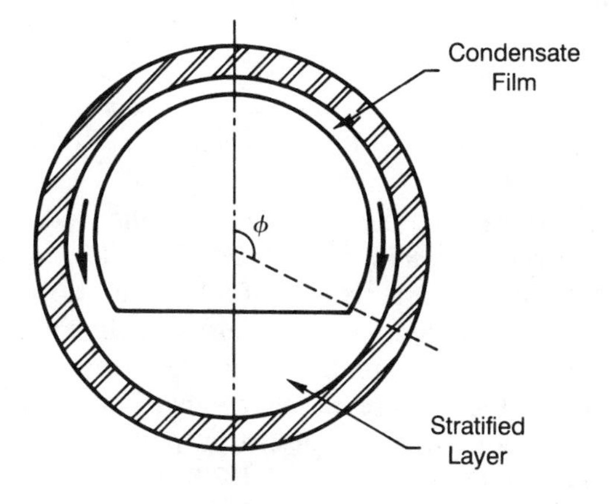

Fig. 10.7. Idealized condensate profile inside a horizontal tube.

important. When the vapor velocity is low (i.e., j_g^* is less than 0.5), the flow will be dominated by gravitational forces and stratification of the condensate will occur. At high vapor velocities (i.e., $j_g^* > 1.5$) where interfacial shear forces are large, gravitational forces may be neglected and the condensate flow will be annular. When the flow is stratified, the condensate forms as a thin film on the top portion of the tube walls. This condensate drains toward the bottom of the tube where a stratified layer exists as shown schematically in Fig. 10.7. The stratified layer flows axially due to vapor shear forces. In this circumstance, the Nusselt theory for laminar flow is generally valid over the top, thin film region of the tube. However, Butterworth [5] points out that if the axial vapor velocity is high, turbulence may occur in this thin film and the Nusselt analysis is no longer valid. In the stratified layer, heat transfer is generally negligible. For laminar flow, the average heat transfer coefficient over the entire perimeter may be expressed by a modified Nusselt result:

$$h_m = \Omega \left\{ \frac{k_l^3 \rho_l (\rho_l - \rho_g) g i_{lg}}{\mu_l D \, \Delta T} \right\}^{1/4} \tag{10.26}$$

where the coefficient Ω depends on the fraction of the tube that is stratified. Jaster and Kosky [39] have shown that Ω is related to the void fraction of the vapor α_g:

$$\Omega = 0.728 \alpha_g^{3/4} \tag{10.27}$$

where

$$\alpha_g = \frac{1}{1 + [(1-x)/x](\rho_g/\rho_l)^{2/3}} \tag{10.28}$$

In the annular flow regime, there are many predictive techniques that are available [40]. Generally, laminar flow models (based on a Nusselt analysis) predict heat transfer coefficients that are too low, and turbulent flow models must be used. These turbulent flow models are either empirically based, dimensionless correlations or are based on the heat transfer–momentum analogy. Some of the models are rather cumbersome to use and result in expressions that are inconvenient for design purposes. Table 10.1 summarizes the most commonly used methods, Eqs, (10.29) through (10.43).

TABLE 10.1 Annular Flow Models

Reference	Method
Akers et al. [41]	$Nu = \frac{hD_i}{k_l} = CRe_e^n Pr_l^{1/3}$ (10.29) where $C = 0.0265 \quad n = 0.8 \quad \text{for } Re_e > 5 \times 10^4$ $C = 5.03 \quad n = \frac{1}{3} \quad \text{for } Re_e < 5 \times 10^4$ $Re_e = \frac{D_i G_e}{\mu_l}$ (10.30) $G_e = G\left[(1-x) + x(\rho_l/\rho_g)^{1/2}\right]$ (10.31)
Cavallini and Zecchin [42]	$Nu = 0.05 Re_e^{0.8} Pr^{0.33}$ (10.32)
Boyko and Kruzhilin [43]	$Nu = 0.021 Re_l^{0.8} Pr_l^{0.43}\left[1 + x(\rho_l/\rho_g - 1)\right]^{1/2}$ (10.33) $Re_l = \frac{GD_i}{\mu_l}$
Shah [44]	$Nu = Nu_l\left[(1-x)^{0.8} + \frac{3.8x^{0.76}(1-x)^{0.04}}{p_r^{0.38}}\right]$ (10.34) where $p_r = \frac{p}{P_c}$ $Nu_l = 0.023 Re_l^{0.8} Pr_l^{0.4}$ (10.35)

TABLE 10.1 *(Continued)*

Reference	Method
Kosky and Staub [45]	$$h = \frac{\rho_l C_{pl} u^*}{T^+} \quad (10.36)$$ where $$u^* = \left(\frac{\tau_w}{\rho_l}\right)^{1/2} \quad (10.37)$$ $$T^+ = Pr_l \delta^+ \qquad \delta^+ \leq 5 \quad (10.38a)$$ $$T^+ = 5\left[Pr_l + \ln(1 + Pr_l(\delta^+/5 - 1))\right] \qquad 5 < \delta^+ \leq 30 \quad (10.38b)$$ $$T^+ = 5\left[Pr_l + \ln(1 + 5Pr_l) + \tfrac{1}{2}\ln(\delta^+/30)\right] \qquad \delta^+ > 30 \quad (10.38c)$$ $$\delta^+ = \frac{\delta u^*}{\nu_l} \quad (10.39a)$$ $$\delta^+ = \left(\frac{Re_l}{2}\right)^{1/2} \qquad Re_l < 1250 \quad (10.39b)$$ $$\delta^+ = 0.0504 Re_l^{0.875} \qquad Re_l > 1250 \quad (10.39c)$$ $$Re_l = \frac{(1-x)GD_i}{\mu_l}$$ $$\tau_w = \frac{-D_i}{4}\left(\frac{dP}{dz}\right)_f \quad (10.40)$$
(Method for calculating the two-phase frictional pressure gradient $(dP/dz)_f$ are presented in Section 10.5.2.)	
Traviss et al. [46]	$$Nu = Pr_l Re_l^{0.9} \frac{F_1(X_{tt})}{F_2(Re_l, Pr_l)} \quad (10.41)$$ where $$Re_l = \frac{(1-x)GD_i}{\mu_l}$$ $$F_1 = 0.15\left[X_{tt}^{-1} + 2.85 X_{tt}^{-0.476}\right] \quad (10.42)$$ $$F_2 = 0.707 Pr_l Re_l^{0.5} \qquad Re_l < 50 \quad (10.43a)$$ $$F_2 = 5Pr_l + 5\ln\left[1 + Pr_l\left(0.0964 Re_l^{0.585} - 1\right)\right] \qquad 50 < Re_l < 1125 \quad (10.43b)$$ $$F_2 = 5Pr_l + 5\ln(1 + 5Pr_l) + 2.5\ln\left(0.0031 Re_l^{0.812}\right) \qquad Re_l > 1125 \quad (10.43c)$$

All of the expressions for the local heat transfer coefficient must be integrated over the length of the tube in order to find an average heat transfer coefficient

$$h_m = \frac{1}{L}\int_0^L h(z)\,dz \tag{10.44}$$

The problem of completing the necessary integration lies with the dependence of the quality x on axial position z. This generally will require subdividing the overall length into a number of subelements of length Δz and following the process from inlet to outlet, using local heat transfer coefficients for each subelement. If the quality is assumed to vary linearly (which unfortunately does not occur in many cases), then an average heat transfer coefficient may be found by using an average quality $x = 0.5$ in the local expressions listed in Table 10.1.

10.4.3 Condensation in Vertical Tubes

Condensation heat transfer in vertical tubes depends on the flow direction and its magnitude. For downward-flowing vapor, at low velocities, the condensate flow is controlled by gravity and the heat transfer coefficient may be calculated by Nusselt theory on a vertical surface

$$Nu_m = \frac{h_m L}{k_l} = 0.943\left\{\frac{\rho_l(\rho_l - \rho_g) g i_{lg} L^3}{\mu_l \,\Delta T k_l}\right\}^{1/4} \tag{10.45}$$

If the condensate film proceeds from laminar, wave-free to wavy conditions, a correction to Eq. (10.45) can be applied

$$\frac{h_{m,c}}{h_m} = 0.8\left(\frac{Re_\Gamma}{4}\right)^{0.11} \tag{10.46}$$

where the film Reynolds number $Re_\Gamma > 30$. If turbulent conditions exist, then the average heat transfer coefficient can be calculated by one of the methods described by Marto [6]. If the vapor velocity is very high, then the flow is controlled by shear forces and the annular flow models outlined in Table 10.1 may be used.

For upward-flowing vapor, interfacial shear will retard the drainage of condensate. As a result, the condensate film will thicken and the heat transfer coefficient will decrease. In this case, Eq. (10.45) may be used with a correction factor of 0.7 to 1.0 applied, depending on the magnitude of the vapor velocity. Care must be exercised to avoid vapor velocities that are high enough to cause "flooding," which occurs when the vapor shear forces prevent the downflow of condensate. One criterion to predict the onset of "flooding" is due to Wallis [47] which is based on air–water systems

$$(v_g^*)^{1/2} + (v_l^*)^{1/2} = C \tag{10.47}$$

where

$$v_g^* = \frac{v_g \rho_g^{1/2}}{\left[g D_i (\rho_l - \rho_g) \right]^{1/2}} \tag{10.48a}$$

$$v_l^* = \frac{v_l \rho_l^{1/2}}{\left[g D_i (\rho_l - \rho_g) \right]^{1/2}} \tag{10.48b}$$

The velocities v_g and v_l should be calculated at the bottom of the tube (where they are at their maximum values). Wallis [47] determined the parameter C to be 0.725 based on his measurements of air and water.

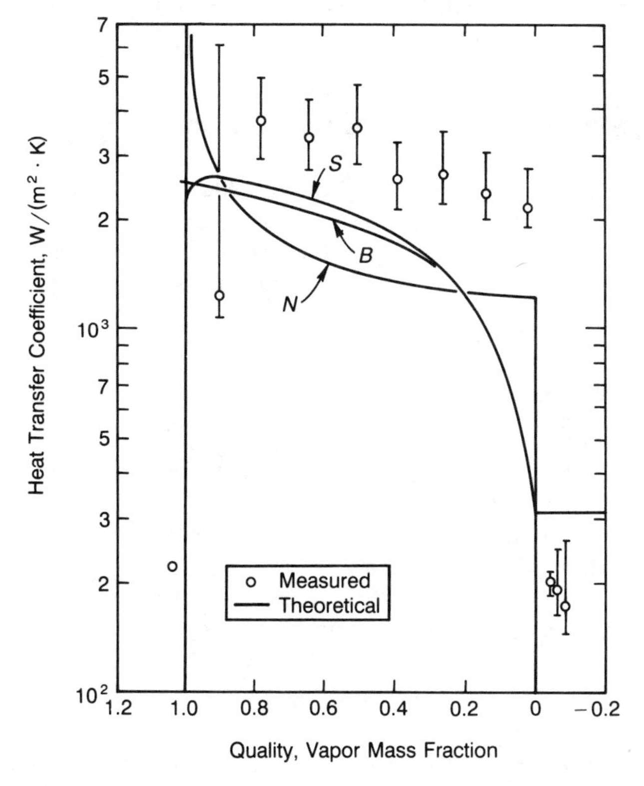

Fig. 10.8. Local heat transfer coefficients for condensation of nitrogen in a plate-fin heat exchanger (from Haseler [48]).

Butterworth [5] suggests that C should be corrected for surface tension and for tube end effects using the relationship:

$$C^2 = 0.53F_\sigma F_g \tag{10.49}$$

where F_σ is a correction factor for surface tension and F_g depends on the geometry of the end of the tube.

10.4.4 Condensation in Noncircular Passages

In recent years, there has been significant interest in condensation within plate-fin heat exchangers [48–51]. In this case, condensation occurs within a noncircular passage which may be rectangular or triangular in shape and the heat transfer is controlled by large vapor shear forces. Figure 10.8 shows a comparison between the local heat transfer coefficients for nitrogen condensing in a plate-fin heat exchanger as measured by Haseler [48], and the in-tube correlations of Nusselt [8], Boyko and Kruzhilin [43], and Shah [44] for circular tubes. None of the correlations give reasonable agreement, and they all underpredict the measured results. This discrepancy may be due to surface tension effects which have been neglected in all of the existing models [50].

10.5 PRESSURE DROP DURING CONDENSATION

The pressure drop during condensation is made up of terms involving wall friction, flow acceleration, and gravity. In general, the gravity term is small and can be neglected. The acceleration term may result in a pressure gain since there may be a deceleration of the vapor as mass is removed by the condensation process.

10.5.1 Shell-Side Pressure Drop

Despite the importance of knowing the shell-side pressure drop in condenser design, especially for steam power plant condensers that operate under vacuum conditions, little research has been performed on shell-side, two-phase flow with condensation. Most of the pressure drop information in the literature pertains to adiabatic flow.

For simplicity, shell-side losses may be calculated by using single-phase (i.e., dry-tube) correlations of the form

$$\Delta P = 4f_m N_t \rho_g \frac{u_m^2}{2} \tag{10.50}$$

where N_t is the number of tube rows, u_m is the maximum vapor velocity in the tube bundle (i.e., based on the minimum flow area), and f_m is the friction factor, which for a limited range of vapor Reynolds numbers may be expressed as $a/Re_g^{0.2}$.

The coefficient a in the friction factor expression depends on the geometry of the tube bank. Equation (10.50) has been used successfully to predict the pressure drop during horizontal flow in condensing tube banks, but the determination of the average velocity u_m to perform the calculation remains uncertain. In recent years, several more accurate methods have been proposed to predict the two-phase frictional pressure drop during shell-side condensation [52–54]. These are discussed more thoroughly by Marto [6].

10.5.2 Pressure Drop Inside Tubes

The local pressure gradient during condensation inside tubes may be written as

$$\frac{dP}{dz} = \left(\frac{dP}{dz}\right)_f + G^2 \frac{d}{dz}\left(\frac{(1-x)^2}{\rho_l(1-\alpha_g)} + \frac{x^2}{\rho_g \alpha_g}\right) \tag{10.51}$$

where the gravity term has been neglected (it may be important for long vertical tubes) and where the second term on the right-hand side represents the acceleration term. In Eq. (10.51), α_g represents the void fraction of the vapor, Eq. (10.28). The two-phase frictional pressure gradient may be related to the single-phase flow of either the liquid or the vapor, assuming that they are flowing at their actual respective mass fluxes (e.g., $G_g = xG$), or to the single-phase flow of either phase, assuming that either phase is flowing at the total mass flux [55]:

$$\left(\frac{dP}{dz}\right)_f = \phi_l^2 \left(\frac{dP}{dz}\right)_l = \phi_g^2 \left(\frac{dP}{dz}\right)_g \tag{10.52}$$

or

$$\left(\frac{dP}{dz}\right)_f = \phi_{\text{lo}}^2 \left(\frac{dP}{dz}\right)_{\text{lo}} = \phi_{\text{go}}^2 \left(\frac{dP}{dz}\right)_{\text{go}} \tag{10.53}$$

where ϕ_l, ϕ_g, ϕ_{lo}, and ϕ_{go} are two-phase frictional multipliers. The respective single-phase pressure gradients are

$$\left(\frac{dP}{dz}\right)_l = \frac{2f_l G^2 (1-x)^2}{\rho_l D_i} \tag{10.54a}$$

$$\left(\frac{dP}{dz}\right)_g = \frac{2f_g G^2 x^2}{\rho_g D_i} \tag{10.54b}$$

$$\left(\frac{dP}{dz}\right)_{\text{lo}} = \frac{2f_{\text{lo}} G^2}{\rho_l D_i} \tag{10.54c}$$

$$\left(\frac{dP}{dz}\right)_{\text{go}} = \frac{2f_{\text{go}} G^2}{\rho_g D_i} \tag{10.54d}$$

where the friction factor f depends on the respective Reynolds number. One such relationship is the Blasius equation:

$$f = 0.079 Re^{-0.25} \quad \text{for } Re > 2000 \tag{10.55}$$

There are many two-phase-flow frictional multiplier correlations in the open literature. Hewitt [55] makes the following tentative recommendations:

1. For $\mu_l/\mu_g < 1000$, the Friedel [56] correlation should be used.
2. For $\mu_l/\mu_g > 1000$, and $G > 100$ kg/(m$^2 \cdot$ s), the Chisholm [57] correlation should be used.
3. For $\mu_l/\mu_g > 1000$ and $G < 100$ kg/(m$^2 \cdot$ s), the Martinelli [58, 59] correlation (as modified by Chisholm [60]) should be used.

These correlations are listed in Table 10.2. The correlations are based on adiabatic, two-phase-flow data. During condensation, because of the mass

TABLE 10.2 Two-Phase Flow Frictional Multiplier Correlations

Reference	Correlation
Friedel [56]	$\phi_{lo}^2 = E + 3.23 FHFr^{0.045} We^{0.035}$ (10.56)
	where
	$E = (1-x)^2 + x^2 \dfrac{\rho_l f_{go}}{\rho_g f_{lo}}$ (10.57)
	$F = x^{0.78}(1-x)^{0.224}$ (10.58)
	$H = \left(\dfrac{\rho_l}{\rho_g}\right)^{0.91} \left(\dfrac{\mu_g}{\mu_l}\right)^{0.19} \left(1 - \dfrac{\mu_g}{\mu_l}\right)^{0.7}$ (10.59)
	$Fr = \dfrac{G^2}{g D_i \rho_h^2}$ (10.60)
	$We = \dfrac{G^2 D_i}{\rho_h \sigma}$ (10.61)
	$\rho_h = \dfrac{\rho_g \rho_l}{x\rho_l + (1-x)\rho_g}$ (10.62)

TABLE 10.2 *(Continued)*

Reference	Correlation
Chisholm [57]	$\phi_{lo}^2 = 1 + (Y^2 - 1)[Bx^{(2-n)/2}(1-x)^{(2-n)/2} + x^{2-n}]$ (10.63a) where $Y = \left[\frac{(dP/dz)_{go}}{(dP/dz)_{lo}}\right]^{1/2}$ (10.63b) n is the Reynolds number exponent in friction factor relationships [e.g., $n = 0.25$, Eq. (10.55)] For $0 < Y < 9.5$, $B = \begin{cases} \frac{55}{G^{1/2}} & G \geq 1900\ \text{kg}/(\text{m}^2 \cdot \text{s}) \\ \frac{2400}{G} & 500 \leq G \leq 1900\ \text{kg}/(\text{m}^2 \cdot \text{s}) \\ 4.8 & G < 500\ \text{kg}/(\text{m}^2 \cdot \text{s}) \end{cases}$ (10.64a) For $9.5 < Y < 28$, $B = \begin{cases} \frac{520}{YG^{1/2}} & G \leq 600\ \text{kg}/(\text{m}^2 \cdot \text{s}) \\ \frac{21}{Y} & G > 600\ \text{kg}/(\text{m}^2 \cdot \text{s}) \end{cases}$ (10.64b) for $Y > 28$, $B = \frac{15,000}{Y^2 G^{1/2}}$ (10.64c)
Martinelli [58, 59] as modified by Chisholm [60]	$\phi_l^2 = 1 + \frac{C}{X_{tt}} + \frac{1}{X_{tt}^2}$ (10.65) $\phi_g^2 = 1 + CX_{tt} + X_{tt}^2$ (10.66) where $X_{tt} = \left[\frac{(dP/dz)_l}{(dP/dz)_g}\right]^{1/2}$ (10.67) [see Eq. (10.22)] $C = \begin{cases} 20 & \text{for turbulent–turbulent flow} \\ 12 & \text{for viscous–turbulent flow} \\ 10 & \text{for turbulent–viscous flow} \\ 5 & \text{viscous–viscous flow} \end{cases}$

transfer across the liquid–vapor interface, a correction must be made to the frictional pressure gradient. Sardesai et al. [61] suggest the following correction:

$$\left(\frac{dP}{dz}\right)_{fc} = \left(\frac{dP}{dz}\right)_{f}\theta \tag{10.68}$$

where

$$\theta = \frac{\phi}{1 - \exp(-\phi)} \tag{10.69a}$$

$$\phi = \frac{G(u_g - u_i)}{\tau_i} \tag{10.69b}$$

The frictional pressure drop must be calculated in a stepwise manner. The tube or channel is divided into a number of short, incremental lengths Δz, over which the conditions change moderately. The pressure drop over one of these lengths would be

$$\Delta P_f = \left(\frac{dP}{dz}\right)_{f}\Delta z \tag{10.70}$$

where the gradient $(dP/dz)_f$ is evaluated using the flow conditions at the midpoint of the length Δz.

10.6 CONDENSATION HEAT TRANSFER AUGMENTATION

Numerous techniques have been proposed to augment condensation heat transfer. During film condensation of a pure vapor, anything that can be done to thin the condensate film will augment the heat transfer process. This can be accomplished on the shell side by using extended surfaces, fluted surfaces, roughness elements, and condensate drainage devices. If dropwise conditions can be promoted, the heat transfer process can be increased by more than an order of magnitude. The most common enhancement techniques for in-tube film condensation involve internal microfins, twisted tapes, and roughened surfaces.

10.6.1 Shell-Side Film Condensation Using Integral-Fin Tubes

During shell-side condensation, the most common technique to enhance the heat transfer is with the use of integral fins. The fins not only increase the

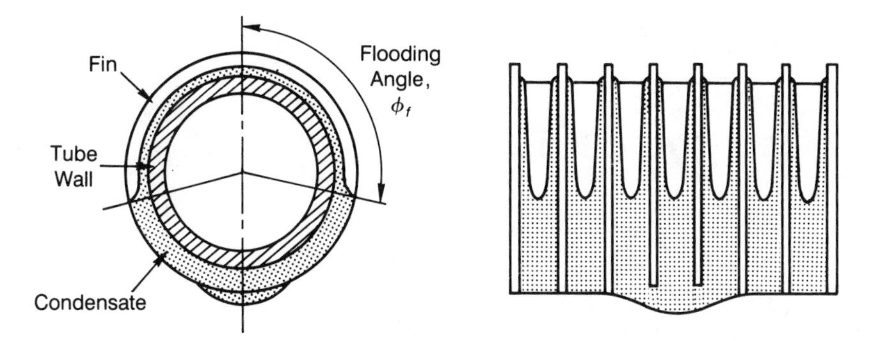

Fig. 10.9. Liquid retention on a finned tube.

surface area but also introduce surface tension forces which can play a significant role in thinning the condensate film. Marto [62] has provided a critical review of the literature pertaining to film condensation on horizontal integral-fin tubes, and has concluded that fin geometry (including spacing, thickness, and height) can have a significant effect in determining the best condensation heat transfer rates.

When a horizontal integral-fin tube is placed in contact with a wetting liquid, surface tension forces cause the liquid to flood the space between fins on the bottom portion of the tube. This phenomenon is known as "flooding," "retention," or "holdup," and is shown schematically in Fig. 10.9 under static (i.e., no condensation) conditions for a tube with rectangular-shaped fins. On the bottom part of the tube, the condensate completely fills the space between fins, whereas on the top part of the tube, only a small liquid "wedge" exists at the intersection of the fins and the tube surface. The flooding angle ϕ_f is defined as the angle from the top of the tube to the circumferential position where the condensate just completely fills the interfin space.

Honda et al. [63] have arrived at an approximate expression for the flooding on a horizontal finned tube with trapezoidal-shaped fins

$$\phi_f = \cos^{-1}\left\{\left(\frac{4(\sigma_l/\rho_l)\cos\theta}{gsD_o}\right) - 1\right\} \tag{10.71}$$

As fin spacing s decreases, more flooding occurs (i.e., ϕ_f decreases) and at a critical fin spacing s_c, it is possible for the entire tube to be flooded (i.e., $\phi_f = 0$). Since heat transfer in the flooded zone is less than in the unflooded zone, it is clear that for a given fluid, an optimum fin spacing must exist which results in the best heat transfer.

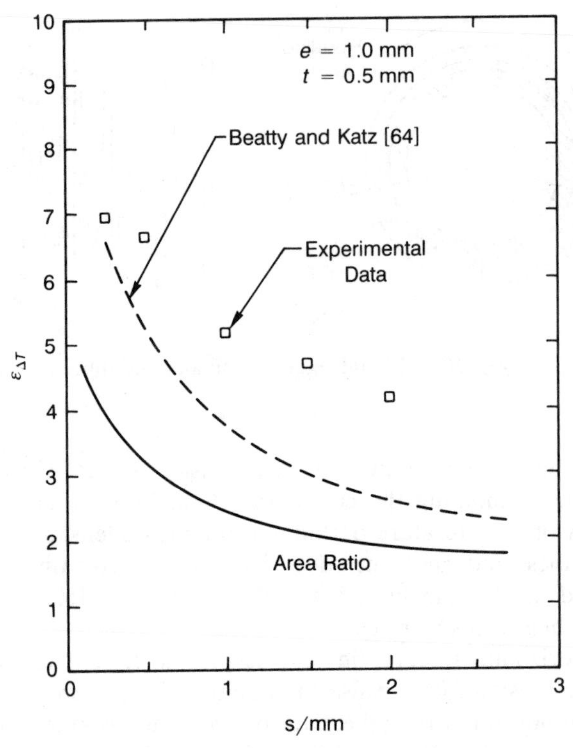

Fig. 10.10. Comparison of measured heat transfer enhancement to Beatty and Katz [64] prediction (adapted from Marto et al. [65]).

The literature contains several theoretical models to predict the heat transfer coefficient on a single integral-fin tube. These models vary in complexity and accuracy. For design purposes, the simple model of Beatty and Katz [64] is still preferred, even though it completely neglects surface tension forces. Because of this simplifying assumption, their model is conservative when compared to data on finned tubes. This is demonstrated in Fig. 10.10 which compares the heat transfer enhancement data of Marto et al. [65] for R-113 condensing on a finned copper tube to the theoretical prediction of Beatty and Katz [64]. At a near-optimum fin spacing of 0.5 mm, the Beatty and Katz [64] model is about 20% less than the data. The Beatty and Katz [64] expression for the average heat transfer coefficient on a single finned tube (based on an effective finned tube surface area A_{ef}) is

$$h_{m,f} = 0.689\left(\frac{k_l^3 \rho_l^2 g i_{lg}}{\mu_l \, \Delta T D_e}\right)^{1/4} \tag{10.72}$$

where

$$\left[\frac{1}{D_e}\right]^{1/4} = 1.30\eta_f \frac{A_{\rm fs}}{A_{\rm ef}} \frac{1}{\overline{L}^{1/4}} + \frac{A_{\rm ft}}{A_{\rm ef}} \frac{1}{D_o^{1/4}} + \frac{A_u}{A_{\rm ef}} \frac{1}{D_r^{1/4}} \tag{10.73}$$

$$\overline{L} = \frac{\pi\left(D_o^2 - D_r^2\right)}{4D_o} \tag{10.74}$$

$$A_{\rm ef} = \eta_f A_{\rm fs} + \eta_f A_{\rm ft} + A_u \tag{10.75}$$

$A_{\rm fs}$ is the surface area of the fin sides, $A_{\rm ft}$ is the surface area of the fin tips, A_u is the interfin surface area of the tube, and η_f is the fin efficiency.

Example 10.5. Consider the condensation of R-22 vapor under the same conditions as Example 10.1, except that the smooth copper tube is to be replaced with an integral-fin copper tube having rectangular-shaped fins and the following dimensions:

Outside diameter	$D_o = 19$ mm
Root diameter	$D_r = 17$ mm
Fin thickness	$t = 0.3$ mm
Fin spacing	$s = 0.5$ mm
Fin height	$e = 1.0$ mm
Fin pitch	$p = 0.8$ mm
Number of fins per unit length	$n_f = 1/p$
	$= 1250$ fins/m (32 fins/in.)

Find the average heat transfer coefficient and the heat transfer augmentation using this tube compared to the smooth tube of Example 10.1.

Solution: We calculate all the surface areas per meter of tube length.

$$\begin{aligned} A_{\rm fs} &= 2n_f\pi\left(D_o^2 - D_r^2\right)/4 \\ &= 2(1250)(\pi)\left[(0.019)^2 - (0.017)^2\right]/4 \\ &= 0.1414 \text{ m}^2/\text{m} \\ A_{\rm ft} &= n_f\pi D_o t \\ &= (1250)(\pi)(0.019)(0.0003) \\ &= 0.0224 \text{ m}^2/\text{m} \\ A_u &= n_f\pi D_r s \\ &= (1250)(\pi)(0.017)(0.0005) \\ &= 0.0334 \text{ m}^2/\text{m} \end{aligned}$$

Since the tube is made of copper and the fins are very short, it is safe to assume that the fin efficiency η_f is 1.0. We now calculate the total effective surface area

$$A_{\text{ef}} = \eta_f A_{\text{fs}} + \eta_f A_{\text{ft}} + A_u$$

$$= (1.0)(0.1414) + (1.0)(0.0224) + (0.0334)$$

$$= 0.1972 \text{ m}^2/\text{m}$$

$$\bar{L} = \pi\left(D_o^2 - D_r^2\right)/4D_o$$

$$= \pi\left[(0.019)^2 - (0.017)^2\right]/(4)(0.019)$$

$$= 0.00298 \text{ m}$$

$$\left[\frac{1}{D_e}\right]^{1/4} = 1.30(1.0)\frac{(0.1414)}{(0.1972)}\frac{1}{(0.00298)^{1/4}}$$

$$+(1.0)\frac{(0.0224)}{(0.1972)}\frac{1}{(0.019)^{1/4}} + \frac{(0.0334)}{(0.1972)}\frac{1}{(0.017)^{1/4}}$$

$$= 4.7646 \text{ m}^{-1/4}$$

Using the properties of R-22 from Example 10.1, the average heat transfer coefficient can be calculated from Eq. (10.72):

$$h_{m,f} = 0.689\left(\frac{(0.077)^3(1099.4)^2(9.81)(1.58 \times 10^5)}{(0.176 \times 10^{-4})(7)}\right)^{1/4}(4.7646)$$

$$= 5329 \text{ W}/(\text{m}^2 \cdot \text{K})$$

The heat transfer rate of the finned tube per meter of tube length is

$$q_f = h_{m,f} A_{\text{ef}}(T_{\text{sat}} - T_w)$$

$$= (5329)(0.1972)(7)$$

$$= 7356 \text{ W/m}$$

The heat transfer rate of the smooth tube per meter of tube length is

$$q_s = h_{m,s} A_s(T_{\text{sat}} - T_w)$$

$$= (3124)(\pi)(0.019)(7)$$

$$= 1305 \text{ W/m}$$

Therefore, by adding integral fins to the smooth copper tube (using the same envelope diameter, D_o), the heat transfer rate can be augmented by the ratio $q_f/q_s = 5.6$.

The average heat transfer coefficient in a bundle of finned tubes may be expressed in a similar way to Eq. (10.11):

$$h_{m,N} = h_{m,f} N^{-a} \tag{10.76}$$

However, the exponent a in Eq. (10.76) is not precisely known for finned tubes although preliminary data in the literature indicate that $a \leq 0.1$. For conservative design purposes, the Kern [12] value of $\frac{1}{6}$ [Eq. (10.13)] can be used. The influence of vapor shear on finned tube condensation is very complex and there is almost no information in the literature. The data of Gogonin and Dorokhov [66] for R-21 indicate that the effect of vapor velocity for finned tubes is very small compared to the effect for smooth tubes. For conservative design purposes, it is recommended that the influence of vapor shear be neglected for finned tubes.

10.6.2 Dropwise Condensation

Dropwise condensation is a complex phenomenon which involves a stochastic process that includes nucleation of microscopic droplets at discrete locations, the rapid growth of these droplets, their coalescence into larger droplets, the formation of large inactive "dead" drops, and the eventual removal of these "dead" drops from the surface either by gravity or vapor shear. The removal of a large drop, with its sweeping action of all condensate in its path, allows fresh microscopic droplets to begin to grow again, continuing the process.

In order for dropwise conditions to occur, the condensate must be prevented from wetting the solid surface. For this to happen, a suitable promoter must be used on the surface. Some common promoters for steam are: oleic acid, benzyl mercaptan, and, generally, any oily or waxy material. However, with most of these organic materials, the promoter gets washed off the surface with time, eventually exposing the bare metal to the vapor and creating filmwise conditions. In recent years, a number of investigations have occurred to arrive at a "permanent" dropwise coating. Either noble metals like gold, silver, and platinum, or organic polymers or inorganic oxides may be used. Figure 10.11 shows the data of Woodruff and Westwater [67] for condensation of steam on different gold-plated, vertical surfaces. The shaded area, labeled DWC, denotes good-quality dropwise condensation. The lower curve, labeled FWC, denotes filmwise condensation conditions. The curves between these two extremes correspond to different gold-plating techniques and plating thicknesses. Woodruff and Westwater [67] found that a strong correlation exists between the heat transfer performance and the chemical composition of the surface. A predominance of gold together with carbon

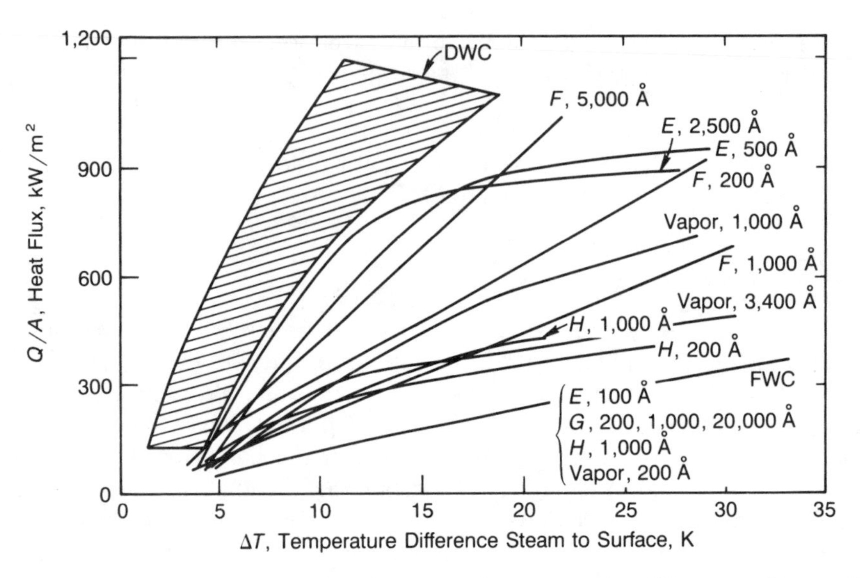

Fig. 10.11. Dropwise condensation heat transfer curves for different gold-coated surfaces (from Woodruff and Westwater [67]).

gave the best dropwise conditions (and a thermal performance improvement over the filmwise case of 5 to 10). Nash and Westwater [68] have recently concluded that pure, organic-free gold will not produce dropwise condensation, and that dropwise condensation will require the presence of some organic material on the surface. Marto et al. [69] studied dropwise condensation of steam on horizontal tubes containing either "permanent" organic coatings or electroplated silver. The silver-plated tube gave an enhancement of about 8 over the smooth tube. The best organic coatings (fluoracrylic and Parylene) gave enhancements of about 3 to 5.

Despite the difficulties of specifying the precise surface conditions of the condensing surface, together with the stochastic nature of the process, there has been good success in predicting dropwise condensation heat transfer rates, Rose [70]. In tube bundles, there is evidence that dropwise condensation heat transfer is not deteriorated by condensate inundation. Because of the sweeping effect of the large drops on the lower tubes in the bundle, the overall bundle performance may actually increase over the single-tube result.

10.7 CONDENSATION OF VAPOR MIXTURES

There are many situations in practice where condensation occurs with a mixture of vapors or in the presence of a noncondensable gas. The condensation process then becomes far more complex than for a pure vapor, involving

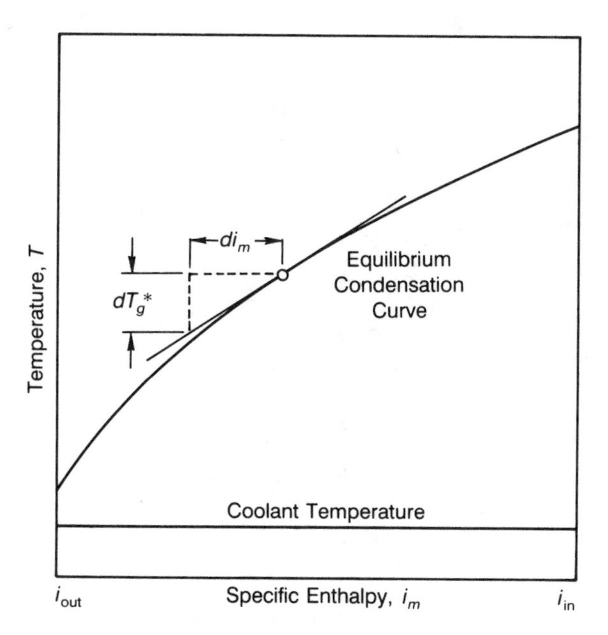

Fig. 10.12. Equilibrium condensation curve.

mass transfer effects that create additional thermal resistances, thus lowering the overall heat transfer coefficient.

Condensation of mixtures differs from condensation of a pure vapor in several important aspects. First of all, the vapor temperature at which condensation occurs can change markedly throughout the condenser. This is illustrated in Fig. 10.12, which shows a typical equilibrium condensation curve for a mixture of vapors, where the equilibrium vapor temperature T_g^* is plotted versus the specific enthalpy of the condensing mixture i_m from inlet to outlet, assuming a constant pressure throughout. Sometimes this curve is plotted versus the cumulative heat release rate $\dot{Q}$ which is related to the specific enthalpy by

$$\dot{Q} = \dot{m}(i_{m,\text{in}} - i_m) \tag{10.77}$$

where $\dot{m}$ is the total mass flow rate of the mixture. The curve clearly indicates that along the path of condensation, as the less volatile components condense out, the equilibrium condensing temperature drops. As a result, the temperature difference between the vapor mixture and the coolant is reduced, leading to a lower heat transfer rate. The real condensing curve may not follow this equilibrium curve closely since condensation is a nonequilibrium process. Nevertheless, this curve shows the correct trend and the implications for design. Equilibrium condensation curves may be of the integral type (where it is assumed that the vapor and the liquid are not

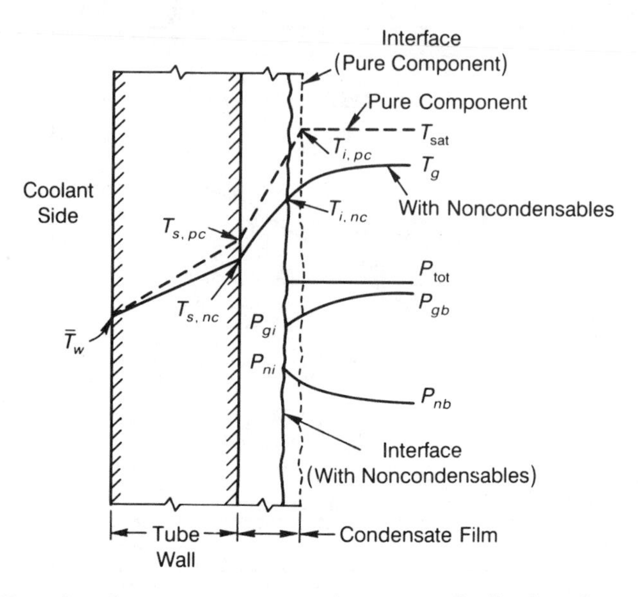

Fig. 10.13. Boundary layer temperature and pressure distributions for pure component condensation (dashed lines) and with noncondensables (solid lines) (from Webb and Wanniarachchi [72]).

separated from one another, as might occur during condensation inside a tube) or of the differential type (where it is assumed that the condensate is separated from the vapor, as might occur on the shell side of a shell-and-tube condenser). Butterworth [71] describes the calculational procedures to obtain these condensation curves.

At a given point along the condensation curve, the local temperatures and pressures vary from the bulk conditions of the mixture toward the interface. This is best illustrated by Fig. 10.13 (Webb and Wanniarachchi [72]), which shows schematically the profiles for a vapor containing a noncondensable gas. The presence of the gas decreases the resulting local heat transfer rate in two ways. First, in the presence of a noncondensable gas, the vapor exists at a partial pressure P_{gb} causing the bulk vapor temperature T_g to be less than T_{sat}. In addition, as the vapor molecules migrate toward the cold wall, they sweep noncondensable gas molecules with them. Since the noncondensable gas does not condense at the prevailing operating conditions in the condenser, these gas molecules accumulate near the liquid–vapor interface. The concentration profile of these gas molecules reaches an equilibrium condition due to a local balance of vapor momentum effects in one direction and back-diffusion effects in the other. As a result, the local partial pressure of the noncondensable gas increases to a maximum at the interface. The vapor molecules must travel through this gas-rich layer and, since the total pressure of the mixture is constant, the vapor partial pressure decreases from P_{gb} to

P_{gi}. This lower vapor pressure at the interface corresponds to a lower vapor temperature T_i, which creates a reduced effective temperature difference across the condensate film. Condensation of vapor mixtures creates similar phenomena to those described previously. The more volatile components will tend to accumulate at the liquid–vapor interface and the less volatile components must diffuse through this layer.

The condensate that collects on the cold wall is usually a completely homogeneous, or miscible, mixture of components. However, there are some applications (involving, for example, steam and some organic vapors) where the condensate forms an immiscible liquid mixture. This liquid mixture can form in several ways that complicate the resulting fluid flow and heat transfer.

Because of the added complexities noted previously and the important role of mass diffusion during condensation of vapor mixtures, the analysis of these processes is more complex than during condensation of a pure vapor. Reviews of these analytical methods have been provided recently by Butterworth [71], Webb and McNaught [73], Sardesai et al. [74], Stephan [75], Jensen [76], and Lee and Rose [77]. The methods described in these reviews vary in both complexity and accuracy, and may be categorized into "equilibrium methods" and "nonequilibrium methods" [78]. These are briefly described in the following sections.

10.7.1 Equilibrium Methods

These methods, as proposed by Silver [79], Ward [80], Bell and Ghaly [81], and Roetzel [82], all assume that there is local equilibrium between the vapor and the condensate throughout the condenser. Thus the gas temperature follows the equilibrium condensation curve (i.e., $T_g = T_g^*$). The method of Silver [79] is described in the following discussion.

The local overall heat transfer coefficient from the bulk vapor mixture to the coolant is written as

$$\frac{1}{U} = \frac{1}{h_c} + R + \frac{1}{h_{\mathrm{ef}}} \tag{10.78}$$

where h_c is the heat transfer coefficient on the coolant side, R is the thermal resistance due to the tube wall (and any fouling), and h_{ef} is an effective condensing-side heat transfer coefficient, which includes the thermal resistance across the condensate film, as well as the sensible cooling of the gas. This effective coefficient is obtained by writing the overall temperature difference from the bulk gas to the wall as

$$(T_g^* - T_w) = (T_g^* - T_i) + (T_i - T_w) \tag{10.79}$$

Since each temperature difference may be written in terms of a heat flux divided by a heat transfer coefficient, Eq. (10.79) may be expressed as

$$\frac{q''}{h_{\text{ef}}} = \frac{q''_g}{h_g} + \frac{q''}{h_l} \tag{10.80}$$

Therefore

$$h_{\text{ef}} = \left[\frac{1}{h_l} + \left(\frac{1}{h_g}\right)\left(\frac{q''_g}{q''}\right)\right]^{-1} \tag{10.81}$$

The ratio q''_g/q'' is usually written as Z where

$$Z = \dot{x}_g c_{pg} \frac{dT_g^*}{di_m} \tag{10.82}$$

In Eq. (10.82), $\dot{x}_g$ is the mass flow fraction of the gas (i.e., $\dot{m}_g/\dot{m}$), c_{pg} is the specific heat of the gas, and dT_g^*/di_m is the local slope of the equilibrium condensation curve (see Fig. 10.12). h_g is calculated for the gas phase flowing along by itself and should be corrected for mass transfer effects as proposed by McNaught [83]:

$$h_g^\circ = h_g\left(\frac{a}{e^a - 1}\right) \tag{10.83}$$

where

$$a = \frac{\sum_{i=1}^{n} \dot{n}_i \tilde{c}_{pgi}}{h_g} \tag{10.84}$$

Therefore, knowing the equilibrium condensation curve, the local conditions of the mixture, and representative values for h_l and h_g (and thus h_g°), h_{ef} can be readily calculated. The total condenser surface area can then be obtained by integration:

$$A_t = \int_{i_{m,\text{out}}}^{i_{m,\text{in}}} \frac{\dot{m}\, di_m}{(T_g^* - T_c)U} \tag{10.85}$$

The preceding methodology can readily be used for condenser design, although there may be some situations where the details of the outlet stream are critical and this method would not provide these details. In these situations, the more complex nonequilibrium methods must be used.

10.7.2 Nonequilibrium Methods

Advanced methods for condensation of multicomponent mixtures include film, penetration, and boundary layer models [76]. These models provide physically realistic formulations of the problem, yielding more accurate local coefficients at the expense of considerable complexity. Colburn and Hougen [84] developed a trial-and-error solution procedure for condensation of a single vapor in the presence of noncondensable gas. Colburn and Drew [85] later extended the method to include condensation of binary vapor mixtures (with no noncondensables). In recent years, considerable progress has been made to further improve upon this method of analysis for application to multicomponent mixtures [86–90]. The procedure of Sardesai et al. [74], which outlines the work of Krishna and Panchal [87], is described in the following discussion.

At any local point along the condenser, the heat flux can be written as

$$q'' = h_g^\circ(T_g - T_i) + \sum_{i=1}^{n} \dot{n}_i \, \Delta\tilde{i}'_{lg,i} \tag{10.86}$$

where

$$\Delta\tilde{i}'_{lg,i} = \Delta\tilde{i}_{lg,i} + \tilde{c}_{pgi}(T_g - T_i) \tag{10.87}$$

The heat flux therefore includes three contributions: (1) sensible cooling of the bulk vapor mixture as it moves through the condenser, (2) sensible cooling of the bulk vapor mixture as it flows from the local bulk conditions to the interface (at a temperature T_i), and (3) latent heat of condensation of the various condensing species. The condensation flux of the ith component n_i is given by

$$\dot{n}_i = \dot{J}_{ib} + \tilde{y}_{ib}\dot{n}_t \tag{10.88}$$

Two mass transfer models exist: (1) interactive models (due to Toor [86] and Krishna and Standart [89]) and (2) noninteractive models known also as effective diffusivity models. For the interactive models, the diffusion flux $\dot{J}_{ib}$ is

$$\dot{J}_{ib} = [B][\zeta](\tilde{y}_{ib} - \tilde{y}_{ii}) \tag{10.89}$$

where $[B]$ is a matrix of binary mass transfer coefficients β_{ij} for all the component pairs and the bulk vapor composition, $[\zeta]$ is a correction matrix that allows for net mass flow on the mass transfer coefficients, and $(\tilde{y}_{ib} - \tilde{y}_{ii})$ is the vapor mole fraction driving force of the ith component. For the noninteractive, or effective diffusivity methods, Eq. (10.89) is simplified to

$$\dot{J}_{ib} = \ulcorner B_{\text{ef}} \lrcorner \ulcorner \zeta_{\text{ef}} \lrcorner (\tilde{y}_{ib} - \tilde{y}_{ii}) \tag{10.90}$$

where $\ulcorner B_{ef} \lrcorner$ and $\ulcorner \zeta_{ef} \lrcorner$ represent diagonal matrices since each species is assumed to have no interaction with the other species involved. Sardesai et al. [74] compared each of these methods to existing experimental data for ternary systems and found that each method agreed with the experimental data to within about $\pm 10\%$. Since the effective diffusivity method is less complex, it requires less computation time and is consequently the preferred method to use. Webb and McNaught [73] provide a comprehensive, step-by-step design example for a multicomponent mixture where the results of the previously outlined methods are compared.

NOMENCLATURE

A	area, m^2
a	constant; parameter defined by Eq. (10.84)
B	parameter defined by Eq. (10.64)
$[B]$	multicomponent mass transfer coefficient matrix, $kmol/(m^2 \cdot s)$
$\ulcorner B_{ef} \lrcorner$	diagonal multicomponent mass transfer coefficient matrix, $kmol/(m^2 \cdot s)$
C	constant
C_N	correction factor for condensate inundation
c_p	specific heat, $J/(kg \cdot K)$
$\tilde{c}_p$	molar specific heat, $J/(kmol \cdot K)$
C_{ug}	correction factor for vapor shear
D	diameter, m
E	dimensionless parameter defined by Eq. (10.57)
e	fin height, m
F	dimensionless parameter defined by Eq. (10.8); dimensionless parameter defined by Eq. (10.58)
F_1	dimensionless parameter defined by Eq. (10.42)
F_2	dimensionless parameter defined by Eq. (10.43)
F_σ	correction factor for surface tension
F_g	correction factor for geometry
Fr	Froude number, $G^2/gD_i\rho_h$
f	friction factor
G	mass velocity, $kg/(m^2 \cdot s)$
g	gravitational acceleration, m/s^2
H	dimensionless parameter defined by Eq. (10.59)
h	heat transfer coefficient, $W/(m^2 \cdot K)$
h_g°	corrected heat transfer coefficient defined by Eq. (10.83), $W/(m^2 \cdot K)$
i	enthalpy per unit mass, J/kg
$\tilde{i}$	molar enthalpy, J/kmol
i_{lg}	latent heat of vaporization, J/kg

$\tilde{i}_{lg}$	molar latent heat of vaporization, J/kmol
$\dot{J}$	diffusive flux, kmol/(m^2 · s)
j	mass flux, kg/(m^2 · s)
k	thermal conductivity, W/(m · K)
L	length, m
$\overline{L}$	average condensing length defined by Eq. (10.74)
M	molecular weight, kg/mol
$\dot{m}$	mass flow rate, kg/s
N	number of tubes in a vertical column
N_t	number of tube rows; number of unit cells
$\dot{n}_i$	molar condensing flux, kmol/(m^2 · s)
n_f	number of fins per unit length, m^{-1}
P	pressure, N/m^2
P_c	critical pressure, N/m^2
ΔP	pressure drop, N/m^2
Pr	Prandtl number
p	pitch, m
p_r	reduced pressure, P/P_c
$\dot{Q}$	cumulative heat release rate, W
q''	heat flux, W/m^2
R	universal gas constant, J/(mol·K); thermal resistance, (m^2 · K)/W
Re	Reynolds number
Re_Γ	film Reynolds number, $4\Gamma/\mu_l$
$\widetilde{Re}$	two-phase Reynolds number, $u_g D/\nu_l$
s	fin spacing, m
T	temperature, K
T^+	dimensionless temperature defined by Eq. (10.38)
T_g^*	equilibrium vapor temperature, K
ΔT	temperature difference, $(T_{sat} - T_w)$, K
t	fin thickness, m
U	overall heat transfer coefficient, W/(m^2 · K)
u	velocity, m/s
u^*	friction velocity defined by Eq. (10.37)
v	velocity, m/s
v^*	dimensionless velocity defined by Eq. (10.48)
We	Weber number, $G^2 D_i/\sigma\rho_h$
w	mean flow width per unit cell, m
X	function defined in Eq. (10.2)
X_{tt}	Lockhart–Martinelli parameter, defined by Eq. (10.22)
x	vapor quality
Y	dimensionless parameter defined by Eq. (10.63b)
$\tilde{y}$	mole fraction of component in gas phase
Z	function defined by Eq. (10.82)
z	axial position, m

Greek Symbols

α	void fraction
Γ	film flow rate per unit length, kg/(m · s)
δ	film thickness, m
δ^+	dimensionless film thickness defined by Eq. (10.39a)
$\varepsilon_{\Delta T}$	heat transfer enhancement ratio, defined as the ratio of finned tube vapor-side coefficient to smooth tube value at the same vapor-to-wall temperature difference and based on the smooth tube surface area of diameter D_r
η	fin efficiency
$[\zeta]$	high flux correction matrix
$\ulcorner \zeta_{ef} \lrcorner$	diagonal high flux correction matrix
θ	function defined by Eq. (10.69a); fin half-angle
μ	dynamic viscosity, Pa · s [kg/(m · s)]
ν	kinematic viscosity, μ/ρ, m^2/s
ρ	density, kg/m^3
σ	surface tension, N/m; condensation coefficient
τ	shear stress, N/m^2
ϕ	function defined by Eq. (10.69b); two-phase frictional multiplier; circumferential angle
ϕ_f	flooding angle defined by Eq. (10.71)
Ω	coefficient defined by Eq. (10.27)

Subscripts

b	bulk
c	coolant; corrected value; critical point value
e	equivalent
ef	effective
f	friction; fin
fs	fin sides
ft	fin tips
g	vapor phase
go	vapor only
G	gravity controlled
h	homogeneous mixture
i	inside; interface
in	inlet
l	liquid phase; longitudinal
lo	liquid only
m	mean or average value; maximum value; mixture value
N	result for N tubes
o	outside
out	outlet

r	root
s	smooth
sat	saturation condition
sh	shear controlled
t	total; transverse
u	unfinned
w	wall
1	single tube
Γ	film

REFERENCES

1. Berman, L. D. (1967) On the effect of molecular kinetic resistance upon heat transfer with condensation. *Int. J. Heat Mass Transfer* **10** 1463.
2. Le Fevre, E. J., and Rose, J. W. (1966) A theory of heat transfer by dropwise condensation. *Proc. Third Int. Heat Transfer Conf., Chicago* **2** 362–375.
3. Niknejad, J., and Rose, J. W. (1981) Interphase matter transfer: An experimental study of condensation of mercury. *Proc. Roy. Soc. London A* **378** 305–327.
4. Rohsenow, W. M. (1985) Condensation. In *Handbook of Heat Transfer Fundamentals*, W. M. Rohsenow, J. P. Hartnett, and E. N. Ganic (eds.), 2nd ed., Chapter 11, pp. 1–50, McGraw-Hill, New York.
5. Butterworth, D. (1983) Film condensation of pure vapor. In *Heat Exchanger Design Handbook*, E. U. Schlünder (ed.), Vol. 2, Section 2.6.2. Hemisphere, New York.
6. Marto, P. J. (1988) Fundamentals of condensation. In *Two-Phase Flow Heat Exchangers: Thermal-Hydraulic Fundamentals and Design*, S. Kakaç, A. E. Bergles, and E. O. Fernandes (eds.), pp. 221–291. Kluwer, Dordrecht.
7. Rose, J. W. (1988) Fundamentals of condensation heat transfer: laminar film condensation. *JSME Int. J.* **31** 357–375.
8. Nusselt, W. (1916) The condensation of steam on cooled surfaces. *Z.d. Ver. Deut. Ing.* **60** 541–546 and 569–575. [Translated into English by D. Fullarton (1982) *Chem. Eng. Fund.* **1**(2) 6–19.]
9. Abramowitz, M. (1951) Tables of the functions $\int \sin^{1/3} x\,dx$ and $\frac{4}{3}\sin^{-4/3} \times \phi \int \sin^{1/3} x\,dx$. *J. Res. National Bureau of Standards* **47** 288–290.
10. Shekriladze, I. G., and Gomelauri, V. I. (1966) Theoretical study of laminar film condensation of flowing vapor. *Int. J. Heat Mass Transfer* **9** 581–591.
11. Butterworth, D. (1977) Developments in the design of shell and tube condensers. ASME Winter Annual Meeting, Atlanta, ASME Preprint 77-WA/HT-24.
12. Kern, D. Q. (1958) Mathematical development of loading in horizontal condensers. *AIChE J.* **4** 157–160.
13. Eissenberg, D. M. (1972) An investigation of the variables affecting steam condensation on the outside of a horizontal tube bundle. PhD Thesis, University of Tennessee, Knoxville.
14. Shklover, G. G., and Buevich, A. V. (1978) Investigation of steam condensation in an inclined bundle of tubes. *Thermal Eng.* **25**(6) 49–52.

15. Nobbs, D. W., and Mayhew, Y. R. (1976) Effect of downward vapor velocity and inundation on condensation rates on horizontal tube banks. *Steam Turbine Condensers*, NEL Report 619, pp. 39–52.
16. Kutateladze, S. S., Gogonin, N. I., Dorokhov, A. R, and Sosunov, V. I. (1979) Film condensation of flowing vapor on a bundle of plain horizontal tubes. *Thermal Eng.* **26** 270–273.
17. Fujii, T., Uehara, H., Hirata, K., and Oda, K. (1972) Heat transfer and flow resistance in condensation of low pressure steam flowing through tube banks. *Int. J. Heat Mass Transfer* **15** 247–260.
18. Fujii, T., Honda, H., and Oda, K. (1979) Condensation of steam on a horizontal tube—the influence of oncoming velocity and thermal condition at the tube wall. In *Condensation Heat Transfer*, P. J. Marto and P. G. Kroeger (eds.), pp. 35–43. ASME, New York.
19. Cavallini, A., Frizzerin, S., and Rossetto, L. (1986) Condensation of R-11 vapor flowing downward outside a horizontal tube bundle. *Proc. Eighth Int. Heat Transfer Conf., San Francisco* **4** 1707–1712.
20. McNaught, J. M. (1982) Two-phase forced convection heat transfer during condensation on horizontal tube bundles. *Proc. Seventh Int. Heat Transfer Conf., Munich* **5** 125–131.
21. Marto, P. J. (1984) Heat transfer and two-phase flow during shell-side condensation. *Heat Transfer Eng.* **5**(1–2) 31–61.
22. Barsness, E. J. (1963) Calculation of the performance of surface condensers by digital computer. ASME Paper 63-PWR-2, National Power Conference, Cincinnati, Ohio.
23. Emerson, W. H. (1969) The application of a digital computer to the design of surface condenser. *The Chemical Engineer* **228**(5) 178–184.
24. Wilson, J. L. (1972) The design of condensers by digital computers. *I. Chem. E. Symp. Ser.*, No. 35, pp. 21–27.
25. Hafford, J. A. (1973) ORCON1: A Fortran code for the calculation of a steam condenser of circular cross section. ORNL-TM-4248, Oak Ridge National Laboratory, Oak Ridge, Tenn.
26. Hopkins, H. L., Loughhead, J., and Monks, C. J. (1983) A computerized analysis of power condenser performance based upon an investigation of condensation. In *Condensers: Theory and Practice, I. Chem. E. Symp. Ser.*, No. 75, pp. 152–170. Pergamon, London.
27. Shida, H., Kuragaska, M., and Adachi, T. (1982) On the numerical analysis method of flow and heat transfer in condensers. *Proc. Seventh Int. Heat Transfer Conf., Munich* **6** 347–352.
28. Al-Sanea, S., Rhodes, N., Tatchell, D. G., and Wilkinson, T. S. (1983) A computer model for detailed calculation of the flow in power station condensers. In *Condensers: Theory and Practice, I. Chem. E. Symp. Ser.*, No. 75, pp. 70–88. Pergamon, London.
29. Caremoli, C. (1983) Numerical computation of steam flow in power plant condensers. In *Condensers: Theory and Practice, I. Chem. E. Symp. Ser.*, No. 75, pp. 89–96. Pergamon, London.

30. Beckett, G., Davidson, B. J., and Ferrison, J. A. (1983) The use of computer programs to improve condenser performance. In *Condensers: Theory and Practice*, *I. Chem. E. Symp. Ser.*, No. 75, pp. 97–110. Pergamon, London.
31. Zinemanas, D., Hasson, D., and Kehat, E. (1984) Simulation of heat exchangers with change of phase. *Computers and Chem. Eng.* **8** 367–375.
32. Breber, G. (1987) In-tube condensation. In *Heat Transfer Equipment Design*, R. K. Shah, E. C. Subbarao, and R. A. Mashelkar (eds.). Hemisphere, New York.
33. Soliman, H. M., and Azer, N. Z. (1974) Visual studies of flow patterns during condensation inside horizontal tubes. *Proc. Fifth Int. Heat Transfer Conf., Tokyo* **3** 241–245.
34. Rahman, M. M., Fathi, A. M., and Soliman, H. M. (1985) Flow pattern boundaries during condensation: new experimental data. *Canadian J. Chem. Eng.* **63** 547–552.
35. Soliman, H. M. (1986) Flow pattern transitions during horizontal in-tube condensation. In *Encyclopedia of Fluid Mechanics*, Chapter 12. Gulf Publishing Co., Houston.
36. Breber, G., Palen, J. W., and Taborek, J. (1980) Prediction of horizontal tubeside condensation of pure components using flow regime criteria. *J. Heat Transfer* **102** 471–476.
37. Tandon, T. N., Varma, H. K., and Gupta, C. P. (1982) A new flow regime map for condensation inside horizontal tubes. *J. Heat Transfer* **104** 763–768.
38. Soliman, H. M. (1986) The mist–annular transition during condensation and its influence on the heat transfer mechanism. *Int. J. Multiphase Flow* **12** 277–288.
39. Jaster, H., and Kosky, P. G. (1976) Condensation heat transfer in a mixed flow regime. *Int. J. Heat Mas Transfer* **19** 95–99.
40. Royal, J. (1975) Augmentation of horizontal in-tube condensation of steam. PhD Thesis, Iowa State University, Ames, Iowa.
41. Akers, W. W., Deans, H. A., and Crosser, O. K. (1959) Condensing heat transfer within horizontal tubes. *Chem. Eng. Prog. Symp. Ser.* **55** 171–176.
42. Cavallini, A., and Zecchin, R. (1971) *Proc. 13th Int. Congress Refrigeration*, Washington, D.C.
43. Boyko, L. D., and Kruzhilin, G. N. (1967) Heat transfer and hydraulic resistance during condensation of steam in a horizontal tube and in a bundle of tubes. *Int. J. Heat Mass Transfer* **10** 361–373.
44. Shah, M. M. A general correlation for heat transfer during film condensation inside pipes. *Int. J. Heat Mass Transfer* **22** 547–556.
45. Kosky, P. G., and Staub, F. W. (1971) Local condensing heat transfer coefficients in the annular flow regime. *AIChE J.* **17** 1037–1043.
46. Traviss, D. P., Rohsenow, W. M., and Baron, A. B. (1972) Forced convection condensation inside tubes: A heat transfer equation for condenser design. *ASHRAE Trans.* **79** 157–165.
47. Wallis, G. B. (1961) Flooding velocities for air and water in vertical tubes. UKAEA Report AEEW-R123.

48. Haseler, L. (1980) Condensation of nitrogen in brazed aluminum plate-fin heat exchangers. 19th National Heat Transfer Conf., Orlando, ASME Paper 80-HT-57.
49. Robertson, J. M. (1980) Review of boiling, condensing and other aspects of two-phase flow in plate-fin heat exchangers. In *Compact Heat Exchangers*, R. K. Shah (ed.), HTD-Vol. 10. ASME, New York.
50. Westwater, J. W. (1986) Compact heat exchangers with phase change. *Proc. Eighth Int. Heat Transfer Conf., San Francisco* **1** 269–278.
51. Robertson, J. M., Blundell, N., and Clarke, R. H. (1986) The condensing characteristics of nitrogen in plain, brazed aluminum, plate-fin heat exchanger passages. *Proc. Eighth Int. Heat Transfer Conf., San Francisco* **4** 1719–1724.
52. Grant, I. D. R., and Chisholm, D. (1979) Two-phase flow on the shell-side of a segmentally baffled shell-and-tube heat exchanger. *J. Heat Transfer* **101** 38–42.
53. Grant, I. D. R, and Chisholm, D. (1980) Horizontal two-phase flow across tube banks. *Int. J. Heat Fluid Flow* **2**(2) 97–100.
54. Ishihara, K., Palen, J. W., and Taborek, J. (1980) Critical review of correlations for predicting two-phase flow pressure drop across tube banks. *Heat Transfer Eng.* **1**(3) 23–32.
55. Hewitt, G. F. (1983) Gas–liquid flow. In *Heat Exchanger Design Handbook* E. U. Schlünder (ed.), Section 2.3.2. Hemisphere, New York.
56. Friedel, L. (1979) Improved friction pressure drop correlations for horizontal and vertical two-phase pipe flow. European two-phase flow group meeting, Ispra, Italy, Paper E2.
57. Chisholm, D. (1973) Pressure gradients due to friction during the flow of evaporating two-phase mixtures in smooth tubes and channels. *Int. J. Heat Mass Transfer* **16** 347–348.
58. Lockhart, R. W., and Martinelli, R. C. (1949) Proposed correlation of data for isothermal two-phase two-component flow in pipes. *Chem. Eng. Prog.* **45**(1) 39–48.
59. Martinelli, R. C., and Nelson, D. B. (1948) Prediction of pressure drop during forced-circulation boiling of water. *Trans. ASME* **70** 695–702.
60. Chisholm, D. (1967) A theoretical basis for the Lockhart–Martinelli correlation for two-phase flow. *Int. J. Heat Mass Transfer* **10** 1767–1778.
61. Sardesai, R. G., Owen, R. G., and Pulling, D. J. (1982) Pressure drop for condensation of a pure vapor in downflow in a vertical tube. *Proc. Seventh Int. Heat Transfer Conf., Munich* **5** 139–145.
62. Marto, P. J. (1988) An evaluation of film condensation on horizontal integral-fin tubes. *J. Heat Transfer* **110**(4B) 1287–1305.
63. Honda, H., Nozu, S., and Mitsumori, K. (1983) Augmentation of condensation on horizontal finned tubes by attaching a porous drainage plate. *Proc. ASME–JSME Thermal Eng. Joint Conf.*, Y. Mori and W.-J. Yang (eds.), **3** 289–296.
64. Beatty, K. O., and Katz, D. L. (1948) Condensation of vapors on outside of finned tubes. *Chem. Eng. Prog.* **44**(1) 55–70.
65. Marto, P. J., Zebrowski, D., Wanniarachchi, A. S., and Rose, J. W. (1988) Film condensation of R-113 on horizontal finned tubes. *ASME Proc. Nat. Heat Trans. Conf.* H. R. Jacobs (ed.) **2** 583–592.

66. Gogonin, I. I., and Dorokhov, A. R. (1981) Enhancement of heat transfer in horizontal shell-and-tube condensers. *Heat Transfer–Soviet Research* **3**(3) 119–126.
67. Woodruff, D. W., and Westwater, J. W. (1981) Steam condensation on various gold surfaces. *J. Heat Transfer* **103** 685–692.
68. Nash, C. A., and Westwater, J. W. (1987) A study of novel surfaces for dropwise condensation, *Proc. Second ASME–JSME Thermal Eng. Joint Conf.*, P. J. Marto and I. Tanasawa (eds.), **2** pp. 485–491.
69. Marto, P. J., Looney, D. J., Rose, J. W., and Wanniarachchi, A. S. (1986) Evaluation of organic coatings for the promotion of dropwise condensation of steam. *Int. J. Heat Mass Transfer* **29** 1109–1117.
70. Rose, J. W. (1988) Some aspects of condensation heat transfer theory. *Int. Comm. Heat Mass Transfer* **15** 449–473.
71. Butterworth, D. (1983) Condensation of vapor mixtures. In *Heat Exchanger Design Handbook*, E. U. Schlünder, (ed.), Vol. 2, Section 2.6.3. Hemisphere, New York.
72. Webb, R. L., and Wanniarachchi, A. S. (1980) The effect of noncondensible gases in water chiller condensers—literature survey and theoretical predictions. *ASHRAE Trans.* **86** 142–159.
73. Webb, D. R., and McNaught, J. M. (1980) Condensers. In *Developments in Heat Exchangers Technology*, D. Chisholm (ed.), 71–126. Applied Science Publishers, London.
74. Sardesai, R. G., Shock, R. A., and Butterworth, D. (1982) Heat and mass transfer in multicomponent condensation and boiling. *Heat Transfer Eng.* **3**(3–4) 104–114.
75. Stephan, K. (1981) Heat transfer with condensation in multicomponent mixtures. In *Heat Exchangers: Thermal-Hydraulic Fundamentals and Design*, S. Kakaç, A. E. Bergles, and F. Mayinger (eds.), pp. 337–355. Hemisphere, New York.
76. Jensen, M. K. (1988) Condensation with noncondensables and in multicomponent mixtures. In *Two-Phase Flow Heat Exchangers: Thermal-Hydraulic Fundamentals and Design*, S. Kakaç, A. E. Bergles, and E. O. Fernandes (eds.), pp. 293–324. Kluwer, Dordrecht.
77. Lee, W. C., and Rose, J. W. (1983) Comparison of calculation methods for non-condensing gas effects in condensation on a horizontal tube. In *Condensers: Theory and Practice*, *I. Chem. E. Symp. Ser.*, No. 75, pp. 342–355. Pergamon, London.
78. Butterworth, D. (1988) Private communication.
79. Silver, L. (1947) Gas cooling with aqueous condensation. *Trans. Inst. Chem. Eng.* **25** 30–42.
80. Ward, D. J. (1960) How to design a multiple component partial condenser. *Petrochem. Eng.* **32**(10) C42–C48.
81. Bell, K. J., and Ghaly, M. A. (1973) An approximate generalized design method for multicomponent/partial condensers. *AIChE Symp. Ser.* **69**(131) 72–79.
82. Roetzel, W. (1975) Approximate design method for mixed vapor condensers. *Wärme Stoffübertrag.* **8** 211–218.
83. McNaught, J. (1979) Mas transfer correction terms in design methods for multicomponent/partial condensers. In *Condensation Heat Transfer*, P. J. Marto and P. G. Kroeger (eds.), pp. 111–118. ASME, New York.

84. Colburn, A. P., and Hougen, O. A. (1934) Design of cooler condensers for mixtures of vapors with non-condensing gases. *Ind. Eng. Chem.* **26** 1178–1184.
85. Colburn, A. P., and Drew, T. B. (1937) The condensation of mixed vapors. *Trans. AIChE* **33** 197–215.
86. Toor, H. L. (1964) Solution of the linearized equations of multicomponent mass transfer. *AIChE J.* **10** 448–460.
87. Krishna, R., and Panchal, C. B. (1977) Condensation of a binary vapor mixture in the presence of an inert gas. *Chem. Eng. Sci.* **32** 741–745.
88. Taylor, R., and Webb, D. R. (1981) Film models for multicomponent mass transfer: computational methods: the exact solution of the Maxwell–Stephan equations. *Computers and Chem. Eng.* **5** 61–73.
89. Krishna, R., and Standart, G. L. (1976) A multicomponent film model incorporating a general matrix method of solution to maxwell–stephan equations. *AIChE J.* **22** 383–389.
90. Taylor, R., and Webb, D. R. (1980) On the relationship between the exact and the linearized solutions of the Maxwell–Stephan equations for the multicomponent film model. *Chem. Eng. Comm.* **7** 287–299.

CHAPTER 11

STEAM POWER PLANT AND PROCESS CONDENSERS

D. BUTTERWORTH
Heat Transfer and Fluid Flow Service (HTFS)
Harwell Laboratory
Oxfordshire, United Kingdom

11.1 INTRODUCTION

The more important types of condensing equipment are described in this chapter including shell-and-tube for power and process applications, plate, spiral, plate-fin, air-cooled, and direct-contact. Advice on the choice between the types is given. Methods available for the design and check rating of these exchangers are given, with particular emphasis on the shell and tube which is the most common type. These methods build upon the detailed theories and correlations for individual condensing processes, which are described in previous chapters. An important calculation step is the determination of the mean temperature difference and mean overall coefficient from their local values. This problem is therefore discussed in some detail. Possible reasons why condensers fail to operate as expected are noted.

Condensers may be classified into two main types: those in which the coolant and condensate streams are separated by a solid surface, usually a tube wall; and those in which the coolant and condensing vapor are brought into direct contact. Each of these two types may be subdivided into further categories, as illustrated in Fig. 11.1. The direct-contact type may consist of a vapor which is bubbled into a pool of liquid, a liquid which is sprayed into a vapor, or a packed-column in which the liquid flows downwards as a film over a packing material against the upward flow of vapor. Those in which the streams are separated may be subdivided into three main types: air-cooled,

Boilers, Evaporators and Condensers, Edited by Sadik Kakaç
ISBN 0-471-62170-6

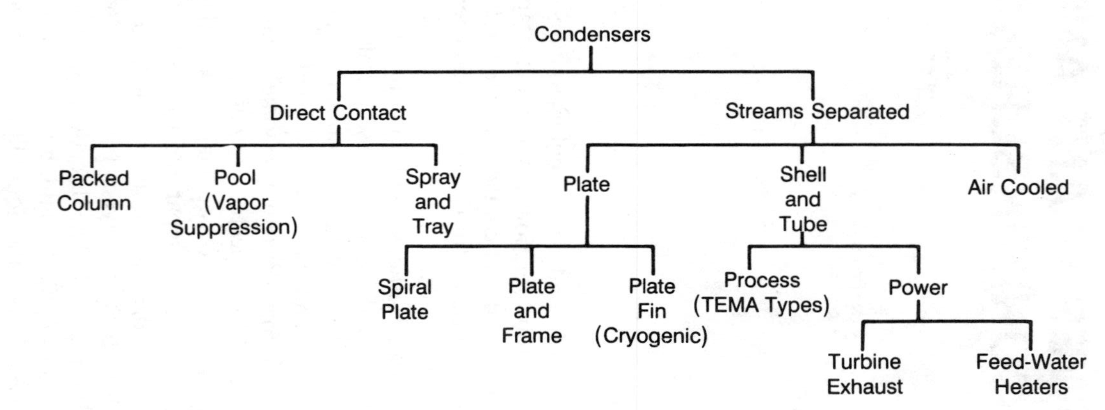

Fig. 11.1. Subdivisions of condenser types.

shell-and-tube, and plate. In the air-cooled type, condensation occurs inside tubes with cooling being provided by air which is blown or sucked across the tubes. Fins are usually provided on the air side to compensate for the low air-side coefficients, by having a large surface area. The shell-and-tube type consists of a large cylindrical shell inside which there is a bundle of tubes. One fluid stream flows inside the tubes while the other flows on the outside of the tubes (known as the "shell side"). The condensation may occur inside or outside the tubes, depending on the design requirements. Although shell-and-tube condensers are used both for process applications and power generation, there are wide differences between the two types, and it is therefore worthwhile considering them separately. There are many different process types which are given letters of designation by the Tubular Exchanger Manufacturers Association (TEMA) [1]. Power station condensers can be subdivided into those used at the exhaust of the low-pressure steam turbines, and those used for heating the boiler feed water. A plate may be used instead of a tube wall to divide the coolant and condensing streams. In one design, the plates are corrugated to give rigidity and also to improve heat transfer. These are then held together in a press or frame, with gaskets between the plates to prevent fluid leakage. In the "plate-fin" design, the plates are flat but corrugated metal sheets are sandwiched between them to act as fins. In the spiral plate type, two plates are rolled into a spiral.

11.2 SHELL-AND-TUBE CONDENSERS FOR PROCESS PLANT

11.2.1 Horizontal Shell-Side Condensers

Perhaps the simplest form of this type of condenser is that shown in Fig 11.2, which is designated by TEMA as an E-type shell with a single tube-side pass. "E-type" means that the shell-side fluid enters the exchanger at one end and then flows in a relatively straightforward way to the other end of the exchanger, where it then leaves. The shell-side flow path is not completely straightforward, though, because baffles are inserted which serve both to direct the flow into a zig-zag path and to support the tubes at regular intervals. The baffles tend to cause the fluid to flow perpendicular to the tubes as the fluid flows between pairs of baffles, and then to flow parallel to the tubes as it flows from one baffle compartment to the next. Figure 11.3 shows the main features of the most common type of baffle which is a single segmental baffle. By "single segmental" we mean that one segment of a circular plate is cut away, as shown. Holes are drilled into or punched out of the baffle through which the tubes pass. The various constructional details of shell-and-tube exchangers are described in TEMA, and this chapter is restricted to those special features which are relevant to condensers.

A very important feature of a condenser, as compared with any other type of heat exchanger, is that it must have a vent for removal of noncondensable

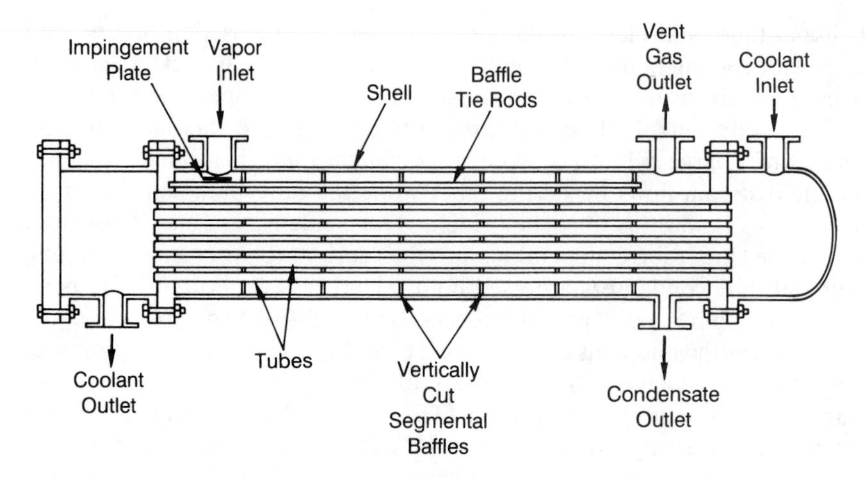

Fig. 11.2. Example of shell-side condenser: TEMA *E*-type shell with single tube-side pass.

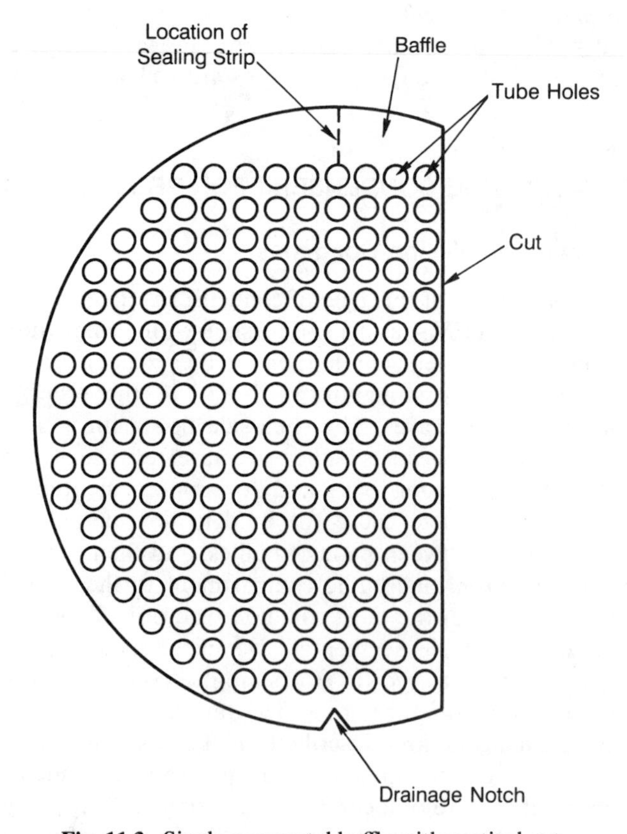

Fig. 11.3. Single segmental baffle with vertical cut.

gas. Thus an *E*-type condenser will have two outlet pipes (known as "nozzles"); one for the vent and the other for the condensate outlet. Noncondensables have the effect both of depressing the condensation temperature, and therefore reducing the temperature difference between the streams, and lowering the effective condensing-side coefficients. It is clear, therefore, that we want to make absolutely sure that any noncondensables present before start-up can be removed, and that there is no chance of noncondensables accumulating during condenser operation. Hence the vent is provided. It is prudent to provide a vent if it is believed that the incoming vapor is pure. This is because even a very pure vapor will contain a small amount of noncondensables which will gradually accumulate in the exchanger over a very long period if no vent is provided. Of course, some of the noncondensables will dissolve in the condensate, but it is not prudent to rely on this. Perhaps one rare occasion when a vent may be justifiably omitted is when the vapor is in a closed cycle, such as a refrigeration cycle, which operates above atmospheric pressure. If, however, such a cycle were operating below atmospheric pressure, leakage of air into the equipment is a possibility. Even with a high-pressure closed cycle, a vent will be needed for start-up.

The vent should be located as near as possible to the coldest part of the condenser which, of course, is where the coolant enters. It should be high enough up that it does not flood with condensate, so that in the simple arrangement shown in Fig. 11.2, it is convenient to have the vent at the top of the shell. The baffles should be positioned so that there is a continuous onward flow of the vapor–gas mixture from the inlet nozzle to the vent nozzle, with no possibility of areas of stagnant or recirculating flow forming where noncondensables could accumulate in a "pocket." By having the vent as far from the inlet as possible, and by having it at the coldest part of the condenser, we ensure that we are venting from the point where the noncondensables have their highest concentration. We are therefore throwing away as little vapor as possible. It is extremely important that there is no short cut which would allow the vapor to go straight to the outlet without being forced across the tubes. This would again cause us to throw away vapor unnecessarily.

Baffles in condensers are usually arranged with the cut vertical, so that the vapor flow is from side to side. This allows the condensate to fall smoothly downwards and away from the tubes and thus collect in the bottom of the shell. Drainage notches must be cut in the baffles as shown in Fig. 11.3 to allow this condensate to run along the shell to the condensate outlet. Condensate will also leak through the clearances between the baffles and the shell and between the tubes and the baffles. In addition, condensate can back up behind baffles and flow around the edge of the baffle cut. It is debatable, therefore, to what extent the drainage notch is a main channel for condensate flow during normal condenser operation, but nevertheless, it is still necessary to have a drainage notch to make absolutely sure that the exchanger may be fully drained during plant shutdown. Also, to ensure such drainage, the exchanger must be inclined toward the condensate outlet.

As has already been mentioned, baffles also serve to support the tubes. It is usual to allow alternate baffles to overlap by at least two tube rows. Some tubes are only supported at alternate baffles and the maximum unsupported length for these tubes is laid down in the TEMA standards. It may be necessary to have the baffles close together in order to prevent the tubes from vibrating. The optimum spacing of the baffles depends on the trade-off between good heat transfer and low pressure drop. As the vapor velocities and noncondensable concentration increase toward the outlet of the condenser, it can sometimes prove very beneficial to have the baffles closer together toward the outlet end. Webb et al. [22] have suggested that an ideal design is one which the pressure drop per row crossed is a constant throughout the exchanger. They therefore start with an ideal design and then try to approximate this in a practical unit by careful arrangement of the tubes and the baffles.

A variety of alternative baffle arrangements are shown in Fig. 11.4. The baffles in this figure are shown shaded, and the W signifies the baffle

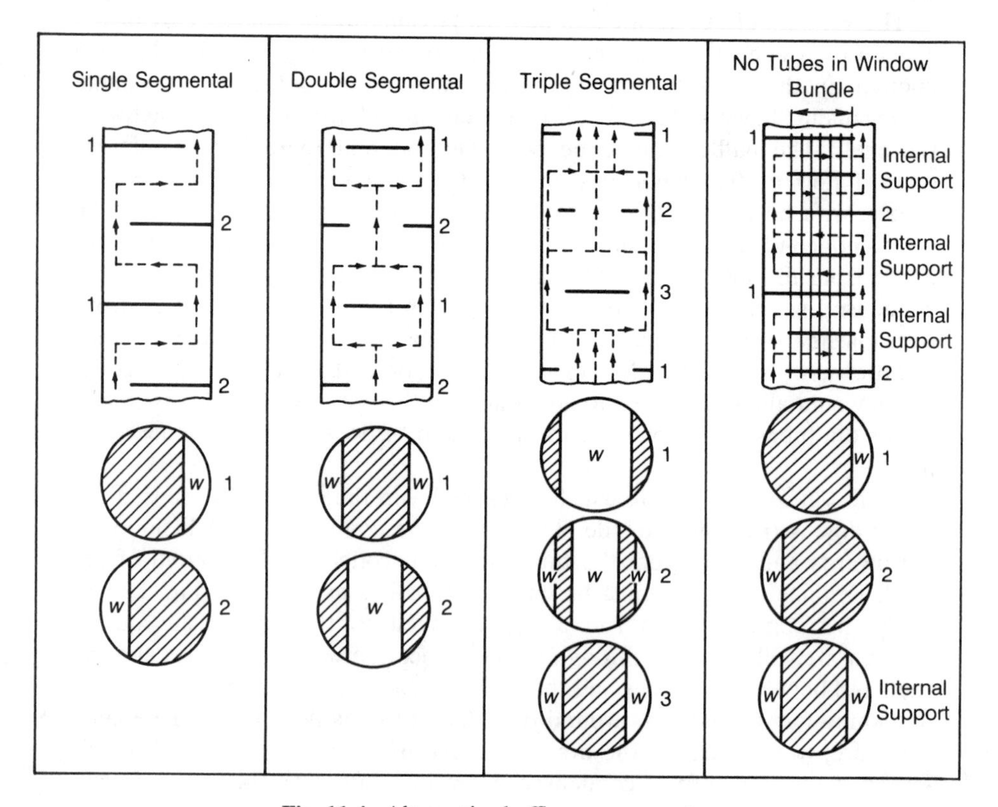

Fig. 11.4. Alternative baffle arrangements.

"window." The window is the part of the baffle which has been cut away to allow the shell-side stream to flow form one baffle compartment to the next. The figure also illustrates the flow paths taken by the fluid stream, as viewed from the top, for the various baffle arrangements. With most of the baffle types, having the baffles closer together increases the shell-side fluid velocities. This would be undesirable if vibration or high pressure drop is a problem. The design with no tubes in the window, however, allows one to have additional intermediate baffles which support the tubes without having any significant effect on the flow. However, such a design is expensive because there is a large volume of empty shell.

Tubes are usually left out of the bundle near the inlet nozzle to prevent excessive flow constriction which could create problems due to high pressure drop, tube erosion, or vibration. Leaving these tubes out may give a path for the vapor to flow along the top of the shell directly from the vapor inlet to the vent without crossing any tube rows. To prevent this happening, a sealing strip should be placed along the top of the shell in the position shown in Fig. 11.3. Tubes may also be left out near the bottom of the shell to allow good condensate drainage.

It is almost universal practice to have an impingement plate under the vapor inlet nozzle to prevent tube erosion. This plate should be located perpendicular to the inlet nozzle, just above the first row of tubes met by the vapor inlet flow (see Fig. 11.5). With small nozzles the plate should be slightly larger than the nozzle; with large nozzles it should extend to the shell, otherwise the outer tubes of the first row could be subject to serious damage due to the very high velocity with which the incoming vapor will hit these tubes. With the latter arrangement, the vapor escapes from the nozzle longitudinally and adequate space must be allowed beyond the edge of the plate and the tube plate, and beyond the other edge of the plate and the baffle, for the vapor to flow into the bundle, as shown in Fig. 11.5. An alternative, though expensive, way of avoiding high velocities is to provide a vapor belt, as shown in Fig. 11.6. When flow-induced vibration is a problem, an extra tube support plate may be inserted near the inlet nozzle, as shown in Fig. 11.7.

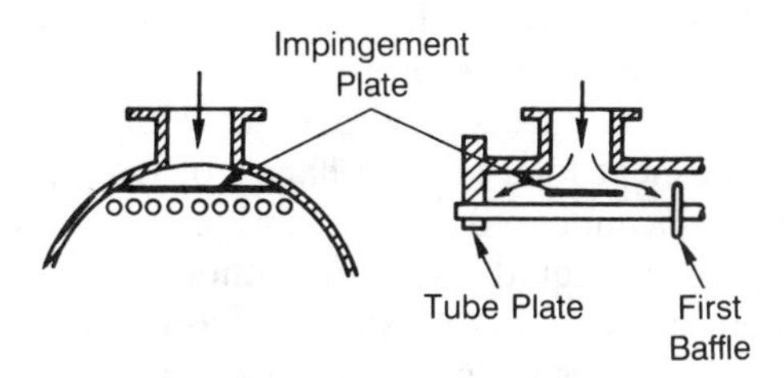

Fig. 11.5. Impingement plate.

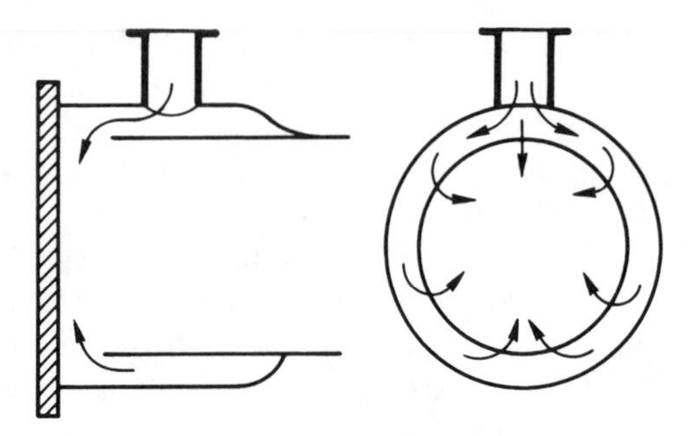

Fig. 11.6. Vapor belt.

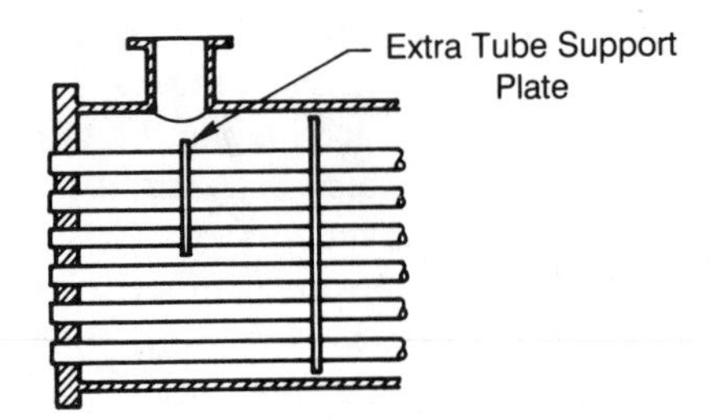

Fig. 11.7. Extra tube support plate to help prevent vibration of tubes near the inlet nozzle.

The tube-side fluid may be made to pass along the exchanger more than once as it flows through the exchanger. The additional tube-side passes are achieved either as a result of having plates in the header or by using U tubes.

So far, we have only described the simplest TEMA shell type, the E type. Figure 11.8 shows the other shell types designated by TEMA. With the exception of the K type, the other types may be used as shell-side condensers, although the F shell is unusual. The broken lines in these figures denote a longitudinal baffle. The J shell has a great advantage over the E shell in that it can be arranged with two nozzles, one at either end, for the vapor inlet, and with one small nozzle in the middle for the condensate outlet. Of course, one would normally have a small nozzle in the middle at the top in order to vent noncondensables. By having these two inlet nozzles, a larger vapor volume coming into the condenser can be accommodated more easily. Also, by splitting the vapor flow into two and by halving the path length for vapor flow, the pressure drop may be reduced substantially over that for a similar size E shell. It is good practice with a J shell to make sure that there is the same heat load in both halves of the exchanger to prevent the possibility that noncondensed vapor coming from one end of the exchanger meets subcooled liquid from the other. This could give rise to periodic violent vapor collapse and possible exchanger damage. This problem usually means that J shells should not be designed with a single tube-side pass if there is a large temperature variation in the tube-side fluid as it flows

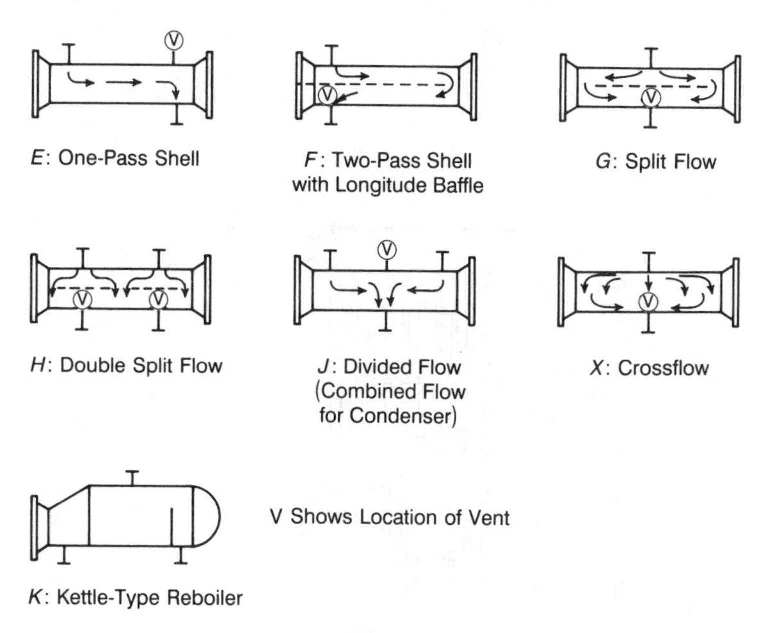

Fig. 11.8. TEMA shell types.

from one end of the exchanger to the other. *J* shells would normally have baffles similar to those found in *E* shells, except that a full-circle tube support plate may be placed in the center of the exchanger.

The *G* shells and *H* shells can also have transverse baffles in addition to the longitudinal baffles. Full-circle tube support plates may be placed in line with the nozzles and, for *H* shells, additional full-circle tube support plates can be placed halfway along the shell. An *H* shell would therefore have three tube support plates along the lengths of the tubes and it may be possible to avoid having further segmented baffles supporting the tubes. In such circumstances, an *H* shell gives a fairly low pressure drop. The vent nozzles in *G* shells and *H* shells have to be placed in the side of the shell above the condensate outlet nozzles but, of course, below the longitudinal baffles. If there are multiple tube-side passes in *G* and *H* shells, these should be arranged so that the coldest pass is at the bottom and warmest at the top, so that there is some degree of countercurrent flow.

The crossflow, or *X*-type exchanger, is a very useful unit for vacuum operation. In such operating conditions, large volumes of vapor must be handled, and it is therefore useful to keep the flow areas in the exchanger as large as possible to avoid the chance of the tube vibration. The large flow area combined with the short flow path also means that pressure drops can be kept low. It is particularly important to keep pressure drops low in vacuum operation so as to avoid reducing the saturation temperature, and therefore losing temperature difference. Figure 11.9 shows a typical crossflow

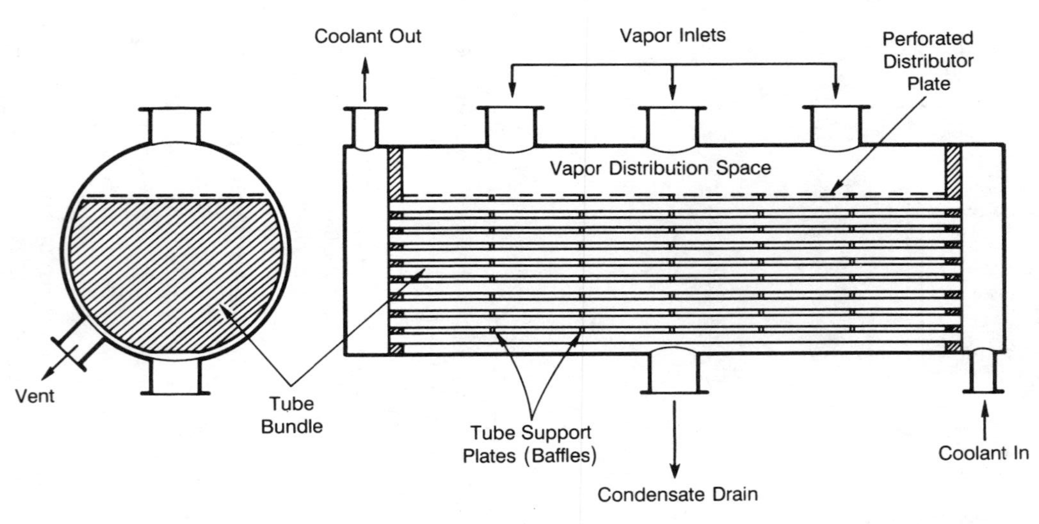

Fig. 11.9. Main features of a crossflow condenser (TEMA X type).

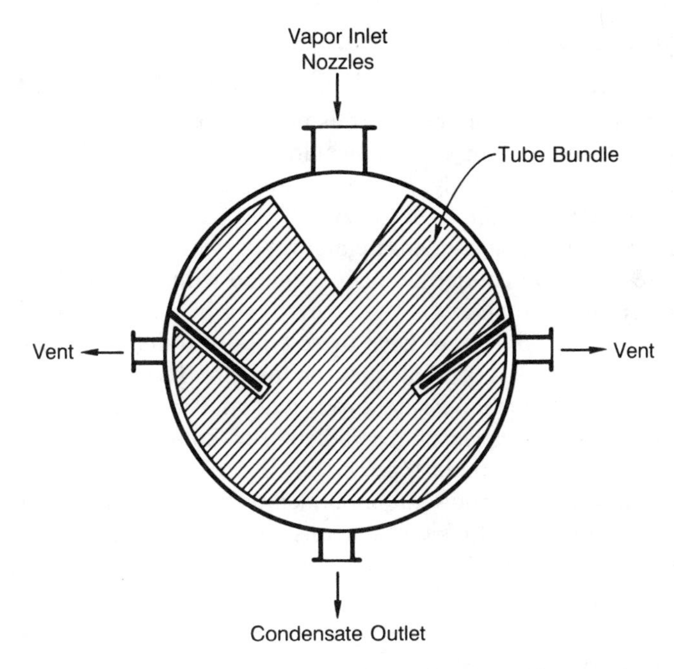

Fig. 11.10. Alternative tube bundle layout for a crossflow condenser.

unit. This particular unit has three inlet nozzles to avoid having a very large single inlet nozzle which may lead to difficulties in mechanical construction. A large space above the top of the bundle is necessary to give good vapor distribution along the exchanger length, and this may be assisted by the introduction of a perforated distributor plate. The large space also helps to prevent tubes encountering high vapor velocities. As many full-circle tube support plates may be inserted as is necessary to give sufficient tube support to avoid tube vibration. Noncondensable gases must be vented from as low as possible in the exchanger, as shown in Fig. 11.9.

Variations on the tube bundle layout are possible in crossflow condensers, and one alternative arrangement is shown in Fig. 11.10. Here, a V-shaped space is left at the top of the bundle for distribution of the incoming vapor and to reduce the vapor velocity across the tubes. In addition, the vents are placed at the side of the condenser and special baffles introduced to prevent the vapor taking a direct path to the vent point. As with G and H shells, X shells with multiple tube-side passes should be arranged with the coldest passes at the bottom.

Webb et al. [22] have proposed the use of a hybrid condenser which is part X shell and E shell. This is done by having a very large baffle space at the vapor inlet to the condenser which acts as a crossflow region. The rest of the condenser has closer baffle spacing more normal to an E shell.

11.2.2 Vertical Shell-Side Condensers

TEMA *E*-type shells may also be installed vertically, as shown in Fig. 11.11. The vent line must, as usual, be as far as possible from the vapor inlet and at the cold end of the exchanger. This would mean having it very close to the bottom tube plate but, clearly, it must be raised a little bit to avoid taking condensate out through the outlet nozzle. These vertical shell-side units give very good mixing between condensate and vapor, which therefore means that, with mixtures, integral condensation is obtained with improved temperature difference between the phases.

It is theoretically possible to have these types of units with the vapor inlet at the bottom and the vent at the top and to have the vapor flowing countercurrent to the condensate produced on the tubes. However, such designs are unusual because of the difficulty of knowing the precise conditions which lead to flooding in these circumstances. These units must not be designed with vertical upflow of both phases because of the almost total impossibility of ensuring that the condensate may be dragged upwards and out of the top of the unit.

11.2.3 Tube-Side Condensers

Condensation inside tubes is often just as convenient as condensation outside them. Indeed, when considering multicomponent mixtures, in-tube condensation can help to ensure integral condensation and better temperature differences than would be the case for, say, condensation outside tubes in a horizontal *E* shell. However, differential condensation may be obtained in horizontal tubes during stratified flow.

Multiple passes with headers must be avoided in tube-side condensers. This is because it is impossible to say how a two-phase mixture discharging

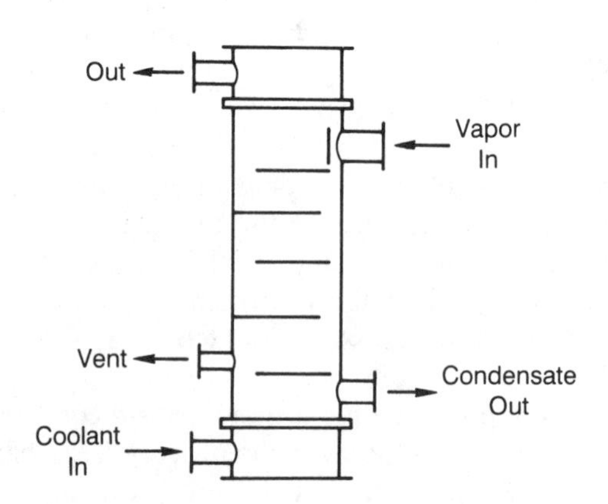

Fig. 11.11. Vertical *E* shell as condenser.

from one pass will distribute itself on entrance to the next pass. An unpredictable amount of separation will occur, making the design of the condenser extremely uncertain. In addition, such separation would often lead to a loss of thermodynamic efficiency of the exchanger. This could occur, say, because some of the tubes are flowing full of liquid and are therefore rapidly cooled to a close approach with the coolant temperature, thus rendering these tubes significantly less useful for heat transfer. In addition, instabilities could arise where the subcooled condensate from these liquid-filled passes meets vapor discharging from other tubes at the end of such a pass. One obvious way of avoiding these problems, and yet still have two passes, is to use U tubes. However, it is dangerous to go to more than two passes.

With vertical exchangers it is even more necessary to keep to a single tube-side pass. Such exchangers are often designed with downward flow of both the vapor and the condensate. Two tube-side passes, using U tubes, would be possible where there is upward flow in the first pass and downward flow in the second. It is then necessary to ensure that the vapor velocities at the top of the first pass are high enough to drive the condensate up and around the bend.

Vertical tube-side condensers may also be designed to operate in the reflux mode, with upward flow of vapor but with a downward counterflow of any condensate forming on the tube walls. Clearly, such units can only operate provided the flooding phenomenon is avoided.

Again, it is possible to design tube-side condensers with vertical upflow of both the condensate and vapor. This is usually rather difficult because there is often insufficient vapor at the top of the tubes to ensure that the condensate is dragged smoothly away under all possible operating conditions. Hence this design is best avoided unless absolutely necessary.

As with the shell-side condensers, tube-side condensers must have adequate venting. In horizontal units, the vent should be placed in the highest convenient position in the outlet header. In vertical units with downflow, the vent line should be placed in the lower header above any possible pool of condensate which forms there. Some sort of shrouding or cover near the vent nozzle may be necessary to avoid entrainment of condensate into the vent line. With condensers operating in the reflux mode the vent should be in the top header.

11.2.4 Sub-cooling in Shell-and-Tube Condensers

It is often desirable to subcool any condensate leaving a condenser to prevent flashing in pipework and equipment downstream of the condenser. A small amount of subcooling in horizontal units can be achieved by having some level control in the condenser which causes some of the tubes in the bottom of the unit to be flooded. However, it should be realized that the baffles have been arranged in order to deal satisfactorily with the condensation heat transfer. Hence the geometry in the bottom of the condenser is not well suited for the liquid cooling requirement. Typically, conditions there are of a very low velocity pool of condensate. The heat transfer coefficients in this pool are both very low and are extremely difficult to predict accurately. Also, it is quite difficult to know what the liquid level is throughout the condenser. Even though the level may be controlled in the outlet baffled compartment, the level elsewhere through the unit will depend on the pressure drop in the vapor phase, the hydraulic gradients in the condensate pool, and the precise nature of any leakage paths from baffle space to baffle space. This, again, adds considerable uncertainty to the prediction of the amount of subcooling in such units. Having said all this, the deliberate flooding of a few tubes is often carried out to provide a little bit of subcooling, and this is fine provided one does not have to rely upon having a precise amount of subcooling.

Where a large and precise amount of subcooling is required from such a unit, a separate, specially designed, subcooling exchanger is preferable as illustrated in Fig. 11.12.

Subcooling with shell-side condensation in vertical tubes is somewhat easier to achieve than with horizontal tubes. Again, some sort of liquid level control is needed. Additional baffling may be provided in this pool to try and give higher liquid velocities and therefore increase the coefficients, but there is usually not much to be gained from this because the shell diameter that has been obtained to get the condensation right is usually rather large to get a sensible design in the subcooling region. Again, therefore, a separate subcooler may be the better solution.

Subcooling during condensation inside horizontal tubes can sometimes be conveniently obtained by having a special pass set aside for the subcooling duty. Hence, for example, the unit may be designed with all the required condensation achieved in the first pass. The vent line would then be placed in the header at the end of this first pass. The liquid level would be maintained in this header and the second pass would be running full of liquid. With this particular method of achieving subcooling, however, a much smaller number of tubes would normally be required in the second pass to give high velocities to ensure good heat transfer.

For condensation inside vertical tubes, subcooling can be achieved by having a liquid level control and running the tubes full of liquid up to a

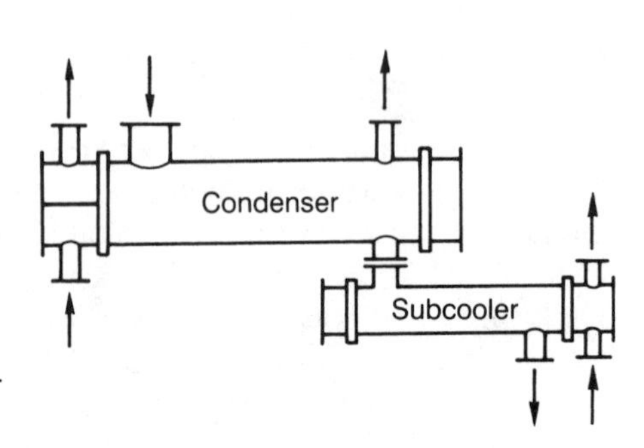

Fig. 11.12. Subcooling in separate condenser.

certain height. However, this is clearly only possible if there is no incondensable present, otherwise there is no convenient way of venting such noncondensables.

At this point, it is worth remarking that noncondensables, which are a nuisance in most other respects, are helpful in achieving subcooling. This is because they suppress the condensing temperature below that for the pure vapor. Hence the condensate drained from a condenser with vapor and noncondensable gas will have effective subcooling unless the condenser layout is such that this condensate can be reheated with fresh vapor. This phenomenon is often used to advantage in order to achieve subcooling when condensing inside vertical tubes.

11.2.5 Choice between Types

Some advantages of particular types of condensers over others have already been mentioned. Further simple rules are given here. When all else is equal, it is best to have the dirtiest fluid in the tubes than on the shell-side because it is usually easier to make provision for cleaning inside tubes than outside them. In addition, it is best to have the highest-pressure stream inside the tubes because, then, only the tubes and the headers need be built to handle the high pressure, while the shell (which is very often a very costly item) need only handle a lower pressure. Also, when special materials are required to handle corrosive fluids, such fluids are better on the tube side in order to avoid having the shell made out of special materials.

Additional guidance on the various types is given in Table 11.1, which is adapted from Bell and Mueller [2]. When there is no obvious advantage in one type of exchanger over another, it may be advantageous to design the alternative types and see which is the cheapest.

11.3. SHELL-AND-TUBE CONDENSERS FOR POWER PLANT

11.3.1 Steam Turbine Exhaust Condensers

For historical reasons, these condensers are often referred to as "surface condensers." In principle, they are no different from the shell-side condensers just described; in particular, the *X* type. In practice, there are certain severe demands placed on these units which have been overcome by special design features. These special demands arise from the large heat duties that they must perform and from the necessity to maintain a low condensing temperature to achieve the highest possible power station efficiency.

The aim is to operate with the condensing temperature only a few degrees above the cooling-water temperature. Typically, the cooling water is about 20°C, with condensation taking place at around 30°C. Saturation pressure of

TABLE 11.1 Guide to the Selection of Type of Shell-and-Tube Condenser

Type	For Subcooled Condensates	For Mixed Vapors and/or Noncondensables	Pressure Range (For Pressures between 0.04 and 10 bar, All Types Are Satisfactory)		If Fouling or Polimerization	For Corrosive Condensates or Vapors
			Below 0.4 bar	Above 10 bar		
Inside tubes						
Vertical upflow (with reflux condensate)	Not applicable	Limited to small quantity of low boiler	Very poor	Good	Not applicable	Fair
Vertical downflow	Good	Excellent	Poor	Excellent	Fair	Good
Horizontal	Very poor but possible with separate sub-cooling pass	Good in annular flow, poor in stratified flow	Poor	Excellent	Fair	Good
Outside tubes						
Vertical	Possible but not recommended	Good	Fair	Poor—high shell cost	Very poor	Poor—alloy shell required
Horizontal	Not recommended since predictions of cooling coefficient and effective area are unreliable	Poor for wide condensing mixtures. Good with noncondensables if baffle space is varied. Heat and mass transfer prediction is unreliable	Fair	Poor—high shell cost	Very poor	Poor

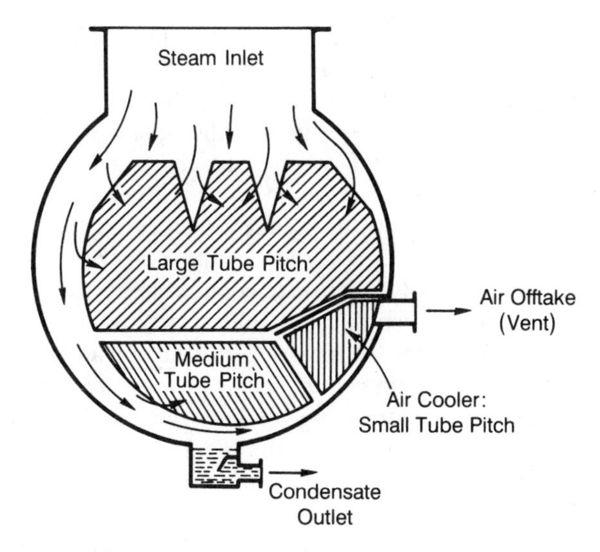

Fig. 11.13. Small turbine exhaust condenser: areas occupied by tubes shown shaded.

water at this temperature is 0.042 bar absolute, which is a typical operating pressure for these condensers. Clearly, there is little pressure available for pressure drop through the unit. There is also little temperature differences to spare in order to overcome the effect of noncondensable gases. Hence the design of surface condensers is governed by the need for good venting and low pressure drop.

Often these condensers are very large. There may, for example, be two condensers serving a single 600 MWe turbine set. Hence each condenser must handle around 200 kg/s of steam, with an approach velocity of up to 60 m/s.

Surface areas are around 25,000 m^2, which may be achieved by having say 1500 tubes of 25 mm diameter with a length of around 20 m. These very large condensers often have box-shaped shells, but the smaller ones, with surface areas less than about 5000 m^2, may have cylindrical shells.

Surface condensers vary widely in their geometric detail and various types are described by the editors of *Power* [3], Simpson [4], and Sebald [5]. Standards for their design are given by the British Electrical and Allied Manufacturers' Association [6] and by the Heat Exchange Institute [7]. Nevertheless, there are many features that are common to most designs, as shown in the diagram of the relatively small surface condenser in Fig. 11.13. This is not an actual unit but a drawing illustrating some of the main features.

The vapor inlet velocity is very high because of the high thermal duty combined with the low pressure. Tubes near the inlet are therefore on a wider pitch than those elsewhere, and tubes are left out in places to provide paths for steam lanes to guide steam into the bundle. The combination of steam lanes and paths around the bundle means that there is a large bundle perimeter allowing the steam to enter the bundle, thereby minimizing the effects of the very large inlet velocity. As the steam passes through the bundle towards the vent line, its flow rate decreases and the air concentration increases. Therefore closer tubes and less superficial flow area are used toward the exit in order to keep the steam velocities up. This improves the gas-phase heat and mass transfer and reduces the danger of stagnant pockets of air forming. There is usually a separate compartment just by the vent line that has the smallest tube pitch and the coldest cooling water in the tubes. Because most of the steam has been extracted from the air by this stage, the compartment is called the "air cooler." The purpose of this section is to extract the last possible moisture from the air, which includes knocking out any entrained condensate. As with any other vented condenser, care is taken to avoid a short-cut path between the steam inlet and the vent line. These condensers are often also used as a condensate deaerator. This is done by allowing some of the steam to flow through condensate dripping from the tubes.

Many of the features just described can be achieved by having a condenser with radial flow paths, as shown in an early Westinghouse design in Fig 11.14. The idea here is that the steam flows radially in from the outside toward the space in the middle of the unit, from which noncondensables are vented axially. Such units cannot, however, be made completely radially symmetric because of the location of the vapor inlet and, of course, gravity causes the condensate to move preferentially downwards. Problems associated with the design of such units are discussed by Coit [8].

As with any other shell-and-tube unit, tubes in surface condensers must be supported at regular intervals along their length with tube support plates. Such support plates also have the advantage of deliberately preventing any axial flow of vapor, thus making it easier for designers to ensure that vapor flow paths through the bundles are relatively straightforward, giving rise to no recirculation pockets where noncondensables can accumulate.

There is such a variety of different surface condenser designs that it is impossible to illustrate them all here, but many examples of modern condensers are described by Sebald [5].

11.3.2 Feed-Water Heaters

Feed-water heaters use steam bled from turbines in order to heat the boiler feed water. The steam used may be at pressures of up to 40 bar, and the water may have pressures of up to about 200 bar. As might be expected from the guidelines given in Section 11.2.5, the higher pressure of water is placed in the tubes with the lower pressure steam condensing on the shell side. The feed water may often pass in series through a number of feed heaters, each operating with condensing steam at a different pressure. The feed water

Fig. 11.14. Radial steam flow, circular bundle design, 1948, Westinghouse (courtesy of Chochrane Environmental Systems).

would then first meet heaters operated with the low pressure condensing and then pass through condensers with successively higher condensing pressures. The condenser with the lowest pressure may well be operating under vacuum conditions at around 0.5 bar.

There is no reason in principle why feed-water heaters should not have the same sort of design as the TEMA types of exchangers already described. However, as with steam turbine condensers, their designs have evolved in order to meet the special requirements of power station operation. It has already been mentioned that conventional shell-side condensers are not normally very good for handling condensing and subcooling in the same unit. With the high-pressure feed-water heaters, however, because of the expense of the high-pressure shell, there is a great incentive to cope with both operations in the same shell. The design has therefore been refined in order to handle these two processes effectively. Another feature of feed-water heaters is that they can be fed with superheated steam. This superheat is

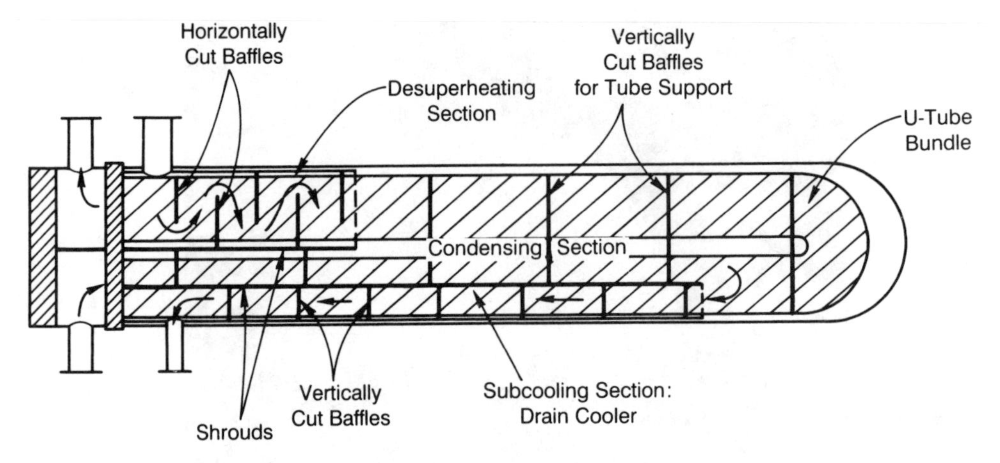

Fig. 11.15. Example of a horizontal feed-water heater.

used to raise the temperature of the feed water leaving the exchanger to a much higher temperature than is achieved in the condensing region. Again, therefore, there are special features in the design to cope with this desuperheating.

Feed-water heaters usually have two tube passes, achieved by having U tubes. The shell may be oriented horizontally or vertically and, with vertical arrangements, the header or channel for the tube-side fluid may be at the top or bottom of the exchanger (often referred to as either "head up" or "head down" designs, respectively).

A feature of the feed-water heater is that the sections for desuperheating and subcooling are usually partitioned off by enclosing the tubes with a shroud of wrapper. The subcooling section is often called the "drain cooler." The desuperheating section has transverse baffles (usually single segmental) which are fairly closely spaced to give fairly high velocities in order to increase the single-phase gas-cooling coefficient. The condensing section has widely spaced baffles whose main purpose is to support the tubes. High velocities are not required in this region because the steam-condensing coefficients are high. The subcooling, or drain-cooling, section again often has closer baffling to increase the coefficients in this region. Figure 11.15 illustrates these main features for a horizontal feed-water heater. The subcooling section in this exchanger only uses some of the tubes in the first pass and is said to have a "split-pass drain cooler." When all the tubes are used, it is known as a "full-pass drain cooler."

An interesting feature of high-pressure units is that they often do not need any vent line. This is in contradiction to the rules elsewhere in this chapter, but is acceptable in these circumstances because the steam at this point in the cycle is free of air because all air has been removed earlier in the circuit by deaeration in the turbine exhaust condenser, and further deaeration in specialized equipment. However, a vent is required for initial start-up and, of course, vents are required on the units using vacuum steam.

Further information on the design and construction of these units is given by the editors of *Power* [3] and by Spence et al. [9]. Standards for the design of such units are provided in the United Kingdom by the British Electrical and Allied Manufacturers' Association [10], and in the United States by the Heat Exchange Institute [11].

11.4 PLATE EXCHANGERS

Plate exchangers consist of a large number of plates, which are sealed round the edge by means of gaskets and are held together in a large press known as a "frame." These plates are corrugated in various patterns, both to improve the heat transfer and to increase the rigidity of the plates. The patterns in the plates are arranged so that successive plates touch one another at many points across the surface but there is still a path, albeit a tortuous one, for the fluid stream to pass from the inlet port to the outlet port in the plate. The gaskets around the parts are arranged so that the two fluid streams are made to flow between alternate plates. Plate exchangers are usually limited to fluid streams with pressures below about 25 bar and temperatures below about 250°C. These limits are dictated by the maximum pressure loadings which can be tolerated by the plates and by the maximum pressure loadings and temperatures which can be withstood by the gaskets. Further details of plate heat exchangers are given by Alfa Laval [12].

Plate heat exchangers have been mainly developed for single-phase duties and have particular advantage when one of the fluid streams has high viscosity. They tend not to be well suited for condensers. One limitation arises because of the size of the ports in these plates, which are often rather small for handling large-volume flows of vapor. Another difficultly is that one cannot conveniently have separate condensate and vent outlets when the design demands this. However, such exchangers are more compact than shell-and-tube exchangers and are frequently used with service steam on one side being used to heat some process stream. Some thought is presently being given to the design of special plate types and exchanger configurations for evaporation and condensation systems.

11.5 SPIRAL EXCHANGER

Figure 11.16 illustrates the main features of a spiral heat exchanger used for condensing duties. It consists of two flat plates which are wound around a central core to form two spiral channels. Studs are normally welded at regular spacing on one side of the plates before rolling to provide a uniform channel separation and to support the plate against internal and external

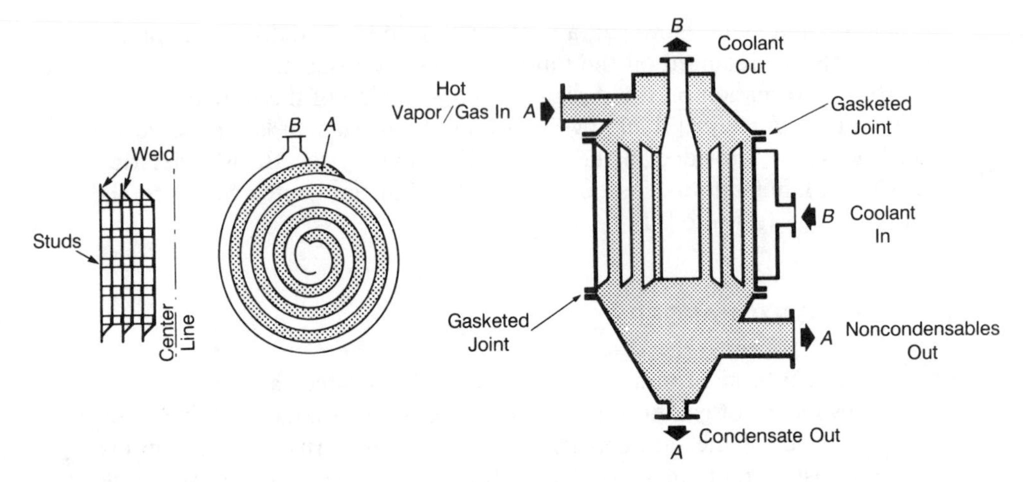

Fig. 11.16. Spiral exchanger for condenser applications (courtesy of APV International, Ltd.).

pressure. Channel widths are typically 4 to 20 mm and the two flow channels can be of different widths if the duty so demands. A pressure shell is achieved by forming the last, outer spiral into a thicker material on which are welded end flanges to receive the end covers. Theses exchangers can be manufactured from stainless steel, high nickel alloys, or any special material that can be cold worked and welded. They can operate at pressures up to about 20 bar and temperatures up to about 400°C.

Like plate exchangers, spiral exchangers are more compact than shell-and-tube exchangers. The configuration shown in Fig. 11.16 has, however, the advantage over plate configurations for handling large vapor volumes with low pressure drop.

11.6 PLATE-FIN HEAT EXCHANGERS

Figure 11.17 shows the general form of a plate-fin heat exchanger. The fluid streams are separated by flat plates between which are sandwiched corrugated fins. A more apt name for this exchanger is therefore "finned-plate" exchanger. Plate-fin heat exchangers are often used in low-temperature (cryogenic) plants and where the temperature differences between the streams are small (1 to 5°C). They are very compact units having a heat transfer area per unit volume of around 2000 m^2/m^3. Special manifold devices are provided at inlet and outlet to these exchangers to provide good flow distributions across the plates and from plate to plate. The plates are typically 0.5 to 1.0 mm thick and the fins 0.15 to 0.75 mm thick. The whole exchanger is

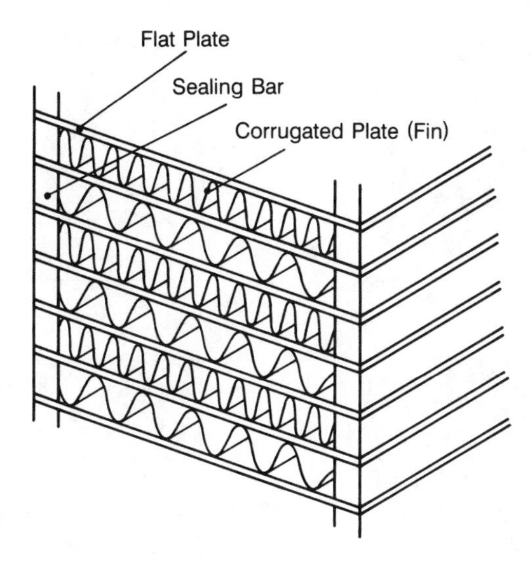

Fig. 11.17. Basic construction of a plate-fin exchanger.

made of aluminum alloy and the various components are brazed together by brazing in a salt bath or now more commonly in a vacuum furnace.

Plate-fin exchangers can be arranged into a variety of configurations with respect to the fluid streams. Figure 11.17 shows the arrangement for co-current or countercurrent flow between the streams. Alternatively, the streams may be arranged in crossflow. While most heat exchangers exchange heat between two streams, plate-fin units may be arranged to distribute heat among a large number of streams. Streams may also be introduced and removed at points along an exchanger.

The corrugated sheets which are sandwiched between the plates serve both to give extra heat transfer area and to give structural support to the flat plates. There are many different forms of corrugated sheets used in these exchangers, but the most common types are:

1. Plain fin
2. Plain-perforated fin
3. Serrated fin (also called "lanced," "interrupted," or "multientry")
4. Herringbone or wavy fin

The plain type is used most frequently for condensing duties. Figure 11.18 shows these four types. The perforated type is essentially the same as the plain type except that it has been formed from a flat sheet with small holes in it.

The flow channels in plate-fin exchangers are small which means that the flows also have to be small [10 to 300 $kg/(m^2 \cdot s)$] to avoid excessive pressure

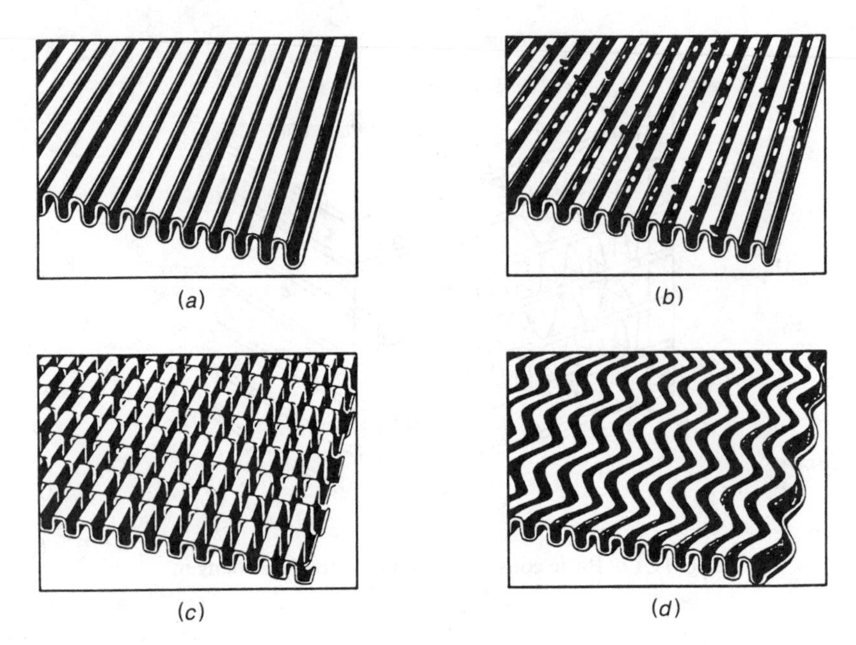

Fig. 11.18. Fin types in plate-fin exchangers: (*a*) plain, (*b*) perforated, (*c*) serrated, (d) herringbone.

drops. This may make the channel prone to fouling which, when combined with the fact that they cannot be mechanically cleaned, means that plate-fin exchangers are restricted to clean fluids. They are frequently used for condensation duties in air liquefaction plants. Further information on these exchangers is given by HTFS [13].

11.7 AIR-COOLED HEAT EXCHANGERS

Many coolants are possible for process condensers: for example, air, cooling-tower water, or a colder process stream which requires heating. In areas where there is a shortage of make-up water, air-cooled condensers may be favored. They can also become economical if condensation is taking place at temperatures which are more than about 20°C above ambient. They suffer the disadvantage, however, of occupying a relatively large ground area and of generating noise from the fans.

Figure 11.19 illustrates a typical air-cooled heat exchanger which may be used as a condenser. It consists of a horizontal bundle of tubes with the air being blown across the tubes on the outside and condensation occurring inside the tubes. The unit shown is a forced-draft unit since the air is blown across. An alternative design is the induced-draft unit which has the fans on

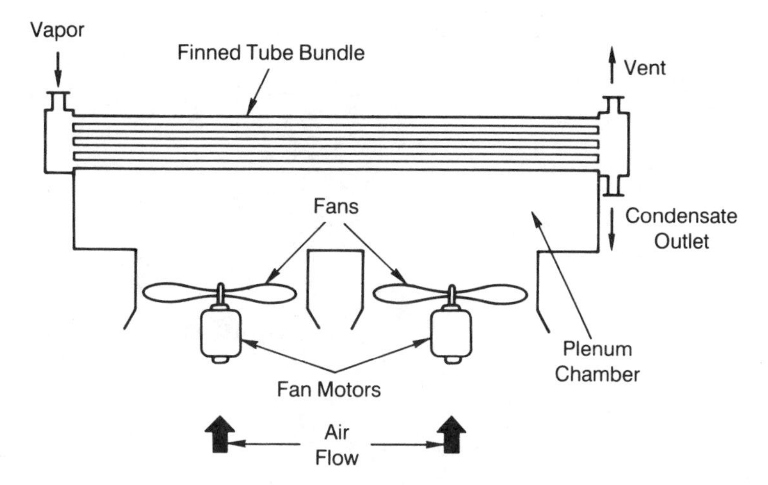

Fig. 11.19. Forced-draft, air-cooled exchanger used as a condenser.

top which sucks the air over the tubes. The tubes are finned with transverse fins on the outside to overcome the effects of the low air-side coefficients. There would normally be a few tube rows and the process stream may take one or more passes through the unit. With multipass condensers, the problem arises with redistributing the two-phase mixture on entry to the next pass. This can be overcome in some cases by using U tubes or by having separate passes just for subcooling or desuperheating duties. In multipass condensers, it is important to have each successive pass below the previous one to enable the condensate to continue downwards. Further information in air-cooled heat exchangers is given by Ludwig [14] and by the American Petroleum Institute [15].

11.8 DIRECT-CONTACT CONDENSERS

Direct-contact condensers are cheap and simple devices but have limited application because the process streams and coolant are mixed. The removal of the intermediate wall means that they are not prone to fouling and very high heat transfer rates per unit volume can be achieved.

Some direct-contact exchangers inject vapor into a pool of liquid. This may be done to heat up a process fluid or to suppress vapor released from a reaction vessel as a result of an accident or malfunction. Two difficulties arise with this method of condensation. The first is that the condensation front may move back into the vapor inlet line, causing the liquid to be periodically ejected, often with some violence. The second is that a very large vapor bubble may form in the liquid pool and this may collapse suddenly, causing damage to the vessel. These problems may be avoided by having the vapor

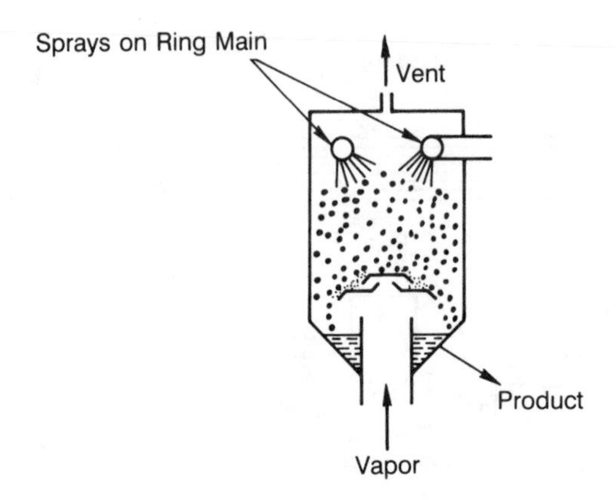

Fig. 11.20. Spray condenser.

injected through a large number of small holes or by using special ejectors which mix the incoming vapor with liquid in a special mixing tube.

The most common type to direct-contact condenser is one in which subcooled liquid is sprayed into the vapor in a large vessel. This arrangement is illustrated in Fig. 11.20. Very often, these units are used for condensing steam using water as a coolant. In these cases, the mixing of water with condensate presents no major problem. When condensing a vapor whose condensate is immiscible with the spray liquid, however, a separator is usually required after the condenser in order to recover the product. Alternatively, the condensate product may be cooled in a single-phase exchanger and some recycled as coolant spray. At first sight, there then seems little benefit in having a direct-contact condenser and a conventional single-phase exchanger instead of using one shell-and-tube condenser. The advantage appears, however, when the condenser is operating under vacuum. As has already been seen, tubular condensers for vacuum operation are large and complex. It can therefore be sometimes be economical to replace such a condenser by a simple spray condenser and a compact single-phase cooler.

Spray condensers cannot be used with dirty coolants since the spray nozzles may become blocked. In these circumstances, a tray condenser may be used, as illustrated in Fig. 11.21. The trays may be sloped slightly to prevent dirt accumulating on them. The tray arrangement can have a slight thermodynamic advantage over a spray unit because some degree of countercurrent flow may be achieved between the falling liquid and upward flowing vapor–gas mixture. In spray units, by contrast, the gas phase is usually thoroughly mixed throughout the vessel, thus making counterflow impossible.

Even better countercurrent flow may be obtained using a packed column but these units give a high pressure drop on the gas–vapor side and are more expensive than spray and tray exchangers. The packing usually consists of

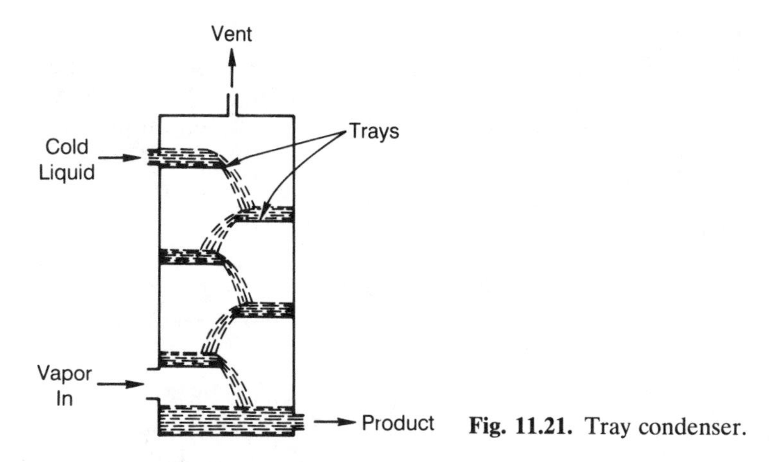

Fig. 11.21. Tray condenser.

randomly packed rings, but a lower pressure drop may be obtained by using stacked rings or grid packing.

11.9 THERMAL EVALUATION METHODS FOR SHELL-AND-TUBE CONDENSERS

11.9.1 Introduction and Definition of Terms

The term *thermal evaluation method* is used to signify the calculation process by which, for a known exchanger geometry, the thermal duty may be calculated or, alternatively, the required heat transfer area determined to suit the duty. In the latter case, the calculated heat transfer area may be incompatible with the assumed geometry. These calculations fall short of a full design calculation which involved calculating stream pressure drops as well as repeating the calculations for many different assumed geometries in order to find ones which satisfy all the imposed constraints. The final stage of design is to choose the best design on the basis of, say, capital cost.

The basis of thermal evaluation methods is an equation of the form

$$\frac{dQ}{dA} = U\theta \tag{11.1}$$

where Q is the heat transfer rate, A is the heat transfer area, U is the overall heat transfer coefficient, and θ is the temperature difference. It is important to appreciate that both U and θ can vary significantly throughout a condenser and hence Eq. (11.1) is based on the local values. The temperature difference θ may be defined in a number of ways provided that it is consistent with the definition of U. One definition often used for θ is that it is the difference between the equilibrium temperature of the two streams if each were well mixed at the point in question. This would, for example, be the right definition of θ to use when determining U by the Silver [16] and Bell and Ghaly [17] method described in Chapter 10. It is the definition used in Sections 11.9.2 to 11.9.4.

The overall coefficient, U, may be regarded as being made up of a number of components as follows:

$$\frac{1}{U} = \frac{1}{h_{\text{hot}}} + r_{\text{hot}} + \frac{s_w}{k_w} + \frac{1}{h_{\text{cold}}} + r_{\text{cold}} \tag{11.2}$$

where h_{hot} and h_{cold} are, respectively, the "film" coefficients for the hot and cold streams, r_{hot} and r_{cold} are the respective fouling layer thermal resistances, s_w is the tube wall thickness, and k_w is the tube wall thermal conductivity. For thick-walled tubes, corrections are necessary to allow for the different surface areas inside and outside the tubes. A simplified form of Eq. (11.2) will be used in subsequent calculations:

$$\frac{1}{U} = \frac{1}{h_{\text{hot}}} + r + \frac{1}{h_{\text{cold}}} \tag{11.3}$$

where r is the combined thermal resistance of the tube wall and fouling. This resistance will be taken as a constant throughout the exchanger, whereas h_{hot} and h_{cold}, and consequently U, may vary considerably. The determination of h_{hot} and h_{cold} is discussed elsewhere in this book, and it is therefore assumed in this chapter that we know how to calculate them locally. This chapter therefore concentrates on how to use this information in thermal evaluation. When considering pure vapor, h_{hot} is the coefficient for the condensate layer and when condensing a mixture it is the effective condensing-side coefficient which combines the gas phase and condensate film coefficient.

Equation (11.1) can be rearranged and written in an integral form as follows:

$$\int_{Q_T} \frac{dQ}{\theta} = \int_{A_T} U\,dA \tag{11.4}$$

where the subscript T refers to the total value for the exchangers. Heat exchanger designs are usually summarized in terms of mean quantities which are related by an equation as follows:

$$Q_T = U_m A_T \theta_m \tag{11.5}$$

where U_m is the mean overall coefficient and θ_m is the mean temperature difference. Comparing Eqs. (11.4) and (11.5) suggests the following definitions for the mean quantities:

$$\frac{1}{\theta_m} = \frac{1}{Q_T}\int_{Q_T}\frac{dQ}{\theta} \tag{11.6}$$

$$U_m = \frac{1}{A_T}\int_{A_T} U\,dA \tag{11.7}$$

In practice, it is unnecessary to evaluate both Eqs. (11.6) and (11.7), since when either θ_m or U_m has been determined the other may be calculated from Eq. (11.5).

Equation (11.1) may also be written as

$$A_T = \int_{Q_T}\frac{dQ}{U\theta} \tag{11.8}$$

Combining this with Eq. (11.3) gives

$$A_T = \int_{Q_T}\frac{dQ}{h_{\text{hot}}\theta} + r\int_{Q_T}\frac{dQ}{\theta} + \int_{Q_T}\frac{dQ}{h_{\text{cold}}\theta} \tag{11.9}$$

Dividing through by $A_T U_m$ and using Eqs. (11.5) and (11.6) gives

$$\frac{1}{U_m} = \frac{1}{U_m A_T}\int_{Q_T}\frac{dQ}{h_{\text{hot}}\theta} + r + \frac{1}{U_m A_T}\int_{Q_T}\frac{dQ}{h_{\text{cold}}\theta} \tag{11.10}$$

which, on comparing with Eq. (11.3), suggests the following definitions for the mean "film" coefficients:

$$\frac{1}{h_m} = \frac{1}{U_m A_T}\int_{Q_T}\frac{dQ}{h\theta} \tag{11.11}$$

The derivation leading to this last equation was proposed by Smith [18].

Using Eq. (11.1), an alternative form of this definition is

$$\frac{1}{h_m} = \frac{1}{U_m A_T}\int_{A_T}\left[\frac{U}{h}\right]dA \tag{11.12}$$

Some special cases of Eqs. (11.6) and (11.7) are useful. If θ varies linearly with Q, Eq. (11.6) can be integrated to give

$$\theta_m = \theta_{\text{LM}} = \frac{\theta_a - \theta_b}{\ln(\theta_a/\theta_b)} \tag{11.13}$$

where θ_{LM} is the well-known logarithmic-mean temperature difference and θ_a and θ_b are the end values of θ. It is most unusual in a condenser for θ to vary linearly with Q over the whole exchanger but small portions of the exchanger can often be identified over which this assumption is well approximated. Examples of this will be seen below.

If U varies linearly with A, Eq. (11.7) may be integrated between U_a and U_b to give

$$U_m = \tfrac{1}{2}(U_a + U_b) \tag{11.14}$$

If both U and θ vary linearly with Q. Eq. (11.7) may be integrated with the aid of (11.1) to give

$$U_m = \frac{U_a\theta_b - U_b\theta_a}{\theta_{LM}\ln(U_a\theta_b/U_b\theta_a)} \tag{11.15}$$

This result was first obtained by Colburn [19]. If both $1/U$ and θ vary linearly with Q, Eq. (11.7) may be integrated with the aid of (11.1) to give

$$\frac{1}{U_m} = \frac{1}{U_a}\frac{\theta_{LM} - \theta_b}{\theta_a - \theta_b} + \frac{1}{U_b}\frac{\theta_a - \theta_{LM}}{\theta_a - \theta_b} \tag{11.16}$$

Again, these equations will not usually be valid over the whole of the condenser but may apply to small portions of it. It is not always clear which of the preceding equations is valid for a given set of circumstances. However, if U_a and U_b vary only by a small amount, Eq. (11.14) is preferred because of its simplicity. There is a long tradition in the use of Eq. (11.15) but with little justification. Equation (11.16) seems more in line with the variations observed in condensers and is hence recommended in those situations when Eq. (11.14) cannot be used due to the large difference between U_a and U_b. Of course, any question about which equation is more accurate can always be avoided by dividing the exchanger into a large number of sections.

11.9.2 Co-current and Countercurrent Condensers

The procedure given here applies to TEMA E-type shells with a single tube-side pass. It also applies to a J shell which can be divided down the middle and treated as two exchangers, one with co-current flow and the other with countercurrent flow. Countercurrent flow is more usual in E shells since it makes best use of the temperature difference between the streams. Indeed, some duties are not possible in co-current flow but can be handled without difficulty in countercurrent flow. The following description is in terms of countercurrent flow but the same approach can be used for co-current flow, and the differences in the results obtained are noted.

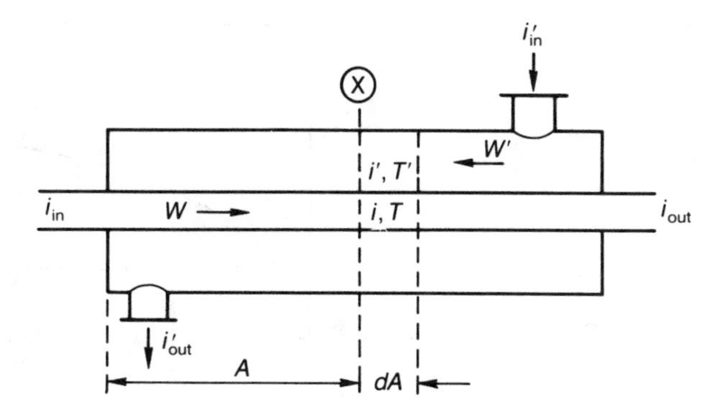

Fig. 11.22. Counterflow heat exchanger.

Figure 11.22 illustrates a countercurrent flow exchanger. In this diagram and the subsequent discussion, the shell-side stream is denoted by a prime ($'$). Hence the shell-side stream enters with specific enthalpy i'_{in} and leaves with specific enthalpy i'_{out}. The tube-side specific enthalpy changes from i_{in} to i_{out}. The shell-side and tube-side mass flows are, respectively, W' and W. A heat balance over area A of the exchanger gives

$$i = i_{\text{in}} + \frac{W'}{W}(i' - i'_{\text{out}}) \tag{11.17}$$

where i' and i are the shell-side and tube-side specific enthalpy, respectively, at position X on Fig. 11.22. The corresponding equation for a co-current flow exchanger is obtained by replacing i'_{out} by i'_{in} and W' by $(-W')$.

The first step in the thermal evaluation is to plot the equilibrium temperature, T', against i' for the shell-side stream. The equilibrium temperature is used in accordance with our definition of the overall coefficient, U. Such a plot is shown in Fig. 11.23. Using Eq. (11.17) and the temperature–specific enthalpy relationship for the tube-side fluid, the corresponding tube-side temperature, T, may be plotted on Fig. 11.23 as shown. This diagram is extremely useful in condenser design and will be called here the "exchanger operating diagram." Figure 11.23 is typical of a condenser with a desuperheating zone and where condensation is occurring in the presence of noncondensable gas. The tube-side curve shown would occur, say, if a pure liquid were being heated up and then boiled. The design is impossible if the two curves cross or touch anywhere.

The next step in the thermal evaluation is to divide this diagram into zones for which both the T curve and the T' curve are linear. This is shown by the vertical broken lines in Fig. 11.23. Now, over each zone, θ (i.e., $T' - T$) varies linearly with i'. This is the same as saying θ varies linearly with Q, since

$$Q = W'(i'_{\text{in}} - i') \tag{11.18}$$

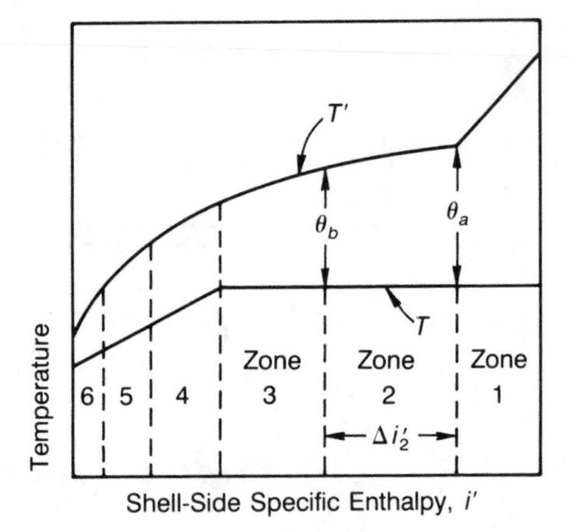

Fig. 11.23. Example of exchanger operating diagram for a counterflow exchanger.

Thus the logarithmic-mean temperature difference, as defined by Eq. (11.13), applies for each zone and can be evaluated. The appropriate θ_a and θ_b for zone 2 are illustrated in Fig. 11.23. Also, the overall coefficients at the zone boundaries may be calculated and a mean overall coefficient for each zone calculated using Eq. (11.14), (11.15), or (11.16), whichever is most appropriate. Equation (11.5) may then be applied to each zone in the form

$$A_j = \frac{W'\Delta i'_j}{U_{m,j}\theta_{\mathrm{LM},j}} \tag{11.19}$$

where the subscript j refers to zone number and $\Delta i'_j$ is the specific enthalpy change of the shell-side fluid in the jth zone. Clearly, the total heat transfer area is given by

$$A_T = \sum_j A_j \tag{11.20}$$

Equations (11.6) and (11.7) may be expressed in summation form to give U_m and θ_m for the whole exchanger:

$$\frac{1}{\theta_m} = \frac{1}{i'_{\mathrm{in}} - i'_{\mathrm{out}}} \sum_j \frac{\Delta i'_j}{\theta_{\mathrm{LM},j}} \tag{11.21}$$

and

$$U_m = \frac{1}{A_T} \sum_j U_{m,j} A_j \tag{11.22}$$

In the preceding calculation, the heat load on the exchanger is known and is given by

$$Q_T = W'(i'_{\text{in}} - i'_{\text{out}}) = W(i_{\text{out}} - i_{\text{in}}) \tag{11.23}$$

Therefore it is not necessary to evaluate both Eqs. (11.21) and (11.22) since U_m, θ_m, and Q_T are related via Eq. (11.5). However, it is useful to evaluate Eqs. (11.21) and (11.22) and substitute the results into Eq. (11.5) in order to cross-check the arithmetic.

A convenient feature of the preceding calculation procedure is that Fig. 11.23 does not depend on the heat transfer coefficient and hence is independent of details of the geometry like the number of tubes, baffles, and so forth. The same applies to the zonal and exchanger mean temperature differences. These quantities may therefore need only to be recalculated when the number of passes is changed.

11.9.3 Shell-Side, *E*-Type Condenser with Two Tube-Side Passes

Figure 11.24 illustrates the operation of an *E* shell with two tube-side passes. A heat balance over area *A* gives

$$W'(i' - i'_{\text{out}}) = W(i^{\text{I}} - i_{\text{in}} + i_{\text{out}} - i^{\text{II}}) \tag{11.24}$$

where the superscripts I and II refer to the first and second tube-side pass, respectively. Combining this with Eq. (11.23) gives

$$i' = i'_{\text{in}} - \frac{W}{W'}(i^{\text{II}} - i^{\text{I}}) \tag{11.25}$$

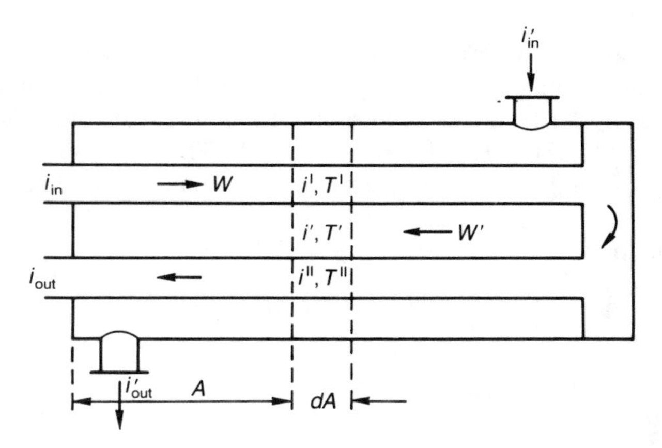

Fig. 11.24. Exchanger with one shell-side pass and two tube passes.

This result is for the first tube pass countercurrent to the shell-side flow. If, however, it is co-current, the result is

$$i' = i'_{out} + \frac{W}{W'}(i^{II} - i^{I}) \tag{11.26}$$

Heat balances over area dA for passes I and II give, respectively,

$$W\,di^{I} = +U^{I}(T' - T^{I})\frac{dA}{2} \tag{11.27}$$

and

$$W\,di^{II} = -U^{II}(T' - T^{II})\frac{dA}{2} \tag{11.28}$$

Dividing Eq. (11.28) by Eq. (11.27) gives

$$\frac{di^{II}}{di^{I}} = -\frac{U^{II}(T' - T^{II})}{U^{I}(T' - T^{I})} \tag{11.29}$$

The same result is obtained if the shell-side stream is co-current to the first tube-side pass.

It is very convenient to simplify Eq. (11.29) by assuming U^{II}/U^{I} is 1. This is often a reasonable approximation for the shell-side condensers provided that the tube-side coefficient is constant or not controlling. With this assumption, Eq. (11.29) becomes

$$\frac{di^{II}}{di^{I}} = -\frac{T' - T^{II}}{T' - T^{I}} \tag{11.30}$$

The right-hand side of Eq. (11.30) is a known function of i^{II} and i^{I}, as becomes evident when one realizes that T' is a known function of i', T a known function of i (whether superscripted I or II), and i' is related to i^{II} and i^{I} by Eq. (11.25) or (11.26). Hence Eq. (11.30) can be integrated along the exchanger with the initial boundary conditions that $i^{I} = i_{in}$ when $i^{II} = i_{out}$.

For example, a simple numerical integration can be done by updating i^{I} and i^{II} as follows:

$$i^{I}_{new} = i^{I} + \delta i^{I} \tag{11.31}$$

and

$$i^{II}_{new} = i^{II} - \frac{T' - T^{II}}{T' - T^{I}}\delta i^{I} \tag{11.32}$$

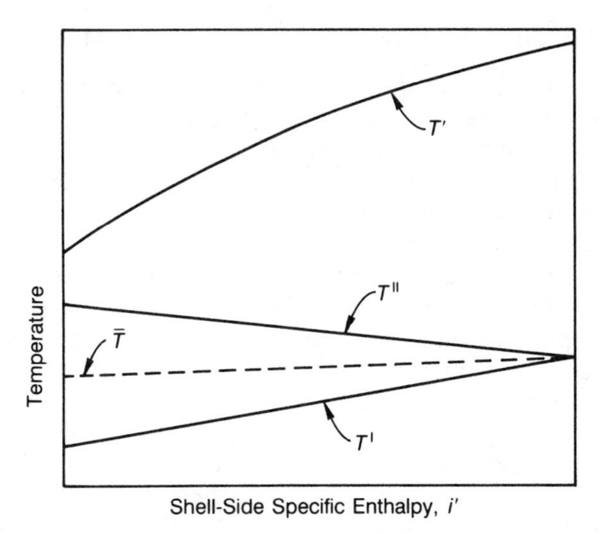

Fig. 11.25. Example of operating diagram for shell-side condenser (E type) with two tube passes.

where δi^{I} is a small change in i^{I}. There are, of course, more sophisticated integration methods for use with computers. This sort of integration may be used to construct the operating diagram shown in Fig. 11.25. As with the counterflow exchanger, this operating diagram is independent of detailed geometrical features and, therefore, applies to all two-pass E shells.

The heat leaving the shell side in area dA is

$$dQ = U^{\mathrm{I}}(T' - T^{\mathrm{I}})\frac{dA}{2} + U^{\mathrm{II}}(T' - T^{\mathrm{II}})\frac{dA}{2} \tag{11.33}$$

which, since $U^{\mathrm{I}} = U^{\mathrm{II}} = U$ (say), gives

$$dQ = (T' - \bar{T})U\,dA \tag{11.34}$$

where

$$\bar{T} = \frac{T^{\mathrm{I}} + T^{\mathrm{II}}}{2} \tag{11.35}$$

that is, $\bar{T}$ is the average temperature between the passes at a given point along the shell.

The curve for $\bar{T}$ can be plotted on the operating diagram as illustrated by the broken line in Fig. 11.25. Equation (11.34) is now identical to Eq. (11.1), except that θ is replaced by $T' - \bar{T}$. Therefore the remainder of the thermal evaluation is now the same as for the counterflow exchanger except T is replaced by $\bar{T}$.

An example using the preceding calculation procedure is given in Section 11.12.1.

11.9.4 Shell-Side, *E*-Type Condenser with Four or More Tube Passes

An analysis of the type described previously can be extended to exchangers with 4, 6, 8, or more even-number passes. The calculations become progressively more complicated, however, as the number of passes are increased. Furthermore, it is found in practice that the average tube-side temperature $\overline{T}$ (now averaged over however many passes there are) does not change significantly as the number of passes are increased beyond 4 (see Butterworth [20]).

Since the objective of the first part of the thermal evaluation is to construct an operating diagram containing the $\overline{T}$ curve, we could construct this for a four-pass exchanger and use this for any number of passes. There is, however, a convenient method of finding $\overline{T}$ for an infinite number of passes which we can use instead. This method is slightly less general than that given previously since it only applies when there is a linear temperature–enthalpy curve for the tube-side stream. Nevertheless, this covers the most important practical case of a single-phase coolant.

The method is due to Emerson [21] who presents a more rigorous deriviation than that given here although he did not spot some minor algebraic manipulations which are used here to simplify the calculation procedure. It is reasonable to postulate that the tube-side stream sees a constant shell-side temperature, T'_{eff} say, as it traverses the length of the exchanger an infinite number of times. The logarithmic-mean temperature difference then applies for heating the tube-side since it has a linear temperature–enthalpy curve:

$$\theta_m = \frac{T_{\text{out}} - T_{\text{in}}}{\ln \dfrac{T'_{\text{eff}} - T_{\text{in}}}{T'_{\text{eff}} - T_{\text{out}}}} \tag{11.36}$$

We know also, for a large number of tube passes, that the mean tube-side temperature, $\overline{T}$, is a constant, independent of i'. We can therefore write θ_m as

$$\theta_m = T'_{\text{eff}} - \overline{T} \tag{11.37}$$

Hence combining Eqs. (11.36) and (11.37) gives

$$\overline{T} = T'_{\text{eff}} - \frac{T_{\text{out}} - T_{\text{in}}}{\ln \dfrac{T'_{\text{eff}} - T_{\text{in}}}{T'_{\text{eff}} - T_{\text{out}}}} \tag{11.38}$$

But θ_m is also given by integrating Eq. (11.6)

The procedure for obtaining the operating diagram is therefore as follows:

1. Plot the temperature versus specific enthalpy for the shell-side stream.
2. Guess a value of T'_{eff} (between T'_{in} and T'_{out}).

3. Calculate $\overline{T}$ from Eq. (11.38) and plot this as a horizontal line on the operating diagram.
4. Determine θ_m by the methods already described for a counterflow exchanger, that is, divide the diagram into linear zones, determine θ_{LM} for each zone, and combine these using Eq. (11.21).
5. Recalculate T'_{eff} from Eq. (11.37) using the previously calculated θ_m and $\overline{T}$.
6. Repeat the calculation from step 3 and continue the process until convergence is obtained. This usually takes two to three iterations.

The procedure described here will give reasonable results also for a multipass *J* shell.

11.9.5 Crossflow Condensers

Let us first consider a single-pass condenser as illustrated in Fig. 11.26. A heat balance over area *dA* of this condenser gives

$$dQ = U\theta \, dA \tag{11.39}$$

For single-phase coolant, and because the shell-side temperature is constant in area dA, θ is given as

$$\theta = \frac{T_x - T_{in}}{\ln \dfrac{T' - T_{in}}{T' - T_x}} \tag{11.40}$$

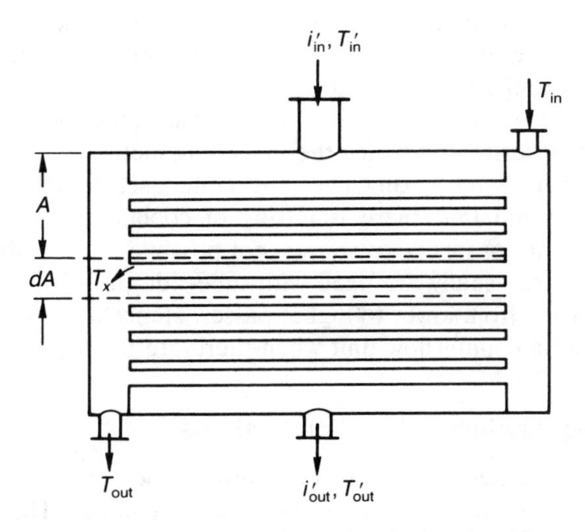

Fig. 11.26. Crossflow condenser with single tube-side pass.

where T_x is the temperature at exit to the tubes in question. Note that this is not the same on the coolant outlet temperature to the exchanger which is obtained after mixing the coolant from each tube. A heat balance over dN tubes sitting in area dA gives

$$G_c c_{pc} \frac{\pi D_i^2}{4} (T_x - T_{in})\, dN = \pi D_o L U \theta\, dN \tag{11.41}$$

where G_c is the coolant mass flux, c_{pc} is the coolant specific heat, D_i and D_o are the tube inside and outside diameters, respectively, and L is the tube length. This equation can be simplified to

$$T_x - T_{in} = \theta U B \tag{11.42}$$

where

$$B = \frac{4 D_o L}{G_c c_{pc} D_i^2} \tag{11.43}$$

Combining Eqs. (11.40) and (11.41) gives (after some manipulation)

$$\theta = (T' - T_{in}) \frac{1 - e^{-UB}}{UB} \tag{11.44}$$

Substituting this into Eq. (11.39) gives

$$\frac{dQ}{dA} = \frac{1}{B}(T' - T_{in})(1 - e^{-UB}) \tag{11.45}$$

The heat transfer area, or the heat load for a given area, can thus be determined by integrating Eq. (11.45). It must be borne in mind when doing this that both U and T' vary with Q.

A crossflow condenser with two tube passes may be treated as two crossflow units in series and hence the preceding method is used again. This assumes, however, good mixing over the condenser length (which is not always the case) and that there is mixing of coolant between the passes. Mixing between the passes is, of course, not possible in a U-tube condenser. Multipass units are usually arranged with tube-side passes layered and with the coolant flowing from lower to higher passes. Hence a crossflow exchanger will approximate a counterflow unit when there are many passes.

11.9.6 Nonequilibrium Calculation Methods

The methods presented so far are equilibrium ones in so far as the local steam temperatures are taken as their equilibrium values. The overall coefficient is then calculated in a way which is consistent with this. However, it becomes very difficult to estimate a suitable overall coefficient in some situations: particularly when condensing mixtures of fluids with large relative volatilities. In addition, the equilibrium temperature method hides the actual liquid- and gas-phase temperatures which may be crucial parts of the design. For example, the method will not tell one whether the gas phase is superheated or supersaturated. If the latter, there is danger of fog formation. The methods given so far, therefore, have severe limitations. Removing these limitations, unfortunately, makes the analysis much more complicated.

In order to illustrate some of the complexities of nonequilibrium methods, and how to deal with these, we will take the case of condensing vapor in the presence of a noncondensable gas. The equations for cooling the gas phase and removing vapor from the gas phase are, respectively,

$$\frac{dT_g}{dA} = -\frac{h_g(T_g - T_I)}{W_g c_{pg}}\left(\frac{a}{e^a - 1}\right) \tag{11.46}$$

$$\frac{dW_v}{dA} = -G_v \tag{11.47}$$

where T_g is the gas-phase temperature, T_I is the gas–liquid interface temperature, h_g is the gas-phase heat transfer coefficient (not corrected for mass transfer), W_g is the gas-phase mass flow rate (vapor plus noncondensables), W_v is the vapor mass flow rate, G_v is the mass flux of condensing vapor toward the interface, and $a = G_v c_{pv}/h_g$ (where c_{pv} is the vapor specific heat).

Let us assume for the moment that the coolant temperature, T_c, is constant and that the condensing stream follows a single path (or identical parallel paths). The full thermal analysis therefore consists of integrating Eqs. (11.46) and (11.47) along the condensing path. In doing this, it is necessary at each integration step to solve iteratively nonlinear equations in order to obtain values T_I, G_v, and a to use in the preceding equations.

If now we let the coolant temperature vary, we have to integrate a further equation to obtain this temperature as the integration proceeds. This integration is straightforward if the coolant and condensing streams follow parallel paths which are either co-current or countercurrent. The appropriate equation is then

$$\frac{dT_c}{dA} = \pm\frac{h_{cI}(T_I - T_c)}{W_c c_{pc}} \tag{11.48}$$

where W_c is the coolant mass flow rate and h_{cI} is the heat transfer coefficient between the interface and coolant. The positive sign is for co-current flow and the negative sign for countercurrent flow.

The integration of these equations, while not trivial, is certainly a feasible design approach given computers with standard library subroutines for integrating equations. A slight additional complication arises with countercurrent flow because the coolant temperature may not be known at the start of the integration (because it is the coolant outlet temperature). However, in such cases, the outlet temperature may be guessed and the inlet temperature then determined by integration. If the calculated coolant inlet temperature is different from its known value, a new guess has to be made for the inlet temperature.

So far we have only outlined the nonequilibrium calculation for a relatively trivial case. A slightly more complicated case is that of a TEMA *E* shell with shell-side condensation and with two equal tube-side passes. If we assume good radial mixing on the shell side, Eq. (11.46) now becomes

$$\frac{dT_g}{dA} = -\frac{1}{2}\left[\frac{h_g(T_g - T_I)}{W_g c_{pg}}\left(\frac{a}{e^a - 1}\right)\right]^{\mathrm{I}} - \frac{1}{2}\left[\frac{h_g(T_g - T_I)}{W_g c_{pg}}\left(\frac{a}{e^a - 1}\right)\right]^{\mathrm{II}} \tag{11.49}$$

where the superscripts I and II refer to conditions pertaining to the tubes in passes I and II, respectively. We also have to integrate separate equations for the pass I and II coolant temperatures:

$$\frac{dT_c^{\mathrm{I}}}{dA} = \left[\frac{h_{cI}(T_I - T_c)}{W_c c_{pc}}\right]^{\mathrm{I}} \tag{11.50}$$

$$\frac{dT_c^{\mathrm{II}}}{dA} = -\left[\frac{h_{cI}(T_I - T_c)}{W_c c_{pc}}\right]^{\mathrm{II}} \tag{11.51}$$

In setting out the signs of the right-hand sides of Eqs. (11.49) to (11.51), it has been assumed that the integration is proceeding from the front-end header and that the shell-side inlet nozzle is at the front end. In addition to these integrations, some iteration is normally required to match the calculated tube-side temperatures, T_c^{I} and T_c^{II}, at the rear-end header.

Clearly, the calculations just outlined for a two-pass exchanger become even more involved when one goes to more passes or if some of the tubes are submerged in condensate. Nevertheless, such calculations can be handled economically with good computer programs.

The paper by Webb et al. [22] gives a useful discussion of the advantages of the nonequilibrium method when applied to the design of condensers.

11.9.7 Multidimensional Shell-Side Flows

The calculation method given in the previous sections, although quite complex, still contains a major simplifying assumption. This is that the flow paths

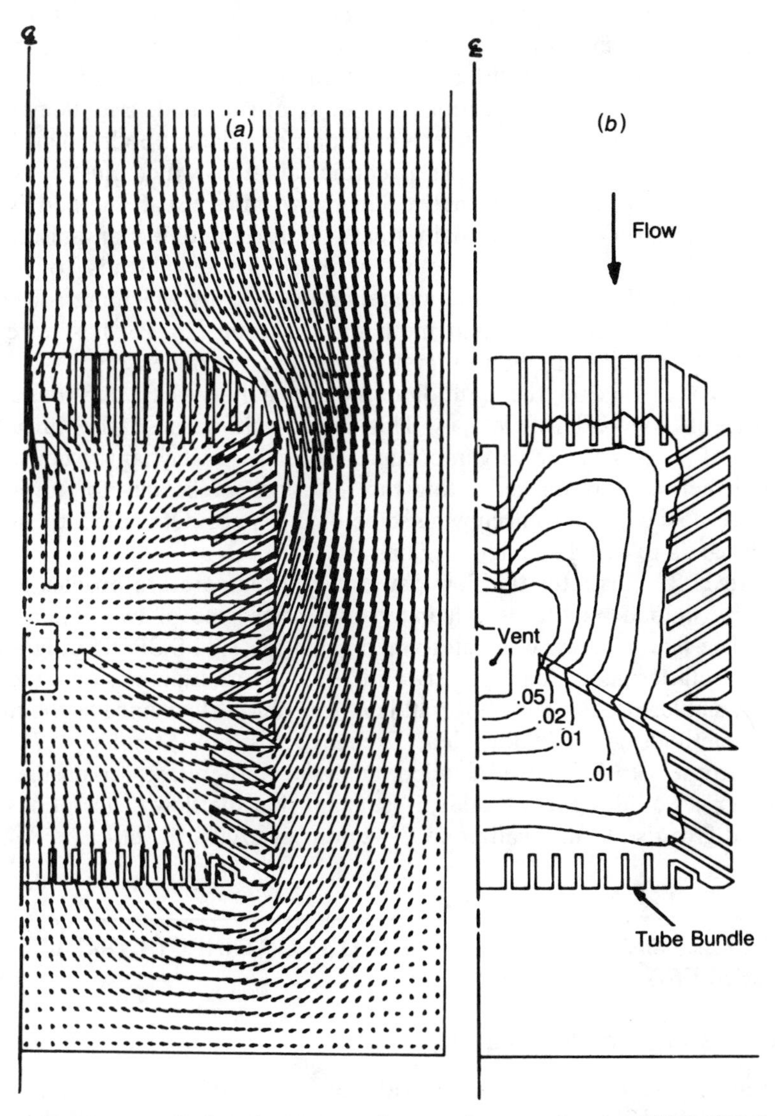

Fig. 11.27. Flow predictions in a power plant condenser using the HTFS SPOC code: (*a*) illustrates the velocity distribution and (*b*) the corresponding air concentration contours.

are one dimensional. In some situations, though, particularly that of steam power condensers, the multidimensional nature of the shell-side flow is a key feature of the design. It is beyond the scope of this chapter to deal with this problem in depth. An outline only is therefore given, together with references to papers where further information of relevance to condenser design is given.

The first step in multidimensional analyses is to determine the flow pattern. This has been attempted by two methods. The first is a subchannel method [23] in which a flow network is set up with the nodes in that network being the regions between three adjacent tubes (in an equilateral triangle tube layout). Mass balance equations are set up at each node and momentum balance equations are set up between each pair of adjacent nodes. This leads to a very large number of nonlinear algebraic equations which must be solved for the flow pattern. The alternative, and now more usual, method is the continuum method which essentially treats the rod bundle as an anisotropic porous medium with flow-dependent permeability. This leads to partial differential equations of mass and momentum continuity which are solved by finite-difference or finite-element methods. Davidson [24, 25] has reviewed the application of this method to large power station condensers.

The second stage in the multidimensional analysis is to determine the transport of noncondensable gas. This is difficult because the turbulent diffusion of the gas can be very important, especially in stagnant regions. The problem then centers on estimating the diffusion coefficients for turbulent two-phase flow with tubes occupying some of the space. Of course, the art of good design is to avoid stagnant regions.

Figure 11.27 illustrates the flow patterns and temperature profiles for an irregularly shaped bundle as calculated using this type of two-dimensional model. The figure shown was determined using the HTFS SPOC program.

For a well-designed steam power condenser, the multidimensional models give very similar predictions of overall heat transfer performance as the much simpler methods given in the Heat Exchange Institute Standards [7]. The main purpose of the more detailed analysis is therefore to ensure the condenser is well designed. By this, it is meant that the flow distributions are good and give rise to no dead areas of regions of excessive velocity.

11.10 THERMAL EVALUATION METHOD FOR DIRECT-CONTACT CONDENSERS

11.10.1 Spray Condensers

These are rather difficult to design with any degree of precision because of the uncertainties in droplet size, droplet trajectory, droplet coalescence, and flow patterns in the gas phase. However, the simplicity of the basic geometry means that they can be oversized without much additional cost.

A preliminary step in spray condenser design is the selection of nozzle type and the determination of the nozzle behavior for the liquid being used. Steinmeyer [26] describes various types of spray nozzles and discusses their method of operation, advantages, and disadvantages. Spray condensers would usually operate with the simplest type of nozzle in which the source of atomization is the pressure loss in the nozzle.

Nozzle types vary so widely from one manufacturer to another that it is necessary to use the manufacturers' data to determine such quantities as the pressure drop in the nozzle, the mean droplet size, and the initial droplet velocity. Unfortunately, the data given by manufacturers are usually limited to water at around 20°C spraying into ambient air. Often, however, there is enough information given to calculate the pressure drop and, from this, the inlet velocity, u_1, may be determined as

$$u_1 = K\left(\frac{2\,\Delta p}{\rho_l}\right)^{1/2} \tag{11.52}$$

where Δp is the nozzle pressure drop and ρ_l is the liquid density. K is a coefficient which would be 1 if there were no energy losses in the nozzle. A reasonable value of K for estimation purposes is 0.8. The mean droplet diameter for water can be determined from manufacturers' data, and Steinmeyer [26] suggests an approximate equation for correcting this for other liquids:

$$\frac{d}{d_{\mathrm{wat}}} = \left(\frac{\sigma}{\sigma_{\mathrm{wat}}}\right)^{0.5}\left(\frac{\mu_l}{\mu_{\mathrm{wat}}}\right)^{0.2}\left(\frac{\rho_{\mathrm{wat}}}{\rho_l}\right)^{0.3} \tag{11.53}$$

where d is the droplet diameter, d_{wat} is the diameter for water sprayed with the same volumetric flow through the nozzle, μ_l is the liquid viscosity, μ_{wat} is the viscosity of water at 20°C, ρ_l is the density of the liquid, and ρ_{wat} is the density of water at 20°C. In reality, the dependence of droplet size on fluid properties is very complex, and hence the preceding equation should be used with caution.

The problem of subcooled droplets injected into saturated vapor has been analyzed by Brown [27]. He treated the droplets as solid spheres and solved the transient conduction equation in order to give the temperature rise in the droplet as a function of exposure time to vapor. The droplet diameter is assumed to be independent of time, which is reasonable since very little vapor can condense before the droplet is heated up to, or close to, saturation.

The droplet temperature rise is given by Brown as

$$\frac{T_{\mathrm{out}} - T_{\mathrm{in}}}{T_{\mathrm{sat}} - T_{\mathrm{in}}} = 1 - \frac{6}{\pi^2}\sum_{n=1}^{\infty}\frac{1}{n^2}\exp\left(-\frac{4n^2\pi^2\alpha_l t_c}{d^2}\right) \tag{11.54}$$

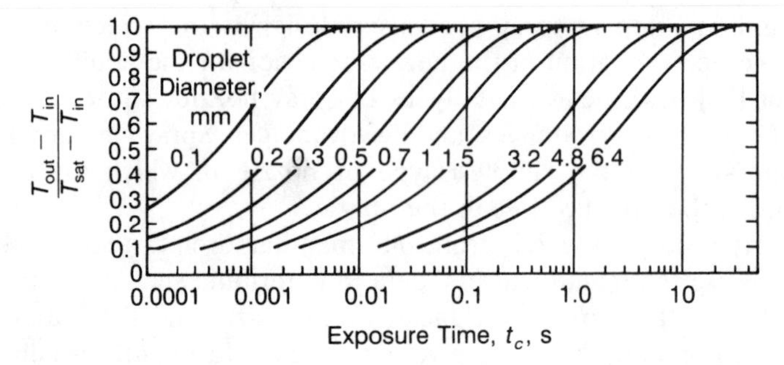

Fig. 11.28. Temperature rise as a function of time for water droplets in saturated steam (Brown [27]).

where T_{out} is the mean outlet temperature, T_{in} is the inlet temperature, T_{sat} is the saturation temperature, α_l is the liquid thermal diffusivity, and t_c is the contact time. Figure 11.28 shows the results of this equation plotted for low-pressure water. These results may be used to determine an effective mean coefficient $\bar{h}$ for a given temperature rise. The results of such a calculation are given in Fig. 11.29. This figure may be used in conjunction with the following equation in order to determine the desired contact time, t_c:

$$t_c = \frac{d\rho_l c_{pl}}{6\bar{h}} \ln\left(1 - \frac{T_{out} - T_{in}}{T_{sat} - T_{in}}\right) \tag{11.55}$$

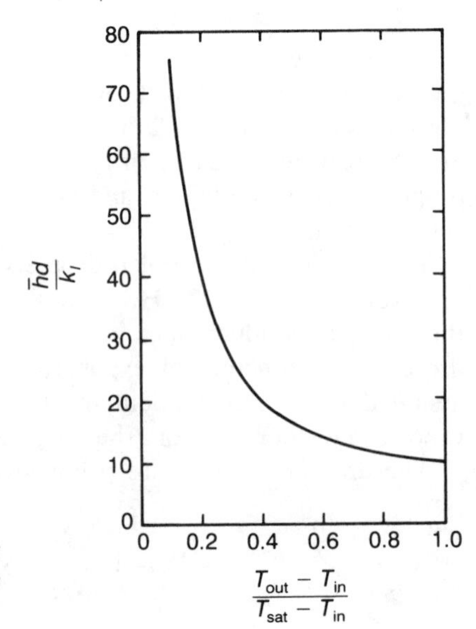

Fig. 11.29. Effective mean heat transfer coefficient for droplets in pure vapor (Brown [27]).

The coefficient $\bar{h}$ can be very large and, indeed, comparable with the interfacial or molecular-kinetic coefficient. This is particularly so when, as is often the case, the condenser is operating at high vacuum. In such circumstances, Eq. (11.55) should be evaluated with $\bar{h}$ replaced by h_{eff} which combines $\bar{h}$ and the interfacial coefficient, h_I, as follows:

$$\frac{1}{h_{eff}} = \frac{1}{\bar{h}} + \frac{1}{h_I} \tag{11.56}$$

Having determined the contact time t_c, it is necessary to estimate how far the droplets will travel in this time, thus enabling one to estimate the vessel size. A force balance on the droplets for vertical downward motion yields

$$Mu\frac{du}{dz} = M\frac{du}{dt} = Mg - C_D\frac{\pi d^2}{4}\frac{\rho_g u^2}{2} \tag{11.57}$$

where u is the droplet velocity, M is the droplet mass, z is the distance, t is the time, g is the gravitational acceleration, C_D is the drag coefficient, and ρ_g is the gas-phase density. For horizontal flow, the same equation applies but g is 0. Also, g is often small compared with the other terms for high-velocity droplets. Pita and John [28] have integrated this equation analytically for the case when g is 0. In order to do this, they used the Ingebo [29] equation for C_D which applies for $6 < Re < 400$:

$$C_D = 27Re^{-0.84} \tag{11.58}$$

where

$$Re = \frac{\rho_g u d}{\mu_g} \tag{11.59}$$

and where μ_g is the gas-phase viscosity. Peta and John obtained the total distance traveled, L, in time t_c, as

$$L = 0.06\frac{d^{1.84}}{\Gamma}\left(u_1^{0.84} - u_2^{0.84}\right) \tag{11.60}$$

where u_2 is the velocity at time t_c given by

$$u_2 = \left(u_1^{-0.16} - 3.23\frac{\Gamma t_c}{d^{1.84}}\right)^{-6.25} \tag{11.61}$$

and where Γ is a physical property grouping given by

$$\Gamma = \frac{\rho_g}{\rho_l} \nu_g^{0.84} \tag{11.62}$$

As an alternative to the preceding calculation method, Fair [30] gives empirical calculation procedures based on the volumetric heat transfer coefficient.

The estimation of the effects of noncondensables is quite difficult and involves step-by-step calculations of the type described by Peta and John for droplet evaporation.

11.10.2 Tray Condensers

The uncertainties involved in the design of spray condensers become even more severe with tray condensers. It is possible to use the preceding heat transfer calculation method while using a large droplet diameter (say greater than 5 mm) but this can lead to error if the sheet does not break up into droplets before falling onto the next tray. The vertical distance traveled by the droplets can be determined by assuming that they fall freely under gravity. Alternatively, empirical calculation methods have been devised by Fair [30] which use volumetric coefficients.

11.11 REASONS FOR FAILURE OF CONDENSER OPERATION

Steinmeyer and Mueller [31] chaired a panel discussion session on why condensers do not operate as they are supposed to. Some of the main points arising from this discussion are noted here:

1. The tubes may be fouled more than expected—a problem not unique to condensers.
2. The condensate may not be drained properly causing tubes to be flooded. This could mean that the condensate outlet is too small, too high, or blocked.
3. Venting of noncondensables may be inadequate. Remarks on the proper arrangement of vents were given in Section 11.2.1.
4. The condenser was designed on the basis of end temperatures without noticing that the design duty would involve a temperature cross in the middle of the range (see Section 11.9).
5. Flooding limits have been exceeded for condensers with backflow of liquid against upward vapor flow.
6. Excessive fogging may be occurring. This can be a problem when condensing high molecular weight vapors in the presence of noncondensable gas.

An additional problem not mentioned in this panel discussion is the possibility of severe maldistribution in parallel condensing paths particularly with vacuum operation. This occurs because there can be two flow rates which satisfy the imposed pressure drops. An example might be that in which the pressure drop is 0. One channel may have a very high vapor inlet flow but achieve a zero pressure drop because the momentum pressure recovery cancels out the frictional pressure loss. The next channel may achieve the zero pressure drop by having no vapor inlet flow. To be stable in this case, the channel would be full of noncondensables. This problem may occur with parallel tubes in a tube-side condenser or with whole condensers when they are arranged in parallel.

11.12 EXAMPLES

11.12.1 Process Condenser

A steam–hydrocarbon–air mixture, whose condensation curve is given in Fig. 11.30, has to be condensed in a TEMA *E*-type condenser with cooling water available at 25°C. The process data are as follows:

$$Q_T = 1820\ k\text{W}$$

$$T_{\text{in}} = 25°\text{C} \qquad T'_{\text{in}} = 172°\text{C}$$

$$T_{\text{out}} = 55°\text{C} \qquad T'_{\text{out}} = 54°\text{C}$$

$$W = 14.51\ \text{kg/s} \qquad W' = 2.72\ \text{kg/s}$$

The shell-side enthalpy curve has been approximated by a number of straight lines for the purposes of this calculation. The differences between these straight lines and the detailed curves are small.

The tube-side enthalpy can be determined from the tube-side specific heat c_p as follows:

$$i = c_p T$$

which, if $c_p = 4.18$ kJ/(kg · K), gives

$$i = 4.18 \times T\ \text{kJ/kg} \tag{11.63a}$$

$$T = i/4.18°\text{C} \tag{11.63b}$$

We will consider the shell-side flow being co-current to the first tube-side pass. Hence Eq. (11.26) is used with

$$\frac{W}{W'} = \frac{14.51}{2.72} = 5.34$$

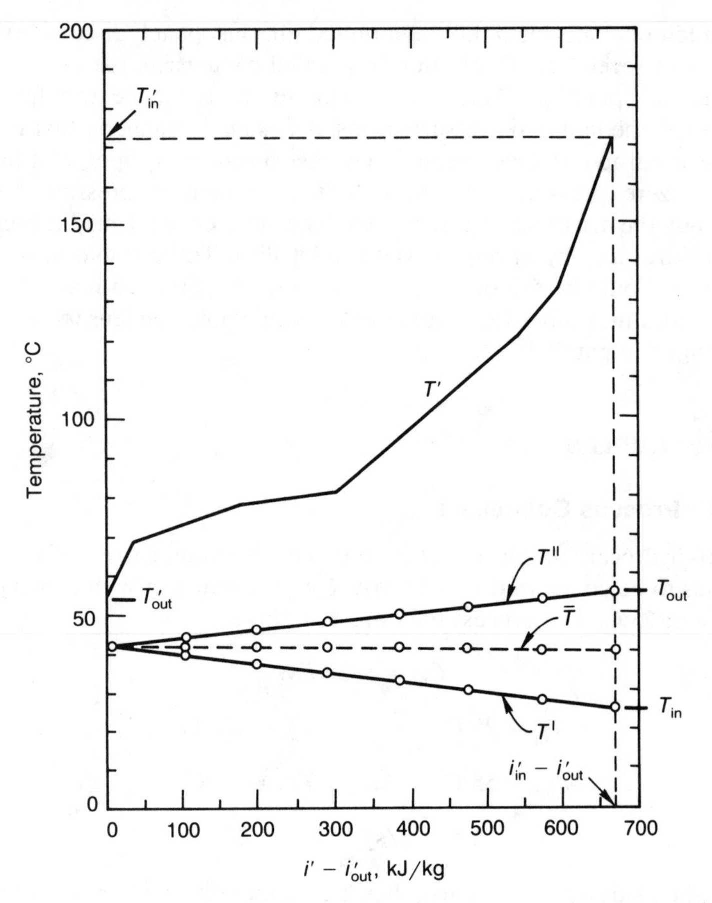

Fig. 11.30. Temperature distributions for the process condenser example.

giving

$$i' - i_{out} = 5.34\,(i^{II} - i^{I}) \tag{11.64}$$

We will choose steps δi^{I} of 10 kJ/kg. Hence putting $\delta i^{I} = 10$ in Eq. (11.32) gives

$$i^{II}_{new} = i^{II} - 10\frac{T' - T^{II}}{T' - T^{I}}\ \text{kJ/kg} \tag{11.65}$$

Note that new i^{I} values for each calculation step are determined as follows:

$$i^{I}_{new} = i^{I} + 10 \tag{11.66}$$

With this information, the calculations shown in Table 11.2 may proceed by going left to right on each line and hence from line to line. The calculation is started by

$$i^{\mathrm{I}} = i_{\mathrm{in}} = 4.18 \times 25 = 104.5 \text{ kJ/kg}$$

when

$$i^{\mathrm{II}} = i_{\mathrm{out}} = 4.18 \times 55 = 229.9 \text{ kJ/kg}$$

The method for calculating each item in the table is indicated above the appropriate column. The calculation is stopped when $i' - i'_{\mathrm{out}}$ becomes 0 or where $i^{\mathrm{I}} = i^{\mathrm{II}}$.

The results of this calculation are shown plotted in Fig. 11.30. The calculation for this example could have been simplified considerably since a linear temperature–enthalpy curve is used on the tube side. Equations (11.64) to (11.66) could have been written directly in terms of tube-side temperature, thus avoiding the step of converting enthalpies to temperatures. However, the calculation was done in full here to show how we would set about dealing with nonlinear, temperature–enthalpy curves on the tube side.

To calculate the heat transfer area, we must be able to calculate local values of the overall heat transfer coefficient, U. Since this is discussed elsewhere is this book, the coefficients are taken here as those given in Fig. 11.31.

Equation (11.19) can then be used to determine the heat transfer area. The equation was applied to give A_j for each of the straight-line sections of the shell-side, temperature–enthalpy curve given in Fig. 11.30. The required $T' - \bar{T}$ values were read off from this figure and the corresponding U values from Fig. 11.31. The steps in the calculation are shown in Table 11.3.

If required, the mean temperature difference and the mean overall coefficient for the whole exchanger may be calculated using Eqs. (11.21) and (11.22), respectively. These calculations are summarized in Table 11.4. Hence the mean temperature difference is determined from Eq. (11.21) as

$$\frac{1}{\theta_m} = \frac{1}{669} \times 15.08$$

$$= 0.0225$$

Hence

$$\theta_m = 44.4°\text{C}$$

TABLE 11.2 Worksheet for Temperature Distribution Calculation

Step	i^{I}, kJ/kg	i^{II}, kJ/kg	$i' - i'_{out}$, kJ/kg	T^{I}, °C	T^{II}, °C	T', °C	$T' - T^{I}$, °C	$T' - T^{II}$, °C	$\frac{T' - T^{I}}{T' - T^{I}}$,	$\frac{T^{I} + T^{II}}{2}$, °C
	Eq. (11.66)	Eq. (11.65)	Eq. (11.64)	Eq. (11.63)	Eq. (11.63)	Fig. 11.30				
1	104.5	229.9	669	25.0	55.0	172.0	117.0	147.0	0.796	40.0
2	114.5	221.9	573.5	27.4	53.1	128.6	101.2	75.5	0.746	40.2
3	124.5	213.6	475.8	29.8	51.1	109.9	80.1	58.8	0.734	40.4
4	134.5	206.3	383.2	32.2	49.3	94.5	62.3	45.1	0.724	40.7
5	144.5	199.1	291.4	34.6	47.6	80.7	46.1	33.1	0.718	41.1
6	154.5	191.9	199.8	37.0	45.9	78.3	41.3	32.4	0.784	41.4
7	164.5	184.1	104.7	39.4	44.0	73.2	33.8	29.2	0.864	41.7
8	174.5	175.5	5.3	41.7	41.9					41.8

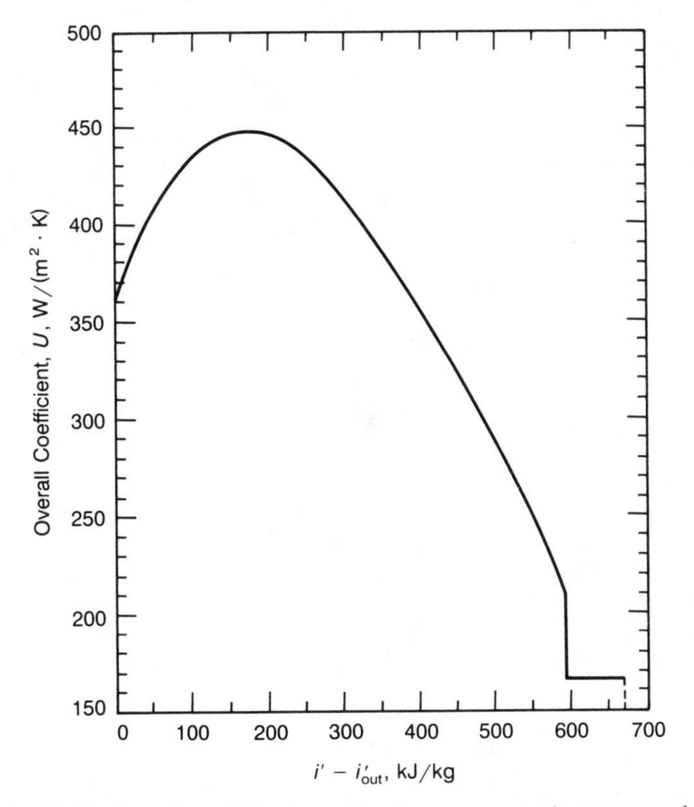

Fig. 11.31. Overall coefficients used in the process condenser example.

The mean overall coefficient is determined from Eq. (11.22):

$$U_m = \frac{1}{115.5} \times 41{,}060 = 355 \text{ W/(m}^2 \cdot \text{K)}$$

We can cross-check the arithmetic by using Eq. (11.5):

$$\begin{aligned} Q_T &= U_m A_T \theta_m \\ &= 355 \times 155.5 \times 44.4 \\ &= 1.82 \times 10^6 \text{ W (1.82 MW)} \end{aligned}$$

The heat duty is also given by

$$\begin{aligned} Q_T &= W'(i'_{\text{in}} - i'_{\text{out}}) \\ &= 2.72(669 \times 10^3) \\ &= 1.82 \times 10^6 \text{ W (1.82 MW)} \end{aligned}$$

TABLE 11.3 Worksheet for Heat-Transfer Area Calculation

$i'_a - i_{out}$, kJ/kg	$\Delta i'_j$, kJ/kg	U_a, W/(m$^2 \cdot$ K)	U_b, W/(m$^2 \cdot$ K)	$U_{m,j}$, W/(m$^2 \cdot$ K), [Eq. (11.14)]	θ_a, °C	θ_b, °C	$\theta_{LM,j}$, °C [Eq. (11.13)]	A_j, m^2 [Eq. (11.19)]
0	37	360	397	378	12.2	27.0	18.8	14.2
37	139	397	447	422	27.0	36.8	31.6	28.4
176	126	447	413	430	36.8	40.0	38.4	20.8
302	238	413	253	333	40.0	80.0	57.7	33.7
540	55	253	209	231	80.0	92.5	86.1	7.5
595	74	166	166	166	92.5	132.0	111.1	10.9
							Total A_T =	115.5

TABLE 11.4 Worksheet for Calculation of Mean Temperature Difference and Mean Overall Coefficient

$i'_a - i'_{out}$, kJ/kg	$\Delta i'_j$, kJ/kg	$U_{m,j}$, W/(m² · K)	$\theta_{LM,j}$, °C	A_j, m²	$A_j U_{m,j}$, W/K	$\Delta i'_j / \theta_{LM,j}$, kJ/(kg · K)
0	37	378	18.8	14.2	5,368	1.968
37	139	422	31.6	28.4	11,985	4.398
176	126	430	38.4	20.8	8,944	3.281
302	238	333	57.7	33.7	11,222	4.125
540	55	231	86.1	7.5	1,732	0.639
595	74	166	111.1	10.9	1,809	0.666
				Totals	41,060	15.08

11.12.2 Power Condenser

Wet steam at 0.1 bar with (mixture) specific enthalpy 2426.3 kJ/kg enters the condenser at a flow rate, W, of 245.34 kg/s. It is designed to condense the vapor without subcooling. A single tube-side pass is used and the cooling-water velocity is selected as 2 m/s as a good compromise between fouling and erosion. Cooling water is available at 15°C and can exit the condenser at 25°C.

The condenser tube details are as follows:

Outside diameter, $D_o = 0.0254$ m
Inside diameter, $d_i = 0.02291$ m (i.e., 18 BWG)
Wall thermal conductivity, $k_2 = 111$ W/(m · K) (admirality metal)

The fouling resistances inside and outside the tube are, respectively,

$$r_i = 0.00018 \ (\mathrm{m}^2 \cdot \mathrm{K})/\mathrm{W}$$

$$r_o = 0.00009 \ (\mathrm{m}^2 \cdot \mathrm{K})/\mathrm{W}$$

The required fluid properties may be obtained from steam tables as follows:

Cooling water (at the mean temperature of 20°C)

$$\rho_c = 997 \ \mathrm{kg/m^3}$$

$$c_{pc} = 4180 \ \mathrm{J/(kg \cdot K)}$$

$$\mu_c = 0.00101 \ (\mathrm{N \cdot s})/\mathrm{m}^2$$

$$k_c = 0.602 \ \mathrm{W/(m \cdot K)}$$

$$Pr_c = 6.96$$

Saturated liquid properties of condensate (at 0.1 bar)

$$T_{sat} = 45.8°\text{C}$$

$$\rho_l = 990\ \text{kg/m}^2$$

$$i_{lg} = 2392\ \text{kJ/kg}$$

$$k_l = 0.635\ \text{W/(m} \cdot \text{K)}$$

$$\mu_l = 5.88 \times 10^{-4}(\text{N} \cdot \text{s})/\text{m}^2$$

$$i_l = 191.8\ \text{kJ/kg}$$

The condenser heat load, Q_T, is calculated from

$$Q_T = W(i_{in} - i_l)$$

$$= 245.34(2426.3 - 191.8)$$

$$= 5.412 \times 10^5\ \text{kW (548.2 MW)}$$

Hence the cooling water mass flow rate, W_c, can be determined from a heat balance

$$W_c = \frac{Q_T}{(T_{c,out} - T_{c,in})c_{pc}}$$

$$= \frac{(5.482 \times 10^5) \times 10^3}{(25 - 15)4180}$$

$$= 1.311 \times 10^4\ \text{kg/s}$$

The number of tubes, N_T, is then determined from the fixed cooling-water velocity, u_c, as follows:

$$W_c = u_c\rho_c\left(\frac{\pi D_i^2}{4}\right)N_T$$

or

$$N_T = \frac{4W_c}{\rho_c u_c \pi D_i^2}$$

$$= \frac{4 \times 1.311 \times 10^4}{997 \times 2 \times \pi \times (0.02291)^2}$$

$$= 15{,}950$$

In order to calculate the condensing-side heat transfer coefficient, we need an estimate of the number of tubes in a vertical column. From typical condenser tube layouts, this was estimated as 70.

The coolant-side heat transfer coefficient, h_c, can be estimated by single-phase, in-tube heat transfer coefficient methods (see Chapter 3). The tube-side Reynolds number is first calculated:

$$Re = \frac{D_i u_c \rho_c}{\mu_c}$$

$$= \frac{0.02291 \times 2 \times 997}{0.00101}$$

$$= 45{,}230$$

The Petukhov–Kirillov correlation can then be used to determine the heat transfer coefficient (see Chapter 3):

$$Nu = \frac{(f/2) Re Pr}{1.07 + 12.7(f/2)^{1/2}(Pr^{2/3} - 1)}$$

where

$$f = (1.58 \ln Re - 3.28)^{-2}$$

$$= [1.58 \ln(45{,}230) - 3.28)^{-2}$$

$$= 0.00536$$

Hence

$$\frac{f}{2} = 0.00268$$

and

$$Nu = \frac{0.00268 \times 45230 \times 6.96}{1.07 + 12.7(0.00268)^{1/2}\left[(6.98)^{2/3} - 1\right]}$$

$$= 300.4$$

Hence

$$h_c = \frac{Nu k_c}{D_i}$$

$$= \frac{300.4 \times 0.602}{0.02291}$$

$$= 7890\ \mathrm{W/(m^2 \cdot K)}$$

The mean temperature difference for the exchanger may be taken as the logarithmic mean because the temperature difference between the streams varies linearly with the amount of heat transferred to the tube-side fluid.

$$\theta_{in} = (45.8 - 15)$$
$$= 30.8°C$$

and

$$\theta_{out} = (45.8 - 25)$$
$$= 20.8°C$$

Hence

$$\theta_{LM} = \frac{\theta_{in} - \theta_{out}}{\ln(\theta_{in}/\theta_{out})}$$
$$= \frac{30.8 - 20.8}{\ln(30.8/20.8)}$$
$$= 25.5°C$$

The next step in the calculation is to determine the shell-side heat transfer coefficient in order to determine the overall coefficient. Unfortunately, this coefficient depends on the local heat flux and hence an iteration is necessary. The equations required in this iteration are developed first.

The overall heat transfer coefficient, U, based on the tube outside diameter, is given by

$$\frac{1}{U} = R + \frac{1}{h_o}$$

where h_o is the coefficient outside the tubes and R is the sum of all the other thermal resistances given by

$$R = r_o + \left(\frac{1}{h_c} + r_i\right)\frac{D_o}{D_i} + \frac{s_w}{k_w}\frac{D_o}{D_w}$$

where

$$D_W = \frac{D_o - D_i}{\ln(D_o/D_i)} \simeq \frac{1}{2}(D_o + D_i)$$
$$= \frac{1}{2}(0.0254 + 0.0229) = 0.0242 \text{ m}$$

The wall thickness, s_w, is given by

$$s_w = \tfrac{1}{2}(D_o - D_i)$$

$$= \tfrac{1}{2}(0.0254 - 0.0229)$$

$$= 0.0013 \text{ m}$$

Hence

$$R = 0.00009 + \left(\frac{1}{7890} + 0.00018\right)\frac{0.0254}{0.0229} + \left(\frac{0.0013}{111}\right)\left(\frac{0.0254}{0.0242}\right)$$

$$= 4.42 \times 10^{-4}$$

Hence

$$\frac{1}{U} = 4.42 \times 10^{-4} + \frac{1}{h_o} \tag{11.67}$$

The condensing-side heat transfer coefficient may be calculated by the Nusselt method with the Kern correction for condensate inundation (see Chapter 10). Hence

$$h_o = 0.728\left\{\frac{\rho_l^2 g i_{lg} k_l^3}{\mu_l \,\Delta T_w\, D_o}\right\}\frac{1}{N^{1/6}}$$

where ΔT_w is the difference between the saturation temperature and the temperature at the surface of the fouling. This equation has been simplified because $\rho_l \gg \rho_g$. Hence

$$h_o = 0.728\left\{\frac{(990)^2(9.81)(2392 \times 10^3)(0.635)^3}{(5.88 \times 10^{-4})\,\Delta T_w(0.0254)}\right\}^{1/4}\frac{1}{(70)^{1/6}}$$

$$= \frac{8990}{\Delta T_w^{1/4}} \tag{11.68}$$

Now, the temperature difference, ΔT_w, is given by

$$\Delta T_w = \theta - Rq$$

where q is the heat flux. But

$$q = U\theta$$

TABLE 11.5 Iteration for Overall Coefficient at the Inlet of the Power Condenser

ΔT_w, °C [Eq. (11.69)]	h_o, W/(m^2 · K) [Eq. (11.68)]	U, W/(m^2 · K) [Eq. (11.67)]
10	5055	1563
9.52	5118	1569
9.44	5129	1570
9.43	5130	1570

Hence

$$\Delta T_w = \theta(1 - RU)$$

$$= \theta(1 - 4.42 \times 10^{-4} U) \tag{11.69}$$

A suggested iteration is therefore to

1. Guess ΔT_w.
2. Calculate h_o from Eq. (11.68).
3. Calculate U from Eq. (11.67).
4. Recalculate ΔT_w from Eq. (11.69).
5. Repeat the calculations from step 2 and continue the iteration until U converges.

Table 11.5 summarizes the results of this iteration for the inlet of the condenser when $\theta = 30.8$°C. The initial guess of ΔT_w is 10°C.

The process is repeated for the outlet end of the condenser, where θ is 20.8°C. The overall coefficient obtained is 1626 W/(m^2 · K). The mean overall coefficient can then be determined from Eq. (11.14) as follows:

$$U_m = \tfrac{1}{2}(1570 + 1626)$$

$$= 1598 \text{ W/(m}^2 \cdot \text{K)}$$

The required surface area for the exchanger is therefore given by Eq. (11.5) as follows:

$$A_T = \frac{Q_T}{U_m \theta_{\text{LM}}}$$

$$= \frac{548.2 \times 10^6}{1598 \times 25.5}$$

$$= 1.345 \times 10^4 \text{ m}^2$$

TABLE 11.6 Comparison of Thermal Resistances at the Inlet of the Condenser

Item	Given by	Value, $(m^2 \cdot K)/kW$	%
Tube-side fluid	$\frac{1}{h_i}\frac{D_o}{D_i}$	0.141	22
Tube-side fouling	$r_i\frac{D_o}{D_i}$	0.200	31
Wall	$\frac{s_w}{k_w}\frac{D_o}{D_w}$	0.012	2
Shell-side fouling	r_o	0.090	14
Shell-side fluid	$\frac{1}{h_o}$	0.195	31

Hence the required tube length is determined using

$$A_T = N_T \pi D_o L$$

or

$$L = \frac{A_T}{N_T \pi D_o}$$

$$= \frac{1.345 \times 10^4}{15{,}950 \times \pi \times 0.0254}$$

$$= 10.6 \text{ m}$$

It is instructive to compare the various thermal resistances in this condenser. Table 11.6 does this for the inlet end of the condenser.

It can be seen that considerable resistance is due to the fouling, particularly the tube-side fouling. In practice, therefore, much effort often goes into keeping the tube-side clean. This is done by careful control of the cooling-water chemistry and by use of mechanical cleaning methods such as balls which are passed down the tubes.

ACKNOWLEDGMENT

This chapter is an extended and modified version of a chapter previously published in *Two-Phase Flow Heat Exchangers*, by Kluwer Academic Publishers (1988). Thanks are due the publisher for permission to use the material here.

NOMENCLATURE

a	$G_v c_{pl}/h_g$
A	heat transfer area, m^2
B	parameter defined by Eq. (11.43)
c_p	specific heat capacity at constant pressure, $J/(kg \cdot K)$
C_D	drag coefficient
d	droplet diameter, m
D	tube diameter, m
f	friction factor
g	gravitational acceleration, m/s^2
h	heat transfer coefficient, $W/(m^2 \cdot K)$
i	specific enthalpy, J/kg
k	thermal conductivity, $W/(m \cdot K)$
K	parameter in Eq. (11.52)
L	length, m
G	mass flow per unit area, $kg/(m^2 \cdot s)$
M	mass of a droplet, kg
N	number of tubes
Nu	Nusselt number, hD/k
p	pressure, Pa
Pr	Prandtl number, $c_p\mu/k$
Q	heat flow through exchanger surface, W
r	thermal resistance, $(m^2 \cdot K)/W$
Re	Reynolds number
s	thickness, m
t	time, s
t_c	contact time, s
T	temperature, K
u	velocity, m/s
U	overall heat transfer coefficient, $W/(m^2 \cdot K)$
W	mass flow rate, kg/s
z	distance, m

Greek Symbols

α	thermal diffusivity, m^2/s
Γ	physical property grouping defined by Eq. (11.62), $m^{1.68}/s^{0.84}$
Δi	enthalpy change, J/kg
Δp	pressure drop, Pa
θ	temperature difference between streams, K
μ	viscosity, $(N \cdot s)/m^2$
ν	kinematic viscosity, m^2/s
ρ	density, kg/m^3
σ	surface tension, N/m

Subscripts

a, b zone boundaries
c coolant
cI between coolant and interface
cold cold stream
eff effective value
g gas phase (including vapor or gas–vapor mixtures)
hot hot stream
i inside of tube
in inlet to exchanger
j for jth zone
l liquid phase
LM logarithmic mean
m mean value for exchanger (or for zone in exchanger)
o outside tube
sat saturated
T total value for exchanger
v vapor
wat ambient water
1 initial value
2 final value

Superscripts

I first pass
II second pass
$'$ shell side

REFERENCES

1. TEMA (1988) *Standard of Tubular Exchanger Manufacturers' Association*, 7th ed. Tarrytown, New York.
2. Bell. K. J., and Mueller, A. C. (1971) *Condensation Heat Transfer and Condenser Design*, AIChE Today Series. American Institute of Chemical Engineers, New York.
3. Editors of *Power* (1967) *Power Generation Systems*, pp. 265–296. McGraw-Hill, New York.
4. Simpson, H. C. (1969) Outline of current problems in condenser design. *Proc. Symp. to Celebrate the Bicentenary of the James Watt Patent*, University of Glasgow, September 1–2, pp. 91–134.
5. Sebald, J. F. (1979) A developmental history of steam surface condensers in the electrical utility industry. ASME/AIChE Nat. Heat Transfer Conf., San Diego, August 6–8, 1979; also published in *Heat Transfer Eng*. **1**(3) 80–87; **1**(4) 76–81.

6. British Electrical and Allied Manufacturers' Association (1967) *Recommended Practice for the Design of Surface Type Steam Condensing Plant*, BEAMA Publication 222. London.
7. Heat Exchange Institute (1984) *Standard for Steam Surface Condensers*, 8th ed. Cleveland, Ohio.
8. Coit, R. L. (1984) A designer's approach to surface condenser venting and deaeration. In *Power Condenser Heat Transfer Technology*, P. J. Marto and R. H. Nunn (eds.), pp. 163–180. Hemisphere, Washington, D.C.
9. Spence, J. R., Rydall, M. L., and McConnell, A. (1967–1968) The development and production of high pressure feed heaters for modern central power stations. *Proc. Inst. Mech. Eng.* **182**(36) 735–756.
10. British Electrical and Allied Manufacturers' Association (1968) *Guide to the Design of Feedwater Heating Plant*, BEAMA Publication 226. London.
11. Heat Exchange Institute (1984) *Standards for Closed Feedwater Heaters*. Cleveland, Ohio.
12. Alfa Laval (1969) *Thermal Handbook*. Sweden.
13. HTFS Plate-Fin Study Group (1987) *Plate-Fin Heat Exchangers—Guide to Their Specification and Use*, M. A. Taylor, (ed). HTFS, Oxfordshire, UK.
14. Ludwig, E. E. (1965) *Applied Process Design for Chemical and Petrochemical Plants*, Vol. 3, pp. 146–131. Gulf Publishing, Houston.
15. American Petroleum Institute (1968) *Air Cooled Heat Exchangers for General Refinery Service*, API Standard 661, Washington, D.C.
16. Silver, L. (1947) Gas Cooling with Aqueous Condensation. *Trans. Inst. Chem. Eng.* **25** 30–42.
17. Bell, K. J., and Ghaly, M. A. (1973) An approximate generalized design method for multicomponent/partial condensers. *Amer. Inst. Chem. Eng. Symp. Ser.* **69**(131) 72–79.
18. Smith, R. A. (1976) Private Communication.
19. Colburn, A. P. (1933) Mean temperature difference and heat transfer coefficient in liquid heat exchangers. *Ind. Eng. Chem.* **25** 873–877.
20. Butterworth, D. (1973) A calculation method for shell-and-tube heat exchangers in which the overall coefficient varies along the length. *Conf. on Advances in Thermal and Mechanical Design of Shell-and-Tube Heat Exchangers*, NEL Report 590, pp. 56–71. National Engineering Laboratory, East Kilbride, Scotland.
21. Emerson, W. H. (1973) Effective tube-side temperatures in multi-pass heat exchangers with non-uniform heat-transfer coefficients and specific heats. *Conf. on Advances in Thermal and Mechanical Design of Shell-and-Tube Heat Exchangers*, NEL Report 590, pp. 32–55. National Engineering Laboratory, East Kilbride, Scotland.
22. Webb, D. R., Bird, R., and Mangnall, K. (1988) A new approach to the design of vacuum condensers. Second UK Nat. Conf. on Heat Transfer, University of Strathclyde, Glasgow, Vol. 2, Paper C206/83, pp. 949–968.
23. Wilson, J. L. (1976) NEL two dimensional condenser computer program. *Meeting on Steam Turbine Condensers*, NEL Report 619, pp. 152–159. National Engineering Laboratory, East Kilbride, Scotland.

24. Davidson, B. J. (1976) Computational methods for evaluating the performance of condensers. *Meeting on Steam Turbine Condensers*, NEL Report 619, pp. 152–159. National Engineering Laboratory, East Kilbride, Scotland.

25. Davidson, B. J. (1981) Simulation of power plant condenser performance by computational methods: an overview. In *Power Condenser Heat Transfer Technology*, P. J. Marta and R. H. Nunn (eds.), pp. 17–49. Hemisphere, Washington, D.C.

26. Steinmeyer, D. E. (1973) Phase dispersions: liquid in gas dispersions. *Chemical Engineers Handbook*, 5th ed. R. H. Perry and C. H. Chilton (eds.), p. **18**–62. McGraw-Hill, New York.

27. Brown, G. (1951) Heat transmission by condensation of steam on spray of water drops. *Inst. Mech. Eng. Proc. General Discussion on Heat Transfer* 49–52.

28. Pita, E. G., and John, J. E. A. (1970) The effect of forced convection on evaporative cooling of sprays in air. *Proc. Fourth Int. Heat Transfer Conf. Versailles*, Vol. 7, Paper CT3.12. Elsevier, Amsterdam.

29. Ingebo, R. D. (1951) Vaporization rates and heat transfer coefficients for pure liquid drops. NACA, TN2368.

30. Fair, J. R. (1972) Design of direct contact coolers/condensers. *Chem. Eng.* **79**, 91–100.

31. Steinmeyer, D. E., and Mueller, A. C. (1974) Why Condensers Don't Operate as They Are Supposed To. *Chem. Eng. Progress* **70**(7), 78–82.

CHAPTER 12

EVAPORATORS AND CONDENSERS FOR REFRIGERATION AND AIR-CONDITIONING SYSTEMS

M. B. PATE

Department of Mechanical Engineering
Iowa State University
Ames, Iowa 50011

12.1 INTRODUCTION

12.1.1 Background

Every refrigeration and air-conditioning system based on a vapor–compression cycle contains an evaporator and a condenser. A schematic diagram of a typical vapor–compression cycle showing these heat exchangers in relationship to other major components, such as the expansion device and the compressor, is shown in Fig. 12.1. The thermodynamic processes and states for the working fluid (i.e., refrigerant) during a vapor–compression cycle are demonstrated in the pressure–enthalpy (p–h) diagram shown in Fig. 12.2. Both an ideal cycle and an actual cycle, which contains nonideal processes such as pressure drops in the evaporator and condenser, are shown. The thermodynamic states at the inlets and exits of the heat exchangers are also marked on both figures. For example, entering the evaporator is a saturated mixture (4) while either a superheated or saturated vapor (1a) exits. For the condenser, a superheated vapor (2a) enters while a subcooled liquid (3) exits. Another important observation that can be made from the pressure–enthalpy diagram is that condensers operate at high pressures and temperatures while

Boilers, Evaporators and Condensers, Edited by Sadik Kakaç
ISBN 0-471-62170-6

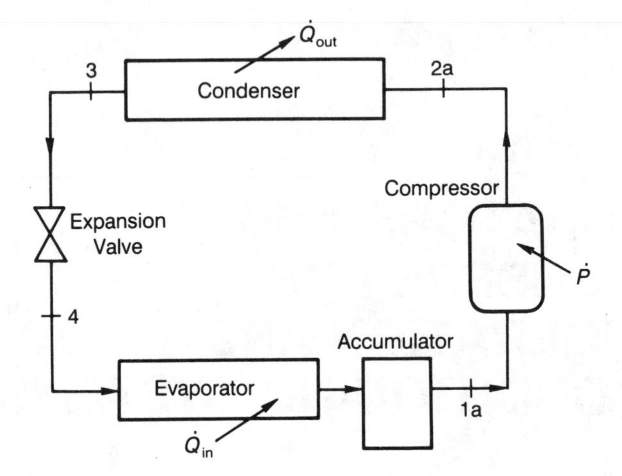

Fig. 12.1. Schematic of typical refrigeration system.

evaporators operate at low pressures and temperatures. The additional pressure drops shown for the actual cycle at the inlet and exit of the compressor occur in the compression intake and discharge valves.

Figure 12.2 also shows that the refrigerant drops in pressure as it flows through the evaporator and condenser. This drop in pressure in the evaporator is undesirable for the performance of a refrigeration system because as

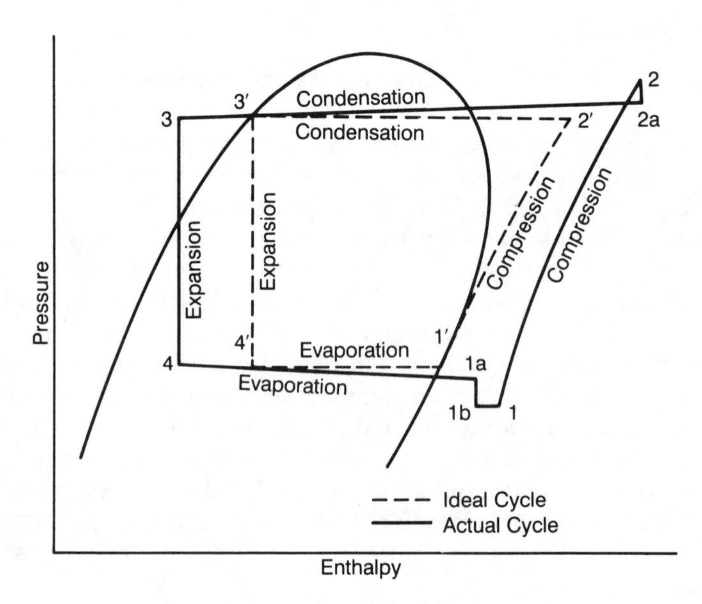

Fig. 12.2. Pressure–enthalpy diagram with vapor–compression cycle.

the low-side pressure decreases, more specific compressor work is required. Thus the coefficient of performance (COP) of the system decreases. For the condenser, the drop in pressure also has an undesirable effect that can be better understood in terms of saturation temperature. Because the state of the refrigerant in most of the condenser is saturated, the pressure and temperature are dependent, which means that the refrigerant temperature also drops. As a result, the temperature difference between the refrigerant and the cooling fluid (e.g., air or water) decreases, thus reducing heat transfer in the condenser. The desire to minimize pressure drops in both the condenser and evaporator for the reasons discussed previously is an important consideration for a designer who is considering the installation of enhanced tubes on the refrigerant side.

12.1.2 Typical Evaporator Behavior

The refrigerant flowing through an evaporator absorbs energy as it cools a fluid (usually water or air). The fluid-cooling process is the reason for the existence of the refrigeration system in most applications. Insights into the behavior of an evaporator can be gained by analyzing sample temperature profiles as a function of position shown in Fig. 12.3*a* (for a counterflow arrangement) and Fig. 12.3*b* (for a parallel-flow arrangement). The coordinates on these figures are not exact, especially the position coordinate, because they depend on the type of heat exchanger and the tube's circuiting arrangement. As mentioned previously, in the two-phase or saturated refrigerant region the temperature trend follows the pressure trend. The refrigerant entering the evaporator is at a saturated state, generally at a quality of approximately 10%. The liquid phase of the refrigerant is then vaporized as it flows through the evaporator, thus increasing the quality of the refrigerant. In some types of evaporators (e.g., flooded evaporators), the refrigerant exits as a saturated vapor at 100% quality. In other types of evaporators (e.g., direct expansion evaporators), the refrigerant superheats before it exits. Superheating is shown in Fig. 12.3, as evidenced by the sharp temperature increase near the exit. Superheated vapor exiting the evaporator will prevent liquid slugging of the compressor and ensure the maximum refrigerating effect.

Both flow arrangements, counterflow and parallel flow, are used in evaporator designs. The advantages of a counterflow arrangement in terms of heat exchanger performance are discussed in numerous texts. However, the fact that the refrigerant temperature drops even though heat is being added also suggests some advantages for parallel flow. In fact, achieving a relatively constant temperature difference between the two fluids may in some cases be easier to accomplish in a parallel-flow arrangement than in a counterflow arrangement. The profiles in Fig. 12.3 are examples only, and therefore a designer must either perform detailed calculations or conform to industry practice when selecting a flow arrangement.

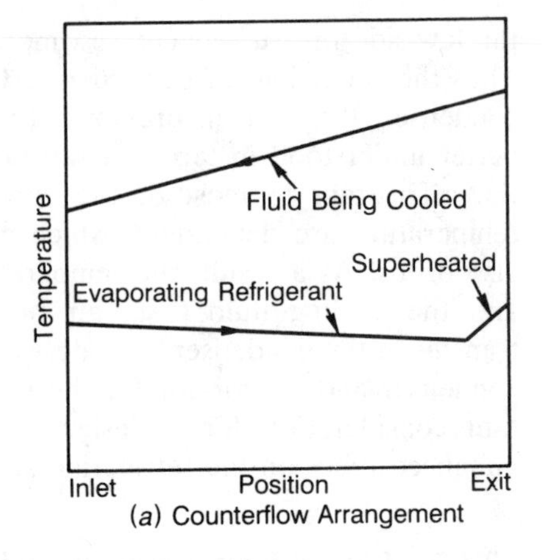

(*a*) Counterflow Arrangement

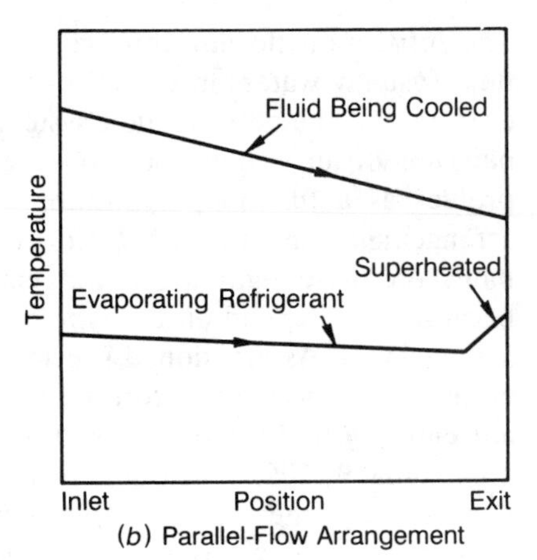

(*b*) Parallel-Flow Arrangement

Fig. 12.3. Typical temperature profiles for an evaporator.

12.1.3 Typical Condenser Behavior

The condenser is used to reject both the work of compression and the heat absorbed by the evaporator. To reject this heat, the condenser's refrigerant temperature must be higher than that of the fluid, usually air or water, cooling the condenser. As with the evaporator, temperature profiles can be used to demonstrate the behavior of a typical condenser.

As shown in Fig. 12.4 the refrigerant state entering the condenser is superheated and in many cases close to the state exiting the compressor, depending on the heat transfer and pressure drops in the piping connecting the two components. A short distance after entering the condenser the refrigerant is cooled to the saturation point. Condensation then occurs over most of the heat exchanger length as the refrigerant goes from 100% to 0% quality. The temperature decrease in this two-phase region is the result of the refrigerant pressure drop. Beyond the point where all the vapor is

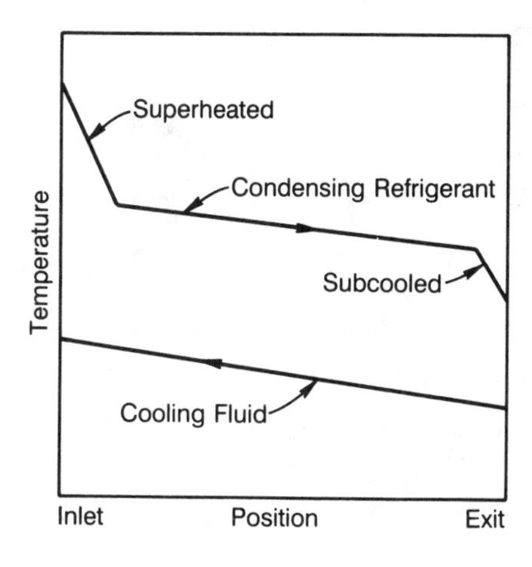

Fig. 12.4. Typical temperature profiles for a condenser.

condensed, the refrigerant is subcooled until it exits the condenser and enters the expansion device. As the cooling fluid absorbs energy, its temperature increases as shown in Fig. 12.4. This cooling fluid is single phase except in the case of cascade refrigeration systems and evaporative condensers.

Most condensers for refrigeration and air-conditioning applications are manufactured in counterflow arrangements as shown in Fig. 12.4. The reason for this arrangement is that counterflow heat exchangers have higher heat exchanger performance compared to parallel flow.

12.1.4 Types of Heat Exchangers in Refrigeration and Air-Conditioning Applications

Two-phase heat exchangers in refrigeration and air-conditioning applications can be categorized according to whether they are coils or shell-and-tube heat exchangers. Evaporator and condenser coils are used when the second fluid is air because a much larger surface area is required on the air side. Shell-and-tube evaporators and condensers are used when the second fluid is a liquid, such as water or a brine. The five major types of heat exchangers used in refrigeration and air-conditioning industries are shown in Fig. 12.5 under the major categories of coils or shell-and-tube heat exchangers.

Some general characteristics, such as whether the refrigerant or fluid flows inside or outside of the tubes, are also listed in Fig. 12.5. Two types of shell-and-tube heat exchangers shown in the figure (namely flooded evaporators and shell-and-tube condensers) have the refrigerant flowing on the shell side, while the liquid being cooled or heated flows through the inside of the tubes. The three remaining heat exchanger types (evaporator and condenser coils and DX evaporators) have the refrigerant flowing inside the tubes. It should be noted that a shell-and-tube heat exchanger with in-tube condensation, analogous to a DX evaporator, is not shown. The reason is that this condenser type is not as commonly used as the other types listed.

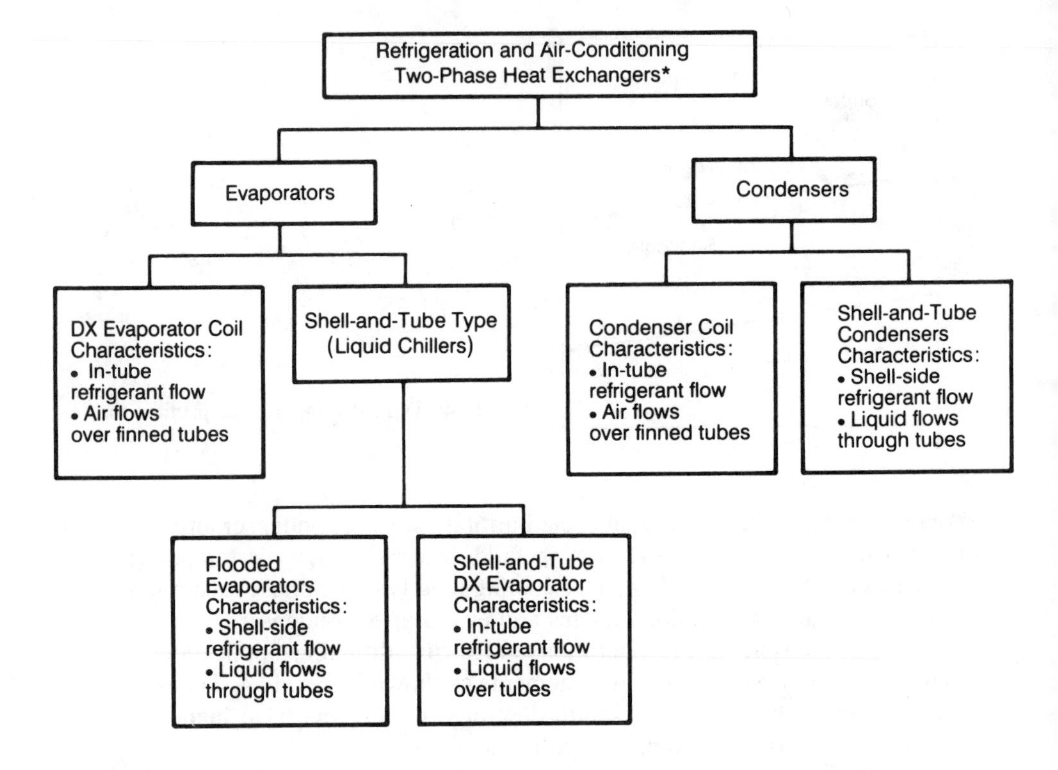

Fig. 12.5. Common heat exchanger types used in refrigeration and air-conditioning applications.

Flooded evaporators and direct expansion (DX) evaporators perform similar functions; however, they differ in that one is based on shell-side evaporation and the other is based on in-tube evaporation, as noted previously. The advantages of a DX evaporator over a flooded evaporator is that better oil circulation can be achieved inside tubes and that a superheated refrigerant can be returned to the compressor. The advantage of flooded evaporators is that flow distribution problems on both the refrigerant and liquid sides are minimized.

The five types of heat exchangers mentioned previously are described in detail in this chapter because they represent the majority of heat exchanger types found in refrigeration and air-conditioning applications. However, the reader should be aware that other types of heat exchangers are also used, but on a more limited basis. For example, plate-fin heat exchangers and double-pipe (i.e., tube-in-tube) heat exchangers are used for some automotive air-conditioning systems and low-tonnage liquid-cooling and heating systems, respectively.

Nearly all of the tubing used in refrigeration and air-conditioning applications is round tubing made from copper or aluminum. The exception to the use of round tubing is oval tubing, which is used in some coils manufactured for automotive applications. In addition, the exception to the use of copper or aluminum tubing is when the refrigerant is ammonia, in which case carbon steel tubing is generally used. Also, steel tubing is used in refrigerator applications for the condenser.

12.2 HEAT EXCHANGER ANALYSIS

12.2.1 General Equations

Heat exchangers for refrigeration and air-conditioning applications can be analyzed and designed by general techniques that are used in single-phase applications and in other two-phase applications such as power plants and chemical processing. Specifically, the energy equations and the methods used to solve them are similar for these different applications. The major differences between these different applications are in the correlations used and in the types of heat transfer enhancement installed. The correlations used to describe evaporation and condensation of refrigerants are often different from other fluids in that they are derived from a refrigerant database. Also, the methods of enhancing heat transfer reflect in some cases special requirements for refrigeration systems, such as low pressure drops. Other differences also exist. Much of this chapter deals with the presentation of methods for quantifying the effects that are different for refrigeration and air conditioning. However, this section is an overview of general heat exchanger analysis and design techniques along with special considerations required for refrigeration and air-conditioning applications.

Two possible approaches are used for analyzing heat exchangers. The first approach is a lumped analysis, in which the heat exchanger is analyzed as a single control volume with two inlets and two outlets. The second approach is a local analysis, in which the heat exchanger is divided into segments or multiple control volumes, with the outlet of one control volume being the inlet to an adjacent control volume, and vice versa. In the local analysis approach, the heat transfer rate for the heat exchanger is obtained by integrating the local values. The first approach is more common and simpler, and it often times provides reasonably accurate results. Efforts to develop computer models based on the second approach are ongoing; however, most of this work is proprietary because of the competitive nature of refrigeration and air-conditioning industries. To date, no widely used computer models based on the local approach are available in the public domain, even for a specific type of heat exchanger, such as those listed in Fig. 12.5.

For both of the heat exchanger analysis approaches, lumped and local, the basic working equation is the relationship for calculating the overall heat transfer coefficient. This equation can be derived by treating the heat flow path between the two fluids (e.g., refrigerant and air or water) as a series of thermal resistances. The following thermal resistances are applicable to most refrigeration and air-conditioning heat exchangers:

$$R_1 = \frac{1}{h_i A_{pi}} \qquad \text{convection inside the tube}$$

$$R_2 = \frac{1}{h_{di} A_{pi}} \qquad \text{inside deposit}$$

$$R_3 = \frac{t_p}{A_{pm} k_p} \qquad \text{tube wall}$$

$$R_4 = \frac{1}{h_c A_{po}} \qquad \text{contact between fin and tube wall}$$

$$R_5 = \frac{1}{h_{do} A_o} \qquad \text{outside deposit}$$

$$R_6 = (\text{see discussion}) \qquad \text{fin}$$

$$R_7 = \frac{1}{h_o A_{po}} \qquad \text{convection outside the tube}$$

The subscripts on the areas, A, represent only the outside area of the tube in the case of "po" and the total area including fins in the case of "o." The final overall heat transfer coefficient, which is obtained by combining thermal resistances and is based on the outside area, is

$$U_o = \cfrac{1}{\cfrac{A_o}{A_{pi} h_i} + \cfrac{A_o t_p}{A_{pm} k_p} + \cfrac{1 - \phi}{h_o \left(\cfrac{A_{po}}{A_f} + \phi \right)} + \cfrac{1}{h_o} + \cfrac{A_o}{A_{pi} h_{di}} + \cfrac{1}{h_{do}} + \cfrac{A_o}{h_c A_{po}}} \tag{12.1}$$

For clean tube surfaces, the values of h_{di} and h_{do} for the inside and outside deposits, respectively, approach ∞, signifying negligible thermal resistance. It is important to note that refrigerants are normally free of contaminants, and therefore the deposit thermal resistance on the refrigerant side is 0. In addition, the walls of tubes are quite often made of high thermal conductivity metals, such as copper and aluminum, and as a result, the thermal resistance for the tube wall, R_3, is negligible. The exceptions are when carbon steel tubing is used for those applications described previously.

As was noted at the beginning of this section, much of this chapter is devoted to showing the designer how to calculate various parameters in the preceding equation. Because heat exchanger types differ, there may be several ways to calculate each parameter.

The differential heat transfer rate can be defined for three differential control volumes defined on the refrigerant side, the nonrefrigerant fluid side, and in the wall–fluid interface region, respectively, as follows:

$$\delta Q_r = m_r \, di_r \tag{12.2}$$

$$\delta Q_f = m_f C_p \, dT_f \tag{12.3}$$

$$\delta Q = U_o \, dA_o \, \Delta T \tag{12.4}$$

where ΔT is the local temperature difference between the refrigerant and fluid. These heat transfer rates are equal for steady state when heat losses from the heat exchanger to the surroundings are negligible.

It is important to note that the form of the energy equation presented previously assumes that the nonrefrigerant fluid is both single phase and single component, such as air or water. However, for cooling air in evaporator coils, water vapor is frequently condensed out so that either a wet film or frost layer builds up on the air-side surface. In this case, Eq. (12.3) is modified to include the other phase, and this equation is written in terms of enthalpies instead of temperatures.

The difference between the lumped and local approaches depends on how the preceding heat transfer equations are solved. For the lumped analysis, these equations are solved by treating the heat exchanger as a single control volume characterized by average properties and parameters (e.g., heat transfer coefficients, temperatures, quality, etc.) with the boundaries being the inlets and outlets to the heat exchanger. For the local analysis, the equations are solved over incremental heat exchanger lengths from one end of the heat exchanger to the other end. The heat transfer rates for each increment are then integrated over the heat exchanger to obtain the total heat transfer. Both approaches are described in the following sections.

12.2.2 Lumped Heat Exchanger Analysis Approach

The lumped analysis approach is based on the assumption that the heat exchanger can be defined by average characteristics, that is, properties and parameters. As a result, the preceding heat transfer equations can be integrated over the length of the heat exchanger as follows:

$$Q_r = \dot{m}_r(i_{r_{\text{in}}} - i_{r_{\text{out}}}) \tag{12.5}$$

$$Q_f = \dot{m}_f(T_{f_{\text{out}}} - T_{f_{\text{in}}}) \tag{12.6}$$

$$Q = U_o A_o \,\Delta T_{\text{av}} \tag{12.7}$$

The average temperature difference can be defined as the log-mean temperature difference, providing the refrigerant is two phase over the length of the heat exchanger and the other fluid is single phase. The log-mean temperature difference is

$$\Delta T_{\text{LMTD}} = \frac{\Delta T_1 - \Delta T_2}{\ln(\Delta T_1/\Delta T_2)} \tag{12.8}$$

where ΔT_1 and ΔT_2 are the differences in temperature for the refrigerant and the fluid at adjacent positions. For example, for a counterflow heat exchanger (see Fig. 12.3*a*) the refrigerant inlet is adjacent to the outlet of the fluid, while for a parallel-flow heat exchanger (see Fig. 12.3*b*) both inlets and exits are adjacent to each other.

The complicating factor in refrigeration and air-conditioning heat exchangers, is that the refrigerant is not two phase over all its length. For example, evaporators (except flooded evaporators) operate with superheated outlets, while condensers have superheated inlets. In addition, most condensers operate with liquid subcooling at the outlet.

Because of the existence of a two-phase region and multiple single-phase regions, the assumption of constant specific heat that is used to derive the log-mean temperature difference is violated. Insight into the errors in the heat exchanger analysis resulting from basing ΔT_1 and ΔT_2 on the refrigerant inlet and outlet temperatures can be gained by observing the evaporator and condenser temperature profiles presented earlier in Figs. 12.3 and 12.4. For example, if the superheating at the evaporator outlet is large, then ΔT_2 can approach 0. This can significantly reduce the average temperature difference (i.e., log-mean temperature difference) calculated from Eq. (12.8). One can easily observe from Fig. 12.3 that the average temperature difference is actually much higher, being influenced much more by the temperature difference in the two-phase region than by the superheat region at the exit. For the case of the condenser in Fig. 12.4, the superheat region at the inlet results in an overestimation of the average temperature difference and, hence, an overestimation of the rate of heat transfer. However, the temperature difference at the subcooled exit offsets some of this overestimation.

Because of the complicating factors described previously, the lumped analysis of refrigeration and air-conditioning heat exchangers can be approached in two ways: (1) The two-phase region determines the characteristics (including temperature differences) of the whole heat exchanger, or (2) the heat exchanger is divided into single-phase and two-phase regions that are analyzed separately. The basis for the first option is that the superheated region, whether at the inlet of a condenser or the exit of an evaporator, makes up a smaller percentage compared to the two-phase region of the total heat transfer in a refrigeration-type heat exchanger.

For the first approach, the refrigerant temperatures used in the calculation of the log-mean temperature difference in Eq. (12.8) are the refrigerant saturation temperatures closest to the inlet or exit. This guideline allows for pressure drops varying the saturation temperatures through the heat exchanger. These saturation temperatures can be calculated from a combination of pressures and pressure drops or from the amount of superheating and subcooling. It should be noted that the assumption of the two-phase region determining the average characteristics affects the calculated log-mean temperature difference; hence the calculated heat transfer rate is higher than actual for evaporators and lower than actual for condensers. In addition, the calculated overall heat transfer coefficients may be slightly higher than actual ones. However, since the refrigerant-side thermal resistance is only part of the overall heat transfer coefficient, this effect is secondary compared to the effect on temperature differences.

The second approach accounts for heat transfer in the single-phase regions by separating the heat exchanger into a single-phase region and a two-phase region. The two-phase region is treated as it was before; however, only the heat exchanger surface area that corresponds to the two-phase region is used in the calculation. The heat transfer in the single phase, whether superheated or subcooled, is calculated by treating it as a separate region with its own surface area, log-mean temperature difference, and overall heat transfer coefficient. This approach may be difficult to implement, depending on the heat exchanger type (e.g., coil, shell and tube, etc.), flow arrangement (e.g., parallel flow, counterflow, etc.), and tube configuration. A specific problem may be the difficulty in associating nonrefrigerant fluid temperatures with the refrigerant-side two-phase and single-phase flow regions of the heat exchanger. This association is necessary for calculating the log-mean temperature difference for each region. An iteration procedure may also be required for the heat exchanger calculation depending on the conditions that are known to a designer or analyzer.

It should be noted that the log-mean temperature difference is often multiplied by a corrective factor, F, to account for the fact that a counterflow heat exchanger assumption is used to derive the log-mean temperature difference. In reality, few refrigeration and air-conditioning heat exchangers are strictly counterflow. However, for two-phase flow heat exchangers where the heat capacity approaches ∞, resulting in a constant temperature, this correction factor is 1. Complications can arise when the temperature of the two-phase refrigerant changes because of pressure drops or when single-phase regions are analyzed separately from the two-phase region as described in the second approach.

12.2.3 Local Heat Transfer Integration Approach

As mentioned previously, the local approach is based on dividing the heat exchanger into control volume elements and then solving the governing differential heat transfer equations for these elements. Depending on the type of evaporator or condenser, these control volume elements can be whole tube rows, an approach used in the past with flooded evaporators, or tube length increments, an approach used in the past with coils. Typically, the solution is marched through the heat exchanger by starting from an element where temperatures and other fluid properties are known. However, the heat exchanger configuration is often such that conditions (e.g., temperature) for one of the two fluids are unknown at the starting element. In this case, a fluid condition is assumed at the starting elements. If the fluid conditions are given at the last element, then the known and calculated conditions are compared after the solution is complete to verify the accuracy of the assumed conditions at the starting elements. The total heat transfer rate for the heat exchanger is calculated by integrating the incremental heat transfer rates over the heat exchanger.

The three energy equations [Eqs. (12.2) to (12.4)] can be written in a variety of ways since the equation form is dependent on how the control volume element is defined (e.g., boundaries) and on the method used to identify the element (e.g., grid system). Therefore the governing equations for the local analysis approach are not presented herein as they were for the lumped analysis approach. However, two considerations are important for the third energy equation, containing the overall heat transfer coefficient and the temperature difference between the two fluids. First, because local conditions are calculated for each element, it is possible to consider the effects of temperature-dependent properties and refrigerant quality on evaporation and condensation heat transfer coefficients. As a result, the overall heat transfer coefficient varies for each element. The second consideration deals with the method of defining the temperature difference that drives heat transfer through the tube wall. If the control volume elements are small enough, then the driving temperature difference can be based on simple arithmetic average temperatures on each side rather than log-mean temperature differences.

Several heat exchanger models based on the local analysis approach have been reported in the open literature for refrigeration and air-conditioning applications. For example, Webb, Choi, and Apparao [1] reported a model of a flooded refrigerant evaporator where each tube row was treated as a control volume element. Payvar [2] used a similar approach in a full-bundle submerged boiler, which closely resembles a flooded evaporator. An evaporator and condenser coil model based on a local analysis approach was reported by Huang and Pate [3]. In this latter study, control volume elements consisted of tube length increments. For example, in one case they divided a 3-m long tube in a coil into 28 increments.

12.3 EVAPORATOR COILS

12.3.1 Description and Special Considerations

Evaporator coils consist of plate-finned tubes with refrigerant flowing on the inside of the tubes and air flowing over the outer tube surface. A simplified sketch of a coil consisting of plate fins and round tubes is shown in Fig. 12.6. Evaporator coils are used for many applications that require a wide range of configurations and sizes. These sizes and configurations also vary from manufacturer to manufacturer even for the same application. For example, the capacity of evaporator coils can be quite small—fractions of a ton—when used for small refrigerator applications, or quite large—hundreds of tons—when used with large building air-conditioning systems. Different applications can also result in different methods of moving air through the

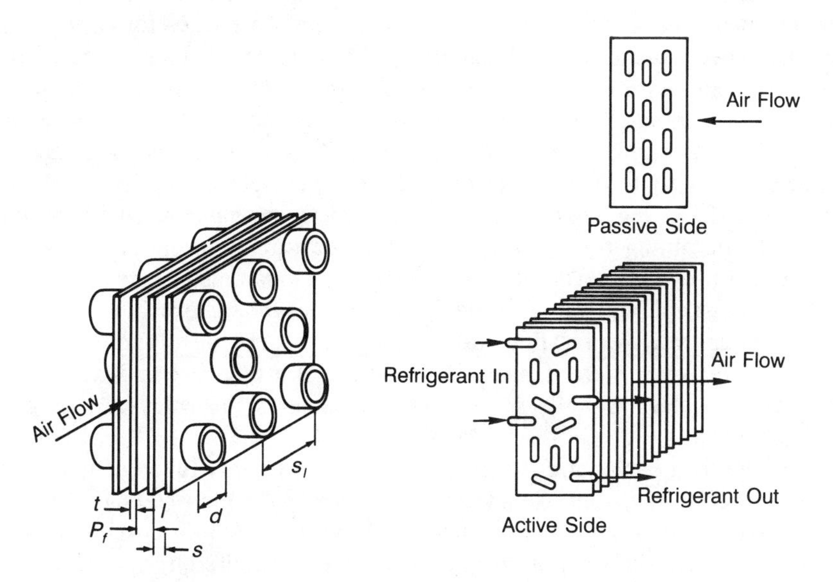

Fig. 12.6. Typical plate-finned-tube heat exchanger including end views of the active and passive sides.

coil; forced circulation is used in most cases, but natural circulation is also used.

Evaporator coils use round tubes for the most part; however, rectangular and oval tubes are used for special applications. Typical tube sizes representing a wide range of applications are outside diameters of 5/16, 3/8, 1/2, 5/8, 3/4, and 1 in. (7.9, 9.5, 12.7, 15.9, 19.1, and 25.4 mm). The tubes are arranged either in staggered arrangements, forming equilateral triangles, or inline arrangements, forming rectangles. The former is the preferred method. The spacing of tubes ranges from 0.6 to 2.5 in. (16 to 64 mm).

The tubes are finned in a continuous manner by using flat plates installed with typical densities of 4 to 14 fins per inch (1.8 to 6.4 mm apart). The lower values of fin density (e.g., fewer than 10 fins per inch) are used in low-temperature applications such as refrigeration, where frosting and thus channel blockage may occur. The higher fin densities are used in higher-temperature applications such as air conditioning, where the water vapor in the air stream is removed by moisture condensation rather than frosting. Considerations other than frosting and moisture accumulation to be used when fin spacing is decided are frictional pressure drop and coil contamination due to lint, dust, and so on. The thickness of the plates that are used to form the fins range from 0.004 to 0.017 in. (0.09 to 0.42 mm). Note that thinner fins produce less gripping of the fin to the tube and, therefore, less surface contact area for heat transfer.

Evaporator coils consist of multipath tubing arranged in a serpentine fashion running perpendicular to the air flow. Because of the wide range of applications for coils and the practices used by manufacturers, an unlimited number of refrigerant circuit arrangements exist. Some rules for configuring coils have been suggested by Hogan [4]. For example, an active side of the coil should contain soldered connections and the inlet and outlet tubes, while a passive side should contain only tube bends. (An example of a passive and active side was shown earlier in Fig. 12.6.) Tubes should be arranged in staggered rows to form equilateral triangles. Parallel-flow paths should be arranged to form regions of symmetry in the heat exchanger. This is important for maintaining relatively constant temperature differences in the coil that in turn prevent uneven heat transfer rates. Hogan also suggests that the heat exchanger should be divisible into unique subheat exchangers, which contain equal refrigerant flow rates, and that each must end at a face other than the one where it starts. An example of a tube circuit that satisfies the previous rules, as well as a circuit that violates the rules, is shown in Fig. 12.7 [4].

The inlet to the evaporator coil can be from a capillary tube in the case of small refrigeration systems (less than a ton), short tube restricters, orifices, or thermostatic expansion valves. The capillary tube is low cost and most applicable when the evaporator operates over a limited range of design conditions. Most room and window air conditioners, household refrigerators,

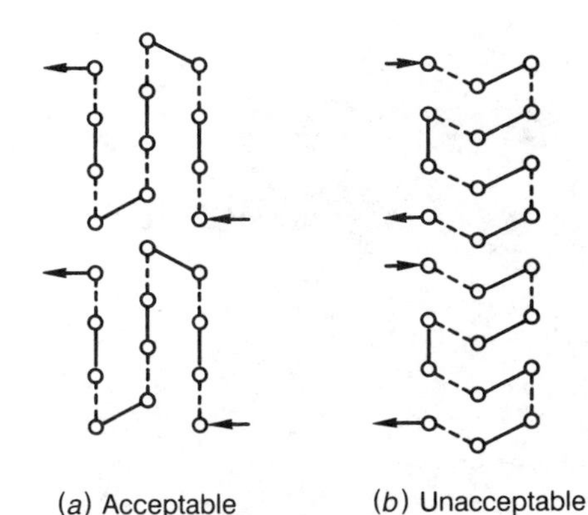

Fig. 12.7. Examples of acceptable and unacceptable tube circuits for plate-finned-tube heat exchangers [4].

and freezers operate with capillary tubes. Thermostatic expansion values have the advantage of being able to control the exit superheat of the evaporator by controlling the flow rate through it. Typically, the thermostatic expansion valve controls the superheat at the evaporator exit to 6 to 12°F (3.3 to 6.7°C). It is used with larger refrigeration and air-conditioning systems.

Larger coils supplied by thermostatic expansion valves have multiple circuits within the same coil. A distributor is placed between the expansion valve and coil for the purpose of supplying equal amounts of refrigerant to each circuit. These distributors consist of small-diameter tubes of equal size and length to ensure equal flow restriction for the saturated mixture of vapor and liquid refrigerant passing through the distributor.

Since evaporator coils frequently operate before the dew-point temperature of the air, moisture collects and drains from the coil. To collect the moisture that drains from the coil, a pan should be located at the bottom of the coil with the drain connection located on the downstream side of the coil. For large coils several pans in the vertical plane may be required. Excessive air velocities that might entrain the moisture and carry it downstream into the ductwork should be avoided. Therefore design air velocities should be in the range of 400 to 500 fpm (2 to 2 m/s) to prevent condensate carry-over.

Because of the diversity of evaporator coils, it would be impossible to describe a single coil to represent all applications; however, an evaporator coil showing some of the basic elements—tubes, plates, and distribution device—is shown in Fig. 12.8.

The information on evaporator coil design presented in the following sections deals primarily with methods for calculating the thermal resistances that exist in the heat flow path between the refrigerant and air. These are necessary for determining the overall heat transfer coefficient by using Eq. (12.1). The thermal resistances discussed here are associated with the in-tube heat transfer coefficients, fin bonding to the tube, and air-side heat transfer coefficients. An additional thermal resistance that may or may not exist is

Fig. 12.8. Typical evaporator coil with expansion device and distributor.

associated with layers of frost and films of water that condense out of the air-flow stream.

12.3.2 In-Tube Refrigerant Evaporation Heat Transfer

The in-tube evaporation heat transfer coefficient must be calculated before one determines the overall heat transfer coefficient. Prior to selecting a correlation for use, the designer must consider several characteristics of the evaporator that affect heat transfer:

1. Evaporator tubes are horizontal.
2. Return bends are adiabatic because they are located outside the air-flow stream.
3. The refrigerant covers the full range of boiling mechanisms including nucleate boiling at low qualities and convective boiling at high qualities.
4. Common refrigerants are R-11, R-12, R-22, R-123, R-134a, R-502, and ammonia.
5. Both smooth and enhanced tubes are used.
6. Lubricants are mixed in small concentrations with refrigerants in the evaporator.

Several correlations for local evaporation heat transfer in smooth, horizontal tubes are presented here. These correlations have been verified for most of the common refrigerants; in fact, a wide range of refrigerant data was used in the original derivation. These correlations are by Shah [5, 6], Kandlikar [7], and Gungor and Winterton [8] for pure refrigerants. Results for refrigerant–lubricant fixtures are also presented here, while in-tube evaporation heat transfer for enhanced tubes is presented in a separate section. Since heat transfer data for enhanced tubes are usually referenced to smooth-tube heat transfer, the correlations presented here are also indirectly applicable to enhanced tubes.

Single-Phase Heat Transfer Since the exit of an evaporator coil is superheated vapor, single-phase heat transfer correlations are also required for design calculations. Two single-phase correlations that have been extensively verified with experimental data for refrigerants are those by Dittus and Boelter (McAdams [9]) and by Petukhov and Popov [10]. Because of their common usage, they are not presented here in equation form. However, in the author's experience with several different refrigerants, including R-12, R-22, and R-113, the Petukhov–Popov equation is slightly more accurate, predicting experimental data to within 5%. However, the Dittus–Boelter equation routinely predicts the same data bank to within 10%.

Evaporation Heat Transfer for Pure Refrigerants In 1982, Shah developed a correlation equation from an earlier chart-based correlation [5, 6]. This correlation is applicable to nucleate, convection, and stratified boiling regions as evidenced by its functional relationship with several nondimensional numbers including the boiling number, Froude number, and convection number. As expected, at low qualities, nucleate boiling dominates, while at high qualities, convection dominates. For horizontal flow, Shah's correlation is

$$\frac{h_{\mathrm{TP}}}{h_l} \equiv \Psi \tag{12.9}$$

where h_l is the all-liquid convection heat transfer coefficient which can be calculated from correlations presented in Chapter 3. The term Ψ is evaluated by the following procedure:

$$N \equiv 0.38 Fr^{-0.3} Co \tag{12.10}$$

$$\Psi_{\mathrm{cb}} = \frac{1.8}{N^{0.8}} \tag{12.11}$$

For $N > 1$:

$$\Psi_{\mathrm{nb}} = 230 Bo^{0.5} \qquad Bo > 3.0 \times 10^{-5} \tag{12.12}$$

$$= 1 + 46 Bo^{0.5} \qquad Bo < 3.0 \times 10^{-5} \tag{12.13}$$

For $0.1 < N \leq 1.0$:

$$\Psi_{\mathrm{bs}} = E Bo^{0.5} e^{2.74 N^{-0.1}} \tag{12.14}$$

For $N \leq 0.1$:

$$\Psi_{\mathrm{bs}} = E Bo^{0.5} e^{2.74 N^{-0.15}} \tag{12.15}$$

The value of Ψ is the larger of Ψ_{cb} and Ψ_{nb} or Ψ_{bs}.

The constant E in the preceding expressions depends on the boiling number, Bo:

$$E = 14.7 \qquad Bo \geq 11 \times 10^{-4} \tag{12.16}$$

$$E = 15.43 \qquad Bo < 11 \times 10^{-4} \tag{12.17}$$

which can be calculated as

$$Bo = \frac{q}{G i_{fg}} \tag{12.18}$$

The convection, Co, and Froude number, Fr, also used in the preceding expressions, can be calculated as follows:

$$Co = \left(\frac{1-x}{x}\right)^{0.8}\left(\frac{\rho_g}{\rho_l}\right)^{0.5} \tag{12.19}$$

and

$$Fr = \frac{G^2}{\rho_l^2 g D} \tag{12.20}$$

The preceding equations presented for the Shah correlation have been used extensively for refrigerants such as R-11, R-12, and R-22. For these three refrigerants, Shah reports mean deviations of 23% when comparing the correlation with experimental data. The correlation proposed by Kandlikar [7] also considers nucleate, convection, and stratified boiling flow. It is

$$\frac{h_{\mathrm{TP}}}{h_l} = C_1(Co)^{C_2}(25Fr_l)^{C_5} + C_3(Bo)^{C_4}F_{fl} \tag{12.21}$$

where if $C_o < 0.65$ then

$$C_1 = 1.1360 \qquad C_4 = 0.7$$

$$C_2 = -0.9 \qquad C_5 = 0.3$$

$$C_3 = 667.2$$

and if $C_o > 0.65$ then

$$C_1 = 0.6683 \qquad C_4 = 0.7$$

$$C_2 = -0.2 \qquad C_5 = 0.3$$

$$C_3 = 1058.0.$$

and h_l is a liquid-only convective heat transfer coefficient based on a form of the Dittus–Boelter equation,

$$h_l = 0.023\left(\frac{G(1-x)D}{\mu_l}\right)^{0.8}\frac{Pr_l^{0.4}k_l}{D} \tag{12.22}$$

and possible effects of flow stratification in the horizontal flow are taken into account by the Froude number, Fr. The coefficients D_1 through D_6 are defined in Kandlikar [7]. Of all the in-tube evaporation correlations, this correlation is the easiest to use. In addition, it is the author's experience, on the basis of numerous comparisons with experimental data, that this correlation is reliable.

The correlation of Gungor and Winterton [8] was derived from a large database that included halocarbon refrigerants such as R-11, R-12, R-22, R-113, and R-114. The basic form of the correlation is

$$h_{\mathrm{TP}} = Eh_l + Sh_{\mathrm{pool}} \tag{12.23}$$

where the all-liquid convection heat transfer coefficient, h_l, was defined in Eq. (12.22). The other parameters are an enhancement factor, E,

$$E = 1 + 2.4 \times 10^4 Bo^{1.16} + 1.37\left(\frac{1}{X_{tt}}\right)^{0.86} \tag{12.24}$$

a suppression factor, S, is given as

$$S = \frac{1}{1 + 1.15 \times 10^{-6} E^2 Re_l^{1.17}} \tag{12.25}$$

and a pool boiling term, h_{pool}, is given as

$$h_{\mathrm{pool}} = 55P_r^{0.12}(-\log_{10}P_r)^{-0.55} m^{-0.5} q^{0.67}$$

Also present in these equations is the well-known Martinelli parameter, X_{tt},

$$X_{tt} = \left(\frac{1-x}{x}\right)^{0.9}\left(\frac{\rho_g}{\rho_l}\right)^{0.5}\left(\frac{\mu_l}{\mu_g}\right)^{0.1} \tag{12.26}$$

The Martinelli parameter is similar to the convection number, Co, used in the Shah and Kandlikar correlation except that the vapor viscosity effects are also considered.

For low Froude number flow in horizontal tubes, $Fr < 0.05$, Gungor and Winterton recommend multiplying E and S by the following factors:

$$E_2 = Fr^{(0.1-2Fr)} \tag{12.27}$$

$$S_2 = \sqrt{Fr} \tag{12.28}$$

Local heat transfer coefficients for the three correlations presented before are compared in Fig. 12.9. This comparison is for in-tube flow of R-134a, an alternative refrigerant with an HFC designation, which does not affect the ozone layer as do the CFC refrigerants such as R-12. The quality range in Fig. 12.9 is from 0% to 95%, which covers most of the conditions present in

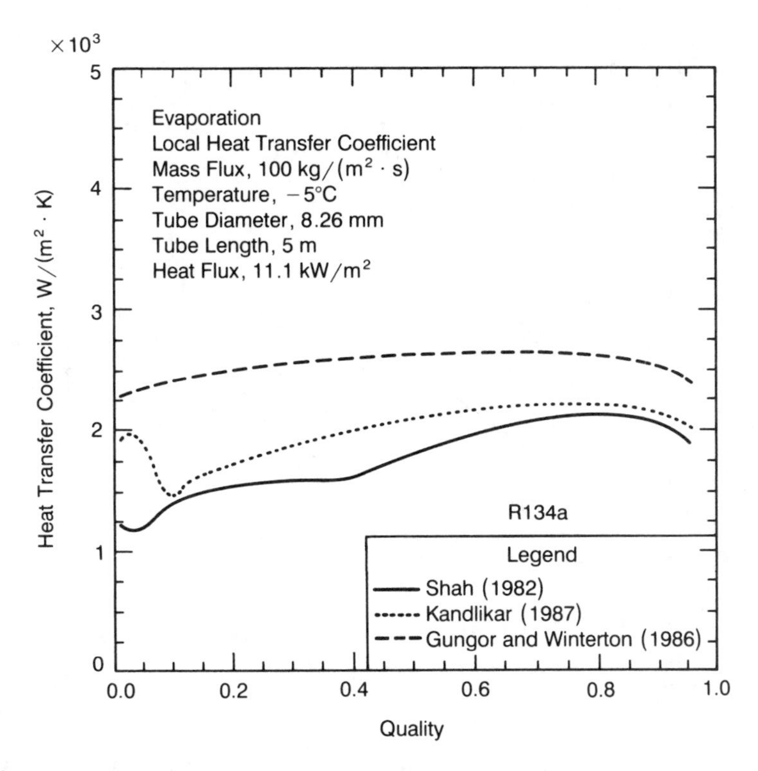

Fig. 12.9. Local evaporation heat transfer coefficients for in-tube flow of R-134a.

an evaporator. A typical evaporator flow rate, temperature, tube diameter, and tube length were selected for the comparison. The Gungor–Winterton and Shah correlations result in the highest and lowest heat transfer coefficients, respectively, over the full quality range. The results for the Kandlikar correlation are somewhere between the other two correlations. Differences in the local heat transfer correlations for the nucleate and convective boiling regions can also be observed in Fig. 12.9. Also apparent is the fact that convective boiling coefficients increase with quality because of higher flow velocities.

Another comparison can be made by integrating the local heat transfer coefficients over the tube length to obtain average heat transfer coefficients. The average heat transfer coefficients as a function of flow rates are shown in Fig. 12.10 for the same size tube. The relative magnitudes of the three correlations follow closely with the local heat transfer coefficients that were presented in Fig. 12.9 for a flow rate of 100 kg/(m^2 · s). Considering the uncertainties associated with two-phase flow and heat transfer, the three correlations produce results that are fairly close. In fact, several past studies have shown that these correlations can consistently predict heat transfer coefficients to within $\pm 20\%$ of experimental values [11, 12].

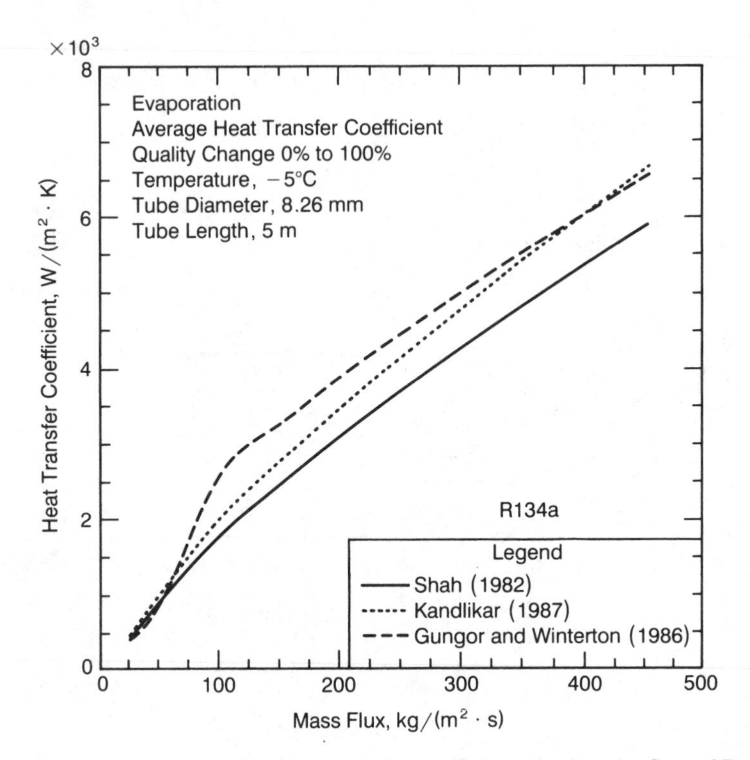

Fig. 12.10. Average evaporation heat transfer coefficients for in-tube flow of R-134a.

Lubricant Effects on Heat Transfer The lubricant used in the compressor circulates with the refrigerant through all components that make up a refrigeration system, including the evaporator and condenser. The circulating concentration of lubricant mixed with refrigerant probably varies from 0.2% to 10%, depending on the type of compressor (e.g., reciprocating, rotary, screw, or scroll) and on whether an oil separator exists. However, the actual lubricant concentration in the evaporator and condenser may be higher than the circulating concentration. This higher concentration is the result of high-viscosity, lubricant-rich films traveling along the tube wall in an annular flow pattern at a slow velocity.

A detailed study of lubricant effects on heat transfer by Schlager et al. [11, 13, 14] showed that in a smooth tube, the heat transfer coefficients were higher when oil was added to the refrigerant. For example, heat transfer coefficients increased by as much as 35% over pure refrigerant values as oil concentrations were increased up to 2.5%. At higher oil concentrations, the heat transfer coefficients began to decrease, with the value being only 15% higher than the pure refrigerant value at 5% oil concentrations. The following correlations can be used by the designer to account for the effects of oil on smooth-tube evaporation heat transfer. For 150 SUS oil and R-22 [14],

$$EF_{s'/s} = 1.03e^{(17.7W_o - 286W_o^2 - 0.0496G')} \tag{12.29}$$

and for 300 SUS oil and R-22 [14],

$$EF_{s'/s} = 1.03e^{W_o(4.98G' - 8.77)} \tag{12.30}$$

The EF in the preceding equation is a type of enhancement factor that reflects changes in heat transfer coefficients due to the presence of oil. The subscripts identify the heat transfer coefficients used to calculate EF. For example, the subscript s is the smooth-tube heat transfer coefficient for the pure refrigerant, while the prime on the subscript s represents lubricant mixed with the refrigerant. The lubricant concentration W_o is the fraction of lubricant mixed with the refrigerant in the evaporator. Experiments for R-22 showed that this value is approximately three times the concentration of the flowing lubricant–refrigerant mixture during evaporation. The mass flux, G', used in the preceding equation is normalized to 300 $kg/(m^2 \cdot s)$.

Equations that account for the effects of lubricants have also been reported for other refrigerants such as R-12. Because these equations are quite complicated and because of the phaseout of CFCs such as R-12, they are not presented here.

12.3.3 In-Tube Heat Transfer Augmentation

A number of techniques have been suggested for in-tube enhancement during evaporation, including rough surfaces, extended surfaces, and several swirl flow devices. Figure 12.11 shows several examples for each of these enhancement techniques. Rough surfaces shown are helical wire inserts, internal threads, and corrugated tubes, while the twisted tape insert is presented as a typical example of a swirl flow device. These rough surfaces and swirl flow techniques have not found widespread usage in refrigeration applications because pressure drop enhancements have exceeded heat transfer enhancements. Pressure drops in evaporator tubes should be kept to a minimum because of the detrimental effect they have on the system coefficient of performance (COP).

In contrast to the preceding two techniques, extended surfaces have found widespread usage in refrigeration applications. Examples of extended surfaces shown in Fig. 12.11 that have been used in refrigeration applications are high-profile fins, microfins, annular offset strip ribbon fins, and intersecting fins. Of these examples of extended surface techniques, the microfin tube is the most commonly used. In addition, it shows potential for even more widespread use in the future. The microfin tube has gained in popularity for

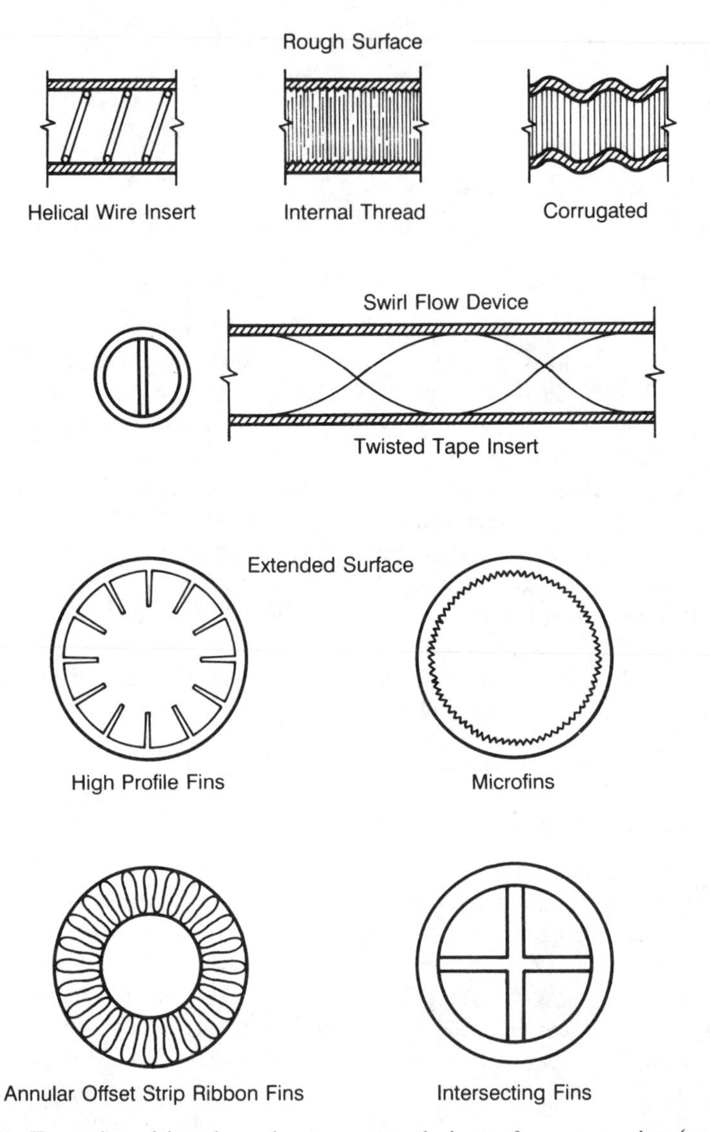

Fig. 12.11. Examples of in-tube enhancement techniques for evaporating (and condensing) refrigerants.

two reasons. First, it can increase evaporation heat transfer coefficients by factors of 2 to 3 over smooth-tube values. The increase in pressure drop is significantly less, on the order of 1 to 2. Second, most enhancement techniques require extra tubing material (e.g., copper), which then raises the cost of the enhanced tube. The microfin tube, however, requires little, if any, additional tube material.

The microfin tube is characterized by having 60 to 70 fins with heights of 0.10 to 0.20 mm and spiral angles of 10 to 30°. The shape of the fin tip and valley can be either flat, round, or sharp. Three different views of a typical microfin tube are shown in Fig. 12.12. The extended surface enhancement technique closest to the microfin tube is the high-profile fin tube; however, these two tubes differ greatly in the number of fins and the height of the fins. For example, most high-profile fin tubes have fewer than 30 fins and fin heights greater than 0.4 mm.

Performance data for evaporation heat transfer are presented here for microfin tubes so that a designer can perform heat exchanger design calculations. The performance information is also applicable to shell-and-tube DX evaporators, which are presented in a later section.

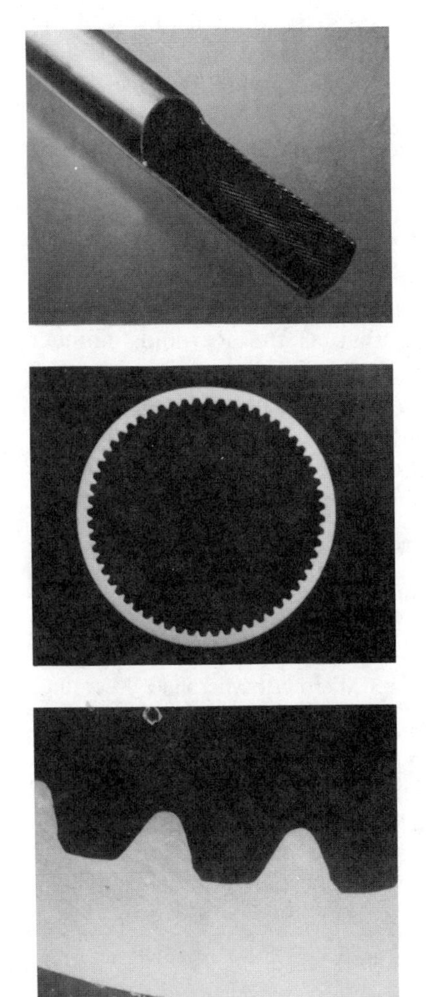

Fig. 12.12. Photograph of microfin tube.

The presentation of performance data is divided into several areas. First, plots of enhancement factors, defined as the enhanced-tube heat transfer coefficient divided by an equivalent-diameter smooth tube measured at similar flow conditions, are presented. These plots are for different microfin tube geometries so that a designer can understand how heat exchanger performance might possibly differ for different types of microfin tubes. The effect of tube diameter on the performance of a microfin tube is also presented. Finally, the effect on microfin tube performance of lubricants mixed with refrigerants is covered.

A detailed comparison of three different microfin tube geometries for 9.52-mm (3.8 in.) and 12.7-mm (1/2 in.) OD tubes with R-22 was reported by Schlager et al. [15, 16]. The tube dimensions are listed in Table 12.1. Enhancement factors, EF, for the three geometries and two different diameter tubes are shown in Fig. 12.13 for evaporation heat transfer. The numbers on the figure correspond to the tube numbers listed in Table 12.1. In addition, it is important to note that the curves shown are spline fits of several data points that contain experimental uncertainties. Because of the magnitude of the experimental uncertainties, one can conclude that the effects of tube diameter and also tube geometries are minor. For example, at a mass flux of 250 kg/(m^2 · s), which is a region where the two different tube diameters overlap, the EF for all the 9.52-mm (3/8-in.) tubes ranged from 1.6 to 1.9, while the EFs for the 12.7-mm (1/2-in.) tubes fall in a tighter band of 1.7 to 1.8.

Figure 12.13 also shows that EF decreases as the mass flux increases. A possible reason for this behavior may be that as the Reynolds number increases, the turbulence induced by the microfins is less important relative to the turbulence level in the smooth tube, with a resulting decrease in the heat transfer augmentation due to turbulence.

The preceding evaporation performance data can be used by a designer in two ways. First, if a heat exchanger is being designed with either 3/8-in. or

TABLE 12.1 Dimensions of Microfin Tubes [15, 16]

	12.7-mm Tubes				9.52-mm Tubes		
	Smooth	Microfin 1	Microfin 2	Microfin 3	Microfin 1	Microfin 2	Microfin 3
d_o, mm	12.7	12.7	12.7	12.7	9.52	9.52	9.52
$d_{i_{max}}$, mm	10.9	11.7	11.7	11.7	8.92	8.92	8.92
t, mm	0–.90	0.50	0.50	0.50	0.30	0.30	0.30
f, mm	—	0.30	0.20	0.15	0.20	0.16	0.15
n	—	60	70	60	60	60	60
β, °	—	18	15	25	18	15	25
A_{i_S}/A_{i_M} [a]	—	1.51	1.33	1.39	1.55	1.38	1.43

[a] Ratio of inside surface area of the microfin tube to the inside area of a smooth tube having the same maximum inside diameter.

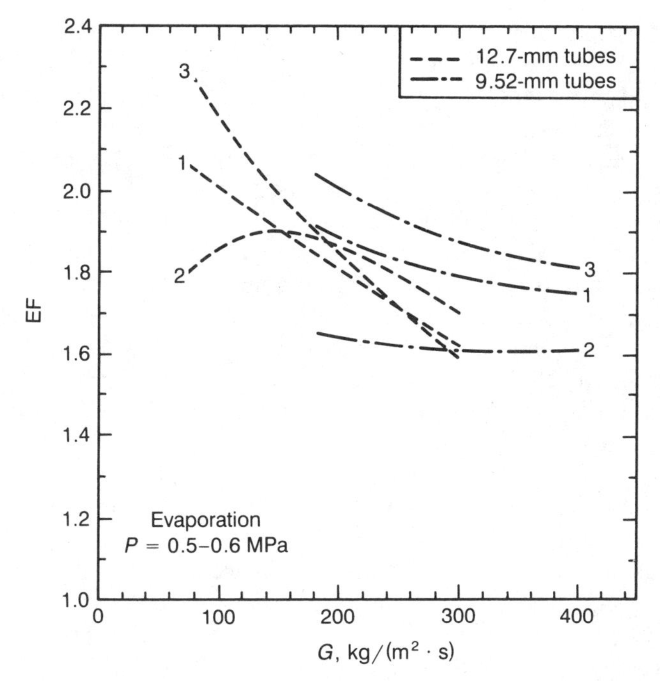

Fig. 12.13. Evaporation heat transfer enhancement factors for three different microfin tube geometries and two different diameters [15, 16].

1/2-in. microfin tubes with geometrical parameters similar to those listed in Table 12.1, then enhancement factors, EF, can be taken directly from Fig. 12.13 at the desired mass flux. Second, a designer can extrapolate enhancement factors plotted for the different diameter tubes to either smaller or larger diameter tubes. Since EF is only a weak function of tube diameter, an extrapolation technique should work satisfactorily. It is important to note that even though EF is not a strong function of diameter, the in-tube heat transfer coefficient for the microfin tube is a function of tube diameter. The effect of diameter on the microfin tube is accounted for when the smooth-tube heat transfer coefficient, which varies with tube diameter, is multiplied by EF.

Even though the previous study, along with several other studies, has compared the heat transfer performance of several different microfin tube geometries, there has not been a study that systematically varied microfin tube geometrical parameters so that general correlations could be developed. These correlations would be extremely useful to the designer in that one could select those microfin tube dimensions (e.g., spiral angle, tip and valley shape, number of fins, etc.) that would optimize the performance of a heat exchanger for a particular application. Using these same equations, one could also perform heat exchanger calculations for each of the several different microfin tubes available from different manufacturers to determine which tube should be installed.

Of special importance for heat exchanger design with microfin tubes are the results of a study performed by Schlager et al. [17] on a microfin tube geometry typical of those available from several manufacturers. This typical microfin tube was evaluated for pure R-22 and for lubricant–refrigerant mixtures for two different oil viscosities (150 SUS and 300 SUS). The 9.52-mm (3/8-in.) OD microfin tube used in this study had 60 fins with heights of 0.2 mm (0.008 in.) and spiral angles of 18°. The area increase of the microfin tube over an equivalent-diameter smooth tube was 1.5. The results for a single mass flux of 300 kg/(m^2 · s) are shown in Table 12.2 for four different oil concentrations–namely, 0%, 1.25%, 2.5%, and 5%. The heat transfer enhancement factors, EF, shown in the table are based on referencing the microfin tube data to smooth-tube data measured at similar conditions, including the same oil viscosity and concentration.

Several observations regarding the trends in Table 12.2 will assist the designer. For example, enhancement factors decrease with the addition of lubricant to the refrigerant. In addition, enhancement factors are not strong functions of oil viscosity, though they are slightly higher for the higher-viscosity 300 SUS oil. The designer can either use the microfin tube performance data in Table 12.2 or obtain additional data from [11, 13, 17]. Also, a correlation is presented in the following discussion for extending the enhancement factors in Table 12.2 to other mass fluxes.

A general one-parameter correlation, namely an enhancement factor as a function of mass flux, has been reported [14]. This correlation is based on a best-fit curve of enhancement factor data for several different microfin tube geometries as reported by several different investigators. In all cases the refrigerant was R-22 and the outside diameter of the tubes was 3/8 in. (9.52 mm). The resulting empirical expression is

$$\frac{\mathrm{EF}_1}{\mathrm{EF}_2} = \left(\frac{G_1}{G_2}\right)^{-0.32} \tag{12.31}$$

TABLE 12.2 Microfin Tube Evaporation Performance Data for Pure R-22 and Lubricant Mixtures at 300 kg / (m^2 · s) [17]

Parameter	Oil Concentration, %	150 SUS	300 SUS
Heat transfer enhancement factor, EF	0	2.05	2.05
	1.25	2.00	1.95
	2.5	1.7	1.90
	5.0	1.75	1.85

Because this equation was able to correlate data from several diverse studies, it should be satisfactory for design in the absence of more applicable data. However, it should be emphasized again that the accuracy of this correlation is limited in that it does not account for the effects of different microfin tube geometries. To use the preceding equation, one needs to know at least one value of enhancement factor at a given mass flux. Any one of several different enhancement factors presented earlier, including those in Table 12.2, can be used as a given value in the previous equation.

All of the preceding studies are for R-22 only. The reason is that several common refrigerants that are CFCs, such as R-11 and R-12, are being rapidly phased out because they have a destructive effect on the protective ozone layer. As a result, no new heat exchanger designs are expected for these two refrigerants. To date, the replacement refrigerants for R-11 and R-12, often referred to as alternative refrigerants, have not been tested with microfin tubes. Until this information is available, it is recommended that designers use the enhancement factors presented previously when designing for alternative refrigerants.

12.3.4 Air-Side Heat Transfer

Most refrigeration and air-conditioning air coils, whether evaporators or condensers, use circular-finned tubes rather than plate-finned tubes. (The exception is evaporators in automotive air conditioning.) The fin plates used on the air side can be flat or plain surfaces, standard or sine-wave corrugated surfaces, or louvered surfaces. Examples of some of these fin surfaces are sketched in Fig. 12.14. The order presented represents increasing heat transfer performance; however, additional manufacturing effort is required to either deform and shape the surface or, in the case of louvering, to cut slits in the surface. It should also be noted that even though the heat transfer performance increases, the pressure drop on the air side, and, hence, required fan power, also increases.

The discussion of air-side heat transfer is organized around the three types of fin surfaces. In addition, moisture buildup and frosting are also discussed because in both cases an additional thermal resistance to heat transfer results. Additional information on air-side heat transfer can be found in articles by Webb [18, 19].

Dry Coils Correlations and/or data plots for air-side heat transfer in dry coils are presented for the three types of air-side fins introduced previously. Plain fins have been studied the most because they were the first fin type to be used. In addition, the development of general correlations for plain-fin tubes has been successful because they can be defined by only a few variables. In contrast, there are no general correlations for either corrugated surfaces or louvered fins, because of the large number of parameters re-

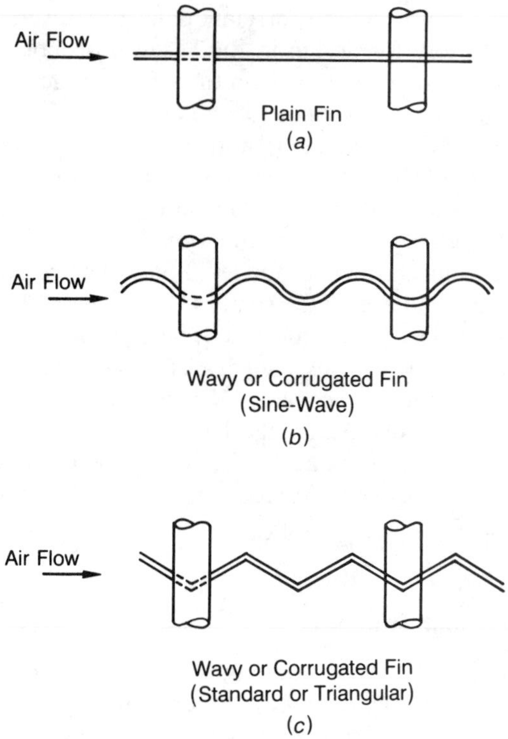

Fig. 12.14. Comparison of several fin plates used on the air side.

quired to define these surfaces. Several flat or plain surfaces used in refrigeration and air-conditioning applications are shown in Fig. 12.15.

Air-side heat transfer for air-conditioning coils with plain fins has been reported by McQuiston and Tree [20], Rich [21, 22], McQuiston [23], and Kays and London [24]. Most of these studies were for four-row heat exchangers with coil parameters as shown in Table 12.3. These experimental studies are important because nearly all of the correlations available in the literature have either been derived directly from this data bank or at least verified against it.

Several air-side heat transfer equations for dry coils are presented in the following discussion. A correlation by McQuiston [25] for a four-row plate-finned-tube heat exchanger using plain fins in terms of the j factor is

$$j_{4\text{-row}} = 0.0014 + 0.2618 Re_d^{-0.4} \left(\frac{A}{A_t} \right)^{-0.15} \tag{12.32}$$

where the Reynolds number is based on the tube outside diameter and the

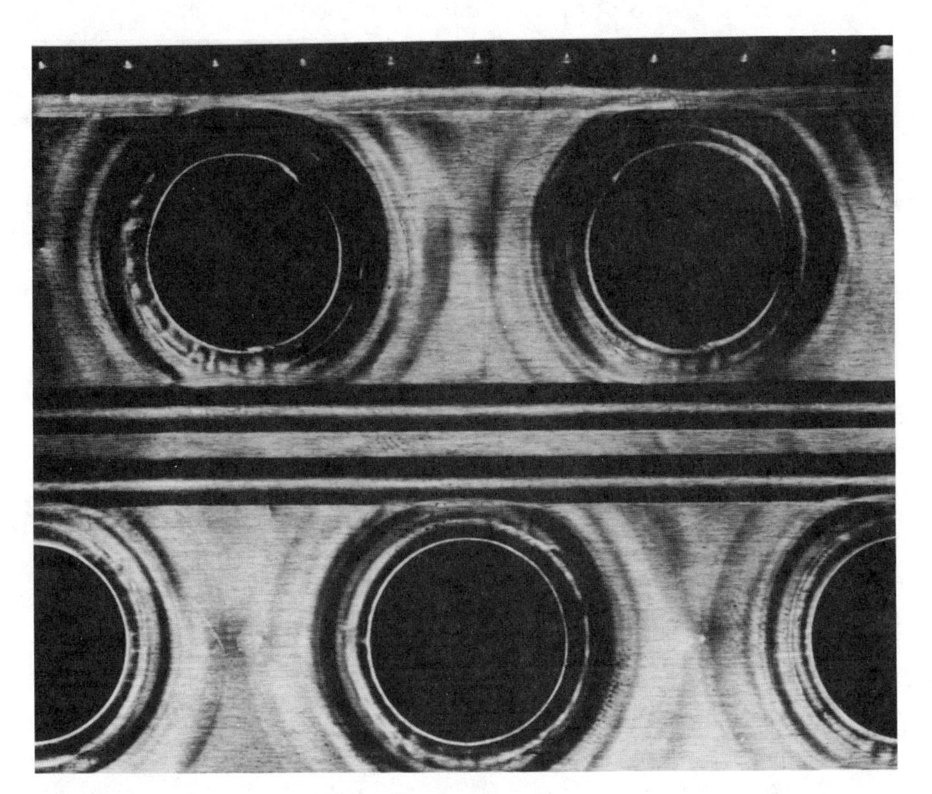

Fig. 12.15. Photograph of plain (flat) fin surfaces used in refrigeration and air-conditioning applications.

TABLE 12.3 Experimental Data for Four-Row Heat Exchangers (Listed in Chronological Order)

d, mm	Spacing, fins/m	t, mm	S_t/d	S_l/d	Number of Coils	Reference
10.2	315	0.33	2.49	2.16	1	Kays and London [24]
17.2	305	0.40	2.22	2.58	1	Kays and London [24]
10.4	157, 551	0.16	1.69	1.95	2	McQuiston and Tree [20]
13.3	115–811	0.15	2.39	2.07	8	Rich [21]
9.96	157–551	0.15	2.21	2.55	5	McQuiston [23]

area ratio is the ratio of the total surface area to the area of a bare tube bank. Another working equation derived by using data from the studies in Table 12.3 for a four-row coil is by Gray and Webb [26] as follows:

$$j_{4\text{-row}} = 0.14Re^{-0.328}\left(\frac{S_t}{S_l}\right)^{-0.502}\left(\frac{s}{D}\right)^{0.0312} \tag{12.33}$$

When the previous two equations are compared to the experimental data for some of the studies shown in Table 12.3, the rms deviation is less than 10%.

For coils with other than four rows of tubes, different procedures are followed depending on whether the number of tube rows is greater than or less than four rows. For example, the preceding equation for four rows is also applicable if the number of tube rows is greater than 4. This observation was verified by Gray and Webb [26] by comparing their equation results with experimental data for five-row and eight-row heat exchangers. It should be noted that these data were not used in the original derivation of their equation. When there are less than four tube rows, Gray and Webb recommend modifying their four-row equation by a factor calculated as follows:

$$\frac{j_N}{j_{4\text{-row}}} = 0.992\left[2.24Re^{-0.092}\left(\frac{N}{4}\right)^{-0.031}\right]^{0.607(4-N)} \tag{12.34}$$

where N is the number of tube rows. Figure 12.16 shows a comparison of the

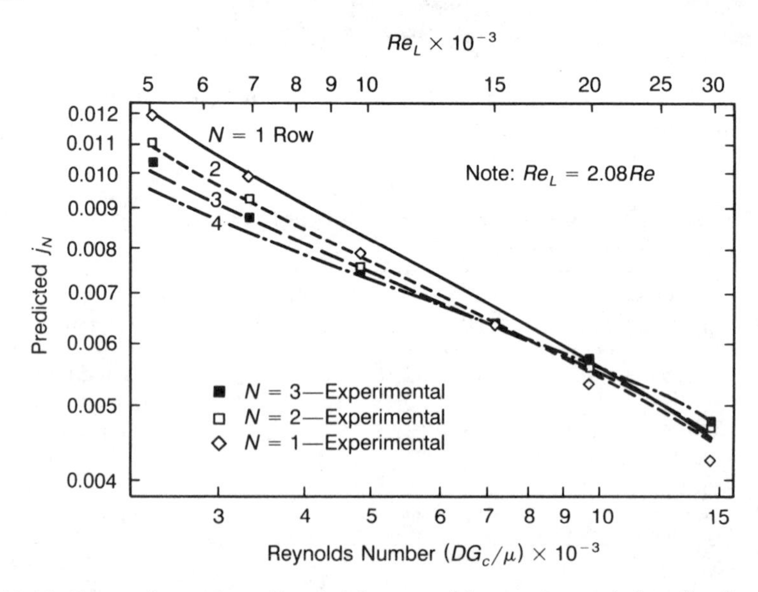

Fig. 12.16. Comparison of predicted j factors with experimental data for four-tube rows or less [26].

preceding equation with experimental data from Rich [22] that were used to originally derive Eq. (12.33).

The plain fins discussed previously can be modified slightly by metal stamping processes to form wavy fin patterns, also called corrugated or ripple fin patterns. Both sine-wave [27, 28] and triangular-shaped [29] (also called wedge-shaped) fins are found in air coils.

Several wavy (corrugated and ripple) fin surfaces used in refrigeration and air-conditioning applications are shown in Fig. 12.17. Because geometries vary slightly from manufacturer to manufacturer, nearly all of the past studies on wavy fins are proprietary. In fact, there are only three reports of experimental heat transfer and pressure drop data readily available in the literature [27–29]. The availability of correlations that account for geometric parameters is even more scarce, consisting only of correlations by Beecher and Fagan [29] for wedge-shaped fins in staggered tube arrangements. In the absence of working equations, some designers have applied enhancement factors to the plain-fin equations presented earlier [Eqs. (12.31) to (12.34)].

Fig. 12.17. Photograph of wavy (corrugated and ripple) fin surfaces used in refrigeration and air-conditioning applications.

Assuming that the factor is based either on experience or actual experiments, it can provide a method of accounting for the increased heat transfer caused by surface alterations and flow passage disruptions. An example of this approach is by Fisher and Rice [30] who used a factor of 1.45 increase in heat transfer over the plain fin.

The working equation for wedge-shaped fins developed by Beecher and Fagan [29] was derived from experiments performed on ripple fins with the following geometric parameters:

Fin patterns per longitudinal tube row (N_P)—2, 3, and 4
Fin pattern depths (P_d)—0.457 to 3.175 mm
Fin spacings (W_f)—1.956, 2.388, and 2.794 mm
Fin densities (for a fin thickness of 0.127 mm)—236, 343, 398, and 480
Transverse tube spacing (P_t)—25.4 and 31.75 mm
Tube diameter (D)—7.94, 9.53, and 12.7 mm
Number of tube rows (N_r)—3

For the preceding fin geometry, Beecher and Fagan used coil face velocities of 3.0 to 4.6 m/s, which correspond to maximum air velocities between the fins of 0.9 to 6.6 m/s. Nusselt numbers, Nu, and Graetz numbers, Gz, as a function of the previous parameters and flow conditions were used to correlate the experimental data. This approach is different from the j-factor approach used to correlate plain fins. Because of the large number of parameters, it was necessary for Beecher and Fagan to develop several different sets of fairly complicated equations. The reader is referred to Beecher and Fagan [29] for the details of the conditions, which can be used for design calculations if the fins and flow conditions are in the applicable range.

Louvered fins can achieve a higher heat transfer performance than either plain or wavy fins. The trade-off, however, is higher pressure drops and the potential for contamination by foreign matter when the fins are used in dirty environments. Contamination occurs as a result of slits and slots that have sharp corners where large particles of foreign matter (e.g., lint) can get caught. Louvered fins are made by cutting and then offsetting strips from a plain fin of the type described earlier. The resulting surfaces differ in the size, shape, and location of the strip, including the distance the strip is lifted above the plate and whether all strips are lifted uniformly. Louvered fins are also known as strip fins, slot fins, or offset strip fins. Several types of louvered and strip fins are shown in Fig. 12.18. Although the use of multiple terminology can become confusing, Webb [31] suggests that lifting uniformity distinguishes louvered fins from offset strip fins because in louvered fins only the leading and trailing edges are bent.

Heat transfer is enhanced as a result of the boundary layer being disrupted by the presence of relatively short slits and strips. Each time a strip is

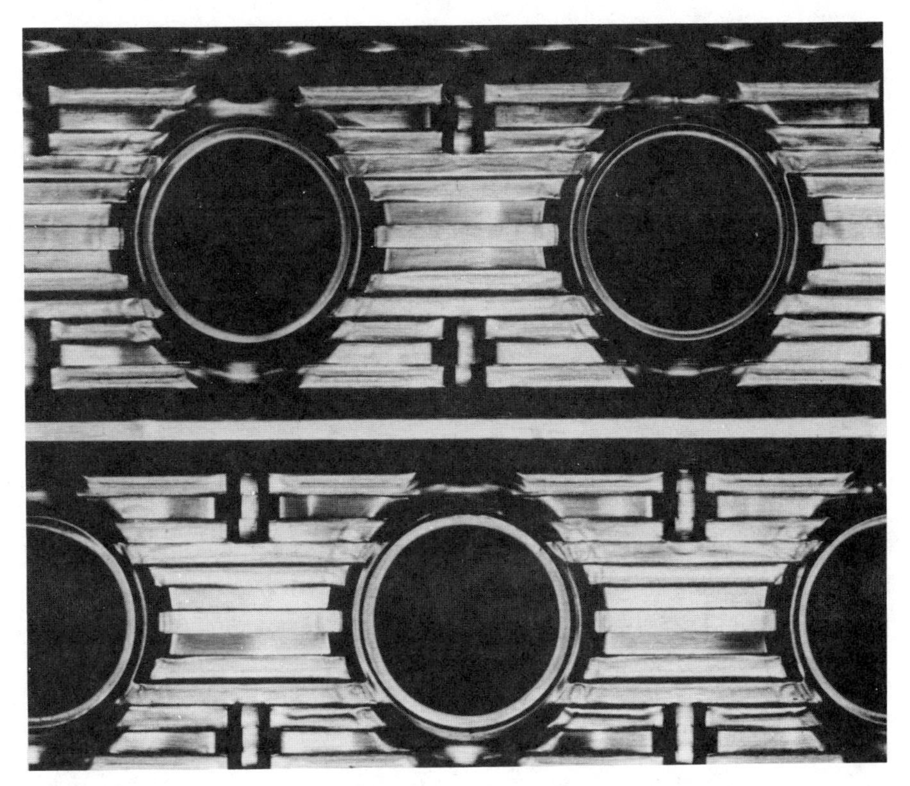

Fig. 12.18. Photograph of louvered fins (also referred to as strip fins, slot fins, and offset strip fins) used in refrigeration and air-conditioning applications.

encountered, the boundary layer is broken up and then formed again. The overall result is a thinning of the boundary layer that results in an increase in the local heat transfer coefficient. The boundary layer thinning also increases the friction factor, f, which in turn increases pressure drops as the air flows through the coil.

Several studies evaluating the performance of strip fins have been reported in the literature [31–36]. For example, Hosada et al. [32] report 60% increases in heat transfer coefficients for louvered fins compared to wavy fins. However, only the study by Nakayama and Xu [37] reports a predictive correlation that can be used in heat exchanger design. They developed correlations for j factors and friction factors by modeling local variations in heat transfer coefficients and then obtaining average coefficients for the entire fin surface.

The correlation developed by Nakayama and Xu in terms of a j factor is

$$j = 0.479Re^{-0.644}F_j \tag{12.35}$$

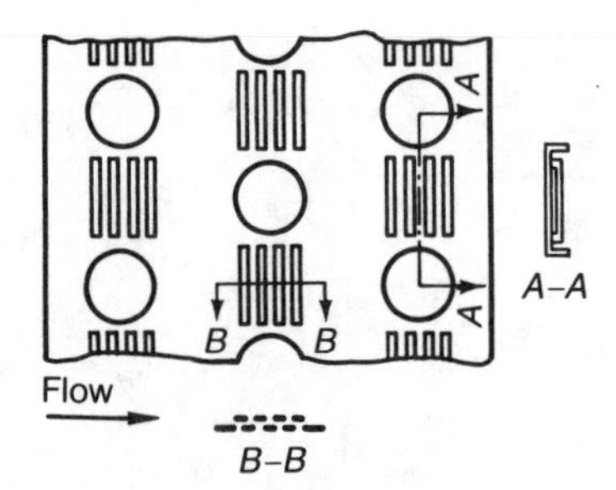

Fig. 12.19. Geometry of an enhanced fin [37].

where

$$F_j = 1 + 1.093 \times 10^3 \left(\frac{\delta_f}{\delta_a} \right)^{1.24} \phi_s^{0.944} Re^{-0.58} + 1.097 \left(\frac{\delta_f}{\delta_a} \right)^{2.09} \phi_s^{2.26} Re^{0.88} \tag{12.36}$$

The strip-fin arrangement, including tube locations, applicable to this correlation is shown in Fig. 12.19. Figure 12.20 shows good agreement between the

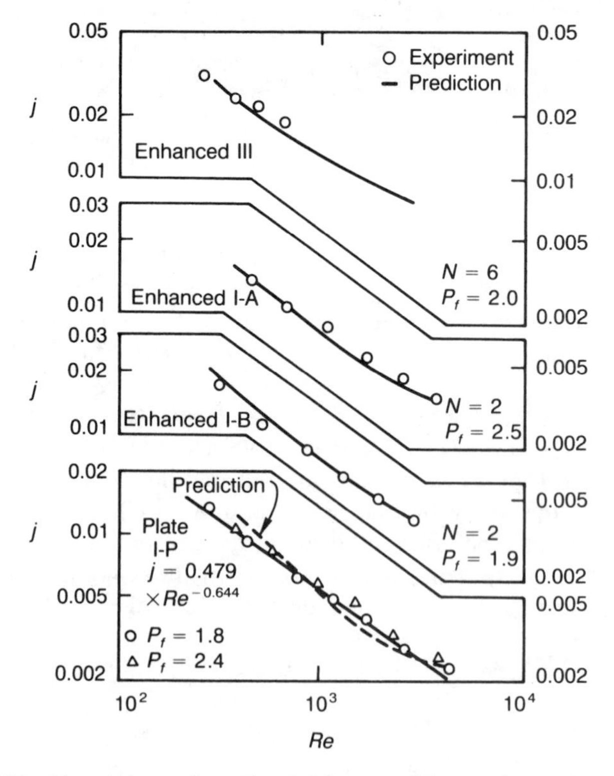

Fig. 12.20. Comparison of predicted j factors with experimental data [37].

preceding correlation and the experimental data for three different strip-fin surfaces and a plain-fin surface. Surface 1 has three strips in each enhanced zone and two rows of tubes while surface 111 has four strips and six tube rows. Surface 1 is further divided into two surfaces with different fin pitches, namely a fin pitch of 2.5 mm for 1-A and 1.9 mm for 1-B. Also shown in Fig. 12.20 is the enhancement capability of strip-fin surfaces over plain-fin surfaces. For example, the j factor for surface 111 is 150% higher than that of a plain-fin surface at a Reynolds number of 1000.

Other investigators have used numerical techniques to model heat transfer from louvered and strip fins. These studies have provided designers with valuable information regarding flow patterns and flow phenomena [38, 39]. However, because of the complicated nature of heat transfer and fluid flows in louvered fins, additional work is necessary before numerical models can be used to design louvered-fin heat exchangers.

12.3.5 Wet-Coil Heat Transfer

Where the evaporator coil surface temperature is reduced below the dew-point temperature of the air, water is condensed out of the air. Under these circumstances the air-side surface of the coil becomes wetted and the evaporator coil is referred to as a wet coil or moist coil. The evaporator coil now serves the dual function of both cooling and dehumidifying the air. Depending on local air-side temperature variations and on the entering humidity, either part or all of the coil is wetted.

Because of the complicated nature of moist coils, a wide range of assumptions can be made when deriving correlations and equations for analysis. As such, the approaches used in the past differ from each other considerably. Some of these approaches are by Fischer and Rice [30], Threlkeld [40], and Stoecker and Jones [41]. Rather than a detailed presentation of equations, which are complicated and vary considerably depending on assumptions made, only a general discussion of moisture removal and dehumidification will be presented.

The presence of moisture of the air-side surfaces affects heat transfer in three ways. First, the condensing moisture enhances convective heat transfer from the air to the fin surfaces. This enhancement is most likely due to the moisture layer creating an uneven or rough surface on the smooth fin surface. Two similar equations available in the literature that account for this heat transfer enhancement are [42]:

$$h_{\text{wet}} = h_{\text{dry}} 0.367 V_a^{0.101} \tag{12.37}$$

and [43]:

$$h_{\text{wet}} = h_{\text{dry}} 0.555 V_a^{0.101} \tag{12.38}$$

Second, the presence of a water layer on the fin surface adds another thermal resistance to the heat transfer path. The thermal conductivity and thickness of the water layer can be used to approximate this resistance. How this resistance gets incorporated into the heat flow path varies from study to study. For example, one approach by Threlkeld [40] uses the water thermal conductivity and the film thickness to modify the air-side heat transfer coefficient. The water film thickness may be difficult to determine. However, since the thermal resistance due to the water film is small, errors in the estimate may have only a minor effect on that transfer calculation.

Third, because heat transfer also occurs as a result of moisture migration and subsequent condensation on the coil surface, the water vapor concentration is also a driving potential for heat transfer. As with other phenomena in moist coils, a range of assumptions can be made and, hence, different approaches can be used. For example, Threlkeld [40] defines the overall heat transfer coefficient in terms of an air enthalpy difference rather than a temperature difference. The mean enthalpy difference is calculated from a logarithmic-mean enthalpy difference, which is a function of a combination of true air-stream enthalpies and fictitious saturated air-stream enthalpies.

It should be noted that if only part of the coil is wet, then it can be divided into two parts. The wet portion of the coil can be analyzed by accounting for the phenomenon described previously, while the dry portion can be analyzed by using conventional approaches.

12.3.6 Frosted-Coil Heat Transfer

If the air-side surface temperature of an evaporator coil is below the freezing point of water, then the potential for air-side frosting exists. The buildup of frost lowers the capacity of the evaporator by adding an extra thermal resistance due to conduction through the frost layer. Since the air-flow channel area between the fins is reduced, the flow of air through the coil is restricted. In turn, the air-side convection coefficient is lowered, the pressure drop through the coil increases, or both. Coil frosting is a transient phenomenon in that frost builds up until a defrosting mechanism is initiated.

Kondepudi and O'Neal [44] have performed a detailed search and evaluation of the literature dealing with frosting of air coils. They made several observations:

1. The air-side heat transfer coefficient increases during initial frosting because of increased surface roughness. However, an increase in the thermal resistance of the frost layer soon offsets the surface roughness effect, and the overall heat transfer coefficient decreases.
2. The frost buildup is more severe near the front face.
3. A wider fin spacing is better than a narrow fin spacing for heat transfer performance during frosting. Coils with variable fin spacings, with the fin spacing increasing downstream, are best.
4. The fin efficiency increases initially with frost growth because of a more uniform temperature distribution over the fin. A constant value is then approached.

Because of the lack of correlations or even raw data in the open literature dealing with the frosting of air coils, a designer is often forced to use simple geometries, such as cylinders and plates. One of the few correlations available in the literature for frost buildup is by O'Neal [45]. For $Re < 15{,}900$, the frost growth in millimeters is

$$x_f = 0.015 Re^{0.393} t^{0.663} (-T_p)^{0.705} (w_a - w_o)^{0.098} \tag{12.39}$$

while for $Re > 15{,}900$ it is

$$x_f = 0.712 t^{0.582} (-T_p)^{0.705} (w_a - w_o)^{0.098} \tag{12.40}$$

The preceding correlation can be used by a designer to approximate the thickness of the frost layer as a function of time. The added thermal resistance due to conduction heat transfer through the frost layer, or changes in air-flow rate because of channel restrictions, can then be calculated.

As mentioned previously, frosting also affects the air-side heat transfer coefficient because the initial frost formation increases roughening of the fin-plate surface. Experimental studies have shown that this increase can last for several minutes until other factors, such as flow restrictions, cause the heat transfer coefficient to decrease. A study by O'Neal [45] for parallel plates with frost initiation showed that a correlation similar to the well-known Dittus—Boelter equation with the coefficient changed from 0.023 to 0.034 can correlate experimental data. These results suggest that in the absence of additional data or correlation, a possible approach for a designer is to increase a frostless air-side heat transfer coefficient by a factor of 1.5, or 50%, to account for surface roughening during frost initiation.

12.3.7 Fin Bonding and Thermal Contact Resistance

In air-conditioning and refrigeration coils, plate fins are attached to tubes by expanding the tube either mechanically or hydraulically into the fin collar. As a result, a discontinuity in the region where the plate fins attach to the round tube surface introduces a thermal resistance in the heat path between the air and refrigerant. Compared to the overall thermal resistance, the thermal contact resistance at this discontinuity can range from 5% to 50%. This wide range is the result of the different applications (e.g., refrigerator coils, air-conditioning coils, etc.) and different manufacturers who apply different tolerances and use different processes for attaching the plate fin to the round tubes.

Most heat exchanger manufacturers apply "rules of thumb" for their particular coil to account for this contact resistance. This estimated value can be in terms of a percentage of overall thermal resistance (e.g., 15%), or in terms of a thermal contact resistance or a thermal contact conductance [e.g., 2000 $\mathrm{Btu/(hr \cdot ft^2 \cdot {}^\circ F)}$]. Because of the proprietary nature of coil design and manufacturing, there is no published information on design rules for various applications. However, for cost reasons home refrigeration coils generally have a higher value of thermal contact resistance than air coils found in industrial air-conditioning systems.

One of the few general correlations available in the literature which do not require information regarding the manufacturing process was developed by Wood et al. [46] by using a procedure proposed by Eckels [47] and experimental data from 31 coils. The resulting equation in terms of contact conductance, h_c, which is the reciprocal of thermal contact resistance, is

$$h_c = \exp[a + bx] \tag{12.41}$$

where $a = 7.247$, $b = 3861.00$, and the value of x is

$$x = \frac{t}{(t_{\text{fpi}} - 1)^2 \text{OD}} \tag{12.42}$$

The units of h_c are Btu/(hr · ft^2 · °F).

Using the same data bank from 31 coils, Wood et al. [46] also derived a more general equation that requires knowledge of the tube diameter before and after the expansion process. The resulting equation is

$$h_c = \exp\left\{6.092 + 2.889\left[\left(I \cdot \text{fpi}\frac{d}{\text{OD}}\right)^{0.75} (t \cdot \text{fpi})^{1.25}\right]\right\} \tag{12.43}$$

where t is in units of inches and the tube-to-collar interference, I, is defined as

$$I = \frac{d_e + 2t_w - d_h}{2} \tag{12.44}$$

An indicator of the accuracy of calculating the thermal contact resistance by using Eq. (12.43) can be seen in Fig. 12.21. In this figure, about 50% of the data points can be predicted within the ±20% limits shown. In addition, one can observe in Fig. 12.21 that of the 31 coils tested, more than half of the coils had thermal conductances of less than 1000 Btu/(hr · ft^2 · °F), signifying higher thermal contact resistances.

If the equations presented previously are not usable because certain parameters are unknown, then on the basis of the data plotted in Fig. 12.21, a value of 2500 Btu/(hr · ft^2 · °F) represents an average thermal conductance which can be used in calculations. It should be noted again that the preceding equations and figure are for larger air-conditioning coils, which are

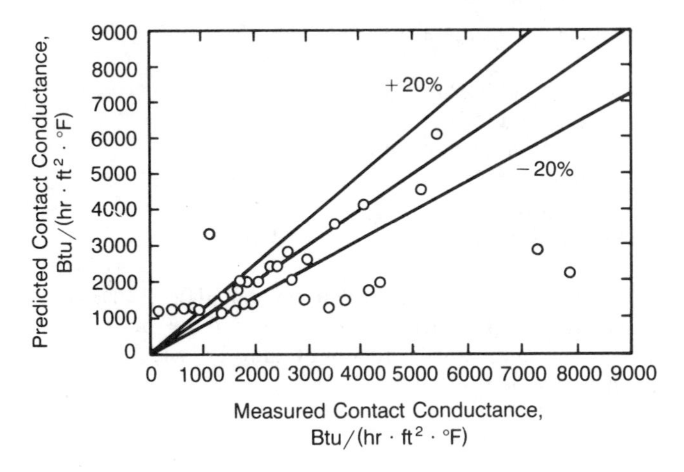

Fig. 12.21. Comparison of predicted and measured contact conductance [46].

generally built with tighter contact between the fin and tube, at least compared to smaller, lower priced (household) refrigerator coils.

Example 12.1. Overall Heat Transfer Coefficient Calculation for an Evaporator Coil. A calculation is performed to demonstrate the use of the equations and methodology presented in Chapter 12 with emphasis on Section 12.3. Many of the equations used in this example are also applicable to Section 12.4.

The dimensions of the coil and fluid conditions, both refrigerant and air, are as follows.

The geometrical parameters are as follows:

Tube inside diameter	0.00826 m
Tube outside diameter	0.00952 m
Width of heat exchanger	1.25 m
Number of rows	4
Number of independent circuits	4
Horizontal tube spacing	0.0222 m
Vertical tube spacing	0.025 m
Number of tubes vertical to air-flow direction	12
Number of tubes along air flow direction	4
Fin thickness	1.6154×10^{-4} m
Fin pitch	551.18 m
Fin surface area/total outside surface area	0.948
Free air-flow area/frontal area	0.5466

The fluid conditions are as follows:

Refrigerant flow rate	19.3 kg/hr (per tube path)
Refrigerant inlet temperature	−5°C
Refrigerant inlet quality	0.1
Refrigerant exit condition	0°C superheat
Air velocity	2 m/s
Air inlet temperature	15°C

The overall heat transfer coefficient equation, (Eq. 12.1), can be simplified by assuming a fin efficiency, ϕ, of 100% and by neglecting the thermal resistances associated with the tube wall and the inside and outside deposits. The resulting equation is

$$U_o = \frac{1}{\dfrac{A_o}{A_{pi}h_i} + \dfrac{1}{h_o} + \dfrac{A_o}{h_c A_{po}}}$$

The area ratios in the preceding equation are calculated as follows:

$$\frac{A_o}{A_{po}} = \frac{1}{1 - 0.948} = 19.23$$

and

$$\frac{A_o}{A_{pi}} = \frac{A_o}{A_{po}}\frac{A_{po}}{A_{pi}} = \frac{A_o}{A_{po}}\frac{D_o}{D_i} = 19.23\left(\frac{0.00952}{0.00826}\right) = 22.16$$

The in-tube evaporation heat transfer coefficient, h_j, is calculated from the equation developed by Kandlikar [7], (Eq. 12.21). It should be noted that any of several other in-tube evaporation equations presented in Chapter 12 could also have been used. The properties for R-134a at −5°C are listed in Table 12.7.

Several nondimensional numbers used in the Kandlikar equation are calculated as follows:

Boiling number:

$$Bo = \frac{q}{Gh_{fg}} = \frac{7.4\ \text{kW/m}^2\ [\text{kJ/(s} \cdot \text{kW)}]}{(100\ \text{kg/(m}^2 \cdot \text{s)})(202.3\ \text{kJ/kg})}$$

$$B_o = 3.64 \times 10^{-4}$$

Convection number:

$$Co = \left(\frac{1-x}{x}\right)^{0.8}\left(\frac{\rho_g}{\rho_l}\right)^{0.5} = \left(\frac{1-0.5}{0.5}\right)^{0.8}\left(\frac{12.2\ \text{kg/m}^3}{1308\ \text{kg/m}^3}\right)^{0.5}$$

$$= 0.0966$$

Froude number:

$$Fr = \frac{G^2}{(\rho_l^2 g D)} = \frac{[100\ \text{kg}/(\text{m}^2 \cdot \text{s})]^2}{(1308\ \text{kg}/\text{m}^3)^2 (9.8\ \text{m}/\text{s}^2)(0.00826\ \text{m})}$$

$$= 0.0722$$

All-liquid Reynolds number

$$Re_l = \frac{G(1-x)D}{\mu_l} = \frac{[100\ \text{kg}/(\text{m}^2 \cdot \text{s})](1-0.5)(0.00826\ \text{m})}{301 \times 10^{-6}\ \text{Pa} \cdot \text{s}\left(\dfrac{\text{kg} \cdot \text{m}}{\text{s}^2 \cdot \text{m}^2 \cdot \text{Pa}}\right)}$$

$$= 1372$$

The liquid-only convective heat transfer coefficient is calculated from Eq. (12.22) as follows:

$$h_l = 0.023 Re_l^{0.8} Pr_l^{0.4} k_l / D$$

$$= 0.023(1372)^{0.8} 3.98^{0.4} [0.0981\ \text{W}/(\text{m} \cdot \text{K})]\ (0.00826\ \text{m})$$

$$= 153.6\ \text{W}/(\text{m}^3 \cdot \text{K})$$

The evaporation heat transfer coefficient calculated from Eq. (12.21), where constants C_1 through C_6 were presented earlier and $F_{fl} = 1.5$ for R-134a, is

$$h_{TP} = h_l \left[C_1 Co^{C_2} (25 Fr_l)^{C_5} + C_3 (Bo)^{C_4} F_{fl} \right]$$

$$= 153.6\ \text{W}/(\text{m}^2 \cdot \text{k})(1.136)(0.0966)^{-0.9} + (667.2)(3.64 \times 10^{-4})^{0.7}(1.3)$$

$$= 2048\ \text{W}/(\text{m}^2 \cdot \text{K})$$

It should be noted that the preceding value in the evaporation heat transfer coefficient is in agreement with the $X = 0.5$ value taken from Fig. 12.9. The outside heat transfer coefficient, h_o, can be calculated from Eq. (12.32) where the properties of the air are based on an average air temperature through the coil. Because this average air temperature requires an iterative procedure which is part of the heat exchange design problem, the average air temperature is assumed to be equal to the inlet air temperature for this initial calculation.

The j factor from Eq. (12.32) is

$$j_{4\text{-row}} = 0.0014 + 0.2618 Re_d^{-0.4} \left(\frac{A}{A_t} \right)^{-0.5}$$

$$= 0.0014 + 0.2618 \left[\frac{(1.1054 \text{ kg/m}^3)(2 \text{ m/s})(0.00952 \text{ m})}{(178.6 \times 10^{-7} \text{ Kg} \cdot \text{m/m}^2 \cdot \text{s})} \right]^{-0.4}$$

$$\times (19.23)^{-0.15}$$

$$= 0.001133$$

The outside heat transfer coefficient can be calculated from the j factor as follows:

$$j = StPr^{2/3} = \frac{h}{\rho V c_p} Pr^{2/3}$$

by rearranging

$$h_o = \frac{j \rho V c_p}{Pr^{2/3}} = (0.001133)(1.1054 \text{ kg/m}^3)(2 \text{ m/s})[1.0068 \text{ kJ/(kg} \cdot \text{K)}]$$

$$= 540$$

The final variable required to calculate the overall heat transfer coefficient is the thermal contact resistance for the region where the fins are mechanically bonded to the tube wall.

A value of 2500 Btu/(hr · ft^2 · °F), which is recommended in Section 12.3, is used. In SI units, this contact conductance is

$$h_c = 14{,}195 \text{ W/(m}^2 \cdot \text{K)}$$

The overall heat transfer coefficient is then calculated as follows:

$$U_o = \frac{1}{\dfrac{A_o}{Ap_i h_i} + \dfrac{1}{h_o} + \dfrac{A_o}{h_c Ap_o}}$$

$$= \frac{1}{\dfrac{22.16}{2190 \text{ W/(m}^2 \cdot \text{K)}} + \dfrac{1}{540 \text{ W/(m}^2 \cdot \text{K)}} + \dfrac{19.23}{14{,}195 \text{ W/(m}^2 \cdot \text{K)}}}$$

$$= 71.3 \text{ W/(m}^2 \cdot \text{K)}$$

12.4 CONDENSER COILS

12.4.1 Description and Special Considerations

Condenser coils usually consist of arrays of copper tubes that penetrate at right angles closely stacked aluminum plates. These plates provide a system of continuous fins for enhancing air-side heat transfer. This approach is similar to that used to make evaporator coils that were described in the previous section. Variations on this arrangement can also be found, such as tube circuits consisting of integral spiny fins on an aluminum base that is wrapped around the copper tube. In this arrangement, there are no thermal bridges between tubes because the spines in one tube do not connect with others. For plate-fin coils, the fins are usually spaced with a density of 8 to 18 fins per inch (3.2 to 1.4 mm spacing). Typical tube diameters range from 1/4 to 3/4 in. OD (6 to 20 mm). The advantage of smaller-diameter tubes is that they reduce the refrigerant charge.

Even though air-cooled condenser coils can be cooled by natural convection, they are usually installed with fans. Propeller fans are usually used if the condenser is installed outdoors while centrifuged fans are used when air is directed to the condenser through a duct. Air-flow requirements vary from 600 to 1200 cfm per ton (80 to 160 L/s per kW). Single-unit condenser coils are designed up to several hundred tons. Larger-tonnage refrigeration and air-conditioning systems use multiple condensers connected in parallel.

Coil circuits are arranged so that the superheated vapor enters the top and the subcooled liquid exits near the bottom. This approach assists with oil circulation. Even more importantly, it maintains a liquid seal at the coil exit while using gravity to aid in refrigerant circulation. The air-side arrangement is such that air flows across the tubes in crossflow. The overall tube circuiting is usually arranged so that the refrigerant enters on the downstream of the air side and exits on the upstream of the air side. This approach produces a type of counterflow arrangement for the air and refrigerant.

The number of separate tube circuits and the number of passes for each circuit are based on having sufficient flow rate in each circuit to obtain satisfactory in-tube refrigerant heat transfer while limiting the pressure drop of the refrigerant. Refrigerant-side pressure drops are detrimental because they decrease the temperature difference between the two fluids. An optimum coil arrangement may consist of multiple circuits that join downstream in the low-quality region closer to the exit. In this arrangement, the maximum number of circuits exist at the superheated inlet, and the minimum number of circuits exist at the subcooled outlet.

12.4.2 Similarities between Condenser and Evaporator Coils

Several similarities exist between evaporator coils and condenser coils used in refrigeration and air-conditioning applications. The coils are constructed from similar materials, namely copper tubing and aluminum plates, and similar techniques are used to fasten the plates and tubes together. Similar heat transfer enhancement techniques are used for both the refrigerant inside the tubes, namely microfinning, and for the air flowing through the plates attached to the outside of the tubes, namely ripple fins and louvered fins.

Because of these similarities, several sections presented during the discussion of evaporator coils are directly applicable to condenser coils. Specific sections that are applicable to condenser coils are the sections on air-side heat transfer (Section 12.3.4) and fin bonding and thermal contact resistance (Section 12.3.7). On the refrigerant side, discussion presented earlier on single-phase flow and oil circulation is also applicable. Even though the same microfin tubes are used for condensation and evaporation, the effect on heat transfer of microfinning is completely different.

12.4.3 In-Tube Refrigerant Condensation Heat Transfer

In-tube condensation heat transfer coefficients must be calculated before one determines the overall heat transfer coefficients. Several correlations available in the literature have been verified for use with refrigerants. Some of these correlations are described in the following discussion. They are the correlation by Traviss et al. [48], the correlation by Cavallini and Zecchin correlation [49], and the Shah correlation [50]. For heat transfer in the superheated and subcooled regions of the condenser, the single-phase correlations discussed earlier during in-tube evaporation are applicable.

The correlation by Traviss et al. [48] was originally developed as part of an extensive study of condensation of R-12 and R-22, two refrigerants widely used today. The correlation was also verified successfully by using these same R-12 and R-22 data. This in-tube condensation equation was derived by applying the momentum and heat transfer analogy to an annular flow model. The velocity distribution in the annular film was described by the von Kármán universal velocity distribution. Radial temperature gradients in the vapor core were neglected, and a saturation temperature was assumed at the liquid–vapor interface. The resulting two-phase heat transfer coefficient is

$$h_{\mathrm{TP}} = \frac{k_l Pr_l Re_l^{0.9} F_1}{D F_2} \tag{12.45}$$

where the liquid Reynolds number is

$$Re_l = \frac{G(1-x)D}{\mu_l} \tag{12.46}$$

and the nondimensional parameter, F_1, is

$$F_1 = 0.15\left[\frac{1}{X_{tt}} + \frac{2.85}{X_{tt}^{0.476}}\right] \tag{12.47}$$

Three functions are given for F_2, with the choice of which function to use in the correlation being dependent on the Reynolds number range. The three functions are

$$F_2 = 0.707 Pr_l Re_l^{0.5} \quad \text{for } Re_l \leq 50 \tag{12.48}$$

$$= 5Pr_l + 5\ln\left[1 + Pr_l\left(0.09636 Re_l^{0.585} - 1.0\right)\right]$$

$$\text{for } 50 < Re_l \leq 1125 \tag{12.49}$$

$$= 5Pr_l + 5\ln(1 + 5Pr_l) + 2.5\ln\left(0.00313 Re_l^{0.812}\right)$$

$$\text{for } Re_l > 1125 \tag{12.50}$$

where Reynolds number is defined in Eq. (12.46).

The preceding correlation has been used extensively in the past. However, two more recent correlations, described next, are simpler to implement, and in addition, they have been shown to correlate experimental data just as well [11].

Cavallini and Zecchin [49] developed a semiempirical equation that is simple in form and correlates refrigerant data quite well. Data for several refrigerants, including R-11, R-12, R-21, R-22, R-113, and R-114, were used to derive and verify the correlation. The basic form of the correlation was developed from a theoretical analysis similar to that used by Traviss et al. [48].

The working equation suggested by Cavallini and Zecchin is

$$h_{\text{TP}} = \frac{k_l}{D} 0.05 Re_{\text{eq}}^{0.8} Pr_l^{0.33} \tag{12.51}$$

where the equivalent Reynolds number, Re_{eq}, is defined by

$$Re_{\text{eq}} = Re_v\left(\frac{\mu_v}{\mu_l}\right)\left(\frac{\rho_l}{\rho_v}\right)^{0.5} + Re_l \tag{12.52}$$

The equation for Re_l was presented earlier in Eq. (12.46), and Re_v is defined similarly as

$$Re_v = \frac{GxD}{\mu_v} \tag{12.53}$$

The Cavallini–Zecchin correlation is very similar in form to any one of several single-phase turbulent correlations (e.g., the well-known Dittus–Boelter equation). Cavallini and Zecchin also suggest that their equation can be used to calculate the average heat transfer coefficients between the condenser inlet and outlet, providing the thermophysical properties and the temperature difference between the wall and fluid do not vary considerably along the tube.

The Shah [50] correlation was developed from a larger group of fluids, including water, than the previous correlations. It was developed by establishing a connection between condensing and Shah's earlier correlations for boiling heat transfer without nucleate boiling. The resulting correlation in terms of h_l, as defined earlier in Eq. (12.22), is

$$h_{\mathrm{TP}} = h_l\left(1 + \frac{3.8}{Z^{0.95}}\right) \tag{12.54}$$

where

$$Z = \left(\frac{1-x}{x}\right)^{0.8} P_r^{0.4} \tag{12.55}$$

Shah also suggested integrating these equations over a length of tubing to obtain the mean heat transfer coefficient in the condensing region:

$$h_{\mathrm{TP}m} = \frac{1}{L}\int_0^L h_{\mathrm{TP}}\, dL \tag{12.56}$$

For the case of a linear quality variation over a 100% to 0% range, the result is

$$h_{\mathrm{TP}m} = 0.55h_{1\phi} + \frac{2.09}{P_r^{0.38}} \tag{12.57}$$

The results from this equation differ by only 5% from the value obtained when a mean quality 50% is used in the local heat transfer correlation, Eq. (12.54).

The local heat transfer coefficients for the previous three correlations are compared in Fig. 12.22. As with the comparisons of in-tube evaporation correlations presented earlier, an ozone-safe refrigerant, namely R-134a, is used. Flow rate, temperature, tube diameter, and tube length conditions

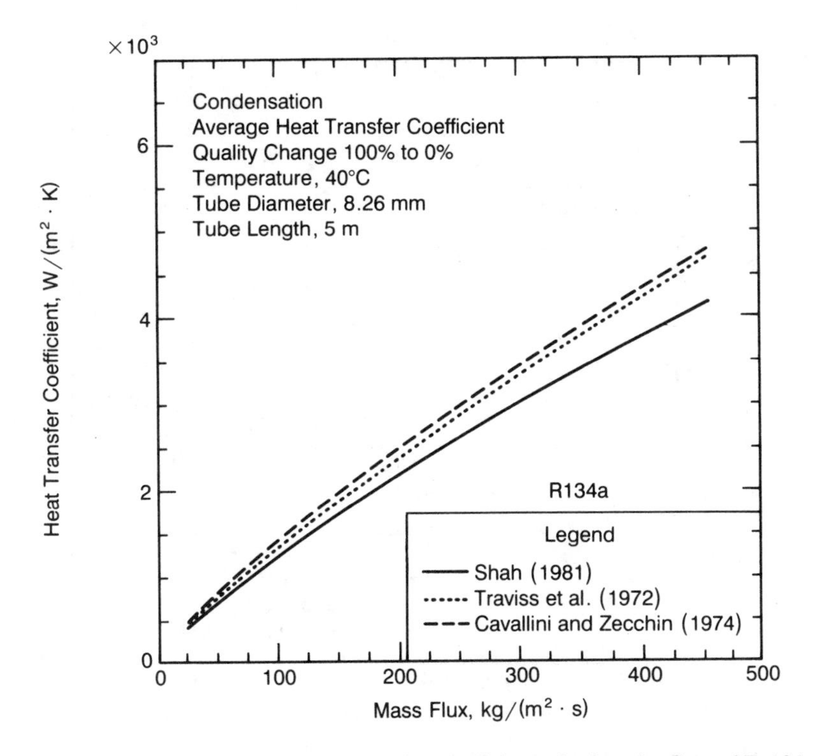

Fig. 12.22. Local condensation heat transfer coefficients for in-tube flow of R-134a.

similar to a typical condenser were selected for this comparison. Over most of the quality range, the correlations agree with each other to within 20%. The local heat transfer coefficients decrease as the quality decreases, which is the result of the annular film thickness increasing as condensation proceeds from the high-quality inlet of the condenser to the low-quality exit.

Average heat transfer coefficients, obtained by using Eq. (12.56) to integrate the local value over tube length, are plotted as a function of mass flux in Fig. 12.23. All three correlations show good agreement, however two of the correlations agree to within 10% of each other. Average coefficient data from several past experimental studied have been predicted to within ±20% by these three correlations [13, 14].

As with evaporation, condensation heat transfer coefficients calculated from the preceding correlations can be modified to account for the presence of lubricants. As before, an equation is presented in terms of a type of enhancement factor, EF, defined as the ratio of heat transfer coefficients for refrigerant–oil mixtures to coefficients for pure refrigerant. For condensation of R-22 mixed with SUS oil, the applicable equation is [14]:

$$\mathrm{EF}_{s'/s} = e^{-3.2W_o} \tag{12.58}$$

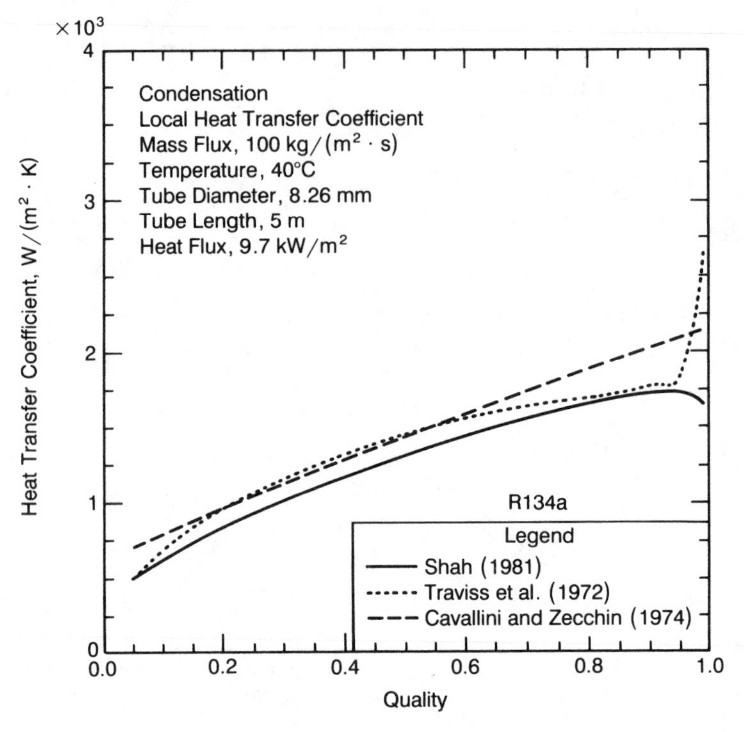

Fig. 12.23. Average condensation heat transfer coefficients for in-tube flow of R-134a.

The subscripts s and s' and the lubricant concentration W_o, were defined earlier during the discussion on in-tube evaporation. Unlike the equation for evaporation, Eq. (12.58) is not a function of mass flux, and, in addition, the in-tube condensation heat transfer coefficient always decreases with oil addition.

A similar correlation for R-22 and 300 SUS oil is [14]:

$$\mathrm{EF}_{s'/s} = e^{-5.0W_o} \tag{12.59}$$

As mentioned previously Eqs. (12.58) and (12.59) are convenient in that refrigerant properties do not have to be modified to account for changes due to the presence of lubricants.

12.4.4 In-Tube Heat Transfer Augmentation

The most popular method for providing heat transfer enhancement for in-tube condensation of refrigerants is the microfin tube. This type of tube was described in detail in Section 12.3.3. As with evaporation, in-tube condensation heat transfer coefficients have been increased by factors of 2 to 3 over smooth-tube values. Again, pressure drop increases are significantly less than heat transfer increases, unlike other types of in-tube enhancement.

The following design information for microfin tubes is for different microfin tube geometries and for different tube diameters. The effects on microfin tubes of lubricants mixed with refrigerants are also described.

The performance of three different 3/8-in. (9.52-mm) microfin tubes are compared to the performance of 1/2-in. (12.7-mm) tubes in Fig. 12.24 [15, 16]. This comparison is for R-22; however, in the absence of additional information, the results can be applied to other refrigerants. These same six tubes were described previously in Table 12.1 during the discussion on evaporation heat transfer in microfin tubes. One can see that the enhancement factors for the larger-diameter tube are slightly larger than they are for the smaller-diameter tube. Specifically, at a mass flux of 300 kg/(m^2 · s), which represents a region where data for the two different diameters overlap, the enhancement factor, EF, for the 1/2-in. (12.7-mm) tubes range from 1.6 to 1.8, while the smaller 3/8-in. (9.52-mm) tube ranges from 1.5 to 1.7. Considering the fact that the authors report experimental uncertainties of ±14% for EF, the effects of both tube diameter and microfin geometry are minor to the point of being negligible. If a designer is performing heat exchanger calculations for one of the preceding tube geometries and diameters, then values of EF for given mass fluxes can be selected from Fig. 12.24 directly. However, if design calculations are required for other tube diame-

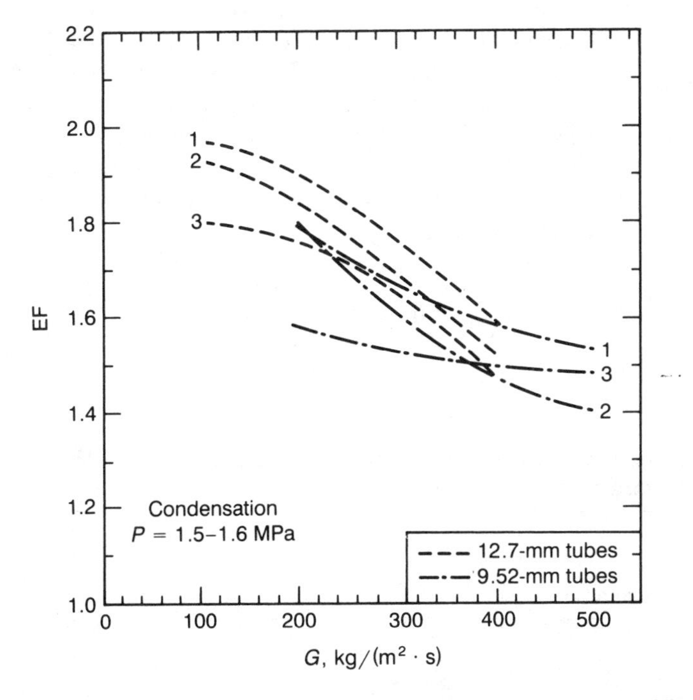

Fig. 12.24. Condensation heat transfer enhancement factors for three different microfin tube geometries and two different diameters [15, 16].

ters, such a 3/4-in. or 1-in. tubes, or other microfin tube geometries, then Fig. 12.24 can also be used in the absence of more applicable heat transfer performance data.

As with the evaporation design obtained by Schlager et al. [17], data for a commonly used microfin tube is also presented for in-tube condensation. As described previously, the 9.52-mm (3.8 in.) OD tube had 60 fins with fin heights of 0.2 mm (0.008 in.) and spiral angles of 18°. The in-tube condensation results for a mass flux of 300 kg (in.^{2} · s) are shown in Table 12.4 for pure R-22 and for oil concentrations of 1.25%, 2.5% and 5%. This table is analogous to Table 12.2 for evaporation. As shown in Table 12.4, there is some slight decrease in the enhancement factor as oil is added to the refrigerant; however, this change is minor. Not so obvious in Table 12.4 is the fact that the in-tube condensation heat transfer coefficient decreases as oil is added to the refrigerant. For example, decreases in heat transfer coefficients can be as high as 20% at 5% oil concentrations. However, oil affects heat transfer in both smooth tubes and microfin tubes similarly, and as a result the enhancement factor changes only slightly.

A generalized equation for condensation heat transfer as a function of geometrical parameters in a microfin tube would be extremely useful to the designer. However, as for evaporators, no such equation is available in the literature. An attempt was made by Schlager et al. [14] to correlate data for several different microfin tube geometries from several different studies. However, they observed that much of the data available from the literature showed inconsistent trends such that it was not possible to develop a generalized equation. Instead, they devised a one-parameter enhancement factor correlation as a function mass flux for their own data taken on a microfin tube with R-22. The microfin tube used was described previously in the discussion on heat transfer of lubricant and refrigerant mixtures. This empirical equation is as follows:

$$\frac{\text{EF}_1}{\text{EF}_2} = \left(\frac{G_1}{G_2}\right)^{-0.21} \tag{12.60}$$

TABLE 12.4 Microfin Tube Condensation Performance Data for Pure R-22 and Lubricant Mixtures at 300 kg / (m^2 · s) [17]

Parameter	Oil Concentration, %	150 SUS	300 SUS
	0	2.10	2.10
Heat transfer	1.25	2.45	2.05
enhancement factor, EF	2.5	2.00	2.00
	5.0	2.05	1.95

To use this equation a designer needs the value of an enhancement factor, such as those listed in Table 12.4 or Fig. 12.24, for at least one mass flow rate.

12.5 FLOODED EVAPORATORS

12.5.1 Description and Special Considerations

As mentioned at the beginning of the chapter, flooded shell-and-tube evaporators cool liquids flowing through tubes by transferring heat to the evaporating refrigerant on the shell side. The "flooded" terminology refers to the fact that the tubes are covered with a saturated mixture of liquid and vapor. The refrigerant at a low quality (e.g., 10%) enters the shell side through a distributor that evenly distributes the refrigerant over all tubes. As boiling occurs and the bubbles rise, the quality of the refrigerant increases from bottom to top. Unlike other types of boilers, such as kettle reboilers, which rely on internal recirculation of liquids to achieve convective boiling enhancement, flooded evaporators use high mass fluxes that pass over the tubes only once.

To prevent liquid carry-over to the compressor, either impingement separators, coalescing filters, gravity dropout, or a combination of these devices are located at the topmost region of the evaporator. An example of a gravity dropout area consisting of an open space between the top tube row and the suction outlet is shown in Fig. 12.25.

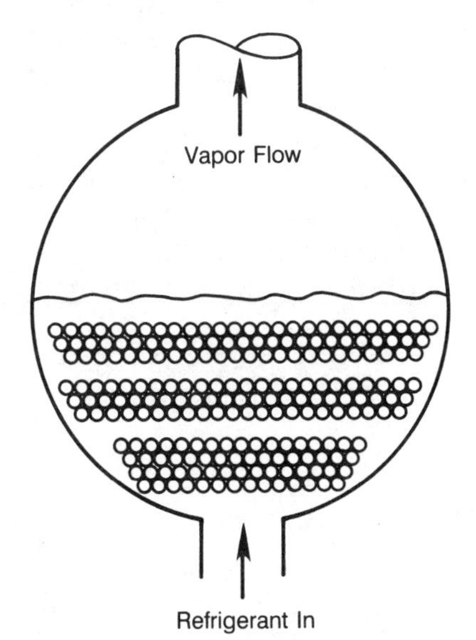

Fig. 12.25. Schematic of typical flooded evaporator.

Two important considerations that affect flooded evaporator design are oil accumulation and refrigerant level control. Oil enters the evaporator when small amounts of oil are mixed with the refrigerant leaving the compressor. Unlike the liquid refrigerant entering the evaporator, oil cannot be evaporated and, in addition, the vapor velocity at the top is not high enough to obtain oil carry-over back to the compressors. As a result, oil builds up on the shell side of the evaporator. Adverse effects of this buildup are that shell-side evaporation heat transfer coefficients are reduced while the compressor can be starved of oil needed for proper lubrication. Oil accumulation in the evaporator is reduced by pumping oil-rich liquid refrigerant back to the compressor through an oil return line.

Level control is needed in the flooded evaporator to ensure that the top row of tubes is completely covered with refrigerant. This control is achieved primarily by controlling the refrigerant inventory in the system or by a low-side float valve. For example, the initial refrigerant charge in the system can be sized so that during normal operation the refrigerant level in the evaporator just covers the tubes. A high-side float valve can be used to control the inventory in the condenser, which indirectly sets the level in the evaporator, providing the refrigerant inventory in the overall system has been sized properly. The second method of control is a low-side float valve that sets the evaporator level by controlling the flow rate into the evaporator. The use of a low-side float valve is more accurate, and it requires less precise knowledge of the overall refrigerant inventory.

Convective boiling is maximized on the shell side by proper distributor design and tight bundle construction. For example, the two-phase distributor at the bottom of the shell is important not only to obtain an even distribution of refrigerant but also to maximize the velocity of the incoming refrigerant. The refrigerant velocity and, hence, convective boiling component of heat transfer is also increased by using a tight bundle construction with no opportunity for vapor bypass, especially in the regions between the tubes and shell.

Methods for calculating shell-side heat transfer are presented for smooth, finned, and high-performance surfaces. However, heat transfer coefficient calculation methods are not presented for the liquid flowing inside the tubes. The reason is that the fluids and tube configurations used are common to many other applications and, therefore, heat transfer calculation methods have been covered in other publications. It is sufficient to state that both smooth tubes and enhanced tubes are routinely used. An overview of the types of enhanced tubes used along with heat transfer results for these tubes has been presented by Kohler and Starner [51].

12.5.2 Shell-Side Refrigerant Heat Transfer

Heat transfer on the outside of tubes in a flooded evaporator is a combination of nucleate and convective boiling phenomena. The different tube surfaces available for augmenting heat transfer over equivalent smooth tubes are based on improving heat transfer for either of these two mechanisms. For example, low fin tubes rely on improved convective boiling through increased surface area and mixing while nucleate boiling surfaces are modified to contain an increased number of small cavities.

Of the four types of two-phase heat transfer mechanisms typically found in refrigeration applications, namely, evaporation and condensation inside and outside the tubes, the calculation of accurate shell-side evaporation heat transfer coefficients is the most difficult for a designer. The reason is multifold: the heat transfer mechanism is complicated by the presence of convective and nucleate boiling, the presence of the bundle made up of multiple tubes, and the fact that the refrigerant quality and temperature changes from the tube row to tube row.

There are several levels of sophistication for calculating the evaporation heat transfer coefficient on the shell side of a flooded-shell-and-tube evaporator. At one extreme is the use of pool boiling or natural convection boiling correlations for a single smooth tube while at the other extreme is the use of evaporation heat transfer coefficients that are applicable to the complete tube bundle as found in a flooded evaporator for refrigeration applications. Correlations for the first case, a single tube in pool boiling, are available but they have limitations when applied to real heat exchangers because flooded evaporators use either finned surfaces or high-performance enhanced surfaces and, in addition, the evaporator is made up of an array of tubes. The latter, a shell-side correlation for a full array of tubes, is applicable but is not available in that there are no published (and experimentally verified) correlations for shell-side evaporation of a refrigerant in a flooded evaporator. However, in the absence of accurate correlations for the shell side, attempts have been made to derive general correlations by making various simplifying assumptions regarding heat transfer mechanisms and flow regimes. Applicability to a full tube array can thus be obtained; however, accuracy is sacrificed as a result of the many simplifying assumptions made.

Because of the advantages and disadvantages of each of the previous extremes, shell-side heat transfer coefficients for flooded evaporators are presented in three categories:

1. Generalized correlations for single, plain tubes applicable to pool boiling of a wide range of refrigerants
2. Correlations developed for two-phase evaporation heat transfer that account for both nucleate boiling and convective boiling effects but based on simplifying assumptions
3. Correlations and experimental data for both finned surfaces and high-performance enhanced surfaces usually limited to a few specific types of refrigerants

The designer can use any of these heat transfer approaches, depending on the trade-offs between accuracy and applicability that the designer considers important.

As mentioned previously, general correlations are based on experimental data for a range of refrigerants taken on single tubes with smooth or plain surfaces. Even so, these equations can be useful to the designer in the following situations:

1. Performing simple, straightforward calculations to obtain order-of-magnitude values
2. Using nondimensional numbers and functional relationships found in the simplified correlation to scale experimental data for one refrigerant to other refrigerants
3. Calculating heat transfer coefficients for enhanced tubes from heat transfer enhancement factor data when the enhancement factor is based on a reference smooth tube
4. Superimposing pool boiling correlations and forced convection correlations to obtain two-phase evaporation heat transfer coefficients that are more applicable to actual flooded shell-and-tube evaporators

Of the many correlations available for pool or natural convection boiling from a single tube, only a few are based on a large refrigerant database. One of these correlations is by Stephen and Abdelsalam [52]. Their experimental database contains more than 15 different refrigerants. Included in this base are common refrigerants such as R-11, R-12, R-22, and ammonia. Using this database, they developed two equations that differ in their simplicity. The more general of the two equations is

$$Nu = 207X_1^{0.745}X_5^{0.581}X_6^{0.533} \tag{12.61}$$

where X_1, X_5, and X_6 are dimensionless parameters defined as follows:

$$X_1 = \frac{qd}{k_f T_w} \tag{12.62}$$

$$X_5 = \frac{\rho_g}{\rho_f} \tag{12.63}$$

$$X_6 = Pr_f \tag{12.64}$$

It is important to note that the previous equation is only applicable to refrigerants. Other forms of the equation are presented for water, hydrocarbons, and cryogenic fluids.

An even simpler form of the preceding equation in terms of the heat transfer coefficient in units of $W/(m^2 \cdot K)$ is

$$h = C_4 q^{0.745} \tag{12.65}$$

where the C_4 parameter is a function of the refrigerant type and the saturation pressure. Stephan and Abdelsalam [52] present a figure, with coordinates of C_4 versus pressure in bars, which contains curves for each refrigerant.

Unlike other equations for pool boiling, the previous equations are based less on a theoretical foundation than they are on empirical curve fits of large amounts of data. However, as evidenced by the dimensionless numbers X_1, X_5, and X_6, these correlations do contain functional relationships that are common to many other correlations. Stephan and Abdelsalam also compared their correlation to the same large database used to derive it and found a mean absolute error of 10.57%. A plot of shell-side evaporation heat transfer coefficients for R-134a obtained from the Stephan–Abdelsalam correlation is shown in Fig. 12.26.

One weakness of the Stephan–Abdelsalam equation might appear to be the absence of a parameter to account for surface roughness effects or fluid-to-surface interactions. A survey of the database used to derive the equations shows that most of the studies, at least for the common refrigerants, was performed by using copper tubes with some of the studies being performed with nickel and stainless steel tubes. Heat transfer for surfaces with roughnesses different than the assumed average of $R_P = 1\ \mu m$ can be corrected on the basis of Stephan and Abdelsalam's suggestion that the heat

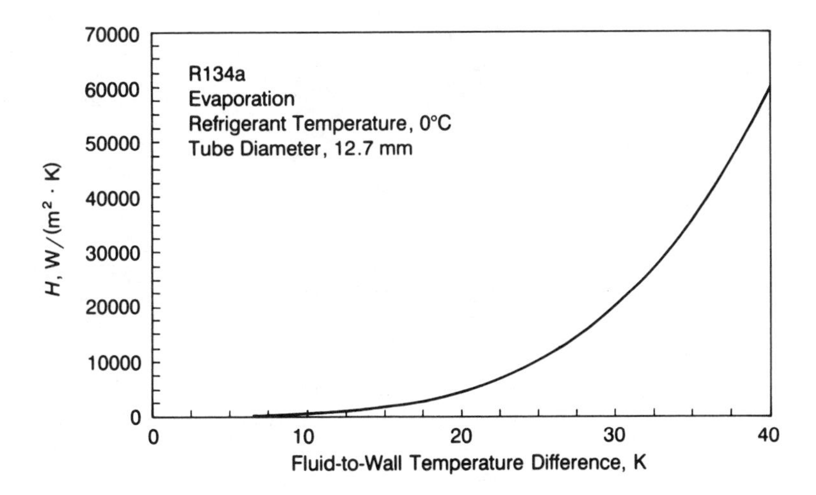

Fig. 12.26. Shell-side evaporator heat transfer coefficient for R-134a based on the Stephan–Abdelsalam correlation.

transfer coefficient, h, is proportional to $R_P^{0.133}$ so that multiplying by this factor with R_P in units of μm one can correct for other roughnesses at least in the range of

$$0.1 \le R_p \le 10 \ \mu\text{m} \tag{12.66}$$

Another approach for calculating evaporation coefficients is presented by Collier [53]. This approach is based on previous work by Borishanski [54]. The equation is

$$h_{\text{nb}} = A^* q^{0.7} F(P_r) \tag{12.67}$$

where $P_r = P/P_c$ and

$$F(P_r) = 0.7 + 2P_r\left(4 + \frac{1}{1 - P_r}\right) \tag{12.68}$$

Values of P_c and A^* for common refrigerants have been tabulated by Wolverine [55]; they are listed in Table 12.5.

Another approach was proposed by Webb et al. [1], who used a modified Chen [56] forced convection boiling model as follows:

$$h_s = Sh_{\text{snb}} + Fh_{\text{sfc}} \tag{12.69}$$

The suppression factor, S, which accounts for nucleate boiling suppression due to temperature gradients in the liquid near the surface, can be calculated from a correlation proposed by Bennett et al. [57]:

$$S = w(1 - e^{-w}) \tag{12.70}$$

and

$$w = \frac{k_f l}{F h_{fe} \beta} \tag{12.71}$$

TABLE 12.5 Critical Pressures and Values of A^* [55]

Refrigerant	Pressure Range, atm	Critical Pressure, atm	A^*
R-11	1–3	42.9	0.681
R-12	1–4.9	40.3	0.956
R-12	6–40.5	40.3	1.01
R-22	0.4–2.15	48.4	0.941
R-113	1–3	33.4	0.488

and

$$\beta = \frac{0.041}{\{\sigma/[g(\rho_f - \rho_g)]\}^{0.5}} \tag{12.72}$$

Webb et al. compared several correlations suggested by several investigators for the F factor in Chen's correlation. Because it produced values for F that were in the midrange compared to the other correlations, they recommend using a correlation by Bennett and Chen [58]. This equation, which includes a dependency on the Prandtl number, is

$$F = Pr_f^{0.296}\left[\phi_f^{z^3}\right]^{0.445} \tag{12.73}$$

Webb et al. discuss several approaches to calculating the frictional multiplier, ϕ_f^2, defined as the ratio of the two-phase frictional pressure gradient to the gradient for the liquid phase flowing alone. An approach that appears to be a balance between simplicity and accuracy is one by Ishihara et al. [59] who modified Chisholm's [60] equation to obtain

$$\phi_f^2 = 1 + \frac{C}{X} + \frac{1}{X^2} \tag{12.74}$$

where

$$X^2 = \left(\frac{u_f}{u_g}\right)^n \left(\frac{\rho_g}{\rho_f}\right)\left[\frac{1-x}{x}\right]^{2-n} \tag{12.75}$$

and $C = 8.0$, as suggested by Ishihara et al. for tube banks. The exponent n is based on the functional relationship between the friction factor and the Reynolds number, $f = g(Re^{-n})$. A more detailed correlation for the frictional multiplier, which accounts for flow regimes, has been developed by Schrage et al. [61] from data for R-113.

The use of Chen's correlation as suggested by Webb et al. requires correlations for calculating the nucleate boiling heat transfer coefficient, h_{nb}, and the forced convection heat transfer coefficient, h_{fc}. Nucleate boiling heat transfer coefficients have been presented previously for either plain tubes or enhanced tubes. The correlations for the enhanced tubes were presented either in the form of heat transfer enhancement factors, EF, which also requires semiempirical correlations for plain tubes such as that by Stefan and Abdelsalam [52] for integral finned tubes (e.g., 19 fins/inch) or in the form of data plots for high-performance nucleate boiling surfaces (e.g., Turbo-B, GEWA-T, etc.).

The single-phase convection heat transfer used in the Chen correlation can be approximated by using enhancement factors that are referenced to plain tubes. If this information is not available the designer can approximate it by assuming that in addition to the area increase caused by the addition of fins, which will be accounted for when the heat flux is calculated, the fins also break up the flow and increase turbulence. A conservative calculation can be made by neglecting the flow turbulence caused by the fins and then accounting for the area increase only.

For high-performance surfaces that rely on enhancing nucleate boiling, the forced convection heat transfer coefficient for a plain tube can be used. The reason is that the presence of either micropores or microcavities, which provide increased nucleation sites, do not result in an increased area for the flowing refrigerant to contact the tube.

12.5.3 Shell-Side Heat Transfer Augmentation

Finned-tube surfaces are used to enhance convective boiling heat transfer, while high-performance heat transfer surfaces, which are optimized to increase nucleation sites, increase the nucleate boiling component. Most of the data for integral fins and enhanced boiling surfaces are for refrigerants used in laboratory experiments, such as R-113. Very few studies have been reported on commonly used refrigerants, such as R-11, R-12, and R-22. In the absence of applicable data information, an alternative approach is to use enhancement factors obtained from existing data and then assume that the enhancement factor remains unchanged for the refrigerant of interest. For example,

$$\mathrm{EF}_{\mathrm{R\text{-}22}} = \mathrm{EF}_{\mathrm{R\text{-}113}} = \frac{h_{\mathrm{enhanced}}}{h_{\mathrm{reference}}} \tag{12.76}$$

can be rearranged to calculate the enhanced-tube heat transfer coefficient as follows:

$$h_{\mathrm{enhanced(R\text{-}22)}} = \mathrm{EF}_{\mathrm{R\text{-}113}} h_{\mathrm{reference(R\text{-}22)}} \tag{12.77}$$

where, the "enhanced" subscript can refer to either a finned tube or a high-performance enhanced surface. The "reference" subscript can refer to either a plain tube or, if a high-performance enhanced tube is being investigated, to an equivalent-diameter finned surface.

The use of the integral finned tubes shown in Fig. 12.27 is quite common in refrigeration and air-conditioning industries. Their sizes vary from manufacturer to manufacturer. For example, one manufacturer, Wolverine Tube Company, makes integral finned tubes with fin densities of 16 to 40 fins per inch that are approximately 1/16 in. high. Area increases over the smooth tube due to the presence of the fins vary from factors of 2.2 to 6.7. The fins are manufactured in a screw-type pattern by an extrusion process. Since most tubes used in refrigeration and air-condition applications (with the exception

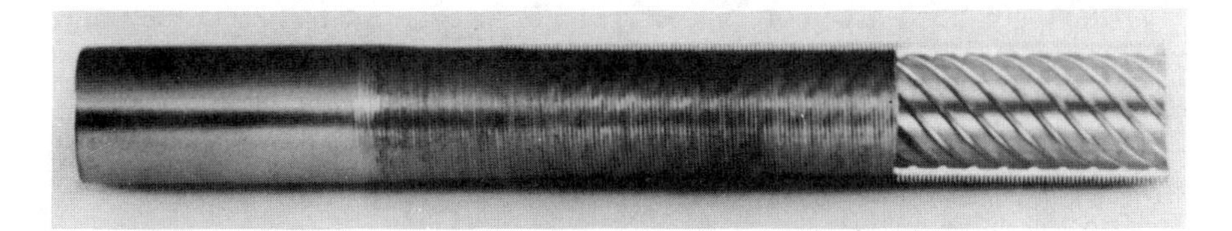

Fig. 12.27. Photograph of finned-tube surface for evaporation and condensation applications (courtesy of Wolverine Tube Inc.).

of ammonia which requires the use of aluminum or steel) are made of copper, the fin efficiencies approach 100%.

Even though low-finned tubes are commonly used in refrigeration and air-conditioning applications, very little design information is available. One of the few studies is published for R-12 on finned tubes, 19 fins per inch, and on plain tubes by Katz et al. [62]. Figure 12.28 shows a plot of heat flux versus wall superheat for the two tubes. From data on this plot, one can find heat transfer coefficients by using Newton's law of cooling. For data plots such as Fig. 12.28 it is often more meaningful to define enhancement factors on the basis of ratio of the wall superheat for the plain surface to the wall superheat for the enhanced surface. For example, the finned tube in Fig. 12.28 has an enhancement factor of 1.8 at a heat flux of 80,000 W/m^2. It should be noted that the area that Katz et al. used in Fig. 12.28 was based on the outside projected area, namely $\pi D_o L$, where D_o is the fin outside diameter. The enhancement factor based on the actual outside area is 1.25.

High-performance boiling surfaces have superior performance over low-fin surfaces. However, because additional manufacturing processes are required

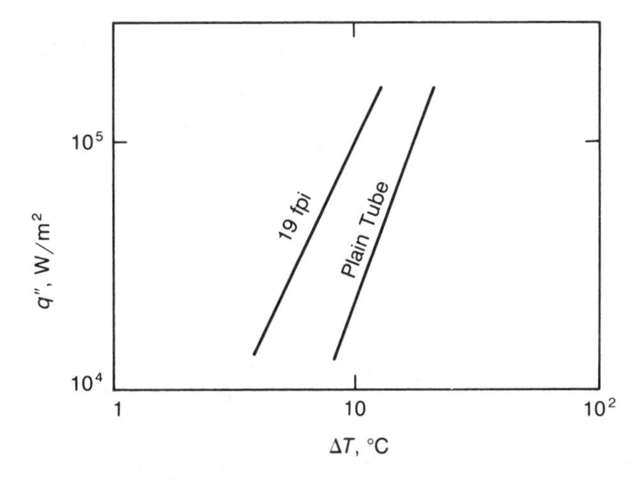

Fig. 12.28. Heat transfer data for 740 fins per meter–low-fin tube for R-12 [62].

to make them, they are more expensive. These performance trade-offs and tube cost makes the designer's calculations and analyses—to determine whether high-performance tubes should be used—that much more important. The high-performance tubes have one thing in common—they have all been designed to increase nucleate boiling by creating numerous nucleation sites on the tube surface.

High-performance boiling surfaces have superior performance over finned surfaces. However, they cost more because of additional manufacturing processes required. Design calculations are important for determining whether the economics for any given situation justify the installation of high-performance tubes.

High-performance boiling surfaces for refrigeration applications are available under commercial names such as High Flux (Union Carbide), Thermoexcel-E (Hitachi), GEWA-T, GEWA-TX, GEWA-TXY (Wieland), and Turbo-B (Wolverine). These tubes are shown in Figs. 12.29*a* through *g*, in the order listed. The performance of some of these tubes is such that it has been possible to achieve boiling at wall superheats below 1°C (1.8°F). Two variations on the GEWA-T are also available from Wieland. They are the GEWA-TX tube, which differs from the GEWA-T in that it has grooved

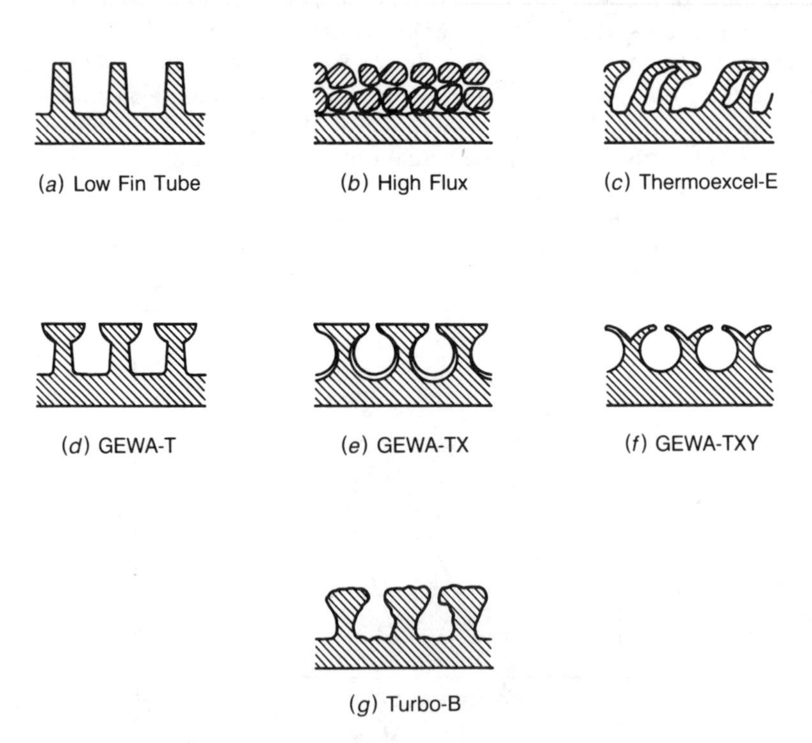

Fig. 12.29. Example of convective and nucleate boiling surfaces used in refrigeration and air-conditioning applications.

TABLE 12.6 Sample Enhancement Factor for High-Performance Boiling Surfaces

			Enhancement Factors		
Tube Type	Reference	Refrigerant	4 kW/m^2	10 kW/m^2	20 kW/m^2
High Flux Tube	Marto and Lepere [63]	113	8	6.5	6
	O'Neill et al. [64]	11	—	16.5	17.5
	Chyu [65]	113	—	—	9
Thermoexcel-E	Hitachi [66]	113	—	7	6
	Nakayama et al. [67]	11	—	11	13
	Kuwahara [68]	11	—	10	9
	Marto and Lepere [63]	113	8	7	5
GEWA-T	Marto and Lepere [63]	113	2	2.5	2.5
	Ayub and Bergles [69]	113	2.5	2	2

channels, and the GEWA-TXY tube, which has the fin modified to a Y shape. There are no experimental data available in the open literature for these two tubes with refrigerants as the test fluid.

Four high-performance boiling surfaces have been presented. Sample results from past studies have also been presented for various refrigerants. To aid the designer, a compilation of enhancement factors taken from these plots is presented in Table 12.6 for three heat fluxes, namely, 4, 10, and 20 kW/m^2. These values represent a typical heat flux for shell-side evaporation in refrigeration applications, 10 kW/m^2, and then extreme high and low values.

The enhancement factors shown are limited to these refrigerants for which experimental data are available. If enhancement factors for other refrigerants are needed, the designer can either use an enhancement factor for a refrigerant listed or apply some corrections for trends that might exist. For example, one can see in Table 12.6 that enhancement factors for R-11 are higher than enhancement factors for R-113 for the same tube. A similar dilemma exists when enhancement factors are available for some refrigerant saturation temperatures but not others for which a design calculation is being performed. It is also important to note that the enhancement factors shown in Table 12.6 are based on the ratios of the wall superheat temperature for the plain tube divided by the wall superheat for the high-performance boiling surface.

12.6 SHELL-AND-TUBE DIRECT EXPANSION EVAPORATORS

12.6.1 Description and Special Considerations

The second type of liquid cooler, is the direct (or dry) expansion (DX) evaporator shown in Fig. 12.30. This type of liquid cooler is also a shell-and-

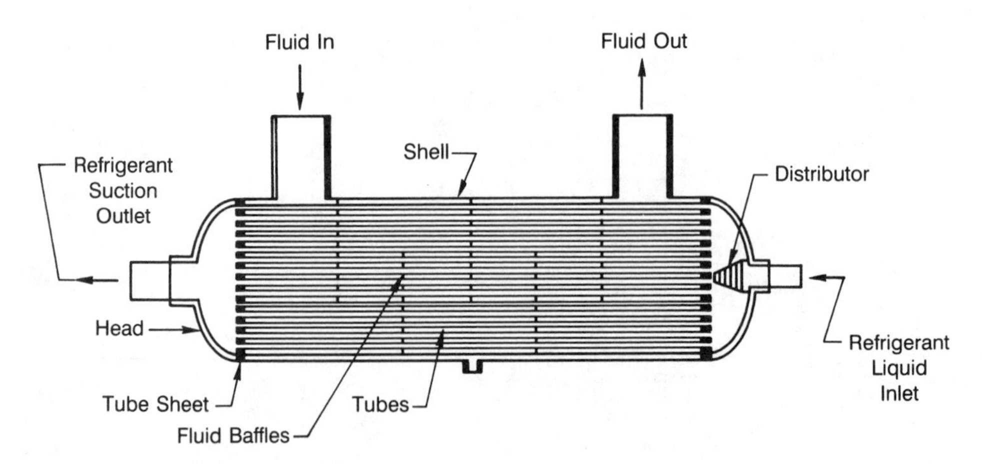

Fig. 12.30. Schematic of shell-and-tube direct expansion evaporator.

tube heat exchanger. However, unlike the flooded evaporator where refrigerant flows on the shell side, the refrigerant evaporates on the inside of tubes in DX evaporators.

The direct expansion evaporator has several advantages over the flooded evaporator type of liquid cooler. First, the DX evaporator assures that oil is returned to the compressor, which is important where substantial compressor lubricating oil is present. Second, a superheated vapor can be obtained at the exit while in the case of the flooded evaporator only saturated vapor is obtained.

Shell-and-tube DX evaporators, commonly referred to as direct expansion coolers, are used with positive displacement compressors such as reciprocating, screw, or scroll compressors. In addition, these heat exchangers are usually designed for horizontal evaporation. The refrigerant enters the heat exchanger through an inlet port, a distributor located inside the head, and then goes into the tubes. The design of the distributor is important for ensuring that refrigerant is evenly supplied to each tube. Uneven distribution can easily occur because the refrigerant entering the evaporator head from the expansion valve is a mixture of vapor and liquid that can be easily stratified in an oversized head or an improperly designed flow distributor.

The undesirable consequence of an oversupply to some tubes is that liquid exits the tube rather than a properly superheated vapor. Because the evaporator exit is controlled to a fixed superheat condition, the liquid is then evaporated by mixing with excessively superheated refrigerant that exits undersupplied tubes. The net consequence of this maldistribution of refrigerant is that the in-tube heat transfer coefficients for some tubes, especially those with low flow rates and high superheats, are too low.

Because low temperature differences result in low thermal expansion forces, economical, fixed-tube heat designs can be used. Direct expansion evaporators can be designed in either a single-phase arrangement, multiphase arrangements, or a U-bend arrangement.

Because of the increasing specific volume of the refrigerant as it progresses through the heat exchanger, the number of tubes can change per pass in multipass applications. A potential problem with multipass arrangements is that it may be difficult to obtain even refrigerant distribution after the first pass. Likewise, a potential problem with single-pass arrrangements is that it may be difficult to evaporate fully the refrigerant unless either long tubes or enhanced tubes are used.

12.6.2 In-Tube and Shell-Side Heat Transfer

Because the in-tube flow of refrigerant is the same as that described in the evaporator coil section, the reader is referred to Section 12.3.2 for design heat transfer information. In-tube enhancement is used to improve the heat transfer performance of the evaporating refrigerant inside the tubes. The most popular type of in-tube enhancement is the microfin tube which was described earlier in Section 12.3.3. The shell side is single-phase liquid across a tube bank; this type of flow has been described extensively in the heat transfer literature.

12.7 SHELL-AND-TUBE CONDENSERS

12.7.1 Description and Special Considerations

The shell-and-tube condenser with superheated refrigerant vapor entering and subcooled liquid-refrigerant exiting is the most common type of water-cooled condenser. The advantage of a shell-and-tube condenser is that a large condensing surface area can be installed in a small space, at least compared to other heat exchanger types. Shell-and-tube condensers are constructed so that refrigerant condenses on the shell side or the outside of tubes while cooling water is pumped through the inside of the tubes as shown in Fig. 12.31. They are normally mounted in the horizontal position to facilitate condensate drainage from the tube surface. Water flow on the inside instead of on the outside of tubes allows the designer to achieve high water velocities that are necessary for good water-side heat transfer. These water velocities always are maintained above 1 m/s (3 fps) and are normally in the range of 3 to 4 m/s (10 to 13 fps). The desire to obtain high water velocities also determines the number of tube passes designed into a condenser. For example, in order to maintain high water velocities in a condenser, the number of tubes in each pass must be increased as the number of tube passes is decreased. The only real limitation to water flow velocities are pressure drop considerations. It should also be noted that the water-side pressure drop decreases as the number of passes are reduced.

Copper tubes are usually used in shell-and-tube condensers with sizes ranging from 5/8 to 1 in. OD, with 3/4 in. OD being the most popular. Ammonia requires carbon steel tubes. In addition, the diameters are generally larger than those described previously. Integral fins spaced 19 to 40 fins per inch (1.33 to 0.64 mm spacing) with fin heights of 0.035 to 0.061 in. (0.9 to 1.5 mm) are routinely used for shell-side enhancement. As will be discussed

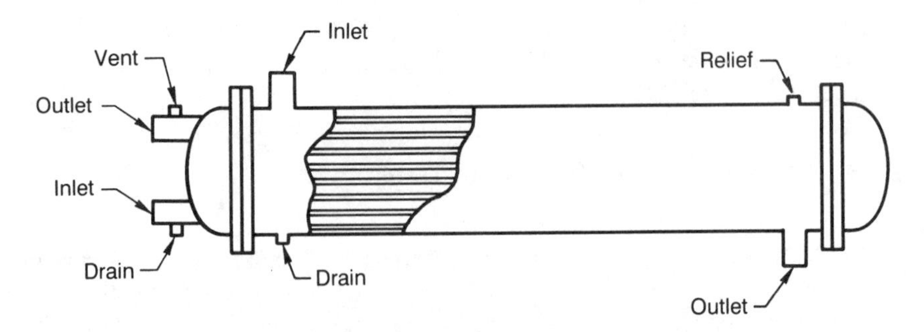

Fig. 12.31. Schematic of shell-and-tube condenser.

later, high performance condensing surfaces are also available. Equilateral triangular pitches are normally used for installing tubes. Tube clearances vary; however, a typical spacing is about 3/16 in. (4.7 mm).

Shell-and-tube condensers are designed so that the inlet and outlet connections are located far enough apart to allow the entering superheated vapor to be exposed to the maximum tube surface area. Baffling located between the inlet and outlet can also aid the distribution of refrigerant vapor. Of special importance for proper distribution of the vapor is a clearance space between the top row of tubes and the shell. In addition to ensuring that refrigerant vapor reaches all of the heat transfer surfaces, the internal construction should also be such that excessive pressure drops are avoided. These pressure drops reduce the saturation temperature of the refrigerant which in turn reduces the temperature difference during heat transfer.

As the refrigerant vapor is condensed on the outside of the tubes, it drips down to lower tubes and collects at the bottom of the condenser. The bottom of the condenser can be designed so that tube rows are located either above or below the condensate liquid level located at the bottom. More refrigerant subcooling can be achieved if the liquid level is above the bottom tube rows. In some cases the condenser bottom is designed to act as a receiver, a reservoir for storing refrigerant. Regardless of which approach is used, a liquid seal should always be maintained over the outlet nozzle.

The shell-and-tube condenser is basically arranged in a counterflow arrangement in that the cooling water first passes through the lower rows of tubes while the upper tube rows correspond to later tube passes.

The tubes in water-cooled condensers are installed in a fixed tube sheet arrangement. Straight tubes are usually used. However, U tubes with a single tube sheet are also manufactured.

Noncondensable gases—either air or water vapor—are a problem in condensers in that they raise the operating pressure of the condenser, which in turn increases the required compressor power. These gases also blanket the condensing surfaces, decreasing the condensation heat transfer coefficient, which in turn decreases the heat transfer rate for a given driving temperature difference. Noncondensable gases should be eliminated through a purge line and/or system. A designer can account for purge gas effects on a design calculation by treating them as either an added thermal resistance or a fouling factor on the refrigerant side. Noncondensable gases enter a refrigerant system during the refrigerant charging process, or they are present as a dissolved gas in the lubricating oil that is mixed with the refrigerant in the compressor. For refrigerants that operate below atmospheric pressure, air can leak into the system from the surroundings.

12.7.2 Shell-Side Refrigerant Condensation Heat Transfer

As with shell-side evaporation, shell-side condensation can be divided up into several different groups of correlations depending on tube type. Specifically, the types of tubes used in condensation are smooth (or plain), finned, and high-performance enhanced tubes.

Surface tension is the dominant force controlling condensate thickness and causing condensate holdup on the bottom of the tubes. Heat transfer to the tube wall from the surrounding vapor is, in turn, inversely proportional to this condensate thickness. Because the surface tension of refrigerants is low, fins have been used successfully for shell-side condensation applications. Finned tubes for refrigeration applications are manufactured with typical fin densities of 620 to 1560 fins per meter with heights of 0.76 to 15.24 mm. If the surface tension is too high, the condensate collects between the fins so that the effective area for heat transfer from the vapor to the tube wall is reduced.

This phenomenon is known as condensate flooding. Tube manufacturers, recognizing the importance of surface tension, have designed special high-performance tubes with surface profiles that produce thin condensate films through improved condensate drainage.

Shell-side condensation heat transfer coefficient correlations are presented for smooth tubes, finned tubes, and high-performance enhanced tubes. The smooth-tube equations are generalized in that their development and verification are based on experimental data for several refrigerants. In addition, because of the ease of modeling heat transfer and fluid flow for a simple tube geometry, smooth-tube correlations are more fundamentally based while more complicated geometries require empirical curve fits of data. The correlations for finned tubes are extensions of smooth-tube correlations to account for fins. Few correlations exist for high-performance tubes; rather, the limited experimental data available have been presented in the form of performance curves.

Several correlations are available in the literature for predicting smooth-tube condensation heat transfer coefficients. Even though these correlations have been verified with experimental data for refrigerants, there is considerable disagreement between them.

The Nusselt [70] equation has been shown to predict condensation heat transfer coefficients for smooth tubes. For example, Williams and Sauer [71] predicted experimental data for R-11 by using this equation. Nusselt's equation for an average heat transfer coefficient is

$$h = 0.729\left[\frac{g\rho_f(\rho_f - \rho_v)k_f^3 h_{fg}}{D\mu_f(T_{\text{sat}} - T_w)}\right]^{1/4} \tag{12.78}$$

This equation is based on several assumptions that may or may not be violated in a particular design situation. Some of these assumptions are that gravity is the only force acting on the liquid film (vapor shear is neglected), the condensate film is laminar, the surrounding vapor is saturated, and, finally, the liquid film is at a constant temperature. Since superheated vapor enters a condenser, one can implement a correlation for this by modifying the heat of vaporization to account for vapor superheat as follows:

$$h'_{fg} = h_{fg} + C_p(T_v - T_s) \qquad (12.79)$$

An example of condensation heat transfer coefficients calculated from Eq. (12.78) is shown in Fig. 12.32.

Nusselt's equation is for a single tube and therefore it does not consider the effects on heat transfer of condensate dripping from row to row. To account for row effects, several investigators have suggested modifications. One such modification is by Nusselt [70] and Short and Brown [72], and it can be used to calculate the condensation heat transfer coefficient on the Nth row:

$$h_N = 1.24 h_{1\text{-row}} \left[N^{0.75} - (N-1)^{0.75} \right] \qquad (12.80)$$

where $h_{1\text{-row}}$ is calculated from the Nusselt equation. It should be noted that Eq. (12.80) is for heat exchangers with more than 10 rows and, in addition, it still does not account for vapor shear.

A correction factor for vapor shear has been developed by Webb [73] at least for the first row of tubes as follows:

$$h = h_{1\text{-row}} 1.32 F^{-0.05} \qquad (12.81)$$

where F is a function of velocity as follows:

$$F = \frac{Prh_{fg}}{C_p Fr\, \Delta T} \qquad (12.82)$$

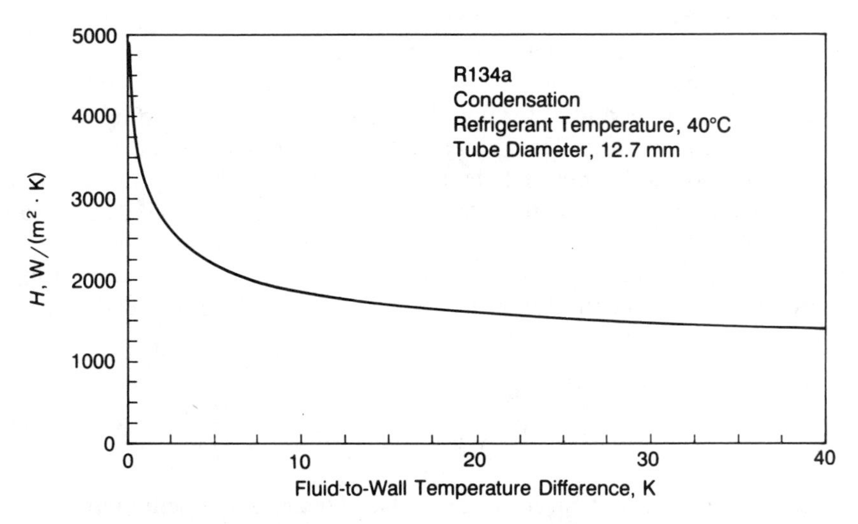

Fig. 12.32. Shell-side condensation heat transfer coefficient for R-134a based on Nusselt correlation.

Beyond the first row of tubes, the vapor velocity decreases because of condensate formation. Therefore, for the overall heat exchanger, the question arises of whether heat exchanger design calculations in refrigeration and air-conditioning applications will be in error if vapor shear is neglected altogether. Webb [73] contends from calculations and experimental data that the velocities are insufficient to affect the condensate film thickness, but they may in fact affect the condensate flow pattern from tube to tube. The latter phenomenon could therefore result in an increased heat transfer coefficient.

An equation for an average heat transfer coefficient for a tube bundle of N tubes has been suggested for Short and Brown [72], Webb [73], and Kern [74] as follows:

$$\bar{h}_N = h_{1\text{-row}} N^{-1/6} \tag{12.83}$$

It should be noted that originally Nusselt suggested a coefficient of 1/4 instead of 1/6. However, this resulted in severe penalties when the number of tube rows was greater than 10, which is common. It is also suggested that if vapor shear effects are important, one may want to decrease the 1/6 coefficient even further. In fact, if the preceding coefficient is decreased to its limit of 0 then one concludes that the heat transfer decrease due to condensate buildup on lower rows is offset by potential heat transfer increases due to excessive liquid motion and film breakup as it drips from tube to tube. In this case, the value of the average heat transfer coefficient for a tube bundle can be approximated by the value for a single tube.

12.7.3 Shell-Side Heat Transfer Augmentation

As mentioned previously finned tubes are common in refrigeration and air-conditioning applications because of the low surface tension of most common refrigerants. The one exception is ammonia when plain tubes are used because of the potential for flooding caused by the high surface tension of ammonia condensation. Heat transfer from horizontal low fin tubes was correlated from experimental data by Beatty and Katz [75]. They also derived an equation similar to Nusselt in terms of fin efficiencies; however, it was more complicated. Since fin efficiencies for integral fins made from copper are close to 100%, similar results for finned tubes can be obtained by using Nusselt's equation and the following equation for the overall outside area including fins and the exposed tube surface:

$$A_o = \pi L \left[d_r \left(\frac{S}{s + y} \right) + \frac{1}{2} (d_o^2 - d_r^2) N_f \right] \tag{12.84}$$

High-performance enhanced surfaces have been developed to produce thin condensate films and to facilitate drainage of the condensate from the tube. Two examples of this type of tube are the Turbo-C tube manufactured

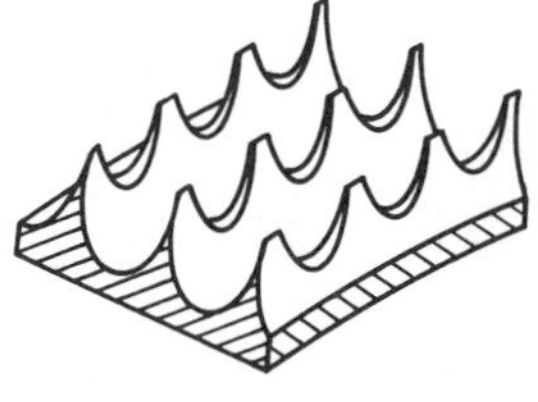

(*a*) Thermoexcel-C
(Similar to Turbo-C)

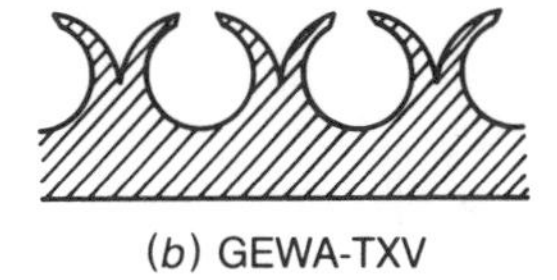

(*b*) GEWA-TXV

Fig. 12.33. Examples of high-performance condensing surfaces used in refrigeration and air-conditioning applications.

by Wolverine Tube Company and Thermoexcel-C manufactured by Hitachi. As shown in Fig. 12.33 these tubes use fins in a spine arrangement, which provides a path for condensate drainage. Another tube is the GEWA-TXV, which has steep-angle, V-shaped fins as shown in Fig. 12.33 to help in maintaining a thin refrigerant film.

12.8 HEAT EXCHANGER DESIGN WITH ALTERNATIVE REFRIGERANTS

As of this writing, nearly all household refrigerators and automotive air conditioners use R-12 as the working fluid. In addition, many industrial refrigeration systems use R-12, while many large chiller systems use R-11 and R-12 as the working fluid. However, R-11 and R-12 are chlorofluorocarbons (called CFCs) and are being phased out because their leakage into the atmosphere destroys the ozone layer surrounding the earth. These refrigerants are stable and have long lives such that they migrate to the upper atmosphere where chlorine atoms are released which then catalyze the destruction of ozone. Over the next decade severe reductions in R-11 and R-12 use will be mandated by governments throughout the world. It is expected that non-CFC refrigerants, formed by replacing chlorine atoms with hydrogen atoms to form an HFC, will replace R-11 and R-12 on a large scale. It is expected that HFC-134a will replace CFC-12 and HFC-123 will replace CFC-11, at least in the immediate future (note that the "R" and the designation of the refrigerant, such as HFC and CFC, can be used interchangeably). The selection of these refrigerants as replacements is based on

TABLE 12.7 Comparison of R-134a and R-12 Properties at a Typical Evaporation Temperature of −5°C [76]

Property	R-134a	R-12	Difference, %	Effect of Heat Transfer
Liquid density, kg/m^3	1308	1417	−7.7	↑ slightly
Vapor density, kg/m^3	12.2	15.4	−20.8	↓ slightly
Enthalpy of vaporization, kJ/kg	202.3	153.9	+31.4	↑ moderately
Saturation pressure, MPa	0.243	0.261	−6.9	≈ 0
Liquid viscosity, $\mu Pa \cdot s$	301	284	+6.0	↓ slightly
Vapor viscosity, $\mu Pa \cdot s$	12.2	11.3	+7.9	↓ slightly
Vapor thermal conductivity, mW/mK	11.77	8.01	+46.9	↑ slightly
Liquid thermal conductivity, mW/mK	98.1	80.8	+21.4	↑ strongly
Liquid specific heat, $kJ/(kg \cdot K)$	1.297	0.922	+40.6	↑ moderately
Vapor specific heat, $kJ/(kg \cdot K)$	0.868	0629	38.0	↑ slightly
Liquid Prandtl number	3.98	3.24	+22.6	↑ slightly
Vapor Prandtl number	0.99	0.89	+11.2	↑ slightly

TABLE 12.8 Comparison of R-134a and R-12 Properties at a Typical Condensation Temperature of 40°C [76]

Property	R-134a	R-12	Difference, %	Effect on Heat Transfer
Liquid density, kg/m^3	1147	1253	−8.5	↑ slightly
Vapor density, kg/m^3	50.0	55.0	−9.1	↓ slightly
Enthalpy of vaporization, kJ/kg	163.1	128.6	+26.8	↑ moderate
Saturation pressure, MPa	1.017	0.9607	+5.9	≈ 0
Liquid viscosity, $\mu Pa \cdot s$	163.4	195	−16.2	↑ slightly
Vapor viscosity, $\mu Pa \cdot s$	14.31	13.78	+3.8	↓ slightly
Vapor thermal conductivity, mW/mK	15.56	11.0	+41.5	↑ slightly
Liquid thermal conductivity, mW/mK	74.6	63.8	+16.9	↑ strong
Liquid specific heat, kJ (kg · K)	1.514	1.01	+49.9	↑ slightly
Vapor specific heat, kJ (kg · K)	1.130	0.7857	+43.8	↑ slightly
Liquid Prandtl number	3.32	3.09	+7.4	↑ slightly
Vapor Prandtl number	1.04	1.03	+0.9	↑ slightly

the fact that their thermodynamic properties are similar to the refrigerants that they are replacing.

To date only R-134a heat transfer coefficients have been measured, and these measurements have only been performed for evaporation and condensation occurring inside tubes. There are no reports of shell-side evaporation and condensation of either R-134a or R-123. A comparison of the properties for R-12 and its replacement R-134a are shown in Tables 12.7 and 12.8 for typical evaporation and condensation temperatures, respectively [76]. If a designer is performing heat exchanger calculations for the purposes of comparing the two refrigerants at the temperatures shown, properties can be taken and used directly from these tables. These tables also provide qualitative information when the designer considers design changes to implement alternative refrigerants. For example, one can observe that the liquid thermal conductivity, enthalpy of vaporization, and liquid specific heat are all significantly higher for R-134a compared to R-12, and therefore all three of these properties contribute to higher heat transfer coefficients.

As mentioned previously, in-tube heat transfer coefficients have been measured for R-134a during condensation and evaporation of R-134a [77]. These coefficients were measured for a 3.67-m-long smooth tube with an inner diameter of 8.0 mm. Average evaporation heat transfer coefficients for almost a full-quality range are shown in Fig. 12.34 for temperatures of 5, 10, and 15°C. For similar mass fluxes the heat transfer coefficients for R-134a are about 30% to 40% higher than values for R-12. Part of this increased heat transfer coefficient is due to the fact that to obtain similar exit qualities for

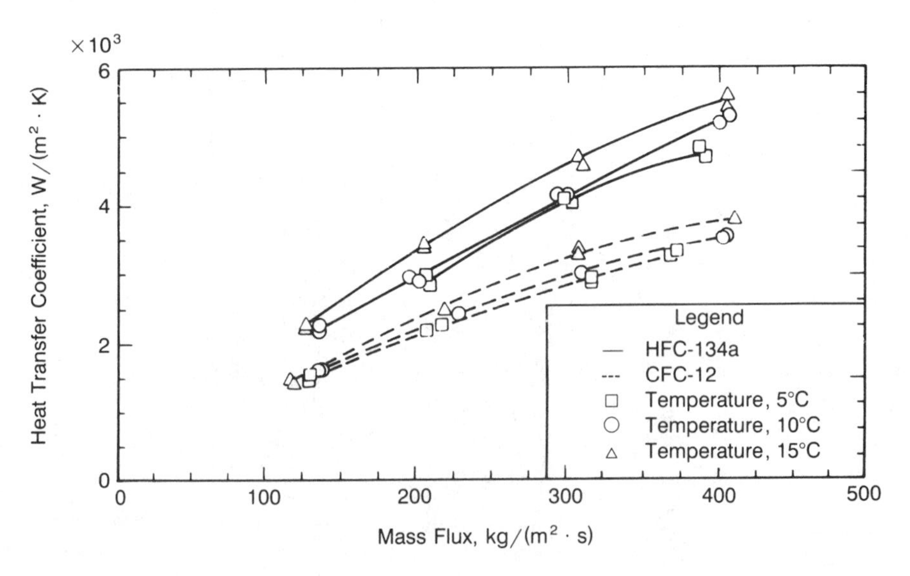

Fig. 12.34. Measured evaporation heat transfer coefficients for HFC-134a and CFC-12 at three temperatures [12].

the same tube length, it was necessary to increase the heat flux for R-134a. The reason for this increase is that the enthalpy of vaporization is higher by R-134a. For example, at 200 kg/(m^2 · s) and 10°C temperature, the heat flux is 12.1 kW/m^2 for R-134a and 9.1 kW/m^2. The increased heat flux probably accounts for about a 10% increase in the heat transfer coefficient for R-134a.

A comparison of R-134a and R-12 at temperatures of 30, 40, and 50°C during a condensation is shown in Fig. 12.35. The in-tube condensation heat transfer coefficients are about 25% to 35% higher for R-134a. The differences in heat flux for the two refrigerants does not have the same effect for condensation as it does for evaporation.

When designing heat exchangers for alternative refrigerants, one must consider differences in enthalpies of vaporization of the alternative refrigerants compared to the refrigerants that they are replacing. For example, for a typical evaporation temperature of −5°C, the enthalpy of vaporization of R-134a is about 31% higher than for R-12. As a result, an evaporator using R-134a would require 31% less mass flow rate of refrigerant to obtain the same heat capacity. The decrease in mass flow rate for R-134a will result in a decrease in the in-tube heat transfer coefficient if the tube diameter stays the same. However, one should note that even when this decrease in mass flow rate compared to R-12 is taken into account, the heat transfer coefficients for R-134a are still about 5% to 15% higher compared to R-12. For condensation, a similar situation can occur, and for these temperature conditions the heat transfer coefficients for R-134a are about 10% to 20% higher than values for R-12 even when the flow rate is reduced.

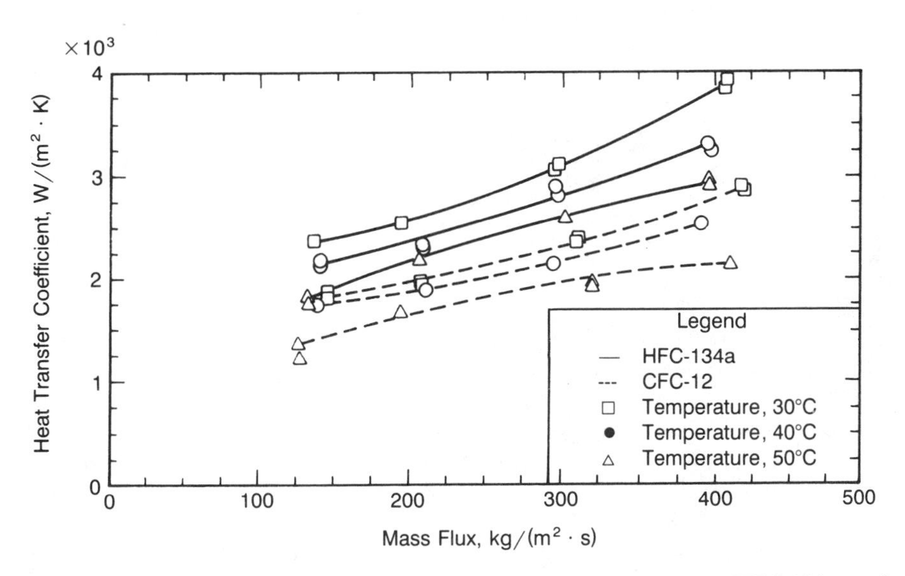

Fig. 12.35. Measured condensation heat transfer coefficients for HFC-134a and CFC-12 at three temperatures [12].

NOMENCLATURE

A	area, m^2
b	width, mm
Bo	boiling number, dimensionless, $q/(Gi_{fg})$
Co	convection number, dimensionless, $((1-x)/x)^{0.8}(\rho_g/\rho_f)^{0.5}$
d	diameter, mm
D	tube diameter, indentation diameter, mm
EF	enhancement factor, dimensionless
f	friction factor, dimensionless
fpi	fins per inch, 1/mm
F_1, F_2	nondimensional parameters
F_{fl}	fluid-dependent parameter, dimensionless
Fr	Froude number, dimensionless, $G^2/(\rho_f^2 gD)$
G	mass flux, $kg/(m^2 \cdot s)$
Gz	Graetz number, dimensionless, $RePrD/x$
h	heat transfer coefficient, $W/(m^2 \cdot K)$
i	enthalpy, J/kg
j	j factor, dimensionless, $StPr^{2/3}$
k	thermal conductivity, W/mK
l	tube-to-collar interference, mm
L	tube length, m
m	molecular weight, kg/mol; mass flow rate, kg/s
N	number of tube rows, dimensionless
N_p	fin patterns per longitudinal row, dimensionless
Nu	Nusselt number, dimensionless, hL/K
OD	outside diameter, mm
p	fin pitch, mm
P	pressure, kPa
P_d	fin pattern depths, mm
Pr	Prandtl number, dimensionless
Q	heat transfer, J
q	heat flux, W/m^2
R	resistance, $m^2 \cdot K/W$
Re	Reynolds number, dimensionless, $\rho VL/\mu$ or $\rho VD/\mu$
R_p	roughness, μm
s	fin spacing, fin height, mm
S	suppression factor, tube spacing, mm
St	Stanton number, dimensionless, $Nu/RePr$
t	time, fin thickness, wall thickness, mm
T	temperature, °C
V	velocity, m/s
W_o	oil concentration
w_a	humidity ratio for saturated air at 0°C, dimensionless
w_o	humidity ratio for entering air, dimensionless

W_f	fin spacings, mm
x	quality, frost height, mm
X_{tt}	Martinelli parameters, dimensionless, $(x/1-x)^{0.8}(\rho_f/\rho_q)^{0.5} \times(\mu_f/\mu_g)$
Z	parameter in Shah's correlation, dimensionless

Greek Symbols

β	spiral angle, °
Ψ	parameter in the Shah correlation
δ	fin height, mm
δ_a	gap distance between fins, mm
δ_f	fin thickness, mm
ε	fin pattern depth parameters, dimensionless
Y	lead angle, °
μ	viscosity, $(N \cdot s)/m^2$
ϕ	fin efficiency
ϕ_s	ratio of the enhanced area to total fin area, dimensionless
1ϕ	one phase
ρ	density, kg/m^3
σ	surface tension, N/m

Subscripts

a	based on arithmetic mean temperature difference, air
av	average
bs	stratified boiling
c	contact conductance
cb	convective boiling
conv	convective boiling region
d	diameter; depth; deposit
dry	dry
e	expanded
eq	equivalent
f	liquid; fin; frost; fluid
g	vapor
h	hydraulic; hole
i	inside
in	in
l	liquid; longitudinal; all-liquid
L	local
LMTD	log mean temperature difference
m	mean
N	number
nb	nucleate boiling

nucl	nucleate region
o	outside
out	out
p	plate; perimeter; pipe
P	pattern or wavy
pool	pool boiling
r	reduced; rows; refrigerant; root
ref	reference
s	smooth tube; solid wall; saturated
sfc	saturated forced convection
snb	saturated nucleate boiling
t	tube; transverse; tip
TP	two phase
v	vapor; valley
w	wall
wet	wet

REFERENCES

1. Webb, R. L., Choi, K., and Apparao, T. R. (1989) A theoretical model for prediction of the heat load in flooded evaporators. *ASHRAE Trans.* **95** (1).
2. Payvar, P. (1985) Analysis of performance of full bundle submerged boilers. ASME HTD-44, Denver, Colo., pp. 11–18. ASME, New York.
3. Huang, K., and Pate, M. B. (1988) A model for air-conditioning condensers and evaporators with emphasis on in-tube enhancement. IIR Conference on Refrigeration Machinery, Purdue University, July 18–21, pp. 266–276.
4. Hogan, M. R. (1980) The development of a low-temperature heat pump grain dryer. Ph.D. thesis, Purdue University.
5. Shah, M. M. (1976) A new correlation for heat transfer during boiling flow through pipes. *ASHRAE Trans.* **82** 66–86.
6. Shah, M. M. (1982) Chart correlation for saturated boiling heat transfer: equations and further study. *ASHRAE Trans.* **88** 185–196.
7. Kandlikar, S. S. (1987) A general correlation for saturated two-phase flow boiling heat transfer inside horizontal and vertical tubes. 1987 ASME Winter Annual Meeting, December 14–18.
8. Gungor, K. E., and Winteron, R. H. S. (1986) A general correlation for flow boiling in tubes and annuli. *Int. J. Heat Mass Transfer* **19**(3) 351–358.
9. McAdams, W. H. (1942) *Heat Transmission*, 2nd ed. McGraw-Hill, New York.
10. Petukhov, B. S. (1970) Heat transfer and friction in turbulent pipe flow with variable physical properties. *Advances in Heat Transfer*, Vol. 6. Academic, New York.
11. Schlager, L. M., Pate, M. B., and Bergles, A. E. (1988) Evaporation and condensation of refrigerant–oil mixture in a smooth tube and a micro-fin tube. *ASHRAE Trans.* **94**(1) 149–166.

12. Eckels, S. J. and Pate, M. B. (1990) An experimental comparison of evaporation and condensation heat transfer coefficients for HFC-134a and CFC-12. *Int. J. of Refrigeration*. To appear.

13. Schlager, L. M., Pate, M. B., and Bergles, A. E. (1989) A comparison of 150 and 300 SUS oil effects on refrigerant evaporation and condensation in a smooth tube and a micro-fin tube. *ASHRAE Trans.* **95**(1).

14. Schlager, L. M., Pate, M. B., and Bergles, A. E. (1990) Performance predictions of refrigerant–oil mixtures in smooth and internally finned tubes. II: Design equations. *ASHRAE Trans.* **96**(1).

15. Schlager, L. M., Pate, M. B., and Bergles, A. E. (1989) Heat transfer and pressure drop during evaporation and condensation of R-22 in horizontal micro-fin tubes. *Int. J. Refrigeration* **12** 6–14.

16. Schlager, L. M., Pate, M. B., and Bergles, A. E. (1989) Evaporation and condensation heat transfer and pressure drop in horizontal, 12.7 mm micro-fin tubes with refrigerant 22. *Proc. 1989 National Heat Transfer Conf.*, August 6–9, Philadelphia.

17. Schlager, L. M., Pate, M. B., and Bergles, A. E. (1989) Performance of micro-fin tubes with refrigerant-22 and oil mixtures. *ASHRAE J.* November 17–28.

18. Webb, R. L. (1980) Air-side heat transfer in finned tube heat exchangers. *Heat Transfer Eng.* **1**(3) 33–49.

19. Webb, R. L. (1983) Heat transfer and friction characteristics for finned tubes having plain fins. In *Low Reynolds Number Flow Heat Exchangers*, S. Kakaç, R. K. Shah, and A. E. Bergles (eds.), pp. 431–450. Hemisphere, Washington, D.C.

20. McQuiston, F. C., and Tree, D. R. (1971) Heat transfer and flow friction data for two fin-tube surfaces. *J. Heat Transfer* **93** 249–250.

21. Rich, D. G. (1973) The effect of fin spacing in the heat transfer and friction performance of multi-row, smooth plate fin-and-tube heat exchangers. *ASHRAE Trans.* **79**(2) 137–145.

22. Rich, D. G. (1975) The effect of the number of tube rows on heat transfer performance of smooth plate-fin-tube heat exchangers. *ASHRAE Trans.* **81**(1) 307–317.

23. McQuiston, F. C. (1978) Correlation of heat, mass, and momentum transport coefficients for plate-fin-tube heat transfer surfaces with staggered tube. *ASHRAE Trans.* **84**(1) 290–308.

24. Kays, W. M., and London, A. L. (1984) *Compact Heat Exchangers*, 3rd ed. McGraw-Hill, New York.

25. McQuiston, F. C. (1978) Heat, mass and momentum transfer data for five plate-fin-tube heat transfer surfaces. *ASHRAE Trans.* **84**(1) 266–293.

26. Gray, D. L., and Webb, R. L. (1986) Heat transfer and friction correlations for plate finned-tube heat exchangers having plain fins. *Proc. Eighth Int. Heat Transfer Conf.* August 17–22, San Francisco, pp. 2745–2750.

27. Hauser, S. G., Kreid, D. K., and Johnson, B. M. (1983) Investigation of combined heat and mass transfer from a wet heat exchanger. II. Experimental results. ASME–JSME Thermal Engineering Joint Conference, March 20–24, Honolulu, Hawaii, pp. 525–535.

28. Johnson, B. M., Kreid, D. K., and Hansen, S. G. (1983) A method of comparing performance of extended-surface heat exchangers. *Heat Transfer Eng*. **4**(1) 32–42.
29. Beecher, D. T., and Fagan, T. J. (1987) Fin-patternation effects in plate finned tube heat exchangers. ASHRAE Annual Meeting, June 27–July 1, Nashville, Tenn.
30. Fisher, S. K., and Rice, C. K. (1981) A steady-state computer design model for air-to-air heat pumps. ORNL/CON-80, Oak Ridge National Laboratories, Oak Ridge, Tenn.
31. Webb, R. L. (1983) Enhancement for extended surface geometries used in air-cooled heat exchangers. In *Low Reynolds Number Flow Heat Exchangers*, S. Kakaç, R. K. Shah, and A. E. Bergles, (eds.), pp. 721–734. Hemisphere, Washington, D.C.
32. Hosada, T., Uzuhashi, H., and Kobayashi, N. (1977) Louver fin type heat exchangers. *Heat Transfer–Japanese Research* **6**(2) 69–74.
33. Itoh, M., Kimura, H., Tanaka, T., and Musah, M. (19xx) Development of air-cooling heat exchangers with rough-surface louver fins. *ASHRAE Trans*. No. 2712, pp. 218–227.
34. Sensku, T., Hatada, T., and Ishibane, K. (1979) Surface heat transfer coefficients of fins used in air cooled heat exchangers, *Heat Transfer–Japanese Research* **8**(8) 16–26.
35. Mori, Y., and Nakayama, W. (1980) Recent advances in compact heat exchangers in Japan. HTD-10, pp. 5–16. ASME, New York.
36. Hatada, T., and Senshu, T. (1984) Experimental study on heat transfer characteristics on convex louver fins for air conditioning heat exchangers. ASME Paper 84-HT-74, pp. 1–8. ASME, New York.
37. Nakayama, W., and Xu, L. P. (1983) Enhanced fins for air-cooled heat exchanger —Heat transfer and friction factor correlations. ASME–JSME Thermal Engineering Joint Conference Proceedings, March 20–24, Honolulu, Hawaii, pp. 495–502.
38. Kadambi, V., and Giansante, J. H. (1982) The effect of lances on finned-tube heat exchanger performance. *ASHRAE Trans*. No. 2741, pp. 85–95.
39. Bemisderfer, C. H. (1987) Heat transfer: A contemporary analytical tool for developing improved heat transfer surfaces. *ASHRAE Trans*. **93**(1).
40. Threlkeld, J. L. (1970) *Thermal Environmental Engineering*. Prentice-Hall, Englewood Cliffs, N.J.
41. Stoecker, W. F., and Jones J. W. (1986) *Refrigeration and Air Conditioning*. McGraw-Hill, New York.
42. Myers, R. J. (1967) The effect of dehumidification on the air-side heat transfer coefficient for a finned tube coil. Master's thesis, University of Minnesota.
43. Hiller, C. C., and Glicksman, L. R. (1979) Improving heat pump performance via compressor capacity control—Analysis and test, Vols. 1 and 2, MIT Energy Laboratory Report MIT-EL 76-0001 and MIT-EL 76-002, Cambridge, Mass.
44. Kondepudi, S. N., and O'Neal, D. L. (1987) The effects of frost growth on extended surface heat exchanger performance: a review. ASHRAE Annual Meeting, June 27–July 1, Nashville, Tenn.

45. O'Neal, D. L. (1982) The effect of frost formation on the performance of a parallel plate heat exchanger. Ph.D. thesis, Purdue University.
46. Wood, R. A., Sheffield, J. W., and Sauer, H. J., Jr. (1987) Thermal contact conductance of finned tubes: A generalized correlation. ASHRAE Annual Meeting, June 27–July 1, Nashville, Tenn.
47. Eckels, P. W. (19xx) Contact conductance of mechanical expanded plate finned tube heat exchangers. Westinghouse Research Laboratories, Scientific Paper 77-1E9, SURCO-P1.
48. Traviss, D. P., Rohsenow, W. M., and Baron, A. B. (1972) Forced convection condensation inside tubes: a heat transfer equation for condenser design. *ASHRAE Trans.* **79** 157–165.
49. Cavallini, A., and Zecchin, R. (1974) A dimensionless correlation for heat transfer in forced convection condensation. *Proc. Fifth Int. Heat Transfer Conf.*, September 3–7, pp. 309–313.
50. Shah, M. M. (1979) A general correlation for heat transfer during film condensation inside pipes. *Int. J. Heat Mass Transfer* **22** 547–556.
51. Kohler, J. A., and Starner, K. E. (1989) High performance heat-transfer surfaces. In *Handbook of Applied Thermal Design*, E. C. Guyer (ed.). McGraw-Hill, New York.
52. Stephan, K., and Abdelsalam, M. (1980) Heat transfer correlations for natural convection boiling. *Int. J. Heat Mass Transfer* **23** 73–87.
53. Collier, J. G. (1981) In *Heat Exchanger Thermal-Hydraulic Fundamentals and Design*, S. Kakaç, A. E. Bergles, and F. Mayinger (eds.). Hemisphere Washington, D.C.
54. Borishanski, V. M. (1969) Correlation of the effect of pressure on the critical heat flux and heat transfer rates using the theory of thermodynamic similarity. *Problems of Heat Transfer and Hydraulics at Two-Phase Media*, pp. 16–37. Pergamon.
55. Wolverine Tube, Inc., Engineering Data Book, Decatur, Ala.
56. Chen, J. C. (1963) A correlation for boiling heat transfer to saturated fluids in convective flow. ASME Paper 63-HT-34, presented at Sixth National Heat Transfer Conference, Boston.
57. Bennett, D. L., Davis, M. W., and Hertzler, B. L. (1980) The suppression of saturated nucleate boiling by forced convective flow. *AIChE Symp. Ser.* **76**(199) 91–103.
58. Bennett, D. L., and Chen, J. C. (1980) Forced convective boiling in vertical tubes for saturated pure components and binary mixtures. *AIChE J.* **26**(3) 454–461.
59. Ishihara, K., Palen, J. W., and Taborek, J. (1980) Critical review of correlations for predicting two-phase flow pressure drop across tube banks. *Heat Transfer Eng.* **1**(3) 23–32.
60. Chisholm, D. (1967) A theoretical basis for the Lockhart–Martinelli correlation for two-phase flow. *Int. J. Heat Mass Transfer* **10** 1767–1778.
61. Schrage, D. S., Hsu, J. T., and Jensen, M. K. (1987) Void fractions and two-phase multipliers in a horizontal tube bundle. *Heat Transfer–Pittsburgh*, *AIChE Symp. Ser.* **83**(257) 1–8.
62. Katz, D. L. Myers, J. E., Young, E. H., and Balekjian, G. (1955) Boiling outside finned tubes. *Petroleum Refiner* **34**(2) 113–116.

63. Marto, P. J., and Lepere, V. J. (1981) Pool boiling heat transfer from enhanced surfaces to dielectric fluids. In *Advances in Enhanced Heat Transfer*. HTD-18, pp. 93–102. ASME, New York.
64. O'Neill, P. S., Gottzmann, C. G., and Terbot, J. W. (1972) Novel heat exchanger increases cascade cycle efficiency for natural gas liquifacture. *Adv. in Cryogenic Eng.* **17** 420–437.
65. Chyu, M. C. (1979) Boiling heat transfer from a structured surface. M.S. thesis, Iowa State University, Ames, Iowa.
66. Hitachi Cable, Ltd. (1978) High flux boiling and condensation heat transfer tube–hitachi thermoexel (Catalog). Tokyo, Japan.
67. Nakayama, W., Daikoku, T., Kawahara, H., and Nakajima, T. (1988) Dynamic model of enhanced boiling heat transfer on porous surfaces. II. Analytic modeling. *J. Heat Transfer* **102** 451–456.
68. Kuwahara, H., Nakayama, W., and Daikoku, T. (1977) Boiling heat transfer from a surface with numerous tiny pores linked by small tunnels running below the surfaces. *14th Symp. on Heat Transfer*, Japan, Paper B104.
69. Ayub, Z. H., and Bergles, A. E. (1987) Pool boiling from GEWA surfaces in water and R-113. *Warme-und Stoffubertragung* **21** 209–219.
70. Nusselt, W. (1916) Die Oberflachen-Kondesation des Wasserdampffs. *Zeitsh. R. Ver. Deutsch. Ing.*, pp. 60, 541, and 569.
71. Williams, P. E., and Sauer, H. J. (1981) Condensation of refrigerant–oil mixtures on horizontal tubes. *ASHRAE Trans.* **87**(1) 52–69.
72. Short, B. E., and Brown, H. E. (1951) Condensation of vapor in vertical banks of horizontal tubes. *Proc. General Discussion on Heat Transfer, Institute Mechanical Engineers*, pp. 27–31.
73. Webb, R. L. (1984) Shell-side condensation in refrigerant condensers. *ASHRAE Trans.* No. AT-84-01, pp. 5–25.
74. Kern, D. Q. (1950) *Process Heat Transfer*. McGraw-Hill, New York.
75. Beatty, K. O., and Katz, D. L. (1948) Condensation of vapors on outside of finned tubes. *Chem. Eng. Progress* **44**(1) 55–70.
76. Eckels, S. J., and Pate, M. B. (1990) A comparison of R-134a and R-12 In-tube Heat Transfer Coefficient, Based on Existing Correlations. *ASHRAE Transactions* **96**(1).

CHAPTER 13

EVAPORATORS AND REBOILERS IN THE PROCESS AND CHEMICAL INDUSTRIES

P. B. WHALLEY

Department of Engineering Science
University of Oxford
Oxford OX1 3PJ, United Kingdom

13.1 INTRODUCTION

This chapter describes the main features of the different classes of evaporator (Section 13.3) and reboiler (Section 13.4) used in the process and chemical industries. The main applications of the various types are briefly discussed. Energy efficiency in evaporation is briefly reviewed in Section 13.5, and the possible energy-saving arrangements of multiple-effect evaporation, vapor recompression, and multistage flash evaporation are introduced briefly. The main problems in the heat transfer and pressure drop in evaporators and reboilers are reviewed in Section 13.6. Finally, Section 13.7 looks at some of the problems encountered during their operation.

Evaporation is a very common industrial process: the solvent in a solution is vaporized to give a concentrated solution. This concentrated stream may be a product or a waste stream. If, as is very often the case, the solvent is water then the steam can be rejected or, more economically, the heat in the steam can be reused as described in Section 13.5. However, if the solvent is not water it is almost always valuable and is therefore recovered for reuse. The particular geometry of the reboiler can vary widely; the various types are described in Section 13.3.

Reboilers are used to vaporize the liquid at the bottom of a distillation column to provide the vapor flow up the column. The heat is removed in the

Boilers, Evaporators and Condensers, Edited by Sadik Kakaç
ISBN 0-471-62170-6

condenser at the top of the column, but of course the heat is removed at a lower temperature than the input temperature in the reboiler. As will be seen in Section 13.4, there is considerable overlap in application and type between evaporators and reboilers.

Before looking at the different types of evaporators and reboilers, it is useful to consider the various classifications into which the different units can be placed. First, the units can be divided into those where the evaporating stream and the heating stream are kept apart by a tube or plate wall, and those where the streams are deliberately allowed to mix. The former category is much more commonly encountered. The latter category are direct-contact units and are known as "submerged-combustion evaporators" (see Section 13.3.10). In these units there is no heat transfer surface. This is especially useful when evaporating highly corrosive or highly fouling solutions. Flash evaporators (see Sections 13.3.3 and 13.5.4) also need no heating surface, relying on the reduction in system pressure from one region of the evaporator to another to produce the required evaporation.

The units in which the evaporating stream and the heating stream are separated may be divided into two main types: shell-and-tube units and plate units. The shell-and-tube designs consist of a large cylindrical shell inside which is a bundle of tubes. The plate evaporator uses a plate instead of a tube to divide the heating and evaporating streams.

In a shell-and-tube unit, the shell may be horizontal or vertical, and the fluid to be evaporated may be introduced into the shell (a "shell-side" evaporator) or into the tubes (a "tube-side" evaporator).

In a plate unit the individual plates have corrugations or ribs to improve the heat transfer rate. The plates are mounted together and held in a frame with gaskets separating and sealing the plates. This arrangement allows alternate heating and evaporating streams in the space between successive plates. In other plate-type designs of evaporators, flat aluminum plates are separated by corrugated metal sheets which acts as fins. The sandwich of plates and fins is brazed together to form an integral unit. Such units are often used in cryogenic applications.

In all types of reboilers and evaporators, to get a good heat transfer rate, it is most advantageous to see that the liquid to be evaporated flows over the heated surface with as large a velocity as possible. This flow or circulation of liquid in the unit may be caused by the density differences between the evaporating mixture of liquid and vapor, and the liquid returning after vapor–liquid separation. These units are known as "natural-circulation" units. Alternatively, the flow can be provided by a pump: these are "forced-circulation" units. In some cases both processes may contribute to the circulation: these are "assisted-circulation" units. If the liquid being evaporated is very viscous, none of these circulation methods may be sufficient. In this case direct agitation at the heat transfer surface must be provided; evaporators with this feature include stirred or agitated evaporators. However, more commonly the liquid is mechanically spread onto the heated

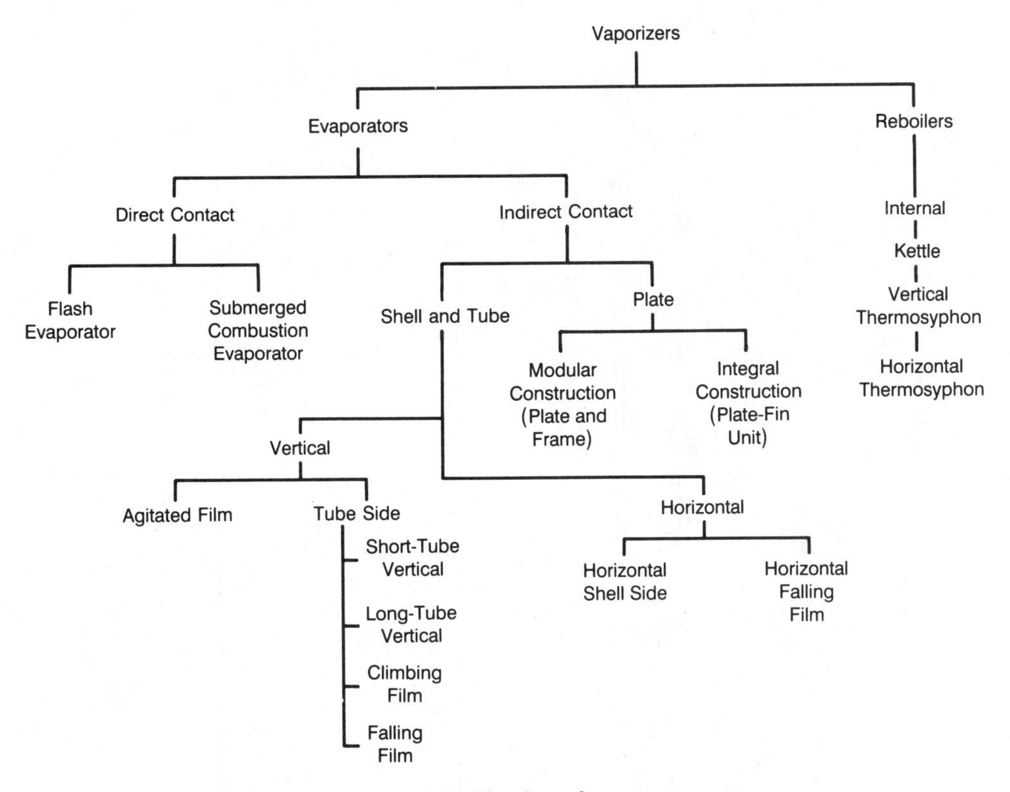

Fig. 13.1. General classification of evaporator types.

surface as a thin film: this can be done by a rotating assembly of mechanical wipers or scrapers. This unit is the agitated thin film evaporator and is described further in Section 13.3.8.

The main categories of evaporators and reboilers are summarized in Fig. 13.1. Further details about the types of unit can be found in [1–5].

13.2 RELEVANCE OF UPFLOW AND DOWNFLOW IN VERTICAL UNITS

An important design and operating consideration in all vertical evaporators or reboilers is whether the liquid feed enters at the bottom or the top of the unit. If the solution flows upwards, the boiling point of the liquid falls as it flows through the unit because of the decreasing hydrostatic pressure. If the liquid enters well below its boiling point, there is no boiling at the base of the unit. Rising through the unit, the temperature of the liquid increases as it is heated, and at the same time the boiling point is falling. Once boiling begins, the temperature then falls corresponding to the fall in pressure, because the liquid is saturated. Evaporation is occurring therefore both because of

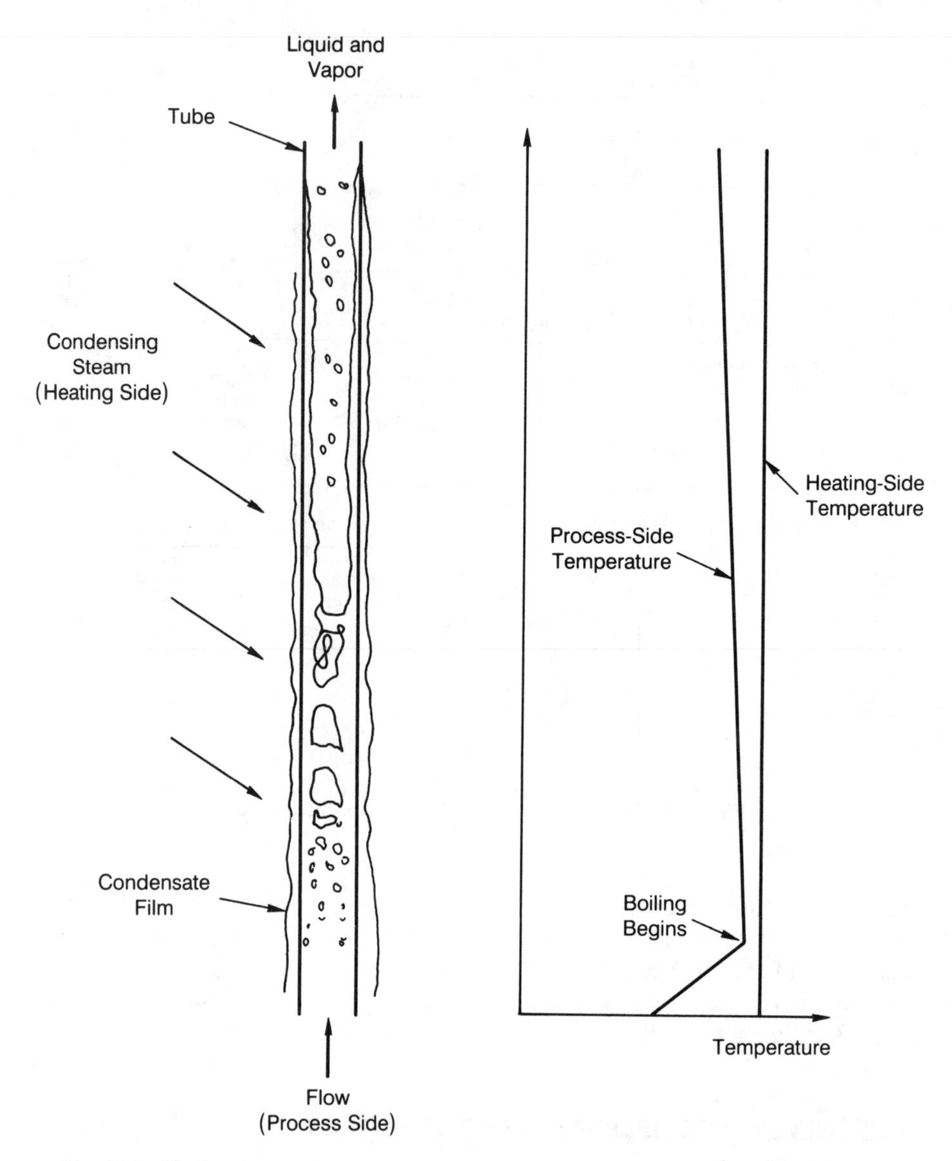

Fig. 13.2. Upflow in an evaporator: typical temperature profile in the boiling liquid.

flashing of the liquid due to the fall in pressure and because of the heat transfer. Hence the temperature difference between the evaporating fluid and the heating process stream passes through a minimum at the start of boiling (see Fig. 13.2). The consequence of this is that a larger steam pressure and a larger heat transfer area are required than if the evaporating fluid were flowing down through the unit as shown in Fig. 13.3. Thus if there is a choice, downflow units are preferable to those where there is an upflow.

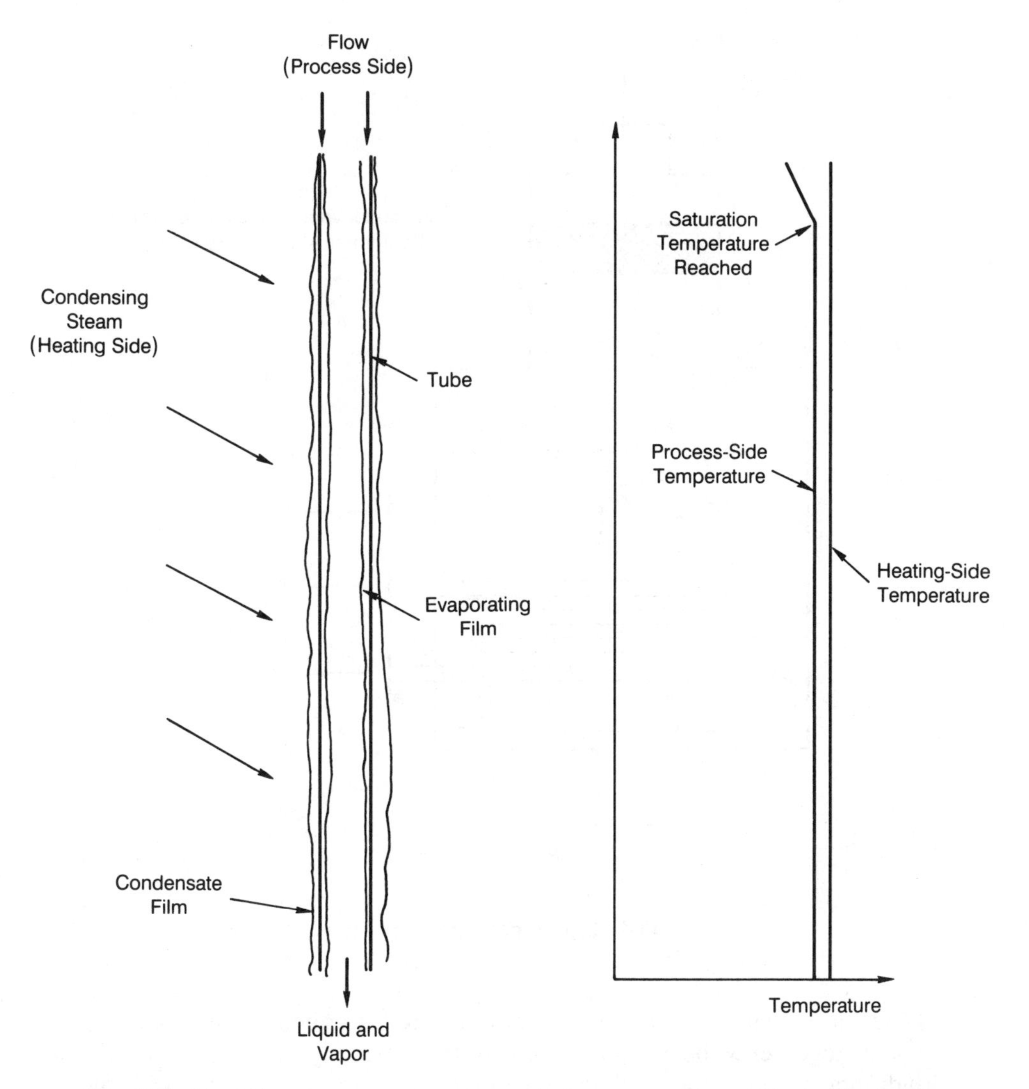

Fig. 13.3. Downflow in an evaporator: typical temperature profile in the boiling liquid.

13.3 EVAPORATOR TYPES

In some parts of the world today, common salt is recovered from seawater by direct solar evaporation of seawater in shallow ponds. A variant of this is to heat the liquid in a pan over a fire. However, the simplest form of industrial evaporator is shown in Fig. 13.4: it is a simple steam-heated evaporator (sometimes called a "still"). In industrial evaporators the heating medium is very commonly process steam (as in Fig. 13.4), but sometimes hot oil or special heating liquid is used.

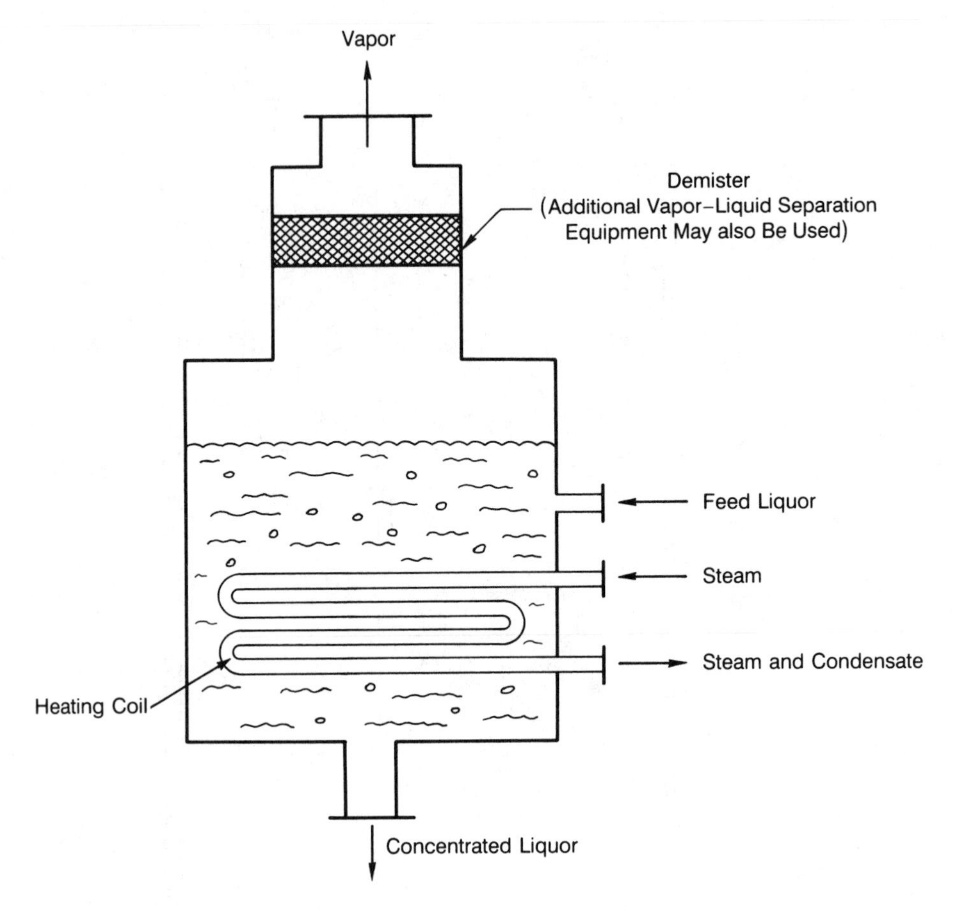

Fig. 13.4. Typical pot-type evaporator.

The most common types of evaporators are described in the following sections, together with some indications of their uses or their advantages and disadvantages. First, the shell-and-tube units are described, starting with the shell-side units (Sections 13.3.1 and 13.3.2), followed by the tube-side units (Sections 13.3.3 to 13.3.8). Then plate units are described (Section 13.3.9), and finally direct-contact units (the submerged-combustion evaporator, see Section 13.3.10).

13.3.1 Horizontal Shell-Side Evaporator

The horizontal shell-side evaporator and the kettle reboiler (see Section 13.4.2) are identical in construction. The horizontal shell-side evaporator is illustrated in detail in Fig. 13.5. The heating fluid (usually steam) is supplied to the inside of the tubes in the bundle, and the liquid to be evaporated is on

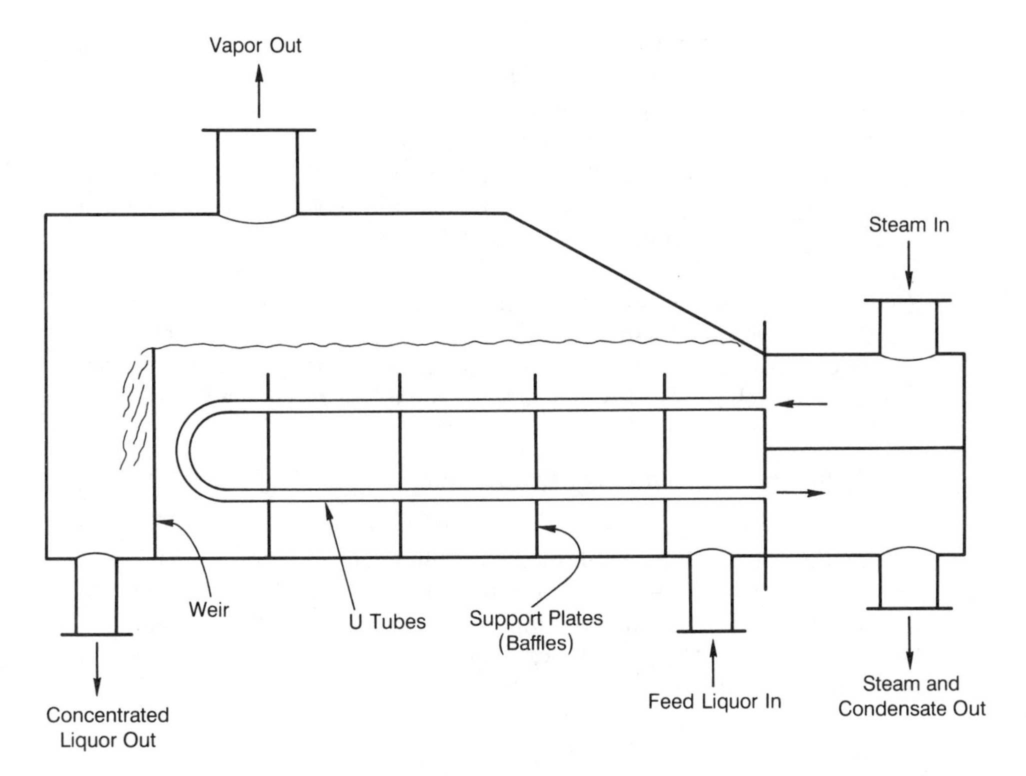

Fig. 13.5. Horizontal shell-side evaporator.

the outside of the tubes. These units are often used as boiler feed-water make-up evaporators, but are otherwise not much used as evaporators. Low entrainment of liquid drops into the vapor is a primary requirement for these evaporators, and the horizontal unit provides a large vapor–liquid surface area for separation in relation to the shell diameter. For economic reasons the shell diameter must be kept as small as possible because these evaporator units often operate at high pressures both on the tube side and the shell side.

The advantages of horizontal shell-side evaporators are:

1. As explained previously they give a relatively large vapor–liquid separation area.
2. They have a very low headroom requirement and can thus be fitted into confined spaces.
3. They are relatively cheap to construct, though the cost increases fairly rapidly with the shell-side pressure as the shell is relatively large. The large shell is necessary to provide the space for the disengagement of the liquid and the vapor.

4. They provide good heat transfer performance because there is a strong circulation induced within the shell and through the tube bundle. This aspect is discussed further in Section 13.4.2 on kettle reboilers.
5. As evaporators they can be used with hard waters: it has been found that the scale formed on the tubes can be removed by draining the shell and rapidly filling the hot tube bundle with cold water to create a thermal shock to crack off the scale.

In general, however, in spite of the comment about hard-water scale, these units are not best suited to the evaporation of fouling liquids. It is common for the most fouling stream to be placed inside the tubes as this surface can be mechanically cleaned. Also they are not suitable for foaming liquids.

13.3.2 Horizontal Falling-Film Evaporator

Falling-film evaporators are usually vertical units, where the liquid falls down inside of the tubes. However, horizontal falling-film evaporators have also been suggested [6] (see Fig. 13.6). Here the liquid to be evaporated is sprayed on the outside of a horizontal tube bundle. The heating fluid flows inside the tubes. The liquid outside the tubes forms a thin film flowing around the tubes. When it reaches the bottom it drips off, either as drops or possibly as a

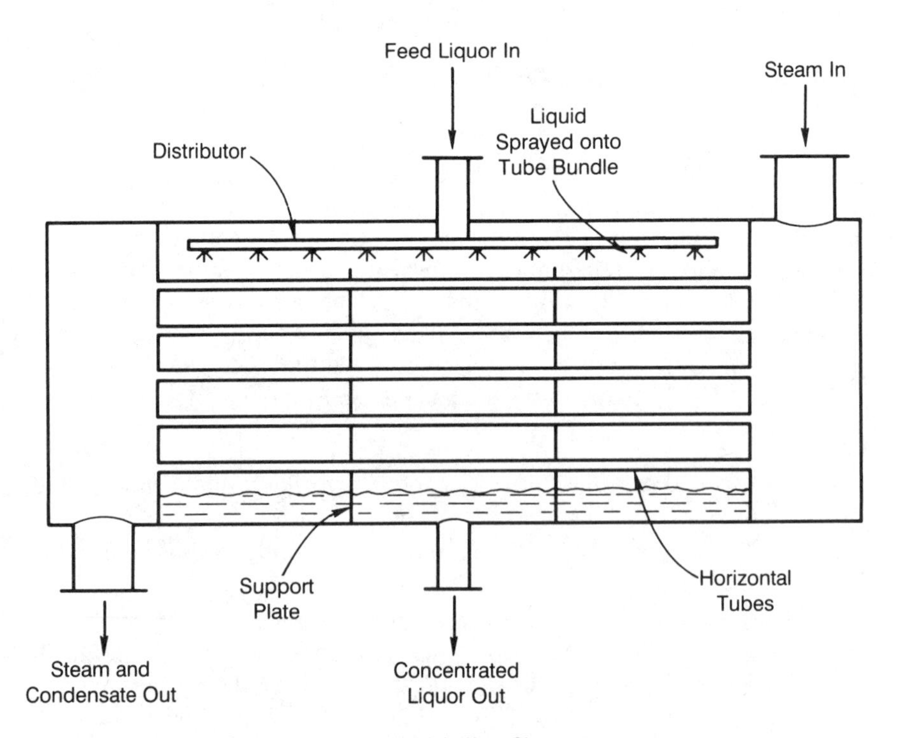

Fig. 13.6. Horizontal falling-film evaporator.

sheet of liquid, onto the tubes below. The flow is very much like the traditional view of condensation in a horizontal tube bundle: indeed the main difference is that the flow rate of liquid falls going down through the bundle instead of rising.

One particular advantage of these units over conventional film evaporators is that the problem of distributing the liquid is much simpler. In a vertical unit, ideally the same amount of liquid is fed to each tube and distributed uniformly around the periphery of each tube. In the horizontal unit, the distribution can be done as a spray onto the top face of the bundle. Like vertical falling-film evaporators, horizontal units give high heat transfer coefficients because the liquid is in the form of a very thin film.

13.3.3 Horizontal Tube-Side Evaporator

This type of evaporator is illustrated in Fig. 13.7; it is also sometimes called the submerged-tube forced-circulation evaporator. It consists of a horizontal shell-and-tube heat exchanger with steam or other heating fluid on the shell

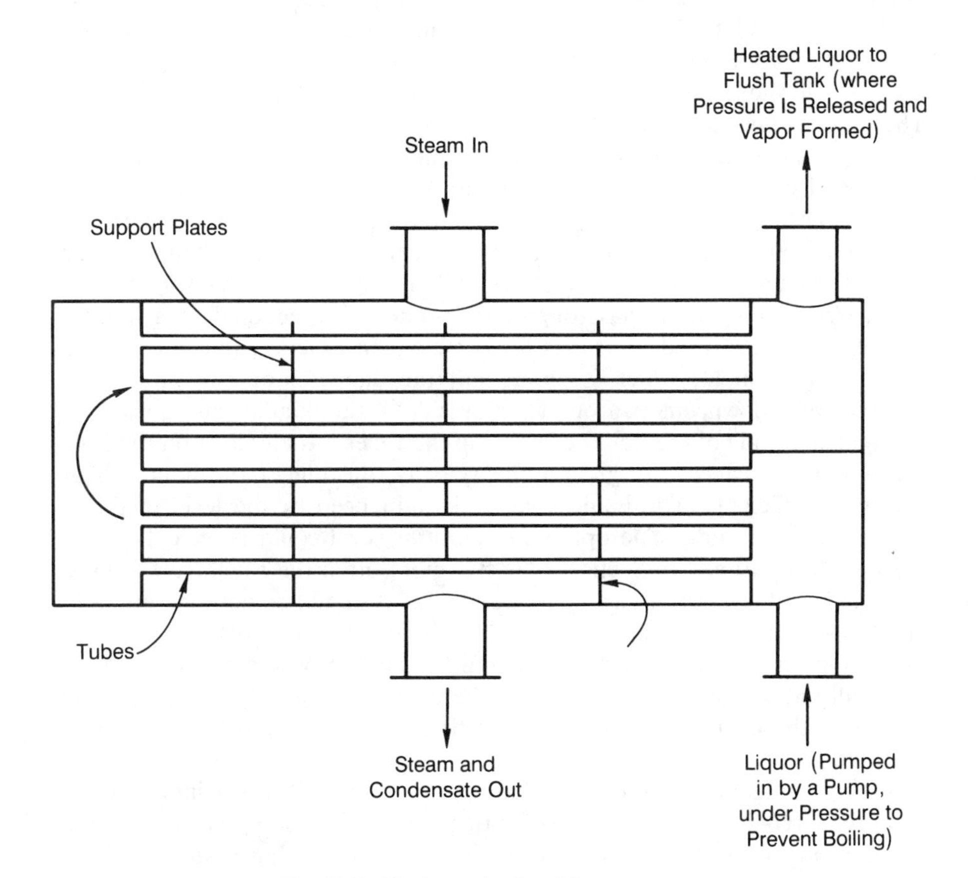

Fig. 13.7. Horizontal tube-side evaporator.

side. The liquid feed flows on the tube side, it is heated but bulk boiling does not occur because the pressure is high. It should be noted that the prevention of boiling inevitably means that the heat transfer is relatively poor, though possibly some subcooled boiling may occur if the temperature is only a few degrees below the boiling point. The heated solution flows through a throttle valve and into a separator. As the liquid flows into the separator, some of its flashes to form vapor. The liquid from the separator is recirculated via a pump to the evaporator. A pump is inevitably necessary because of the large pressure loss across the throttle valve. These evaporators are suitable for crystallization of, for example, common salt, and similar duties. They are suitable for such applications because the fact that boiling is prevented in the heat exchanger prevents crystallization or large-scale fouling occurring on the heat transfer surfaces.

13.3.4 Short-Tube Vertical Evaporator

This type of unit is illustrated in Fig. 13.8. It was one of the first types of evaporator to be developed, it is also called the "calandria" evaporator. It consists of a relatively squat vertical cylinder, and it has horizontal tube sheets that go right across the shell. The tubes are relatively large in diameter, in the range 25 to 75 mm, but relatively short (only 1.5 to 2 m long). The tubes are expanded into tube sheets. The larger tube diameters are used for crystallizing evaporators. The evaporating liquid fills the lower part of the vessel and comes part way up the tubes. The tubes are heated from the outside, usually by condensing steam. The liquid boils in the tubes. As the liquid boils it is carried upward by the steam, the liquid is then returned to the lower part of the evaporator through a large central hole (or "well"). This large hole typically has roughly the same cross-sectional area as that available for flow in all the tubes. This means, in practice, that the central hole has a diameter which is about half the diameter of the tube sheet. Alternatively, downcomers around the outside of the bundle can be used.

The magnitude of the liquid velocity up the tubes has an effect on the heat transfer performance: the greater the velocity, the higher the boiling heat transfer coefficients. The liquid velocity is influenced by the level of liquid within the whole unit. The optimum operating level could be calculated by the methods outlined in Section 13.6, but more often the rough rule is used that the liquid level (as indicated by a sight glass) should be between one-half and two-thirds the way up the calandria. If the level is too high, then the saturation temperature will be increased by the hydrostatic pressure effect. This will reduce the heat transfer and the circulation rate. If the level is too low, then there may be incomplete wetting of the upper part of the tube surfaces.

If these units are used as crystallizing evaporators, then it is important to keep the circulation rate high. Assisted circulation is used by putting a large impeller in the central downcomer. Such a unit is shown in Fig. 13.9. Crystallization is not wanted in the tubes, and so the liquid level is increased,

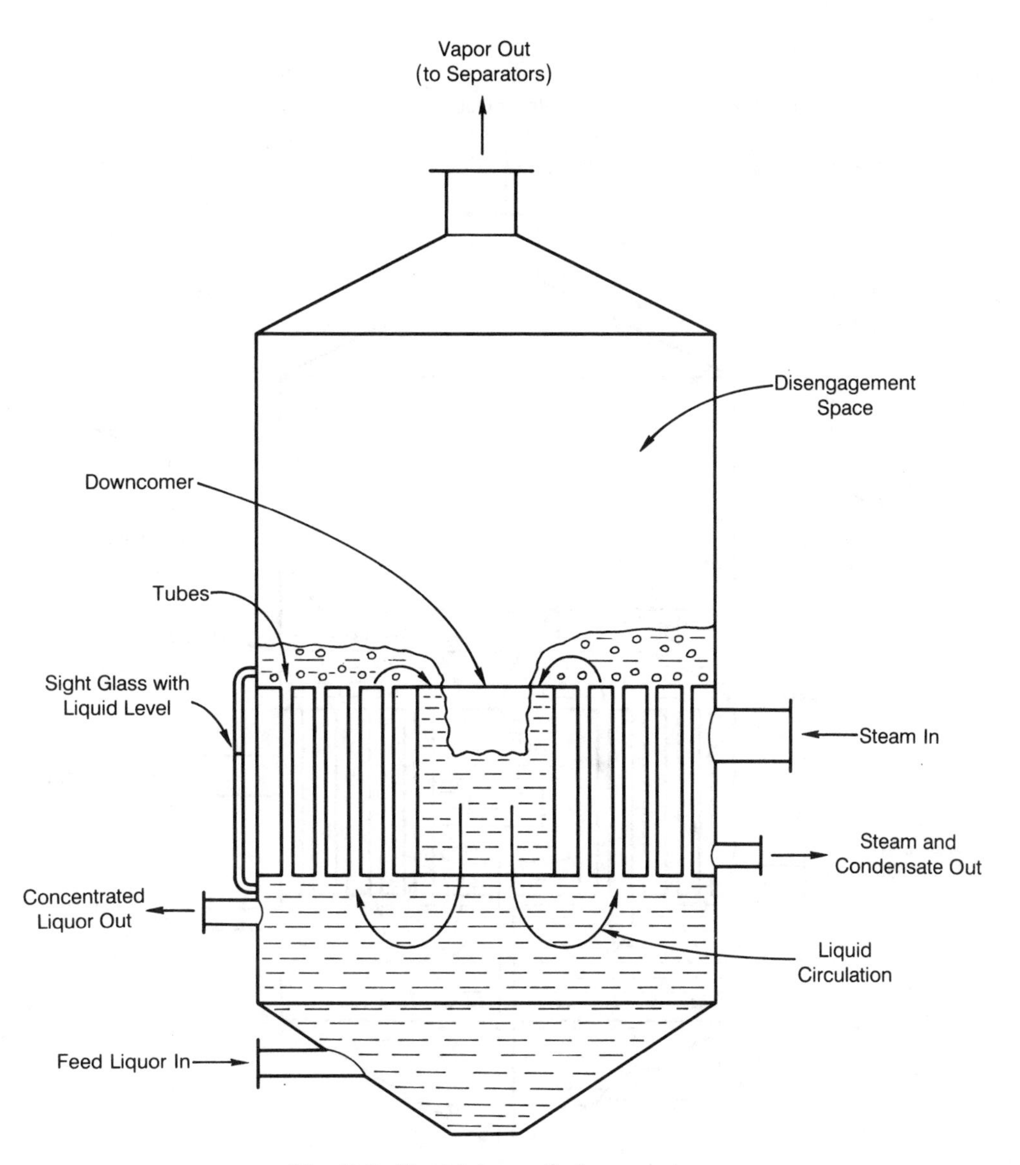

Fig. 13.8. Short-tube vertical evaporator.

as shown in Fig. 13.9. The pumping unit, if used to increase the liquid flow rate, should be installed at the bottom of the downcomer to minimize the risk of cavitation.

The advantages of short-tube vertical evaporators are:

1. They give good heat transfer performance particularly at large temperature differences. This is because large temperature differences lead to high liquid velocities through the tubes.

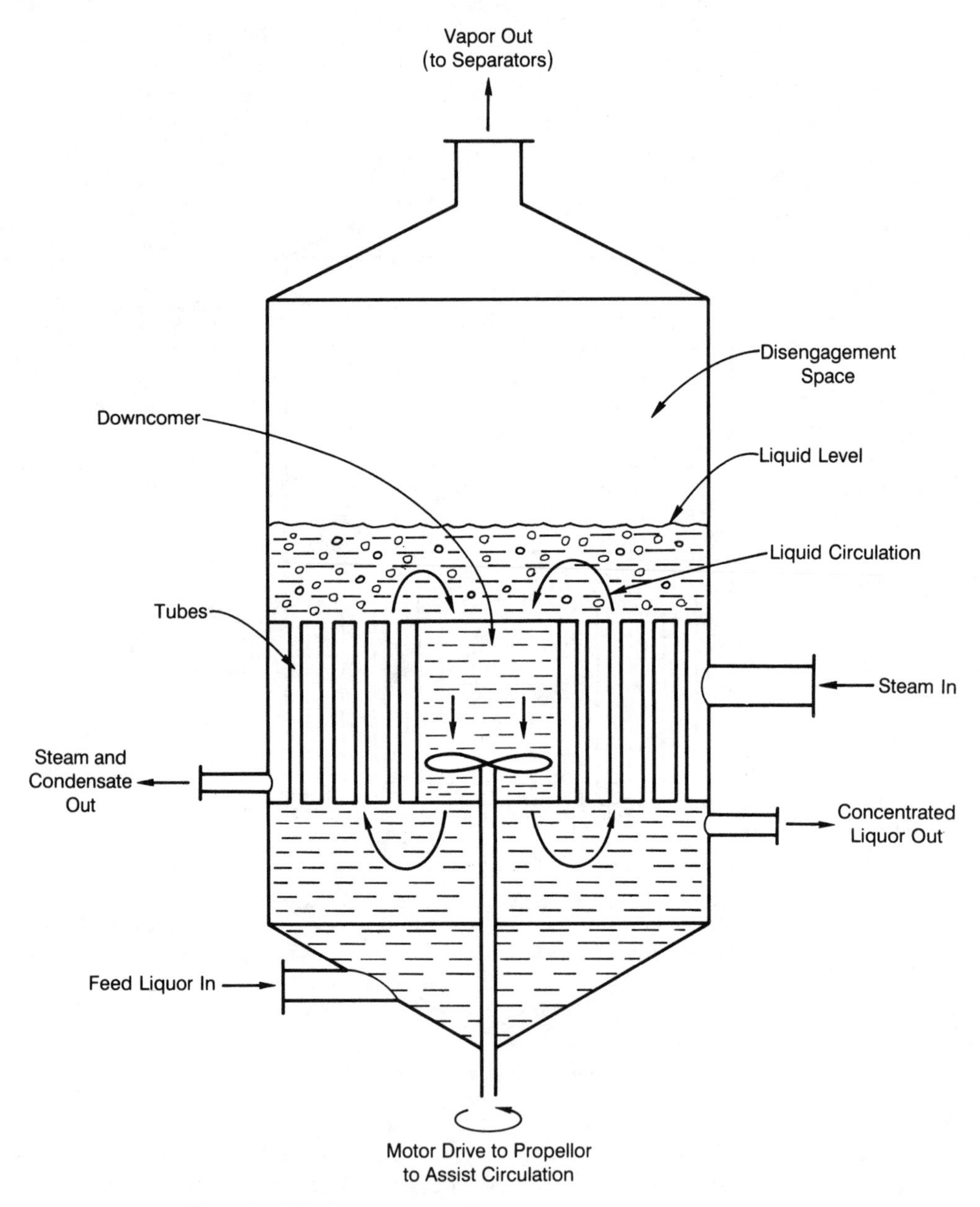

Fig. 13.9. Short-tube vertical evaporator used as a crystallizer.

2. They are relatively cheap to construct and can be built as large units.
3. They are suitable for crystallizing liquids if assisted circulation is used, as in Fig. 13.9.
4. They require low headroom.
5. They often have large-diameter tubes and therefore the inside surface, in particular, is relatively easy to clean mechanically. They are thus suitable for fouling liquids.

13.3.5 Long-Tube Vertical Evaporator

The long-tube vertical evaporator is illustrated in Fig. 13.10. It is very similar to the vertical thermosyphon reboiler (see Section 13.4.3). The main difference is that the vertical thermosyphon reboiler is, as its name implies, a natural-circulation device. The long-tube vertical evaporator is often equipped with a pump in the feed line as shown in Fig. 13.10. The unit consists of a

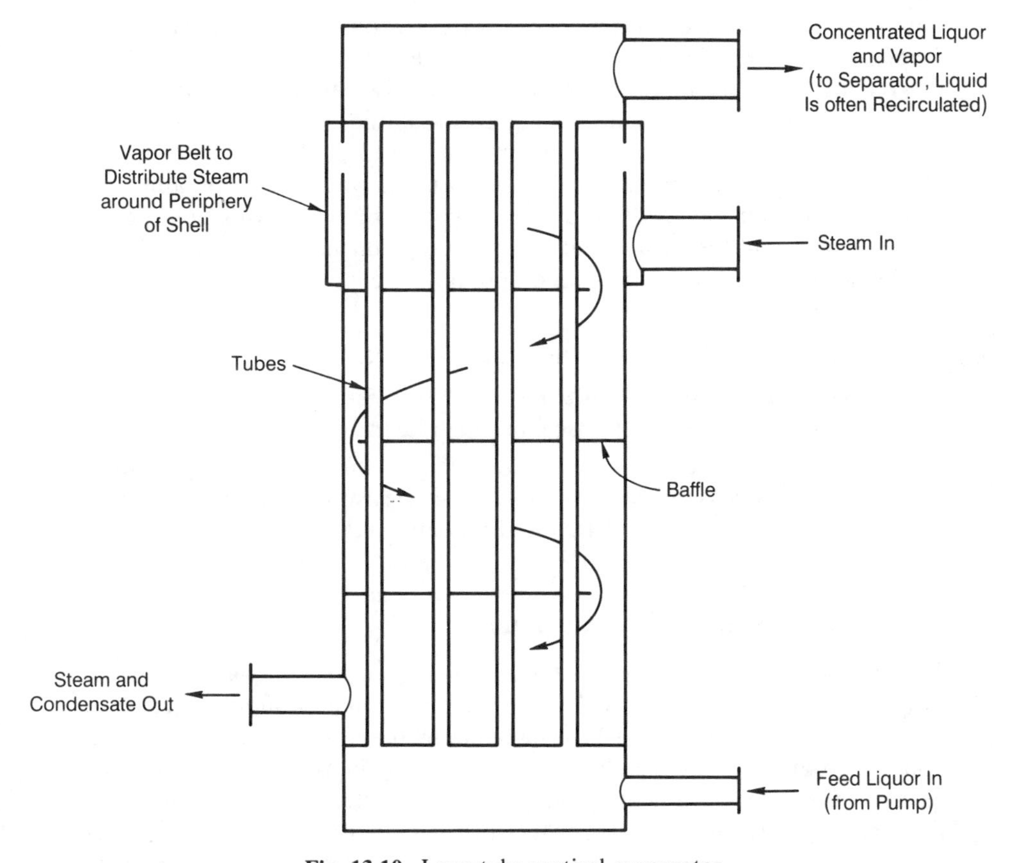

Fig. 13.10. Long-tube vertical evaporator.

vertical shell-and-tube heat exchanger. The tubes are usually smaller in diameter, longer, and fewer in number than in the short-tube version (see Section 13.3.4). Here the tube length may be up to 6 m. Boiling takes place inside the tubes, which are usually heated by steam condensing on the shell side. The vapor–liquid separator may be integral, but is more usually carried out in a different vessel as shown in Fig. 13.10. From the separator the liquid phase recirculates to the evaporator.

13.3.6 Climbing-Film Evaporator

A slight variant on the long-tube evaporator described in Section 13.3.5 is the so-called "climbing-film evaporator." Here the flow into the tubes is relatively low and so a large fraction of the flow is vaporized fairly quickly. This has the result of producing a high vapor velocity which causes the "annular" vapor–liquid flow pattern to be formed. Here most of the liquid flows as a thin film on the walls of the tube, and the vapor flows up the center of the tube. The advantage of the annular flow pattern is that the liquid film is very thin and so high heat transfer coefficients are obtained. A climbing-film evaporator can be used, in some applications, as a once-through device. Here the required concentration is reached in one pass through the evaporator, and so no recirculation of the feed liquid is required.

The change in the saturation temperature due to the hydrostatic pressure effect can make this evaporator type difficult to analyze. If it is operating in a natural-circulation mode (as in a thermosyphon reboiler) without a pump, then calculation of the throughput is particularly difficult (see Section 13.6.3).

The advantages of climbing-film evaporators are:

1. They give excellent heat transfer performance.
2. They are generally inexpensive to manufacture.
3. They take up little floor space, but high headroom is needed.
4. The liquid hold-up and the liquid residence time are low. These units are therefore useful for evaporating heat-sensitive liquids, or where a low inventory of liquid is required for safety reasons.

13.3.7 Vertical Falling-Film Evaporator

This type of unit is illustrated in Fig. 13.11; it is not used as a reboiler. As explained in Section 13.2.1, a downflow evaporator is desirable because it does not suffer from problems brought about by the hydrostatic head effect in upflow evaporators. The feed liquid is arranged to fall as a thin film down the inside of the tubes. The heating fluid (condensing steam or hot liquid) is on the shell side of the exchanger. Because the film is very thin, the heat transfer coefficient for the evaporating film can be very high. The vapor usually flows

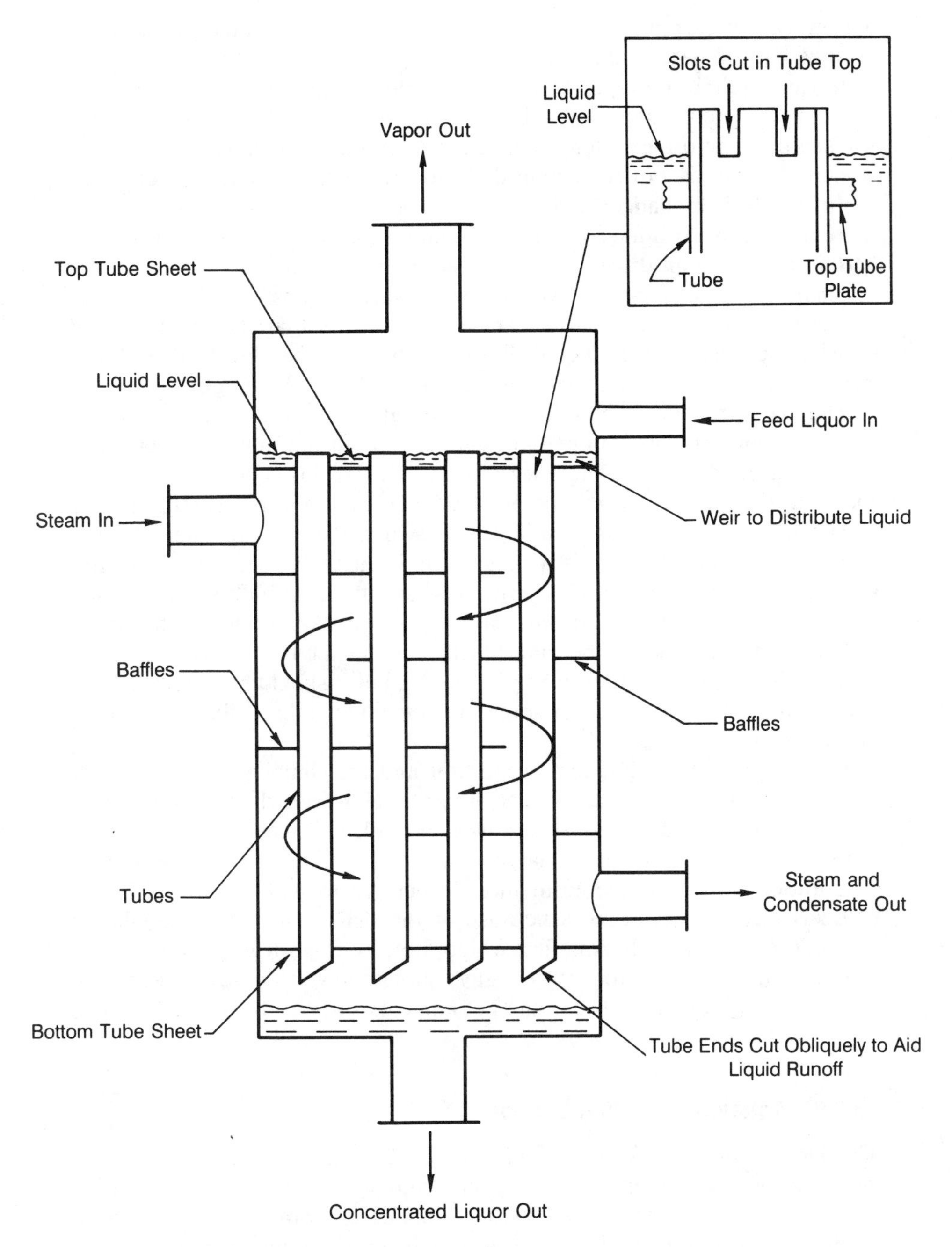

Fig. 13.11. Vertical falling-film evaporator.

downwards co-currently with the liquid and the vapor–liquid separator is arranged at the base of the unit.

If it is required to evaporate the liquid below its normal boiling point and vacuum operation is undesirable because the liquid would be oxidized by any leakage of air into the system, then the evaporation may be carried out in a stream of inert gas. For co-current downflow the inert gas would enter at the top of the heat exchanger and be taken off the bottom. In this type of unit there may be an advantage in having countercurrent operation as this gives a better concentration driving force for the mass transfer. This means having the inert gas and evaporated vapor flow upwards against the liquid flowing downwards. One danger of such an arrangement is that the flooding limit may be exceeded and the liquid film may no longer be able to flow downwards against the upflow of gas (see Section 13.7.7 on flooding).

The main problem with falling-film evaporators is that it is very important to have a good distribution of the liquid feed both uniformly to every tube and also around the circumference of each tube. If this is not achieved then some of the tubes will not be wetted over their entire length. A weir arrangement can work reasonably well provided that the top tube sheet is accurately flat and level: this arrangement is, however, sensitive to hydrostatic pressure gradients across large tube sheets. Distributors with orifices are often preferable to weirs in that they give a better feed distribution; however, the orifices can become blocked if the feed liquid is dirty. This problem is discussed further in Section 13.7.3, but it should be noted that the distribution problem is not so severe in horizontal falling-film evaporators (see Section 13.3.2).

Both the vertical falling-film evaporator and the climbing-film evaporator produce very good heat transfer performance. However, the falling film unit has the inherent advantage that it is a downflow device. The falling-film evaporator is, however, slightly larger than the climbing-film unit for a given evaporation duty, and therefore more expensive to manufacture. The expense is increased even more because of the delicate nature of the liquid distributors required. Falling-film evaporators are, however, probably the type of evaporator able to operator most successfully at the lowest values of temperature difference between the hot and cold sides of the heat exchanger.

13.3.8 Agitated Thin Film Evaporator

This type of unit is illustrated in Fig. 13.12. The falling-film evaporator uses gravity to produce a thin liquid film, the climbing-film evaporator uses shear stress produced by the vapor flow, and in this evaporator the thin film is produced mechanically; this is done by an integral rotor equipped with blades that spread out the liquid into a thin film [7]. Thin films are of course desirable because they lead to high heat transfer coefficients. Such evapora-

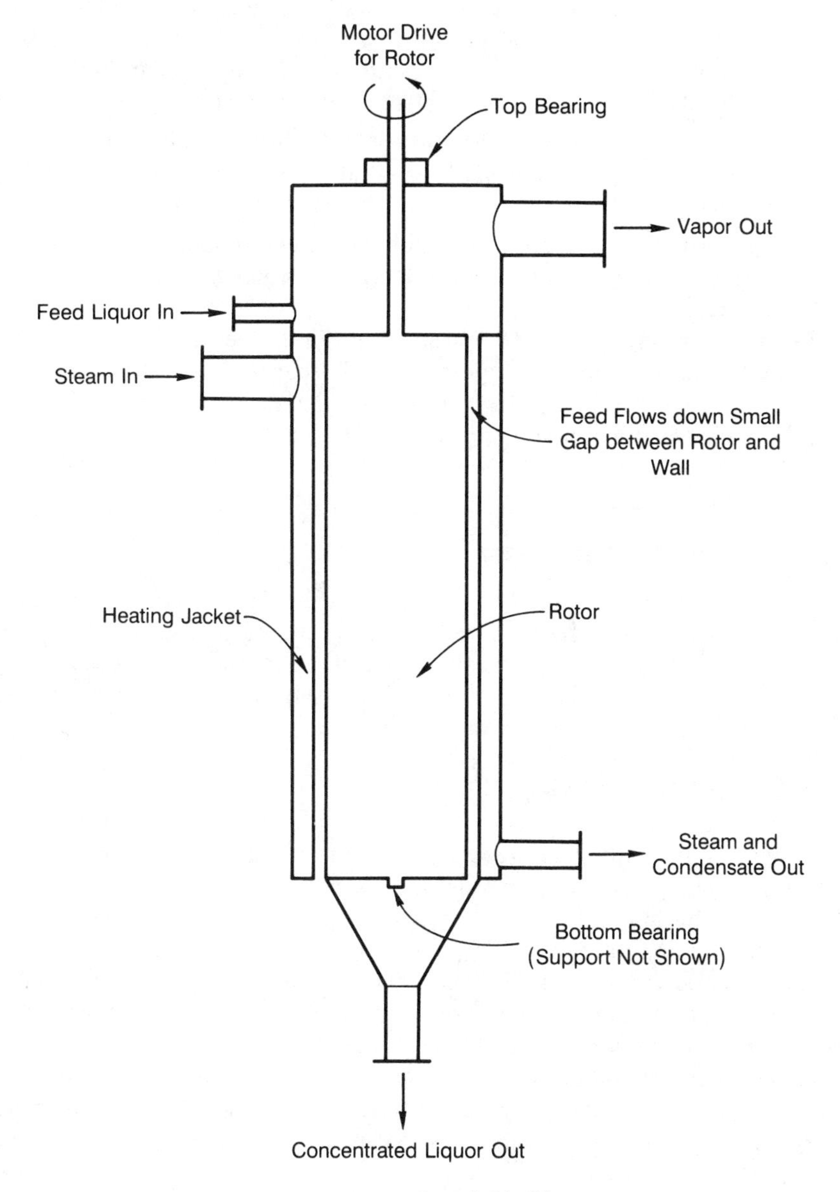

Fig. 13.12. LUWA-type agitated thin film evaporator.

tors are used when the product is:

1. heat sensitive, for example, food products because the residence time on the heated surface is very low, typically in the range 6 to 40 s. In addition, the residence time distribution is very narrow so that there is only a small probability of a heat-sensitive product being damaged.
2. known to have a tendency to foam or to foul the heat transfer surface.
3. highly viscous, as the mechanical action produces the thin film.

As well as producing high heat transfer coefficients under difficult circumstances, these evaporators are able to operate at high evaporation ratios (see Section 13.6.10).

Typical operating parameters for agitated thin film evaporators have been given by Salden [7]. They can operate up to 3 bar and 400°C with fluids of viscosity up to 10^4 $(N \cdot s)/m^2$. The size of the unit can be up to a heat transfer area of 40 m^2, a volumetric flow of 35 m^3/hr, and an evaporation rate of 12,000 kg/hr.

13.3.9 Plate-Type Evaporator

This type of unit is illustrated in an exploded view in Fig. 13.13 [8, 9]. Plate heat exchangers can often be used as an alternative to shell-and-tube exchangers. They are very commonly used in heat transfer to single-phase liquids but their use with boiling and condensing flows is less well known. Normal evaporators usually involve condensation of the heating steam, as well as partial vaporization of the feed liquid. There are two approaches to using a plate heat exchanger as an evaporator.

The first is to use the heat exchanger as a conventional single-phase heat exchanger by preventing the feed liquid from evaporating by increasing the pressure. When the pressure is released after the heat exchanger, the vapor is formed by flashing.

The second approach is to accept that changes of phase will take place in the plate heat exchanger and to modify the exchanger accordingly. Inevitably, a large change of volume occurs, so in the evaporation, for example, provision must be made for the large extra volume of vapor produced. Hence a novel arrangement and design of plates is needed (see Fig. 13.13). The plates are relatively widely spaced to allow sufficient flow area for the two-phase mixtures. The plates are grouped in sections of four plates. Referring to Fig. 13.13 and considering the plates from the left:

Plate 1 is a condensing-steam plate. Steam condenses on this plate and heats the plates on each side. This plate has downwards flow, and the condensate is removed at the bottom of the plate.

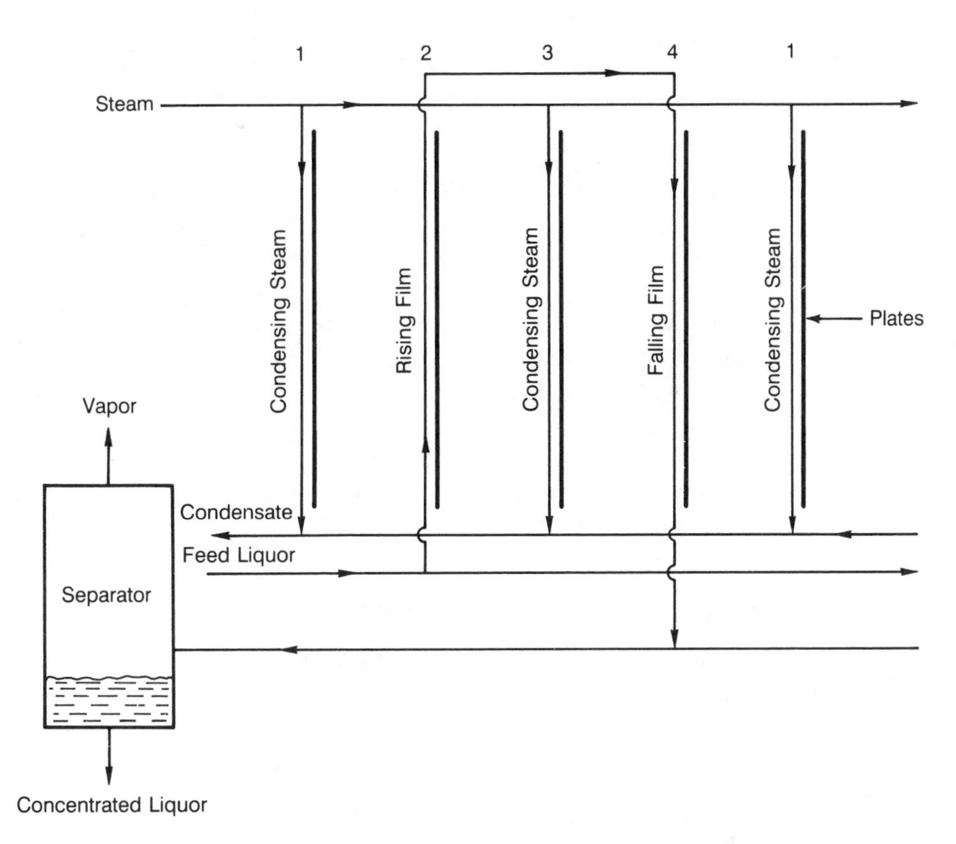

Fig. 13.13. APV plate-and-frame evaporator.

Plate 2 is an evaporation plate. The feed liquid enters at the bottom, and the liquid is partially evaporated as it rises. The plate is heated by the two adjacent condensing-steam plates. The liquid and the vapor are transferred to plate 4.

Plate 3 is another condensing-steam plate. Again the steam condenses in downwards flow, and the condensate is removed at the bottom of the plate.

Plate 4 is another evaporation plate that receives the liquid and vapor from plate 2. The liquid flows downwards on this plate, further evaporation taking place as it falls. The remaining liquid and the vapor are removed through the large rectangular port at the bottom of the plate and flow into the separation vessel. Again this plate is heated by the two surrounding steam plates.

Plate 5 is the first plate of another group of four.

The advantages of these plate-type evaporators are:

1. They can be used with very heat sensitive products as the residence time is short. In addition, the liquid hold-up on the plates is low.
2. They do not require much headroom, typically 3 m.
3. They give high heat transfer rates. Once again this is a consequence of the liquid being in the form of a thin film.
4. They can be arranged in very flexible groups. For example, plate evaporators can be arranged as multiple-effect evaporators. Another consequence of this flexibility is that a number of units can easily be arranged to handle large flow rates.

The second main class of evaporator indicated in Fig. 13.1 consists of those utilizing direct-contact heat transfer. The primary example of a direct-contact evaporator is the subemerged-combustion evaporator.

13.3.10 Submerged-Combustion Evaporator

This type of unit is illustrated in Figure 13.14. Submerged combustion is the combustion of hydrocarbon fuel so that either the combustion itself takes place under liquid, or the hot combustion product gases are released under the surface of the liquid. In either case the energy released by the combustion process is transferred by direct contact with the liquid. Submerged-combustion systems are unusually efficient in that the maximum possible energy is transferred to the liquid.

As indicated previously the burner itself can be submerged in the liquid; however, this is relatively unusual. It has the advantage, however, that the burner is very efficiently cooled by the surrounding liquid. This may mean that refractory material or special steel is not needed for the burner.

The exhaust gas to the stack from any type of submerged-combustion unit will be unusually cool and wet. This of course is a necessary consequence of the efficiency of the combustion process. It does mean, however, that the stack gas will be unusually corrosive, and this fact leads to a tendency to use very clean fuels in this type of evaporator.

In the submerged exhaust system shown in Fig. 13.14, the exhaust gas is released into the annulus between the downcomer and a draft tube. This has the effect of producing a strong circulation of the liquid in the tank and so gives good mixing between the hot exhaust gas and the liquid.

Submerged-combustion evaporators have found two main uses. They are used to revaporize liquified gases. This is done in a submerged-combustion unit in two stages. First, the combustion product gases are used to heat water in a tank. Then the hot water in the tank is used to vaporize the liquid that flows through coiled tubes immersed in the liquid. This can be done safely when the direct vaporization of the liquified gases might be hazardous [10].

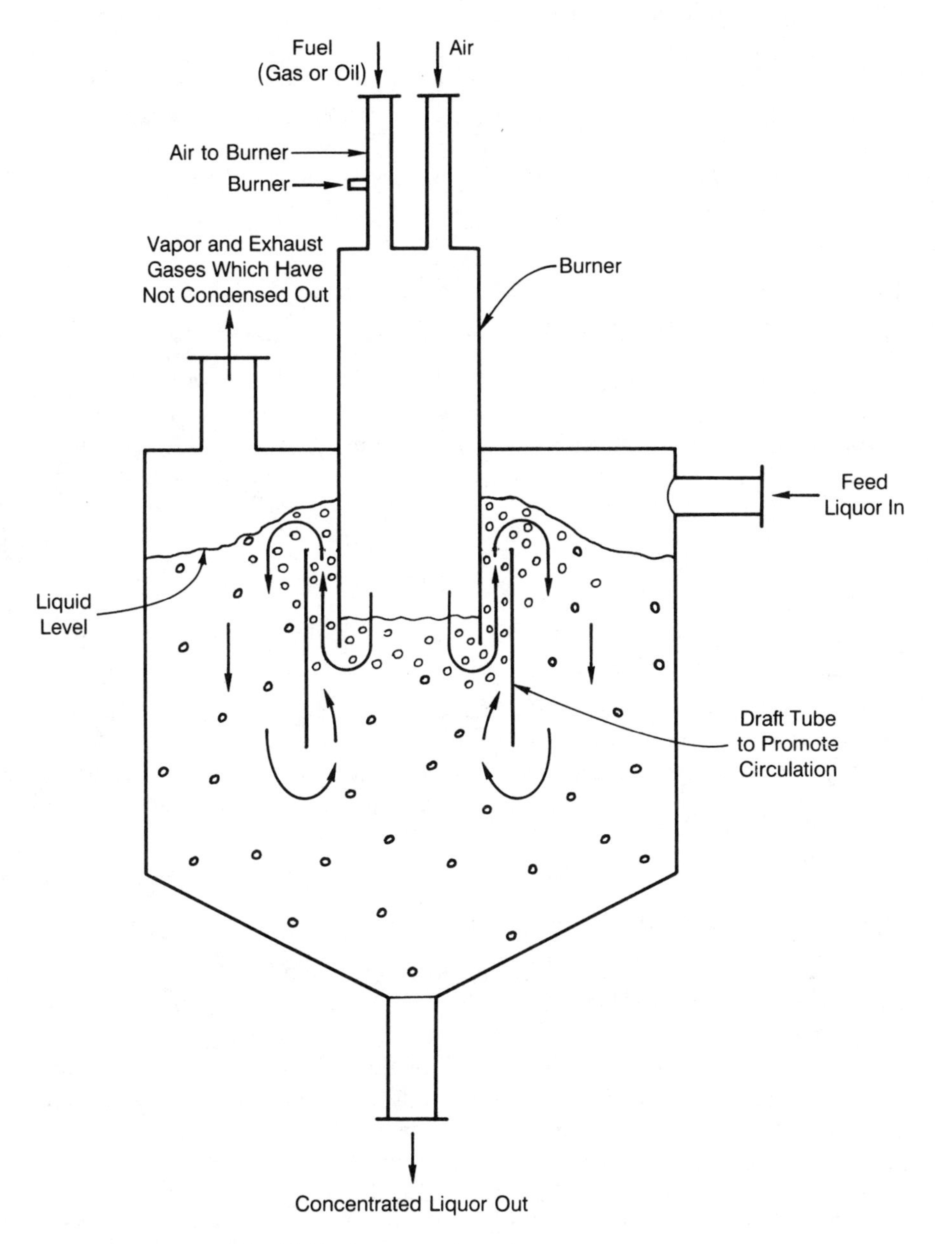

Fig. 13.14. Submerged-combustion evaporator.

Submerged-combustion evaporators are also commonly used in the concentration of corrosive chemicals such as sulfuric acid, phosphoric acid, or various forms of chemical waste.

The main advantages of these submerged-combustion units come from the absence of any fixed heat transfer surface, so there can be no corrosion or fouling. They can also handle viscous liquids, very corrosive liquids, and liquids with dissolved solids. The disadvantage, as mentioned before, is the contamination of the liquid by the combustion products.

13.4 REBOILER TYPES

Here the main types of reboiler are briefly described. They are the internal reboiler (Section 13.4.1), the kettle reboiler (Section 13.4.2), the vertical thermosyphon reboiler (Section 13.4.2), and the horizontal thermosyphon reboiler (Section 13.4.4). Some of these types are very similar to evaporator types already considered.

13.4.1 Internal Reboiler

The simplest type of reboiler, the internal reboiler, is shown in Fig. 13.15. It is simply a tube bundle placed in the base of the distillation column. Although the idea is very attractive in principle because there is no separate reboiler at all and the distillation column would only have to be modified slightly, it is rarely practicable. This is because it is often not possible to fit enough heat transfer area into the bundle to reach the required vaporization rate.

13.4.2 Kettle Reboiler

The kettle reboiler is illustrated in Fig. 13.16. It is similar to the horizontal shell-side evaporator (see Section 13.3.1, Fig. 13.5). The heating fluid, usually condensing steam flows insides the tubes which are commonly U tubes. The tube bundle occupies only the lower part of the enlarged *K*-type shell, and the upper part of the shell provides space for the vapor and the liquid to disengage. The liquid level is fixed by a weir, and the level is such that the top of the tube bundle is only just submerged. The liquid usually enters the reboiler by gravity feed, controlled by a valve if necessary, and the overflow from the weir is the bottom product from the distillation column. If necessary a pump can be installed in the pipe between the distillation column and the reboiler. The return pipe to the distillation column contains, if the vapor–liquid separation is efficient, only vapor. The thermosyphon reboilers described in Sections 13.4.3 and 13.4.4 both return a vapor–liquid mixture to the column. For information on the vapor–liquid separation, see Smith [1].

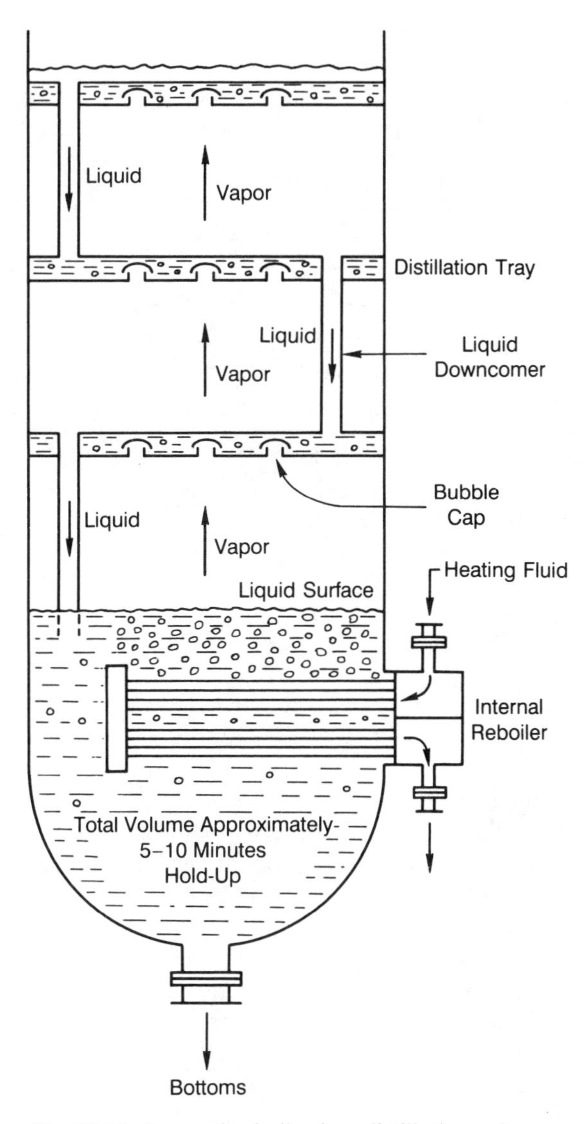

Fig. 13.15. Internal reboiler in a distillation column.

The liquid from the distillation column forms a large pool surrounding the tubes; thus the inventory of liquid in this type of reboiler is large. This can be a disadvantage in some circumstances. The liquid, however, is certainly not stagnant, neither is the boiling pure "pool" boiling. The vaporization of part of the liquid in the bundle causes a strong upflow through the bundle because of the density differences. This strong upflow increases the heat transfer coefficients to well above the pure "pool" boiling values. However,

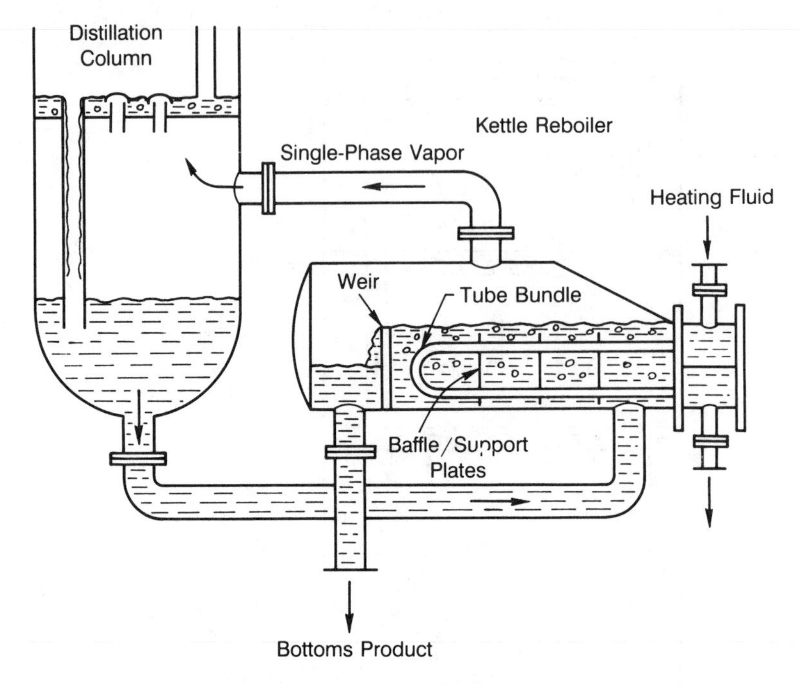

Fig. 13.16. Kettle reboiler showing connections to the distillation column.

the inflow of liquid is hindered by the fairly closely packed tubes in the bundle, so at high heat fluxes, and therefore vaporization rates, insufficient liquid may be able to enter the bundle. For this reason, in a large bundle, the maximum permissible heat flux might be rather low.

Kettle reboilers have the advantages that they are relatively easy to design, and that there is a reasonable amount of design information available. There are few control problems or stability problems. The disadvantages are that they are not economical for high-pressure operation, the liquid inventory is high, and they are not suitable for foaming liquids. They are also not suitable for fouling liquids as it is not easy to clean the outside of the tube bundle.

13.4.3 Vertical Thermosyphon Reboiler

This reboiler type is illustrated in Fig. 13.17. It is very similar to the long-tube evaporator and the climbing-film evaporator (see Sections 13.3.5 and 13.3.6 and Fig. 13.10). The liquid being evaporated is on the tube side, and the heating fluid is on the shell side. The flow of the process liquid is upwards; this is a disadvantage, but not often a serious one. This is because, in contrast to evaporators, there is usually more than sufficient temperature difference available to dwarf any effect of the hydrostatic head on the saturation temperature of the liquid. The liquid residence time is low as most of the

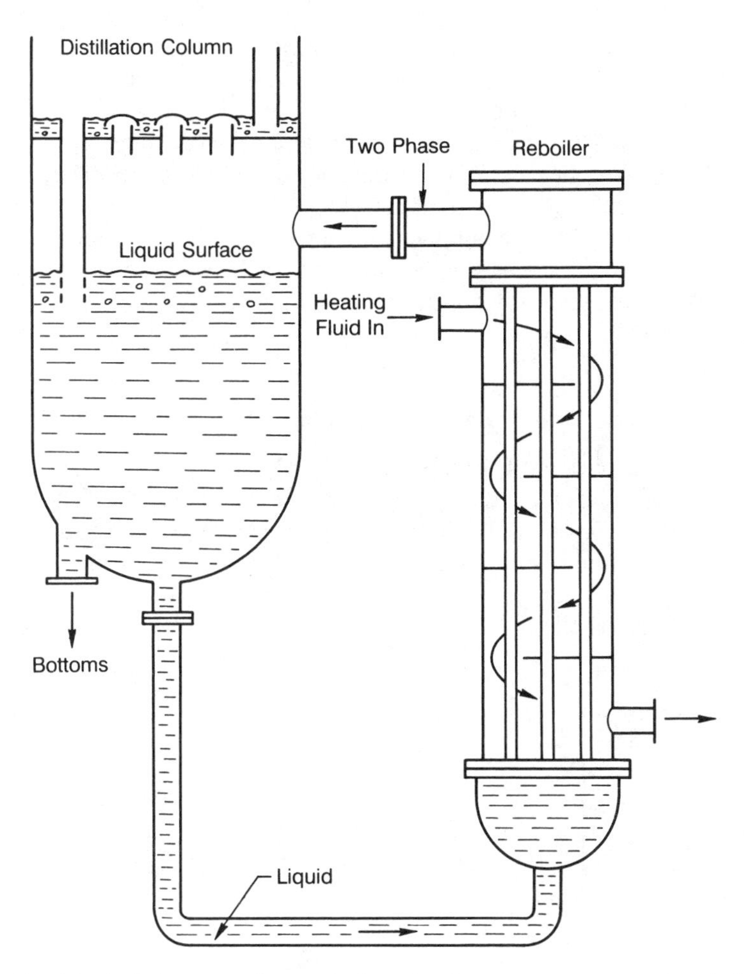

Fig. 13.17. Vertical thermosyphon reboiler showing connections to the distillation column.

tube length is occupied by a flow with a thin film of liquid on the walls of the tube and with the vapor flowing in the center of the tube. Heat transfer coefficients are therefore high. The return pipe to the distillation column carries both liquid and vapor, and separation of the liquid and the vapor occurs in the space above the liquid pool at the bottom of the distillation column.

The flow through the reboiler is determined by the natural-circulation flow which is controlled by the density differences and the height of the liquid in the bottom of the distillation column. The surface of the liquid is commonly at the top tube plate level. Vacuum conditions usually mean that the liquid level has to be lowered to give satisfactory operation.

The advantages of the vertical thermosyphon reboiler are the low residence time of the process liquid, the low liquid inventory, the high heat transfer coefficients, and the low floor area required. Vertical thermosyphon reboilers are usually the smallest and cheapest for a given duty. They can also be used for fouling liquids as the inside of the tubes is relatively easy to clean. The disadvantages are that they require a high headroom (this can mean that the distillation column has to be raised up to accommodate the reboiler below it), and there can be stability problems. The design process, because of the thermosyphon action, is not easy. However, there is some design information available [5].

13.4.4 Horizontal Thermosyphon Reboiler

This type of reboiler unit is illustrated in Fig. 13.18. This consists of a horizontal shell-and-tube exchanger with a single, large, horizontal baffle. The process fluid flows on the shell-side along the length of the tube bundle from its point of entry midway along the shell to the ends. The fluid then turns through 180° and flows back to the midpoint of the shell along the upper part of the shell. Boiling takes place over most of this flow path. The

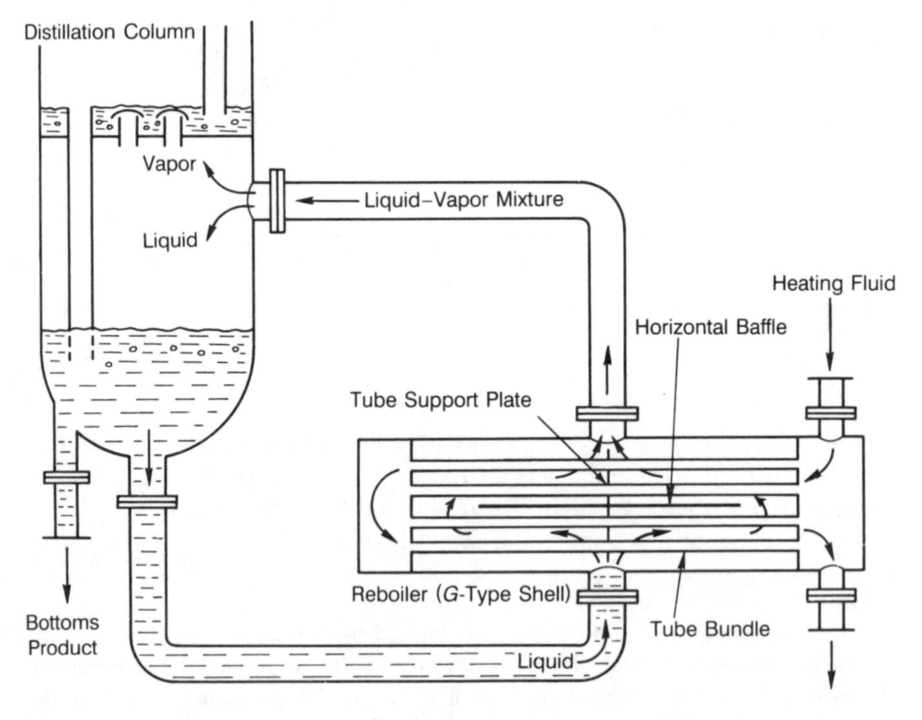

Fig. 13.18. Horizontal thermosyphon reboiler showing connections to the distillation column.

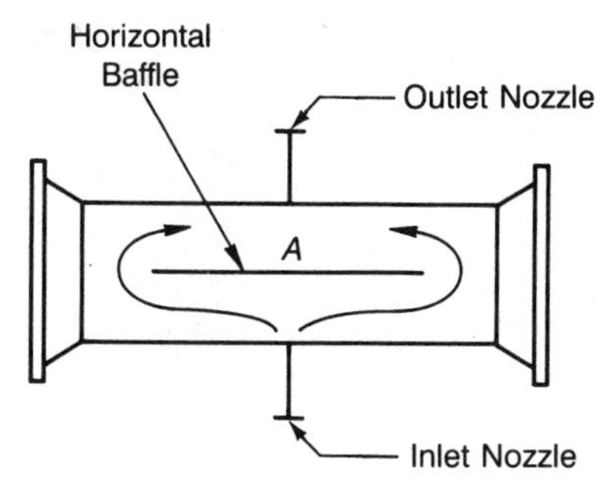

G-Type Shell (Tube Support Position Is Labeled *A*)

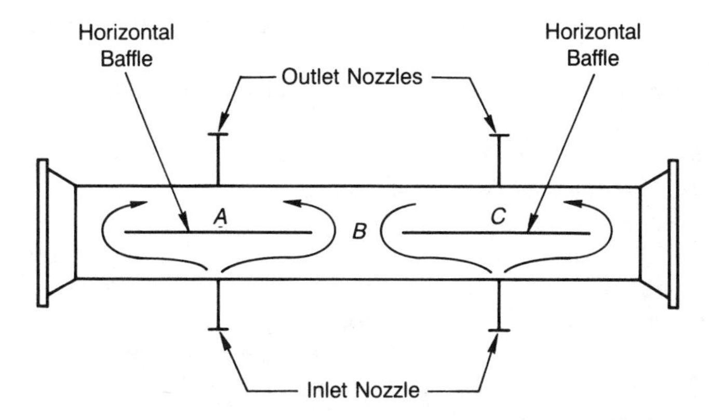

H-Type Shell (Tube Support Positions Are Labeled *A*, *B*, and *C*)

Fig. 13.19. Horizontal thermosyphon reboiler: *G*-type and *H*-type shells.

heating fluid (usually condensing steam) flows inside the tubes usually in two passes. The steam enters along the upper pass and leaves along the lower pass, allowing the condensate to drain naturally out of the bundle. The flow of the process fluid through the reboiler is again governed by a thermosyphon action, although a pump could be installed in the inlet pipe if necessary. Because the process fluid flow is controlled by density differences, calculation of the flow rate through the reboiler is difficult. If the thermosyphon reboiler requires a high heat transfer area, it may be convenient to have two liquid inlets and two mixture outlets (an *H* shell), rather than a single inlet and outlet (a *G* shell), see Fig. 13.19.

These units give high heat transfer coefficients, normally need no pump, and need relatively low headroom. They do not therefore suffer from the disadvantages of the vertical thermosyphon reboiler that the distillation column may have to be lifted and the saturation temperature changes little along the flow path. There are, however, no experimental results published on horizontal thermosyphon reboilers and no tested design information.

13.5 ENERGY EFFICIENCY IN EVAPORATION

13.5.1 Introduction

In the design of evaporators, care must be taken if an economical use of energy is to be achieved. In reboiler design there is comparatively little scope for improvement on the simple use of steam for the process heating. In the overall design of the distillation column, it may be possible to use the heat released in the condenser for other purposes in the plant. However, in an evaporator, because the designer has some freedom about the pressure, and therefore the temperature, at which the evaporation takes place, various energy-saving arrangements are possible. This aspect is particularly important when evaporating water as the latent heat of water is unusually large. Note that in a reboiler the designer does not have this freedom to alter the temperature and pressure as the reboiler operating pressure and temperature are fixed by external constraints.

Here three commonly used variants on the simple evaporation process are considered. Multiple-effect evaporation (Section 13.5.2) uses the vapor produced in one evaporator to be the heating medium for an evaporator operating at lower pressure and therefore temperature. Vapor recompression (Section 13.5.3) uses the vapor produced in the evaporation to be the heating medium in that evaporator after its temperature has been increased by mechanical compression. Multistage flash evaporation is a process where evaporation occurs in a series of flashing stages that occur at successively lower pressures and temperatures. The vapor released during the flashing is used to reheat the liquid back up to near its original temperature so that it can be used as a recycle stream to which the feed is added.

13.5.2 Multiple-Effect Evaporators

As indicated previously in multiple-effect evaporation, the basic idea is to make use of the steam generated in one evaporator to act as the heating medium in a second evaporator, and so on. Of course, it is obvious that the second evaporator must be at a lower pressure than the first so that the saturation temperature of the solvent is reduced. Because over the working pressure range of a multiple-effect system the latent heat of vaporization varies comparatively little, the steam usage can be reduced by a factor that is almost equal to the number of "effects" in series. The word "effect" in this context simply means an evaporation stage.

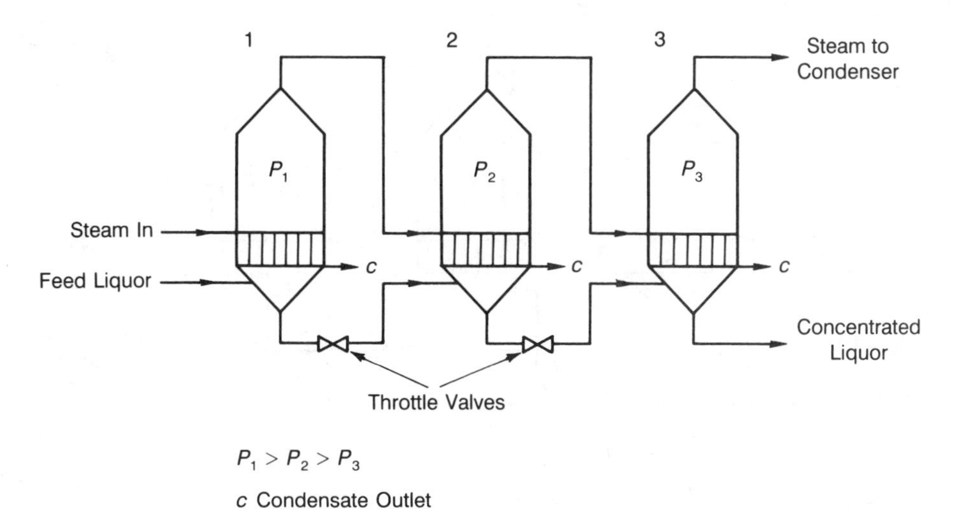

Fig. 13.20. Multiple-effect evaporator: forward feed arrangement.

Three feed arrangements can be used in multiple-effect evaporation, see Figs. 13.20, 13.21, and 13.22:

1. *Forward feed* (see Fig. 13.20). Here the solution being concentrated and the steam flow in the same direction between effects. This arrangement is used when the feed liquid is hot or when pumping the liquid between effects would be inconvenient or difficult. Here the pressure in the liquid is reduced between effects by throttle valves.

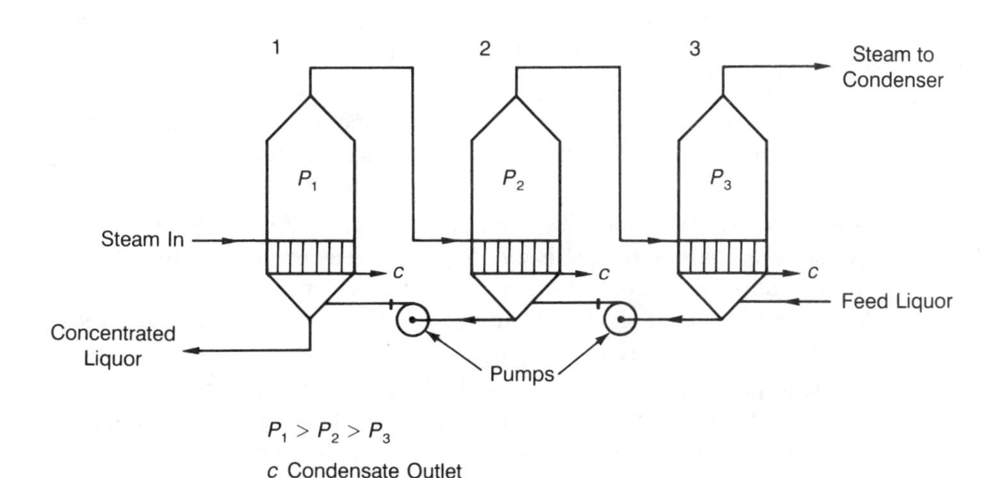

Fig. 13.21. Multiple-effect evaporator: backward feed arrangement.

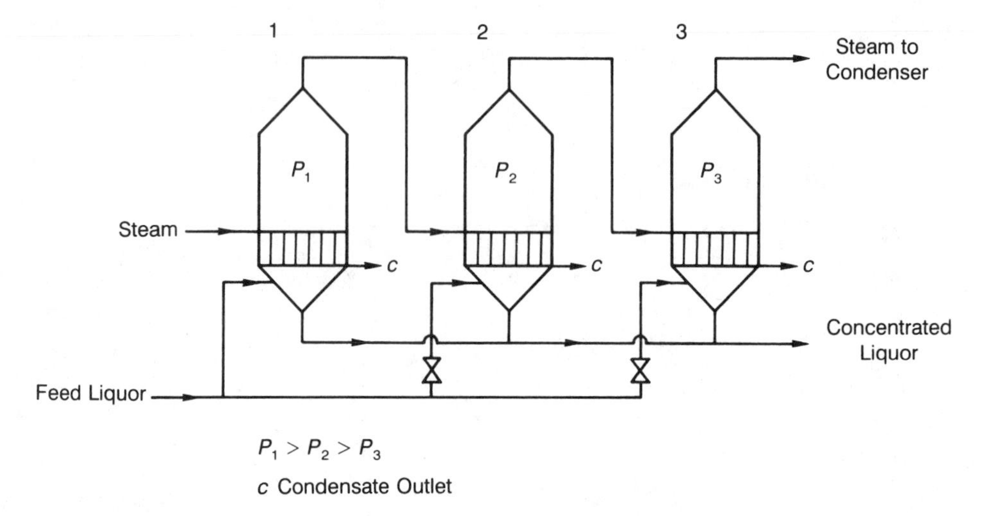

Fig. 13.22. Multiple-effect evaporator: parallel feed arrangement.

2. *Backward feed* (see Fig. 13.21). Here the solution being concentrated and the steam flow in opposite directions between effects. This arrangement is used when the feed liquid is relatively cool or when there is a significant boiling point elevation during the process. Pumps are necessary to raise the pressure of the liquid between effects.
3. *Parallel feed* (see Fig. 13.22). Here the feed solution is fed directly in parallel to each evaporator. The feeds to the later effects drop significantly in pressure across throttle valves. If the feed is hot, some flashing may occur when the pressure is reduced. Parallel feed is used when the concentration in each effect is to be the same or when it is desired to concentrate different solutions in the different effects.

Smith [1] identifies a number of important points that need careful attention if multiple-effect evaporation is to produce satisfactory results:

1. Noncondensible gases may be present due to gas coming out of solution as the evaporation proceeds or due to in-leakage of air to effects working at below atmospheric pressure. Such gas means that there is no longer the expected simple relationship between the pressure and the saturation temperature of the liquid.
2. The feed must be heated to as near its initial saturation temperature as possible. If this is not done, then some of the heating in the multiple-effect system will be required merely to do this heating.
3. The vapor must be efficiently separated from the vapor. If this is not done then the liquid droplet carry-over will occur, and, as the liquid is

certainly not pure solvent, scale may be formed on the heating surface of the next effect.

In a multiple-effect evaporator, the actual reduction in energy input, R, and the number of stages, n, are almost equal. R is defined as the energy input required per kilogram of product in a single-stage evaporator divided by the energy input required per kilogram of product in the multistage evaporator. It is also known as the performance ratio as it defines the energy advantage to be gained by using a multistage process. What then limits the number of stages since it appears that this should be as large as possible? The important variable is the minimum temperature difference between the evaporating and the condensing liquids. In a short-tube vertical evaporator, a minimum temperature difference of about 8°C in each effect is needed. In addition, in large units the effects of hydrostatic head increasing the boiling point can be relevant. In practice, the number of stages, n, and the performance ratio, R, are limited to about 6. An improvement can be made if an evaporator capable of operating at a lower value of temperature difference is used. A falling-film evaporator, in which there is also no hydrostatic head effect, can mean that n and R be increased to 10 or even 12.

For some applications, although a multiple-effect evaporator gives a very useful energy saving, a multistage flash evaporator can be better because there is no relation between the number of stages, n, and the energy saving, R, and the number of stages can be very large.

13.5.3 Vapor Recompression in Evaporation

Instead of using the vapor released in the evaporation in a different evaporator, vapor recompression provides a way of using it in the same effect; one possible arrangement is shown in Fig. 13.23. Steam produced in the evaporator is compressed either by a compressor, as shown in Fig. 13.23, or by a steam injector. The compression increases the temperature of the steam and, more importantly, its saturation temperature. Thus the condensing steam is now hot enough to evaporate the liquid in the evaporator. It is important to realize that the system shown in Fig. 13.23 uses mechanical energy in the compressor to enable the steam to be reused. As mechanical energy is generally expensive compared to thermal energy, vapor recompression tends to be used when mechanical energy is for some reason available or unusually cheap. Possible sources of cheap mechanical energy are hydropower and process plants, where a turbine can be used to let the steam down to the desired pressure from the site main pressure to the required pressure for a particular process. A disadvantage is that for an industrial scale process, the compressor required turns out to be a very large and expensive machine.

Clearly, vapor recompression can be applied to many evaporator types, though it will be most useful when comparatively little energy has to be put into the steam in the compressor. It will thus tend to be used with evapora-

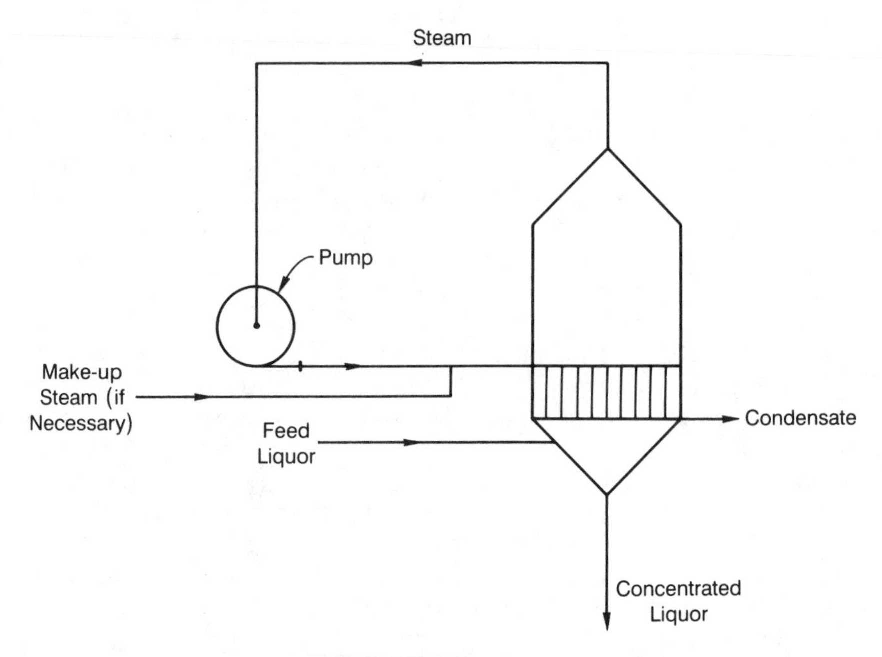

Fig. 13.23. Vapor recompression evaporation.

tors that can operate with small values of the temperature difference between the heating streams and the evaporating liquid, for example, a falling-film evaporator [6, 11].

13.5.4 Multistage Flash Evaporator

Multistage flash evaporators are used extensively for the desalination of seawater. There is no heat transfer surface for the evaporation and hence no fouling problem. The evaporation takes place as a flashing process as the pressure in a hot liquid is reduced. The high latent heat of vaporization of water (2256 kJ/kg at 1 bar) is partially compensated by the high specific heat of liquid water [4.2 kJ/(kg · K) at around the normal boiling point]. Thus, to evaporate 1 kg of water to form steam, the temperature of 10 kg of water or aqueous solution must fall by 54°C (e.g., from 92 to 38°C).

The following numerical example is taken from Silver [12]. Figure 13.24 ([12], modified) shows a four-stage flash process that has been arranged to reduce the necessary heat input by a factor of about 3 (down to approximately 753 kJ/kg of product) compared with the single-stage flash process or with a simple single-stage evaporator. The system is of course a steady-flow one; but instead of considering the flow rates and the power inputs and outputs, we will look at a reference mass of fluid and consider heat transfers. The reference flows are 10 kg of brine flowing around a circuit and a feed of 2 kg of seawater at 15°C. The outputs from the process are 1 kg of distilled

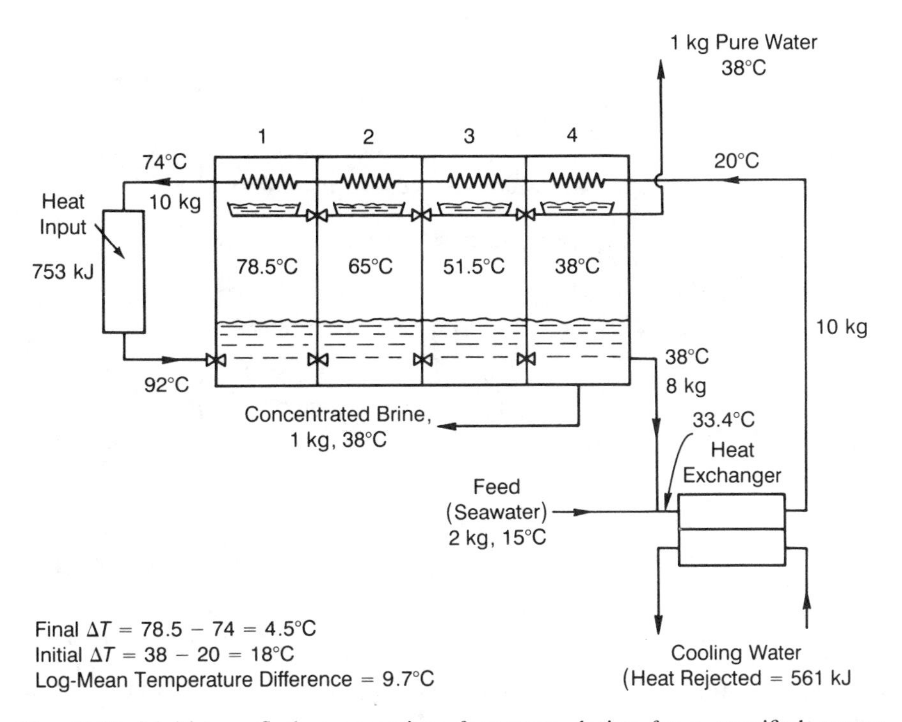

Fig. 13.24. Multistage flash evaporation: four-stage design for a specified energy consumption.

water at 38°C and 1 kg of concentrated brine at 38°C. Again we take a top temperature of 92°C; but now since the heat input is only 753 kJ, a temperature rise of only 18°C can be produced in 10 kg of aqueous solution. Hence the temperature can rise only from 74 to 92°C. The steam produced in the flashing must therefore be used to heat up the solution, in the four stages, to 74°C. At the end of the flashing the aqueous solution is compressed back to 1 bar and cooled, here to 20°C. This situation is exactly the same no matter how many stages there are: Fig. 13.25 ([12], modified) shows a nine-stage process. With four stages the highest temperature at which vapor is available is 78.5°C, the terminal temperature differences are 4.5 and 18°C, giving a logarithmic-mean temperature difference of 9.7°C. With nine stages the highest temperature at which vapor is available is 86°C, the terminal temperature differences are 12 and 18°C, giving a logarithmic-mean temperature difference of 14.8°C. Thus, because of this higher logarithmic-mean temperature difference, the nine-stage process requires only 65% of the heat transfer area of the four-stage process, assuming that the heat transfer coefficients are the same. This improvement has been brought about merely by installing five extra partition walls.

From this example it can be seen that the number of effects does not, on its own, have any influence on the economy of the process. For the four stage

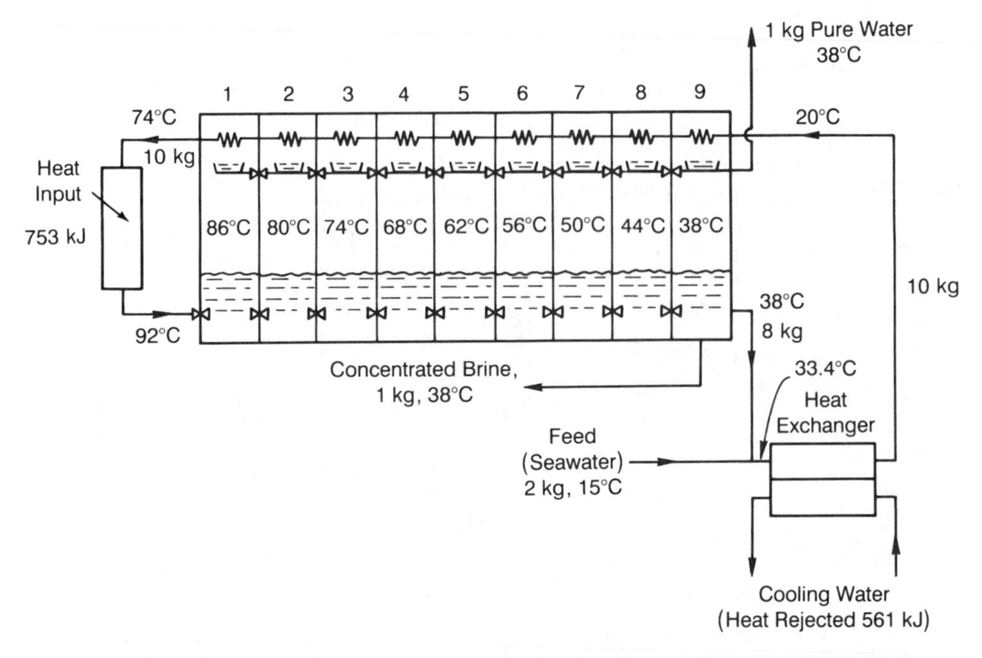

Fig. 13.25. Multistage flash evaporation: nine-stage design for a specified energy consumption.

and the nine-stage process, the same heat input (753 kJ/kg of product) was required. A reduction in the heat input will mean that the temperature difference available for the heat transfer will rapidly decrease. For the four-stage design the temperature difference becomes 0 when the heat input is 564 kJ. The corresponding figure for the nine-stage process is 251 kJ. It is simple to show that these minimum heat input values per kilogram of product are $\Delta h_v/n$, where Δh_v is the latent heat of vaporization, J/kg, and n is the number of stages. Heat inputs approaching these minimum values would have very low values of the logarithmic-mean temperature difference, and therefore very large surface areas would be necessary for the heat transfer.

The main variables in these multistage flash evaporators were the reduction in energy input required (here 3), the number of stages (here 4 or 9), and the ratios of the product flow to the recirculating flow (here 1 kg/10 kg) and the reject brine flow to the recirculating flow (here again 1 kg/10 kg). There is no simple way of deciding suitable values of these variables, the design process is a trial-and-error one. A good design will have a large reduction in energy input, not too many stages, and a large logarithmic temperature difference so that the heat transfer area required is small.

In contrast to multistage flash evaporation, in multiple-effect evaporation the number of stages and the actual reduction in energy input are directly related and are, in fact, almost equal. In multistage flash evaporation, however, the number of stages can be greater than the reduction in energy input in order to reduce the heat transfer area. In practice, it is found that a multistage flash evaporator will always require more heat transfer area than the equivalent multiple-effect evaporator. In addition, the pumping duty in the multistage flash process is larger because the whole of the recycle stream has to be pumped around the circuit and raised in pressure. In the multiple-effect evaporator, there is no recycle stream and the volumes of liquid are much lower.

However, in spite of these disadvantages, the multistage flash evaporator does have some definite advantages which mean that it is often the preferred type of unit for the desalination of seawater:

1. The multistage flash evaporator is simpler and cheaper to construct than a multiple-effect evaporator for the same heat duty.
2. In a multistage flash evaporator the temperature drops between each stage can be very small; figures of under 2°C have been quoted [12]. This can mean that a very large number of stages can be used, possibly up to 40. Reductions of energy input of about a factor of 10 compared with a single-stage flashing process or with a simple evaporator can be achieved. As has been seen in Section 13.5.2, this can be achieved in multiple-effect evaporation by using falling-film evaporators which can operate, when equipped with enhanced heat transfer surfaces, at very low values of temperature differences. More commonly, multiple-effect evaporators reach a reduction in energy input of about 6.

It is important in the design of a multistage flash evaporator to allow sufficient residence time in each stage for the flashing to occur, and for the liquid and the vapor to attain thermodynamic equilibrium. Equilibration will proceed much more quickly if a large amount of surface area can be created at which the flashing can take place. This means that the more chaotic the flow with splashing and bubble formation, the more quickly will equilibrium be reached. However, this is at variance with another practical requirement —to avoid any liquid reaching the heat transfer surface in the upper part of the flash chambers. This is because the liquid is brine, and the dissolved salt will foul the heat transfer surface. Careful design of the flashing chambers, with the installation of wire mesh separators [1], are therefore necessary.

13.6 HEAT TRANSFER AND PRESSURE DROP PROBLEMS

13.6.1 Initial Sizing of the Unit

Of course, even before any initial sizing of a unit can be performed, the first design step is the selection of the type of evaporator or reboiler appropriate for the duty required. Some comments have been made earlier with regard to particular types of equipment, and further advice is provided in [1–5]. Reviewing the experience of operators with similar requirements and successful designs can also be very helpful. However, it must be said that for a given duty, different designers, however experienced, will often select different types. Often there is no single "correct" answer.

Once the type of evaporator or reboiler has been decided upon, it is necessary to arrive at a preliminary "design" giving rough estimates of the dimensions of the unit. This rough design can then be refined in the light of detailed calculations of the pressure drop and heat transfer characteristics. An initial estimate of the surface area (A, m^2) required can be estimated from the normal heat exchanger equation:

$$A = \frac{Q}{U(T_{cond} - T_{evap})} \tag{13.1}$$

where Q is the total heat or duty W; U is the overall heat transfer coefficient, $W/(m^2 \cdot K)$; T_{cond} is the condensing steam temperature, K; and T_{evap} is the evaporating liquid temperature, K.

The major difficulty in Eq. (13.1) is estimating the value of U, the overall heat transfer coefficient. Values are given for some circumstances [1, 2, 3, and 5]. In particular, Smith [1] suggests that:

1. For an aqueous solution in a forced-flow evaporator, U ranges from 2000 $W/(m^2 \cdot K)$ for clean fluids, down to 700 $W/(m^2 \cdot K)$ for fairly dirty fluids.
2. A value of U of 2000 $W/(m^2 \cdot K)$ may be assumed for a clean aqueous solution in a falling film. Injection of an inert gas to perform the evaporation at a lower temperature will give a higher heat transfer coefficient because of the increased vapor shear on the liquid film.
3. For an aqueous solution in a natural-circulation evaporator, it may be assumed for a first approximation that U ranges from 1000 $W/(m^2 \cdot K)$ for clean fluids to 500 $W/(m^2 \cdot K)$ for fairly dirty fluids. Natural circulation gives lower heat transfer coefficients than forced flow because the flow velocities are generally lower.

Values of the overall heat transfer coefficient for the vaporization of organic liquids are usually lower, by up to a factor of 2 (see Whalley and Hewitt [5]).

Typical liquid velocities in evaporators are in the range 0.05 to 0.1 m/s. The diameter and length of the tubes are largely determined by the choice of evaporator or reboiler type. At this stage typical values for the type chosen should be assumed and so the number of tubes can be calculated from the approximate value of the surface area. The diameter and length may have to be changed later in the design, depending on the outcome of more accurate calculations.

The problems encountered in the detailed simulation of an evaporator or reboiler can be divided into two main categories:

1. *Pressure drop problems.* The flow can be single-phase liquid or a mixture of vapor and liquid. The single-phase problem is relatively simple and is not discussed here. The vapor–liquid two-phase pressure drop problem is much more complicated (see Section 13.6.2). If the unit is a natural-circulation unit, then the flow rate will have to be calculated; this problem is discussed in Section 13.6.3. Even in a forced-flow unit, the pressure drop is important because it will partly determine the size and type of pump required.
2. *Heat transfer problems.* The problem here is to calculate the various parts of the total heat transfer coefficient (see Section 13.5.4 and following sections).

There are no simple rules available for refining the design in terms of the actual heat transfer coefficient or the actual pressure drop that will result. What can be said is that each case has to be treated separately and the calculation performed as an integral calculation for the heat transfer coefficient and the pressure gradient as the conditions change through the evaporator. The methods referred to in the rest of Section 13.6 will indicate the methods that can be used, but no attempt has been made here to show the entire calculation. Indeed such a complete calculation will be different in important respects for the various types of evaporators and reboilers.

13.6.2 Two-Phase Vapor – Liquid Pressure Drop

Calculation of the pressure drop and the two-phase pressure drop is necessary to size the pump in a forced-flow unit and to calculate the flow in a natural-circulation unit (see Section 13.6.3).

It is most convenient in analyzing the two-phase pressure gradient to start with a simple analysis of single-phase flow (see Fig. 13.26). Application of the normal momentum equation to an element of length δz of this pipe gives

$$-\frac{dp}{dz}\delta z\frac{\pi d^2}{4} - \tau\,\delta z\,\pi d - \frac{\pi d^2}{4}\,\delta z\,\rho g \sin\theta = \frac{\pi d^2}{4}\frac{d}{dz}(\rho u^2) \quad (13.2)$$

where $-(dp/dz)$ is the total pressure gradient, N/m^3; δz is the element length, m; d is the tube diameter, m; ρ is the fluid density, kg/m^3; τ is the wall shear stress, N/m^2; g is the acceleration due to gravity, m/s^2; u is the velocity of the flow, m/s; and θ is the angle of inclination to the horizontal as shown in Fig. 13.26.

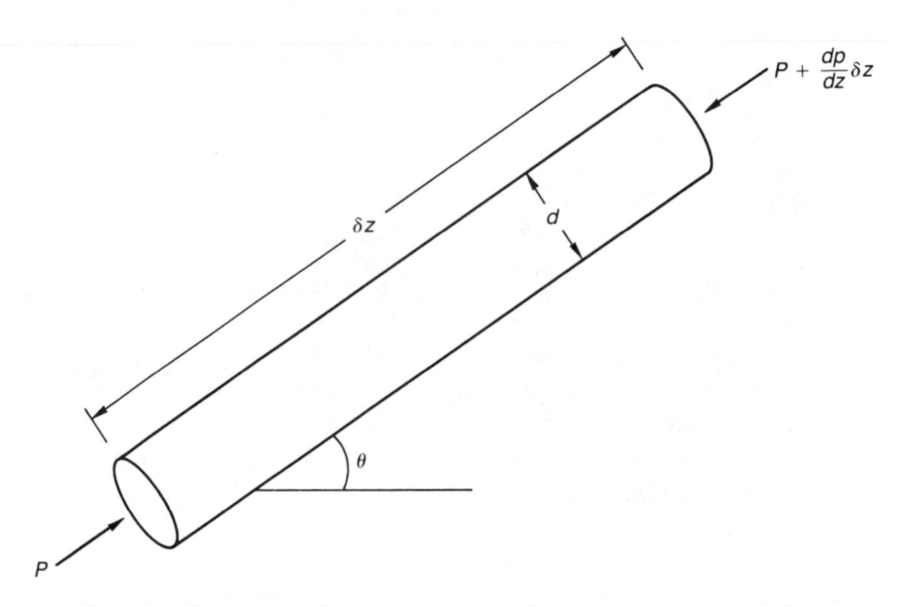

Fig. 13.26. Single-phase flow: control volume for the momentum equation.

This equation can be easily rearranged into a much more simple form:

$$-\frac{dp}{dz} = \frac{4\tau}{d} + \rho g \sin\theta + G^2 \frac{d}{dz}\left[\frac{1}{\rho}\right] \tag{13.3}$$

Total pressure gradient = Frictional pressure gradient + Gravitational pressure gradient + Accelerational pressure gradient

where G is the fluid mass flux, $\text{kg}/(\text{m}^2 \cdot \text{s})$; this is the mass flow rate divided by the tube cross-sectional area and is given by $G = \rho u$.

From Eq. (13.3) it can be seen that the total pressure gradient divides naturally into three components: the frictional part caused by the wall shear stress, $-(dp/dz)_F$; the gravitational part caused by the weight of the fluid, $-(dp/dz)_G$; and the accelerational part caused by the change of velocity of the fluid, $-(dp/dz)_A$.

The simplest way of treating a two-phase flow is to assume that it behaves as a single-phase flow with some mean fluid properties somewhere between the gas properties and the liquid properties. The relevant properties are the density, which appears directly in Eq. (13.3), and the viscosity, which helps to determine the wall shear stress. This is the "homogeneous model" of two-phase flow: the phases are assumed implicitly to behave as a homogeneous mixture and thus have equal velocities. Appropriate values for the mean properties are discussed by Collier [13] and Whalley [14]. Reasonable equations are as follows:

for the homogeneous density, ρ_h, kg/m^3:

$$\frac{1}{\rho_h} = \frac{x}{\rho_g} + \frac{1-x}{\rho_l} \tag{13.4}$$

where ρ_g is the vapor density, kg/m^3; ρ_l is the liquid density, kg/m^3; and x is the quality, that is, the fraction of the mass flow rate that is vapor.

for the homogeneous viscosity μ_h, (N · s)/m^2:

$$\frac{1}{\mu_h} = \frac{x}{\mu_g} + \frac{1-x}{\mu_l} \tag{13.5}$$

where μ_g is the vapor viscosity, (N · s)/m^2; and μ_l is the liquid viscosity, (N · s)/m^2.

The homogeneous model gives good results at high pressures (where the phase densities are not too different) or at high velocities (where the phases really do have about the same velocity). At more moderate conditions, however, this model gives rather poor results, and the separated model, which recognizes that the phases flow at different velocities, must be used. If once again, the momentum equation is applied to the flow in Fig. 13.27,

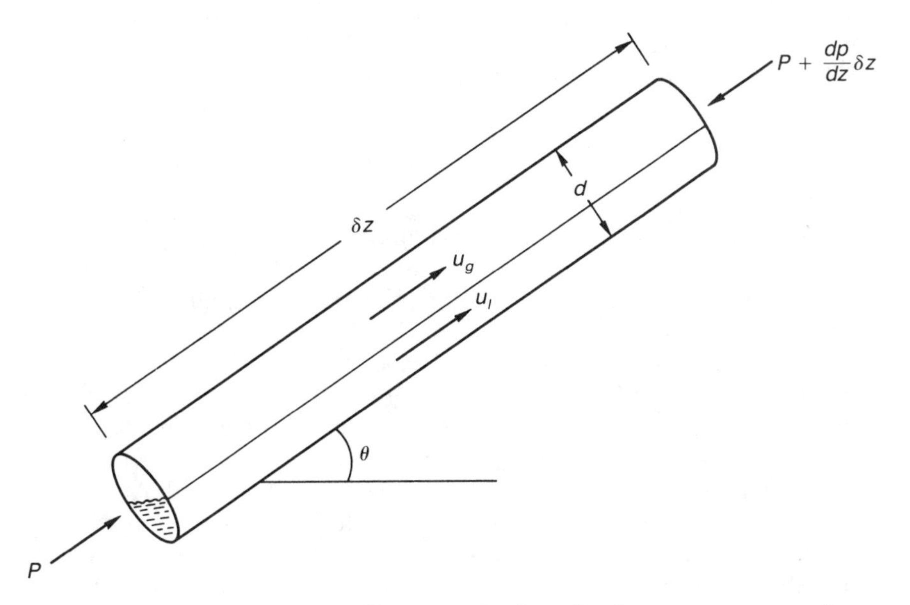

Fig. 13.27. Separated two-phase flow: control volume for the momentum equation.

where now the phases are shown separately, the result is

$$-\underset{\substack{\text{Total}\\\text{pressure}\\\text{gradient}}}{\frac{dp}{dz}} = \underset{\substack{\text{Frictional}\\\text{pressure}\\\text{gradient}}}{\frac{4\tau}{d}} + \underset{\substack{\text{Gravitational}\\\text{pressure}\\\text{gradient}}}{\left[\alpha\rho_g + (1-\alpha)\rho_l\right] g \sin\theta}$$

$$+ \underset{\substack{\text{Accelerational}\\\text{pressure}\\\text{gradient}}}{G^2 \frac{d}{dz}\left[\frac{x^2}{\alpha\rho_g} + \frac{(1-x)^2}{(1-\alpha)\rho_l}\right]} \tag{13.6}$$

$$\left[-\frac{dp}{dz}\right] = \left[-\frac{dp}{dz}\right]_F + \left[-\frac{dp}{dz}\right]_G + \left[-\frac{dp}{dz}\right]_A \tag{13.7}$$

Here the variables have the same meaning as before, and α is the void fraction, that is, the fraction of the flow cross-sectional area that is occupied by the vapor.

If Eq. (13.6) is to be used to calculate the actual total pressure gradient in a two-phase flow, then we must have methods of calculating the frictional pressure gradient and the void fraction. A very large number of correlations have been proposed for these quantities [13, 14]. A number of workers have carried out systematic comparisons between these various correlations and data banks containing large numbers of experimental void fraction and pressure drop measurements.

The frictional pressure gradient can be written as

$$-\left[\frac{dp}{dz}\right]_F = -\left[\frac{dp}{dz}\right]_{fo} \phi_{fo}^2 \tag{13.8}$$

where ϕ_{fo}^2 is the two-phase frictional multiplier and $-[dp/dz]_{\text{fo}}$ is the frictional pressure gradient is the total mass flow rate (liquid plus vapor) were flowing in single-phase flow and had the properties of the liquid.

Values of ϕ_{fo}^2 must be obtained from correlations (see [13, 14]). Probably the most accurate calculation method is that of Friedel [15]. It is quite complicated algebraically, but is straightforward in application. It does not perform very well for viscous liquids where the ratio of the liquid viscosity to the vapor viscosity exceeds 1000. The Friedel correlation for frictional pressure gradient in gas–liquid flow is given in detail in Chapter 10. Alternatives are the graphical correlation of Baroczy [16], which is applicable to any fluid, and the Thom [17] correlation, which works well for steam–water systems where the pressure exceeds 17 bar. It should always be noted that even the best correlations produce results with a root mean square error of around 35%.

It can be seen from Eq. (13.6) that to calculate the accelerational and gravitational terms, it is necessary to be able to calculate the void fraction α. Once again a large number of correlations have been proposed for the void fraction. The correlation is often expressed in terms of the slip velocity ratio, S, which is defined as the mean velocity of the vapor phase divided by the mean velocity of the liquid phase. The slip ratio is related to the void fraction by the equation:

$$\alpha = \frac{1}{1 + \left[S \frac{1-x}{x} \frac{\rho_g}{\rho_l} \right]} \tag{13.9}$$

Systematic comparisons have also been carried out between the various void fraction correlations and date banks containing large numbers of experimental measurements of either void fraction, or equivalently, mean fluid density. One of the simplest correlations which is reasonably accurate is that of Chisholm [18]:

$$S = \left[\frac{x \rho_l}{\rho_g} + (1 - x) \right]^{1/2} \tag{13.10}$$

In order to obtain better accuracy, it is necessary to have considerably more complicated methods. The best correlation which is applicable to all fluids is generally termed the CISE correlation [19]. The CISE correlation is shown in detail in Table 13.1. A more accurate correlation which is only applicable to steam–water systems is that of Bryce [20]. The root mean square error in the calculation of mean fluid density from these correlations is typically around 25%.

In practice, it is usual for the thermohydraulic conditions within the evaporator or reboiler to be incorporated into a computer-based model; this enables the more complicated and accurate separated flow model of two-phase flow to be conveniently used.

13.6.3 Calculation of Natural-Circulation Units

The calculation of the flow rate in a natural-circulation evaporator or a thermosyphon reboiler is an iterative process. The procedure is to assume a value for the circulation rate and then to calculate the pressure loss around the whole circulation loop. It must be remembered that the pressure loss will be negative in some regions; that is, the pressure rises in the flow direction. The circulation flow rate must then be adjusted until the total pressure loss is 0. More details are given by Whalley and Hewitt [5] who also give analytic expressions for the flow rate in a vertical thermosyphon reboiler and for the internal circulation inside the shell and the bundle of a kettle reboiler. These expressions are based on a homogeneous two-phase flow theory which,

TABLE 13.1 CISE Correlation for Void Fraction in Gas–Liquid Flow

The correlation of Premoli et al. (1970), usually known as the CISE correlation is a correlation in terms of the slip ratio, S. The void fraction, α, is then given by

$$\alpha = \frac{1}{1 + \left(S \frac{1 - x}{x} \frac{\rho_g}{\rho_l} \right)}$$

The slip ratio is then given by

$$S = 1 + E_1 \left(\frac{y}{1 + yE_2} - yE_2 \right)^{0.5}$$

where

$$y = \frac{\beta}{1 - \beta}$$

$$\beta = \frac{\rho_l x}{\rho_l x + \rho_g (1 - x)}$$

$$E_1 = 1.578 Re^{-0.19} \left(\frac{\rho_l}{\rho_g} \right)^{0.22}$$

$$E_2 = 0.0273 We Re^{-0.51} \left(\frac{\rho_l}{\rho_g} \right)^{-0.08}$$

$$Re = \frac{Gd}{\mu_l}$$

and

$$We = \frac{G^2 d}{\sigma \rho_l}$$

where x is the quality; ρ_g is the gas density, kg/m^3; ρ_l is the liquid density, kg/m^3; d is the tube diameter, m; μ_l is the liquid viscosity $N \cdot s/m^2$; G is the total (liquid + gas) mass flux, $kg/(m^2 \cdot s)$; and σ is the surface tension, N/m.

although certainly not the most accurate, has the advantage that analytical expressions can be obtained. These analytical expressions are shown in Tables 13.2 (vertical thermosyphon reboiler) and 13.3 (kettle reboiler bundle).

One important check that must always be made in a natural-circulation evaporator or reboiler is to see if the downward velocities at any point are sufficiently low to allow any vapor bubbles to rise out of the downflowing

TABLE 13.2 Approximate Method for Calculating the Flow Rate in a Natural-Circulation Vertical Thermosyphon Reboiler

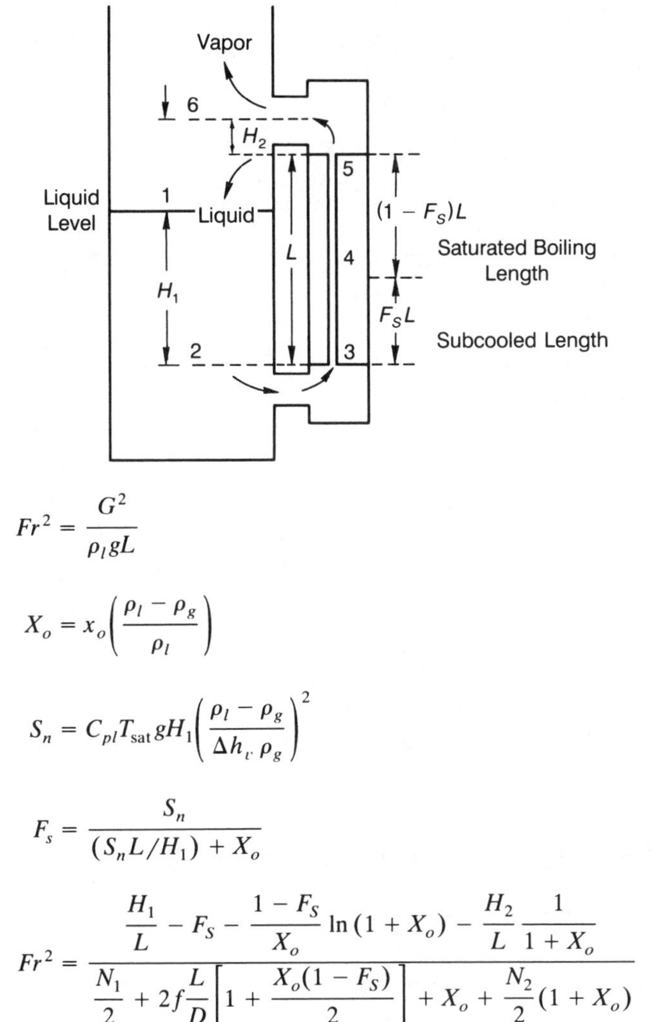

$$Fr^2 = \frac{G^2}{\rho_l g L}$$

$$X_o = x_o\left(\frac{\rho_l - \rho_g}{\rho_l}\right)$$

$$S_n = C_{pl} T_{\text{sat}}\, g H_1 \left(\frac{\rho_l - \rho_g}{\Delta h_v\, \rho_g}\right)^2$$

$$F_s = \frac{S_n}{(S_n L/H_1) + X_o}$$

$$Fr^2 = \frac{\dfrac{H_1}{L} - F_S - \dfrac{1 - F_S}{X_o}\ln(1 + X_o) - \dfrac{H_2}{L}\dfrac{1}{1 + X_o}}{\dfrac{N_1}{2} + 2f\dfrac{L}{D}\left[1 + \dfrac{X_o(1 - F_S)}{2}\right] + X_o + \dfrac{N_2}{2}(1 + X_o)}$$

where G is the mass flux through the tubes, kg/(m$^2 \cdot$ s); x_o is the quality at the tube exit; f is the Fanning friction factor for flow in the tubes; N_1 is the number of velocity heads (based on the mass flux inside the tubes) lost in the inlet pipework; and N_2 is the number of velocity heads (based on the mass flux inside the tubes) lost in the outlet pipework.

TABLE 13.3 Approximate Method for Calculating the Recirculation Flow Rate in the Bundle of a Kettle Reboiler

$$Fr^2 = \frac{G^2}{\rho_l g D_B}$$

$$X_o = x_o\left(\frac{\rho_l - \rho_g}{\rho_l}\right)$$

$$S_n = C_{pl} T_{sat} g D_B \left(\frac{\rho_l - \rho_g}{\Delta h_v \rho_g}\right)^2$$

$$Ga = \frac{\rho_l^2 g D_o^2 D_B}{\mu_l^2}$$

$$B_f = \frac{D_o D_B D_v}{(P_T - D_o)^3}$$

$$D_v = \frac{4P_T^2 - \pi D_o^2}{\pi D_o}$$

$$A_c = 0.061 \qquad n_c = 0.267$$

$$Fr^2 = \frac{1 - \dfrac{S_n}{S_n + X_o} - \dfrac{1}{S_n + X_o}\ln(1 + X_o)}{X_o + 2(A_c B_F)^{2/2-n_c}\left[1 + \dfrac{X_o^2}{2(S_n + X_o)}\right]\left[\left\{X_o + \dfrac{4(S_n + X_o)}{X_o^2}\right\}\Big/ Ga\right]^{\frac{n_c}{(2-n_c)}}}$$

where D_B is the bundle diameter, m; x_o is the quality at the top of the bundle; D_o is the tube outside diameter, m; μ_l is the liquid viscosity, $(\mathrm{N \cdot s})/\mathrm{m}^2$; and P_T is the tube pitch, m.

Note that these equations are valid only for square and rotated square tube layouts.

---

liquid and not be "carried-under" with the liquid and so reduce the effective head which is driving the natural-circulation flow. If the downward velocities are high, then it is extremely difficult to prevent carry-under of vapor. Such a check is also relevant in a forced-flow unit, but it easier to obtain substantial carry-under in a natural-circulation unit as the flow rates are not initially known.

13.6.4 Heat Transfer Rates

If information is available about the flow at every point, then it is possible to make more realistic calculations of the overall heat transfer coefficient. In a natural-circulation unit, such as a thermosyphon reboiler, the value of the heat transfer coefficients will of course affect the vapor production rate and so in turn affect the driving head for the circulation. This means that in a natural-circulation evaporator or reboiler, the iteration loop for the flow rate (see Section 13.6.3) must include the heat transfer calculations.

The overall heat transfer resistance, the reciprocal of the overall heat transfer coefficient, U, is the sum of the five thermal resistances through which the heat must pass:

1. The heating fluid, usually condensing steam
2. The fouling layer deposited by the heating fluid
3. The material of the tube wall
4. The fouling layer deposited by the evaporating fluid and
5. The evaporating fluid

Each of these heat transfer resistances is the reciprocal of the individual film heat transfer coefficients, h, corrected to refer to the same surface area, either inside or outside the tube. In the case of the tube wall material, at least for a thin-walled tube, the thermal resistance is the tube wall thickness divided by the thermal conductivity of the material of the tube wall.

Often in evaporator calculations, a knowledge of single-phase heat transfer coefficients is needed. These are relevant not only in the single-phase regions of the flow, but also in the two-phase regions. This is because the two-phase heat transfer coefficients are often calculated by means of some kind of equivalent single-phase flow and a multiplication factor to take account of the extra heat transfer in the two-phase region. Single-phase heat transfer coefficients can be calculated by reference to Chapter 3.

Evaporating-side heat transfer coefficients (see Sections 13.6.7 to 13.6.10) depend on:

1. Whether evaporation is occurring on the shell side or the tube side of the bundle
2. The configuration of the tube bundle, for example, whether it is horizontal or vertical
3. Whether the evaporation is occurring in a climbing film or in a falling film
4. The flow rate and the quality of the flow

13.6.5 Heat Transfer on the Heating Side

Most evaporators and reboilers use condensing steam as the heating medium. The stem condensation may occur on the shell side or the tube side, and the steam flow may be horizontal or vertical [1]. If the flow is vertical, it is invariably downwards so that the condensate easily drains out of the system.

As with all condensation systems, it is very important to see that the condenser is vented regularly to prevent the build up of incondensable gas in the condenser. Incondensable gas in condensers is the most common reason why the condensation heat transfer coefficients are less than expected. Smith [1] suggests that a heat transfer coefficient of 10,000 $W/(m^2 \cdot K)$ can safely be used for general purposes for the condensation of steam. Much more detailed consideration of condensing heat transfer coefficients is given in Chapter 10. Information relevant to shell-side heat transfer coefficients is also given in Chapter 5.

13.6.6 Fouling

As always in heat transfer, the fouling resistance is one of the most difficult problems. Previous experience in similar equipment and similar fluids is probably the best and safest guide.

TEMA [21] suggest that for condensing steam taken directly from a steam main, a fouling coefficient of 10,000 $W/(m^2 \cdot K)$ should be used. In multiple-effect evaporators the fouling coefficients are usually lower; a figure of 3000 $W/(m^2 \cdot K)$ is suggested.

Fouling on the boiling side is more variable and more uncertain, and it is one of the most difficult quantities to establish. Again previous experience with similar types of evaporators and reboilers and similar liquids is without a doubt the best guide. In the absence of other information, it may be necessary to undertake test work on a tube pilot scale unit to establish the propensity of a particular solution to foul the particular heat transfer surface. Smith [1] recommends a fouling coefficient of 1000 $W/(m^2 \cdot K)$ for heavy scaling deposits and 10,000 $W/(m^2 \cdot K)$ for slight scale. More detailed consideration of fouling heat transfer coefficients is given in Chapter 4, where the TEMA [21] values for the fouling heat transfer coefficient are also given. These values should be treated with considerable caution and used only to indicate the magnitude of the fouling heat transfer coefficient that may occur.

In general, it is probably best not to overdesign the evaporator by allowing for heavy fouling. In a natural-circulation unit, this may result in inadequate circulation when the evaporator is clean. The inadequate circulation may then itself be the cause of rapid fouling. Palen [22] recommends the use of realistic (rather than conservative) design fouling coefficients. Two further measures are advocated in an attempt to see that fouling does not occur very rapidly: designs with high exit qualities should be avoided, and designs should always aim for the best possible flow distribution.

13.6.7 Boiling inside Tubes

As the liquid enters a tube and heat is added, vapor is gradually formed and the flow passes through a series of flow patterns (see Fig. 13.28 for vertical upwards flow). Obviously, the mechanisms of heat transfer and the factors

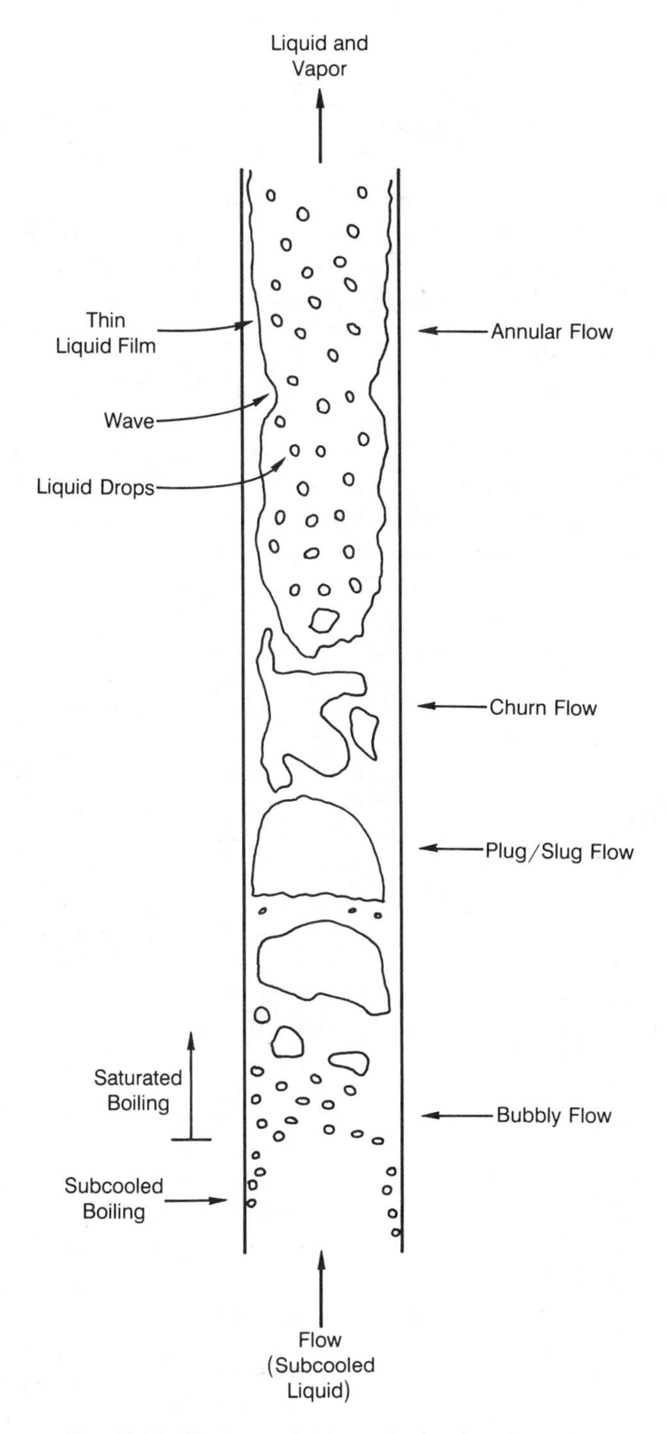

Fig. 13.28. Flow patterns in vertical upflow in a tube.

influencing the heat transfer will change greatly over the length of the tube. At the bottom of the tube where there is no thin film of liquid, the boiling tends to nucleate boiling. Here there is bubble nucleation at the walls of the tube at "nucleation sites" or imperfections in the tube wall surface. In this region the heat transfer coefficient is a strong function of the heat flux. In the upper region of the tube, there is a thin film, and convective boiling tends to occur. Here the heat is conducted through the liquid film and the liquid evaporates at the vapor–liquid interface. There is no nucleation or bubble formation, and the heat transfer coefficient is now almost independent of the heat flux. Ideally, of course, different calculation methods would be used in these two regions and possibly in each flow pattern. However, it is usual to use one method for the entire region of the boiling flow.

The most well-known correlation is that of Chen [23], which is described in detail by Collier [13] and Whalley [14]. It also has the advantage that it has been widely tested. The method assumes that nucleate and convective boiling occur to some degree over the entire length of the evaporator. However, the form of the correlation is such that the nucleate boiling part tends to dominate in the lower parts of the tube where the quality is low, and the convective part dominates when the quality is high.

The form of the Chen correlation is complicated; probably more complicated than is actually necessary. The nucleate boiling method used within the correlation is particularly complicated and tedious to use. In addition, the assumption that there is a very gradual shift from nucleate boiling at low quality to convective boiling at high quality is probably also over-elaborate. It is likely that just as satisfactory results could be obtained as assuming an abrupt change in the boiling mechanism. However, the Chen correlation has been tested in many circumstances and been found to work reasonably well overall, though some of the individual data points can show large errors. Because of this extensive background of testing, it is recommended that the Chen correlation be used for the calculation of boiling heat transfer coefficients inside tubes. The details of the Chen correlation are shown in Table 13.4.

For boiling within horizontal tubes an alternative graphical correlation by Shah [24] takes into account the stratification of the liquid and vapor that occurs in horizontal flow. Chapter 12 gives more details of the Shah correlation, as well as some other alternative correlations.

These correlations, which produce heat transfer coefficients with typical accuracies of about 30%, remain valid only in the region where the wall is fully wetted by the liquid. If the tube is long enough there will come a point where the liquid film is evaporated away, and the tube wall becomes dry. Heat transfer to vapor is very much less efficient than heat transfer to a boiling liquid, and so the heat transfer coefficient drops very dramatically. This phenomenon is "dry-out" or "critical heat flux," and is explained in detail in [13] and [14]. Estimation of the quality at which this occurs is complicated. Fair [25] gives a simple method, which works reasonably well for

TABLE 13.4 Chen Correlation for Boiling Heat Transfer

$$h = h_{NB} + h_{FC}$$

$$h_{NB} = S h_{FZ}$$

$$h_{FZ} = \frac{0.00122 \Delta T_{sat}^{0.24} \Delta p_{sat}^{0.75} C_{pl}^{0.45} \rho_l^{0.49} k_l^{0.79}}{\sigma^{0.5} \lambda^{0.24} \mu_l^{0.29} \rho_g^{0.24}}$$

Note that C_{pl} is the liquid specific heat, J/(kg · K); k_l is the liquid thermal conductivity, W/(m · K); $\Delta T_{sat} = T_w - T_{sat}$; and Δp_{sat} is the difference in saturation pressure corresponding to ΔT_{sat}. Δp_{sat} is shown on the vapor pressure curve.

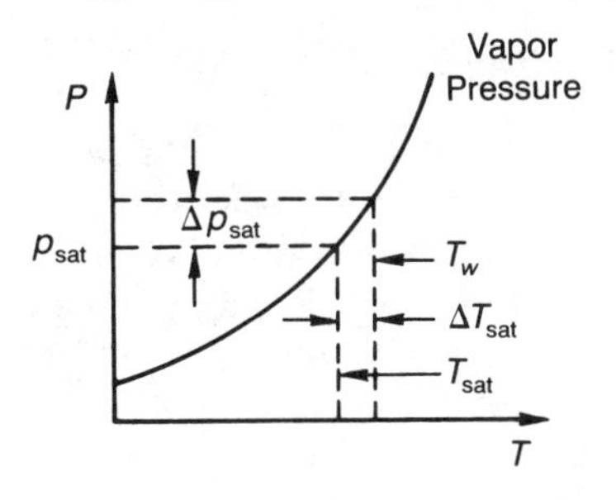

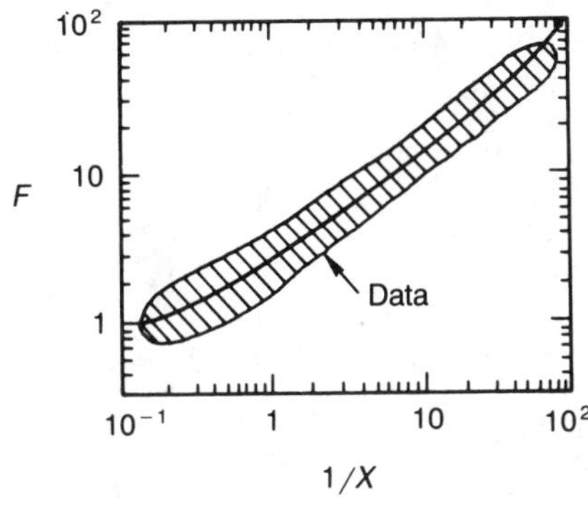

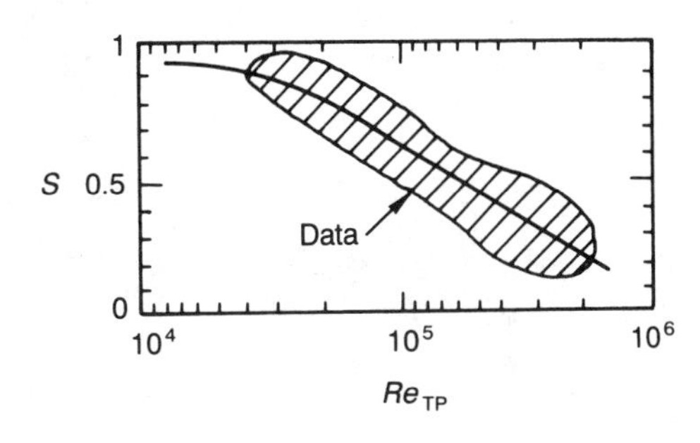

$$h_{FC} = h_l F$$

$$Nu_l = 0.023 Re_l^{0.8} Pr_l^{0.4}$$

$$Nu_l = \frac{h_l d}{k_l}$$

$$Re_l = \frac{G(1-x)d}{\mu_l}$$

$$Pr_l = \frac{\mu_l C_{pl}}{k_l}$$

$$X = \left(\frac{1-x}{x}\right)^{0.9} \left(\frac{\rho_g}{\rho_l}\right)^{0.5} \left(\frac{\mu_l}{\mu_g}\right)^{0.1}$$

$$Re_{TP} = Re_l F^{1.25}$$

tubes of length 4 to 5 m and at low pressures, but there is considerable doubt about the physical basis of the correlation. Other calculation methods are also available, which are certainly better tested but which are far more complicated to use. For steam–water there is the correlation of Bowring [26], and for fluids there is a general correlation of Katto and Ohne [27]. Details of both these methods are given by Whalley [14]. The root mean square error in the calculated critical heat flux of the Bowring method is about 7%, and for the Katto and Ohne method typical errors are about 20%.

13.6.8 Boiling outside Tubes

Available methods for the prediction of boiling heat transfer coefficients on the outside of tube bundles are not in general very satisfactory. One conservative approach is to assume that the boiling is nucleate pool boiling, so that the heat transfer coefficient can be calculated by an appropriate correlation, such as that as of Cooper [28]. This procedure neglects entirely any contribution of convective effects of the total boiling heat transfer coefficient.

However, boiling in large tube bundles is not very similar to pool boiling on an isolated tube, wire, or plate. The production of vapor in the bundle produces a large natural-circulation flow of liquid up through the bundle (see Fig. 13.29), then the liquid recirculates to the bottom of the bundle. The

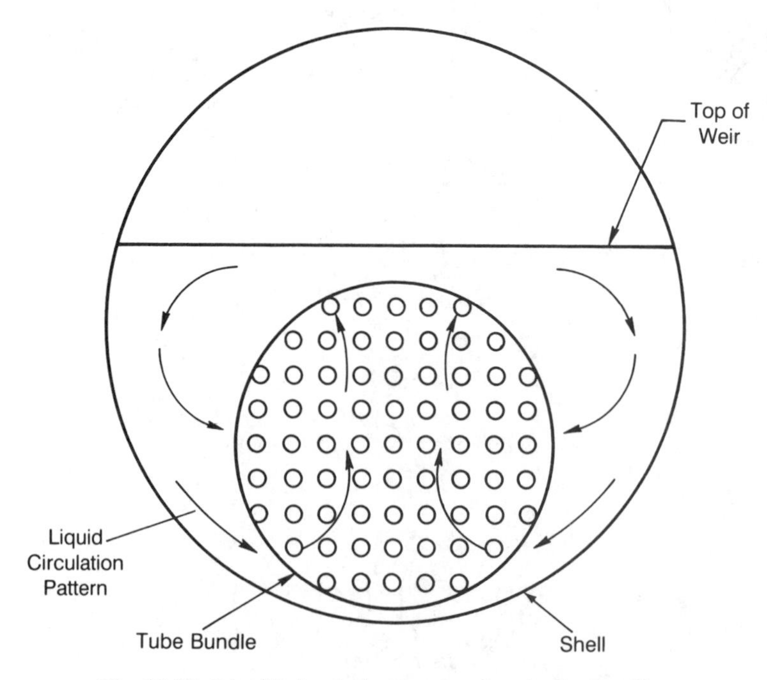

Fig. 13.29. Liquid circulation in a kettle reboiler bundle.

nucleate boiling at the tube surface is enhanced by the convective effects of the flow through the bundle. These circulation patterns and the calculation of the circulation strength are reviewed in detail by Whalley and Hewitt [5].

Even if the flow rate across or along the tube bundle is known, there still remains the problem of calculating the total boiling heat transfer coefficient. One approach that has been tried is to use an in-tube correlation for boiling. The modifications that have to be made are:

1. The "equivalent diameter" is used in place of any actual, physical dimension. The equivalent diameter is four times the wetted perimeter of the flow divided by the cross-sectional area available for the flow.
2. The mass flux in the equivalent tube is assumed to be the same as the mass flux in the bundle.

For large tube bundles the overall maximum heat flux may have to be reduced considerably from that possible with boiling on the outside of a single tube bundle, because of the restriction on the access of the liquid feed to the innermost parts of the bundle. Palen [22] has given in some detail the various methods for estimating heat transfer coefficients and maximum heat fluxes. Indeed the maximum heat flux is one of the greatest unknowns for boiling on the outside bundles of horizontal tubes. Palen and Small [29] give a calculation method that gives values for the critical heat flux which are very much lower than for a single tube. For large bundles this can be an extreme effect, and the calculated critical heat flux can only be a few percent of the single tube value. This is almost certainly very pessimistic.

13.6.9 Falling-Film Evaporation

Heat transfer coefficients for evaporating liquid films falling inside or outside vertical tubes are generally higher than in the corresponding climbing-film situation. This is because, particularly when the vapor flows co-currently with the falling film, the falling film is thinner than the climbing film for the same film mass flow rate. Reviews on heat transfer in falling-film evaporation have been given by Ganic [30] and by Sideman [6, 11].

In a convective boiling region where the film is very thin, experimental values of the heat transfer coefficient tend to be somewhat lower than the values predicted from laminar or turbulent flow theory. It can be noted that the theoretical equations, not surprisingly, are very similar to those for condensation on a vertical surface. Nucleate boiling does not usually occur in a falling film; however, when it does occur the film flow rate and film thickness are no longer important parameters in determining the heat transfer coefficient, and the heat transfer rates are similar to ordinary submerged nucleate boiling.

The major practical difficulty in the design and indeed the operation of falling-film evaporators is ensuring an even distribution of the liquid film, both between tubes and around the circumference of each tube. The distribution will determine when the liquid film breaks down, as all films will do when the flow rate is sufficiently low. The film spontaneously breaks into rivulets that wet only a small portion of the tube surface. The minimum film flow rate depends on the physical properties of the liquid [1], but for water the film flow rate per unit periphery of tube wall at which film breakdown occurs, Γ, kg/(m · s), appears to be related also to the heat flux. Heating a falling film appears to destablize the film flow. If a significant fraction of the feed is to be evaporated, then to keep safely above the minimum film flow rate it will probably be necessary to have a substantial recycle stream. In this way only a small portion of the falling film can be evaporated, but still the output can be very concentrated. The disadvantage of this is that the whole of the recycle stream and the falling film are very concentrated and thus possibly viscous or liable to crystallization.

Falling films with evaporation across bundles of horizontal tubes have been examined by Sideman [6, 11]. Good heat transfer performance was obtained, and the flow distribution problem is not so severe.

13.6.10 Agitated-Film Evaporation

In an agitated-film evaporator the liquid film thicknesses is determined by the flow rate and viscosity of the liquid. Typical film thickness are 1 mm for a fairly low viscosity liquid to 15 mm at a viscosity of 10 $(N \cdot s)/m^2$. Very high viscosity liquids do not flow readily under the action of gravity, and so they have be moved through the evaporator by an appropriate design of rotor and wiper. General correlations for the heat transfer coefficient are not available, but it is found that the heat transfer coefficient increase with flow rate, rotor speed, and heat flux, and decreases with viscosity, surface tension, and rotor to gap wall [7]. Some of these trends are simply a reflection of the fact the thinner films give higher heat transfer coefficients.

Burrows and Beveridge [31] have tested agitated-film evaporators with a range of materials. They found that product concentrations up to 90% could be obtained with increases in concentration through the evaporator up to a factor of 25. Typical values of the overall heat transfer coefficient varied widely—from 290 to 2200 $W/(m^2 \cdot K)$. It is not easy to see any definite trends in the data for heat transfer coefficients.

13.6.11 Mixture Effects

In an evaporator there will always be a boiling-point elevation effect because of the presence of the solute in the solvent. Of course, the maximum value of this elevation will be expected to occur where the solute concentration is highest. As a safety margin it is a good idea to add this temperature difference to that required for the boiling process itself. In many circumstances the solute does not alter very much the value of the heat transfer coefficient to the solvent after the boiling-point elevation effect has been taken into account. For example, experiments by Chun and Seban [32] on falling-film evaporation of water and 14% brine showed that the evaporation rates with the brine were about 10% less than with water at the same flow rate and with the same temperature difference from the wall to the liquid. A 10% change with this fairly strong solution is not detectable with the rather poor level of accuracy expected from most correlations for boiling heat transfer.

In reboilers it is usually a mixture of hydrocarbons that is being boiled. Typically, all the components of the mixture undergo some vaporization, though of course some of them are more easily vaporized than others. In a two-component mixture undergoing nucleate boiling, the growing bubble tends to be richer in the lighter component. This relative excess (compared to the bulk mixture) of light component must have come from the liquid immediately surrounding the bubble. This layer is then relatively rich in heavy component, and its saturation temperature will be greater than the bulk liquid. Thus the effective wall superheat (the wall temperature minus the saturation temperature) is less than would be expected for the bulk fluid. Nucleate boiling heat transfer coefficients depend strongly on the wall superheat—typically to the power 3 or 3.33. Thus for a mixture, the nucleate boiling heat transfer coefficients are less than expected. Calculation methods include those of Stephan and Korner [33] and Stephan and Preusser [34]. Shock [35] gives a full review.

The position in convective boiling is not so clear. Mixtures do give lower boiling heat transfer coefficients, but the reduction seems not to be so marked as in nucleate boiling.

13.6.12 Enhanced Surfaces

Enhanced surfaces are specially shaped surfaces that give higher evaporating-side or condensing-side heat transfer coefficients than are obtained with flat surfaces. They have been used extensively in evaporators, especially those used for desalination. It is probably true to say that without the use of enhanced heat transfer surfaces, there would be no chance that in the foreseeable future seawater desalination would be economical. Production of even more effective surfaces is probably the key to the large-scale adoption of the desalination process. Bergles [36, 37] has reviewed the various techniques of heat transfer enhancement and the application of these special surfaces to industrial equipment.

Enhanced heat transfer surfaces have been used in horizontal evaporators [38], vertical evaporators [39], and most particularly in falling-film evaporators [40]. It is for falling-film evaporators that the most impressive heat transfer improvements have been reported: the use of internally fluted tubes can increase the evaporating-side heat transfer coefficient 10 times. "Roped" tubes are tubes where both the internal and external surfaces are modified in a spiral pattern. They can increase both the condensing-side coefficient and the evaporating-side coefficient [41]. They therefore have a particularly large effect on the overall heat transfer coefficient.

In multistage flash evaporators, enhanced surfaces are often used in the condenser. "Roped" tubes increase the condensing coefficient on the outside of the tube and also the single-phase heat transfer coefficient on the inside of the tube: area reductions up to 30% are possible. Of course, there is a price to pay for the increase in performance: the tube-side pressure drop is increased and these special tubes are more expensive than plain tubes [42].

Because there are no satisfactory design correlations available, these special surfaces needed to be tested under realistic conditions using the actual proposed fluids. The mechanism for the large increases in heat transfer coefficients observed is a combination of a number of factors. First, of course, the shape of the surface corrugations increases the area available for heat transfer. Then the surface corrugations increase the turbulence of the fluid flowing over the surface, thus promoting heat transfer. Both these effects apply to single- and two-phase flows. The remaining factor applies only to two-phase flow; that is, the surface shape produces areas where the liquid film is very thin, thus giving high heat transfer coefficients. This latter effect is the most significant. This is why the enhancement effect is so dramatic in falling-film flow with steam condensing on the other side of the tube wall.

These detailed effects are very dependent on the exact surface geometry: designing surfaces that are effective over a range of conditions and that can easily be manufactured (and are therefore not prohibitively expensive) is by no means simple.

13.7 POSSIBLE PROBLEMS IN THE OPERATION OF EVAPORATORS AND REBOILERS

13.7.1 Introduction

The operational problems encountered in evaporators and reboilers are not very different from those encountered in most types of heat exchangers. The problem considered here are: corrosion and erosion (Section 13.7.2), maldistribution (Section 13.7.3), fouling (Section 13.7.4), flow instability (Section 13.7.5), tube vibration (Section 13.7.6), and flooding (Section 13.7.7).

Like other heat exchangers, evaporators and reboilers can be damaged or suffer impaired performance because of maloperation. Smith [1] outlines the detailed tests and routine measurements that should be carried out. Like all heat exchangers, these units should be equipped with a reasonable amount of instrumentation both for control and to establish performance as a function of time. Operators should be encouraged to keep careful logs to detect any change in operating characteristics. From these records, for example, the progress of fouling can be watched, so that rational decisions about the frequency of stripping down the evaporator and cleaning can be made. Regrettably, many evaporators and reboilers are not equipped with sufficient instrumentation to allow the cause of poor performance to be ascertained.

13.7.2 Corrosion and Erosion

When a solution containing aggressive salts is being concentrated, corrosion should always be expected. It is most likely to occur between the tube wall and the fouling layer. In this region salts can concentrate locally to the extent that the elevation of the boiling point matches the local tube wall temperature. The tubes in an evaporator are not normally expected to last the lifetime of the evaporator, and so the tube bundle or the individual tubes should be designed for easy replacement.

In the usual case where the evaporator or reboiler is heated by condensing steam, some steps should be taken to prevent the high-velocity steam hitting the tubes near the steam inlet nozzle. This can be done in one of two ways. An impingement baffle can be installed opposite the inlet nozzle to protect the tubes (see also Chapter 11). Alternatively, the steam can be introduced into the unit in an annular space around the shell: this is a "vapor belt" (see Fig. 13.30).

Corrosion problems in evaporators and reboilers are reviewed by Dunmore [43].

13.7.3 Maldistribution

Maldistribution is only a problem with vertical falling-film evaporators: hence it is a very serious problem. It is necessary to see that the top tube plate is flat and horizontal. The aim is to have the liquid level on the tube plate as uniform as possible. To achieve this, the liquid pool has to be shielded from the disturbance effects of the liquid entering the evaporator. To get a film that will not break up prematurely, it is also necessary to obtain good flow distribution around the circumference of each tube. This can be done by means of an insert fitted into the top of each tube to spread the film out. Alternatively, several weirs can be cut around the circumference of the tube (see Fig. 13.30). The tubes in a vertical falling-film evaporator also need to be vertical to within a fraction of a tube diameter between top and bottom. Great care installing the evaporator is time wasted if the top tube plate is not flat or if the tubes are not straight!

13.7.4 Fouling

Fouling has been referred to on a number of occasions, in the initial selection of a suitable equipment type and at the design stage (see Section 13.6.6) where a fouling coefficient has to be assumed. It is very likely that fouling will

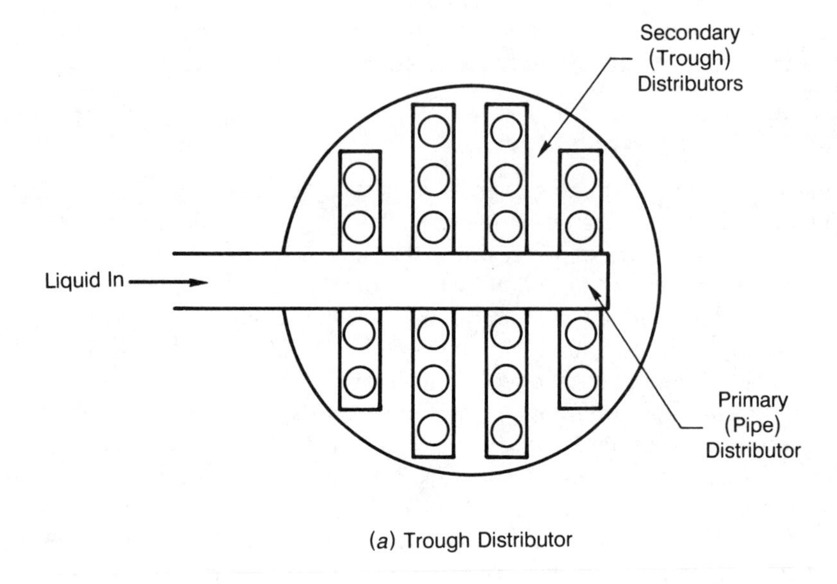

(*a*) Trough Distributor

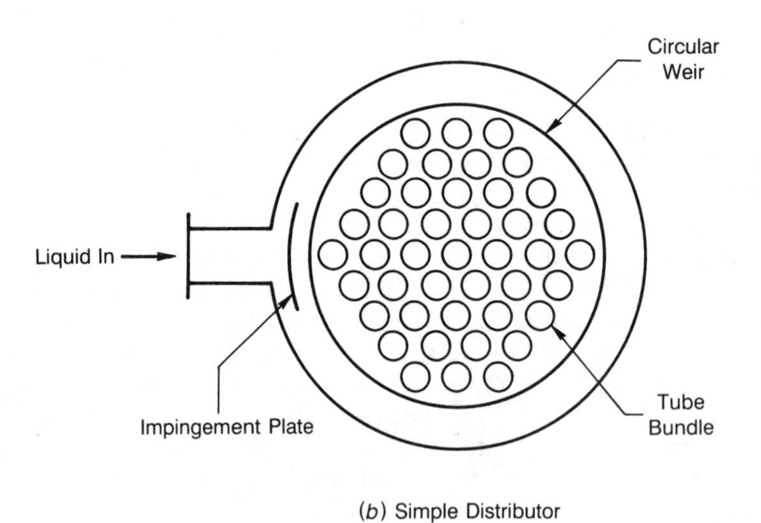

(*b*) Simple Distributor

Fig. 13.30. Alternative types of distribution for a vertical falling-film evaporator.

occur and plant measurements may indicate that the performance of the evaporator or reboiler is unsatisfactory. This may be because, to reach the desired performance, excessively high steam temperatures are becoming necessary, a situation that will only aggravate the fouling. A proper choice of equipment type for the specified duty together with the adoption of design features like plate units or large-diameter tubes (50 to 60 mm) for easy cleaning can help. It may be necessary to have standby units available which can be piped in while the unit is mechanically or chemically cleaned.

For further details on fouling see Chapter 4.

13.7.5 Flow Instability

Fluid flow instabilities are a common problem in boiling systems, particularly in natural-circulation units. A number of different types of instabilities can occur for a number of reasons. General reviews of the very complicated subject of flow instabilities have been provided by Kakaç and Veziroglu [44], and more recently by Lahey and Podowski [45].

A dynamic (or density-wave) instability is caused by a pressure drop feedback mechanism and can be stabilized by increasing the proportion of the total pressure drop that is caused by single-phase friction. The special relevance of the single-phase frictional pressure drop is that it is in phase with the velocity fluctuations, whereas there can be a phase difference with the vapor–liquid region pressure drop. If the single-phase pressure drop is greater than the two-phase pressure drop, the flow is usually stable. This is most easily achieved by inserting a throttle or flow restriction at the entrance to the boiling section. Conversely, extra pressure drop at the end of the boiling section will destabilize the flow. The instability can take the form of a gross instability of the entire flow or an instability between vaious tubes.

The other important type of instability is "bumping" or "geysering." This occurs if the temperature difference and therefore the heat flux is too low. At low values of the temperature difference, the temperature of the tube wall and the surrounding fluid is not hot enough to cause nucleation of bubbles at the wall. The result of this is that the bulk liquid becomes superheated, and eventually when a vapor bubble does form it grows very rapidly. This rapid, almost explosive growth, results in a large change in the local pressure which can stop or even reverse the flow. A large amount of liquid is expelled from the system quickly, and refilling with liquid from the feed then takes place. Hence evaporators should always be designed to exceed a minimum heat flux (see [13]):

$$q''_{\min} = \frac{8\sigma T_{\text{sat}}(v_g - v_l)h_l^2}{\Delta h_v\, k_l} \tag{13.11}$$

where $q''_{\min}$ is the minimum heat flux, W/m^2; σ is the surface tension, N/m;

T_{sat} is the saturation temperature in absolute degrees, K; h_l is the heat transfer coefficient to the single-phase liquid flow, W/(m^2 · K); k_l is the thermal conductivity of the liquid, W/(m · K); v_g is the vapor specific volume, m^3/kg (1/density); and v_l is the liquid specific volume, m^3/kg (1/density).

13.7.6 Tube Vibration

In heat exchangers, the motion of the fluid past the tubes can excite the tubes into severe vibration. This vibration can sometimes be so severe that the noise generated is painful and damaging, and also the vibration can physically damage the tubes. The subject of flow-induced vibration is reviewed by Saunders [46]. He gives calculation methods that can be used to check if vibration is likely to be a problem.

Evaporators and reboilers are not normally prone to fluid-excited tube vibration. However, the one area where care is needed in the design is close to the heating steam inlet. The steam velocities can be large in this region. The tube natural frequency decreases rapidly with increasing span length, so care should be taken so that there are no long unsupported tube lengths in this region. Provision of an impingement baffle or a vapor belt (see Section 13.7.2) will also improve the situation.

13.7.7 Flooding

In countercurrent falling-film evaporators flooding can occur. When the upward gas flow rate is too large, the liquid can no longer travel downwards in the film. Flooding is the start of this phenomenon and is thus a limit of countercurrent flow. Correlations for flooding in tubes and in other geometries have been derived by many workers; brief details of the available methods are given by Smith [1], Whalley [14], and McQuillan and Whalley [47]. A longer review is given by Bankoff and Lee [48].

13.8 DESIGN EXAMPLE

It is required to design in outline a short-tube vertical evaporator (see Section 13.3.4) to evaporate 0.5 kg/s of water from a dilute aqueous solution which is at a pressure of 1 bar. The evaporator tubes are to be 2 m long, 50 mm inside diameter, and 55 mm outside diameter. The tube metal thermal conductivity is 50 W/(m · K).

Solution:

1. As the solution is dilute and aqueous, the liquid can be assumed to have the properties of pure water at 1 bar. In particular, the latent heat of vaporization will be 2256×10^3 kJ/kg, and so, assuming the feed liquid to the

evaporator is saturated, the heat load for this evaporator will be given by

$$\text{Heat load} = \text{evaporation rate} \times \text{latent heat} = 0.5 \times 2256 \times 10^3$$

$$= 1.13 \times 10^6 \text{ W} = 1.13 \text{ MW}$$

2. If the heat is supplied by condensing steam at 110°C (pressure, 1.45 bar; latent heat, 2229×10^3 kJ/kg), then the steam flow necessary is given by

$$\frac{\text{Steam flow} = \text{heat load}}{\text{latent heat}} = \frac{1.13 \times 10^6}{2229 \times 10^3}$$

$$= 0.507 \text{ kg/s}$$

3. It is a reasonable assumption for the moment that the evaporating liquid temperature is constant at 100°C and the condensing vapor temperature is constant at 110°C. The logarithmic-mean temperature difference is therefore 10 K.

4. Turning now to the heat transfer coefficient, the tube is fairly thin walled and so the curvature of the tube wall will be neglected. The difference in area between the inside and outside of the tube will also be neglected. Reasonable estimates of the various heat transfer coefficients are then as follows:

Condensing heat transfer coefficient = 10,000 W/(m^2 · K)
Outside fouling heat transfer coefficient = 10,000 W/(m^2 · K)
Tube metal heat transfer coefficient = tube thickness/metal thermal conductivity = 50/0.0025 = 20,000 W/(m^2 · K)
Inside fouling heat transfer coefficient = 1000 W/(m^2 · K)
Evaporating heat transfer coefficient = 8000 W/(m^2 · K)

The overall heat transfer coefficient, U, is then given by

$$\frac{1}{U} = \frac{1}{10{,}000} + \frac{1}{10{,}000} + \frac{1}{20{,}000} + \frac{1}{1000} + \frac{1}{8000}$$

$$= 1.375 \times 10^{-3}$$

Hence $U = 730$ W/(m^2 · K). Note here how the overall heat transfer coefficient is dominated by the individual heat transfer coefficient which is by far the lowest—the inside fouling coefficient.

5. The required total heat transfer area can now be calculated from

$$\text{Total area} = \frac{\text{heat load}}{(U \times \text{log-mean temperature difference})}$$

$$= \frac{1.129 \times 10^6}{730 \times 10} = 155 \text{ m}^2$$

6. The area provided by each tube is

$$\text{Area per tube} = \pi \times \text{inside diameter} \times \text{length}$$

$$= \pi \times 0.05 \times 2 = 0.314 \text{ m}^2$$

Note here the inside area is calculated because the dominant heat transfer coefficient is an inside heat transfer coefficient.

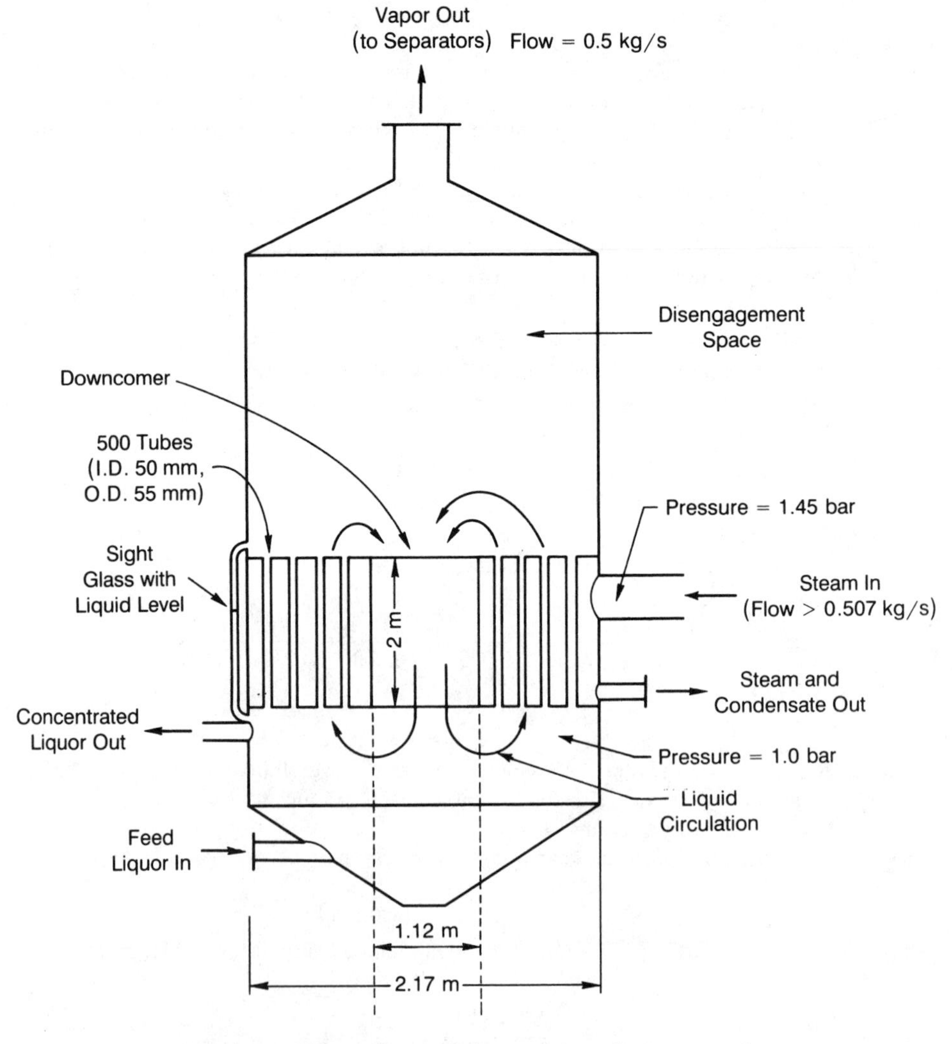

Fig. 13.31. Short-tube vertical evaporator: design example.

7. The required number of tubes is now

$$\text{Number of tubes} = \frac{\text{total area}}{\text{area per tube}}$$

$$= \frac{155}{0.314} = 493$$

A reasonable number of tubes is therefore 500.

8. The cross-sectional flow area of these 500 tubes is

$$\text{Flow area} = \frac{500 \times \pi \times (\text{inside diameter})^2}{4}$$

$$= \frac{500 \times \pi \times 0.05^2}{4} = 0.98\ \text{m}^2$$

The downcomer area (see Section 13.3.4) should be equal to the total tube cross-sectional area. Hence the downcomer diameter should be 1.12 m.

9. If the tubes are on a square pitch with a pitch to outside diameter ratio of 1.33, then each tube takes up an area of tube sheet of $(1.33 \times 0.055)^2 = 5.4 \times 10^{-3}$ m^2. The total tube sheet area for 500 tubes is thus approximately $5.4 \times 10^{-3} \times 500 = 2.7$ m^2. Allowing for the central downcomer, the necessary outside diameter of the calandria is 2.17 m. The tubes will not pack conveniently into an annular space, and so the outer diameter will certainly have to be somewhat larger than this.

The main dimensions of the completed initial design are shown in Fig. 13.31.

13.8.1 Further Refinements in the Design

The greatest uncertainty in the calculation, as often in heat exchanger design, is the magnitude of the fouling coefficients. As the problem stands at the present, the effort involved in producing more accurate condensing and evaporating heat transfer coefficients would be wasted.

The tube sheet could be laid out accurately so that the outer diameter of the calandria could be more accurately found.

The natural circulation through the tubes and the downcomer could be estimated by a simple adaptation of the information in Table 13.2. Once the circulation is known, the pressure in the liquid at the bottom of the tubes can be calculated. This will be greater than 1 bar and the corresponding saturation temperature will be greater than 100°C. This effect will reduce the logarithmic-mean temperature difference below the previously assumed value of 10 K.

NOMENCLATURE

A	heat transfer area, m^2
d	tube diameter, m
$[dp/dz]_A$	accelerational or momentum pressure gradient, N/m^3
$[dp/dz]_F$	frictional pressure gradient, N/m^3
$[dp/dz]_{fo}$	frictional pressure gradient if the total mass flow rate (liquid and vapor) were flowing in single-phase flow and had the properties of the liquid, N/m^3
$[dp/dz]_G$	gravitational or static head pressure gradient, N/m^3
$[dp/dz]$	total single-phase or two-phase pressure gradient, N/m^3
G	mass flux (mass flow rate/flow cross-sectional area), $kg/(m^2 \cdot s)$
g	acceleration due to gravity, m/s^2
h	heat transfer coefficient, $W/(m^2 \cdot K)$
h_l	single-phase liquid heat transfer coefficient, $W/(m^2 \cdot K)$
k_l	liquid thermal conductivity, $W/(m \cdot K)$
n	number of effects in a multistage process
Q	total heat duty, W
q''_{min}	minimum heat flux to avoid geysering, W/m^2
R	reduction factor in energy input
T_{cond}	condensing-steam saturation temperature, K
T_{evap}	evaporating-fluid saturation temperature, K
T_{sat}	saturation temperature (absolute degrees), K
U	overall heat transfer coefficient, $W/(m^2 \cdot K)$
u	velocity of flow, m/s
v_g	vapor specific volume, m/kg^3
v_l	liquid specific volume, m/kg^3
x	quality (mass fraction of the two-phase flow which is vapor)

Greek Symbols

α	void fraction
Γ	liquid film flow rate per unit periphery, $kg/(m \cdot s)$
Δh_v	latent heat of vaporization, J/kg
ΔT	temperature difference between the heating fluid and the evaporating fluid, K
δz	length of momentum equation control volume, m
θ	angle of inclination of the flow to the horizontal (θ positive for upflow)
μ_g	vapor viscosity, $(N \cdot s)/m^2$
μ_h	homogeneous mixture viscosity, $(N \cdot s)/m^2$
μ_l	liquid viscosity, $(N \cdot s)/m^2$
ρ_g	vapor density, kg/m^3
ρ_h	homogeneous mixture density, kg/m^3
ρ_l	liquid density, kg/m^3

σ	surface tension, N/m
τ	wall shear stress, N/m^2
ϕ^2_{fo}	two-phase pressure drop multiplier

REFERENCES

1. Smith, R. A. (1986) *Vaporisers—Selection, Design and Operation*. Longman Scientific and Technical/Wiley, New York.
2. Minton, P. E. (1986) *Handbook of Evaporation Technology*. Noyes Publications, Park Ridge, N. J.
3. Bell, K. J. (1973) Thermal design of heat transfer equipment. In *Perry's Chemical Engineers Handbook*, R. H. Perry and C. H. Chilton (eds.), 5th ed., Section 10, pp. 10-22–10-47. McGraw-Hill, New York.
4. Standiford, F. C. (1973) Evaporators. In *Perry's Chemical Engineers Handbook*, R. H. Perry and C. H. Chilton (eds.), 5th ed., Section 11, pp. 11-27–11-37. McGraw-Hill, New York.
5. Whalley, P. B., and Hewitt, G. F. (1986) Reboilers. In *Multiphase Science and Technology*, G. F. Hewitt, J. M. Delhaye, and N. Zuber, (eds.), Vol. 2, pp. 275–331. Hemisphere, Washington, D.C.
6. Sideman, S. (1981) Film evaporation and condensation in desalination. In *Heat Exchangers Thermal Hydraulic Fundamentals and Design*, S. Kakaç, A. E. Bergles, and F. Mayinger (eds.), pp. 357–375. Hemisphere, Washington, D.C.
7. Salden, D. M. (1987) Agitated thin film evaporators. *The Chemical Engineer Supplement* September, 17–19.
8. Gray, R. M. (1984) The design and use of plate heat exchangers in boiling and condensing applications. *First National U.K. Conference on Heat Transfer* (*Inst. Chem. Eng. Symp. Ser.* 86) **2** 684–694.
9. Usher, J. D. (1970) Evaluating plate heat exchangers. *Chemical Engineering* **72** 90–94.
10. Edwards, R. M. (1967) Efficient new heat exchanger suited to LNG vaporisation. *The Oil and Gas Journal* October, 96–99.
11. Sideman, S. (1981) Some aspects of thin film evaporators and condensers in water desalination. In *Heat Exchangers Thermal Hydraulic Fundamentals and Design*, S. Kakaç, A. E. Bergles, and F. Mayinger (eds.), pp. 681–703. Hemisphere, Washington, D.C.
12. Silver, R. S. (1968) The physics of desalination by multi-stage flash distillation. *Journal of B.N.E.S.* **7** 36–42.
13. Collier, J. G. (1981) *Convective Boiling and Condensation* 2nd ed. McGraw-Hill, New York.
14. Whalley, P. B. (1987) *Boiling, Condensation and Gas–Liquid Flow*. Oxford University Press, Oxford.
15. Friedel, L. (1979) Improved friction pressure drop correlations for horizontal and vertical two-phase flow. *European Two-Phase Flow Group Meeting*, Ispra, Italy (calculation method quoted in full by Whalley [14]).

16. Baroczy, C. J. (1966) A systematic correlation for two-phase pressure drop. *Chem. Eng. Prog. Symp. Ser.*, **62**(64) 232–249.
17. Thom, J. R. S. (1964) Prediction of pressure drop during forced circulation boiling of water. *Int. J. Heat Mass Transfer*, **7** 709–724.
18. Chisholm, D. (1973) Void fraction during two-phase flow. *J. Mech. Eng. Sci.*, **15** 235–236.
19. Premoli, A., Francesco, D., and Prina, A. (1970) An empirical correlation for evaluating two-phase mixture density under adiabatic conditions. *European Two-Phase Flow Group Meeting*, Milan (calculation method quoted in full be Whalley [14]).
20. Bryce, W. M. (1977) A new flow dependent slip correlation which gives hyperbolic steam–water mixture equations. A.E.E.W.-R1099.
21. TEMA (1978) *Standards of the Tubular Exchanger Manufacturers Association* 6th ed. TEMA, New York.
22. Palen, J. W. (1983) Shell and tube reboilers. In *Heat Exchanger Design Handbook*, Section 3.6. Hemisphere, Washington, D.C.
23. Chen, J. C. (1966) A correlation for boiling heat transfer to saturated fluids in convective flow. *Ind. Eng. Chem. Proc. Des. Dev.* **5**(3) 322–333.
24. Shah, M. (1976) A new correlation for heat transfer during boiling flow through pipes. *ASHRAE Trans.* **82**(2) 66–86.
25. Fair, J. R. (1960) What you need to design thermosyphon reboilers. *Petroleum Refiner* **39**(2) 105.
26. Bowring, R. W. (1972) A simple but accurate round tube uniform heat flux dryout correlation over the pressure range 0.7 to 17 MN/m^2. A.E.E.W.-R789.
27. Katto, Y., and Ohne, H. (1984) An improved version of the generalised correlation of critical heat flux for convective boiling in uniformly heated vertical tubes. *Int. J. Heat Mass Transfer* **27** 1641–1648.
28. Cooper, M. G. (1984) Saturation nucleate pool boiling—a simple correlation. *First National U.K. Heat Transfer Conference* (*Inst. Chem. Eng. Symp. Ser.* 86) **2** 785–793.
29. Palen, J. W., and Small, W. M. (1964) A new way to design kettle and internal reboilers. *Hydrocarbon Processing* **43** 199–208.
30. Ganic, E. N. (1981) On the heat transfer and fluid flow in falling film shell and tube evaporators. In *Heat Exchangers Thermal Hydraulic Fundamentals and Design*, *edited by* S. Kakaç, A. E. Bergles, and F. Mayinger (eds.), pp. 705–719. Hemisphere Washington, D.C.
31. Burrows, M. J., and Beveridge, G. S. G. (1979) The centrifugally agitated wiped film evaporator. *The Chemical Engineer*, **343** 229–232.
32. Chun, K. R., and Seban, R. A. (1972) Performance prediction of falling film evaporators. *J. Heat Transfer* **94** 432–436.
33. Stephan, K., and Korner, M. (1969) Calculation of heat transfer in evaporating binary mixtures. *Chem. Ing. Tech.* **41** 409–417.
34. Stephan, K., and Preusser, P. (1978) Heat transfer in natural convection boiling of polynary mixtures. *Sixth International Heat Transfer Conference, Toronto*, Vol. 1, pp. 187–192.

35. Shock, R. A. W. (1982) Multicomponent boiling. In *Multiphase Science and Technology* G. F. Hewitt, J. M. Delhaye, and N. Zuber (eds.), Vol. 1, pp. 281–386. Hemisphere, Washington, D.C.

36. Bergles, A. E. (1983) Augmentation of boiling and evaporation. In *Heat Exchanger Design Handbook*, Section 2.7.9. Hemisphere, Washington, D.C.

37. Bergles, A. E. (1981) Augmentation of two-phase heat transfer and Industrial applications of heat transfer augmentation. In *Two-Phase Flow and Heat Transfer in the Power and Process Industries*, Chapters 12 and 21. McGraw-Hill/Hemisphere, New York.

38. Oates, H. S. (1982) The thermal performance of horizontal falling film evaporator tubes with enhanced heat transfer surfaces. A.E.E.W.-R932.

39. Satone, H., Funahashi, S., and Inoue, S. (1977) Experience on VTE distillation plants in Japan, *Desalination* **22** 169–179.

40. Thomas, D. G., and Young, G. (1970) Thin film evaporation enhancement by finned surfaces. *Ind. Eng. Chem. Proc. Des. Dev.* **9** 317–323.

41. Johnson, B. M., Jansen, G., and Otwzarski, P. C. (1971) Enhanced evaporating film heat transfer from corrugated surfaces. A.S.M.E. paper 71-HT-33.

42. Newson, I. H. (1976) Enhanced heat transfer condenser tubing for advanced multi-stage flash distillation plants. *Proc. 5th Int. Symp. on Fresh Water from the Sea* **2** 107–115.

43. Dunmore, O. J. (1986) Materials of construction for vaporisers. In *Vaporisers—Selection, Design and Operation*, R. A. Smith (ed.), Chapter 12. Longman Scientific and Technical Wiley, New York.

44. Kakaç, S., and Veziroglu, T. N. (1983) Two-phase flow instabilities in boiling systems: summary and review. In *Advances in Two-Phase Flow and Heat Transfer*, S. Kakaç and M. Ishii (eds.), Vol. 2, pp. 577–667 Kluwer, Dordrecht.

45. Lahey, R. T., and Podowski, M. Z. (1989) On the analysis of various instabilities in two-phase flows. In *Multiphase Science and Technology*, G. F. Hewitt, J. M. Delhaye, and N. Zuber (eds.), Vol. 4, pp. 183–370. Hemisphere, Washington, D.C.

46. Saunders, E. A. D. (1988) *Heat Exchangers: Selection, Design and Construction.* Longman Scientific and Technical Harlow, U.K.

47. McQuillan, K. W., and Whalley, P. B. (1985) A comparison between flooding correlations and experimental flooding data. *Chem. Eng. Sci.* **40** 1425–1440.

48. Bankoff, S. G., and Lee, S. C. (1985) A critical review of the flooding literature. In *Multiphase Science and Technology*, G. F. Hewitt, J. M. Delhaye, and N. Zuber (eds.), Vol. 1, pp. 95–195. Hemisphere, Washington, D.C.

APPENDIX A

THERMOPHYSICAL PROPERTIES

P. E. LILEY

School of Mechanical Engineering and
Center for Information and Numerical Data
Analysis and Synthesis (CINDAS)
Purdue University
West Lafayette, Indiana 47907

NOMENCLATURE

b	normal boiling point
c	critical point
c_p	specific heat at constant pressure, kJ/(kg · K)
c_v	specific heat at constant volume, kJ/(kg · K)
c_{pf}	specific heat at constant pressure of the saturated liquid, kJ/(kg · K)
c_{pg}	specific heat at constant pressure of the saturated vapor, kJ/(kg · K)
h	specific enthalpy, kJ/(kg · K)
h_f	specific enthalpy of the saturated liquid, kJ/(kg · K)
h_g	specific enthalpy of the saturated vapor, kJ/(kg · K)
k	thermal conductivity, W/(m · K)
k_f	thermal conductivity of the saturated liquid, W/(m · K)
k_g	thermal conductivity of the saturated vapor, W/(m · K)
M	molecular weight, kg/k mol
P	pressure, bar
P_f	saturation vapor pressure, liquid, bar
P_g	saturation vapor pressure, gas, bar
Pr	Prandtl number, $c_p\mu/k$
Pr_f	Prandtl number of the saturated liquid, $(c_p\mu)_f/k_f$
Pr_g	Prandtl number of the saturated vapor, $(c_p\mu)_g/k_g$
s	specific entropy, kJ/(kg · K)
s_f	specific entropy of the saturated liquid, kJ/(kg · K)
s_g	specific entropy of the saturated vapor, kJ/(kg · K)
t	triple point
T	temperature, °C, K
T_b	normal boiling point temperature, °C, K

T_m normal melting point temperature, °C, K
v specific volume, m^3/kg
v_f specific volume of the saturated liquid, m^3/kg
v_g specific volume of the saturated vapor, m^3/kg
$\bar{v}_s$ velocity of sound, m/s
Z compressibility factor

Greek Symbols

α thermal expansion coefficient
β isothermal compressibility coefficient
γ ratio of principal specific heat, c_p/c_v
μ viscosity, Pa · s
μ_f viscosity of the saturated liquid, Pa · s
μ_g viscosity of the saturated vapor, Pa · s
σ surface tension, N/m
ρ density, kg/m^3

LIST OF SOURCES OF TABLES IN APPENDIX MATERIAL

Table	Source
A1	Table 22.4, Kakaç [1]
A2	Table 22.5, Kakaç [1]
A3	Table 22.6, Kakaç [1]
A8	Table 22.8, Kakaç [1]
A9	Table 22.9, Kakaç [1]
A14	Table 22.12, Kakaç [1]
A15	Table 52.14, Kutz [2]
A18	Table 22.18, Kakaç [1]
A21	Table 22.21, Kakaç [1]
A22	Table 22.22, Kakaç [1]
A27	Table 22.29, Kakaç [1]
A28	Table 22.30, Kakaç [1]
A29	Table 22.31, Kakaç [1]
A30	Table 22.32, Kakaç [1]
A31	Table 22.33, Kakaç [1]
A32	Table 22.34, Kakaç [1]
A33	Inside front cover, Kakaç [1]

REFERENCES

1. Liley, P. E. (1987) Thermophysical properties. In *Handbook of Single-Phase Convective Heat Transfer*, S. Kakaç, R. K. Shah, and W. Aung (eds.), Chapter 22. Wiley, New York.
2. Liley, P. E. (1986) Thermophysical properties of fluids. In *Mechanical Engineers Handbook*, M. Kutz (ed.), Chapter 52. Wiley, New York.

TABLE A1 Thermophysical Properties of 113 Fluids at 1 bar, 300 K[a]

Name	Formula	M	T_m, K	T_b, K	P_g, bar	ρ, kg/m^3
Acetaldehyde	C_2H_4O	44.053	149.7	293.7	2.94	
Acetic acid	$C_2H_4O_2$	60.053		391.1	0.0225	1046
Acetone	C_3H_6O	58.080	178.5	329.3	0.0318	782.5
Acetylene	C_2H_2	26.038	179.0	189.2	50.66	1.0508
Air (R729)	mix	28.966	60.0	var	—	1.1614
Ammonia (R717)	NH_3	17.031	195.4	239.7	10.614	
Aniline	C_6H_7N	93.129	266.8	457.5	0.00072	
Argon (R740)	Ar	39.948	83.8	87.5	—	
Benzene	C_6H_6	78.114	278.7	353.3	0.1382	871
Bromine	Br	159.81	265.4	331.5	0.310	3094
Butadiene, 1,3-	C_4H_6	54.088	164.2	268.7		
Butane, *iso-*	C_4H_{10}	58.124	113.7	261.5	3.696	
Butane, *n-*	C_4H_{10}	58.124	134.9	272.6	2.581	
Butanol	$C_4H_{10}O$	74.124	183.9	390.8	0.01064	804.3
Butylene	C_4H_8	56.108	87.8	266.9		
Carbon dioxide	CO_2	44.010	216.6	194.7	67.10	1.7734
Carbon disulfide	CS_2	76.131	161.1	319.4		1262
Carbon tetrachloride	CCl_4	153.82	250.3	349.8		1580.8
Carbon tetrafluoride	CF_4	88.005	89.5	145.2	—	
Chlorine	Cl_2	70.906	172.2	238.6	8.	
Chlorine trifluoride	ClF_3	92.449	196.8	284.9	1.8	1.820
Chlorine pentafluoride	ClF_5	130.446	178.2	260.1	4.0	1.788
Chloroform	$CHCl_3$	119.377	209.7	334.5	0.2771	1530
Cresol, *o-*	C_7H_8O	108.134	303.8	464.1		
Cresol, *m-*	C_7H_8O	108.134				
Cresol, *p-*	C_7H_8O	108.134				
Cyclobutane	C_4H_8	56.104	182.4	285.7	1.6	
Cyclohexane	C_6H_{12}	84.156	279.8	353.9		772.1
Cyclopentane	C_5H_{10}	70.130	179.3	322.5		
Cyclopropane	C_3H_6	42.081	145.5	240.3	6.	
Decane	$C_{10}H_{22}$	142.276	243.4	447.3	0.0017	724.1
Deuterium	D_2	4.028	18.7	23.7	—	
Diphenyl	$C_{12}H_{10}$	154.200	342.4	527.6		
Ethane (R170)	C_2H_6	30.070	89.9	184.6	43.541	1.2157
Ethanol	C_2H_6O	46.069	159.0	351.5		1247
Ethyl acetate	$C_4H_8O_2$	88.106	189.4	350.3	0.1361	892.3
Ethyl bromide	C_2H_5Br	108.97	153.5	311.5		1450
Ethyl chloride (R160)	C_2H_5Cl	64.515	134.9	285.4	1.68	
Ethyl ether	$C_4H_{10}O$	74.123	150.0	307.8		705.6
Ethyl fluoride (R161)	C_2H_5F	48.060	130.0	236.1	9.1	
Ethyl formate	$C_3H_6O_2$	74.080	193.8	327.4		
Ethylene	C_2H_4	28.054	104.0	169.5	—	0.9035
Ethylene oxide	C_2H_4O	44.054	160.6	283.6	1.82	
Fluorine	F_2	37.997	53.5	85.1	—	1.5442
Glycerol						1284.3
Helium	He	4.003	—	4.3	—	0.1625
Heptane	C_7H_{16}	100.21	182.5	371.6	0.06719	678.0
Hexane	C_6H_{14}	86.178	177.8	341.9	0.06674	652.5
Hydrazine	N_2H_4	32.045	274.7	386.7		
Hydrogen, *n-*	H_2	2.016	14.0	28.4	—	0.0808

TABLE A1 (*Continued*)

c_p, kJ/(kg · K)	c_v, kJ/(kg · K)	γ	μ, 10^{-4} Pa · s	k, W/(m · K)	Pr —	σ, N/m	$\bar{v}_s$, m/s
1.49							
2.066			11.15	0.1635	14.1		1130
2.13			3.08	0.1595	4.11	0.0229	1160
1.7032	1.380	1.234	0.1039	0.0220	0.804	—	342
1.005	0.718	1.400	0.184	0.0261	0.711	—	347.3
2.10	1.63	1.33	0.102	0.246	0.870	—	434
2.079			34.7	0.173	41.7		1615
0.522	0.313	1.667	0.2271	0.0177	0.670	—	322.6
1.73			5.88	0.1444	7.04	0.0279	1275
				0.122		0.041	
1.475			0.0876	0.0181	0.714		
			0.0760	0.0164	0.814		
1.731	1.569	1.103	0.0757	0.0160	0.811	0.0116	211
2.39			24.9	0.152	39.2	0.0241	1215
			0.0780			—	
0.845	0.657	1.288	0.150	0.0166	0.763	—	269.6
1.016				0.161			1143
0.8637			8.81	0.1026	7.42	0.0261	918
0.7071	0.6095	1.150	0.1740	0.01711	0.719	—	
0.4728			0.1371	0.00889	0.729	—	216
0.695			0.1421	0.0138	0.716		
0.745			0.1431	0.0146	0.730		
0.975			5.39	0.117			988
			70.2	0.153	96.0		
			112	0.149	174		
			126	0.144			
1.289			0.0812	0.0148	0.707		
1.865			8.69	0.123	13.2	0.0242	1230
1.81			3.97	0.132	5.46		
1.331			0.0890	0.0163	0.727	—	
2.213							
			0.126	0.141			
1.586							
1.769	1.482	1.192	0.094	0.0218	0.765	—	314
2.456			10.46	0.167	15.4	0.0222	11.37
1.947			4.37	0.143	5.95	0.0234	1120
0.880			3.63	0.101	3.16		905
0.973			0.1002	0.0126	0.774	—	
2.21			2.25	0.130	3.83		968
1.243			0.0954	0.0163	0.727		
				0.159			
1.560	1.262	1.237	0.104	0.0203	0.796	—	332
1.096			0.0879	0.0130	0.742		
0.827	0.607	1.362	0.227	0.0279	0.673	—	299
2.38			7.922	0.280	6730	0.0587	1875
5.193	3.115	1.667	0.1976	0.165	0.668		1020
2.252			3.85	0.1242	6.96	0.0197	1120
2.270			2.87	0.123	5.30	0.0177	1130
14.27	10.18	1.405	0.0894	0.182	0.701	—	1350

TABLE A1 (*Continued*)

Name	Formula	M	T_m, K	T_b, K	P_g, bar	ρ, kg/m^3
Hydrogen, *p*-	H_2	2.016			—	
Hydrogen bromide	HBr	80.912	186.3	206.4	25.5	
Hydrogen chloride	HCl	36.461	160.0	188.1	48.7	
Hydrogen fluoride	HF	20.006	181.8	272.7	1.3	
Hydrogen iodide	HI	127.91	222.4	237.8	8.5	
Hydrogen peroxide	H_2O_2				0.0031	1449
Hydrogen sulfide	H_2S	34.076	187.5	213.0	20.9	
Krypton	Kr	83.80	116.0	121.4	—	3.3659
Mercury	Hg	200.59	234.3	630.1		
Methane	CH_4	16.043	90.7	111.5	—	0.6443
Methanol	CH_4O	32.042	175.5	337.7	0.1860	784.9
Methyl acetate	$C_3H_6O_2$	74.080	175	330.3		
Methyl bromide (R40B1)	CH_2Br	94.939	179.5	276.7	2.38	
Methyl chloride (R40)	CH_3Cl	50.487	175.4	249.4	6.189	
Methyl fluoride (R41)	CH_3F	34.033	131.3	194.7	39.5	
Methyl formate (R611)	$C_2H_4O_2$	60.052	173.4	304.7		
Methylene chloride (R30)	CH_2Cl_2	84.922	176.5	312.9		
Neon (R720)	Ne	20.179	24.5	27.3	—	0.8091
Neopentane	C_5H_{12}	72.151	256.6	282.7		
Nitric oxide	NO	30.006		121.4	—	
Nitrogen (R728)	N_2	28.013	63.1	77.3	—	
Nitrogen peroxide	NO_2	46.006	263	294.5		
Nitrous oxide (R744a)	N_2O	44.013	176	184.7	57.5	
Nonane	C_9H_{20}	128.250			0.0064	712.3
Octane	C_8H_{18}	114.220	216.4	398.9	0.0207	704.2
Oxygen	O_2	31.999	54.4	90.0	—	
Pentane, *iso*-	C_5H_{12}	72.151	113.7	301.1		
Pentane, *n*-	C_5H_{12}	72.151	143.7	309.2		
Propadiene	C_3H_4	40.062	136.9	238.8	8.4	
Propane	C_3H_8	44.097	86.0	231.1	9.9973	
Propanol	C_3H_8O	60.096	147.0	370.4		
Propylene	C_3H_6	42.081	87.9	225.5	12.118	
Refrigerant 11	$CFCl_3$	137.37	162.2	296.9	1.1341	
Refrigerant 12	CF_2Cl_2	120.91	115.4	243.4	6.8491	1311
Refrigerant 12B1	CF_2BrCl	165.37	113.7	269.2	2.8	
Refrigerant 12B2	CF_2Br_2	209.82	131.6	295.9	1.4	
Refrigerant 13	CF_2Cl	104.46	92.1	191.7	37.05	
Refrigerant 13B1	CF_3Br	148.91	105.4	215.4	16.914	
Refrigerant 21	$CHFCl_2$	102.91	138.2	282.1		
Refrigerant 22	CHF_2Cl	86.469	113.2	232.4	10.96	1194
Refrigerant 23	CHF_3	70.014	118.0	191.2	—	
Refrigerant 32	CH_2F_2	52.023		221.5		
Refrigerant 113	$C_2F_3Cl_3$	187.38	238.2	320.8	0.4817	1557
Refrigerant 114	$C_2F_4Cl_2$	170.92	179.2	276.7	2.2788	
Refrigerant 115	C_2F_5Cl	154.47		234.0		
Refrigerant 116	C_2F_6	138.02				
Refrigerant 142b	$C_2F_2H_3Cl$	100.50	142.4	263.9	3.5825	
Refrigerant 152a	$C_2F_4H_4$	66.051	156	248	6.3132	

TABLE A1 (*Continued*)

c_p, kJ/(kg · K)	c_v, kJ/(kg · K)	γ	μ, 10^{-4} Pa · s	k, W/(m · K)	Pr	σ, N/m	$\bar{v}_s$, m/s
14.84	10.72	1.384	0.0814	0.176	0.686	—	1310
0.360			0.1888	0.0098	0.693	—	210
0.845			0.1461	0.0170	0.728	—	310
1.456			0.1251	0.0260	0.702	—	
0.228			0.1872	0.00624	0.684	—	165
1.48			11.3	0.481	3.47		
1.00			0.1294	0.01461	0.886	—	300
0.249	0.149	1.667	0.256	0.0094	0.678	—	
0.139			15.3				1450
2.235	1.711	1.306	0.112	0.0344	0.733	—	450
2.528			5.35	0.200	6.76	0.0221	1097
			3:56	0.1555	4.89		1125
0.449			0.136	0.00832	0.734	—	905
0.808			0.1088	0.012	0.732	—	
1.100			0.1161	0.01809	0.706		
			3.26	0.186			
			4.08	0.1388	3.57		
1.030	0.618	1.667	0.317	0.0493	0.649	—	455
1.77			0.0726	0.0146	0.880		
0.996			0.1919	0.0257	0.743	—	
1.040	0.743	1.400	0.180	0.0260	0.715	—	352
0.879			0.1490	0.0166	0.790	—	275
2.22			6.53	0.129	11.2	0.022	1205
2.231			5.04	0.128	8.78	0.0210	1163
0.920	0.660	1.394	0.2072	0.0267	0.714	—	332
2.28			2.125	0.146	3.17	0.0153	976
2.41			2.24	0.152	3.55		1050
1.473			0.0844	0.0170	0.733		
1.693	1.492	1.135	0.0826	0.01844	0.762		248
2.446			19.17	0.138	34.0	0.0232	1195
1.536	1.336	1.150	0.081	0.0168	0.797		
0.573	0.509	1.126	0.1095	0.0079	0.814	—	139
0.602	0.542	1.111	0.1260	0.0097	0.781		
0.460	0.395	1.165	0.1381	0.0078	0.811		
0.367	0.316	1.160	0.1340				
0.644	0.562	1.146	0.1449	0.0122	0.766	—	
0.468	0.409	1.143	0.1575	0.0099	0.745	—	
0.594	0.504	1.179	0.115	0.0089	0.768	—	
0.647	0.544	1.190	0.1299	0.0110	0.767	—	
0.732	0.611	1.198	0.1488	0.0148	0.736	—	
0.958			6.64	0.0747	8.52		
0.715	0.660	1.083	0.1157	0.0105	0.775	—	
0.686	0.635	1.081	0.1289	0.0119	0.745	—	
0.762	0.698	1.092	0.1453	0.0150	0.740		
				0.0117			
1.029			0.1037	0.0148	0.721		
				0.0619			

TABLE A1 *(Continued)*

Name	Formula	M	T_m, K	T_b, K	P_g, bar	ρ, kg/m^3
Refrigerant 216	$C_3F_6Cl_2$	220.93		308		
Refrigerant 245					4.888	
Refrigerant C318	C_4F_8	200.031	232.7	267.2	3.325	
Refrigerant 500	mix	99.303	114.3	239.7	8.081	
Refrigerant 502	mix	111.63		237	12.186	
Refrigerant 503	mix	87.267		184		
Refrigerant 504	mix	79.240		216	—	
Refrigerant 505	mix	103.43		243.6		
Refrigerant 506	mix	93.69		260.7		
Silane	SiH_4	32.12	86.8	161.8		
Sulfur dioxide	SO_2	64.059	197.8	268.4	4.168	
Sulfur hexafluoride	SF_6	146.05	222.4	209.4	—	
Toluene	C_7H_8	92.141	178.2	383.8	0.0418	859
Water	H_2O	18.015	273.2	373.2	0.0353	
Xenon	Xe	131.36	161.5	165.0	—	5.291

[a]A blank entry means no information is available; a dash means not applicable at 300 K.

TABLE A2 Thermophysical Properties of Liquid and Saturated-Vapor Air[a]

T, K	P_f, bar	P_g, bar	v_f, m^3/kg	v_g, m^3/kg	h_f, kJ/kg	h_g, kJ/kg	s_f, kJ/(kg · K)	s_g, kJ/(kg · K)
60	0.066	0.025	0.001018	6.8752	−437.6	−243.8	2.862	6.302
65	0.159	0.076	0.001068	2.4144	−436.7	−239.0	2.872	6.057
70	0.340	0.194	0.001093	1.0205	−434.9	−234.4	2.891	5.862
75	0.658	0.424	0.001121	0.4958	−430.6	−230.1	2.950	5.701
80	1.174	0.827	0.001150	0.2677	−423.8	−226.0	3.036	5.567
85	1.955	1.473	0.001182	0.1570	−415.6	−222.2	3.134	5.455
90	3.080	2.439	0.001218	0.0981	−406.5	−218.9	3.237	5.357
95	4.627	3.808	0.001257	0.0645	−396.7	−216.2	3.341	5.270
100	6.679	5.673	0.001303	0.0440	−386.3	−214.1	3.444	5.189
105	9.317	8.128	0.001355	0.0309	−375.4	−212.9	3.548	5.113
110	12.626	11.280	0.001418	0.0220	−363.8	−212.9	3.651	5.037
115	16.687	15.242	0.001497	0.0159	−351.4	−214.3	3.756	4.959
120	31.584	20.138	0.001602	0.0115	−337.8	−217.7	3.865	4.874
125	27.41	26.10	0.001757	0.0081	−322.2	−224.3	3.985	4.773
130	34.25	33.27	0.002069	0.0054	−301.1	−237.6	4.140	4.631
132.5	38.08	37.36	0.002594	0.0041	−280.4	−250.6	4.291	4.517

(Continued)

TABLE A1 (*Continued*)

c_p, kJ/(kg · K)	c_v, kJ/(kg · K)	γ	μ, 10^{-4} Pa · s	k, W/(m · K)	Pr	σ, N/m	$\bar{v}_s$, m/s
0.821	0.778	1.056	0.1187	0.0127	0.765		
						—	
				0.0117	0.769	—	
						—	
						—	
						—	
						—	
1.338			0.1184	0.0221	0.718		
0.623			0.129	0.0096	6.837	—	
0.667			0.1654	0.0141	0.784		
1.69			5.54	0.133	7.04	0.0275	1285
4.179			8.9	0.609	5.69	0.0717	1501
0.160	0.097	1.655	0.232	0.0056	0.663	—	

TABLE A2 (*Continued*)

T, K	$c_{p,f}$, kJ/(kg · K)	$c_{p,g}$, kJ/(kg · K)	μ_f, 10^{-4} Pa · s	μ_g, 10^{-5} Pa · s	k_f, W/(m · K)	k_g, W/(m · K)	Pr_f	Pr_g
60			3.25		0.180	0.005		
65			2.64		0.171	0.006		
70			2.21	0.47	0.163	0.006	2.54	0.69
75	1.79	1.13	1.89	0.51	0.154	0.007	2.35	0.73
80	1.82	1.17	1.65	0.55	0.145	0.008	2.23	0.77
85	1.85	1.21	1.47	0.60	0.137	0.008	2.17	0.81
90	1.88	1.26	1.32	0.65	0.128	0.009	2.14	0.84
95	2.00	1.31	1.20	0.70	0.119	0.009	2.15	0.87
100	2.12	1.37	1.10	0.75	0.110	0.010	2.21	0.91
105	2.24	1.48	1.02	0.80	0.102	0.011	2.29	0.98
110	2.41	1.64	0.95	0.86	0.093	0.012	2.47	1.06
115	2.65	1.91	0.87	0.93	0.084	0.014	2.66	1.13
120	3.09	2.40	0.75	1.02	0.076	0.015	2.76	1.33
125	4.12	3.53	0.62	1.17	0.067	0.018	2.83	1.65

[a] Since air is a multicomponent mixture, the dew and bubble points vary with composition and there is no unique critical point.

TABLE A3 Thermophysical Properties of Gaseous Air at Atmospheric Pressure[a]

T, K	v, m^3/kg	h, kJ/kg	s, kJ/(kg · K)	c_p, kJ/(kg · K)	γ	$\bar{v}_s$, m/s	μ, 10^{-5} Pa · s	k, W/(m · K)	Pr
100	0.2783	−204.5	5.755	1.030	1.424	198.0	0.71	0.0092	0.795
110	0.3076	−194.1	5.854	1.024	1.420	208.5	0.77	0.0102	0.786
120	0.3367	−183.9	5.943	1.020	1.417	218.3	0.84	0.0111	0.778
130	0.3657	−173.7	6.024	1.016	1.415	227.6	0.91	0.0120	0.770
140	0.3946	−163.6	6.099	1.014	1.413	236.4	0.97	0.0129	0.762
150	0.4233	−153.4	6.169	1.011	1.410	244.9	1.03	0.0139	0.755
160	0.4519	−143.3	6.234	1.010	1.407	253.1	1.09	0.0147	0.749
170	0.4805	−133.3	6.296	1.009	1.404	261.1	1.15	0.0156	0.743
180	0.5091	−123.3	6.353	1.009	1.402	268.7	1.21	0.0166	0.739
190	0.5376	−113.1	6.408	1.008	1.400	276.2	1.27	0.0174	0.736
200	0.5666	−103.0	6.4591	1.008	1.398	283.3	1.33	0.0183	0.734
210	0.5949	−92.9	6.5082	1.007	1.399	290.4	1.39	0.0191	0.732
220	0.6232	−82.8	6.5550	1.006	1.399	297.3	1.44	0.0199	0.730
230	0.6516	−72.8	6.5998	1.006	1.400	304.0	1.50	0.0207	0.728
240	0.6799	−62.7	6.6425	1.005	1.400	310.5	1.55	0.0215	0.726
250	0.7082	−52.7	6.6836	1.005	1.400	317.0	1.60	0.0222	0.725
260	0.7366	−42.6	6.7230	1.005	1.400	323.3	1.65	0.0230	0.723
270	0.7649	−32.6	6.7609	1.004	1.400	329.4	1.70	0.0237	0.722
280	0.7932	−22.5	6.7974	1.004	1.400	335.5	1.75	0.0245	0.721
290	0.8216	−12.5	6.8326	1.005	1.400	341.4	1.80	0.0252	0.720
300	0.8499	−2.4	6.8667	1.005	1.400	347.2	1.85	0.0259	0.719
310	0.8782	7.6	6.8997	1.005	1.400	352.9	1.90	0.0265	0.719
320	0.9065	17.7	6.9316	1.006	1.399	358.5	1.94	0.0272	0.719
330	0.9348	27.7	6.9625	1.006	1.399	364.0	1.99	0.0279	0.719
340	0.9632	37.8	6.9926	1.007	1.399	369.5	2.04	0.0285	0.719
350	0.9916	47.9	7.0218	1.008	1.398	374.8	2.08	0.0292	0.719
360	1.0199	57.9	7.0502	1.009	1.398	380.0	2.12	0.0298	0.719
370	1.0482	68.0	7.0778	1.010	1.397	385.2	2.17	0.0304	0.719
380	1.0765	78.1	7.1048	1.011	1.397	390.3	2.21	0.0311	0.719
390	1.1049	88.3	7.1311	1.012	1.396	395.3	2.25	0.0317	0.719
400	1.1332	98.4	7.1567	1.013	1.395	400.3	2.29	0.0323	0.719
410	1.1615	108.5	7.1817	1.015	1.395	405.1	2.34	0.0330	0.719
420	1.1898	118.7	7.2062	1.016	1.394	409.9	2.38	0.0336	0.719
430	1.2181	128.8	7.2301	1.018	1.393	414.6	2.42	0.0342	0.718
440	1.2465	139.0	7.2535	1.019	1.392	419.3	2.46	0.0348	0.718
450	1.2748	149.2	7.2765	1.021	1.391	423.9	2.50	0.0355	0.718
460	1.3032	159.4	7.2989	1.022	1.390	428.3	2.53	0.0361	0.718
470	1.3315	169.7	7.3209	1.024	1.389	433.0	2.57	0.0367	0.718
480	1.3598	179.9	7.3425	1.026	1.389	437.4	2.61	0.0373	0.718
490	1.3882	190.2	7.3637	1.028	1.388	441.8	2.65	0.0379	0.718
500	1.4165	200.5	7.3845	1.030	1.387	446.1	2.69	0.0385	0.718
520	1.473	221.1	7.4249	1.034	1.385	454.6	2.76	0.0398	0.718
540	1.530	241.8	7.4640	1.038	1.382	462.9	2.83	0.0410	0.718
560	1.586	262.6	7.5018	1.042	1.380	471.0	2.91	0.0422	0.718
580	1.653	283.5	7.5385	1.047	1.378	479.0	2.98	0.0434	0.718
600	1.700	304.5	7.5740	1.051	1.376	486.8	3.04	0.0446	0.718
620	1.756	325.6	7.6086	1.056	1.374	494.4	3.11	0.0458	0.718
640	1.813	346.7	7.6422	1.060	1.371	501.9	3.18	0.0470	0.718
660	1.870	368.0	7.6749	1.065	1.369	509.3	3.25	0.0482	0.717
680	1.926	389.3	7.7067	1.070	1.367	516.5	3.32	0.0495	0.717

TABLE A3 (*Continued*)

T, K	v, m^3/kg	h, kJ/kg	s, kJ/(kg · K)	c_p, kJ/(kg · K)	γ	$\bar{v}_s$, m/s	μ, 10^{-5} Pa · s	k, W/(m · K)	Pr
700	1.983	410.8	7.7378	1.075	1.364	523.6	3.38	0.0507	0.717
720	2.040	432.3	7.7682	1.080	1.362	530.6	3.45	0.0519	0.716
740	2.096	453.9	7.7978	1.084	1.360	537.5	3.51	0.0531	0.716
760	2.153	475.7	7.8268	1.089	1.358	544.3	3.57	0.0544	0.716
780	2.210	497.5	7.8551	1.094	1.356	551.0	3.64	0.0556	0.716
800	2.266	519.4	7.8829	1.099	1.354	557.6	3.70	0.0568	0.716
820	2.323	541.5	7.9101	1.103	1.352	564.1	3.76	0.0580	0.715
840	2.380	563.6	7.9367	1.108	1.350	570.6	3.82	0.0592	0.715
860	2.436	585.8	7.9628	1.112	1.348	576.8	3.88	0.0603	0.715
880	2.493	608.1	7.9885	1.117	1.346	583.1	3.94	0.0615	0.715
900	2.550	630.4	8.0136	1.121	1.344	589.3	4.00	0.0627	0.715
920	2.606	652.9	8.0383	1.125	1.342	595.4	4.05	0.0639	0.715
940	2.663	675.5	8.0625	1.129	1.341	601.5	4.11	0.0650	0.714
960	2.720	698.1	8.0864	1.133	1.339	607.5	4.17	0.0662	0.714
980	2.776	720.8	8.1098	1.137	1.338	613.4	4.23	0.0673	0.714
1000	2.833	743.6	8.1328	1.141	1.336	619.3	4.28	0.0684	0.714
1050	2.975	800.8	8.1887	1.150	1.333	633.8	4.42	0.0711	0.714
1100	3.116	858.5	8.2423	1.158	1.330	648.0	4.55	0.0738	0.715
1150	3.258	916.6	8.2939	1.165	1.327	661.8	4.68	0.0764	0.715
1200	3.400	975.0	8.3437	1.173	1.324	675.4	4.81	0.0789	0.715
1250	3.541	1033.8	8.3917	1.180	1.322	688.6	4.94	0.0814	0.716
1300	3.683	1093.0	8.4381	1.186	1.319	701.6	5.06	0.0839	0.716
1350	3.825	1152.3	8.4830	1.193	1.317	714.4	5.19	0.0863	0.717
1400	3.966	1212.2	8.5265	1.199	1.315	726.9	5.31	0.0887	0.717
1450	4.108	1272.3	8.5686	1.204	1.313	739.2	5.42	0.0911	0.717
1500	4.249	1332.7	8.6096	1.210	1.311	751.3	5.54	0.0934	0.718
1550	4.391	1393.3	8.6493	1.215	1.309	763.2	5.66	0.0958	0.718
1600	4.533	1454.2	8.6880	1.220	1.308	775.0	5.77	0.0981	0.717
1650	4.674	1515.3	8.7256	1.225	1.306	786.5	5.88	0.1004	0.717
1700	4.816	1576.7	8.7622	1.229	1.305	797.9	5.99	0.1027	0.717
1750	4.958	1638.2	8.7979	1.233	1.303	809.1	6.10	0.1050	0.717
1800	5.099	1700.0	8.8327	1.237	1.302	820.2	6.21	0.1072	0.717
1850	5.241	1762.0	8.8667	1.241	1.301	831.1	6.32	0.1094	0.717
1900	5.383	1824.1	8.8998	1.245	1.300	841.9	6.43	0.1116	0.717
1950	5.524	1886.4	8.9322	1.248	1.299	852.6	6.53	0.1138	0.717
2000	5.666	1948.9	8.9638	1.252	1.298	863.1	6.64	0.1159	0.717
2050	5.808	2011.6	8.9948	1.255	1.297	873.5	6.74	0.1180	0.717
2100	5.949	2074.4	9.0251	1.258	1.296	883.8	6.84	0.1200	0.717
2150	6.091	2137.3	9.0547	1.260	1.295	894.0	6.95	0.1220	0.717
2200	6.232	2200.4	9.0837	1.263	1.294	904.0	7.05	0.1240	0.718
2250	6.374	2263.6	9.1121	1.265	1.293	914.0	7.15	0.1260	0.718
2300	6.516	2327.0	9.1399	1.268	1.293	923.8	7.25	0.1279	0.718
2350	6.657	2390.5	9.1672	1.270	1.292	933.5	7.35	0.1298	0.719
2400	6.800	2454.0	9.1940	1.273	1.291	943.2	7.44	0.1317	0.719
2450	6.940	2517.7	9.2203	1.275	1.291	952.7	7.54	0.1336	0.720
2500	7.082	2581.5	9.2460	1.277	1.290	962.2	7.64	0.1354	0.720

[a]These properties are based on constant gaseous composition. The reader is reminded that, at the higher temperatures, the pressure can affect the composition and the thermodynamic properties.

TABLE A4 Thermophysical Properties of Saturated Ammonia (R717)

T, K	P, bar	v_f, m³/kg	v_g, m³/kg	h_f, kJ/kg	h_g, kJ/kg	s_f, kJ/(kg · K)	s_g, kJ/(kg · K)	c_{pf}, kJ/(kg · K)	c_{pg}, kJ/(kg · K)	μ_f, 10^{-4} Pa · s	μ_g, 10^{-4} Pa · s	k_f, W/(m · K)	k_g, W/(m · K)	Pr_f	Pr_g
195.5[a]	0.0608	0.001327	15.648	−1110.1	380.1	4.203	11.827	4.73		4.25		0.715		2.81	
200	0.0865	0.001372	11.237	−1088.8	388.5	4.311	11.698	4.61		4.07		0.709		2.65	
210	0.1775	0.001394	5.729	−1044.1	406.7	4.529	11.438	4.38	2.03	3.69		0.685		2.36	
220	0.3381	0.001417	3.135	−1000.6	424.1	4.731	11.207	4.35	2.08	3.34		0.661		2.20	
230	0.6044	0.001442	1.822	−957.0	440.7	4.925	11.002	4.38	2.15	3.02		0.638		2.07	
240	1.0226	0.001468	1.115	−912.9	456.2	5.113	10.817	4.43	2.24	2.73	0.085	0.615	0.0188	1.97	1.01
250	1.6496	0.001495	0.712	−868.2	470.6	5.294	10.650	4.48	2.34	2.45	0.089	0.592	0.0196	1.86	1.06
260	2.5529	0.001524	0.472	−823.1	483.8	5.471	10.498	4.54	2.47	2.20	0.094	0.569	0.0205	1.76	1.12
270	3.8100	0.001551	0.324	−777.3	495.6	5.643	10.358	4.60	2.61	1.97	0.099	0.546	0.0219	1.66	1.18
280	5.5077	0.001589	0.228	−730.9	506.0	5.811	10.228	4.66	2.77	1.76	0.104	0.523	0.0235	1.57	1.23
290	7.741	0.001626	0.165	−683.8	514.7	5.975	10.108	4.73	2.96	1.58	0.109	0.500	0.0255	1.49	1.27
300	10.61	0.001666	0.121	−636.0	521.5	6.135	9.994	4.82	3.18	1.41	0.114	0.477	0.0279	1.42	1.31
310	14.24	0.001710	0.091	−587.2	526.1	6.293	9.885	4.91	3.43	1.26	0.119	0.454	0.0308	1.36	1.35
320	18.72	0.001760	0.069	−537.5	528.2	6.448	9.779	5.02	3.72	1.13	0.124	0.431	0.0333	1.32	1.39
330	24.20	0.001815	0.053	−486.7	527.5	6.602	9.675	5.17	4.14	1.02	0.129	0.408	0.0374	1.29	1.43
340	30.79	0.001878	0.0410	−434.3	523.3	6.755	9.571	5.37	4.54	0.92	0.134	0.385	0.0417	1.28	1.47
350	38.64	0.001952	0.0319	−380.0	515.1	6.908	9.465	5.64	5.14	0.83	0.139	0.361	0.0472	1.30	1.52
360	47.90	0.002039	0.0249	−323.2	501.8	7.063	9.354	6.04	5.97	0.75	0.144	0.337	0.0536	1.34	1.60
370	58.74	0.002148	0.0194	−262.6	481.9	7.222	9.235	6.68	7.20	0.69	0.151	0.313	0.0608	1.47	1.79
380	71.35	0.002291	0.0149	−196.5	452.7	7.391	9.100	7.80	9.30	0.61	0.160	0.286	0.0690	1.66	2.16
390	85.98	0.002499	0.0113	−120.9	408.1	7.578	8.935	10.3	13.85	0.50	0.172	0.254	0.0780	2.03	3.05
400	103.0	0.002882	0.0077	−23.5	329.0	7.813	8.694	21.0	32.09	0.39	0.19		0.108		5.64
405.4[b]	113.0	0.004255	0.0043	142.7	142.7	8.216	8.216								

[a]Triple point.
[b]Critical point.

TABLE A5 Thermophysical Properties of Ammonia (R717) at 1-bar Pressure

T, K	v, m^3/kg	h, kJ/kg	s, kJ/(kg · K)	c_p, kJ/(kg · K)	Z	$\bar{v}_s$, m/s	λ, W/(m · K)	μ, 10^{-6} Pa · s	Pr
239.6[a]	1.138	455.5	10.825	2.233	0.9729	388	0.0188	8.5	1.010
240	1.140	456.5	10.829	2.232	0.9731	389	0.0188	8.5	1.009
260	1.246	500.7	11.006	2.193	0.9816	404	0.0205	9.1	0.973
280	1.349	544.4	11.168	2.173	0.9868	420	0.0225	9.7	0.937
300	1.450	587.8	11.317	2.169	0.9900	435	0.0246	10.3	0.908
320	1.550	631.2	11.458	2.179	0.9921	450	0.0267	11.0	0.896
340	1.650	675.0	11.590	2.198	0.9940	464	0.0290	11.7	0.886
360	1.749	719.2	11.717	2.224	0.9949	477	0.0314	12.4	0.878
380	1.847	764.0	11.838	2.255	0.9957	491	0.0341	13.2	0.874
400	1.946	809.4	11.954	2.289	0.9964	503	0.0365	13.9	0.871
420	2.044	855.5	12.067	2.325	0.9970	516	0.0388	14.5	0.869
440	2.143	902.4	12.176	2.362	0.9975	528	0.0418	15.4	0.868
460	2.241	950.0	12.282	2.400	0.9979	540	0.0446	16.1	0.867
480	2.339	998.4	12.384	2.438	0.9981	552	0.0475	16.9	0.866
500	2.437	1047.6	12.485	2.476	0.9983	563	0.0504	17.6	0.865

[a]Normal boiling point.

TABLE A6 Thermophysical Properties of Saturated Normal Butane (R600)

T, K	P, bar	v_f, m³/kg	v_g, m³/kg	h_f, kJ/kg	h_g, kJ/kg	s_f, kJ/(kg · K)	s_g, kJ/(kg · K)	c_{pf}, kJ/(kg · K)	c_{pg}, kJ/(kg · K)	μ_f, 10^{-4} Pa · s	μ_g, 10^{-4} Pa · s	k_f, W/(m · K)	k_g, W/(m · K)	Pr_f	Pr_g
134.9[a]	6.7. − 6[b]	0.001360	28,630	−388.8	134.2	2.302	5.968	1.875	1.109	22.80	0.0367	0.208	0.0043	20.55	0.947
140	1.7. − 5	0.001369	11,635	−379.0	123.2	2.372	5.877	1.946	1.127	19.00	0.0378	0.202	0.0046	18.30	0.926
150	8.7. − 5	0.001387	2,470	−359.2	125.5	2.508	5.724	2.001	1.159	13.80	0.0400	0.191	0.0051	14.46	0.909
160	3.5. − 4	0.001405	645	−339.3	135.4	2.639	5.602	2.011	1.190	10.60	0.0422	0.182	0.0056	11.71	0.889
170	1.2. − 3	0.001424	207	−319.1	147.0	2.761	5.502	2.009	1.220	8.40	0.0444	0.173	0.0062	9.75	0.874
180	0.00337	0.001443	76.3	−299.0	159.2	2.875	5.421	2.013	1.250	6.86	0.0467	0.165	0.0067	8.37	0.860
190	0.00853	0.001463	31.7	−278.9	171.7	2.985	5.356	2.022	1.282	5.73	0.0489	0.159	0.0073	7.29	0.850
200	0.0195	0.001484	14.7	−258.6	184.6	3.088	5.304	2.039	1.316	4.87	0.0512	0.152	0.0080	6.53	0.843
210	0.0405	0.001505	7.4	−237.9	197.7	3.190	5.263	2.063	1.352	4.20	0.0535	0.146	0.0086	5.93	0.838
220	0.0781	0.001528	4.0	−217.3	210.9	3.286	5.232	2.094	1.391	3.67	0.0558	0.141	0.0093	5.45	0.834
230	0.1410	0.001551	2.308	−196.1	224.5	3.379	5.208	2.128	1.432	3.24	0.0582	0.136	0.0101	5.07	0.831
240	0.2408	0.001575	1.404	−174.6	238.3	3.470	5.192	2.166	1.476	2.87	0.0606	0.131	0.0108	4.75	0.829
250	0.3915	0.001601	0.894	−152.7	252.2	3.560	5.180	2.207	1.524	2.57	0.0631	0.127	0.0117	4.47	0.827
260	0.6099	0.001628	0.592	−130.4	266.3	3.647	5.173	2.252	1.574	2.31	0.0657	0.122	0.0125	4.26	0.825
270	0.9155	0.001656	0.406	−107.6	280.4	3.733	5.170	2.300	1.628	2.09	0.0683	0.118	0.0135	4.07	0.824
272.3	1.0000	0.001663	0.374	−102.3	283.7	3.752	5.170	2.312	1.641	2.04	0.0689	0.117	0.0137	4.03	0.825
272.6	1.0133	0.001664	0.369	−101.6	284.2	2.756	5.170	2.314	1.643	2.03	0.0690	0.117	0.0137	4.01	0.827
280	1.3297	0.001686	0.286	−84.4	294.7	3.818	5.172	2.350	1.686	1.89	0.0710	0.114	0.0145	3.90	0.829
290	1.8765	0.001718	0.207	−60.5	309.0	3.900	5.175	2.403	1.748	1.72	0.0738	0.110	0.0155	3.76	0.832
300	2.5816	0.001753	0.153	−36.2	323.3	3.983	5.182	2.460	1.813	1.57	0.0767	0.106	0.0166	3.63	0.837
310	3.472	0.001790	0.1155	−11.2	337.5	4.064	5.189	2.520	1.884	1.43	0.0797	0.102	0.0178	3.51	0.844
320	4.575	0.001831	0.0884	14.3	351.7	4.145	5.199	2.586	1.961	1.30	0.0829	0.099	0.0190	3.40	0.854
330	5.920	0.001875	0.0686	40.6	365.6	4.225	5.209	2.656	2.047	1.18	0.0863	0.095	0.0204	3.30	0.866
340	7.537	0.001925	0.0539	67.6	379.2	4.305	5.222	2.735	2.142	1.08	0.0899	0.091	0.0218	3.23	0.882
350	9.459	0.001980	0.0427	95.4	392.6	4.384	5.234	2.825	2.250	0.98	0.0937	0.088	0.0234	3.15	0.901
360	11.72	0.002043	0.0340	124.0	405.3	4.463	5.246	2.930	2.379	0.88	0.098	0.084	0.0251	3.07	0.930
370	14.35	0.002116	0.0273	153.7	417.5	4.544	5.256	3.057	2.538	0.80	0.103	0.081	0.0269	3.02	0.972
380	17.39	0.002204	0.0218	184.6	428.7	4.625	5.266	3.222	2.746	0.71	0.108	0.077	0.0289	2.97	1.026
390	20.89	0.002310	0.0175	217.1	438.5	4.705	5.273	3.456	3.045	0.62	0.114	0.074	0.0312	2.90	1.113
400	24.90	0.002451	0.0138	251.5	445.9	4.791	5.277	3.843	3.542	0.55	0.122	0.070	0.0343	3.02	1.260
410	29.49	0.00265	0.0106	289.5	449.2	4.881	5.271	4.69	4.62	0.46	0.133	0.068	0.040	3.17	1.54
420	34.83	0.00305	0.0075	335.7	441.6	4.989	5.242	8.75	9.60	0.35	0.153	0.075	0.062	4.08	2.37
425.2[c]	37.96	0.00439	0.0044	395.4	395.4	5.127	5.127							20.55	0.947

[a] Triple point.

[b] The notation 6.7. − 6 signifies 6.7×10^{-6}.

[c] Critical point.

TABLE A7 Thermophysical Properties of Normal Butane (R600) at Atmospheric Pressure

T, K	v, m^3/kg	h, kJ(kg · K)	s, kJ/(kg · K)	c_p, kJ/(kg · K)	Z	$\bar{v}_s$, m/s	λ, W/(m · K)	μ, 10^{-4} Pa · s	Pr
272.6[a]	0.369	284.2	5.170	1.643	0.9594	200.1	0.0137	6.90	0.827
280	0.380	296.7	5.214	1.673	0.9613	203.1	0.0142	7.07	0.822
300	0.411	330.5	5.332	1.755	0.9705	211.2	0.0164	7.55	0.802
320	0.441	366.1	5.447	1.820	0.9765	218.7	0.0186	8.03	0.784
340	0.471	403.3	5.560	1.903	0.9809	225.7	0.0208	8.51	0.776
360	0.500	442.2	5.671	1.985	0.9843	232.5	0.0232	9.00	0.768
380	0.529	482.8	5.781	2.071	0.9868	239.0	0.0258	9.48	0.760
400	0.558	525.1	5.889	2.156	0.9888	245.3	0.0284	9.96	0.753
420	0.587	568.9	5.996	2.238	0.9903	251.3	0.0312	10.4	0.747
440	0.616	614.5	6.102	2.321	0.9919	257.2	0.0340	10.9	0.744
460	0.645	661.8	6.207	2.402	0.9928	263.0	0.0369	11.4	0.742
480	0.673	710.5	6.311	2.481	0.9938	268.6	0.0399	11.9	0.740
500	0.702	760.9	6.414	2.558	0.9949	274.0	0.0430	12.3	0.738

[a] Normal boiling point.

TABLE A8 Thermophysical Properties of Solid, Saturated-Liquid, and Saturated-Vapor Carbon Dioxide

Temp. T, K	Absolute Pressure P, bar	Specific Volume, m^3/kg		Specific Enthalpy, kJ/kg		Specific Entropy, kJ/(kg · K)		Specific Heat c_p kJ/(kg · K)		Thermal Conductivity, W/(m · K)		Viscosity 10^{-4} Pa · s		Prandtl Number	
		Condensed[a]	Vapor	Condensed[a]	Vapor	Condensed[a]	Vapor	Condensed[a]	Vapor	Liquid	Vapor	Liquid	Vapor	Liquid	Vapor
200	1.544	0.000644	0.2362	164.8	728.3	1.620	4.439								
205	2.277	0.000649	0.1622	171.5	730.0	1.652	4.379								
210	3.280	0.000654	0.1135	178.2	730.9	1.682	4.319								
215	4.658	0.000659	0.0804	185.0	731.3	1.721	4.264								
216.6	5.180	0.000661	0.0718	187.2	731.5	1.736	4.250								
216.6	5.180	0.000848	0.0178	386.3	731.5	2.656	4.250	1.707	0.958	0.182	0.011	2.10	0.116	1.96	0.96
220	5.996	0.000857	0.0624	392.6	733.1	2.684	4.232	1.761	0.985	0.178	0.012	1.86	0.118	1.93	0.97
225	7.357	0.000871	0.0515	401.8	735.1	2.723	4.204	1.820	1.02	0.171	0.012	1.75	0.120	1.87	0.98
230	8.935	0.000886	0.0428	411.1	736.7	2.763	4.178	1.879	1.06	0.164	0.013	1.64	0.122	1.84	0.99
235	10.75	0.000901	0.0357	420.5	737.9	2.802	4.152	1.906	1.10	0.160	0.013	1.54	0.125	1.82	1.01
240	12.83	0.000918	0.0300	430.2	738.9	2.842	4.128	1.933	1.15	0.156	0.014	1.45	0.128	1.80	1.02
245	15.19	0.000936	0.0253	440.1	739.4	2.882	4.103	1.959	1.20	0.148	0.015	1.36	0.131	1.80	1.04
250	17.86	0.000955	0.0214	450.3	739.6	2.923	4.079	1.992	1.26	0.140	0.016	1.28	0.134	1.82	1.06
255	20.85	0.000977	0.0182	460.8	739.4	2.964	4.056	2.038	1.34	0.134	0.017	1.21	0.137	1.84	1.08
260	24.19	0.001000	0.0155	471.6	738.7	3.005	4.032	2.125	1.43	0.128	0.018	1.14	0.140	1.89	1.12
265	27.89	0.001026	0.0132	482.8	737.4	3.047	4.007	2.237	1.54	0.122	0.019	1.08	0.144	1.98	1.17
270	32.03	0.001056	0.0113	494.4	735.6	3.089	3.981	2.410	1.66	0.116	0.020	1.02	0.150	2.12	1.23
275	36.59	0.001091	0.0097	506.5	732.8	3.132	3.954	2.634	1.81	0.109	0.022	0.96	0.157	2.32	1.32
280	41.60	0.001130	0.0082	519.2	729.1	3.176	3.925	2.887	2.06	0.102	0.024	0.91	0.167	2.57	1.44
285	47.10	0.001176	0.0070	532.7	723.5	3.220	3.891	3.203	2.40	0.095	0.028	0.86	0.178	2.90	1.56
290	53.15	0.001241	0.0058	547.6	716.9	3.271	3.854	3.724	2.90	0.088	0.033	0.79	0.191	3.35	1.68
295	59.83	0.001322	0.0047	562.9	706.3	3.317	3.803	4.68		0.081	0.042	0.71	0.207	4.1	1.8
300	67.10	0.001470	0.0037	585.4	690.2	3.393	3.742			0.074	0.065	0.60	0.226		
304.2[b]	73.83	0.002145	0.0021	636.6	636.6	3.558	3.558								

[a]Above the solid line, the condensed phase is solid; below the line, it is liquid.
[b]Critical point.

Table A9 Thermophysical Properties of Gaseous Carbon Dioxide at 1-bar Pressure

T, K	v, m^3/kg	h, kJ/kg	s, kJ/(kg · K)	c_p, kJ/(kg · K)	k, W/(m · K)	μ, 10^{-4} Pa · s	Pr
300	0.5639	809.3	4.860	0.852	0.0166	0.151	0.778
350	0.6595	853.1	4.996	0.898	0.0204	0.175	0.770
400	0.7543	899.1	5.118	0.941	0.0243	0.198	0.767
450	0.8494	947.1	5.231	0.980	0.0283	0.220	0.762
500	0.9439	997.0	5.337	1.014	0.0325	0.242	0.755
550	1.039	1049	5.435	1.046	0.0364	0.261	0.750
600	1.133	1102	5.527	1.075	0.0407	0.281	0.742
650	1.228	1156	5.615	1.102	0.0445	0.299	0.742
700	1.332	1212	5.697	1.126	0.0481	0.317	0.742
750	1.417	1269	5.775	1.148	0.0517	0.334	0.742
800	1.512	1327	5.850	1.168	0.0551	0.350	0.742
850	1.606	1386	5.922	1.187	0.0585	0.366	0.742
900	1.701	1445	5.990	1.205	0.0618	0.381	0.742
950	1.795	1506	6.055	1.220	0.0650	0.396	0.743
1000	1.889	1567	6.120	1.234	0.0682	0.410	0.743

A10 Thermophysical Properties of Saturated Ethane (R170)

T, K	P, bar	v_f, m³/kg	v_g, m³/kg	h_f, kJ/kg	h_g, kJ/kg	s_f, kJ/(kg · K)	s_g, kJ/(kg · K)	c_{pf}, kJ/(kg · K)	c_{pg}, kJ/(kg · K)	μ_f, 10^{-4} Pa · s	μ_g, 10^{-4} Pa · s	λ_f, W/(m · K)	λ_g, W/(m · K)	Pr_f	Pr_g
90.3[a]	1.131. – 5[b]	0.001534	21945	–494.8	122.8	2.552	9.145	2.249	1.170	12.60	0.0315	0.254	0.0034	11.15	1.084
100	1.110. – 4	0.001559	2490	–472.6	114.8	2.785	8.633	2.305	1.189	8.02	0.0342	0.248	0.0040	7.45	1.024
110	7.467. – 3	0.001586	405.0	–449.6	124.7	3.005	8.221	2.318	1.207	5.68	0.0376	0.241	0.0045	5.46	1.000
120	3.545. – 3	0.001614	93.5	–426.3	136.5	3.208	7.898	2.320	1.227	4.36	0.0403	0.232	0.0051	4.36	0.962
130	0.01291	0.001644	27.7	–403.1	148.7	3.392	7.639	2.323	1.248	3.52	0.0431	0.222	0.0058	3.68	0.934
140	0.03831	0.001675	10.030	–379.8	160.9	3.565	7.429	2.330	1.272	2.95	0.0460	0.212	0.0064	3.24	0.914
150	0.09672	0.001707	4.239	–356.5	173.2	3.728	7.256	2.342	1.299	2.52	0.0490	0.202	0.0071	2.92	0.896
160	0.2146	0.001742	2.026	–332.9	185.3	3.878	7.117	2.358	1.331	2.20	0.0520	0.192	0.0078	2.70	0.886
170	0.4290	0.001778	1.068	–309.3	197.2	4.021	7.000	2.381	1.369	1.93	0.0551	0.182	0.0086	2.52	0.878
180	0.7874	0.001818	0.609	–285.9	208.9	4.157	6.904	2.411	1.414	1.71	0.0580	0.172	0.0094	2.40	0.872
184.3	1.0000	0.001835	0.4885	–275.0	213.8	4.217	6.867	2.426	1.436	1.67	0.0593	0.167	0.0098	2.38	0.865
184.5	1.0133	0.001836	0.4826	–274.4	214.1	4.220	6.864	2.427	1.438	1.63	0.0594	0.167	0.0098	2.37	0.866
190	1.347	0.001859	0.3708	–261.1	220.1	4.290	6.821	2.448	1.468	1.53	0.0610	0.162	0.0103	2.31	0.869
200	2.174	0.001905	0.2376	–236.4	230.9	4.416	6.751	2.495	1.531	1.37	0.0640	0.152	0.0112	2.25	0.875
210	3.340	0.001955	0.1589	–211.1	240.9	4.539	6.691	2.552	1.607	1.22	0.0670	0.143	0.0122	2.18	0.883
220	4.923	0.002011	0.1100	–185.1	250.1	4.656	6.635	2.622	1.698	1.09	0.0701	0.134	0.0133	2.13	0.895
230	7.004	0.002073	0.0782	–158.3	258.3	4.776	6.585	2.710	1.810	0.98	0.0735	0.126	0.0146	2.11	0.911
240	9.67	0.002144	0.0568	–130.6	265.2	4.892	6.538	2.822	1.951	0.88	0.0771	0.117	0.0159	2.11	0.946
250	13.01	0.002226	0.0420	–101.6	270.7	5.005	6.495	2.967	2.138	0.78	0.0813	0.109	0.0174	2.12	0.999
260	17.12	0.002323	0.0313	–71.0	274.2	5.121	6.452	3.164	2.397	0.69	0.0862	0.101	0.0191	2.16	1.082
270	22.10	0.002444	0.0235	–38.3	275.0	5.241	6.402	3.447	2.791	0.60	0.0922	0.093	0.0212	2.24	1.21
280	28.06	0.002603	0.0175	–2.9	272.0	5.364	6.345	3.918	3.479	0.52	0.100	0.084	0.0242	2.43	1.44
290	35.15	0.002834	0.0128	37.3	262.5	5.500	6.275	4.909	5.038	0.44	0.112	0.076	0.031	2.82	1.85
300	43.55	0.003275	0.0086	88.4	237.2	5.663	6.159	9.322	12.61	0.34	0.134	0.079	0.056	3.95	3.02
305.3[c]	48.71	0.004838	0.0048	165.3	165.3	5.910	5.910								

[a] Triple point.
[b] The notation 1.131 · – 5 signifies 1.131×10^{-5}.
[c] Critical point.

TABLE A11 Thermophysical Properties of Ethane at Atmospheric Pressure

T, K	v, m^3/kg	h, kJ/kg	s, kJ/(kg · K)	c_p, kJ/(kg · K)	Z	$\bar{v}_s$, m/s	λ, W/(m · K)	μ 10^{-6} Pa · s	Pr
184.6[a]	0.483	214.1	6.864	1.439	0.9583	247.3	0.0098	5.94	0.872
200	0.531	236.8	6.984	1.458	0.9736	257.8	0.0110	6.42	0.850
220	0.589	266.4	7.123	1.500	0.9804	270.3	0.0127	7.04	0.831
240	0.645	296.9	7.256	1.554	0.9851	281.8	0.0146	7.66	0.815
260	0.701	328.6	7.383	1.618	0.9884	292.5	0.0167	8.28	0.804
280	0.757	361.8	7.506	1.689	0.9907	302.6	0.0189	8.89	0.795
300	0.812	396.1	7.626	1.765	0.9925	312.2	0.0213	9.48	0.788
320	0.868	432.3	7.742	1.845	0.9941	321.4	0.0238	10.1	0.782
340	0.923	469.9	7.855	1.928	0.9952	330.2	0.0265	10.7	0.776
360	0.978	509.5	7.968	2.012	0.9960	338.8	0.0293	11.2	0.770
380	1.033	550.4	8.081	2.097	0.9966	347.1	0.0323	11.8	0.765
400	1.088	593.3	8.191	2.183	0.9971	355.2	0.0354	12.3	0.760
420	1.143	637.8	8.297	2.268	0.9975	363.1	0.0386	12.9	0.756
440	1.198	684.1	8.407	2.352	0.9979	370.8	0.0418	13.4	0.752
460	1.253	732.0	8.513	2.435	0.9982	378.4	0.0452	13.9	0.748
480	1.308	781.5	8.617	2.517	0.9985	385.8	0.0487	14.4	0.744
500	1.363	832.4	8.723	2.597	0.9988	393.0	0.0522	14.9	0.741

[a]Normal boiling point.

TABLE A12 Thermophysical Properties of Saturated Ethylene (R1150)

T, K	P, bar	v_f, m^3/kg	v_g, m^3/kg	h_f, kJ/kg	h_g, kJ/kg	s_f, kJ/(kg · K)	s_g, kJ/(kg · K)	c_{pf}, kJ/(kg · K)	c_{pg}, kJ/(kg · K)	μ_f, 10^{-4} Pa · s	μ_g, 10^{-4} Pa · s	λ_f, W/(m · K)	λ_g, W/(m · K)	Pr_f,	Pr_g
104[a]	0.0012	0.001527	255	235.7	803.9	3.005	8.476	2.197	1.187						
110	0.0033	0.001545	96	250.0	811.0	3.139	8.238	2.488	1.188	5.63		0.263		5.33	
120	0.0138	0.001576	25.8	275.3	822.7	3.359	7.921	2.539	1.192	4.26		0.250		4.33	
130	0.0445	0.001609	8.63	300.4	834.4	3.560	7.667	2.465	1.199	3.38		0.236		3.53	
140	0.1191	0.001644	3.46	324.7	845.8	3.740	7.462	2.405	1.212	2.77		0.233		2.99	
150	0.275	0.001681	1.598	348.6	856.9	3.905	7.293	2.377	1.232	2.33		0.210		2.64	
160	0.564	0.001721	0.8230	372.4	867.5	4.058	7.152	2.376	1.260	2.00		0.198		2.40	
169.1	1.000	0.001760	0.4849	394.1	876.7	4.190	7.043	2.393	1.295	1.76	0.060	0.187	0.0085	2.25	0.91
169.4	1.013	0.001761	0.4790	394.7	876.9	4.193	7.040	2.393	1.296	1.75	0.060	0.187	0.0085	2.24	0.92
170	1.053	0.001763	0.4625	396.3	877.6	4.202	7.033	2.395	1.299	1.74	0.061	0.186	0.0086	2.24	0.93
180	1.821	0.001810	0.2784	420.5	886.9	4.339	6.931	2.428	1.351	1.53	0.064	0.175	0.0090	2.12	0.96
190	2.957	0.001861	0.1771	445.0	895.3	4.471	6.841	2.472	1.417	1.35	0.067	0.164	0.0096	2.03	0.98
200	4.559	0.001918	0.1177	470.2	902.7	4.599	6.761	2.531	1.501	1.20	0.070	0.153	0.0105	1.99	1.00
210	6.728	0.001981	0.0810	496.0	908.9	4.722	6.689	2.608	1.610	1.07	0.073	0.144	0.0116	1.94	1.02
220	9.571	0.002054	0.0573	522.7	913.7	4.844	6.622	2.710	1.751	0.955	0.076	0.134	0.0128	1.93	1.04
230	13.201	0.002139	0.0413	550.5	916.7	4.964	6.557	2.852	1.943	0.848	0.082	0.125	0.0142	1.93	1.12
240	17.734	0.002241	0.0302	579.7	917.4	5.084	6.492	3.055	2.220	0.748	0.091	0.115	0.0157	1.99	1.29
250	23.923	0.002369	0.0222	611.1	951.1	5.207	6.423	3.372	2.660	0.653	0.102	0.106	0.0180	2.08	1.51
260	30.036	0.002540	0.0163	645.4	908.2	5.335	6.346	3.946	3.479	0.559	0.116	0.097	0.0199	2.27	2.03
270	38.126	0.002804	0.0115	685.3	893.3	5.478	6.248	5.397	5.586	0.460	0.132	0.091	0.0239	2.73	3.09
280	47.830	0.003430	0.0072	742.4	853.8	5.674	6.072	18.13	25.07		0.151		0.333		
282.4[b]	50.401	0.004669	0.0047	795.4	795.4	5.859	5.859								

[a] Triple point.
[b] Critical point.

TABLE A13 Thermophysical Properties of Ethylene (R1150) at Atmospheric Pressure

T, K	v, m^3/kg	h, kJ/kg	s, kJ/(kg · K)	c_p, kJ/(kg · K)	Z	$\bar{v}_s$, m/s	λ, W/(m · K)	μ, 10^{-6} Pa · s	Pr
169.4	0.4790	876.9	7.040	1.296	0.9671	252	0.0097	6.03	0.804
180	0.5123	890.7	7.119	1.289	0.9730	261	0.0102	6.38	0.806
200	0.5736	916.5	7.255	1.299	0.9806	275	0.0113	7.02	0.809
220	0.6341	942.7	7.380	1.327	0.9855	288	0.0125	7.67	0.812
240	0.6942	969.7	7.497	1.369	0.9889	300	0.0141	8.34	0.810
260	0.7538	997.5	7.609	1.420	0.9912	311	0.0160	9.02	0.802
280	0.8133	1026.5	7.716	1.479	0.9930	321	0.0181	9.70	0.791
300	0.8726	1056.7	7.820	1.543	0.9944	331	0.0206	10.38	0.779
320	0.9317	1088.3	7.922	1.610	0.9954	340	0.0232	11.05	0.767
340	0.9907	1121.2	8.022	1.680	0.9962	349	0.0259	11.71	0.758
360	1.0496	1155.5	8.120	1.751	0.9968	358	0.0288	12.35	0.751
380	1.1087	1191.2	8.216	1.822	0.9975	366	0.0317	12.98	0.747
400	1.1676	1228.4	8.312	1.893	0.9979	374	0.0347	13.60	0.743
420	1.2262	1266.9	8.406	1.963	0.9981	383	0.0377	14.21	0.740
440	1.2850	1306.9	8.499	2.031	0.9985	390	0.0407	14.80	0.739
460	1.3441	1347.9	8.602	2.098	0.9990	397	0.0437	15.38	0.738
480	1.4031	1389.4	8.686	2.164	0.9994	405	0.0468	15.95	0.737
500	1.4624	1431.4	8.775	2.227	0.9999	412	0.0500	16.51	0.736

TABLE A14 Thermophysical Properties of *n*-Hydrogen (R702) at Atmospheric Pressure

T, K	v, m^3/kg	h, kJ/kg	s, kJ/(kg · K)	c_p, kJ/(kg · K)	Z	$\bar{v}_s$, m/s	k, W/(m · K)	μ, 10^{-4} Pa · s	Pr
250	10.183	3,517	67.98	14.04	1.000	1,209	0.162	0.079	0.685
300	12.218	4,227	70.58	14.31	1.000	1,319	0.187	0.089	0.685
350	14.253	4,945	72.79	14.43	1.000	1,423	0.210	0.099	0.685
400	16.289	5,669	74.72	14.48	1.000	1,520	0.230	0.109	0.684
450	18.324	6,393	74.43	14.50	1.000	1,611	0.250	0.118	0.684
500	20.359	7,118	77.96	14.51	1.000	1,698	0.269	0.127	0.684
550	22.39	7,844	79.34	14.53	1.000	1,780	0.287	0.135	0.684
600	24.48	8,571	80.60	14.55	1.000	1,859	0.305	0.143	0.684
650	26.47	9,299	81.76	14.57	1.000	1,934	0.323	0.151	0.684
700	28.50	10,029	82.85	14.60	1.000	2,006	0.340	0.159	0.684

TABLE A15 Thermodynamic Properties of Saturated Methane (R50)

T, K	P, bar	v_f, m³/kg	v_g, m³/kg	h_f, kJ/kg	h_g, kJ/kg	s_f, kJ/(kg · K)	s_g, kJ/(kg · K)	c_{pf}, kJ/(kg · K)	$\bar{v}_s$, m/s
90.68	0.117	2.215. − 3[a]	3.976	216.4	759.9	4.231	10.255	3.288	1576
92	0.139	2.2263	3.410	220.6	762.4	4.279	10.168	3.294	1564
96	0.223	2.250. − 3	2.203	233.2	769.5	4.419	10.006	3.326	1523
100	0.345	2.278. − 3	1.479	246.3	776.9	4.556	9.862	3.369	1480
104	0.515	2.307. − 3	1.026	259.6	784.0	4.689	9.731	3.415	1437
108	0.743	2.337. − 3	0.732	273.2	791.0	4.818	9.612	3.458	1393
112	1.044	2.369. − 3	0.536	287.0	797.7	4.944	9.504	3.497	1351
116	1.431	2.403. − 3	0.401	301.1	804.2	5.068	9.405	3.534	1308
120	1.919	2.438. − 3	0.306	315.3	801.8	5.187	9.313	3.570	1266
124	2.523	2.475. − 3	0.238	329.7	816.2	5.305	9.228	3.609	1224
128	3.258	2.515. − 3	0.187	344.3	821.6	5.419	9.148	3.654	1181
132	4.142	2.558. − 3	0.150	359.1	826.5	5.531	9.072	3.708	1138
136	5.191	2.603. − 3	0.121	374.2	831.0	5.642	9.001	3.772	1093
140	6.422	2.652. − 3	0.0984	389.5	834.8	5.751	8.931	3.849	1047
144	7.853	2.704. − 3	0.0809	405.2	838.0	5.858	8.864	3.939	999
148	9.502	2.761. − 3	0.0670	421.3	840.6	5.965	8.798	4.044	951
152	11.387	2.824. − 3	0.0558	437.7	842.2	6.072	8.733	4.164	902
156	13.526	2.893. − 3	0.0467	454.7	843.2	6.177	8.667	4.303	852
160	15.939	2.971. − 3	0.0392	472.1	843.0	6.283	8.601	4.470	802
164	18.647	3.059. − 3	0.0326	490.1	841.6	6.390	8.533	4.684	749
168	21.671	3.160. − 3	0.0278	508.9	839.0	6.497	8.462	4.968	695
172	25.034	3.281. − 3	0.0234	528.6	834.6	6.606	8.385	5.390	637
176	28.761	3.428. − 3	0.0196	549.7	827.9	6.720	8.301	6.091	570
180	32.863	3.619. − 3	0.0162	572.9	818.1	6.843	8.205	7.275	500
184	37.435	3.890. − 3	0.0131	599.7	802.9	6.980	8.084	9.831	421
188	42.471	4.361. − 3	0.0101	634.0	776.4	7.154	7.912	19.66	327
190.56	45.988	6.233. − 3	0.0062	704.4	704.4	7.516	7.516		

[a]The notation 2.215. − 3 signifies 2.215×10^{-3}.

TABLE A16 Thermophysical Properties of Methane (R50) at Atmospheric Pressure

T, K	v, m³/kg	h, kJ/kg	s, kJ/(kg · K)	c_p, kJ/(kg · K)	Z	$\bar{v}_s$, m/s	k, W/(m · K)	μ, 10^{-6} Pa · s	Pr
111.6[a]	0.560	225.2	9.518	2.205	0.9675	271.6	0.0119	4.46	0.826
120	0.596	242.5	9.668	2.174	0.9710	282.8	0.0129	4.78	0.806
140	0.703	285.5	9.998	2.131	0.9814	307.5	0.0152	5.55	0.778
160	0.808	327.9	10.279	2.112	0.9878	329.9	0.0175	6.32	0.763
180	0.912	370.1	10.528	2.105	0.9910	350.7	0.0201	7.07	0.743
200	1.016	412.2	10.752	2.106	0.9936	369.9	0.0225	7.81	0.740
220	1.120	454.4	10.952	2.116	0.9952	388.0	0.0244	8.53	0.739
240	1.223	496.9	11.139	2.133	0.9964	404.9	0.0267	9.24	0.738
260	1.326	539.8	11.307	2.159	0.9972	420.8	0.0290	9.92	0.738
280	1.429	583.3	11.469	2.193	0.9978	435.7	0.0315	10.6	0.737
300	1.532	627.7	11.619	2.236	0.9984	449.7	0.0341	11.2	0.736
320	1.635	672.6	11.768	2.285	0.9989	463.0	0.0369	11.9	0.736
340	1.737	719.3	11.906	2.341	0.9991	475.7	0.0398	12.5	0.736
360	1.840	766.7	12.043	2.401	0.9993	487.8	0.0428	13.1	0.736
380	1.943	815.3	12.174	2.466	0.9994	494.4	0.0459	13.7	0.736

TABLE A16 (*Continued*)

T, K	v, m^3/kg	h, kJ/kg	s, kJ/(kg · K)	c_p, kJ/(kg · K)	Z	$\bar{v}_s$, m/s	k, W/(m · K)	μ, 10^{-6} Pa · s	Pr
400	2.045	865.2	12.304	2.534	0.9996	510.5	0.0491	14.3	0.737
420	2.147	916.3	12.429	2.604	0.9997	521.4	0.0524	14.8	0.737
440	2.250	969.3	12.554	2.677	0.9998	531.9	0.0559	15.4	0.737
460	2.352	1023.5	12.672	2.749	0.9999	542.2	0.0594	15.9	0.738
480	2.455	1079.0	12.791	2.821	1.0000	552.2	0.0629	16.5	0.739
500	2.558	1136.3	12.909	2.891	1.0002	562.0	0.0665	17.0	0.739

[a]Normal boiling point.

TABLE A17 Thermophysical Properties of Nitrogen (R728) at Atmospheric Pressure

T, K	v, m^3/kg	h, kJ/kg	s, kJ/(kg · K)	c_p, kJ/(kg · K)	Z	$\bar{v}_s$, m/s	μ 10^{-6} Pa · s	k W/(m · K)	Pr
77.4[a]	0.2164	76.7	5.403	1.341	0.9545	172	5.0	0.0074	0.913
80	0.2252	80.0	5.446	1.196	0.9610	177	5.2	0.0077	0.811
100	0.2871	101.9	5.690	1.067	0.9801	202	6.7	0.0098	0.728
120	0.3474	123.1	5.884	1.056	0.9883	222	8.0	0.0117	0.727
140	0.4071	144.2	6.046	1.050	0.9927	240	9.3	0.0136	0.723
160	0.4664	165.2	6.186	1.047	0.9952	257	10.6	0.0154	0.721
180	0.5255	186.1	6.309	1.045	0.9967	273	11.8	0.0171	0.720
200	0.5845	207.0	6.419	1.043	0.9977	288	12.9	0.0187	0.719
220	0.6434	227.8	6.519	1.043	0.9984	302	14.0	0.0203	0.718
240	0.7023	248.7	6.609	1.042	0.9990	316	15.0	0.0218	0.717
260	0.7611	269.5	6.693	1.042	0.9994	329	16.0	0.0232	0.717
280	0.8199	290.3	6.770	1.041	0.9997	341	17.0	0.0247	0.716
300	0.8786	311.2	6.842	1.041	0.9998	359	17.9	0.0260	0.716
320	0.9371	332.0	6.909	1.042	0.9999	365	18.8	0.0273	0.717
340	0.9960	352.8	6.972	1.042	1.0000	376	19.7	0.0286	0.717
360	1.0546	373.7	7.032	1.043	1.0001	387	20.5	0.0299	0.717
380	1.1134	394.5	7.088	1.044	1.0002	397	21.4	0.0311	0.717
400	1.1719	411.5	7.142	1.045	1.0002	408	22.2	0.0324	0.717
420	1.2305	436.3	7.193	1.047	1.0002	417	23.0	0.0336	0.717
440	1.2892	457.3	7.242	1.048	1.0003	427	23.8	0.0347	0.717
460	1.3481	478.3	7.288	1.051	1.0003	437	24.5	0.0359	0.718
480	1.4065	499.3	7.333	1.053	1.0004	446	25.3	0.0371	0.718
500	1.4654	520.4	7.376	1.056	1.0004	455	26.0	0.0383	0.718
600	1.758	626.9	7.570	1.075	1.000	496	29.5	0.0440	0.722
700	2.052	735.6	7.738	1.098	1.000	534	32.8	0.0496	0.726
800	2.344	846.6	7.886	1.122	1.000	568	35.9	0.0551	0.730
900	2.636	960.0	8.019	1.146	1.000	601	38.8	0.0606	0.734
1000	2.931	1075.7	8.141	1.167	1.001	631	41.6	0.658	0.737
1500	4.396	1680.5	8.630	1.244	1.001	765			
2000	5.862	2313.5	8.995	1.284	1.001	879			

[a]Normal boiling point.

Table A18 Thermophysical Properties of Oxygen (R732) at Atmospheric Pressure

T, K	v, m^3/kg	h, kJ/kg	s, kJ/(kg · K)	c_p, kJ/(kg · K)	Z	k, W/(m · K)	μ, 10^{-4}Pa · s	Pr
250	0.6402	226.9	6.247	0.915	0.9987	0.0226	0.179	0.725
300	0.7688	272.7	6.414	0.920	0.9994	0.0266	0.207	0.716
350	0.9790	318.9	6.557	0.929	0.9996	0.0305	0.234	0.713
400	1.025	365.7	6.682	0.942	0.9998	0.0343	0.258	0.710
450	1.154	413.1	6.973	0.956	1.0000	0.0380	0.281	0.708
500	1.282	461.3	6.895	0.972	1.0000	0.0416	0.303	0.707
550	1.410	510.3	6.988	0.988	1.0001	0.0451	0.324	0.708
600	1.539	560.1	7.075	1.003	1.0002	0.0487	0.344	0.708
650	1.667	610.6	7.156	1.018	1.0002	0.0521	0.363	0.709
700	1.795	661.9	7.232	1.031	1.0002	0.0554	0.381	0.710

Table A19 Thermophysical Properties of Saturated Normal Propane (R290)

T, K	P, bar	v_f, m³/kg	v_g, m³/kg	h_f, kJ/kg	h_g, kJ/kg	s_f, kJ/(kg · K)	s_g, kJ/(kg · K)	c_{pf}, kJ/(kg · K)	c_{pg}, kJ/(kg · K)	μ_f, 10^{-4} Pa · s	μ_g, 10^{-4} Pa · s	λ_f, W/(m · K)	λ_g, W/(m · K)	Pr_f	Pr_g
85.5[a]	3.0. − 9[b]	0.001363	5.37 · + 7	−495.9	95313	1.879	8.463	1.903	0.884	111.0	0.026	0.212	0.0026	99.6	0.898
90	1.5. − 8	0.001372	1.12. + 7	−487.3	17584	1.977	8.180	1.915	0.902	75.1	0.028	0.211	0.0028	68.2	0.889
100	3.2. − 7	0.001391	5.85. + 5	−468.0	830.2	2.180	7.662	1.932	0.940	37.7	0.031	0.207	0.0033	35.2	0.877
110	3.9. − 6	0.001411	5.33. + 4	−448.8	149.8	2.365	7.261	1.944	0.976	22.5	0.033	0.203	0.0037	21.6	0.866
120	3.1. − 5	0.001432	7.35. + 3	−429.0	107.3	2.535	6.944	1.955	1.010	15.0	0.036	0.199	0.0042	14.7	0.856
130	0.00018	0.001453	1420	−409.5	111.3	2.694	6.687	1.966	1.042	10.80	0.038	0.194	0.0047	10.9	0.844
140	0.00077	0.001475	375	−389.8	120.8	2.839	6.483	1.980	1.073	8.27	0.041	0.188	0.0052	8.71	0.837
150	0.00277	0.001497	100	−369.9	131.5	2.977	6.318	1.996	1.105	6.56	0.043	0.182	0.0057	7.19	0.829
160	0.0085	0.001520	35.6	−349.7	142.5	3.107	6.182	2.014	1.137	5.36	0.045	0.176	0.0063	6.13	0.847
170	0.0220	0.001545	14.6	−329.5	153.8	3.229	6.073	2.035	1.172	4.48	0.048	0.169	0.0069	5.39	0.820
180	0.0505	0.001570	6.693	−309.1	165.4	3.347	5.982	2.060	1.209	3.81	0.051	0.163	0.0075	4.82	0.817
190	0.1051	0.001596	3.384	−288.2	177.1	3.458	5.907	2.088	1.249	3.29	0.053	0.156	0.0082	4.40	0.815
200	0.2013	0.001624	1.8515	−267.1	188.9	3.567	5.846	2.120	1.294	2.87	0.056	0.149	0.0089	4.08	0.817
210	0.3593	0.001653	1.0821	−245.8	200.9	3.671	5.798	2.157	1.342	2.53	0.059	0.143	0.0096	3.82	0.820
220	0.6044	0.001684	0.6684	−224.0	212.9	3.771	5.758	2.197	1.395	2.24	0.062	0.136	0.0104	3.62	0.825
230	0.9661	0.001717	0.4325	−201.7	224.9	3.871	5.726	2.242	1.452	2.00	0.064	0.130	0.0113	3.45	0.828
230.8	1.0000	0.001719	0.4189	−200.0	225.8	3.878	5.724	2.246	1.457	1.98	0.065	0.130	0.0113	3.43	0.833
231.1	1.0133	0.001721	0.4139	−199.3	226.1	3.880	5.721	2.247	1.459	1.97	0.065	0.129	0.0114	3.42	0.828
240	1.479	0.001752	0.2911	−179.0	236.7	3.966	5.699	2.293	1.515	1.79	0.067	0.124	0.0122	3.31	0.836
250	2.179	0.001790	0.2025	−155.7	248.5	4.061	5.678	2.349	1.584	1.60	0.070	0.118	0.0132	3.19	0.845
260	3.107	0.001831	0.1448	−131.8	260.1	4.154	5.662	2.413	1.659	1.44	0.074	0.112	0.0143	3.10	0.853
270	4.306	0.001876	0.1060	−107.2	271.4	4.247	5.649	2.485	1.743	1.30	0.077	0.107	0.0154	3.02	0.870
280	5.819	0.001926	0.0791	−81.8	282.6	4.338	5.640	2.565	1.837	1.17	0.080	0.102	0.0166	2.94	0.890
290	7.694	0.001981	0.0600	−55.7	293.0	4.429	5.631	2.653	1.944	1.06	0.084	0.097	0.0180	2.91	0.908
300	9.978	0.002044	0.0461	−28.6	303.0	4.517	5.624	2.767	2.070	0.95	0.088	0.092	0.0195	2.86	0.936
310	12.71	0.002116	0.0357	−0.5	312.0	4.608	5.617	2.898	2.225	0.85	0.093	0.087	0.0211	2.83	0.976
320	15.98	0.002200	0.0279	28.9	320.2	4.699	5.610	3.061	2.424	0.76	0.098	0.083	0.0231	2.82	1.025
330	19.82	0.002302	0.0218	59.9	326.8	4.792	5.601	3.28	2.70	0.68	0.104	0.078	0.026	2.84	1.10
340	24.31	0.002431	0.0169	92.9	331.3	4.887	5.588	3.60	3.13	0.59	0.111	0.074	0.029	2.88	1.20
350	29.54	0.002608	0.0130	128.9	332.0	4.989	5.569	4.20	3.94	0.51	0.121	0.071	0.035	2.99	1.37
360	35.64	0.002894	0.0095	170.7	325.2	5.102	5.531	5.90	6.26	0.42	0.137	0.073	0.048	3.34	1.78
369.9[c]	42.48	0.004535	0.0045	259.6	259.6	5.338	5.338								

[a]Triple point.
[b]The notation 3.0. − 9 signifies 3.0×10^{-9}.
[c]Critical point.

Table A20 Thermophysical Properties of Propane (R290) at Atmospheric Pressure

T, K	v, m^3/kg	h, kJ/kg	s, kJ/kg · K	c_p, kJ/kg · K	Z	$\bar{v}_s$, m/s	λ, W/(m · K)	μ, 10^{-6} Pa · s	Pr
231.1[a]	0.413	226.1	5.721	1.459	0.9626	218.3	0.0114	6.47	0.828
240	0.432	239.2	5.778	1.484	0.9664	222.7	0.0121	6.70	0.822
260	0.471	269.6	5.898	1.549	0.9745	232.2	0.0139	7.23	0.805
280	0.511	301.1	6.016	1.623	0.9803	241.0	0.0159	7.76	0.792
300	0.549	334.5	6.132	1.704	0.9843	249.3	0.0180	8.29	0.786
320	0.588	369.4	6.243	1.789	0.9873	257.2	0.0202	8.82	0.781
340	0.626	406.1	6.354	1.876	0.9891	264.8	0.0226	9.35	0.775
360	0.664	444.5	6.465	1.963	0.9912	272.2	0.0252	9.88	0.770
380	0.702	484.6	6.574	2.051	0.9925	279.3	0.0278	10.4	0.766
400	0.740	526.6	6.681	2.138	0.9937	286.2	0.0306	10.9	0.762
420	0.778	570.1	6.787	2.224	0.9948	292.9	0.0334	11.4	0.759
440	0.815	615.4	6.891	2.308	0.9954	299.5	0.0363	11.9	0.757
460	0.853	662.4	6.996	2.392	0.9960	305.9	0.0393	12.4	0.755
480	0.891	711.1	7.100	2.474	0.9965	312.2	0.0424	12.9	0.753
500	0.928	761.3	7.202	2.553	0.9970	318.3	0.0455	13.4	0.752

[a] Normal boiling point.

TABLE A21 Thermophysical Properties of Saturated Refrigerant 12

P, bar	T, K	v_f, 10^{-4} m^3/kg	v_g, m^3/kg	h_f, kJ/kg	h_g, kJ/kg	s_f, kJ/(kg · K)	s_g, kJ/(kg · K)
0.10	200.1	6.217	1.365	334.8	518.1	3.724	4.640
0.15	206.3	6.282	0.936	340.1	521.0	3.750	4.627
0.20	211.1	6.332	0.716	344.1	523.2	3.769	4.618
0.25	214.9	6.374	0.582	347.4	525.0	3.785	4.611
0.30	218.2	6.411	0.491	350.2	526.5	3.798	4.606
0.4	223.5	6.437	0.376	354.9	529.1	3.819	4.598
0.5	227.9	6.525	0.306	358.8	531.2	3.836	4.592
0.6	231.7	6.570	0.254	362.1	532.9	3.850	4.588
0.8	237.9	6.648	0.198	367.6	535.8	3.874	4.581
1.0	243.0	6.719	0.160	372.1	538.2	3.893	4.576
1.5	253.0	6.859	0.110	381.2	542.9	3.929	4.568
2.0	260.6	6.970	0.0840	388.2	546.4	3.956	4.563
2.5	266.9	7.067	0.0681	394.0	549.2	3.978	4.560
3.0	272.3	7.183	0.0573	399.1	551.6	3.997	4.557
4.0	281.3	7.307	0.0435	407.6	555.6	4.027	4.553
5.0	288.8	7.444	0.0351	414.8	558.8	4.052	4.551
6.0	295.2	7.571	0.0294	421.1	561.5	4.073	4.549
8.0	306.0	7.804	0.0221	431.8	565.7	4.108	4.546
10	314.9	8.022	0.0176	440.8	569.0	4.137	4.544
15	332.6	8.548	0.0114	459.3	574.5	4.193	4.539
20	346.3	9.096	0.0082	474.8	577.5	4.237	4.534
25	357.5	9.715	0.0062	488.7	578.5	4.275	4.527
30	367.2	10.47	0.0048	502.0	577.6	4.311	4.517
35	375.7	11.49	0.0036	515.9	574.1	4.347	4.502
40	383.3	13.45	0.0025	532.7	564.1	4.389	4.471
41.2[a]	385.0	17.92	0.0018	548.3	548.3	4.429	4.429

TABLE A21 (*Continued*)

P, bar	$c_{p,f}$, kJ/(kg · K)	$c_{p,g}$, kJ/(kg · K)	μ_f, 10^{-4} Pa · s	μ_g, 10^{-5} Pa · s	k_f W/(m · K)	k_g, W/(m · K)	Pr_f	Pr_g	σ, N/m
0.10	0.855		6.16		0.105	0.0050	5.01		
0.15	0.861		5.61		0.103	0.0053	4.69		
0.20	0.865		5.28		0.101	0.0055	4.52		
0.25	0.868		4.99		0.099	0.0056	4.38		
0.30	0.872		4.79		0.098	0.0057	4.26		
0.4	0.876		4.48		0.097	0.0060	4.05		0.0189
0.5	0.880	0.545	4.25	1.00	0.095	0.0062	3.94	0.89	0.0182
0.6	0.884	0.552	4.08	1.02	0.094	0.0063	3.84	0.88	0.0176
0.8	0.889	0.564	3.81	1.04	0.091	0.0066	3.72	0.88	0.0167
1.0	0.894	0.574	3.59	1.06	0.089	0.0069	3.61	0.88	0.0159
1.5	0.905	0.600	3.23	1.10	0.086	0.0074	3.40	0.89	0.0145
2.0	0.914	0.613	2.95	1.13	0.083	0.0077	3.25	0.90	0.0134
2.5	0.922	0.626	2.78	1.15	0.081	0.0081	3.16	0.91	0.0125
3.0	0.930	0.640	2.62	1.18	0.079	0.0083	3.08	0.91	0.0118
4.0	0.944	0.663	2.40	1.22	0.075	0.0088	3.02	0.92	0.0106
5.0	0.957	0.683	2.24	1.25	0.073	0.0092	2.94	0.93	0.0096
6.0	0.969	0.702	2.13	1.28	0.070	0.0095	2.95	0.95	0.0087
8.0	0.995	0.737	1.96	1.33	0.066	0.0101	2.95	0.97	0.0074
10	1.023	0.769	1.88	1.38	0.063	0.0107	3.05	1.01	0.0063
15	1.102	0.865	1.67	1.50	0.057	0.0117	3.23	1.11	0.0042
20	1.234	0.969	1.49	1.69	0.053	0.0126	3.47	1.30	0.0029
25	1.36	1.19	1.33		0.047	0.0134	3.84		0.0019
30	1.52	1.60	1.16		0.042	0.014	4.2		0.0009
35	1.73	2.5			0.037	0.016			0.0005
40									0.0001
41.2									0.0000

[a]Critical point.

Table A22 Thermophysical Properties of Refrigerant 12 at 1-bar Pressure

T, K	v, m^3/kg	h, kJ/kg	s, kJ/(kg · K)	μ, 10^{-5} Pa · s	c_p, kJ/(kg · K)	k, W/(m · K)	Pr
300	0.2024	572.1	4.701	1.26	0.614	0.0097	0.798
320	0.2167	584.5	4.741	1.34	0.631	0.0107	0.788
340	0.2309	597.3	4.780	1.42	0.647	0.0118	0.775
360	0.2450	610.3	4.817	1.49	0.661	0.0129	0.760
380	0.2590	623.7	4.853	1.56	0.674	0.0140	0.745
400	0.2730	637.3	4.890	1.62	0.684	0.0151	0.730
420	0.2870	651.2	4.924	1.67	0.694	0.0162	0.715
440	0.3009	665.3	4.956	1.72	0.705	0.0173	0.703
460	0.3148	697.7	4.987	1.78	0.716	0.0184	0.693
480	0.3288	694.3	5.018	1.84	0.727	0.0196	0.683
500	0.3427	709.0	5.048	1.90	0.739	0.0208	0.674

TABLE A23 Thermophysical Properties of Saturated Refrigerant 22

T, K	P, bar	v_f, m^3/kg	v_g, m^3/kg	h_f, kJ/kg	h_g, kJ/kg	s_f, kJ/(kg · K)	s_g, kJ/(kg · K)	c_{pf}, kJ/(kg · K)	c_{pg}, kJ/(kg · K)
150	0.0017	6.209. − 4	83.40	268.2	547.3	3.355	5.215	1.059	
160	0.0054	6.293. − 4	28.20	278.2	552.1	3.430	5.141	1.058	
170	0.0150	6.381. − 4	10.85	288.3	557.0	3.494	5.075	1.057	
180	0.0369	6.474. − 4	4.673	298.7	561.9	3.551	5.013	1.058	
190	0.0821	6.573. − 4	2.225	308.6	566.8	3.605	4.963	1.060	
200	0.1662	6.680. − 4	1.145	318.8	571.6	3.675	4.921	1.065	0.502
210	0.3116	6.794. − 4	0.6370	329.1	576.5	3.707	4.885	1.071	0.544
220	0.5470	6.917. − 4	0.3772	339.7	581.2	3.756	4.854	1.080	0.577
230	0.9076	7.050. − 4	0.2352	350.6	585.9	3.804	4.828	1.091	0.603
240	1.4346	7.195. − 4	0.1532	361.7	590.5	3.852	4.805	1.105	0.626
250	2.174	7.351. − 4	0.1037	373.0	594.9	3.898	4.785	1.122	0.648
260	3.177	7.523. − 4	0.07237	384.5	599.0	3.942	4.768	1.143	0.673
270	4.497	7.733. − 4	0.05187	396.3	603.0	3.986	4.752	1.169	0.703
280	6.192	7.923. − 4	0.03803	408.2	606.6	4.029	4.738	1.193	0.741
290	8.324	8.158. − 4	0.02838	420.4	610.0	4.071	4.725	1.220	0.791
300	10.956	8.426. − 4	0.02148	432.7	612.8	4.113	4.713	1.257	0.854
310	14.17	8.734. − 4	0.01643	445.5	615.1	4.153	4.701	1.305	0.935
320	18.02	9.096. − 4	0.01265	458.6	616.7	4.194	4.688	1.372	1.036
330	22.61	9.535. − 4	9.753. − 3	472.4	617.3	4.235	4.674	1.460	1.159
340	28.03	1.010. − 3	7.479. − 3	487.2	616.5	4.278	4.658	1.573	1.308
350	34.41	1.086. − 3	5.613. − 3	503.7	613.3	4.324	4.637	1.718	1.486
360	41.86	1.212. − 3	4.036. − 3	523.7	605.5	4.378	4.605	1.897	
369.3[a]	49.89	2.015. − 3	2.015. − 3	570.0	570.0	4.501	4.501	∞	∞

[a] Critical point.

TABLE A23 (*Continued*)

T, K	μ_f, 10^{-4} Pa · s	μ_g, 10^{-4} Pa · s	λ_f, W/(m · K)	λ_g, W/(m · K)	$\bar{v}_{sf}$, m/s	$\bar{v}_{sg}$, m/s	Pr_f	Pr_g	σ, N/m
150			0.161						
160			0.156						
170	7.70		0.151			142.6	5.39		
180	6.47		0.146			146.1	4.69		
190	5.54		0.141			149.4	4.16		
200	4.81		0.136		1007	152.6	3.77		0.024
210	4.24		0.131		957	155.2	3.47		0.022
220	3.78		0.126		909	157.6	3.24		0.021
230	3.40	0.100	0.121	0.0067	862	159.7	3.07	0.89	0.019
240	3.09	0.104	0.117	0.0073	814	161.3	2.92	0.89	0.017
250	2.82	0.109	0.112	0.0080	766	162.5	2.83	0.89	0.0155
260	2.60	0.114	0.107	0.0086	716	163.1	2.78	0.89	0.0138
270	2.41	0.118	0.102	0.0092	668	163.4	2.76	0.90	0.0121
280	2.25	0.123	0.097	0.0098	622	162.1	2.77	0.93	0.0104
290	2.11	0.129	0.092	0.0105	578	161.1	2.80	0.97	0.0087
300	1.98	0.135	0.087	0.0111	536	160.1	2.86	1.04	0.0071
310	1.86	0.141	0.082	0.0117	496	157.2	2.96	1.13	0.0055
320	1.76	0.148	0.077	0.0123	458	153.4	3.14	1.25	0.0040
330	1.67	0.157	0.072	0.0130	408	148.5	3.39	1.42	0.0026
340	1.51	0.171	0.067	0.0140	355	142.7	3.55	1.60	0.0014
350	1.30		0.060		290	135.9	3.72		0.0008
360	1.06								
369.3									

Table A24 Thermophysical Properties of Refrigerant R22 at Atmospheric Pressure

T, K	v, m^3/kg	h, kJ/kg	s, kJ/(kg · K)	c_p, kJ/(kg · K)	Z	$\bar{v}_s$, m/s	μ, 10^{-6} Pa · s	k, W/(m · K)	Pr
232.3	0.2126	586.9	4.8230	0.608	0.9644	160.1	10.1	0.0067	0.893
240	0.2205	591.5	4.8673	0.6117	0.9682	163.0	10.4	0.0074	0.860
260	0.2408	604.0	4.8919	0.6255	0.9760	169.9	11.2	0.0084	0.838
280	0.2608	616.8	4.9389	0.6431	0.9815	176.2	12.0	0.0094	0.820
300	0.2806	630.0	4.9840	0.6619	0.9857	182.3	12.8	0.0106	0.804
320	0.3001	643.4	5.0274	0.6816	0.9883	188.0	13.7	0.0118	0.790
340	0.3196	657.3	5.0699	0.7017	0.9906	193.5	14.4	0.0130	0.777
360	0.3390	671.7	5.1111	0.7213	0.9923	198.9	15.1	0.0142	0.767
380	0.3583	686.5	5.1506	0.7406	0.9936	204.1	15.8	0.0154	0.760
400	0.3775	701.5	5.1892	0.7598	0.9945	209.1	16.5	0.0166	0.755
420	0.3967	717.0	5.2267	0.7786	0.9953	214.0	17.2	0.0178	0.753
440	0.4159	732.8	5.2635	0.7971	0.9961	218.8	17.9	0.0190	0.752
460				0.8150		223.5	18.6	0.0202	0.751
480				0.8326		227.9	19.3	0.0214	0.751
500				0.8502			19.9	0.0225	0.750

TABLE A25 Thermophysical Properties of Saturated Refrigerant R134a

T, K	P, bar	v_f, m^3/kg	v_g, m^3/kg	h_f, kJ/kg	h_g, kJ/kg	s_f, kJ/(kg · K)	s_g, kJ/(kg · K)	c_{pf}, kJ/(kg · K)	c_{pg}, kJ/(kg · K)	μ_f, 10^{-4} Pa · s	μ_g, 10^{-4} Pa · s	k_f, W/(m · K)	k_g, W/(m · K)	Pr_f	Pr_g	τ, N/m
200	0.070	0.000661	2.32	−36.0	201.0	−0.1691	1.0153									
210	0.187	0.000674	0.906	−26.5	208.1	−0.1175	0.9941									
220	0.252	0.000687	0.698	−15.3	214.5	−0.0664	0.9758									
230	0.438	0.000701	0.416	−3.7	220.8	−0.0158	0.9602	1.113	0.732							
240	0.728	0.000716	0.258	8.1	227.1	0.0343	0.9471	1.162	0.764	4.25	0.095	0.099	0.008	4.99	0.90	
250	1.159	0.000731	0.167	20.3	233.3	0.0840	0.9363	1.212	0.798	3.70	0.099	0.095	0.008	4.72	0.96	0.0145
260	1.765	0.000748	0.112	32.9	239.4	0.1331	0.9276	1.259	0.835	3.25	0.104	0.091	0.008	4.49	1.02	0.0131
270	2.607	0.000766	0.077	45.4	244.8	0.1817	0.9211	1.306	0.876	2.88	0.108	0.087	0.009	4.31	1.08	0.0117
280	3.721	0.000786	0.055	59.2	251.1	0.2299	0.9155	1.351	0.921	2.56	0.112	0.083	0.009	4.17	1.14	0.0103
290	5.175	0.000806	0.040	72.9	256.6	0.2775	0.9114	1.397	0.972	2.30	0.117	0.079	0.010	4.07	1.20	0.0090
300	7.02	0.000821	0.029	87.0	261.9	0.3248	0.9080	1.446	1.030	2.08	0.121	0.075	0.010	4.00	1.27	
310	9.33	0.000865	0.022	101.5	266.8	0.3718	0.9050	1.497	1.104	1.89	0.125	0.071	0.010	3.98	1.34	
320	12.16	0.000895	0.016	116.6	271.2	0.4189	0.9021	1.559	1.198	1.72	0.129	0.068	0.011	3.94	1.44	
330	15.59	0.000935	0.012	132.3	275.0	0.4663	0.8986	1.638	1.324	1.58	0.133	0.064	0.011	3.98	1.57	
340	19.71	0.000984	0.0094	148.9	277.8	0.5146	0.8937	1.750	1.520	1.45	0.137	0.060	0.012	4.23	1.74	
350	24.60	0.00105	0.0071	166.6	279.1	0.5649	0.8861	1.931	1.795	1.34	0.14	0.056	0.012	4.62	2.09	
360	30.40	0.00115	0.0051	186.5	277.7	0.6194	0.8721	2.304	2.610	1.20	0.16	0.054	0.013	5.16	3.21	
370	37.31	0.00139	0.0035	216.0	270.0	0.6910	0.8370			0.95	0.26					
374.3[a]	40.67	0.00195	0.0020	248.0	248.0	0.7714	0.7714									

[a] Critical point.

Table A26 Properties of Refrigerant 134a at Atmospheric Pressure

T, K	v, m^3/kg	h, kJ/kg	s, kJ/(kg · K)	c_p, kJ/(kg · K)	Z	$\bar{v}_s$, m/s
247	0.1901	231.5	0.940	0.787	0.957	145.9
260	0.2017	241.8	0.980	0.801	0.965	150.0
280	0.2193	258.1	1.041	0.827	0.974	156.3
300	0.2365	274.9	1.099	0.856	0.980	162.1
320	0.2532	292.3	1.155	0.885	0.984	167.6
340	0.2699	310.3	1.209	0.915	0.987	172.8
360	0.2866	328.8	1.263	0.945	0.990	177.6
380	0.3032	347.8	1.313	0.976	0.992	182.0
400	0.3198	367.2	1.361	1.006	0.994	186.0

TABLE A27 Thermophysical Properties of Saturated Ice–Water–Steam

P, bar	T, K	v_f^a, 10^{-3} m^3/kg	v_g, m^3/kg	h_f^a, kJ/kg	h_g, kJ/kg	μ_g, 10^{-4} Pa · s	k_f^a, W/(m · K)	k_g, W/(m · K)	Pr_f^a	Pr_g
0.001	252.84	1.0010	1167	−374.9	2464.1	0.0723	2.40	0.0169		
0.002	260.21	1.0010	600	−360.1	2477.4	0.0751	2.35	0.0174		
0.003	265.11	1.0010	408.5	−350.9	2486.0	0.0771	2.31	0.0177		
0.004	267.95	1.0010	309.1	−344.4	2491.9	0.0780	2.29	0.0179		
0.005	270.74	1.0010	249.6	−337.9	2497.3	0.0789	2.27	0.0180		
0.006	273.06	1.0010	209.7	−333.6	2502	0.0798	2.26	0.0182		
0.0061	273.15	1.0010	206.0	−333.5	2502	0.0802	2.26	0.0182		
0.0061	273.15	1.0002	206.0	0.0	2502	0.0802	0.566	0.0182	13.04	0.817
0.008	276.73	1.0001	159.4	21.9	2508	0.0816	0.568	0.0184	11.66	0.823
0.010	280.13	1.0001	129.2	29.4	2513.4	0.0829	0.578	0.0186	10.39	0.828
0.02	290.66	1.0013	67.00	73.5	2532.7	0.0872	0.595	0.0193	7.51	0.841
0.03	297.24	1.0028	45.66	101.1	2544.8	0.0898	0.605	0.0195	6.29	0.854
0.04	302.13	1.0041	34.80	121.4	2553.6	0.0918	0.612	0.0198	5.57	0.865
0.05	306.04	1.0053	28.19	137.8	2560.6	0.0933	0.618	0.0201	5.08	0.871
0.06	309.33	1.0065	23.74	151.5	2566.6	0.0946	0.622	0.0203	4.62	0.877
0.08	314.68	1.0085	18.10	173.9	2576.2	0.0968	0.629	0.0207	4.22	0.883
0.10	318.98	1.0103	14.67	191.9	2583.9	0.0985	0.635	0.0209	3.87	0.893
0.20	333.23	1.0172	7.65	251.5	2608.9	0.1042	0.651	0.0219	3.00	0.913
0.30	342.27	1.0222	5.23	289.3	2624.6	0.1078	0.660	0.0224	2.60	0.929
0.40	349.04	1.0264	3.99	317.7	2636.2	0.1105	0.666	0.0229	2.36	0.941
0.5	354.50	1.0299	3.24	340.6	2645.4	0.1127	0.669	0.0233	2.19	0.951
0.6	359.11	1.0331	2.73	359.9	2653.0	0.1147	0.673	0.0236	2.06	0.961
0.8	366.66	1.0385	2.09	391.7	2665.3	0.1176	0.677	0.0242	1.88	0.979
1.0	372.78	1.0434	1.6937	417.5	2675.4	0.1202	0.6805	0.0244	1.735	1.009
1.5	384.52	1.0530	1.1590	467.1	2693.4	0.1247	0.6847	0.0259	1.538	1.000

TABLE A27 (*Continued*)

P, bar	T, K	v_f^a, 10^{-3} m^3/kg	v_g, m^3/kg	h_f^a, kJ/kg	h_g, kJ/kg	μ_g, 10^{-4} Pa · s	k_f^a, W/(m · K)	k_g, W/(m · K)	Pr_f^a	Pr_g
2.0	393.38	1.0608	0.8854	504.7	2706.3	0.1280	0.6866	0.0268	1.419	1.013
2.5	400.58	1.0676	0.7184	535.3	2716.4	0.1307	0.6876	0.0275	1.335	1.027
3.0	406.69	1.0735	0.6056	561.4	2724.7	0.1329	0.6879	0.0281	1.273	1.040
3.5	412.02	1.0789	0.5240	584.3	2731.6	0.1349	0.6878	0.0287	1.224	1.050
4.0	416.77	1.0839	0.4622	604.7	2737.6	0.1367	0.6875	0.0293	1.185	1.057
4.5	421.07	1.0885	0.4138	623.2	2742.9	0.1382	0.6869	0.0298	1.152	1.066
5	424.99	1.0928	0.3747	640.1	2747.5	0.1396	0.6863	0.0303	1.124	1.073
6	432.00	1.1009	0.3155	670.4	2755.5	0.1421	0.6847	0.0311	1.079	1.091
7	438.11	1.1082	0.2727	697.1	2762.0	0.1443	0.6828	0.0319	1.044	1.105
8	445.57	1.1150	0.2403	720.9	2767.5	0.1462	0.6809	0.0327	1.016	1.115
9	448.51	1.1214	0.2148	742.6	2772.1	0.1479	0.6788	0.0334	0.992	1.127
10	453.03	1.1274	0.1943	762.6	2776.1	0.1495	0.6767	0.0341	0.973	1.137
12	461.11	1.1386	0.1632	798.4	2782.7	0.1523	0.6723	0.0354	0.943	1.156
14	468.19	1.1489	0.1407	830.1	2787.3	0.1548	0.6680	0.0366	0.920	1.175
16	474.52	1.1586	0.1237	858.6	2791.8	0.1569	0.6636	0.0377	0.902	1.191
18	480.26	1.1678	0.1103	884.6	2794.8	0.1589	0.6593	0.0388	0.889	1.206
20	485.53	1.1766	0.0995	908.6	2797.2	0.1608	0.6550	0.0399	0.877	1.229
25	497.09	1.1972	0.0799	962.0	2800.9	0.1648	0.6447	0.0424	0.859	1.251
30	506.99	1.2163	0.0666	1008.4	2802.3	0.1684	0.6347	0.0499	0.849	1.278
35	515.69	1.2345	0.0570	1049.8	2802.0	0.1716	0.6250	0.0472	0.845	1.306
40	523.48	1.2521	0.0497	1087.4	2800.3	0.1746	0.6158	0.0496	0.845	1.331
45	530.56	1.2691	0.0440	1122.1	2797.7	0.1775	0.6068	0.0519	0.849	1.358
50	537.06	1.2858	0.0394	1154.5	2794.2	0.1802	0.5981	0.0542	0.855	1.386
60	548.70	1.3187	0.0324	1213.7	2785.0	0.1854	0.5813	0.0589	0.874	1.442
70	558.94	1.3515	0.0274	1267.4	2773.5	0.1904	0.5653	0.0638	0.901	1.503

80	568.12	1.3843	0.0235	1317.1	2759.9	0.1954	0.5499	0.0688	0.936	1.573
90	576.46	1.4179	0.0205	1363.7	2744.6	0.2005	0.5352	0.0741	0.978	1.651
100	584.11	1.4526	0.0180	1408.0	2727.7	0.2057	0.5209	0.0798	1.029	1.737
110	591.20	1.4887	0.0160	1450.6	2709.3	0.2110	0.5071	0.0859	1.090	1.837
120	597.80	1.5268	0.0143	1491.8	2689.2	0.2166	0.4936	0.0925	1.163	1.963
130	603.98	1.5672	0.0128	1532.0	2667.0	0.2224	0.4806	0.0998	1.252	2.126
140	609.79	1.6106	0.0115	1571.6	2642.4	0.2286	0.4678	0.1080	1.362	2.343
150	615.28	1.6579	0.0103	1611.0	2615.0	0.2373	0.4554	0.1307	1.502	2.571
160	620.48	1.7103	0.0093	1650.5	2584.9	0.2497	0.4433	0.1280	1.688	3.041
170	625.41	1.7696	0.0084	1691.7	2551.6	0.2627	0.4315	0.1404	2.098	3.344
180	630.11	1.8399	0.0075	1734.8	2513.9	0.2766	0.4200	0.1557	2.360	3.807
190	634.58	1.9260	0.0067	1778.7	2470.6	0.2920	0.4087	0.1749	2.951	8.021
200	638.85	2.0370	0.0059	1826.5	2410.4	0.3094	0.3976	0.2007	4.202	12.16

P, bar	s_f,[a] kJ/(kg · K)	s_g, kJ/(kg · K)	$c_{p,f}$,[a] kJ/(kg · K)	$c_{p,g}$, kJ/(kg · K)	μ_f, 10^{-4} Pa · s	$\gamma_{f'}$	γ_g	$\bar{v}_{s,f'}$, m/s	$\bar{v}_{s,g}$, m/s	σ,[a] N/m
0.001	−1.378	9.848	1.957							
0.002	−1.321	9.585	2.015							
0.003	−1.280	9.456	2.053							
0.004	−1.260	9.339	2.075							
0.005	−1.240	9.250	2.097	1.851						
0.006	−1.222	9.160	2.106	1.854						
0.0061	−1.221	9.159	2.116	1.854						
0.0061	0.0000	9.159	4.217	1.854	17.50					0.0756
0.008	0.0543	9.0379	4.206	1.856	15.75					0.0751
0.010	0.1059	8.9732	4.198	1.858	14.30					0.0747
0.02	0.2605	8.7212	4.183	1.865	10.67					0.0731
0.03	0.3543	8.5756	4.180	1.870	9.09					0.0721
0.04	0.4222	8.4724	4.179	1.874	8.15					0.0714

TABLE A27 (*Continued*)

P, bar	s_f^a, kJ/(kg · K)	s_g, kJ/(kg · K)	$c_{p,f}^a$, kJ/(kg · K)	$c_{p,g}$, kJ/(kg · K)	μ_f, 10^{-4} Pa · s	$\gamma_{f'}$	γ_g	$\bar{v}_{s,f'}$, m/s	$\bar{v}_{s,g}$, m/s	σ^a, N/m
0.05	0.4761	8.3928	4.178	1.878	7.51					0.0707
0.06	0.5208	8.3283	4.178	1.881	7.03					0.0702
0.08	0.5925	8.2266	4.179	1.887	6.35					0.0693
0.10	0.6493	8.1482	4.180	1.894	5.88					0.0686
0.20	0.8321	7.9065	4.184	1.917	4.66					0.0661
0.30	0.9441	7.7670	4.189	1.935	4.09					0.0646
0.40	1.0261	7.6686	4.194	1.953	3.74					0.0634
0.5	1.0912	7.5928	4.198	1.967	3.49					0.0624
0.6	1.1454	7.5309	4.201	1.978	3.30					0.0616
0.8	1.2330	7.4338	4.209	2.015	3.03					0.0605
1.0	1.3027	7.3598	4.222	2.048	2.801	1.136	1.321	438.74	472.98	0.0589
1.5	1.4336	7.2234	4.231	2.077	2.490	1.139	1.318	445.05	478.73	0.0566
2.0	1.5301	7.1268	4.245	2.121	2.295	1.141	1.316	449.51	482.78	0.0548
2.5	1.6071	7.0520	4.258	2.161	2.156	1.142	1.314	452.92	485.88	0.0534
3.0	1.6716	6.9909	4.271	2.198	2.051	1.143	1.313	455.65	488.36	0.0521
3.5	1.7273	6.9392	4.282	2.233	1.966	1.143	1.311	457.91	490.43	0.0510
4.0	1.7764	6.8943	4.294	2.266	1.897	1.144	1.310	459.82	492.18	0.0500
4.5	1.8204	6.8547	4.305	2.298	1.838	1.144	1.309	461.46	493.69	0.0491
5	1.8604	6.8192	4.315	2.329	1.787	1.144	1.308	462.88	495.01	0.0483
6	1.9308	6.7575	4.335	2.387	1.704	1.144	1.306	465.23	497.22	0.0468
7	1.9918	6.7052	4.354	2.442	1.637	1.143	1.304	467.08	498.99	0.0455
8	2.0457	6.6596	4.372	2.495	1.581	1.142	1.303	468.57	500.55	0.0444
9	2.0941	6.6192	4.390	2.546	1.534	1.142	1.302	469.78	501.64	0.0433
10	2.1382	6.5821	4.407	2.594	1.494	1.141	1.300	470.76	502.64	0.0423
12	2.2161	6.5194	4.440	2.688	1.427	1.139	1.298	472.23	504.21	0.0405

14	2.2837	6.4651	4.472	2.777	1.373	1.137	1.296	473.18	505.33	0.0389
16	2.3436	6.4175	4.504	2.862	1.329	1.134	1.294	473.78	506.12	0.0375
18	2.3976	6.3751	4.534	2.944	1.291	1.132	1.293	474.09	506.65	0.0362
20	2.4469	6.3367	4.564	3.025	1.259	1.129	1.291	474.18	506.98	0.0350
25	2.5543	6.2536	4.640	3.219	1.193	1.123	1.288	473.71	507.16	0.0323
30	2.6455	6.1837	4.716	3.407	1.143	1.117	1.284	472.51	506.65	0.0300
35	2.7253	6.1229	4.792	3.593	1.102	1.111	1.281	470.80	505.66	0.0280
40	2.7965	6.0685	4.870	3.781	1.069	1.104	1.278	468.72	504.29	0.0261
45	2.8612	6.0191	4.951	3.972	1.040	1.097	1.275	466.31	502.68	0.0244
50	2.9206	5.9735	5.034	4.168	1.016	1.091	1.272	463.67	500.73	0.0229
60	3.0273	5.8908	5.211	4.582	0.975	1.077	1.266	457.77	496.33	0.0201
70	3.1219	5.8162	5.405	5.035	0.942	1.063	1.260	451.21	491.31	0.0177
80	3.2076	5.7471	5.621	5.588	0.915	1.048	1.254	444.12	485.80	0.0156
90	3.2867	5.6820	5.865	6.100	0.892	1.033	1.249	436.50	479.90	0.0136
100	3.3606	5.6198	6.142	6.738	0.872	1.016	1.244	428.24	473.67	0.0119
110	3.4304	5.5595	6.463	7.480	0.855	0.998	1.239	419.20	467.13	0.0103
120	3.4972	5.5002	6.838	8.384	0.840	0.978	1.236	409.38	460.25	0.0089
130	3.5616	5.4408	7.286	9.539	0.826	0.956	1.234	398.90	453.00	0.0076
140	3.6243	5.3803	7.834	11.07	0.813	0.935	1.232	388.00	445.34	0.0064
150	3.6859	5.3178	8.529	13.06	0.802	0.916	1.233	377.00	437.29	0.0053
160	3.7471	5.2531	9.456	15.59	0.792	0.901	1.235	366.24	428.89	0.0043
170	3.8197	5.1855	11.30	17.87	0.782	0.867	1.240	351.19	420.07	0.0034
180	3.8765	5.1128	12.82	21.43	0.773	0.838	1.248	336.35	410.39	0.0026
190	3.9429	5.0332	15.76	27.47	0.765	0.808	1.260	320.20	399.87	0.0018
200	4.0149	4.9412	22.05	39.31	0.758	0.756	1.280	298.10	387.81	0.0011

[a]Above the solid line, solid phase; below the line, liquid.

TABLE A28 Thermophysical Properties of Steam at 1-bar Pressure

.T, K	v, m^3/kg	h, kJ/kg	s, kJ/(kg · K)	c_p, kJ/(kg · K)	c_v, kJ(kg · K)	γ	Z	$\bar{v}_s$, m/s	μ, 10^{-5}Pa · s	λ, W/(m · K)	Pr
373.15	1.679	2676.2	7.356	2.029	1.510	1.344	0.9750	472.8	1.20	0.0248	0.982
400	1.827	2730.2	7.502	1.996	1.496	1.334	0.9897	490.4	1.32	0.0268	0.980
450	2.063	2829.7	7.741	1.981	1.498	1.322	0.9934	520.6	1.52	0.0311	0.968
500	2.298	2928.7	7.944	1.983	1.510	1.313	0.9959	540.3	1.73	0.0358	0.958
550	2.531	3028	8.134	2.000	1.531	1.306	0.9971	574.2	1.94	0.0410	0.946
600	2.763	3129	8.309	2.024	1.557	1.300	0.9978	598.6	2.15	0.0464	0.938
650	2.995	3231	8.472	2.054	1.589	1.293	0.9988	621.8	2.36	0.0521	0.930
700	3.227	3334	8.625	2.085	1.620	1.287	0.9989	643.9	2.57	0.0581	0.922
750	3.459	3439	8.770	2.118	1.653	1.281	9.9992	665.1	2.77	0.0646	0.913
800	3.690	3546	8.908	2.151	1.687	1.275	0.9995	685.4	2.98	0.0710	0.903
850	3.921	3654	9.039	2.185	1.722	1.269	0.9996	705.1	3.18	0.0776	0.897
900	4.152	3764	9.165	2.219	1.756	1.264	0.9996	723.9	3.39	0.0843	0.892
950	4.383	3876	9.286	2.253	1.791	1.258	0.9997	742.2	3.59	0.0912	0.886
1000	4.614	3990	9.402	2.286	1.823	1.254	0.9998	760.1	3.78	0.0981	0.881
1100	5.076	4223	9.625	2.36			0.9999	794.3	4.13	0.113	0.858
1200	5.538	4463	9.384	2.43			1.0000	826.8	4.48	0.130	0.837
1300	5.999	4711	10.032	2.51			1.0000	857.9	4.77	0.144	0.826
1400	6.461	4965	10.221	2.58			1.0000	887.9	5.06	0.160	0.816
1500	6.924	5227	10.402	2.65			1.0002	916.9	5.35	0.18	0.788
1600	7.386	5497	10.576	2.73			1.0004	945.0	5.65	0.21	0.735
1800	8.316	6068	10.912	3.02			1.0011	999.4	6.19	0.33	0.567
2000	9.263	6706	11.248	3.79			1.0036	1051.0	6.70	0.57	0.445

TABLE A29 Thermophysical Properties of Water–Steam at High Pressures

T, K	v, m^3/kg	h, kJ/kg	s, kJ/(kg · K)	c_p, kJ/(kg · K)	c_v, kJ/(kg · K)	γ	Z	$\bar{v}_s$, m/s	μ, Pa · s	k, W/(m · K)	Pr
					$P = 10$ bar						
300	1.003. − 3	113.4	0.392	4.18	4.13	1.01	0.0072	1500	8.57. − 4	0.615	5.82
350	1.027. − 3	322.5	1.037	4.19	3.89	1.08	0.0064	1552	3.70. − 4	0.668	2.32
400	1.067. − 3	533.4	1.600	4.25	3.65	1.17	0.0058	1509	2.17. − 4	0.689	1.34
450	1.123. − 3	749.0	2.109	4.39	3.44	1.28	0.0054	1399	1.51. − 4	0.677	0.981
500	0.221	2891	6.823	2.29	1.68	1.36	0.957	535.7	1.71. − 5	0.038	1.028
600	0.271	3109	7.223	2.13	1.61	1.32	0.987	592.5	2.15. − 5	0.047	0.963
800	0.367	3537	7.837	2.18	1.70	1.28	0.994	686.2	2.99. − 5	0.072	0.908
1000	0.460	3984	8.336	2.30	1.83	1.26	0.997	759.4	3.78. − 5	0.099	0.881
1500	0.692	5224	9.337	2.66			1.000	917.2	5.35. − 5	0.18	0.80
2000	0.925	6649	10.154	3.29			1.002	1050	6.70. − 5	0.39	0.57
					$P = 50$ bar						
300	1.001. − 3	117.1	0.391	4.16	4.11	1.01	0.0362	1508	8.55. − 4	0.618	5.76
350	1.025. − 3	325.6	1.034	4.18	3.88	1.08	0.0317	1561	3.71. − 4	0.671	2.31
400	1.064. − 3	536.0	1.596	4.24	3.64	1.16	0.0288	1519	2.18. − 4	0.691	1.34
450	1.120. − 3	751.4	2.103	4.37	3.43	1.27	0.0270	1437	1.52. − 4	0.681	0.975
500	1.200. − 3	976.1	2.575	4.64	3.25	1.43	0.0260	1246	1.19. − 4	0.645	0.856
600	0.0490	3013	6.350	2.85	1.94	1.47	0.885	560.5	2.14. − 5	0.054	1.129
800	0.0713	3496	7.049	2.31	1.74	1.32	0.966	674.5	3.03. − 5	0.075	0.929
1000	0.0911	3961	7.575	2.35	1.85	1.27	0.987	756.5	3.81. − 5	0.102	0.880
1500	0.1384	5214	8.589	2.66			1.000	918.8	5.37. − 5	0.18	0.81
2000	0.1850	6626	9.398	3.12			1.002	1053	6.70. − 5	0.33	0.64
					$P = 100$ bar						
300	9.99. − 4	121.8	0.390	4.15	4.09	1.01	0.0722	1516	8.52. − 4	0.622	5.69
350	1.022. − 3	329.6	1.031	4.17	3.87	1.08	0.0633	1571	3.73. − 4	0.675	2.31
400	1.061. − 3	539.6	1.590	4.23	3.64	1.16	0.0575	1532	2.20. − 4	0.694	1.34
450	1.116. − 3	754.1	2.097	4.35	3.43	1.27	0.0537	1452	1.53. − 4	0.685	0.975

500	1.193. − 3	977.3	2.567	4.60	3.24	1.42	0.0517	1269	1.21. − 4	0.651	0.853
600	0.0201	2820	5.775	5.22	2.64	1.97	0.726	502.3	2.14. − 5	0.073	1.74
800	0.0343	3442	6.685	2.52	1.82	1.38	0.929	662.4	3.08. − 5	0.081	0.960
1000	0.0449	3935	7.233	2.44	1.88	1.30	0.973	753.3	3.85. − 5	0.107	0.876
1500	0.0692	5203	8.262	2.68			1.000	921.1	5.37. − 5	0.18	0.82
2000	0.0926	6616	9.073	3.08			1.003	1057	6.70. − 5	0.31	0.67
					P = 250 bar						
300	9.93. − 3	135.3	0.385	4.12	4.06	1.02	0.1792	1542	8.48. − 4	0.634	5.50
350	1.016. − 3	341.7	1.022	4.14	3.84	1.08	0.1572	1599	3.78. − 4	0.686	2.28
400	1.053. − 3	550.1	1.578	4.20	3.62	1.16	0.1426	1568	2.24. − 4	0.704	1.33
450	1.105. − 3	762.4	2.078	4.30	3.41	1.26	0.1330	1496	1.57. − 4	0.696	0.969
500	1.175. − 3	981.9	2.541	4.50			0.1273	1331	1.24. − 4	0.666	0.838
600	1.454. − 3	1479	3.443	5.88	4.22	1.40	0.1313	896.9	8.63. − 5	0.532	0.952
800	0.0120	3261	6.086	3.41			0.813	627.3	3.29. − 5	0.109	1.03
1000	0.0173	3845	6.741	2.69	1.97	1.36	0.935	745.9	3.98. − 5	0.125	0.856
1500	0.0277	5186	7.827	2.73			1.000	929.1	5.40. − 5	0.18	0.819
2000	0.0372	6608	8.642	3.04			1.008	1068			
					P = 500 bar						
300	9.83. − 4	157.7	0.378	4.06	3.98	1.02	0.3549	1583	8.45. − 4	0.650	5.28
350	1.005. − 3	361.8	1.007	4.10	3.81	1.08	0.3112	1644	3.87. − 4	0.700	2.27
400	1.041. − 3	567.8	1.557	4.14	3.59	1.15	0.2820	1623	2.31. − 4	0.719	1.33
450	1.088. − 3	776.9	2.050	4.23	3.39	1.25	0.2618	1561	1.62. − 4	0.714	0.960
500	1.151. − 3	991.5	2.502	4.37			0.2493	1418	1.29. − 4	0.689	0.822
600	1.362. − 3	1456	3.346	5.08	3.72	1.37	0.2459	1080	9.34. − 5	0.588	0.808
800	4.576. − 3	2895	5.937	5.84	2.79	2.10	0.620	597.8	4.04. − 5	0.178	1.33
1000	8.102. − 3	3697	6.302	3.17	1.81	1.76	0.878	742.1	4.28. − 5	0.150	0.905
1500	0.0139	5157	7.484	2.82			1.004	943.6			
2000	0.0188	6595	8.310	3.04			1.018	1086			

TABLE A30 Thermal Expansion Coefficient $\tilde{\alpha}$ of Water

T,[a] K	$\tilde{\alpha}$, 10^{-4} K^{-1}				
	0	2	4	6	8
270	$(-1.298)^b$	$(-0.899)^b$	−0.530	−0.185	0.137
280	0.394	0.717	0.993	1.217	1.491
290	1.723	1.944	2.157	2.361	2.558
300	2.747	2.930	3.107	3.279	3.445
310	3.607	3.764	3.917	4.067	4.213
320	4.356	4.496	4.633	4.767	4.899
330	5.029	5.157	5.287	5.407	5.530
340	5.651	5.770	5.888	6.004	6.120
350	6.234	6.347	6.459	6.570	6.681
360	6.790	6.899	7.008	7.170	7.278
370	7.385	7.492	7.600	7.707	7.814

[a]Column headings (0, 2, . . .) give third digit of T.
[b]Subcooled liquid.

TABLE A31 Isothermal Compressibility Coefficient β_T of Water

T,[a] K	β_T, 10^{-5} bar^{-1}				
	0	2	4	6	8
270	$(5.219)^b$	$(5.135)^b$	5.057	4.986	4.922
280	4.863	4.810	4.738	4.696	4.676
290	4.640	4.607	4.577	4.551	4.527
300	4.506	4.487	4.471	4.457	4.445
310	4.435	4.428	4.422	4.418	4.415
320	4.415	4.416	4.419	4.423	4.428
330	4.436	4.444	4.454	4.465	4.478
340	4.492	4.507	4.524	4.541	4.560
350	4.580	4.602	4.624	4.648	4.673
360	4.699	4.727	4.755	4.785	4.816
370	4.848	4.882	4.916	4.953	4.992

[a]Column headings (0, 2, . . .) give third digit of T.
[b]Subcooled liquid.

TABLE A32 Thermophysical Properties of Unused Engine Oil

T, K	v_f, m^3/kg	$c_{p,f}$, kJ/(kg.K)	μ_f, Pa.s	k_f, W/(m · K)	Pr_f	α_f, m^2/s
250	1.093. − 3[a]	1.72	32.20	0.151	367,000	9.60. − 8
260	1.101. − 3	1.76	12.23	0.149	144,500	9.32. − 8
270	1.109. − 3	1.79	4.99	0.148	60,400	9.17. − 8
280	1.116. − 3	1.83	2.17	0.146	27,200	8.90. − 8
290	1.124. − 3	1.87	1.00	0.145	12,900	8.72. − 8
300	1.131. − 3	1.91	0.486	0.144	6,450	8.53. − 8
310	1.139. − 3	1.95	0.253	0.143	3,450	8.35. − 8
320	1.147. − 3	1.99	0.141	0.141	1,990	8.13. − 8
330	1.155. − 3	2.04	0.084	0.140	1,225	7.93. − 8
340	1.163. − 3	2.08	0.053	0.139	795	7.77. − 8
350	1.171. − 3	2.12	0.036	0.138	550	7.62. − 8
360	1.179. − 3	2.16	0.025	0.137	395	7.48. − 8
370	1.188. − 3	2.20	0.019	0.136	305	7.34. − 8
380	1.196. − 3	2.25	0.014	0.136	230	7.23. − 8
390	1.205. − 3	2.29	0.011	0.135	185	7.10. − 8
400	1.214. − 3	2.34	0.009	0.134	155	6.95. − 8

[a]The notation 1.093. − 3 signifies 1.093×10^{-3}.

TABLE A33 Conversion Factors

Area: 1 m^2 = 1550.0 $in.^2$ = 10.7639 ft^2 = 1.19599 yd^2 = 2.47104×10^{-4} acre = 1×10^{-4} ha = 10^{-6} km^2 = 3.8610×10^{-7} mi^2

Density: 1 kg/m^3 = 0.06243 lb_m/ft^3 = 0.01002 lb_m/U.K. gallon = 8.3454×10^{-3} lb_m/U.S. gallon = 1.9403×10^{-3} $slug/ft^3$ = 10^{-3} g/cm^3

Energy: 1 kJ = 737.56 ft · lb_f = 238.85 cal = 0.94783 Btu = 3.7251×10^{-4} hp · hr = 2.7778×10^{-4} kW · hr

Heat transfer coefficient: 1 W/(m^2 · K) = 0.8598 kcal/(m^2 · hr · °C) = 0.1761 Btu/(ft^2 · hr · °F) = 10^{-4} W/(cm^2 · K) = 0.2388×10^{-4} cal/(cm^2 · s · °C)

Inertia: 1 kg · m^2 = 3.41717×10^3 lb · $in.^2$ = 0.73756 slug · ft^2

Length: 1 m = 10^{10} Angstrom units = 39.370 in. = 3.28084 ft = 4.971 links = 1.0936 yd = 0.54681 fathoms = 0.04971 chain = 4.97097×10^{-3} furlong = 10^{-3} km = 5.3961×10^{-4} U.K. nautical miles = 5.3996×10^{-4} U.S. nautical miles = 6.2137×10^{-4} mi

Mass: 1 kg = 2.20462 lb_m = 0.06852 slug = 1.1023×10^{-3} U.S. ton = 10^{-3} tonne = 9.8421×10^{-4} U.K. ton

Mass flow rate: 1 kg/s = 2.20462 lb/s = 132.28 lb/min = 7936.64 lb/hr = 3.54314 long ton/hr = 3.96832 short ton/hr

Power: 1 W = 44.2537 ft · lb_f/min = 3.41214 Btu/hr = 1 J/s = 0.73756 ft · lb_f/s = 0.23885 cal/s = 0.8598 kcal/hr

TABLE A33 (*Continued*)

Pressure: 1 bar = 10^5 N/m^2 = 10^5 Pa = 750.06 mm Hg at 0°C = 401.47 in. H_2O at 32°F = 29.530 in. Hg at 0°C = 14.504 lb_f/in.^{2} = 14.504 psia = 1.01972 kg/cm^2 = 0.98692 atm = 0.1 MPa

Specific energy: 1 kJ/kg = 334.55 ft · lb_f/lb_m = 0.4299 Btu/lb_m = 0.2388 cal/g

Specific energy per degree: 1 kJ/(kg · K) = 0.23885 Btu/(lb_m · °F) = 0.23885 cal/(g · °C)

Surface tension: 1 N/m = 5.71015 × 10^{-3} lb_f/in.

Temperature: T (K) = T (°C) + 273.15 = [T (°F) + 459.67]/1.8 = T (°R)/1.8

Temperature difference: ΔT (K) = ΔT (°C) = ΔT (°F)/1.8 = ΔT (°R)/1.8

Thermal conductivity: 1 W/(m · K) = 0.8604 kcal/(m · hr · °C) = 0.5782 Btu/(ft · hr · °F) = 0.01 W/(cm · K) = 2.390 × 10^{-3} cal/(cm · s · °C)

Thermal diffusivity: 1 m^2/s = 38,750 ft^2/hr = 3600 m^2/hr = 10.764 ft^2/s

Torque: 1 N · m = 141.61 oz · in. = 8.85073 lb_f · in. = 0.73756 lb_f · ft = 0.10197 kg_f · m

Velocity: 1 m/s = 100 cm/s = 196.85 ft/min = 3.28084 ft/s = 2.23694 mi/hr = 2.23694 mph = 3.6 km/hr = 1.94260 U.K. knot = 1.94384 Int. knot

Viscosity, dynamic: 1 (N · s)/m^2 = 1 Pa · s = 10^7 μP = 2419.1 lb_m/(ft · hr) = 10^3 cP = 75.188 slug/(ft · hr) = 10 P = 0.6720 lb_m/(ft · s) = 0.02089 (lb_f · s)/ft^2

Viscosity, kinematic: (see Thermal diffusivity)

Volume: 1 m^3 = 61,024 in.^{3} = 1000 liters = 219.97 U.K. gallon = 264.17 U.S. gallon = 35.3147 ft^3 = 1.30795 yd^3 = 1 stere = 0.81071 × 10^{-3} acre-foot

Volume flow rate: 1 m^3/s = 35.3147 ft^3/s = 2118.9 ft^3/min = 13198 U.K. gallon/min = 791,891 U.K. gallon/hr = 15,850 U.S. gallon/min = 951,019 U.S. gallon/hr

INDEX

accelerational pressure drop, 756
adiabatic temperature of combustion, 404
agitated thin film evaporator, 732–734
agitation and heat transfer, 718, 768
air flow system, 365
air heater, 340, 365, 394, 396, 397
alternative refrigerants, 705
annular flow, 541, 542, 680, 730
annular offset strip ribbon fins, 657
aspects of fouling, 115
attemperation, 200
attemperator, 200, 388
augmentation, 657

backward feed in multiple-effect evaporation, 745–746
baffles in shell-and-tube exchangers, 575
basic equations of heat exchangers, 11
bituminous coal-fired 740-MW once-through steam generator, 339
bled-steam feed-water heaters, 279, 335
blowdown line, tank, 332, 334
boiler, 363
boiler classification, 187
boiler feed-water evaporators, 723
boiling, flow instability, 773–774
boiling heat transfer, 229, 762–770
  Chen correlation for, 765
  effects of enhanced surfaces, 769–770
  inside tubes, 762–766
  mixture effects, 768–769
  outside tubes, 766–767
boiling on the shell side, 500
breakdown of film in falling-film flow, 768
buckstays, 315, 326
bumping instability, 773
burning equipment, 366, 369
bypass:
  high pressure, low pressure, 334

calandria evaporator, 726–729
capacity, 278, 280, 321
capacity rate ratio, 32
capillary tube, 648
carry-under of vapor, 760
cast-iron economizer, 392
chain- and traveling-grate stoker, 369–371, 378
Chen correlation for boiling heat transfer, 764–765
Chisholm correlation for slip ratio, 757
chlorofluorocarbons, 705
circuit pressure losses, 453
circulation, 204
  assisted, 205
  design, 206
  forced, 205
  natural, 205
  once through, 205
  thermal, 205
circulation circuit, 448
circulation of liquid, 719, 724, 726, 737, 740
  natural, 729
circulation ratio, 440, 445, 448, 450, 451, 456, 465
circulation reliability, 448, 454
circulation stagnation, 448
circulation velocity, 442
CISE correlation for void fraction, 757–758
Clausius–Rankine Cycle, 184, 279
cleanliness factor, 122
climbing-film evaporator, 730
coefficient of performance, 637
combined-circulation steam generator compared with once through, 283
combined flow system, 384
combustion products flow system, 365
comparison of steam generator systems, 283
compressor, 636, 656
computational models, furnace, 211
computer modeling, 539
condensate flooding, 702
condensate inundation, 532
condensate pump, 335
condensation heating of evaporators, flow orientation, 762–763

condensation heat transfer:
  augmentation, 550
  typical heat transfer coefficients, 762
condensation, use of enhanced surfaces, 769–770
condenser(s), 334
  air cooled, 594
  classification, 571
  direct contact, 595, 612
  feed-water heaters, 588, 590
  packed column, 596
  plate and frame, 591
  plate fin, 592, 594
  selection of shell-and-tube types, 586
  shell and tube, 573, 583, 585, 586, 597
  spiral, 573, 591
  spray, 596, 612
  TEMA types, 573, 574–581
  tray, 616
  trouble shooting, 616
  turbine exhaust, 585, 587
control, 317, 337
controlled-circulation steam generator compared with once through, 283
controlled sliding pressure operation, 288
convection heating surface, 406
convective boiling, 764
  mixture effects, 769
convective heat transfer, 229
corrosion in evaporators and reboilers, 771
corrosive liquids, concentration of, 738
corrugated fin, 667
corrugated tubes, 657
cost of fouling, 114
costs, 280
counterflow system, 384, 416
critical heat flux, 229, 494, 496
  in boiling, 764
critical steam quality, 450, 456
critical velocity, 499
crossflow heat exchangers, 22
cryogenic evaporators, 718
crystallization and evaporators, 726, 728–729
cyclones, 328, 331

deaeration, 590
density, homogeneous, 755
density wave instability, 773
departure from nucleate boiling, 233
deposits, 252
desalination, use of enhanced surfaces, 769
design, 321
design methods, 9
direct-contact evaporators, 736–738
direct expansion (DX) evaporators, 640, 698
distillation column, 738, 741
distribution of liquid:
  in evaporators, 725
  in falling-film evaporators, 732
disturbances, 317
DNB (departure for nucleate boiling), 233, 298
DO (dryout), 298
downcomer(s), 294, 365, 442
downflow evaporators, 719–721
drain cooler, 590
droplet heat transfer, 614
droplet motion, 613–615
dropwise condensation, 525, 526, 555
dropwise promoters, 550
drum, 384, 397–398
drum, steam, 200
dryout, 233, 298, 494
  and boiling, 764
dry-patch formation in falling-film flow, 768

economizer, 199, 391–394
economizer steaming, 297
effectiveness versus NTU, 38
effect of fouling, 108
  on heat transfer, 109
  on pressure drop, 113
efficiency, 278, 281
enclosure, furnace or furnace enclosure, 192
energy efficiency:
  and evaporation, 744–751
  and reboilers, 744
enhanced surfaces, effects in boiling, 769–770
enhancement factors, 660, 662, 683
enthalpy–pressure diagram, 306, 343
entrance length, 70
  hydrodynamic, 70
  thermal, 70
equilibrium condensation curve, 557
equilibrium methods, 559
erosion in evaporators and reboilers, 771
evaporating heat transfer, 762–770
evaporation, 717
  energy efficiency, 744–751
  multiple-effect, 744–747
  vapor recompression, 744
evaporation at the inlet of downcomers, 451
evaporator(s), 717–781
  agitated thin film, 732–734
  calandria, 726–729
  climbing-film, 730
  corrosion and erosion, 771
  direct-contact, 736–738

furnace wall design, 314
general types, 718–719
horizontal falling-film, 724–725
horizontal shell, 722–724
horizontal tube side, 725–726
instrumentation, 770
liquid distribution, 725
long-tube vertical, 729–730
operational problems, 770–774
orientation effects, 719–720
plate type, 734–736
short-tube vertical, 726–729
submerged-combustion, 736–738
submerged-tube forced-circulation, 725–726
temperature differences, 309
temperatures, 305
tube design, 298
types, 721–722
vacuum operation, 732
vertical falling-film, 730–732

falling-film evaporation, heat transfer, 767–768
falling-film evaporators:
enhanced surfaces, 769
flooding in, 774
horizontal, 726
liquid distribution, 771–772
vapor recompression, 748
vertical, 730–732
Fanning friction coefficient, 52
feed water, 201, 365
feed-water heater failures, 317
feed-water pump:
failures, 317
power consumption, 302
feed-water quality, 316
film Reynolds number, 527
film thickness, 526
fin bonding, 673
finned wall, 16
fins, 592–594, 595
fixed pressure operation, 288
flashing effects, 725, 734, 751
flooded evaporators, 639, 640, 687
flooding, 544
and falling film evaporators, 774
in two-phase flow, 732
flow balancing, 221
flow diagrams, 335, 339, 341, 342
flow instability, 773–774
flow rate pulsation, 455
foaming liquids, evaporation of, 734, 740
forward feed in multiple-effect evaporation, 745
fossil-fuel boilers, 179
fouling, 107, 115
biofouling, 116
chemical reaction fouling, 116
corrosion fouling, 116
crystallization fouling, 116
effect on design in evaporators, 762
particulate fouling, 116
fouling coefficients, 762
fouling control, 137
additives, 138
cleaning, 137
fouling effects, 762, 771
and evaporators, 724, 738
in evaporators and reboilers, 771–772
and multistage flash evaporator, 748
on overall heat transfer coefficients, 753
and reboilers, 742
in reboilers, 740
fouling factor, 15
fouling resistance, 121
free water level, 448, 450
frictional pressure drop, 756
correlations for, 756
Friedel correlation for friction two-phase pressure drop, 756
frosting, 672
fuel burning rate of a stoker, 377
fuel consumption of a boiler, 401
fuel oil burner, 368
furnace, 208, 369
design, 209
heating surface, 405
heat release rates (cross section and volume), 321
outlet temperature, 306
wall design, 280, 314, 321
wall temperatures, 310

gas burner, 374
gas-cooled reactors, 473
gas temperature at the furnace outlet, 405
GEWA-T, GEWA-TX, GEWA-TXY, 696
geysering instability, 773
gravitational pressure drop, 756
G-type shell-and-tube heat exchanger, 743

hard water, evaporation of, 724
headroom and evaporators, 723, 729, 730, 736, 742
headroom and reboilers, 742, 743

heat exchanger(s) (general):
plate for evaporation, 734–736
thermal effectiveness, 32
types and selection, 143
heat exchanger, shell and tube:
computer program, 167
condenser design, comments, 158
constructional components, 153
design strategy, 145
reboiler design, comments, 159
size estimation, 147
tube bundle, 154
heat flux, 298
furnace, 208
heat release rate:
per unit furnace area, 377–378
per unit furnace volume, 379–380, 405
heat retention coefficient, 401
heat-sensitive liquids, evaporation of, 736
heat transfer:
in agitated-film evaporation, 768
area number, 33
crisis, 450, 454, 456
in evaporators and reboilers, 760–770
in evaporator tubes, 298
falling-film evaporation, 767–768
single phase, 761
heavy-water reactors, 473
helical wire inserts, 657
High Flux tube, 696, 697
high-profile fins, 657
history of once-through boilers, 277
hold-up of liquid, 738
homogeneous model:
calculation of circulation rates, 757
of two-phase flow, 755
horizontal evaporators, enhanced surfaces, 769
horizontal falling-film evaporator, 724–725
horizontal shell evaporator, 722–724
horizontal thermosyphon reboiler, 742–744
horizontal tube-side evaporator, 725–726
H-type shell-and-tube heat exchanger, 743
hydraulic diameter, 76
hydraulic resistance, 445
hydrostatic pressure effects, 719–720, 730

impingement baffles, and tube vibration, 774
impingement plates, 577
industrial boiler, 189
inlet subcooling, 455
instability:
in boiling flows, 773–774
density wave, 227
excursive, 226
flow, 224
Ledinegg, 226
instrumentation of evaporators and reboilers, 770
integral finned tubes, 694
integral-fin tubes, 550
interfacial thermal resistance, 526
internal reboiler, 738
internal threads, 657
intersecting fin, 657
in-tube flow patterns, 540
in-tube pressure drop, 547

kettle reboiler, 738–740, 758, 760
calculation of circulation, 760
K-type shell-and-tube heat exchanger, 738

laminar forced convection, 72
in concentric smooth ducts, 75
simultaneously developing, 74
thermal entrance, 73
laminar wavy film, 544
light-water reactors, 472
lignite-fired once-through steam generator, 338
liquid distribution:
in falling-film evaporators, 732, 771–772
in falling-film flow, 768
by orifices, 732
by weirs, 732
liquid hold-up, 730, 736, 741
liquid-metal-cooled reactors, 473
liquid–vapor phase transition point, 304
liquified gases, revaporization of, 737
LMTD correction factor, 23
LMTD Method, 19
load changes, 317
local analysis, 646
log mean temperature difference, 19, 644
long-tube vertical evaporator, 729–730
lost heat, 399
louvered fins, 668
low circulation ratio boiler, 451, 456
lubricants, 656, 662, 683
lumped analysis, 644

maldistribution effects in falling-film evaporators, 771
Martinelli parameter, 654
mass flux, 298, 302, 314, 321
mean flow width, 534
mean quantities:
temperature difference (MTD), 598–600
overall heat transfer coefficient, 598–600
stream (film) coefficients, 599

mean temperature difference, 416
microfin tube, 659
mixture effects in boiling, 768–769
momentum equation:
  and single-phase flow, 753–754
  and two-phase flow, 755–756
multiple-effect evaporation, 744–747
  feed arrangements, 745–746
  liquid feed temperature, 745
multiple-effect evaporators:
  compared to multistage flash process, 750–751
  performance ratio, 747
multiple heat exchangers, 22
multistage flash evaporator, 748–751
  compared to multiple-effect process, 750–751
multistream heat exchangers, 591
multivalueness of the hydraulic characteristic curve, 455, 456

natural circulation:
  boiler, 364, 367, 441
  calculation of, 759, 760
  effects, 741, 742, 777
  in evaporators and reboilers, calculation of, 757–758
  and flow instability, 773–774
  of liquid, 729
  steam generator compared with once through, 283
noncircular passages, 546
noncondensable gas(es), 556, 701
  effect of condensation heat transfer, 762
  in multiple-effect evaporation, 746
nonequilibrium methods, 561
nozzles:
  exchanger, 574, 577
  spray, 596
ε–NTU method, 29
nuclear steam generators, 472–486
nucleate boiling, 764
  mixture effects, 769
Nusselt theory, 526

offset strip fin, 668
oil, 657, 662, 683
oil fields steam generator for steam soak and steam drive, 345
once-through boiler, 440, 441, 456
once-through with superimposed circulation steam generator compared with once through, 283
once-through steam generators, 335
operating modes, 288
operation:
  of evaporators and reboilers, 770–774
  of heat exchangers with fouling, 133
operational problems, 507
  boilers, 507
  corrosion, 509
  corrosion fatigue, 512
  erosion, 510
  maldistribution, 508
  maloperation, 512
  steam generators, 507
  thermal fatigue, 511
  vibration, 511
  water hammer, 512
orifices for liquid distribution, 732
overall heat transfer coefficient, 14, 432, 434, 437, 597–598, 642
  calculation of, 760–761
  in evaporators, 752–753
ozone, 705

parallel feed in multiple-effect evaporation, 746
parallel-flow system, 384
part-load behavior, 304
percentage over surface, 127
performance ratio in multiple-effect evaporators, 747
plain fin, 667
plate-fin heat exchangers, 640
plate-fin units, 718
platen superheater, 383, 384
plate-type evaporator, 734–736
plate units, 718
ψ–P method, 44
P–NTU_c method, 42
post-CHF heat transfer, 231
prediction of fouling, 119
preheating boiler, 327
preheating feed water, 278, 335
pressure:
  operating pressure range, 286
  operational, 189
pressure drop, 52, 217, 291, 302
  acceleration, 217
  friction, 217
  hydrostatic, 217
  local, 217
  single-phase, 753–754
  two-phase, 753–757
pressure loss, two phase, 217
pressure transient, 291, 327

processes of fouling, 117
  aging, 118
  attachment, 118
  initiation, 117
  removal, 118
  transport, 117
pumping power, 62

Rankine Cycle, 184
rating of heat exchangers, 48
reboiler(s),717
  corrosion and erosion, 771
  general types, 718–719
  horizontal thermosyphon, 742–744
  instrumentation, 770
  internal, 738
  kettle, 738–740
  operational problems, 770–774
  types, 738–744
  vertical thermosyphon, 740–742
recirculation pump:
  arrangement, 296
  power consumption, 302
  suction pipe, optimal design, 294
recirculation ratio, 283
recuperators, 9
refrigeration system, 636
regenerative-type air heater, 394, 396, 397, 410
regenerators, 9
reheater, 195, 278, 280, 335, 339, 340, 342, 382–387
relative heat input, 304, 306
reliability criterion, 449, 450
residence time, 730, 734, 736, 738–739, 740
ribbed bore tubing, 236
ripple fin, 667
ripple formation, 302
risers, 365, 367, 398, 445
rough surfaces, 657

safety valve, 317
saturation temperature and change with pressure, 291, 292
separated model of two-phase flow, 755–756
separation of vapor and liquid, 723, 725, 730, 738, 746–747
separators, steam–water, 244
Shah correlation for boiling heat transfer, 764
shell-and-tube heat exchanger:
  G type, 743
  H type, 743
  K type, 738
shell-and-tube unit, 718, 730
shell-side pressure drop, 58, 546
short-tube vertical evaporator, 726–729
  design example, 774–777
single horizontal tube, 526
single-phase forced convection, 69
single-phase pressure drop, 753–754
sizing of heat exchangers, 48
slagging, 304, 310
slag screens, 374
sliding pressure operation, 288
slip ratio, 473
  CISE correlation, 758
  and void fraction, 757
slot fins, 668
spiral ribbed tubes, 456
spray attemperators, 305, 335
spreader stoker, 369, 370
start-up:
  boiler, 252
  diagrams, 336
  equipment, 296
  period, 288
  problems, 296
  system, 332
steam, 179
steam-generating systems, 277, 283
steam lanes, 588
steam mass velocity, 385
steam quality, 283, 285, 298, 392
steam temperature control, 200, 388
steam–water mixture, 364, 365, 444, 447
steam–water separator, 244
steam–water system, 365, 440
steel tube economizer, 392, 393
storage capacity, 317
stratification, 541
stratified flow, 451
strip fins, 668
subcooled boiling, 726
subcooling, 583–585, 590
  of water, 445
submerged-combustion evaporator, 736–738
submerged-tube forced-circulation evaporator, 725–726
supercritical 475-MW once-through steam generator, 341
superheater, 195, 283, 304, 305, 334, 340, 341, 343, 382–387, 432
  design, 304, 305
suppression factor, 654
surface efficiency, 17
surface tension, 702
swirl flow techniques, 657

temperature differences in evaporator tubes, 309–313
temperature transient, 291–294, 327
thermal contact resistance, 673
thermal evaluation of condensers, 597
co-current and countercurrent, 600–603
crossflow, 607
direct contact, 612–616
multidimensional flow, 611–612
multipass exchangers, 606–608, 617
nonequilibrium, 608–611
spray, 612–616
tray, 616
Thermoexcel-C, 705
Thermoexcel-E, 696, 697
thermostatic expansion valve, 649
thin film evaporator, agitated, 732–734
throttle, and flow instability, 773
tie bars, 315, 327
total thermal resistance, 15
tube bundles, 531
tube-in-tube heat exchangers, 641
tube-side pressure drop, 52
tube-to-collar interference, 674
tube vibration, in evaporators and reboilers, 774
tubular-type air heater, 394, 395
Turbo-B, 696
Turbo-C, 704
turbulent film, 542
turbulent forced convection, 84
circular ducts, 84
noncircular ducts, 88
two-phase-flow heat exchangers, 1
applications, 1
liquid to vapor, 4
vapor to liquid, 5
two-phase pressure drop, 753–757
two-phase Reynolds number, 528

underfeed stoker, 369, 370
unfired boiler design, 490
water-side design, 491
circulation, 491
upflow evaporators, 719–720
useful heat, 399, 404
utility boiler, 189

vacuum operation in reboilers, 741
vapor belt, and tube vibration, 774
vapor-compression cycle, 635, 636
vapor mixtures, 556
vapor recompression in evaporation, 744, 747–748
vapor shear, 534, 535
variable overall heat transfer coefficient, 50
variable physical properties, 77
laminar flow of gases, 82
laminar flow of liquids, 78
turbulent flow, 84
velocity limits, 229
venting of condensers, 573–574
vertical evaporators, enhanced surfaces, 769
vertical falling-film evaporator, 730–732
vertical thermosyphon reboiler, 740–742, 758–759
calculation of circulation, 759
vibrating-grate stoker, 369–371
vibration of tubes in evaporators and reboilers, 774
viscosity, homogeneous, 755
void fraction:
correlations for, 756
in two-phase flow, 756
in two-phase flow, CISE correlation, 758
volumetric quality, 445

wall temperatures, 298, 309–313
waste gas temperature, 399
waste heat boilers, 486
bayonet tube, 487
calendria type, 486
horizontal crossflow unit, 489
horizontal U tube, 488
vertical U tube, 487
water, 179
water chemistry, 252
water-cooled walls, 374
water separation, 328–331
water wall design, 321
wavy fin, 667
weight of fuel fired, 398
weirs for liquid distribution, 732
wet coil, 671